Digital Signal Processing

A Practical Approach

Second edition

Emmanuel C. Ifeachor
University of Plymouth

Barrie W. Jervis
Sheffield Hallam University

Prentice
Hall

An imprint of **Pearson Education**

Harlow, England · London · New York · Reading, Massachusetts · San Francisco · Toronto · Don Mills, Ontario · Sydney
Tokyo · Singapore · Hong Kong · Seoul · Taipei · Cape Town · Madrid · Mexico City · Amsterdam · Munich · Paris · Milan

Pearson Education Limited
Edinburgh Gate
Harlow
Essex CM20 2JE
England

and Associated Companies throughout the world

Visit us on the World Wide Web at:
www.pearsoneduc.com

First published under the Addison Wesley imprint 1993
Second edition 2002

© Pearson Education Limited 1993, 2002

ISBN 0201-59619-9

British Library Cataloguing-in-Publication Data
A catalogue record for this book can be obtained from the British Library

Library of Congress Cataloging-in-Publication Data
Ifeachor, Emmanuel C.
    Digital signal processing : a practical approach / Emmanuel C. Ifeachor, Barrie W. Jervis.
        p. cm.
    Includes bibliographical references and index.
    ISBN 0–201–59619–9
    1. Signal processing–Digital techniques. I. Jervis, Barrie W. II. Title.
    TK5102.9.I34 2001
    621.382'2–dc21
                                                2001021116

10 9 8 7 6 5 4 3 2
06 05 04 03 02

Typeset in 10/12pt Times by 35
Printed and bound in the United States of America

Contents

Preface

In the last five years, digital signal processing (DSP) has continued to have a major and increasing impact in many key areas of technology, including telecommunication, digital television and media, biomedicine, digital audio and instrumentation. DSP is now at the core of many new and emerging digital products and applications in the information society (e.g. digital cellular phones, digital cameras and TV, and digital audio systems). The need and expectations for electronic, computer and communication engineers to be competent in DSP have grown even stronger since the first edition. DSP is now a core subject in most electronic/computer/communication engineering curricula.

This second edition of the book has been modernized by including additional topics of increasing importance, by providing MATLAB-based problems, and by offering a companion handbook and a home page on the web. These additions have been made in response to software developments, the wider availability of information technology, developments in the teaching of signal processing, and reader demand. Universities are increasingly making use of web-based materials and signal processing software tools such as MATLAB. We have consequently found a demand amongst our readers for MATLAB-based material. This high-level language permits sophisticated signal processing and the immediate display of results with relatively few commands. It is possible to have fun in developing signal processing and the solutions to problems without the distraction of having to produce detailed programs. We believe that the MATLAB examples and exercises in the book will enhance the learning experience of the student and increase the teaching resource available to the instructor.

As in the first edition, the second edition aims to bridge the gap between theory and practice. Thus, we have retained the main features of the book, namely coverage of modern topics and the provision of practical examples and applications. As in the first edition, we have mixed practical examples and systems with theory, to keep students' interest and motivation high in order to enhance learning. Many of the chapters have been extensively revised, to bring the materials up to date and to improve clarity. End-of-chapter problems have been extended to test, reinforce and extend understanding.

In revising the book, we have drawn from our experience and feedback from readers, worldwide, over the last eight years since the first edition was published.

The new topics introduced include oversampling and bandpass sampling techniques in analog/digital conversions to exploit the advantages that DSP offers; wavelet transforms used for time–frequency representation and resolution of signals; blind signal deconvolution for identifying input signals from the output of an unknown system; parametric spectrum estimation for greater resolution applicable to shorter signals and with fewer pitfalls; architectures of new DSP processors and practical schemes for round-off noise reduction in fixed-point DSP systems; and computer-based multi-choice questions to aid revision. Throughout the book, MATLAB-based examples and exercises are provided.

This book was born out of our experience in teaching practically oriented courses in digital signal processing to undergraduate students at the University of Plymouth and the Sheffield Hallam University, and to application engineers in industry for many years. It appeared to us that many of the available textbooks were either too elementary or too theoretical to be of practical use for undergraduates or application engineers in industry. As most readers will know from experience, the gap between learning the fundamentals in any subject and actually applying them is quite wide. We therefore decided to write this book which we believe undergraduates will understand and appreciate and which will equip them to undertake practical digital signal processing assignments and projects. We also believe that higher degree students and practising engineers and scientists will find this text most useful.

Our own research work over the last two and a half decades in applied DSP has also inspired the contents, by identifying practical issues for discussion and presentation to bridge the gap between theoretical concepts and practical implementation, and by suggesting application examples, case studies, and problems.

The great interest and developments in DSP in both industry and academia are likely to continue for the foreseeable future. The availability of numerous digital signal processors is a testimony to the commercial importance of DSP. Its major attraction lies in the ability to achieve guaranteed accuracy and perfect reproducibility, and in its inherent flexibility compared with analog signal processing. In industry, many engineers still lack the necessary knowledge and expertise in DSP to utilize the immense potential of the very powerful digital signal processors now available off the shelf. This book provides insight and practical guidance to enable engineers to design and develop practical DSP systems using these devices.

In academia, DSP is generally regarded as one of the more mathematical topics in the electrical engineering curriculum, and based on our experiences of teaching we have reduced the mathematical content to what we consider useful, essential, and interesting; we have also emphasized points of difficulty. Our experiences indicate that students learn best if they are aware of the practical relevance of a subject, and while more theoretical texts are essential for completeness and reference as the student matures in the subject, we believe in producing graduates equipped also with practical knowledge and skills. This book was written with these considerations in mind.

The book is not a comprehensive text on DSP, but it covers most aspects of the subject found in undergraduate electrical, electronic or communication engineering degree courses. A number of DSP techniques which are of particular relevance to industry are also covered and in the last few years are beginning to find their way into undergraduate curricula. These include techniques such as adaptive filtering and multirate processing.

The emphasis throughout the book is on the practical aspects of DSP. An important feature of the second edition is the inclusion of MATLAB examples and exercises for signal processing, analysis, design and exploration in a time-efficient manner. The reader is encouraged to carry out the MATLAB exercises to gain further insight into DSP. We have also provided the C language DSP software tool from the first edition, after minor revisions, as this has proved popular.

MATLAB is now widely used as a generic tool in industry and academia and requires less programming skills than C. It has good graphics and display facilities and provides a good environment for developing DSP. We believe that MATLAB is a useful tool for students to become familiar with, and competence in it is a valuable transferable skill to acquire. All the MATLAB m-files referred to in the book are available electronically via the web. These include MATLAB m-files which may be used to perform similar tasks as several C-language programs in the first edition. In addition, the m-files (as well as the C-language programs from the first edition) are also available on the CD in the companion handbook (see below for details of how to obtain copies of these).

Main features of the book

- Provides an understanding of the fundamentals, implementation and applications of DSP techniques from a practical point of view.
- Clear and easy to read, with mathematical contents reduced to that which is necessary for comprehension.
- DSP techniques and concepts are illustrated with practically oriented, fully worked, real-world examples designed to provide insight into DSP.
- Provides practical guidance to enable readers to design and develop actual DSP systems. Complete design examples and practical implementation details are given, including assembly language programs for DSP processors.
- MATLAB examples and problems to provide hands-on experience.
- Provides C language implementation of many DSP algorithms and functions, including programs for:
  - digital FIR and IIR filter design,
  - finite wordlength effects analysis of user-designed fixed-point IIR filters,
  - converting from cascade to parallel realization structures,
  - correlation computation,
  - discrete and fast Fourier transform algorithms,
  - inverse z-transformation,

- frequency response estimation, and
- multirate processing systems design.

■ PC-based MATLAB m-files are available electronically on the web (the C programs from the first edition are available on the CD that comes with the companion handbook – see the section 'Website, CD and companion handbook for this book' in this preface for details).

■ Contains many end-of-chapter problems and provides multiple-choice questions to assist with revision.

■ Uses realistic examples to illustrate important concepts and to reinforce the knowledge gained.

The intended audience

The book is aimed at engineering, science and computer science students, and application engineers and scientists in industry who wish to gain a working knowledge of DSP. In particular, final year students studying for a degree in electronics, electrical or communication engineering will find the book valuable for both taught courses as well as their project work, as increasingly a greater proportion of student project work involves aspects of DSP. Postgraduates studying for a master's degree or PhD in the above subjects will also find the book useful.

Undergraduate students will find the fundamental topics very attractive and, we believe, the book will be a valuable source of information both throughout their course as well as when they go into industry.

Large commercial or government organizations who undertake their own internal DSP short courses could base them on the book. We believe the book will serve as a good teaching text as well as a valuable self-learning text for undergraduate, graduate and application engineers.

Contents and organization

Chapter 1 contains an overview of DSP and its applications to make the reader aware of the meaning of DSP and its importance. In Chapter 2 we present, from a practical point of view using real-world examples, many fundamental topics which form the cornerstone of DSP, such as sampling and quantization of signals and their implications in real-time DSP. New features include important topics such as oversampling techniques in AD/DA conversion, sampling of bandpass signals, and uniform and non-uniform quantization. Discrete-time signals and systems are introduced in this chapter, and discussed further in Chapter 4.

Discrete transforms, particularly the discrete and fast Fourier transforms (FFT), provide important mathematical tools in DSP as well as relating the time and frequency domains. They are introduced and described in Chapter 3 with a discussion of some applications to put them in context. The derivation of the discrete Fourier transform (DFT) from the Fourier transform and the exponential Fourier series provides a logical justification for the DFT which does not require coverage of the discrete Fourier series which would unnecessarily increase the length of the book (and the amount of work for the student!). The discussion has also been restricted to the description and implementation of the transforms. In particular, the topic of windowing has not been included in this chapter but is more appropriately discussed in detail in Chapter 11 on spectrum analysis. As an important application of the discrete cosine transform the JPEG standard for image compression is described. The wavelet transform has been growing increasingly popular for a variety of applications because of its applicability to non-stationary signals and its ability to resolve signals in both frequency and time. An introduction to the topic has therefore been included. Applications to multi-resolution analysis and singularity detection for the denoising of signals are described.

In Chapter 4 the basics of discrete-time signals and systems are discussed. Important aspects of the z-transform, an invaluable tool for representing and analyzing discrete-time signals and systems, are discussed. Many applications of the z-transform are highlighted, for example its use in the design, analysis and computation of the frequency response of discrete-time signals and systems. As in the rest of the book, the concepts as well as applications of the z-transform are illustrated with fully worked examples.

Correlation and convolution are fundamental and closely related topics in DSP and are covered in depth in Chapter 5. The authors consider an awareness of all the contents of this chapter to be essential for DSP, but after a preliminary scanning of the contents the reader may well be advised to build up his or her detailed knowledge by progressing through the chapter in stages. The contents might well be spread over several years of an undergraduate course. In this second edition the additional topics of system identification, deconvolution and blind deconvolution have been included. Blind deconvolution is especially interesting, since by exploiting information maximization it is possible to determine an unknown input signal measured at the output of a system of unknown impulse response.

Chapters 6, 7 and 8 include detailed practical discussions of digital filter design, one of the most important topics in DSP, being at the core of most DSP systems. Digital filter design is a vast topic and those new to it can find this somewhat overwhelming. Chapter 6 provides a general framework for filter design. A simple but general step-by-step guide for designing digital filters is given.

Techniques for designing FIR (finite impulse response) filters from specifications through to filter implementations are discussed in Chapter 7. Several fully worked examples are given throughout the chapter to consolidate the important concepts. In this edition, additional topics covered include automatic design of frequency FIR filters. A complete filter design example is included to show how all the stages of filter design fit together.

IIR (infinite impulse response) filter design is discussed in detail in Chapter 8, based on the simple step-by-step guide. This chapter has been substantially reorganized and extended. In particular, the sections on coefficient calculation have been reorganized for clarity and new materials added to cover important topics in IIR filter design, in response to feedback from readers. Additionally, fully worked examples have been included to help the reader to design IIR filters from specifications through to implementation. Design examples using MATLAB as well as C language software are given.

We have reduced the overall material on IIR filter design, by moving the material on finite wordlength effects to Chapter 13. Thus, in response to readers' feedback the materials in Chapters 1–8 contain essential materials for most DSP courses. The more advanced DSP topics now appear in later chapters. Detailed treatment of finite wordlength effects in DSP algorithms now appear logically together in Chapter 13.

Multirate processing techniques allow data to be processed at more than one sampling rate and have made possible such novel applications as single-bit ADCs and DACs (digital-to-analog converters), and oversampled digital filtering, which are exploited in a number of modern digital systems, including for example the familiar compact disc player. In Chapter 9, the basic concepts of multirate processing are explained, illustrated with fully worked examples and by the design of actual multirate systems. The materials in this chapter have been extended to include polyphase. More design examples and applications have been integrated into the theory to illustrate both the principles and design issues in practical multirate systems.

In Chapter 10, key aspects of adaptive filters are described, based on the LMS (least-mean-squares) and RLS (recursive least-square) algorithms which are two of the most widely used algorithms in adaptive signal processing. The treatment is practical with only the essential theory included in the main text.

In Chapter 11, the important topic of spectrum estimation and analysis, used to describe and study signals in the frequency domain, is described. With the introduction of software packages for parametric spectrum estimation it seemed appropriate to provide a detailed introduction to these methods. Provided the signals are accurately represented by models of the correct order, parametric spectrum estimation is applicable to shorter signal lengths and provides spectrum estimates of improved resolution compared with non-parametric methods. An application of autoregressive spectrum estimation of evoked response signals in electroencephalogram signals is used to illustrate the method. Readers who are particularly interested in spectral analysis should study both Chapters 11 and 3 as Chapter 11 draws on explanations and worked examples given in Chapter 3. Those who master the contents of these chapters will be well placed to become competent in the analysis of signals in the frequency domain.

In the last decade and a half, tremendous progress has been made in DSP hardware, and this has led to the wide availability of low cost digital signal processors. For a successful application of DSP using these processors, it is necessary to appreciate the underlying concepts of DSP hardware and software. Chapter 12 discusses the key issues underlying general- and special-purpose processors for DSP, the impact of DSP algorithms on the hardware and software architectures of these processors, and the architectural requirements for efficient execution of DSP functions. The materials in this chapter have been brought up to date. In particular, we have discussed new DSP architectures such as very long instruction word and super scalar, and new fixed and floating point DSP processors (including Texas Instruments fixed point processors, e.g. TMS320C54 and TMS320C62, Motorola fixed point processors DSP56300, and Analog Devices TigerSHARC IS0001).

In Chapter 13, a detailed analysis of finite wordlength effects in modern fixed point DSP systems is presented. Solutions are provided, where appropriate, to the degrading effects of using fixed precision arithmetic.

Chapter 14 is new (although some materials from the first edition are retained) and serves as a teaching and learning resource for the instructor and the student. The chapter includes a description of low-cost DSP boards for implementation of DSP algorithms and a description of a number of real-world applications in the form of case studies. Other features include computer-based, multiple-choice questions which cover key aspects of the topics covered in earlier chapters, and are valuable for revision and for assessing large classes. Complete laboratory exercises are described and case study/project ideas provided.

How to use the book

A useful approach for undergraduate teaching will be to cover the materials in Chapters 1 and 2, to provide the understanding of fundamental topics such as the sampling theorem and discrete-time signals and systems, and to establish the benefits and applications of DSP. Then discrete transforms should be introduced, starting with the DFT and FFT (Chapter 3), and the z-transform (Chapter 4). Aspects of Chapters 11 and 5 may be used to illustrate the application of the DFT and FFT. After an introduction to correlation processing using a selection of materials from Chapter 5, a detailed treatment of digital filters should be undertaken.

In our experience students learn more when they are given realistic assignments to carry out. To this end we would encourage substantial assignments on, for example, filter design, the inverse z-transform, the DFT and FFT. Laboratory work should also be designed to demonstrate and reinforce the techniques taught. It is important that students actually participate as well as attend lectures.

For final-year undergraduates and postgraduate students the approach could be the same but the pace will be more brisk, and the more specialist topics of multirate processing and adaptive filters will also be included.

Website, CD and companion handbook for this book

Additional information about this book may be found at the web home page:

www.booksites.net/ifeachor

Readers are strongly encouraged to send feedback to the authors via the publishers using the 'Contact us' button at:

www.booksites.net/ifeachor

Electronic copies of all the MATLAB m-files can be downloaded from the companion web site for this book at:

www.booksites.net/ifeachor

These include a number of MATLAB m-files which the reader can use to perform similar tasks as they would with several C-language programs in the first edition. The MATLAB m-files, C programs and assembly language codes are also available on the CD that comes with the companion handbook. The C-language programs, taken from the first edition (after minor revision), are available in both executable form and as source codes. A C compiler is required to run the source codes, but not to run the executable codes. The programs were written in standard ANSI C under Borland Turbo C version 2.0. The companion handbook *A Practical Guide for MATLAB and C Language Implementations of DSP Algorithms*, published by Pearson, together with the CD, can be purchased separately. The handbook also contains many illustrative examples of the use of the MATLAB m-files and C programs in the main book. You will find an order form at the back of this book.

Acknowledgements

We are fortunate to have received many useful comments and suggestions from many of our present and past students, which have improved the technical content and clarity of the book. We are grateful to all of them, but especially to Nick Outram, Eddie Riddington, Robin Clark, Steve Harris, Brahim Hamadicharef, Ian Scholey, François Amand, Nichola Gater, Robert Ruse and Andrew Paulley. A number of design exercises in the book, especially in Chapter 14, were developed for our DSP courses by Nick Outram, Eddie Riddington, Robin Clark and Brahim Hamadicharef. James Britton is thanked for computation and plots for examples in Chapters 3 and 11. Several of our ex-students have continued to contribute to our DSP courses from industry. We are grateful to all of them, but especially to Robin Clark and Nick Outram, for their stimulating inputs.

The authors would like to thank Mr Mike Fraser, until a few years ago a technical member of staff of the University of Plymouth, and formerly a Chief Engineer with Rank Toshiba, Plymouth. His considerable experience and valuable comments have been most helpful. We would also like to thank him and Paul Smithson for developing and constructing the DSP target boards from our initial design, and developing the environment in which many programs were implemented and tested. We acknowledge the comments and assistance from many other colleagues, especially Mr Peter Van Eetvelt for deriving the mathematical formulae in Appendices 8C and 8D.

We are indebted to many readers in industry and academia, worldwide, for invaluable feedback, and for taking the trouble to draw our attention to errors in the first edition and to let us know what they think of our book. We very much hope that they and others will keep the feedback coming.

The practical nature of the book made it difficult to keep to deadlines. Each chapter took much longer to write than we had imagined or planned for. We thank the first acquisition editor, Tim Pitts, for his patience and encouragement. We are indebted to Anna Faherty, the second acquisition editor, for talking us into writing the second edition, and to Karen Sutherland, Julie Knight and Mary Lince for taking the project forward.

Finally, the authors are especially grateful to their families for their tolerance, patience and support throughout this very time-consuming project.

Emmanuel Ifeachor
Barrie Jervis
March 2001

Publisher's Acknowledgements

We are grateful to the following for permission to reproduce copyright material:

Fig. 1.8 courtesy of Allen & Heath, Cornwall; Figs. 1.12, 1.15 from *Philips Technical Review* Vol. 40(6) published by Konintlyke Philips Electronics N.V.; Figs. 8.37, 8.38, 8.39 from 'Add DTMF generation and decoding to DSP-up designs' from *EDN Magazine* Vol. 30, published by Cahners Business Information (Mock, P., 1985); Fig. 13.2 from *Journal of Audio Engineering Society* Vol. 41(9), published by the Audio Engineering Society, Inc. (Wilson, R., 1993); Tables 7.11, 7.18 adapted from *An Approach to the Approximation Problem for Nonrecursive Digital Filters* published by The Institute of Electrical and Electronics Engineers Inc. (Rabiner, L.R., Gold, B., McGonegal, C.A., 1970); Fig. 14.12 reprinted from *Clinical Neurophysiology Ireland* No. 38, D.G. Girton and A.J. Kamiya, 'Electroencephalography and Clinical Neurophysiology', pp. 623–39, 1973, with permission from Excerpta Medica Inc.

Whilst every effort has been made to trace the owners of copyright material, in a few cases this has proved impossible and we take this opportunity to offer our apologies to any copyright holders whose rights we may have unwittingly infringed.

Introduction

1

The aims of this chapter are to explain the meaning and benefits of digital signal processing (DSP), to introduce basic DSP operations on which much of DSP is founded, and to make the reader aware of the wide range of application areas for DSP. Specific real-world application examples are presented, drawn from areas with which most readers can relate.

1.1 Digital signal processing and its benefits

By a signal we mean any variable that carries or contains some kind of information that can, for example, be conveyed, displayed or manipulated. Examples of the types of signals of particular interest are

- speech, which we encounter for example in telephony, radio and everyday life,
- biomedical signals, such as the electroencephalogram (brain signals),
- sound and music, such as reproduced by the compact disc player,
- video and image, which most people watch on the television, and
- radar signals, which are used to determine the range and bearing of distant targets.

Digital signal processing is concerned with the digital representation of signals and the use of digital processors to analyze, modify, or extract information from signals. Most signals in nature are analog in form, often meaning that they vary continuously with time, and represent the variations of physical quantities such as sound waves. The signals used in most popular forms of DSP are derived from analog signals which have been sampled at regular intervals and converted into a digital form.

The specific reason for processing a digital signal may be, for example, to remove interference or noise from the signal, to obtain the spectrum of the data, or to transform the signal into a more suitable form. DSP is now used in many areas where analog methods were previously used and in entirely new applications which were difficult or impossible with analog methods. The attraction of DSP comes from key advantages such as the following.

- *Guaranteed accuracy.* Accuracy is only determined by the number of bits used.
- *Perfect reproducibility.* Identical performance from unit to unit is obtained since there are no variations due to component tolerances. For example, using DSP techniques, a digital recording can be copied or reproduced several times over without any degradation in the signal quality.
- No drift in performance with temperature or age.
- Advantage is always taken of the tremendous advances in semiconductor technology to achieve greater reliability, smaller size, lower cost, low power consumption, and higher speed.
- *Greater flexibility.* DSP systems can be programmed and reprogrammed to perform a variety of functions, without modifying the hardware. This is perhaps one of the most important features of DSP.
- *Superior performance.* DSP can be used to perform functions not possible with analog signal processing. For example, linear phase response can be achieved, and complex adaptive filtering algorithms can be implemented using DSP techniques.
- In some cases information may already be in a digital form and DSP offers the only viable option.

DSP is not without disadvantages. However, the significance of these disadvantages is being continually diminished by new technology.

- *Speed and cost*. DSP designs can be expensive, especially when large bandwidth signals are involved. At the present, fast ADCs/DACs (analog-to-digital converters/digital-to-analog converters) either are too expensive or do not have sufficient resolution for wide bandwidth DSP applications. Currently, only specialized ICs can be used to process signals in the megahertz range and these are quite expensive. Furthermore, most DSP devices are still not fast enough and can only process signals of moderate bandwidths. Bandwidths in the 100 MHz range are still processed only by analog methods. Nevertheless, DSP devices are becoming faster and faster.

- *Design time*. Unless you are knowledgeable in DSP techniques and have the necessary resources (software packages and so on), DSP designs can be time consuming and in some cases almost impossible. The acute shortage of suitable engineers in this area is widely recognized. However, the situation is changing as many new graduates now possess some knowledge of digital techniques and commercial companies are beginning to exploit the advantages of DSP in their products.

- *Finite wordlength problems*. In real-time situations, economic considerations often mean that DSP algorithms are implemented using only a limited number of bits. In some DSP systems, if an insufficient number of bits is used to represent variables serious degradation in system performance may result.

1.2   Application areas

DSP is one of the fastest growing fields in modern electronics, being used in any area where information is handled in a digital form or controlled by a digital processor. Application areas include the following:

- Image processing
  - pattern recognition
  - robotic vision
  - image enhancement
  - facsimile
  - satellite weather map
  - animation
- Instrumentation/control
  - spectrum analysis
  - position and rate control
  - noise reduction
  - data compression

- ▪ Speech/audio
  - – speech recognition
  - – speech synthesis
  - – text to speech
  - – digital audio
  - – equalization
- ▪ Military
  - – secure communication
  - – radar processing
  - – sonar processing
  - – missile guidance
- ▪ Telecommunications
  - – echo cancellation
  - – adaptive equalization
  - – ADPCM transcoders
  - – spread spectrum
  - – video conferencing
  - – data communication
- ▪ Biomedical
  - – patient monitoring
  - – scanners
  - – EEG brain mappers
  - – ECG analysis
  - – X-ray storage/enhancement
- ▪ Consumer applications
  - – digital, cellular mobile phones
  - – universal mobile telecommunication system
  - – digital television
  - – digital cameras
  - – Internet phones, music and video
  - – digital answer machines, fax and modems
  - – voice mail systems
  - – interactive entertainment systems
  - – active suspension in cars

A look at the list, which is by no means complete, will confirm the importance of DSP. A testimony to the recognition of the importance of DSP is the continual introduction of powerful DSP devices by semiconductor manufacturers. However, there are insufficient engineers with adequate knowledge in this area. An objective of this book is to provide an understanding of DSP techniques and their implementation, to enable the reader to gain a working knowledge of this important subject.

1.3 Key DSP operations

Several DSP algorithms exist and many more are being invented or discovered. However, all these algorithms, including the most complex, require similar basic operations. It is instructive to examine some of these operations at the outset so as to appreciate the implementational simplicity of DSP. The basic DSP operations are convolution, correlation, filtering, transformations, and modulation. Table 1.1 summarizes these operations and a brief description of each is given below. An important point to note in the table is that all the basic DSP operations require only simple arithmetic operations of multiply, add/subtract, and shifts to carry out. Notice also the similarity between most of the operations.

1.3.1 Convolution

Convolution is one of the most frequently used operations in DSP. For example, it is the basic operation in digital filtering. Given two finite and causal sequences, $x(n)$ and $h(n)$, of lengths N_1 and N_2, respectively, their convolution is defined as

$$y(n) = h(n) \circledast x(n) = \sum_{k=-\infty}^{\infty} h(k)x(n-k) = \sum_{k=0}^{\infty} h(k)x(n-k),$$

$$n = 0, 1, \ldots, (M-1)$$

where the symbol $\circledast$ is used to denote convolution and $M = N_1 + N_2 - 1$. As we shall see in later chapters, DSP device manufacturers have developed signal processors that perform efficiently the multiply–accumulate operations involved in convolution. An example of the linear convolution of the two sequences depicted in Figures 1.1(a) and 1.1(b) is given in Figure 1.1(c). In this example, $h(n)$, $n = 0, 1, 2, \ldots$, can be viewed as the impulse response of a digital system, and $y(n)$ the system's response to the input sequence, $x(n)$. The numerical values for the convolution, that is $y(n)$, were obtained by direct evaluation of Equation 1.1. For example, $y(1)$ is obtained as follows:

$$y(1) = h(0)x(1) + h(1)x(0) + h(2)x(-1) + \ldots + h(12)x(-11)$$

$$= 0 \times 1 + (-0.02) \times 1 + 0 \times 0 + \ldots + 0 \times 0$$

$$= -0.02$$

The significance of convolution is more apparent when it is observed in the frequency domain, and use is made of the fact that convolution in the time domain is equivalent to multiplication in the frequency domain. A more detailed discussion of convolution including its properties and graphical interpretation is given in Chapter 5.

Table 1.1   Summary of key DSP operations.

(1) *Convolution.* Given two finite length sequences, $x(k)$ and $h(k)$, of lengths N_1 and N_2, respectively, their linear convolution is

$$y(n) = h(n) \circledast x(n) = \sum_{k=-\infty}^{\infty} h(k)x(n-k) = \sum_{k=0}^{M-1} h(k)x(n-k), \; n = 0, 1, \ldots, M-1$$

(1.1)

where $M = N_1 + N_2 - 1$.

(2) *Correlation.*

  (a) Given two N-length sequences, $x(k)$ and $y(k)$, with zero means, an estimate of their cross-correlation is given by

$$\rho_{xy}(n) = \frac{r_{xy}(n)}{[r_{xx}(0)r_{yy}(0)]^{1/2}} \quad n = 0, \pm 1, \pm 2, \ldots$$

(1.2)

where $r_{xy}(n)$ is an estimate of the cross-covariance and defined as

$$r_{xy}(n) = \begin{cases} \dfrac{1}{N} \displaystyle\sum_{k=0}^{N-n-1} x(k)y(k+n) & n = 0, 1, 2, \ldots \\[4mm] \dfrac{1}{N} \displaystyle\sum_{k=0}^{N+n-1} x(k-n)y(k) & n = 0, -1, -2, \ldots \end{cases}$$

$$r_{xx}(0) = \frac{1}{N}\sum_{k=0}^{N-1}[x(k)]^2, \; r_{yy}(0) = \frac{1}{N}\sum_{k=0}^{N-1}[y(k)]^2$$

  (b) An estimate of the autocorrelation, $\rho_{xx}(n)$, of an N-length sequence, $x(k)$, with zero mean is given by

$$\rho_{xx}(n) = \frac{r_{xx}(n)}{r_{xx}(0)} \quad n = 0, \pm 1, \pm 2, \ldots$$

(1.3)

where $r_{xx}(n)$ is an estimate of the autocovariance and defined as

$$r_{xx}(n) = \frac{1}{N}\sum_{k=0}^{N-n-1} x(k)x(k+n) \quad n = 0, 1, 2, \ldots$$

(3) *Filtering.* The equation for finite impulse response (FIR) filtering is

$$y(n) = \sum_{k=0}^{N-1} h(k)x(n-k)$$

(1.4)

where $x(k)$ and $y(k)$ are the input and output of the filter, respectively, and $h(k)$, $k = 0, 1, \ldots, N-1$, are the filter coefficients.

(4) *Discrete transform.*

$$X(n) = \sum_{k=0}^{N-1} x(k)W^{kn}, \text{ where } W = \exp(-j2\pi/N)$$

(1.5)

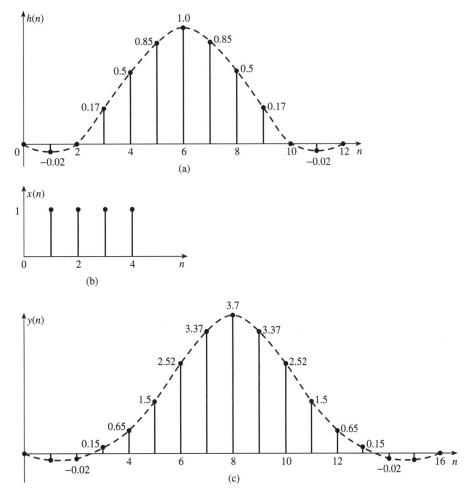

Figure 1.1 An example of the convolution of two sequences. $y(n)$ is the convolution of $h(n)$ and $x(n)$. If $h(n)$ is considered the impulse response of a system, then $y(n)$ is the system's output in response to the input $x(n)$. The values of $y(n)$ above were obtained directly from Equation 1.1.

1.3.2 Correlation

There are two forms of correlations: auto- and cross-correlations.

(1) The cross-correlation function (CCF) is a measure of the similarities or shared properties between two signals. Applications of CCFs include cross-spectral analysis, detection/recovery of signals buried in noise, for example the detection of radar return signals, pattern matching, and delay measurements. CCF is defined in Equation 1.2 in Table 1.1.

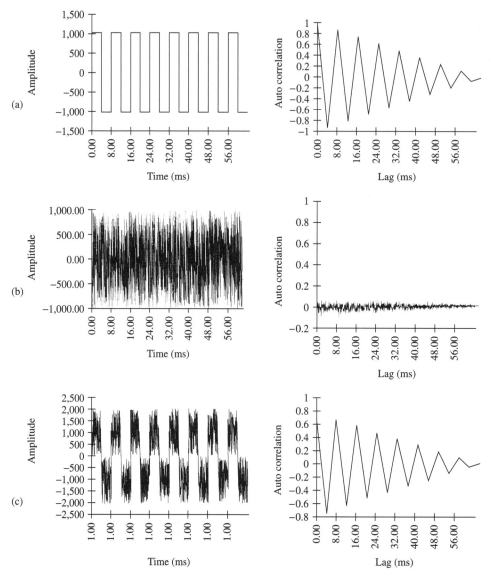

Figure 1.2   Autocorrelations of (a) a periodic signal, (b) noise and (c) periodic signal plus noise. Note that in (c) the periodic nature of the signal buried in noise is still evident, illustrating why autocorrelation is used in detecting hidden periodicity.

(2)   The autocorrelation function (ACF) involves only one signal and provides information about the structure of the signal or its behaviour in the time domain. It is a special form of CCF and is used in similar applications. It is particularly useful in identifying hidden periodicities. The ACF is defined in Equation 1.3 in Table 1.1.

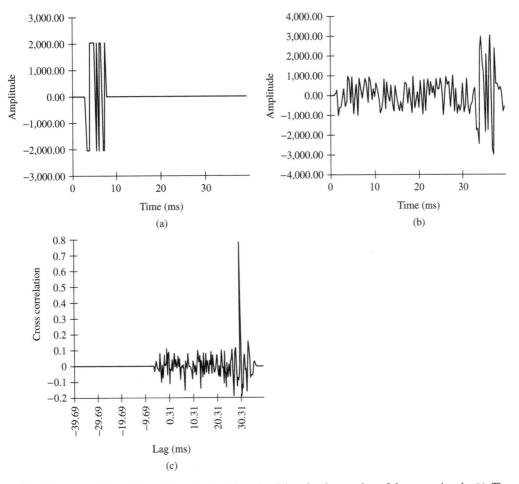

Figure 1.3 Cross-correlation of a random signal, $x(t)$, and a delayed noisy version of the same signal, $y(t)$. The delay between the two signals is the time from the origin to the time where the peak occurred in their cross-correlation in (c).

Examples of CCF and ACF for certain signals are given in Figures 1.2 and 1.3. Notice, for example, that the ACF of the noise-corrupted signal shows clearly that there is a periodic signal buried in noise (Figure 1.2). Figure 1.3 illustrates how to measure delays. The amount of delay introduced by the system is clearly evident from the CCF and can be measured from the time origin to the large peak.

1.3.3 Digital filtering

Digital filtering is one of the most important operations in DSP as will become clear in subsequent chapters. The digital filtering operation for an important class of filters is defined as

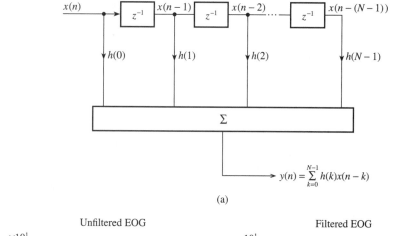

$$y(n) = \sum_{k=0}^{N-1} h(k)x(n-k)$$

(a)

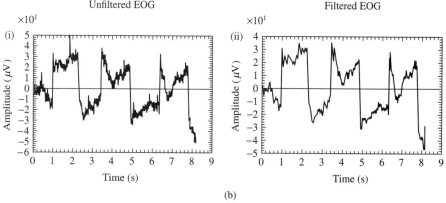

(b)

Figure 1.4   (a) Block diagram representation of the transversal filter. $h(k)$, $k = 0, 1, \ldots,$ $N-1$, are the filter coefficients, and each box containing z^{-1} represents a delay of one sampling period. (b) Digital lowpass filtering of a biomedical signal to remove noise.

$$y(n) = \sum_{k=0}^{N-1} h(k)\,x(n-k)$$

where $h(k)$, $k = 0, 1, \ldots, N-1$, are the coefficients of the filter, and $x(n)$ and $y(n)$, respectively, the input and output of the filter. For a given filter, the values of its coefficients are unique to it and determine the filter's characteristics.

We note that filtering is in fact the convolution of the signal and the filter's impulse response in the time domain, that is $h(k)$. Figure 1.4(a) shows a block diagram representation of the filter defined above. In this form, the filter is popularly known as the transversal filter. In the figure, z^{-1} represents a delay of one sample time.

A common filtering objective is to remove or reduce noise from a wanted signal. For example, Figure 1.4(b) shows the effects of digital lowpass filtering of a certain biomedical signal to remove high frequency distortion. The use of a digital filter in this application was especially important to minimize the distortion of the in-band signal components.

1.3.4 Discrete transformation

Discrete transforms allow the representation of discrete-time signals in the frequency domain or the conversion between time and frequency domain representations. The spectrum of a signal is obtained by decomposing it into its constituent frequency components using a discrete transform. A knowledge of such a spectrum is invaluable in, for example, determining the bandwidth required to transmit the signal. Conversion between time and frequency domains is necessary in many DSP applications. For example, it allows for a more efficient implementation of DSP algorithms, such as those for digital filtering, convolution and correlation.

Many discrete transformations exist, but the discrete Fourier transform (DFT) is the most widely used and is defined as

$$X(k) = \sum_{n=0}^{N-1} x(n)W^{nk}, \text{ where } W = e^{-j2\pi/N}$$

An example of the use of the DFT is given in Figure 1.5. Here, the impulse response of a filter, $h(n)$, $n = 0, 1, \ldots, N-1$, is transformed to give the frequency response of the filter using the DFT. Details of the DFT and its applications are given in Chapters 3, 4 and 11.

1.3.5 Modulation

Digital signals are rarely transmitted over long distances or stored in large quantities in their raw form. The signals are normally modulated to match their frequency characteristics to those of the transmission and/or storage media to minimize signal distortion, to utilize the available bandwidth efficiently, or to ensure that the signals have some desirable properties. Perhaps the two application areas where modulation is extensively employed are telecommunications and digital audio engineering.

The process of modulation often involves varying a property of a high frequency signal, known as the carrier, in sympathy with the signal we wish to transmit or store, called the modulating signal. The three most commonly used digital modulation schemes for transmitting digital data over a bandpass channel (for example a microwave link) are amplitude shift keying (ASK), phase shift keying (PSK), and frequency shift keying (FSK). When digital data is transmitted over an all-digital network, a scheme known as pulse code modulation (PCM) is commonly used (see, for example, Bellamy, 1982). Several other modulation schemes have been developed for digital audio, details of which can be found in Watkinson (1987).

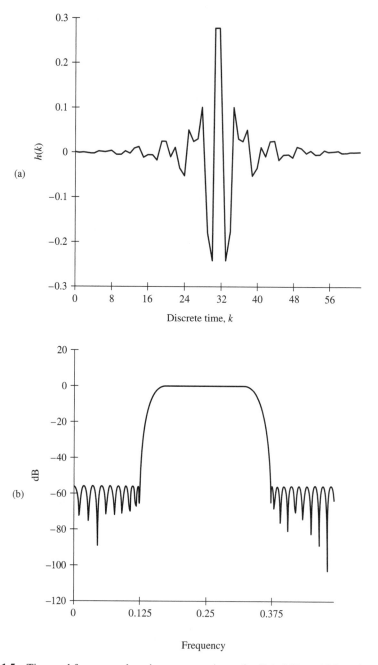

Figure 1.5   Time and frequency domain representations of a digital filter: (a) impulse response; (b) filter spectrum. The filter spectrum was obtained from the discrete transform of $h(n)$, illustrating one of the many uses of the DFT.

1.4 Digital signal processors

DSP systems are characterized by real-time operation, with emphasis on high throughput rate, and the use of algorithms requiring intensive arithmetic operations, notably multiplication and addition or multiply–accumulate. These lead to a heavy flow of data through the processor.

The architectures of standard microprocessors are unsuited to the DSP characteristics and this has led to the development of a new kind of processor whose architecture and instruction set are tailored to DSP operations. The new processors or DSP chips have features that include the following.

- Built-in hardware multiplier(s) to allow fast multiplications. Newer DSP chips incorporate single-cycle multiply–accumulate instructions and some have several multipliers working in parallel.
- Separate busses/memories for program and data – the well-known Harvard architecture, which permits an overlap of instruction fetch and execution.
- Cycle-saving instructions for branching or looping. For example, the following instructions for Texas Instruments' TMS320C25 reduce significantly the number of cycles and program size for a digital filter:

```
RPTK N      ;Repeat the next instruction N times
MACD        ;Move data into memory, multiply and accumulate with delay
```

- Very fast raw speed. For example, the TMS320C25 uses a 40 MHz clock and has a cycle time of 100 ns.
- Use of pipelining which reduces instruction time and increases speed.

Newer DSP chips are faster and more versatile. Some now have floating point arithmetic capabilities, and incorporate features found in standard microprocessors such as serial line, extended memory space, timers, and multilevel interrupts. A more detailed discussion of DSP chips and how to design with them is presented in Chapters 12, 13 and 14.

1.5 Overview of real-world applications of DSP

DSP technology is at the core of many new and emerging digital information products and applications that support the information society. Such products and applications often require the collection, processing, analysis, transmission, display, and/or storage of real-world information, sometimes in real time. The ability of DSP technology to handle real-world signals digitally has made it possible to create affordable,

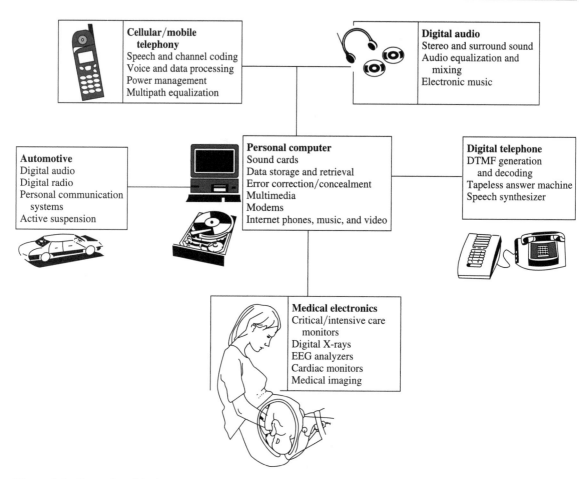

Figure 1.6   Examples of the impact of DSP technology on modern living.

innovative, and high quality products and applications for large consumer markets (e.g. digital cellular mobile phones, digital television, and video games). The impact of DSP is also evident in many other areas, such as medicine and healthcare (e.g. in patient monitors for intensive care, digital X-ray appliances, advanced cardiology and brain mapping systems), digital audio (e.g. CD players, audio mixers, and electronic music), and personal computer systems (e.g. disks for efficient data storage and error correction, modems, sound cards, and video conferencing).

The beneficial impact of DSP technology on the way we live, work and play is illustrated in Figure 1.6. In the next three sections, we will describe some of the innovative applications in more detail, with emphasis on DSP aspects of the applications.

1.6 Audio applications of DSP

1.6.1 Digital audio mixing

The audio mixing system is a prime example of where DSP has been successfully employed to improve audio quality and enhance its functionality.

Audio mixing is used in professional and semi-professional audio applications, e.g. studio recording, broadcasting, sound reinforcement, public address systems and live performance. The mixing console allows the adjustment, mixing and monitoring of the characteristics of multichannel audio signals from a variety of sources to meet the requirements of a particular application.

A digital mixing system includes facilities for audio equalization, audio mixing and post mix processing; see Figure 1.7. The digital audio equalizer (EQ) is a set of digital filters, with adjustable characteristics, that is used to manipulate parts of the frequency bands of the audio inputs to achieve the desired sound (e.g. a boost or cut) similar to a treble and bass control. The equalized audio signals are then mixed using a mix matrix facility which allows any and every audio input to be mixed to every output. After mixing, further signal processing operations may include reverberation and equalization.

A mixing system has interactive controls to adjust mixer parameters, such as fader values (signal level controllers) and EQ control parameters (frequency, Q and gain of the filters), in real time. A challenge in digital audio mixing is how to achieve user control of mixer parameters at a relatively high data rate without audible distortion (Clark *et al.*, 2000). Each time the user adjusts the controls the mixer parameters have to be modified to match the new requirements. Such adjustments can lead to audible distortion that is unacceptable in professional audio mixing systems. As discussed in Clark *et al.* (2000) and in Chapter 12, a careful implementation of the mixing algorithm is critical to achieve professional audio quality.

A typical mixer is shown in Figure 1.8. The mixer features the following:

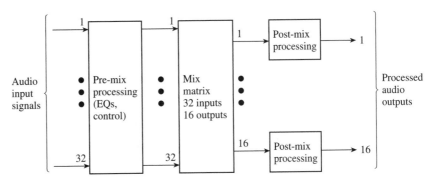

Figure 1.7 A simplified block diagram of a 32-input, 16-output stereo digital mixing system.

Figure 1.8   Controls in a typical 8-channel mixing console (courtesy of Allen & Heath, Cornwall, UK).

- 8 mono inputs (8 microphones or 8 lines);
- 2 pairs of stereo channels (left and right) signal paths;
- master encoder strip for central control, e.g. memory selection, power amp set-up and signal processing.

The digital mixer depicted in Figure 1.8 uses an advanced DSP processor to implement novel audio signal processing algorithms for mixing (e.g. equalization, noise gating, dynamic control), mixing, and post-mixing operations.

1.6.2   Speech synthesis and recognition

1.6.2.1   Speech synthesis

Synthetic speech has in the past been perceived as sounding mechanical. However, advances in semiconductor technology and DSP have made it economically possible to obtain a speech quality that is indistinguishable from human speech.

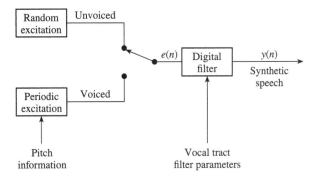

Figure 1.9 Linear predictive coding of speech.

The Speak and Spell (Frantz and Wiggins, 1982) is an example of a successful commercial product with speech output which many readers may be familiar with. It is an electronic learning aid for children and uses the LPC (linear predictive coding) technique, where the actual human speech to be reproduced later is modelled as the response of a time-varying digital filter to a periodic or random excitation signal (Figure 1.9). The periodic excitation is used for voiced sounds (for example vowels), and represents the air flow through the vocal cords as they vibrate. The random excitation is used for unvoiced sounds (for example S, SH), and represents the noise created by forcing air past constrictions in the vocal tract. The filter models the behaviour of the vocal tract. Human speech contains a lot of redundant information. The LPC retains only the relevant information necessary to preserve the characteristics of speech such as intonation, accent, and dialect, allowing minutes of high quality sound to be held in a moderately sized memory.

In the Speak and Spell, use is made of the TMS5100 speech synthesizer chip which incorporates all the components of an LPC model (a digital filter and excitation sources) as well as a decoder and an 8-bit DAC. The synthesizer chip operates in conjunction with a 4-bit microprocessor and two 128-kbit ROMs which together hold a vocabulary of about 300 words and phrases (Figure 1.10). Speech information is stored in the ROMs in frames (representing 25 ms of speech) where each frame is

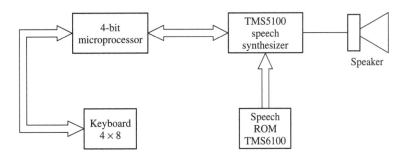

Figure 1.10 A configuration for the Speak and Spell learning aid.

characterized by a set of 10 or 12 LPC parameters. The frame parameters are fed to the synthesizer once every 25 ms and used to update the digital filter coefficients and to select the excitation source and its energy level. The output of the digital filter is converted to analog and applied to the loudspeaker to produce the required sound with a specific pitch, amplitude, and harmonic content. To obtain a smooth transition in speech spectrum, once every 3 ms, the synthesizer updates the LPC parameters by interpolating between the previous and the current frame parameters.

In one mode of operation, the child is asked to spell a word. The child enters the word, one letter at a time via the keyboard. If the spelling is correct, on pressing 'enter', the Speak and Spell responds 'That is right' or 'Correct'. If it is wrong it says 'Wrong, try again'. If the next attempt is wrong, it chides, 'That is incorrect', and adds, 'Correct spelling of . . . is . . .'.

1.6.2.2 Speech recognition

Voice recognition involves inputting of information into a computer using human voice and the computer listening and recognizing the human speech. Voice recognition is still being actively researched as problems posed are more difficult than those of speech synthesis. Thus, successful commercial speech recognition systems are few and far between. The more successful ones are speaker-dependent single-word systems. Such systems operate in one of two modes. In the training mode, the user trains the system to recognize his or her voice by speaking each word to be recognized into a microphone. The system digitizes and creates a template of each word and stores this in its memory. In the recognition mode, each spoken word is again digitized and its template compared with the templates in memory. When a match occurs the word has been recognized and the system informs the user or takes some action. The performance of such systems is affected by speakers not pausing long enough after each word, background noise, and how clearly and carefully the word is spoken. The two important DSP operations in a recognizer are parameter extraction, where distinct patterns are obtained from the spoken word and used to create a template, and pattern matching, where the templates are compared with those stored in memory; see Figure 1.11.

For most people, voice is the most natural form of communication, being faster than writing or typing. Thus, in the office environment, voice systems now exist which allow application programs to be driven by voice commands instead of by keyboard

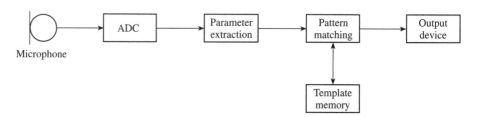

Figure 1.11 Block diagram of a speech recognition system.

entries. Systems which will allow the usual office documents, such as letters and memos, to be created and sent by voice are envisaged. Word recognizers are being incorporated into consumer products, such as voice-operated telephone dialling systems, and are used in voice-activated domestic appliances for disabled people with limited movement. This increases their independence by enabling them to perform simple tasks such as turning on/off lights, radio or TV.

There are of course numerous potential applications of voice recognition. However, it appears that future advances in this area will rely significantly on artificial intelligence (AI) techniques because of the need for machines to understand as well as recognize speech.

1.6.3   The compact disc digital audio system

Most readers are familiar with the undesirable sounds that accompany music reproduced from LP (long play) records when there is damage, a scratch, dirt or fingermarks on the records. The compact disc (CD) system is an advanced audio system which overcomes the disadvantages of the LP record. Table 1.2 compares the important features of the LP and the CD (Bloom, 1985).

In the CD, the information is recorded in digital form as a spiral track that consists of a succession of pits (Figure 1.12) (Carasso et al., 1982). Each bit recorded on the CD occupies an area of only 1 μm^2, that is 10^6 bits per square millimetre, leading to very high density of information on the CD.

A simplified block diagram of the audio signal processing in the CD, during recording, is depicted in Figure 1.13. The analog audio signal in each of the stereo channels is sampled at 44.1 kHz and digitized. Each sample is represented as a 16-bit code, representing a dynamic range of 90 dB. Thus at each sampling instant 32 bits are obtained, 16 bits each from the left and right audio channels. The digital samples are encoded using a two-level Reed–Solomon coding scheme to enable errors to be

Table 1.2   Comparison of features of the LP record and the compact disc (CD).

Features	*LP record*	*Compact disc*
Frequency response	30 Hz to 20 kHz (±3 dB)	20 Hz to 20 kHz (+0.5 to −1 dB)
Dynamic range	70 dB (at 1 kHz)	>90 dB
Signal-to-noise ratio	60 dB	>90 dB
Harmonic distortion	1–2%	0.004%
Separation between stereo channels	25–30 dB	>90 dB
Wow and flutter	0.03%	Not detectable
Effect of dust, scratches and fingermarks	Causes noise	Leads to correctable or concealable errors
Durability	Hf response degrades with playing	Semi-permanent
Stylus life	500–600 h	Semi-permanent
Playing time	40–45 min (both sides)	50–75 min (extendable)

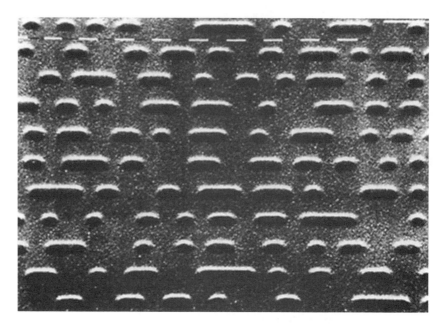

Figure 1.12   Laser-cut pits on a compact disc. Each pit is 0.5 μm wide, 0.8–3.5 μm in length, and has a depth of 0.11 μm. The distance between tracks is 1.6 μm. (Reproduced from *Philips Technical Review*, **40**(6), 1982.)

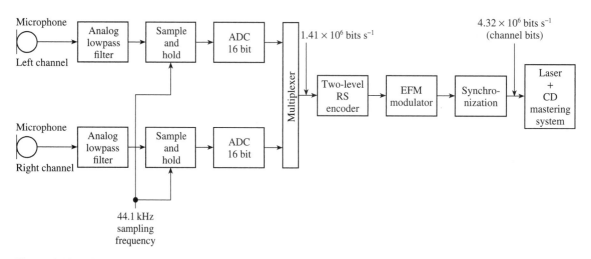

Figure 1.13   Simplified block diagram of the audio signal processing and recording in the compact disc system.

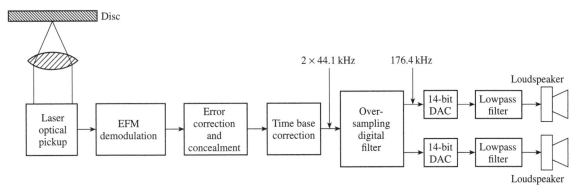

Figure 1.14 Audio signal reproduction in the compact disc system.

detected and corrected or concealed during reproduction of the audio signal. Additional bits are added for control and display information for the listener. The resulting data bit streams are then modulated to translate them into a form more suitable for disc storage. The EFM (eight-to-fourteen modulation) scheme translates each byte in the data stream into a 14-bit code. The resulting channel bit stream, after further processing, is used to control a laser beam, causing the digital information to be recorded onto a light-sensitive layer on a rotating glass disc. A photographic developing process is then used to produce a pattern of pits on the master disc from which users' compact discs are subsequently produced.

In the CD player, during reproduction, the tracks on the disc are optically scanned at a constant velocity of $1.2 \, \text{m s}^{-1}$ while the disc rotates at a speed of between 8 rev s^{-1} and about 3.5 rev s^{-1} to pick up the recorded information (Figure 1.14). The digital signal from the disc is first demodulated, and any errors in the data are detected and, if possible, corrected. The errors may be due to manufacturing defects, damage, fingermarks or dust on the disc. If the errors are not correctable, they are concealed either by replacing the sample in error with a new one obtained by interpolating between adjacent correct samples or, if more than one sample is in error, by zeroing them (muting).

After error correction and/or concealment, the resulting data is a series of 16-bit words, each representing an audio signal sample. These samples could be applied directly to a 16-bit DAC and then analog lowpass filtered. However, this would require analog filters with very tight specifications. In particular, the levels of frequencies above 20 kHz should be reduced by at least 50 dB relative to the maximum audio signal, and the filter should have a linear phase characteristic in the audio band to avoid impairments to the sound waveform. To avoid this, the digital signals are processed further by passing them through a digital filter operating at four times the audio sampling rate of 44.1 kHz. The effect of raising the sampling frequency is to make the output of the DAC smoother, simplifying the analog filtering requirements. It also helps to achieve a 16-bit signal-to-noise ratio performance with a 14-bit DAC. The use of a digital filter allows a linear phase response to be achieved, reduces the chances of intermodulation, and yields a filter with a characteristic that

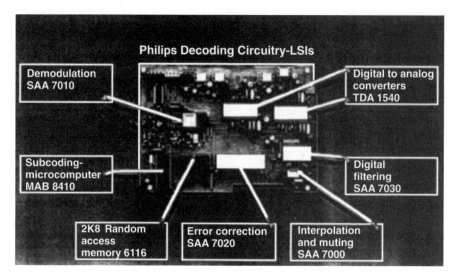

Figure 1.15   The printed circuit board for the decoding circuitry of the Philips CD player. (Reproduced from *Philips Technical Review*, **40**(6), 1982.)

varies with the clock rate, making it insensitive to the speed of rotation of the disc. Figure 1.15 shows the printed circuit board of the decoding circuitry, for the first generation of Philips CD players. The key ICs are clearly identifiable.

Apart from the CD, DSP also plays significant roles in other areas of digital audio engineering. It is widely used in both consumer products and professional audio work such as in the recording studio, transmission and distribution of TV programmes by the broadcasting authorities, film and music industries. Specific uses of DSP in digital audio work, some of which we have already mentioned, include the following:

∎ Use of advanced DSP techniques in encoding, detecting, and correcting or concealing errors caused by dropouts, and in eliminating wow and flutter during replay, ensuring that the limitations imposed by the recording medium (magnetic or optical) no longer dictate the quality attainable in recording and reproduction. Thus the output of a recorder will sound the same for different brands of tapes with similar error rates.

∎ Enhancement of the listening environment and enrichment of the sound. For example, simple digital filter structures have been used to create echoes, natural reverberation, and chorus effects.

∎ Synthesis of sounds that are close imitations of musical instruments and of those no other instruments can produce.

∎ In the creation and use of sound effects, for example gunshots, footsteps, applause, car sounds, punches, in TV commercials, cartoons and motion pictures to enhance the illusion of reality or to add credibility to a scene.

∎ Enhancement of archival recordings or forensic recordings.

1.7   Telecommunication applications of DSP

1.7.1   Digital cellular mobile telephony

1.7.1.1   Introduction

Mobile communications is one of the fastest growing industries in the world today, with the mobile phone now established as an indispensable facility in the information society for keeping in touch. It is estimated that within a few years the number of mobile phone users in the world will exceed the number of fixed line users. This has already been achieved in countries such as Finland. DSP is one of the key technologies that have made the mobile phone revolution possible. DSP is used extensively to process signals and data at radio base stations and in the mobile phones themselves (e.g. for speech coding, multipath equalization, signal strength measurements, voice messaging, error control coding, modulation, and demodulation). DSP chips optimized for wireless communications are now available off the shelf, enabling the mobile communications industry to offer affordable, high quality products for the mass market.

Modern mobile phone systems employ digital cellular radio concepts, but the first generation mobile phone systems which laid the foundation for cellular radio telephony employed analog techniques to process and convey voice signals. Some of the most successful analog mobile phone systems include the Advanced Mobile Phone System (AMPS) used in North America, the Nordic Mobile Telephone (NMT) system developed jointly by Denmark, Finland, Norway and Sweden, and the Total Access Communication Systems (TACS) used in the UK. The earlier systems were incompatible with each other, being mostly country-specific designs, and did not have enough capacity to cope with the rapidly growing demands for mobile communication. Modern digital cellular networks offer greater capacity, coverage, quality, security and reliability. We will use the Global System for Mobile communication (GSM) to illustrate the issues involved in mobile communication.

The GSM was the first all-digital cellular radio telephony system and is now regarded as the *de facto* world standard for mobile digital communication. The GSM was launched in 1992 and by the end of 1998 it was said to have over 130 million users in over 100 countries worldwide. The forecast is that by 2005, the number of mobile phone subscribers will be more than 1 billion. A substantial number of these would be GSM users. Some of the features of the GSM 900 which is used in Europe are summarized in Table 1.3.

1.7.1.2   The architecture of the cellular telephone network

The mobile cellular radio network is a two-way telephony system that allows the mobile phone to send and receive information (speech, data and message) via a radio link. In cellular radio systems, the coverage area is divided up into smaller units known as a cell (radio cell), with each served by a radio base station. The available

Table 1.3   Basic features of the GSM 900.

Parameters for cellular mobile systems	GSM 900 specific parameters
Frequency band	
Mobile transmit/base station receive	890–915 MHz
Mobile receive/base station transmit	935–960 MHz
Duplex frequency separation	45 MHz
Channel frequency spacing	200 kHz
Number of channels	124
Speech rate	13 kbit/s (half rate 5.6 kbit/s)
Overall bit rate	270 kbit/s
Radio transmission method	narrowband CDMA
Typical cell size	300 m–35 km

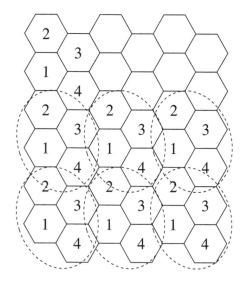

Figure 1.16   Cellular concept showing cell repeat pattern as a result of frequency reuse.

frequency band is split into radio frequencies or channels, and a set of radio frequencies is allocated to each cell. The radio frequencies are reused within the coverage area to make efficient use of the available frequency band, as illustrated in Figure 1.16 (Macario, 1991, 1996). To minimize co-channel interference from other base stations using the same radio frequencies, cells with the same radio frequency set are kept as far apart as possible.

In practice, cell sizes vary according to traffic density. In cities and densely populated areas, cells are small to cater for high traffic levels (micro cells with sizes of less than 300 metres are often used). In rural areas with low traffic levels, cell sizes of

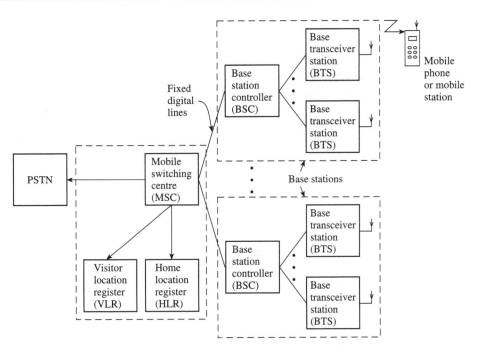

Figure 1.17   A simplified block diagram of a mobile cellular telephone system.

up to 35 kilometres may be used. Such large cells require large transmission power and may contain holes in the coverage area.

Communication takes place between the mobile phone and the network via a radio base station. The cellular network binds the radio base stations together into a single radio telephony system over the coverage area. Each base station has a capacity for a certain number of voice channels and is connected to the mobile switching centre (MSC); see Figure 1.17 (Macario, 1991; Horrocks and Scarr, 1993). Some of the MSCs are themselves connected to the PSTN and provide a gateway between the cellular network and the PSTN. A radio base station consists of base station controllers (for managing the radio channels), each with a set of base transceiver stations. The MSC maintains a record of the locations of mobile phones and manages their mobility.

The two key features of the mobile cellular radio network that enable users to make and receive calls over a wide area are its ability to determine the location of the mobile phone and to hand over radio links with the phone from one base station to another. The network keeps a record of the location of each mobile phone registered with it so that incoming and outgoing calls can be routed to or from the mobile phone wherever it is within the coverage area. This is vital because the mobile phone user is not in a fixed location. When the mobile phone is switched on, it registers with the network and this enables the network to update its location. Each mobile phone has at least two location registers assigned to it in the network – a home location register (HLR) and a visitor location register (VLR). The HLR is in the user's home network and holds management information about the user, such as the services the phone has access to. The VSR contains information about the mobile phone when it is used outside its home network. When a mobile phone registers with a network, it is requested to pass on unique coded information about the phone to the network via the nearest base station. This is then used to obtain the necessary information from the mobile's HLR to authenticate and grant the mobile phone access to the network.

As indicated earlier, the radio link between the mobile phone and the network is via a base station. All the base stations regularly transmit coded control information. A mobile not engaged in a call locks onto the control channel of the nearest base station. As the mobile roams in the coverage area, it will lock onto new base stations as necessary and its location is updated. The signal strength of a mobile engaged in a call is monitored continuously. When the signal strength falls below a certain threshold, the network hands over the radio link to another base station which can provide better reception, if one exists. The ability to hand over the radio link from one base station to another when the signal strength is weak is what allows the mobile user to roam freely within a coverage area and still make and receive calls. The handover process takes a few seconds, although only about a 200–300 ms break in conversation may be noticed. As most mobile phone users know, in some remote areas there is just no other base station nearby that can provide better reception. When the signal strength becomes too weak, the voice channel is cut off and the user is unable to make or receive calls. The signal strength can be inferred by the user by one form of display or another on the mobile phone display screen (e.g. the number of bars).

1.7.1.3   Signal processing aspects

Modern cellular radio telephony systems, such as the GSM, employ digital technology and so DSP is the natural choice for processing and conveying information. In mobile radio telephony, DSP finds use in speech coding, multipath equalization, signal strength and quality measurements, voice messaging, error control coding, modulation, and demodulation (Macario, 1991).

In the GSM, the speech CODEC is based on the Regular Pulse Excitation, Linear Predictive Coding (RPE-LPC). Unlike the PSTN in which speech is encoded at a rate of 64 kbit/s or 32 kbit/s (adaptive ADPCM), in mobile radio telephony, speech is encoded at the relatively low bit rate of 13 kbit/s for efficient use of the radio spectrum. The speech coding algorithm for GSM has been implemented on most popular DSP processors (e.g. Motorola DSP56000, Texas Instrument TMS320C50). The codec has a basic data rate of 13 kbit/s. The codec provides the equivalent of a 13-bit linear ADC and DAC.

In mobile communication, problems associated with multipath propagation are often encountered because of the adverse environment in which the mobile phone operates. At the frequencies used by cellular radiotelephony, transmitted signals are often reflected from high rise buildings, etc. The reflected signals are delayed and arrive at the receiver later than the direct signals, having travelled longer distances. This leads to fluctuations in the amplitude and phase of the combined signal at the receiver, depending on the nature of the multipath and the movement of the mobile phone. The effects of multipath propagation are reduced by the use of digital equalization in the receiver. A known sequence, 26 bits long, is transmitted at regular intervals. At the receiving end, the equalizer uses the training sequence to adjust the coefficients of a digital filter to estimate the characteristics of the radio path and hence to counteract the effects of the multipath on the received data. A knowledge of the transfer function enables the receiver to determine the most likely transmitted bit sequence and hence to demodulate the signal. The GSM equalization algorithm has been implemented on a variety of DSP processors.

Besides speech coding and multipath equalization, DSP techniques have also found use in digital modulation. A variety of techniques are used to modulate the carrier in the GSM system, including multirate DSP techniques. DSP also provides a means of measuring the received signal strength to aid handover and so that the output power levels of the mobile by the base station may be adjusted. In the GSM networks, the mobile phone monitors the signals from surrounding base stations. In cellular radio systems, co-channel interference from cells using the same frequencies is always an issue. This is a function of the transmitted power. In the GSM, the transmitted power may be controlled automatically to keep co-channel interference low to increase the talk and standby times of the battery. Decisions about adjustments to transmitted power are based on the received signal level and quality. DSP techniques are used to analyze the received signals so that these parameters can be assessed.

Mobile radio communication suffers from a variety of errors, e.g. those due to random interference and fading. In the GSM, a convolutional coding technique is used to reduce the effects of these errors.

1.7.2 Set-top box for digital television reception

Large scale digital television broadcasting has recently begun and promises a great deal to the consumer – interactivity, more choice, and better picture and sound quality. Interactivity enables subscribers to play games, access the Internet, shop, have instant replay, etc. TV has re-established itself as an essential part of the information society.

In digital TV, digital information (video, audio, text) may be transmitted to TV sets in homes by satellite (using satellite and existing satellite dishes), cable (cable TV) and terrestrial (using existing TV transmitters and aerials). Most existing TV sets in our homes can only receive analog transmission and so require a digital decoder (set-top box) to receive digital TV; see Figure 1.18. The set-top box converts the digital information into a form suitable for reception by analog TV sets. The latest TV sets have a built-in decoder.

In digital TV, DSP plays a crucial role in the processing, encoding/decoding and modulation/demodulation of video and audio signals from the point of capture to when they are viewed on TV (Benoit, 1997). Without DSP, the remarkable picture and sound quality that we now take for granted would not be possible. For example, DSP is at the heart of the MPEG encoding algorithms which are used to compress video and audio information before transmission (to make good use of the bandwidth). In the set-top box an MPEG decoder is used to recover the information. A key

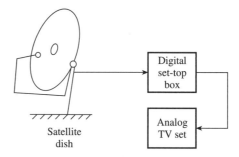

Figure 1.18 An illustration of the use of a set-top box for digital television reception.

component of the MPEG algorithm is the discrete cosine transform which is covered in more detail in Chapter 3.

1.7.3 Adaptive telephone echo cancellation

Echoes arise primarily in communication systems when signals encounter a mismatch in impedance. Figure 1.19 shows a simplified long-distance telephone circuit. The hybrid circuit at the exchange converts the two-wire circuit from the subscriber premises to a four-wire circuit, and provides separate paths for each direction of transmission. This is largely for economic reasons, for example to allow the multiplexing or simultaneous transmission of many calls.

   Ideally, the speech signal originating from customer A travels along the upper transmission path to the hybrid on the right and from there to customer B, while that from B travels along the lower transmission path to A. The hybrid network at each end should ensure that the speech signal from the distant customer is coupled into its two-wire port and none to its output port. However, because of impedance mismatches, the

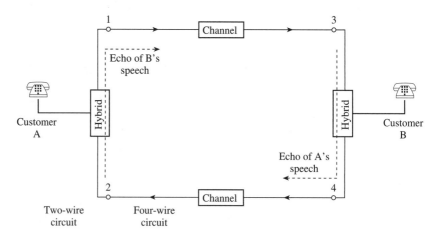

Figure 1.19 Simplified long-distance telephone circuit.

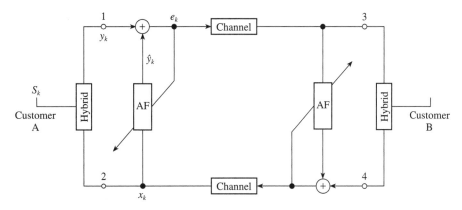

Figure 1.20   Echo cancellation in long-distance voice telephony.

hybrid network allows some of the incoming signals to leak into the output path and return to the talker as an echo. When the telephone call is made over a long distance (for example using geostationary satellites) the echo may be delayed by as much as 540 ms and represents an impairment that can be annoying to users. The impairment increases with distance. To overcome this problem, echo cancellers are installed in the network in pairs, as illustrated in Figure 1.20 (Duttweiller, 1978).

At each end of the communication system (Figure 1.20), the incoming signal, x_k, is applied to both the hybrid and the adaptive filter (AF). The cancellation is achieved by making an estimate of the echo and subtracting it from the return signal, y_k. The estimate of the echo is given by

$$\hat{y}_k = \sum_{i=0}^{N-1} w_{k+1}(i) x_{k-i}$$

where x_k are the samples of the incoming signal from the far-end speaker, and $w_k(i)$, $i = 0, 1, \ldots, N-1$, is the estimate of the impulse response of the echo path at the discrete time k.

1.8   Biomedical applications of DSP

Biomedicine represents an important and very fertile area both for the application of conventional DSP and for the development of new and robust DSP algorithms. Often, medical data are not well behaved and this represents a challenge to the DSP practitioner who must come up with a new way of handling the data. In most cases, medical data are in the audio frequency range. Thus, we find that DSP techniques inspired by problems in biomedicine find use in other areas, such as audio, telecommunication and control, and vice versa.

Many applications of DSP in biomedicine involve signal enhancement and/or the extraction of features of clinical interest. The need for signal enhancement arises from problem of artefacts or signal contamination which is pervasive in biomedicine. Artefacts are caused by sources both external (e.g. mains supply and other medical equipment) and internal (head and body movements, muscle and cardiac activities, and eye movements). Artefacts reduce the clinical usefulness of biomedical signals and make both manual and automatic analysis difficult or, in some cases, impossible because of the similarity between artefacts and signals of clinical interest (Ifeachor *et al.*, 1990).

Signal enhancement tasks are often characterized by the twin problems of low signal levels compared to the interfering noise, and overlap between the signal and noise spectra. Thus, a great deal of care is often required to minimize the distortion of information of clinical interest by the signal enhancement process (Outram *et al.*, 1995, Wu *et al.*, 1997). We describe in the next two sections, two new applications of DSP in biomedicine that involve signal enhancement and/or feature extraction. The basic biomedical signals of interest to us here are physiological signals, in particular electrocardiographic signals – the electrical activity of the heart, and the electroencephalographic signals – the electrical activity of the brain.

1.8.1 Fetal ECG monitoring

The fetal electrocardiogram (ECG) represents the electrical activity of the baby's heart as measured from the body surface (Outram *et al.*, 1995). The fetal heart rate (FHR) is derived from the R-to-R intervals of the ECG (Figure 1.21). A visual analysis of the continuous display of the FHR together with the contraction of the womb (uterine activity), known as the cardiotocogram (CTG), is normally used to assess the condition of the fetus during labour. Difficulties with interpreting the CTG during labour can lead to unnecessary medical intervention (e.g. caesarean section or forceps deliveries), fetal injury or a failure to intervene when needed (Keith *et al.*, 1995).

The correct use of the combined fetal ECG and CTG analysis can significantly reduce unnecessary medical intervention with no adverse effect on neonatal outcome (Westgate *et al.*, 1993). A commercial fetal ECG monitor, the ST Analyser (STAN, Neoventa AB, Sweden) has been developed to exploit this.

A simplified fetal ECG-based monitoring scheme is depicted in Figure 1.22. The ECG is obtained from a scalp electrode to achieve a good signal-to-noise ratio, band limited to about 0.05–100 Hz, and digitized to 12 bits accuracy at a rate of 500 samples per second. The fetal ECG data is processed, for noise reduction and feature extraction, and then analyzed to quantify and display changes in the waveform in association with the CTG.

An important feature of the fetal ECG is the shape of the ST segment. Significant pattern changes in shape associated with stress or distress include persistently rising T wave amplitude, negative T waves and depressed ST segments. Changes in the ST waveform may be quantified as a ratio of the amplitude of the T wave to that of the

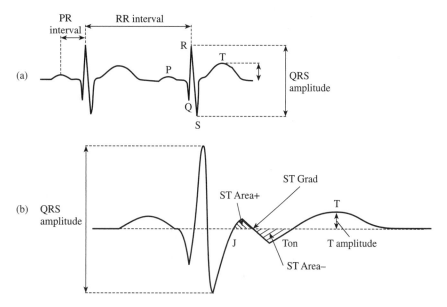

Figure 1.21   Fetal ECG showing key features that are of clinical interest.

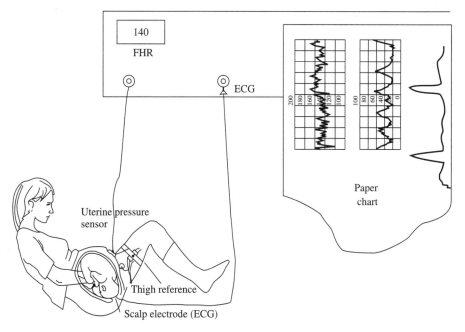

Figure 1.22   Fetal ECG monitoring during labour.

QRS, known as the T/QRS ratio: see Figure 1.21. Other features of interest include the ST area, the variability of the R-to-R intervals, the duration of the P-wave and the width of the QRS complex.

The fetal scalp ECG is susceptible to low frequency noise and other artefacts which may induce false changes in the waveform, such as mains noise, baseline shifts, muscle artefacts, and random noise. Artefacts hinder feature extraction and may lead to inaccurate ECG features and waveform analysis.

A variety of signal processing methods (including curve fitting and multirate digital filtering) are employed to reduce noise and extract key features from the ECG. A summary of a fetal ECG signal processing scheme is shown in Figure 1.23. The first major task is to accurately detect the R-wave. The baseline shift, muscle noise and power line frequencies are then removed from the raw ECG to obtain a waveform suitable for reliable analysis. An example of baseline shift estimation and removal is shown in Figure 1.24.

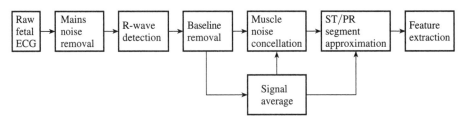

Figure 1.23 Fetal ECG signal processing scheme.

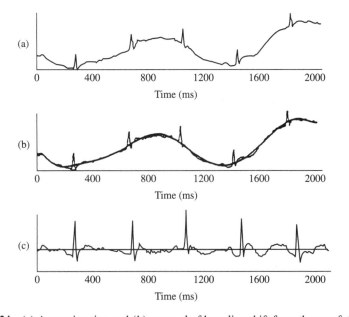

Figure 1.24 (a) Approximation and (b) removal of base line shift from the raw fetal ECG.

It is evident that without DSP, fetal ECG-based monitoring would not be possible. However, although the signal processing problems that previously impeded fetal ECG analysis have been solved, fetal ECG-based monitoring is still limited to a few centres. At present, the analysis and interpretation of changes in the ECG in association with the CTG are carried out by visual inspection. The proper use of the fetal ECG by clinicians requires training, but even with training, it is still possible to miss important ECG patterns without constant expert supervision. A bedside, intelligent monitor for continuous analysis and interpretation of changes in the ECG and CTG to assist clinicians is under development (Keith *et al.*, 1995, Ifeachor and Outram, 1995).

1.8.2 DSP-based closed loop controlled anaesthesia

Evidence of medical applications of DSP abound in the intensive care units of all major hospitals. More advanced techniques are continually being developed for use in anaesthesia. During surgery patients are normally anaesthetized, e.g. by injecting anaesthetic drugs intravenously, so that they do not feel pain and to create a suitable condition for the surgeon to carry out the operation. Anaesthetists aim to deliver just the right amount of drug to induce anaesthesia to the required depth as quickly as possible and to maintain the level until a change is necessary. Injecting too much drug into a patient can lead to complications and other side effects, while inadequate drug leads to intra-operative awareness which may have long-term psychological consequences (Huang *et al.*, 1999). In most cases, the depth of consciousness of the patient is gauged by an experienced anaesthetist by observation of clinical signs. The anaesthetist then makes appropriate changes in the dose of anaesthetic drugs to control anaesthesia. Automated drug delivery using closed loop control techniques offers potential benefits to busy anaesthetists and leads to better patient care and reduced costs. Its use reduces the possibility of excessive dosing and enables the anaesthetist to identify and respond promptly to perturbations which might go unnoticed or be deemed too small to merit manual alterations of drug administration.

However, automated closed loop controlled drug delivery requires a reliable means of monitoring depth of anaesthesia to determine changes to the drug delivery necessary to maintain anaesthesia. State-of-the-art closed-controlled anaesthesia systems use biological signals to measure depth of anaesthesia and as feedback signals to determine the adjustments to be made to drug delivery. In particular, a variety of signal processing methods are used to process the EEG to extract features such as the auditory evoked response (AER) and bispectral index, and to estimate depth of anaesthesia from these (Huang *et al.*, 1999). The EEG is the electrical activity of the brain, measured from electrodes placed on the scalp, and the AER is the electrical response of the brain to external sound stimuli. The AER signals are valuable for assessing the transition from consciousness to unconsciousness, but are difficult to obtain because they are buried in the EEG signals, being several times smaller than the EEG. Signal averaging of responses to successive auditory stimuli is often used to extract them. Thus, AER signals need to be processed, to extract them from the background EEG, and to extract features of clinical interest (e.g. peaks, latencies,

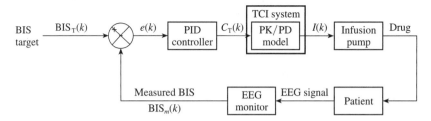

Figure 1.25 DSP-based closed-controlled anaesthesia system.

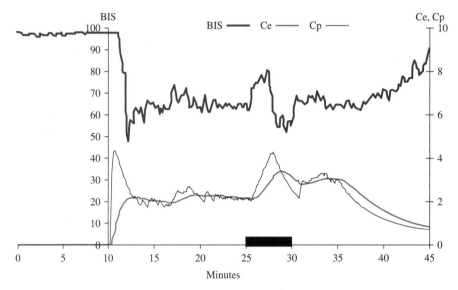

Figure 1.26 An example of the changes in the bispectral indices in a volunteer during closed-controlled anaesthesia.

and shape). The bispectral index is derived from higher order spectrum analysis (Nikias and Raghuveer, 1987) of the EEG. It provides a quantitative measure of the complex changes and interrelationships in the frequency components of the EEG at different levels of consciousness. It is known to be linearly correlated with the blood concentration of propofol, the most common anaesthetic drug.

A simplified block diagram of an EEG-based closed loop controlled anaesthesia (CLAN) system is depicted in Figure 1.25 (Dong *et al.*, 1998, 1999). A key component of the system is the EEG analyzer which is connected to the patient via bifrontal electrodes to collect raw EEG signals. A variety of signal processing methods are used in the analyzer to reduce noise, extract features, analyze changes in the features, and to compute a suitable EEG index. These include wavelet transform, signal averaging, bispectrum analysis and neural networks.

The computed EEG index, such as the bispectral index, gives a measure of the effect of the infused drug on the patient. The computed EEG index then serves as a feedback signal which is compared to a target EEG index to determine changes to be made to the target blood concentration, $C_T(k)$. A commercial EEG analyzer exists for computing the bispectral index and other indices (e.g. A-1000, Aspect Medical System). The pharmacokinetic/pharmacodynamic (PK/PD) model determines the rate, $I(k)$, at which drug is infused into a patient. This rate is a function of the target blood concentration and patient-specific parameters, such as age, weight, and sex. The intravenous drug is normally infused into the patient via a suitable anaesthetic pump under the control of a target controlled infusion program (e.g. the STANPUMP®, S.L. Shafer, Stanford University). An example of the changes in the bispectral indices during anaesthesia is shown in Figure 1.26.

1.9 Summary

In this chapter, the meaning of DSP has been explained, the areas of application have been discussed, and key DSP operations have been identified.

Specific application examples discussed have shown that DSP is already making a significant impact in consumer and professional electronics.

Problems

1.1 State, with justifications, two major advantages and two major disadvantages of DSP compared with analog signal processing system design.

1.2 Describe, with the aid of a block diagram, the audio signal reproduction process in the compact disc player. State and justify four advantages of using DSP techniques in this application.

1.3 What does GSM stand for today? Explain what it stood for in the past.

1.4 Describe, with the aid of a block diagram, how mobile cellular radio telephony works.

1.5 Describe, briefly, the role of DSP in digital cellular radio telephony. Why is DSP the natural choice for processing voice information in a digital radio-telephony system? What are the problems in using DSP in mobile radio telephony?

1.6 Explain what is meant by a radio cell. Sketch a three-cell repeat pattern.

References

Advanced Mobile Phone Service. *Bell System Technical Journal*, **58**(1), January 1979.

Bellamy J.C. (1982) *Digital Telephony*. New York: Wiley.

Benoit H. (1997) *Digital Television: MPEG-1, MPEG-2 and Principles of the DVB System*. London: Arnold.

Bloom P.J. (1985) High-quality digital audio audio in the entertainment industry: an overview of achievements and challenges. *IEEE ASSP Magazine*, October, 2–25.

Carasso M.G., Peek J.B.H. and Sinjou J.P. (1982) The compact disc digital audio system. *Philips Technical Rev.*, **40**(6), 151–6.

Clark R.J., Ifeachor E.C., Rogers G.M. and Van Eetvelt P.W.J. (2000) Techniques for generating digital equaliser coefficients. *Journal of Audio Engineering Society*, **48**(4), 281–98.

Dong C., Kehoe J., Henry J., Ifeachor E.C., Reeve C.D. and Sneyd J.R. (1998) Closed loop computer controlled sedation with propofol. *British Journal of Anaesthesia*, **81**, 631P.

Dong C., Reeve C.D., Sneyd J.R. and Ifeachor E.C. (1999) Closed-loop control of intravenous drug infusion. *IEE Proc. Sci. Meas. Technol.* (submitted).

Duttweiler D.L. (1978) A twelve-channel digital echo canceler. *IEEE Trans. Communications*, **26**, 647–53.

ETSI/GSM Recommendations: ETSI, BP 152, F-06561 Valbonne CEDEX, France.

Frantz G.A. and Wiggins R.H. (1982) Design case history: Speak and Spell learns to talk. *IEEE Spectrum*, February, 45–9.

Horrocks R.J. and Scarr R.W.A. (1993) *Future Trends in Telecommunications*. New York: Wiley.

Huang J.W., Lu Y., Nayak A. and Roy R.J. (1999) Depth of anaesthesia estimation and control. *IEEE Trans. Biomed. Eng.*, **46**(1), 71–81.

Ifeachor E.C., Hellyar M.T., Mapps D.J. and Allen E.M. (1990) Knowledge-based enhancement of EEG signals. *Proceedings of IEEE Radar and Signal Processing*, **37**, 302–10.

Ifeachor E.C. and Outram N.J. (1995) A fuzzy expert system to assist in the management of labour. *Proc. International ICSC Symposium on Fuzzy Logic*, ICSC, C97–102, Zürich, Switzerland.

Keith R.D.F., Beckley S., Garibaldi J.M., Westgate J., Ifeachor E.C. and Greene K.R. (1995) A multicentre comparison study of 17 experts and an intelligent computer system for managing labour using the cardiotocogram. *Brit. J. Obstet. Gynaecol.*, **102**, 688–700.

Macario R.C.V. (ed.) (1991) *Personal and Mobile Radio Systems* (Chapters 4, 9, 13 and 14). Peter Peregrinus Ltd for the Institution of Electrical Engineers.

Macario R.C.V. (ed.) (1996) *Modern Personal Radio Systems* (Chapters 3, 8, 11 and 12). The Institution of Electrical Engineers, London.

Nikias C.L. and Raghuveer M.R. (1987) Bispectrum estimation: a digital signal processing framework. *Proc. IEEE*, **75**, July, 869–91.

Outram N.J., Ifeachor E.C., Van Eetvelt P.W.J. and Curnow J.S.H. (1995) Techniques for optimal enhancement and feature extraction of fetal electrocardiogram. *IEE Proc. Sci. Meas. Technol.*, **142**(6), November, 482–9.

Watkinson J. (1994) *The Art of Digital Audio*. Second edition. Oxford: Butterworth-Heinemann.

Westgate J., Harris M., Curnow J. and Greene K.R. (1993) Plymouth randomised trial of the cardiotocogram only versus ST waveform plus cardiotocogram for intrapartum monitoring in 2400 cases. *Am. J. Obstet. Gynecol.*, **169**, 1151–60.

Wu J., Ifeachor E.C., Allen E.M., Wimalarantna S.K. and Hudson N.R. (1997) Intelligent artefact identification in electroencephalography signal processing. *IEEE Proceedings Science, Measurement and Technology*, **144**(5), 193–201.

Bibliography

Mitra S.K. (1998) *Digital Signal Processing – A Computer-Based Approach*. New York: McGraw-Hill.

Mulgrew B., Grant P. and Thompson J. (1999) *Digital Signal Processing – Concepts and Applications*. Basingstoke: Macmillan.

Oppenheim A.V. and Schafer R.W. (1975) *Digital Signal Processing*. Englewood Cliffs NJ: Prentice-Hall.

Oppenheim A.V. and Schafer R.W. (1989) *Discrete-Time Signal Processing*. Englewood Cliffs NJ: Prentice-Hall.

Orfanidis S.J. (1996) *Introduction to Signal Processing*. Englewood Cliffs NJ: Prentice-Hall.

Papamichalis P. (1987) *Practical Approaches to Speech Coding*. Englewood Cliffs NJ: Prentice-Hall.

Rabiner L.R. and Gold B. (1975) *Theory and Applications of Digital Signal Processing*. Englewood Cliffs NJ: Prentice-Hall.

Analog I/O interface for real-time DSP systems

2

In many real-world applications, the signals are in analog form, but DSP operates on digital data. Thus, to interface DSP systems to the real world, we need an analog input/output (I/O) interface to allow the conversion between analog and digital formats.

In real-time DSP, the analog I/O interface is a weak link because it introduces irreducible errors and imposes speed constraints (see later). A good understanding of the design issues at the analog I/O interface is a key prerequisite for the successful design of real-time DSP systems, with analog input and/or output, and indeed for other areas of DSP (e.g. multirate processing). Many of the design issues are covered in this chapter. Specifically, by the end of this chapter the reader should:

(1)   understand the fundamentals of the design of the analog I/O interface for real-time DSP (e.g. lowpass and bandpass sampling theorems and how to apply them to practical problems and the nature of the errors that arise in the analog I/O interface);

(2)   be able to specify, analyze and determine the basic parameters of the analog I/O interface (e.g. sampling frequency and aliasing error level);

(3)   understand the basic principles of oversampling at the analog I/O interface (e.g. oversampling and noise shaping, and design and analysis of simple over-sampling converters).

We have used applications drawn from audio, telecom and biomedicine to illustrate the principles.

2.1   Typical real-time DSP systems

The block diagram of a typical DSP system operating in real time is depicted in Figure 2.1. The analog input filter is used to bandlimit the analog input signal prior to digitization to reduce aliasing (see later). The ADC converts the analog input signal into a digital form. For wide bandwidth signals or when a slow ADC is used, it is necessary to precede the ADC with a sample and hold circuit, although newer ADCs now have built-in sample and hold circuits. After digital processing in the processor, the DAC converts the processed signal back into analog form. The output filter smooths out the outputs of the DAC and removes unwanted high frequency components.

The heart of the system in Figure 2.1 is the digital processor which may be based on a general purpose microprocessor such as the Motorola MC68000, a digital signal processor chip such as the Texas Instruments TMS320C50, or the Motorola DSP56000 or some other piece of hardware. The digital processor may implement one of several DSP algorithms, for example digital filtering, mapping the input, $x(n)$, into the output, $y(n)$.

Signal processing using a digital processor implies that the input signal must be in a digital form before it can be processed. In some real-time applications the data may already be in a digital form or does not need to be converted to an analog signal. For example, after processing, the signal may be stored in a computer memory for later use or it may be displayed graphically on a display unit. In other applications it may

Figure 2.1   Block diagram of a simplified, generalized real-time digital signal processing system. In some applications, the input filter and the ADC, or the DAC and the output filter, will not be necessary.

be required to generate signals digitally. Examples of this are in speech synthesis, digital frequency synthesis and pseudorandom binary sequence generators. Much of the discussion in this book assumes that the signal is in a digital form or has been adequately digitized as described in the next section.

2.2 Analog-to-digital conversion process

As was mentioned above, before any DSP algorithm can be performed, the signal must be in a digital form. Most signals in nature are in analog form, necessitating an analog-to-digital conversion process, which involves the following steps.

■ The (bandlimited) signal is first sampled, converting the analog signal into a discrete-time continuous amplitude signal.

■ The amplitude of each signal sample is quantized into one of 2^B levels, where B is the number of bits used to represent a sample in the ADC.

■ The discrete amplitude levels are represented or encoded into distinct binary words each of length B bits.

The process is depicted in Figure 2.2. Three distinct types of signals can be identified in the figure.

■ *The analog input signal.* This signal is continuous in both time and amplitude.

■ *The sampled signal.* This signal is continuous in amplitude but defined only at discrete points in time. Thus the signal is zero except at time $t = nT$ (the sampling instants).

■ *The digital signal, $x(n)$ ($n = 0, 1, \ldots$).* This signal exists only at discrete points in time and at each time point can only have one of 2^B values (discrete-time discrete-value signal). This is the type of signal that is of concern to us in this book.

Note that the discrete-time (that is, sampled) signal and the digital signal can each be represented as a sequence of numbers, $x(nT)$, or simply $x(n)$ ($n = 0, 1, 2, \ldots$). Let us now look more closely at the steps in digitizing a signal.

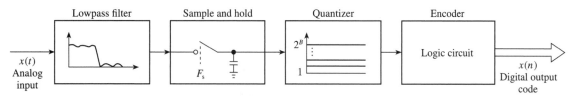

Figure 2.2 A pictorial representation of the analog-to-digital conversion process.

2.3   Sampling – lowpass and bandpass signals

Sampling is the acquisition of a continuous signal (for example, analog) at discrete-time intervals and is a fundamental concept in real-time signal processing. An example of a sampled analog signal is shown in Figure 2.3. Note that after sampling, in this ideal case, the analog signal is now represented only at discrete times, with the values of the samples equal to those of the original analog signal at the discrete times.

In this chapter, we shall give an intuitive presentation of the sampling theorem, which specifies the rate at which an analog signal should be sampled to ensure that all the relevant information contained in the signal is captured or retained by sampling. In practice, we often encounter two forms of sampling – sampling for lowpass signals and sampling for bandpass signals. As you will see later, sampling for bandpass signals can be viewed as a special case of the more general lowpass sampling.

2.3.1   Sampling lowpass signals

2.3.1.1   The sampling theorem

If the highest frequency component in a signal is f_{max}, then the signal should be sampled at the rate of at least $2f_{max}$ for the samples to describe the signal completely:

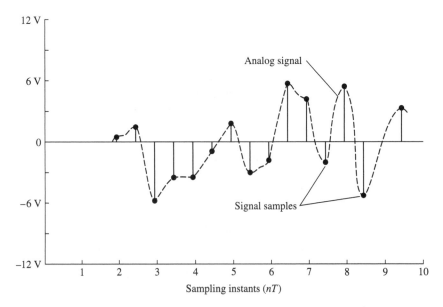

Figure 2.3   An example of a sampled signal (ideal sampling). The values of the signal samples are equal to those of the original analog signal at the sampling instants.

$$F_s \geq 2f_{max} \tag{2.1}$$

where F_s is the sampling frequency or rate. Thus, if the maximum frequency component in an analog signal is 4 kHz, then to preserve or capture all the information in the signal it should be sampled at 8 kHz or more. Sampling at less than the rate specified by the sampling theorem leads to a folding over or 'aliasing' of 'image' frequencies into the desired frequency band so that the original signal cannot be recovered if we were to convert the sampled data back to analog. An important point to remember is that a signal often has significant energy outside the highest frequency of interest and/or contains noise, which invariably has a wide bandwidth. For example, in telephony the highest frequency of interest is about 3.4 kHz but speech signals may extend beyond 10 kHz. Thus, the sampling theorem will be violated if we do not remove the signal or noise outside the band of interest. In practice, this is achieved by first passing the signal through an analog anti-aliasing filter.

2.3.1.2 Aliasing and spectra of sampled signals

Suppose we sampled a time domain signal at intervals of T (seconds) (that is, a sampling frequency of $1/T$ (hertz)). It is seen, Figure 2.4, that another frequency component with the same set of samples as the original signal exists. Thus, the frequency component can be mistaken for the lower frequency component and this is what aliasing is about. In practice, it is more instructive, from the point of view of analyzing the effects or finding the solution to the problem of aliasing, to examine aliasing in the frequency domain.

Figure 2.5 shows the sampling process, which can be regarded as the multiplication of the analog signal $x(t)$ by a sampling function, $p(t)$. $p(t)$ consists of pulses of unit amplitudes, width dt (which is infinitesimally small) and period T. The spectra of $x(t)$, $p(t)$, and their product are shown in Figure 2.5. Note that $X'(f)$ is the convolution of $X(f)$ and $P(f)$ – multiplication in the time domain is equivalent to convolution in the frequency domain.

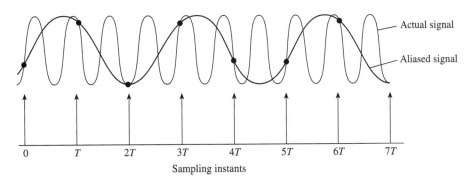

Figure 2.4 An example of aliasing in the time domain. Notice that the two signals have the same values at the sampling instants, although their frequencies are different.

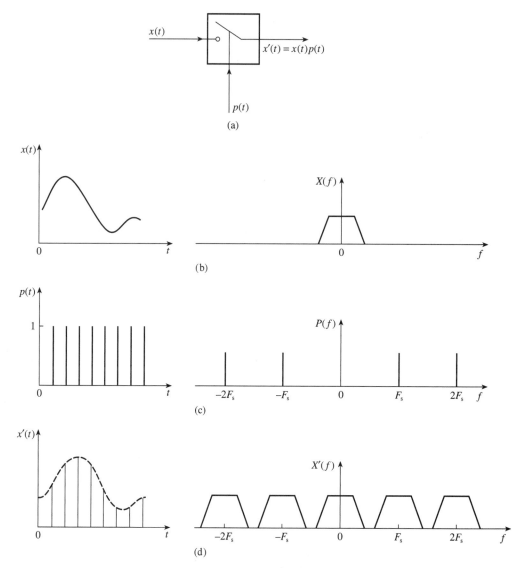

Figure 2.5 Time and frequency domain representations of the sampling process. The spectra of the signal (b) before and (d) after sampling should be compared. (d) Note the changes in the sampled signal and in particular that the spectrum of the sampled signal repeats at multiples of the sampling frequency, F_s.

The following points should be noted for the sampled signal in Figure 2.5(d).

■ The spectrum is the same as the original analog spectrum, but repeats at multiples of the sampling frequency, F_s. The higher order components which are centred on the multiples of F_s are referred to as image frequencies.

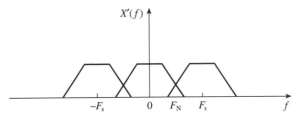

Figure 2.6 Spectrum of an undersampled signal, showing aliasing (foldover region). Signals in the foldover region are not recoverable. F_N is equal to half the sampling frequency and it is often called the Nyquist frequency. To recover all the components of a signal we must sample at a rate greater than (or equal to) twice the highest frequency component.

- If the sampling frequency, F_s, is not sufficiently high the image frequencies centred on F_s, for example, will fold over or alias into the base band frequencies (Figure 2.6). In this case, the information of the desired signal is indistinguishable from its image in the foldover region.

- The overlap or aliasing occurs about the point F_N, that is half the sampling frequency point. This frequency point is variously called the folding frequency, Nyquist frequency, and so on.

In practice, aliasing is always present because of noise and the existence of signal energy outside the band of interest. The problem then is deciding the level of aliasing that is acceptable and then designing a suitable anti-aliasing filter and choosing an appropriate sampling frequency to achieve this.

2.3.1.3 Anti-aliasing filtering

To reduce the effects of aliasing sharp cutoff anti-aliasing filters are normally used to bandlimit the signal and/or the sampling frequency is increased so as to widen the separation between the signal and image spectra. Ideally, the anti-aliasing filter should remove all frequency components above the foldover frequency, that is it should have a frequency response similar to that depicted in Figure 2.7(a). A more practical response is given in Figure 2.7(b), where f_c and f_s are the cutoff and stopband frequencies, respectively. We note from Figures 2.7(b) and 2.7(c) that the practical response introduces an amplitude distortion into the signal as it is not flat in the passband. Also, the signal components greater than f_s will be attenuated by A_{min}, but those between f_c and f_s, the transition width, will have their amplitudes reduced monotonically.

The anti-aliasing filter should provide sufficient attenuation at frequencies above the Nyquist frequency. Because of the non-ideal response of practical filters, the effective Nyquist frequency is taken as f_s (the stopband edge frequency). In specifying the anti-aliasing filter it is useful to take the ADC resolution requirements into account. Thus a filter would be designed to attenuate the frequencies above the

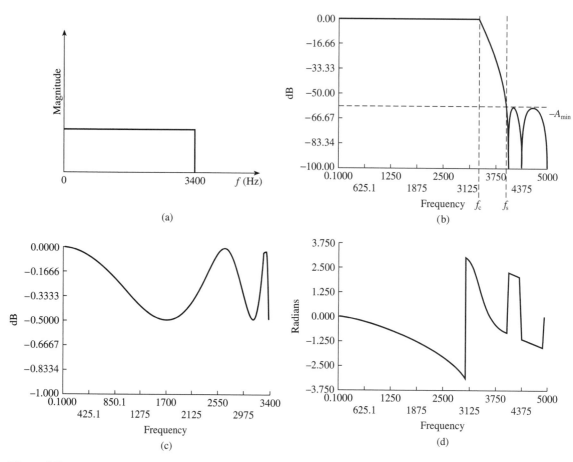

Figure 2.7   Ideal and practical frequency responses of an anti-aliasing filter, showing the errors introduced by the practical responses: (a) ideal response; (b) practical amplitude response; (c) practical passband amplitude response; (d) practical phase response. Compare, for example, the flat passband response in (a) and the practical passband response in (c): the ripples in the practical response would introduce an amplitude distortion of the in-band signal components.

Nyquist frequency to a level not detectable by the ADC, for example to less than the quantization noise level (see later). Thus, for a system using a B-bit linear ADC, the minimum stopband attenuation of the filter would typically be

$$A_{min} = 20 \log (\sqrt{1.5} \times 2^B) \tag{2.2}$$

where B is the number of bits in the ADC (see Example 2.3 for further details). Table 2.1 gives the values for A_{min} for various values of B.

The use of an analog filter at the front end of a DSP system also introduces other constraints, such as phase distortion. Figure 2.7(d) depicts the phase response of the

Table 2.1 Estimates of minimum lowpass filter stopband attenuation, A_{min}, for various ADC resolutions, B.

B	A_{min}(dB)
8	50
10	62
12	74
16	98

anti-aliasing filter whose amplitude response is given in Figure 2.7(c), and it shows that the phase response is not linear with frequency, so that the components of the desired signal will be shifted in phase or delayed by amounts which are not in proportion to their frequencies. The amount of distortion depends on the characteristics of the filter including how steep its roll-off is. In many cases, the steeper the roll-off (that is, the narrower the transition width) the worse the phase distortion introduced by the filters, and the more difficult it is to achieve a good match in amplitude and group delay between channels in a multichannel system. However, the use of steep roll-off filters allows the use of a low sample rate and a slower, cheaper ADC.

The trend in real-time signal processing is to use a high sampling frequency, that is to oversample the signal, even though it may be at the risk of using a fast and expensive ADC. The reasons for this are many-fold. Firstly, it leads to the use of simple anti-aliasing filters which minimizes the phase distortion and, for a multi-channel system, to lower costs. Secondly, oversampling combined with additional digital signal processing leads to an improved signal-to-noise ratio (see Chapter 9). For a DSP system with an analog front end to be usable in different applications the filter cutoff frequency needs to be variable. Programmable analog filters are coming into use, for example MF10, but their performance is not completely satisfactory and, for multichannel systems, may be expensive. Oversampling of the analog signal permits the use of digital sample rate conversion techniques (see Chapter 9) to achieve easily the requirements of variable cutoff frequency.

2.3.1.4   Illustrative examples on choice of sampling frequency and aliasing control

Factors that influence the choice of the sampling frequency include the following:

- the frequency content of the input signals;
- anti-aliasing filtering requirements;
- acceptable aliasing error level;
- resolution of the ADC;
- storage requirements.

We have already discussed the impact of the first two on the sampling frequency. There are a number of ways of specifying acceptable aliasing error level. In most cases, the interrelationships between the sampling frequency, aliasing error level and filter parameters are exploited. For example, we can specify an acceptable aliasing error, for a given anti-aliasing filter characteristic, and then determine the sampling frequency necessary to achieve this. Alternatively, for a given sampling frequency, we can work out the minimum stopband attenuation to be provided by the anti-aliasing filter to give the specified aliasing error level. In practice, the ADC resolution should be taken into account as it establishes the system noise floor.

The following examples illustrate the issues involved in the choice of the sampling frequency and aliasing control.

Example 2.1

Illustrating the effects of sampling and the interrelationship between aliasing error level and the sampling frequency   The front end of a real-time DSP system is depicted in Figure 2.8. Assume a wideband input signal.

(a)   Sketch the spectrum of the signal before sampling (point A) and after sampling (point B) between the range $\pm F_s/2$.

(b)   Determine the signal and aliasing error levels at 10 kHz and the Nyquist frequency (i.e. 20 kHz).

(c)   Determine the minimum sampling frequency, F_s (min), to give a signal-to-aliasing error level of 10 : 1 at 10 kHz. State any other assumptions made.

Solution

(a)   Sketches of the spectrum of the signal before and after sampling are shown in Figure 2.9. We note that the shape of each spectrum component is governed by the equation of the response of the Butterworth filter, i.e.

$$|H(f)| = \frac{1}{\sqrt{1 + \left(\dfrac{f}{f_c}\right)^8}}$$

(b)   Signal spectrum at the output of the filter is equal to the product of the signal spectrum and the response of the filter, i.e. $X(f)|H(f)|$. For a wideband input, the spectrum $X(f)$ is essentially flat. If we assume that both $X(f)$ and $H(f)$ have a maximum value of 1 (i.e. normalized), then the signal levels before sampling (at the output of the filter) and after sampling (at the output of the sample and hold) are governed by the shape of the analog filter.

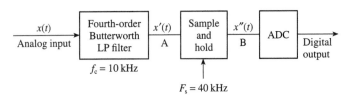

Figure 2.8   The front end of a real-time DSP system.

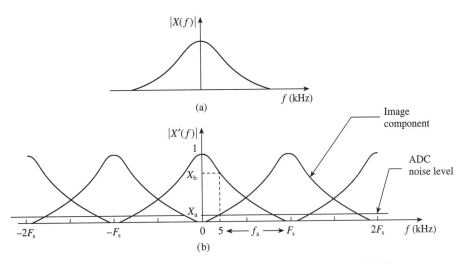

Figure 2.9   Spectrum of sampled signal showing error due to aliasing and ADC quantization.

Thus, at 10 kHz, with $f_c = 10$ kHz, the normalized signal level (from the equation above) is simply 0.707 (i.e. $1/\sqrt{2}$). The aliasing error level (from Figure 2.9(b)) is given by

$$\text{Aliasing level, } X_a = \cfrac{1}{\sqrt{1 + \left(\dfrac{30}{10}\right)^8}} = 0.012$$

The Nyquist frequency is 20 kHz (i.e. half the sampling frequency). This is the crossover point in Figure 2.9(b) and so the signal and aliasing error levels are the same. Signal and aliasing levels at 20 kHz are each (using the Butterworth equation, with $f = 20$ kHz and $f_c = 10$ kHz) equal to 0.062.

(c)   At 10 kHz, the signal level is 0.707. A ratio of 10 : 1 between the signal level and aliasing level implies an aliasing level of 0.0707. The image component that causes aliasing is governed by the Butterworth equation.
   Thus, from

$$\cfrac{1}{\sqrt{1 + \left(\dfrac{f}{10}\right)^8}} = 0.0707$$

we find that $f = 19.39$ kHz.
   This corresponds to the aliasing frequency at 10 kHz, i.e. f_a in Figure 2.9(b) above. Thus, the sampling frequency, $F_s = f_a + 10 = 29.39$ kHz.

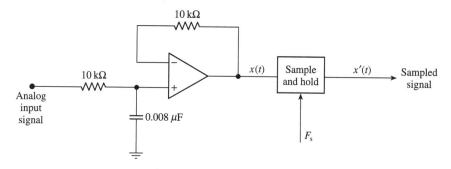

Figure 2.10 Front end of a simple data acquisition system. The simple active filter is used to bandlimit the signal before it is sampled at the rate of F_s.

Example 2.2	Figure 2.10 depicts the front end of a simple data acquisition system. Determine the minimum sampling frequency, F_s, to give an aliasing error of less than 2% of the signal level in the passband.

Solution

The amplitude response of the active filter is given by

$$|H(f)| = \frac{1}{[1 + (f/f_c)^2]^{1/2}} \quad \text{where } f_c = 1/2\pi RC = 2 \text{ kHz}$$

The spectrum of the bandlimited input signal and the sampled signal are depicted in Figure 2.11, where we have assumed a wideband analog input.

We note from the figure that the spectrum of the sampled signal repeats at multiples of the sampling frequency. The foldover of the image frequencies into the desired frequency band (0 to 2 kHz) is aliasing.

At 2 kHz, the signal level, $X_b = 0.7071$, so that

desired aliasing level $< 0.7071 \times 2/100 = 0.014\,14$

Thus,

$$0.014\,14 < \frac{1}{[1 + (f_a/2)^2]^{1/2}}$$

where f_a is the aliasing frequency. Solving for f_a, we have $f_a < 141.4$ kHz. Thus,

$$F_s(\text{min}) > f_c + f_a = 2 \text{ kHz} + 141.4 \text{ kHz} = 143.4 \text{ kHz}$$

To meet the specification and to account for the effects of the image frequencies centred on $2F_s$, $3F_s$ and so on (ignored above) then $F_s(\text{min}) > 143.4$ kHz. Let $F_s(\text{min}) = 150$ kHz.

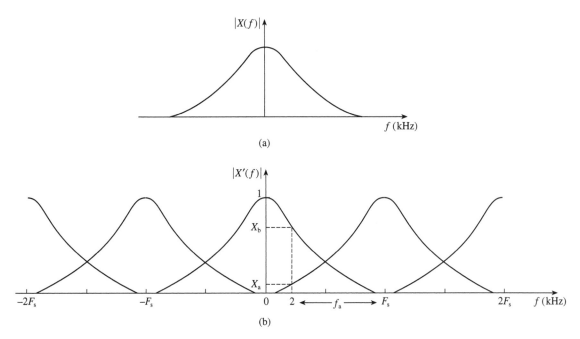

(a)

(b)

Figure 2.11   Spectrum of signal (a) at the output of the analog filter and (b) after sampling (Example 2.2).

Example 2.3	*Illustrating the interrelationship between the ADC resolution and filter parameters*

Figure 2.12 depicts a real-time DSP system. Assuming that the band of interest extends from 0 to 4 kHz and that a 12-bit, *bipolar*, ADC is used, estimate:

(1)   the minimum stopband attenuation, A_{min}, for the anti-aliasing filter,

(2)   minimum sampling frequency, F_s, and

(3)   the level of the aliasing error relative to signal level in the passband for the estimated A_{min} and F_s.

Sketch and label the spectrum of the signal at the output of the analog filter, assuming a wideband signal at the input, and that of the signal after sampling.

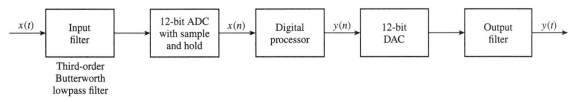

Figure 2.12   Real-time digital signal processing system.

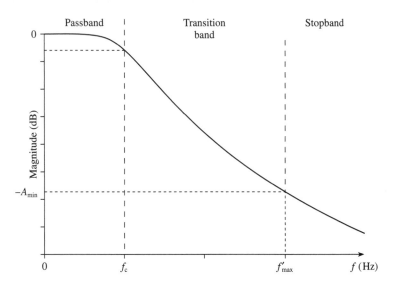

Figure 2.13   A typical magnitude frequency response for a practical anti-aliasing filter.

Solution   To satisfy the sampling theorem, the anti-aliasing filter bandlimits the input signal spectrum such that frequency components above the Nyquist frequency are removed to avoid aliasing.

In practice, because we cannot have an ideal filter, the anti-aliasing filter is normally required to attenuate frequency components above the Nyquist frequency to less than the rms quantization noise level for the ADC so that they are not detectable by the ADC.

A typical magnitude frequency response for the anti-aliasing filter is depicted in Figure 2.13 and shows the pass, transition and stopbands. The anti-aliasing filter is designed to attenuate the levels of the frequency components in the stopband, i.e. frequencies $\geq f'_{max}$, to less than the rms quantization noise level for the ADC.

Thus, the effective Nyquist frequency is f'_{max} and the effective sampling rate can be defined as

$$F_s \geq 2 f'_{max} \tag{2.3}$$

Now, the quantization step size, q, is given by

$$q = \frac{V_{fs}}{2^B - 1} \approx \frac{V_{fs}}{2^B}$$

where B is the number of bits of the ADC and V_{fs} is the full-scale input range. Thus

$$V_{fs} \approx q \times 2^B$$

The rms quantization noise level is given by

$$\sqrt{\frac{q^2}{12}} = \frac{q}{2\sqrt{3}}$$

If we assume a sine wave input, for simplicity, with a peak amplitude A (which just fills the ADC range), then the maximum passband signal level is

$$V_{fs} = 2A = q \times 2^B$$

Therefore

$$A = \frac{q \times 2^B}{2}$$

The ratio of the maximum passband signal level to the stopband signal level gives a measure of the maximum stopband attenuation of the filter.

$$\frac{\text{maximum passband signal level}}{\text{stopband signal level}} = \frac{q \times 2^B/2\sqrt{2}}{q/2\sqrt{3}}$$

$$= \sqrt{1.5} \times 2^B$$

(1) Thus, for the DSP system the minimum stopband attenuation, A_{min}, is given by (for a sine wave input)

$$A_{min} = 20 \log (\sqrt{1.5} \times 2^B) \text{ dB}$$

$$= 74 \text{ dB}$$

(2) The signal spectrum, before and after sampling (ignoring higher order image frequencies), is shown in Figure 2.14.
   From

$$A_{min} = 74 = 20 \log \left[1 + \left(\frac{f'_{max}}{f_c} \right)^6 \right]^{\frac{1}{2}}$$

we obtain

$$\left(\frac{f'_{max}}{f_c} \right)^6 = (5011.87)^2 - 1$$

and thus f'_{max} equals 68.45 kHz.
   From Equation 2.3, we have $F_s = 2f'_{max} = 136.9$ kHz.

(3) The aliasing level at 4 kHz is

$$\frac{1}{\left[1 + \left(\dfrac{136.9 - 4}{4} \right)^6 \right]^{\frac{1}{2}}} = 2.73 \times 10^{-5}$$

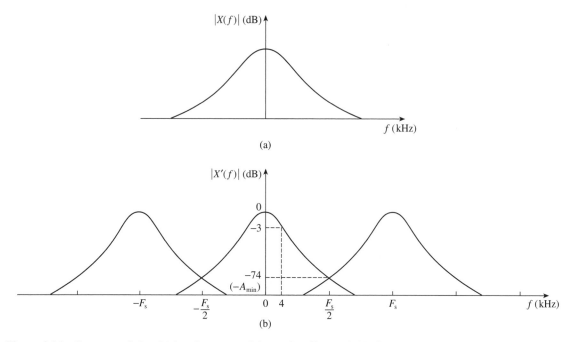

Figure 2.14   Spectrum of signal (a) at the output of the analog filter and (b) after sampling (Example 2.3).

The aliasing level relative to signal level at 4 kHz is

$$\frac{2.73 \times 10^{-5}}{0.7071}$$

If we wish the bandedge frequency to be attenuated to just below the quantization noise level, the sampling frequency can be reduced to 68.4 kHz + 4 kHz = 72.4 kHz.

Example 2.4   *Using the ADC noise floor as the reference*   An analog signal with uniform power spectrum density is bandlimited by an anti-aliasing filter with the following amplitude response:

$$|H(f)| = \frac{1}{\left[1 + \left(\dfrac{f}{f_c}\right)^8\right]^{\frac{1}{2}}}$$

where $f_c = 5$ kHz. The signal is digitized with a linear 12-bit, bipolar, ADC. Determine the:

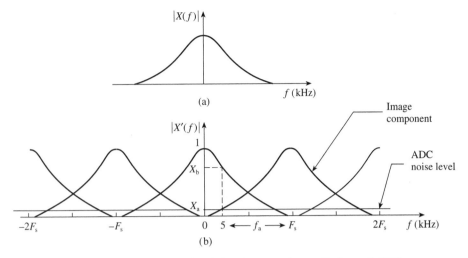

Figure 2.15   Spectrum of sampled signal showing error due to aliasing and ADC quantization (Example 2.4).

(1)   minimum sampling frequency to keep the maximum aliasing error in the passband to no greater than the quantization error level;

(2)   maximum passband signal level, in dB, relative to the ADC quantization noise floor.

Solution   (1)   The sampling frequency should be chosen such that the anti-aliasing filter attenuates the aliasing error folded back into the passband to less than the maximum rms quantization level of the ADC so that they are not detectable by the ADC (Figure 2.15).

   Assume a sine wave input, with peak amplitude of A (which just fills the ADC input range), then

RMS input signal: $\dfrac{A}{\sqrt{2}}$

Quantization step size: $q \approx \dfrac{2A}{2^B}$

RMS quantization noise: $\dfrac{q}{2\sqrt{3}} = \dfrac{A}{\sqrt{3} \times 2^B}$

At 5 kHz, the maximum aliasing error folded back is:

$$\frac{A}{\sqrt{2}} \times \frac{1}{\left[1 + \left(\dfrac{f_a}{5}\right)^6\right]^{\frac{1}{2}}} = \frac{A}{\sqrt{3} \times 2^B} \tag{2.4}$$

With $B = 12$ bits, we can solve for f_a, the aliasing frequency, and hence the sampling frequency, F_s:

$$f_a = 85.59 \text{ kHz}$$

$$F_s = f_a + 5 = 90.59 \text{ kHz}$$

(2)   Maximum signal relative to ADC noise floor:

$$\frac{\text{maximum rms signal level}}{\text{maximum quantization error}} = \frac{A/\sqrt{2}}{A/(\sqrt{3} \times 2^B)} = \sqrt{1.5} \times 2^B$$

$$\text{Signal : ADC noise floor} = 20 \log (\sqrt{1.5} \times 2^B) \tag{2.5}$$

Note also that the signal-to-ADC noise floor in this case can be obtained from (see Equation 2.4)

$$\text{Signal : noise floor} = 20 \log \left[1 + \left(\frac{f_a}{5} \right)^6 \right]^{\frac{1}{2}} \text{dB}$$

2.3.1.5   Other practical issues associated with sampling: accuracy and bandwidth limitations

In practical systems, instantaneous sampling shown in Figure 2.5(d) is not possible; instead, the sampling function has a finite width. This leads to a problem termed the aperture effect, to indicate that the signal is measured over a finite time interval instead of instantaneously. Non-zero aperture time limits the accuracy and maximum signal frequency that can be digitized because the signal may be changing while it is being sampled. A measure of the effects of aperture can be obtained if we assume that the input voltage can only change during the aperture interval by a maximum of $\frac{1}{2}$ LSB (least significant bit) (say). Thus, for a sine wave input, the maximum frequency that can be digitized to $\frac{1}{2}$ LSB accuracy for a system using a B-bit ADC is given by

$$f_{max} = \frac{1}{\pi 2^{B+1} \tau} \tag{2.6}$$

where τ is the aperture time (see Example 2.5 for the proof).

Example 2.5   A real-time DSP system uses a 12-bit ADC with a conversion time of 35 μs and no sample and hold. What is the highest frequency that can be digitized to within $\frac{1}{2}$ LSB accuracy, assuming a binary system with uniform quantization? Comment on the result.

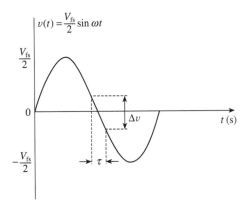

Figure 2.16   Sine wave signal for Example 2.5.

Solution   Consider a sine wave signal with a peak amplitude equal to half the full-scale range of the ADC, $V_{fs}/2$ (Figure 2.16). In the figure, τ is the aperture time and Δv is the change in $v(t)$ during τ. The point of greatest change is at $t = 0$ and the ADC must handle this to measure the signal with desired accuracy. At this point

$$\left. \frac{dv(t)}{dt} \right|_{t=0} = (V_{fs}/2)w \cos wt = \pi f V_{fs} \quad (\text{V s}^{-1}) = \frac{\Delta v}{\tau}$$

For $\frac{1}{2}$ LSB accuracy, $\Delta v = \alpha/2$, where $\alpha = (V_{fs}/2^B)$. Thus $\Delta v/\tau = \pi f V_{fs}$. Substituting for V_{fs} and Δv, and simplifying,

$$f_{max} = \frac{1}{\pi 2^{B+1} \tau}$$

For the DSP system, $B = 12$ and $\tau = 35$ μs. Thus $f_{max} = 1.11$ Hz.

An ADC which can only convert a maximum frequency of 1.11 Hz is clearly of little use. In practice, the ADC is often preceded by a sample and hold which freezes the signal sample during conversion, enabling signals in the kilohertz range to be accurately digitized. For example, if the ADC above is preceded by a sample and hold with an aperture time of 25 ns, and an acquisition time of 2 μs, then the maximum frequency that can be converted becomes

$$2f_{max} \leqslant F_s = 1/(35 + 2 + 0.025) \times 10^{-6} \text{ kHz, that is } f_{max} = 13.5 \text{ kHz}$$

Thus the signal with a maximum frequency of 13.5 kHz would be sampled at a rate of 27 kHz, or at intervals of $(35 + 2 + 0.025)$ μs $= 37.025$ μs.

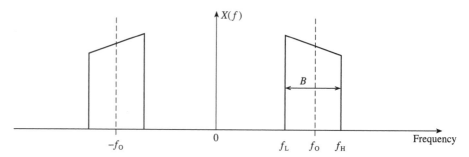

Figure 2.17   A bandpass signal.

Sampling bandpass signals

2.3.2.1   Introduction and basic principles

In some applications, such as communication systems, the signal of interest occupies a narrow part of the available band only; see Figure 2.17.

In such cases, the bandwidth of the signal, B, is often very small compared to the lower and upper bandedge frequencies (f_L and f_H), and it is uneconomical to use the lowpass sampling theorem. A way to deal with this is to use the bandpass sampling theorem (Equation 2.7):

$$\frac{2f_H}{n} \leqslant F_S \leqslant \frac{2f_L}{n-1} \tag{2.7}$$

where

$$n = \frac{f_H}{B} \quad (n \text{ is an integer, rounded up to largest integer})$$

The bandpass sampling theorem allows us to sample narrowband HF signals at a much reduced rate and still avoid aliasing (Vaughan *et al.*, 1991; Del Re, 1978). There are two common approaches for alias-free undersampling of bandpass signals. One approach is the so-called integer-band sampling and the other employs quadrature modulation techniques. The second method is outside the scope of this book.

2.3.2.2   Undersampling techniques for integer bands

Given a bandpass signal, if the bandedge frequencies, f_L and f_H, are integer multiples of the signal bandwidth, then the signal can be sampled at a theoretical minimum rate of $2B$ without aliasing:

$$F_S(\min) = 2B \tag{2.8a}$$

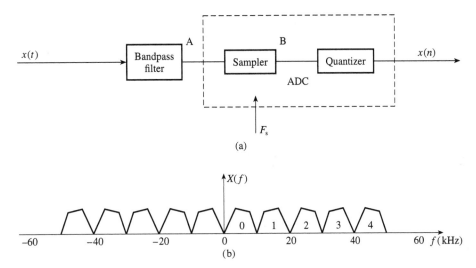

Figure 2.18   (a) Front end of system. (b) Spectrum of received signal. (Example 2.6)

Equation 2.8a is valid provided the ratios of the lower bandedge to the signal bandwidth and/or the upper bandedge to the signal bandwidth are integers:

$$n = \frac{f_H}{B} \text{ or } n = \frac{f_L}{B} \tag{2.8b}$$

When the conditions in Equations 2.8b are satisfied, then the signal band is said to be integer positioned. If the signal band is not integer positioned, the bandedge frequencies can be extended such that the effective band becomes integer positioned.

Example 2.6

Illustrating principles of bandpass undersampling   The front end of the receiver for a multichannel communication system is depicted in Figure 2.18(a). The received signal has the spectrum shown in Figure 2.18(b), with the channel numbers indicated. A bandpass filter is used to isolate the signal in the desired channel before the signal is digitized at the lowest possible rate.

Assume an ideal bandpass filter with the following characteristics:

$$H(f) = 1 \quad 40 \text{ kHz} \leqslant f \leqslant 50 \text{ kHz}$$
$$0 \quad \text{otherwise}$$

(a) (i)   Determine the minimum theoretical sampling frequency.
     (ii)   Sketch the spectrum of the signal before sampling (point A) and after sampling (point B).
(b) Repeat parts (i) and (ii) for a bandpass filter that passes channel no. 3.

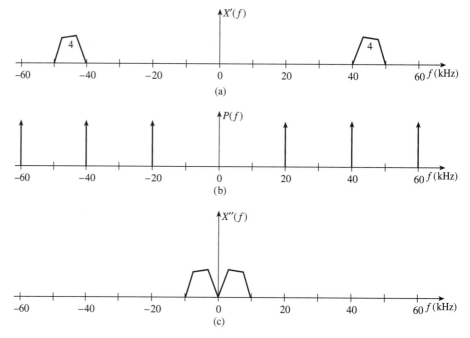

Figure 2.19 (a) Output of the bandpass filter. (b) The sampling function. (c) Output of sampler. (Example 2.6)

Solution   (a)   (i)   The minimum theoretical sampling frequency is 2×10 kHz, i.e. 20 kHz.

(ii)   The spectrum at point A (output of the bandpass filter) is simply the signal spectrum for channel 4 (Figure 2.19(a)).

The spectrum at point B (i.e. after sampling) may be obtained by convolving the spectrum of the signal at the output of the bandpass filter (Figure 2.19(a)) and the spectrum of the sampling function (Figure 2.19(b)). This gives Figure 2.19(c).

(b)   (i)   The sampling frequency remains at 20 kHz.

(ii)   Proceeding as in part (a), the spectra at points A and B in this case are shown in Figures 2.20(a) and 2.20(c), respectively.

Example 2.7   *Illustrating the requirements for alias-free bandpass undersampling technique*   The spectrum of a narrowband signal is depicted in Figure 2.21. Obtain and sketch the spectrum of the sample signal, in the range $\pm F_s/2$, for each of the following three cases:

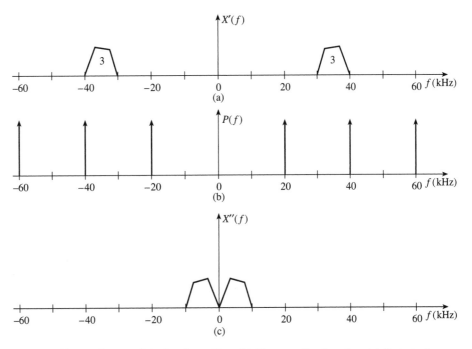

Figure 2.20   (a) Output of the bandpass filter. (b) The sampling function. (c) Output of sampler for channel no. 3. (Example 2.6)

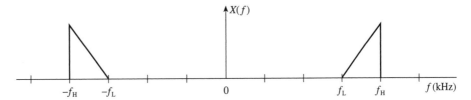

Figure 2.21   Spectrum of narrowband signal (Example 2.7).

(1) $\dfrac{f_{\mathrm{H}}}{B} = 4$

(2) $\dfrac{f_{\mathrm{H}}}{B} = 5$

(3) $\dfrac{f_{\mathrm{H}}}{B} = 6.5$

Assume that the bandwidth of the signal $B = 4$ kHz, and that the signal is sampled at the rate of $2B$ in each case.

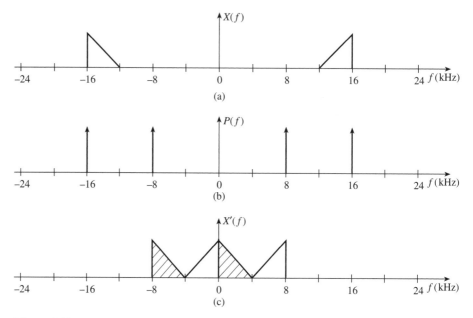

Figure 2.22   Spectrum of sampled signal for n an even integer ($n = f_H/B = 4$) (Example 2.7, case 1).

Solution   (1)   In this case, the spectrum of the signal is shown in Figure 2.22(a). Sampling at $2B$ gives a sampling frequency of 8 kHz. The spectrum of the sampled signal may be obtained by graphically convolving the spectrum of the signal, Figure 2.22(a), and that of the sampling function, Figure 2.22(b).

We will keep the spectrum of the signal fixed and shift the spectrum of the sampling function to perform the convolution. In graphical convolution, we would normally rotate the waveform to be shifted about the vertical axis first before the shifting process starts. However, as the sampling function is symmetrical about the frequency axis, this step is unnecessary as we will end up with the same wave shape.

We note that the frequency point at −16 kHz in Figure 2.22(b) is just lined up with the negative portion of the signal spectrum. Thus, as we shift the spectrum of the sampling function to the right, the frequency point at −16 kHz convolves with the portion of the signal spectrum in the negative frequency. This produces the spectrum from 0 to 4 kHz in Figure 2.22(c). The spectrum point at 8 kHz in Figure 2.22(b) then starts to convolve with the portion of signal band between 12 and 16 kHz, generating the spectrum between 4 and 8 kHz in Figure 2.22c.

A mirror image of the spectrum of the sampled signal is obtained by shifting the sampling function to the left. The portions of the spectrum of the sampled signal produced by the negative frequency components are shown with a hatch.

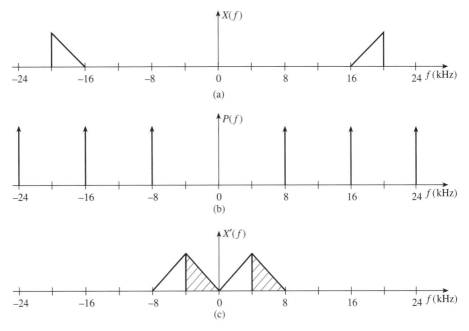

Figure 2.23 Spectrum of sampled signal for n an odd integer ($n = f_H/B = 5$) (Example 2.7, case 2).

> **Comment** *The portion of the spectrum of the sampled signal between 0 and 4 kHz is reversed compared to the original signal spectrum between 12 and 16 kHz. Apart from this the signal band is not aliased and so may be recovered by an appropriate spectrum reversal algorithm.*

(2) Again the sampling frequency is 8 kHz and the spectrum of the signal and that of the sampling function are shown in Figures 2.23(a) and (b).

As before, keeping the spectrum of the signal fixed and shifting the spectrum of the sampling function first to the right and then to the left produces the spectrum of the sampled signal shown in Figure 2.23(c).

As before, the portions of the spectrum of the sampled signal produced by the negative frequency components are shown with a hatch.

> **Comment** *The portion of the spectrum of the sampled signal between 0 and 4 kHz is the right way round compared to the original signal spectrum between 16 and 20 kHz and the signal band is not aliased and so may be recovered.*

(3) As in the previous case, the sampling frequency is 8 kHz and the spectrum of the signal and that of the sampling function are shown in Figures 2.24(a) and (b).

As before, we will keep the spectrum of the signal fixed and shift the spectrum of the sampling function first to the right and then to the left. This produces the spectrum of the sampled signal shown in Figure 2.24(c).

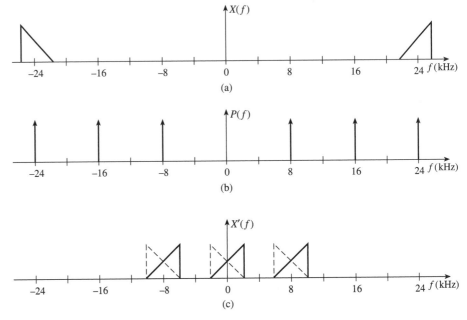

Figure 2.24   Spectrum of sampled signal ($n = f_H/B = 6.5$) (Example 2.7, case 3).

We note that the frequency point at −24 kHz in Figure 2.24(b) is in the middle of the negative portion of the signal spectrum, and the point at 24 kHz in the middle of the spectrum portion between 22 and 26 kHz. Thus, as we shift the spectrum of the sampling function to the right, the frequency point at −24 kHz convolves with the portion of the signal spectrum in the negative frequency and that at 24 kHz convolves with positive portion of the signal spectrum. This produces the spectrum from 0 to 2 kHz in Figure 2.24(c). The portion of the spectrum of the sampled signal produced by the negative frequency components is shown with dotted lines and that produced by the positive frequency by solid lines.

After a further shift of 4 kHz, the spectrum point at 16 kHz in Figure 2.24(b) then starts to convolve with the positive signal band to generate the spectrum centred at 8 kHz in Figure 2.24(c) shown with a solid line. The dotted line is produced by the −16 kHz spectrum point. A mirror image of the spectrum of the sampled signal is obtained by shifting the sampling function to the left.

Comment   *The overlap of the spectrum produced by the positive and negative spectrum points is indicative of aliasing, thus the signal component cannot be recovered by sampling at 8 kHz in this case.*

2.3.2.3   Extending the bandwidth of the signal to achieve alias-free bandpass undersampling

As we have seen in integer bandpass sampling, provided one of the bandedge frequencies is an integer multiple of the bandwidth we can sample a narrow band HF signal at a much reduced rate ($2B$) and still avoid aliasing errors.

Thus, an important parameter in integer bandpass sampling is the ratio of the upper bandedge to the bandwidth, B (or equivalently, the ratio of the lower bandedge, f_L, to the bandwidth):

$$n = \frac{f_H}{B} \tag{2.9a}$$

or

$$n = \frac{f_L}{B} \tag{2.9b}$$

In either case, we can sample at the rate of $2B$ without aliasing, i.e.:

$$F_s = 2B \tag{2.10}$$

For the case where the ratio is an even integer, the spectrum of the sampled waveform is reversed in the baseband region.

When the value of n in Equation 2.9a or 2.9b is not an integer, we find that there is aliasing. We can avoid aliasing by extending the bandedge frequencies or the centre frequency such that n becomes an integer. For example, we can extend the lower bandedge frequency, f_L, to f_1 such that

$$f_1 \leqslant f_L \tag{2.11a}$$

$$f_H = n(f_H - f_1) = nB' \tag{2.11b}$$

From Equation 2.11b we can write

$$f_1 = \left(\frac{n-1}{n}\right) f_H \tag{2.12}$$

and from Equations 2.11a and 2.12 we can write

$$\left(\frac{n-1}{n}\right) f_H \leqslant f_L$$

from which we can obtain an expression for n:

$$n \leqslant \frac{f_H}{f_H - f_L} = \frac{f_H}{B} \tag{2.13}$$

Thus, we can extend the lower bandedge frequency to achieve the desired relationship between the bandedge frequencies and the bandwidth by extending the lower bandedge frequency as indicated in Equation 2.12, where n is the least nearest integer obtained from Equation 2.13.

It can be shown that we can also achieve the desired goal by extending the upper bandedge frequency as follows:

$$f_2 = \left(\frac{n}{n-1}\right)f_L \tag{2.14}$$

where n is given by Equation 2.13 above. The proof of Equation 2.14 is left as an exercise for the reader.

Example 2.8   *Illustrating how to extend the bandwidth to avoid aliasing*   Determine the minimum sampling frequency to avoid aliasing in Example 2.7 (case 3) by extending the lower bandedge frequency accordingly.

Sketch the spectrum of the modified signal before and after sampling at the new rate.

Solution   Now the ratio of the upper bandedge frequency and the bandwidth is

$$\frac{f_H}{B} = \frac{26}{4} = 6.5$$

If we let $n = 6$ (the least nearest integer), then we can reduce the lower bandedge frequency to a new value given by

$$f_L' = \left(\frac{n-1}{n}\right)f_H = 21.66 \text{ kHz}$$

The new bandwidth, B', and sampling frequency, F_s', then become

$$B' = f_H - f_L' = 4.34 \text{ kHz}$$
$$F_s' = 2B' = 8.68 \text{ kHz}$$

It is left as an exercise for the reader to sketch the spectrum of the signal before and after sampling and hence show that aliasing has been avoided.

In some applications, it may be possible to achieve the conditions for alias-free integer-band sampling by altering the centre frequency and not the bandwidth. Such may be the case in digital radio communication where the local IF (intermediate frequency) may be selectable by the designer.

2.4 Uniform and non-uniform quantization and encoding

After sampling, the amplitudes of the analog samples may be quantized and encoded using either uniform or non-uniform quantization and encoding, depending on the application. In biomedicine and audio systems, uniform quantization and encoding are often used, whereas in communications systems, non-uniform quantization and encoding are used extensively because of the need to compress speech signals (see later).

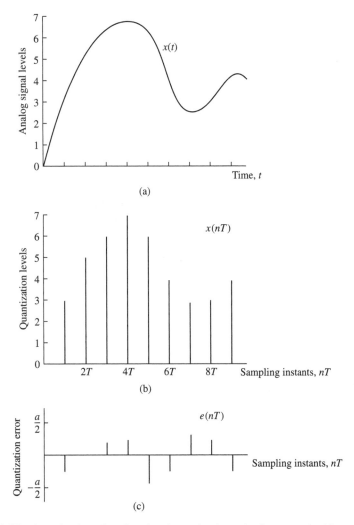

Figure 2.25 Quantization of analog signal samples (quantization errors in (c) are obtained by subtracting the signal samples in (a) from the quantized samples (3-bit quantizer) in (b)).

2.4.1 Uniform quantization and encoding (linear pulse code modulation (PCM))

In uniform quantization and encoding, each analog sample is assigned to one of 2^B values (see Figure 2.25), where B is the number of ADC bits. This process, termed quantization, introduces an error which cannot be removed. The level of the error is a function of the number of bits of the ADC, being approximately equal to one-half of an LSB (assuming rounding). For example, a 12-bit ADC with an input voltage range of ± 10 V will have an LSB of $20/2^{12}$ V, that is 4.9 mV, and a quantization error of 2.45 mV.

For an ADC with B binary digits the number of quantization levels is 2^B and the interval between the levels, that is the quantization step size, q, is given by

$$q = V_{fs}/(2^B - 1) \approx V_{fs}/2^B \qquad (2.15)$$

where V_{fs} is the full-scale range of the ADC with bipolar signal inputs. The maximum quantization error, for the case where the values are rounded up or down, is $\pm q/2$. For a sine wave input of amplitude A (such that the peak-to-peak amplitude of the signal just fills the ADC input range), the quantization step size becomes

$$q = 2A/2^B \qquad (2.16)$$

The quantization error for each sample, e, is normally assumed to be random and uniformly distributed in the interval $\pm q/2$ with zero mean. In this case, the quantization noise power, or variance, is given by

$$\sigma_e^2 = \int_{-q/2}^{q/2} e^2 P(e) \, de$$

$$= \frac{1}{q} \int_{-q/2}^{q/2} e^2 \, de = \frac{q^2}{12} \qquad (2.17)$$

For the sine wave input, the average signal power is $A^2/2$. The signal-to-quantization noise power ratio (SQNR), in decibels, is

$$SQNR = 10 \log\left(\frac{A^2/2}{q^2/12}\right) = 10 \log\left(\frac{3 \times 2^{2B}}{2}\right)$$

$$= 6.02B + 1.76 \text{ dB} \qquad (2.18)$$

This is a theoretical maximum. In practice, when real-world input signals are used, achievable SQNR is less than this value. However, the SQNR increases with the number of bits, B. Practical factors such as speed, inherent signal-to-noise ratio (SNR) of the analog signal, and costs limit the number of bits used. For example, it is unnecessary to use a converter that yields an accuracy better than the SNR of the analog signal to be converted, since this will merely give a more accurate representation of the noise. In many DSP applications, an ADC resolution between 12 and 16 bits is adequate.

The digital samples, $x(n)$, which in many cases are in a binary form, are next encoded into a form suitable for further manipulation. Encoding means assigning discrete codes to the quantized samples. In DSP, the commonest forms are fixed point (two's complement), floating point and block floating point representations. Note that it is possible to perform the three operations of sampling, quantization, and encoding simultaneously. This is the case when we use the ADC without a sample and hold.

Example 2.9

Explain the meaning of dynamic range and aperture time in relation to the analog-to-digital conversion process.

If, in Example 2.2, the dynamic range of the ADC is to be greater than 70 dB and the samples are to be digitized to $\frac{1}{2}$ LSB accuracy, determine

(1)   the minimum resolution of the ADC in bits, and

(2)   the maximum allowable aperture time, assuming the highest frequency of interest to be digitized is 20 kHz.

Solution

The dynamic range is the ratio of the maximum to minimum signal levels that an analog-to-digital conversion system can handle. The dynamic range is often expressed in decibels in terms of the number of bits in the converter:

$$D = 20 \log_{10} 2^B \tag{2.19}$$

In some applications, the dynamic range is defined in terms of the signal power. For example, in digital audio it may be defined as the ratio of the maximum signal power to the minimum power which can be discerned from the noise power.

For the ADC when used alone, the aperture time is essentially the conversion time of the ADC and refers to the period of time over which the analog input must remain stable so that accurate conversion may be made. In relation to the sample and hold, it is the time required to achieve a hold following the hold command.

(1)   Using the expression for D, we have

$$70 = 20 \log_{10} 2^B$$

from which $B = 11.62$. Let $B = 12$ bits (the nearest integer).

(2)   The maximum allowable aperture τ is given by

$$\tau = 1/2^{B+1} \pi f_{max} = 1/(2^{13} \times \pi \times 20 \times 10^3) \text{ s} = 1.94 \text{ ns}$$

This small aperture time calls for the use of a sample and hold ahead of the ADC.

2.4.2   Non-uniform quantization and encoding (nonlinear PCM)

The linear analog-to-digital (A/D) conversion process we have discussed so far is sometimes referred to as linear PCM. In such converters, the level of quantization noise is directly related to the number of bits of the ADC. Such converters are well suited to applications where the signal amplitudes are reasonably uniformly distributed or where the size of the A/D converter wordlength is not a major issue. In applications where the signal amplitudes are not uniformly distributed (e.g. telephony) a large number of bits will be required to represent the data accurately and this may not be efficient.

Some signals, e.g. speech, contain both low and large amplitudes, but the small amplitudes are more likely. Thus, uniform quantization is inappropriate for speech. Non-uniform quantization can provide more quantization levels for low level signals than uniform quantization with the same number of bits and in telephony this means that both quiet and loud talkers can be more readily accommodated.

The standard non-uniform quantization used in telephony (for both public and private telephone networks) is determined from a knowledge of the amplitude distribution of speech. The speech waveform is sampled at 8 kHz rate, with each sample quantized and encoded into 8 bits giving a 64 kbit/s system. The amplitudes of the speech samples are logarithmically compressed into 8 bits before transmission. The compressed data is expanded at the receiving end.

The process of compressing speech signal and expanding it is referred to as companding (an acronym from the words COMpressing and exPANDING). The process is depicted in Figure 2.26. In practice, companding is performed by codecs or combo-codecs (combined PCM codec and anti-aliasing and anti-imaging filters) which are fitted to each speech channel in digital telephony.

In modern digital telephony with codecs, a reverse companding process is necessary in the chain to allow the data to be processed using DSP. The compressed PCM is converted into linear PCM data and after DSP the data is converted back into non-uniform PCM. The reverse companding operations can be performed by DSP chips (e.g. Texas Instrument TMS320 or Motorola 56000 series) as a table look-up or by performing the compression and expansion algorithms in real time.

2.4.2.1   Companding methods: the μ-law and A-law PCM

Two international standards (see CCITT recommendation G.711, 1998) are used to achieve non-uniform quantization in telephony: the μ-law standard (used in the US and Japan) and the A-law standard (used mainly in Europe). Both standards compress the speech into 8 bits which is equivalent to about 14 bits in linear ADC.

For the μ-law, the companding characteristics and equations are defined in Figure 2.27. Eight-bit signed magnitude words are used to represent each sample. The companding characteristics are approximated by a set of eight straight line segments (see Figure 2.27). It is seen that successively larger input signal samples are compressed into a uniform output scale. The step size of the input signal is successively doubled between segments. This makes it easier to convert between linear

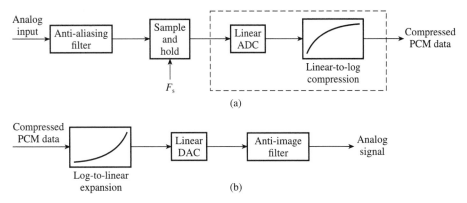

(a)

(b)

Figure 2.26 Non-uniform analog-to-digital and digital-to-analog conversions.

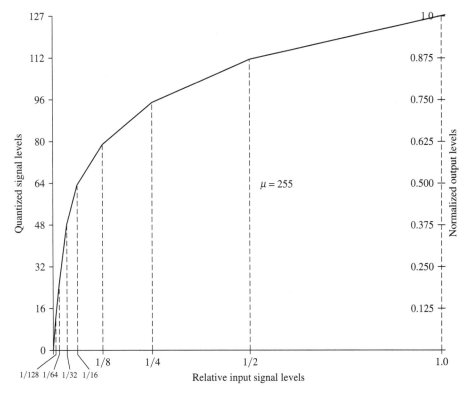

Figure 2.27 Companding characteristics for the μ-255 law (ITU, 1998). The characteristics are governed by the equation

$$F(x) = \text{sgn}(x)\frac{\ln(1 + \mu|x|)}{\ln(1 + \mu)}$$

where $\mu = 255$, x is the normalized input signal, sgn its sign, and $F(x)$ is the compressed output signal.

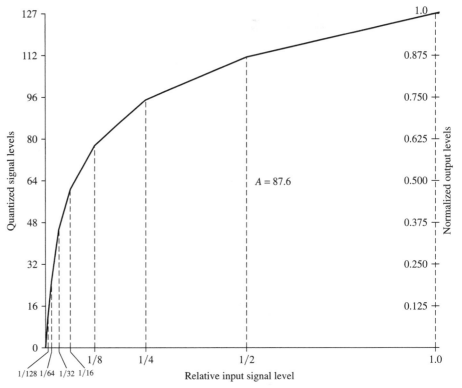

Figure 2.28   Companding characteristics for the A-255 law (ITU, 1998). The characteristics
are governed by the equation

$$F(x) = \operatorname{sgn}(x)\frac{A|x|}{1 + \ln(A)},\ 0 \le |x| < 1/A;\ F(x) = \operatorname{sgn}(x)\frac{1 + \ln(A|x|)}{1 + \ln(A)},\ 1/A \le |x| < 1$$

where $A = 87.6$, x is the normalized input signal, sgn its sign, and $F(x)$ is the
compressed output signal.

and non-uniform PCM. The μ-law PCM consists of 8-bit words. The MSB is the sign
bit, the next 3 bits represent the segment number and the last 4 bits the position within
the segment. In practice, the bits are inverted before transmission to increase the
density of 1's to aid clock recovery and error correction. This is necessary because
speech is a low energy signal. Typically, the signal to quantization noise ratio for
non-uniform quantization is comparable to those of a 14-bit linear ADC.

The A-law characteristics are shown in Figure 2.28. They are similar to the μ-law in
attributes. However, the A-law characteristics give less fidelity for small signals but
have superior dynamic range characteristics.

2.4.2.2   Adaptive differential pulse code modulation (ADPCM)

Non-uniform PCM quantizes the speech signal samples into 8 bits for each data
sample. This approach does not take advantage of the redundancies in speech signals.

In ADPCM (CCITT recommendation G.726, 1990; CCITT, 1989), the value of each sample is adjusted depending on the value of the sample. This reduces the number of bits used to represent each sample from 8 bits to 4 bits (8 kHz × 4 bits = 32 kbits/s). ADPCM transmits the difference between the predicted sample value and the actual sample value. In practice, an ADPCM transcoder may be inserted into a PCM system to increase its voice channel capacity. The ADPCM encoder accepts PCM values as input and the ADPCM decoder outputs PCM values.

A variety of standards now exists for speech coding to serve as benchmarks for the many services and applications in the communications industry. In most cases, the emphasis is on the reduction of data rates. For example, the GSM mobile phone system has a speech rate of 13.2 kbits/s. The CCITT G721 ADPCM standard operates at 32 kbits/s.

2.5 Oversampling in A/D conversion

2.5.1 Introduction

In practice, oversampling means sampling the input signal at a rate much greater than the Nyquist rate. The ratio between the actual sampling rate and the Nyquist rate is referred to as the oversampling ratio (assuming lowpass signals):

$$\text{Oversampling ratio} = F_s/2f_{\max} \tag{2.20}$$

The trend in many areas of modern DSP is always to oversample to exploit the practical implications of the sampling theorem. At the A/D interface, the main benefits of oversampling are (1) simplification of the anti-aliasing filter, (2) support for anti-aliasing filtering with variable cutoff frequency (each cutoff frequency would require a different sampling frequency), and (3) a reduction in the ADC noise floor by spreading the quantization noise over a wider bandwidth. This makes it possible to use an ADC with fewer bits to achieve the same SNR performance as a higher resolution ADC.

2.5.2 Oversampling and anti-aliasing filtering

In high fidelity digital systems, the need to keep aliasing error levels low often dictates the use of relatively complex analog anti-aliasing filters. In a multichannel system, each analog channel must be fitted with a separate anti-aliasing filter as such filters cannot be readily multiplexed. This becomes expensive where the number of analog channels is large (e.g. in biomedicine as many as 64 channels may be required). Further, phase matching of the tightly specified analog filters (to retain the relationship between signal components across all channels) may be difficult to achieve.

Oversampling techniques allow us to overcome many of these problems. As should be evident by examining the spectrum of a sampled signal, the higher the sampling frequency, the wider the separation between the image components and the base-band (see, for example, Figure 2.29). In Chapter 9, we shall see how oversampling techniques are combined with multirate DSP to meet the requirements of hi-fidelity systems.

Example 2.10

(a) A requirement exists for a general purpose, multichannel (up to 64 channels) data acquisition system for collecting neurophysiological data. Each analog channel is to be individually configured, by the user, to have a passband edge frequency between 0.5 Hz and 500 Hz, and a selectable sampling frequency in the range of 1 Hz to 5 kHz. In the passband, the maximum permissible ripple is 0.5 dB and the image components must be at least 40 dB below the signal components.

Explain the strategy you would use to satisfy the above requirement. Your answer should address the following points:

(i)   considerations for application-specific issues;
(ii) how oversampling techniques may be used in this application to satisfy the requirement in an efficient and economical way (in terms of cost/component count).

(b) Assume that identical anti-aliasing filters are used for all the channels in the system in (a), each with the following Butterworth characteristics.

$$A(f) = \frac{1}{\sqrt{1 + \left(\dfrac{f}{f_c}\right)^6}}$$

where f_c = 3 dB cutoff frequency of filter.

(c) Determine, with the aid of sketches of the spectrum of the data before and after sampling:

(i)   the cutoff frequency, f_c;
(ii) a suitable common sampling frequency, F_s.

Comment on your answer.

Solution

(a) High resolution ADC/DAC are slow and impose a limit on the maximum sampling frequency achievable – this is a major bottleneck in a number of real-time applications. To overcome this, it may be necessary to use multiple ADC/DAC devices and/or multirate DSP techniques.

The $\sin x/x$ effect at the output which progressively reduces the high frequency component of the signal may be compensated for by the use of a post processing digital filter with an $x/\sin x$ response. Other reasonable answers are acceptable.

(b) (i) To preserve information of clinical interest in the signals, both amplitude and phase distortions should be kept as low as possible. The time relationships between features across the channels should be preserved. The use of identical anti-aliasing filters with reasonably good amplitude/phase responses is desirable.

  (ii) To reduce component count/cost and size of the PCB for the system, all 64 channels should be fitted with identical, simple anti-aliasing filters. The channels should then be oversampled at a common fixed rate. The high, common sample rate can be reduced to the desired rates using multirate techniques. At least a second-order Butterworth filter should be used to avoid an excessive common sampling rate.

(c) From a consideration of the specifications and the spectrum of the data before and after sampling we find that:

  (i) To keep within the specifications, the amplitude error between 0 and 500 Hz should satisfy the following criterion:

$$20 \log \left[1 + \left(\frac{500}{f_c} \right)^6 \right]^{\frac{1}{2}} \leq 0.5 \, \text{dB}$$

where we have assumed a third-order Butterworth filter with a cutoff frequency of f_c.

Solving for f_c, we find that

$$f_c = \frac{500^6}{0.122} \approx 710 \, \text{Hz}$$

To allow for additional errors in subsequent stages and for convenience, let $f_c = 1000$ Hz (this is equivalent to a maximum ripple of 0.26 dB).

  (ii) After bandlimiting each channel, the spectrum of the sampled data has the form shown in Figure 2.29.

Now, F_s is chosen such that the aliasing error level is down by at least 40 dB at 500 Hz, i.e.

$$20 \log \left[1 + \left(\frac{F_s - 500}{1000} \right)^6 \right]^{\frac{1}{2}} \geq 40 \, \text{dB}$$

Solving for F_s gives $F_s = 5141.5$ Hz.

Some care is required in the choice of the common sampling frequency, F_s, to allow for an efficient reduction in the sample rate. A possible choice is 8192 Hz which allows the common rate to be reduced by simple integer factors.

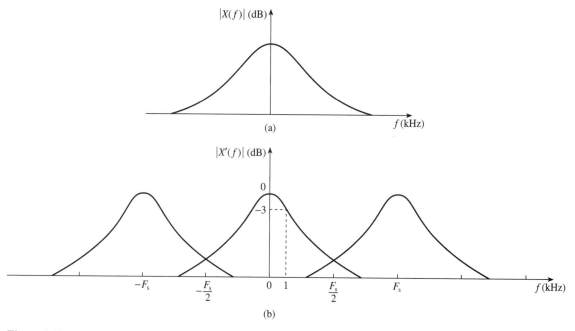

Figure 2.29   Spectrum of the input signal (a) before sampling and (b) after sampling.

2.5.3   Oversampling and ADC resolution

Oversampling the input signal spreads the quantization noise energy over a much wider frequency range, thereby reducing the noise level in the band of interest and extending the resolution of the ADC. This has been exploited in digital audio to achieve the so-called single-bit or oversampled ADCs. First, let us review the basic concepts of quantization and quantization errors.

2.5.3.1   Quantization and quantization errors

In the traditional A/D process, each signal sample is quantized into one of 2^B levels and represented by B binary bits, where B is the number of bits of the ADC. Quantization introduces an error which is a function of the number of bits of the ADC.

The quantization noise power (for uniformly distributed error with zero mean) is given by

$$\sigma_e^2 = \frac{q^2}{12}$$

where q is the quantization step size. The theoretical maximum signal-to-quantization noise ratio (SQNR) for a linear ADC is given by

$$\text{SQNR} = 6.02B + 4.77 - 20 \log (A/\sigma_x) \text{ dB} \tag{2.21}$$

where $\pm A$ is the input range of the ADC and σ_x is the rms value of the input signal. For a sine wave input with a peak amplitude, A, that just fills the ADC range $\sigma_x = \dfrac{A}{\sqrt{2}}$, $20 \log \left(\dfrac{A}{\sigma_x} \right) = 3.01$ dB, and so Equation 2.21 reduces to the familiar form:

$$\text{SQNR} = 6.02B + 1.7 \text{ dB} \tag{2.22}$$

A bipolar, linear 16-bit ADC, for example, with an input range of ± 5 V has a quantization step size, $q = \dfrac{10 \text{ V}}{2^{16} - 1} = 0.152$ mV, a maximum quantization error of $\dfrac{q}{2} = 76 \ \mu\text{V}$, and SQNR = 98 dB.

Drill

A sinusoidal signal with a peak-to-peak amplitude of 10 V is digitized with a 12-bit ADC. Assuming linear quantization, determine

(1)   the quantization step size;
(2)   the quantization noise power;
(3)   the theoretical maximum signal-to-quantization noise ratio.

2.5.3.2   Oversampling and quantization noise power

The intrinsic quantization noise power, introduced by the process of A/D conversion, is given by

$$\sigma_e^2 = \frac{q^2}{12} = \frac{2^{-2(B-1)}}{12} \quad \text{(normalized)} \tag{2.23}$$

where B is the ADC wordlength (including the single bit).

   For sufficiently large or random analog input signals, the energy of the quantization noise is spread evenly over the available spectrum, i.e. from 0 to $F_s/2$, where F_s is the sampling frequency. In this case, the power spectral density of the quantization noise, $P_e(f)$, is given by (see Figure 2.30)

$$P_e(f) = \frac{\sigma_e^2}{F_s} \tag{2.24}$$

Thus, the effective resolution of the ADC can be increased by sampling the input data at a high rate to spread the quantization noise energy over a wider frequency band, thereby reducing the levels of noise in the band of interest. This is what is meant by oversampling.

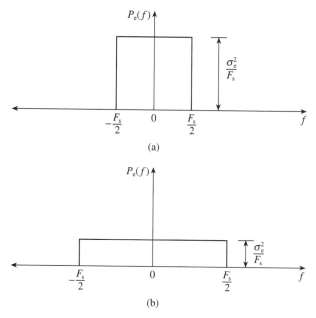

Figure 2.30   Quantization noise power spectral density for (a) Nyquist rate converter and (b) oversampled converter. (*The total noise power is the same for both converters, but for the oversampled converters the noise power is distributed over a much wider frequency range leading to smaller in-band noise power levels.*)

Referring to Equation 2.20, for Nyquist rate converters $f_{max} = F_s/2$, so the total in-band noise power is given by the area in Figure 2.30(a), i.e. σ_e^2. For oversampled converters (Figure 2.30(b)), some of the quantization noise power falls outside the desired band (since $f_{max} < F_s/2$) and the in-band noise is less than for the Nyquist converters.

The in-band noise power for the oversampled converter is given by

$$P_e = \int_{-f_{max}}^{f_{max}} P_e(f)\,\mathrm{d}f = \frac{2f_{max}}{F_s}\sigma_e^2 \tag{2.25}$$

Thus, when we oversample a bandlimited signal, the quantization noise energy in the signal band is lowered by the oversampling ratio. In practice, the oversampling ratio is chosen to be an integer power of 2 for ease of implementation.

Example 2.11   (a)   An audio system handles signals with a baseband that extends from 0 to 20 kHz. Determine the oversampling ratio and the minimum sampling frequency that will be necessary to achieve a performance that would be obtained with a 16-bit ADC using a 12-bit converter.

(b) A digital audio system uses oversampling techniques and an 8-bit bipolar Nyquist rate converter to digitize an analog input signal which has frequency components in the range 0–4 kHz. Estimate the effective resolution, in bits, of the converter if the sampling rate is 40 MHz. Comment on the practical problems associated with this approach.

Solution   (a) At the Nyquist rate (i.e. $F_s = 2f_{max}$), the normalized in-band quantization noise power for the 12-bit and 16-bit converters are, respectively,

$$\sigma_1^2 = 2^{\frac{-2(B_1-1)}{12}} \quad \text{(where } B_1 = 12)$$

$$\sigma_2^2 = 2^{\frac{-2(B_2-1)}{12}} \quad \text{(where } B_2 = 16)$$

To achieve 16-bit performance with a 12-bit ADC, we will need to oversample the input to the 12-bit converter to reduce the in-band quantization noise power. The in-band quantization noise power is reduced by the oversampling factor

$$\sigma_1'^2 = \frac{2f_{max}}{F_s} \sigma_1^2$$

Equating the new in-band quantization noise to that for the 16-bit ADC, we have

$$\frac{2f_{max}}{F_s} \sigma_1^2 = \sigma_2^2$$

Thus

$$\frac{2f_{max}}{F_s} = \frac{\sigma_2^2}{\sigma_1^2} = \frac{2^{-2(B_2-1)}}{2^{-2(B_1-1)}} = 2^{-2(B_2-B_1)} = \frac{1}{256}$$

Thus, the oversampling ratio is given by:

$$F_s/(2f_{max}) = 256, \text{ i.e. } F_s = 10.24 \text{ MHz.}$$

(b) The in-band quantization noise is reduced by the oversampling ratio, i.e. by

$$\frac{40\ 000}{2 \times 4} = 5000$$

From $\dfrac{\sigma_1^2}{\sigma_2^2} = \dfrac{2^{-2(B_1-1)}}{2^{-2(B_2-1)}} = 2^{2(B_2-B_1)} = 5000$ and with $B_1 = 8$ bits, we find that the resolution of the ADC, B_2 is about 14 bits.

As is evident in the previous examples, oversampling techniques on their own may not be economically feasible for achieving the desired resolution using low resolution ADCs, because often such resolution requires very high sampling frequencies which may not be supported by current technology.

In practice, oversampling is combined with noise shaping to shift the quantization noise to higher frequencies well outside the signal band, where it can be filtered out. The principles of oversampled ADCs are covered in the next section.

2.5.4   An application of oversampling – single-bit (oversampling) ADC

The requirements in high fidelity DSP systems, such as digital audio for high quality, high resolution and very fast (and cheap!) ADC, are difficult to satisfy with conventional successive approximation or dual slope converters because of errors associated with the analog parts of such converters (e.g. trimming networks for the ADC, antialiasing filters and sample and hold circuits).

Single-bit, or more appropriately, oversampling ADCs do not require sample and hold amplifiers and use simple or no anti-aliasing filters and so are free from most of the above errors.

The two techniques that make single-bit ADCs possible are as follows.

- Oversampling, to spread the quantization noise energy over a much wider frequency range thereby reducing the noise level in the band of interest.

- Noise shaping, to push most of the noise to higher frequencies, well outside the desired signal band where they can be filtered out digitally.

The concepts of oversampling ADC are depicted in Figure 2.31. The analog input signal is oversampled (e.g. 64 times) to spread the quantization noise power over a wide frequency band. The oversampled data is then noise shaped to push the noise power to higher frequencies away from the signal band. The sample rate is then reduced to the Nyquist rate by decimation. This process also serves to transform the data from a single-bit stream into a multibit data stream. Decimation will be covered in detail in Chapter 9.

One of the most effective ways of achieving noise shaping is via sigma delta modulation. Figure 2.32 depicts a first-order sigma delta modulator (SDM). It consists of an integrator, a single-bit quantizer (i.e. 1-bit ADC implemented as a comparator) and a single-bit DAC in a feedback arrangement. The analog input signal, $x(t)$, is

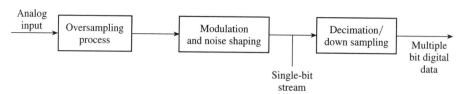

Figure 2.31   Principles of oversampled (single-bit) A/D conversion.

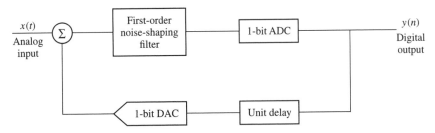

Figure 2.32 A first-order sigma delta modulator.

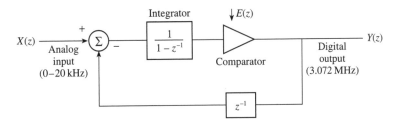

Figure 2.33 z-plane model of first-order SDM.

sampled at a very high rate and then quantized to a single-bit stream which contains very high quantization noise. By a proper choice of the characteristics of the integrator, the noise spectrum is shaped such that most of the noise energy is pushed up and outside the signal band.

To appreciate how the modulator shapes the quantization noise, consider a z-plane model of the first-order sigma delta modulator in Figure 2.33, where we have assumed that the noise samples are uncorrelated.

From the figure, the z-transform of the output is given by

$$Y(z) = E(z) + [X(z) - Y(z)z^{-1}]\left(\frac{1}{1 - z^{-1}}\right)$$

$$= X(z) + E(z)(1 - z^{-1})$$

$$= X(z) + E(z)H_n(z) \tag{2.26}$$

where

$X(z)$ = z transform of the input signal
$Y(z)$ = z transform of the bit stream output
$E(z)$ = z transform of the quantization noise
$H_n(z) = (1 - z^{-1})$ is the noise transfer function

Equation 2.26 shows clearly that the output transform is the same as the input transform plus the quantization noise, modified by the noise transfer function. The

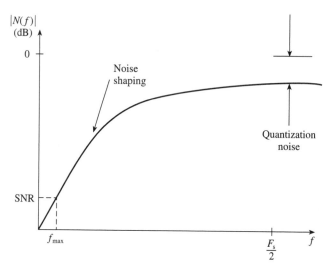

Figure 2.34   Effects of noise shaping on quantization noise.

noise transfer function, $(1 - z^{-1})$, is essentially a highpass filter with zero at d.c. Its effect is to push the quantization noise energy up into the higher frequency spectrum: see Figure 2.34.

For a system with the input bandlimited to f_{max}, the in-band noise power after noise shaping is given by

$$\sigma_n^2 = \int_{-f_{max}/F_s}^{f_{max}/F_s} |H_n(f)|^2 P_d \, df \tag{2.27}$$

Clearly, the performance of the SDM is dependent on the oversampling ratio and on the SDM's ability to shape the noise spectrum. For the first-order SDM, doubling the sampling rate increases the SNR by 9 dB, of which 6 dB is attributable to the noise shaping and the other 3 dB to oversampling. Further reduction in quantization can be achieved by increasing the order of the noise transfer function (i.e. the integrator). It can be shown that for an Nth order SDM, the output transform is given by

$$Y(z) = X(z) + E(z)(1 - z^{-1})^N \tag{2.28}$$

This provides a noise filter with a $6N$ dB/octave roll-off characteristic. Unfortunately, for $N > 3$, the stability of the modulator cannot be guaranteed due to large phase shifts. For SDMs with order higher than two, special configurations are used to avoid instability. One such arrangement is known as the MASH. Figure 2.35 shows a MASH arrangement for a third-order SDM.

The output of the third-order MASH SDM is given by

$$Y(z) = X(z) + E_3(z)(1 - z^{-1})^3 \tag{2.29}$$

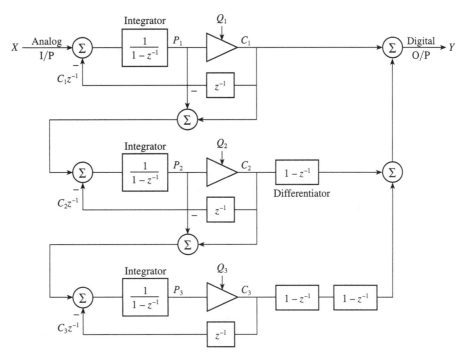

Figure 2.35 Third-order MASH sigma delta modulator z-plane model.

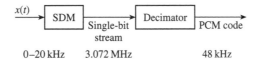

Figure 2.36 A simplified single-bit (oversampled) ADC.

We note that only the quantization noise, $E_3(z)$, from the last stage affects the output, the noise from the first two stages having been suppressed.

Regardless of the order of the SDM, its output contains very small in-band quantization noise, but very large out-of-band noise. The out-of-band noise is removed by lowpass digital filtering. Because of the high sampling rate, direct use of a digital filter is impractical. Instead filtering is achieved by decimation which also serves to reduce the rate to the desired value. After filtering the resulting signal is B-bit quantized data. The filtering serves to average out the high quantization noise. Typically, the FIR coefficients of the decimating filter are represented by 16–24 bits.

A simplified block diagram of a fast single-bit ADC process is shown in Figure 2.36. The analog audio signal is first converted into a single-bit stream, using sigma delta modulation at 3.071 MHz rate. The single-bit stream is then downsampled to 48 kHz, using a multistage decimator (see Chapter 9), and converted into 16-bit linear PCM words.

For a given signal type, the effective wordlength of the ADC is determined by the signal-to-noise ratio achieved through oversampling, noise shaping and decimation. For example, if we desire a 16-bit ADC, then the SNR must be at least 96 dB. Commercial oversampled ADCs now exist and can be purchased off the shelf.

Example 2.12   A digital signal processing system, with analog audio signal input in the range 0–20 kHz, uses oversampling techniques and a second-order sigma delta modulator to convert the analog signal into a digital bit stream at a rate of 3.072 MHz. The z-plane model of the sigma delta modulator is depicted in Figure 2.37.

(1)   Explain how the digital bit stream may be converted into a digital multibit stream at 48 kHz rate.

(2)   Determine the overall improvement in signal-to-quantization noise ratio made possible by oversampling and noise shaping and hence estimate the effective resolution, in bits, of the digitizer.

Solution   (1)   The single-bit stream is converted to multibit words by decimation (down sampling process). The output of the SDM contains very small in-band quantization noise, but very large out-of-band noise. The out-of-band noise is removed by lowpass digital filtering. Because of the high sampling rate, direct use of a digital filter is impractical. Instead, filtering is achieved by decimation which also serves to reduce the rate to the desired value. Typically, a two-stage decimator would be used (factors of 16 and 4). After filtering, the resulting signal is B-bit quantized data. The filtering serves to average out the high quantization noise. Typically, the FIR coefficients of the decimating filter are represented by 16–24 bits.

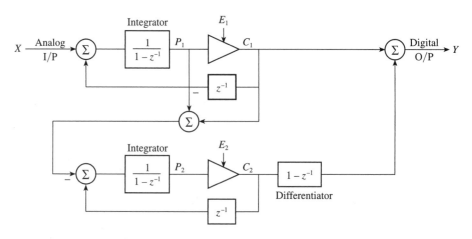

Figure 2.37   z-plane model of second-order SDM (Example 2.12).

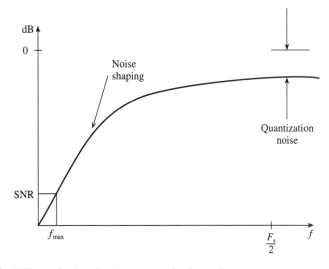

Figure 2.38 Effects of noise shaping on quantization noise.

(2) An estimate of the effective resolution can be obtained by a simplified analysis as follows.

The noise power is reduced by oversampling and noise shaping. The reduction in noise power, due to oversampling, is given by the oversampling ratio.

The oversampling ratio is

$$\frac{F_s}{2f_{max}} = \frac{3.072 \times 10^6}{2 \times 24 \times 10^3} = 64$$

That is a reduction of 18 dB in the quantization noise power.

From the z-plane model of the sigma delta modulator, the transfer function seen by the quantization noise is given by

$$N(z) = (1 - z^{-1})^2$$

This is essentially a highpass filter, with a double zero at d.c. It attenuates the noise component at the low frequency end, Figure 2.38. The magnitude response is given by

$$|N(z)|^2_{z=e^{j\omega T}} = |(1 - e^{-j\omega T})^2|^2$$

At $f = 24$ kHz (the bandedge) and $F_s = 3.072$ MHz, $\omega T = 2.8125°$ and

$$|N(e^{j\omega T})|^2 = 2.412 \times 10^{-3}$$

This offers a reduction in SQNR of 52.35 dB. The effective wordlength of the ADC is determined mainly by the signal-to-noise ratio achieved through

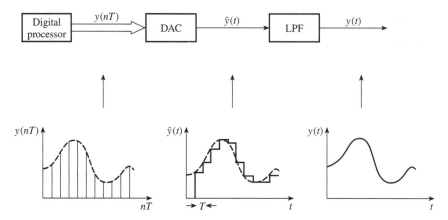

Figure 2.39   Digital-to-analog conversion process used to recover the analog signal after digital processing. Note that the inputs to the DAC are a series of impulses, while the output of the DAC has a staircase shape as each impulse is held for a time T(s).

oversampling and noise shaping. The overall reduction in SQNR is 70.41 dB. This corresponds to an effective ADC resolution of 11.4 bits (from SQNR = 6.02B + 1.77 dB).

2.6   Digital-to-analog conversion process: signal recovery

The digital-to-analog conversion process is employed to convert the digital signal into an analog form after it has been digitally processed, transmitted or stored. The reason for such a conversion may be, for example, to generate an audio signal to drive a loudspeaker (as in the compact disc system) or to sound an alarm. The most commonly used arrangement is shown in Figure 2.39, and can be seen to consist of two main components: the DAC (digital-to-analog converter) and a lowpass filter sometimes called a reconstruction, smoothing or anti-image filter.

2.7   The DAC

The basic DAC accepts parallel digital data and produces an analog output signal which is related to the digital code at its input. A register is used to buffer the DAC's

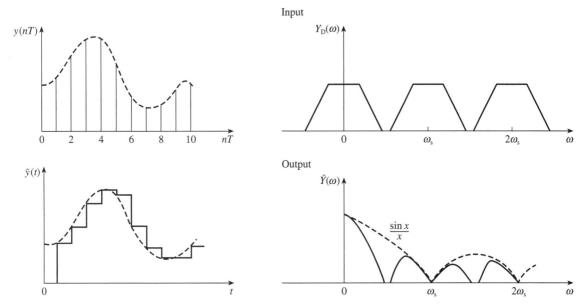

Figure 2.40   Time and frequency domain representation of the DAC input and output signals. Notice that the sin x/x effect manifests itself as a lowpass filter.

input to ensure that its output remains the same until the DAC is fed the next digital input. The register may be external to the DAC or it may form part of the DAC chip, as in Figure 2.39. In some applications, additional circuitry may be required to prevent false digital code from generating transient spikes at the DAC output.

The DAC shown in the figure is referred to as a zero-order hold. By comparing its output, $\bar{y}(t)$, and its input, $y(nT)$, it is evident that for each digital code fed into the DAC, its output is held for a time T. The result is the characteristic staircase shape at the DAC output. In the frequency domain, the holding action of the DAC introduces a type of distortion known as the sin x/x or aperture distortion, where $x = \omega T/2$.

Figure 2.40 shows the time and frequency domain representations of the signals at the input and output of the zero-order hold DAC, from which the following may be noted.

- The input and output signals of the DAC are both wideband signals. Each consists of the signal spectrum (which had been digitized) plus an infinite number of images of the original spectrum centred at the multiples of the sampling frequency.

- The amplitude of the output signal spectrum is multiplied by the sin x/x function, which acts like a lowpass filter, with the image frequencies heavily attenuated.

The $\sin x/x$ effect is due to the holding action of the DAC and, in signal recovery, introduces an amplitude distortion. The average error due to the effect at a given frequency may be expressed as a percentage deviation from unity:

$$(1 - \sin x/x) \times 100\% \tag{2.30}$$

For the zero-order hold, the function $\sin x/x$ falls to about 4 dB at half the sampling frequency $(F_s/2)$ giving an average error of about 36.4%. Aperture error can be eliminated by equalization. In practice this can be achieved by first applying the signal, before converting it to analog, through a digital filter whose amplitude–frequency response has an $x/\sin x$ shape.

In some applications, the digital processor may be used to insert or interpolate between the actual sample points applied to the DAC. This helps to smooth out the analog signal and gives a substantially better result than the simple zero-order hold. Another approach that is becoming popular where high quality audio signals are required is to perform the digital-to-analog conversion at a much higher rate than specified by the sampling theorem (see, for example, Goedhart *et al.*, 1982) using multirate techniques. This has the advantage of improved signal-to-noise performance and simplifies the anti-image filter. Further details are given in Section 2.9 and Chapter 9.

2.8 Anti-imaging filtering

The output of the DAC contains unwanted high frequency or image components centred at multiples of the sampling (that is, the update) frequency as well as the desired frequency components. Depending on the application, the h.f. components may cause undesirable side effects. For example, in the compact disc player, although the image frequencies are not audible they could overload the player amplifier and cause intermodulation products with the desired baseband frequency components. The result is an unacceptable degradation in the audio signal quality.

The role of the output (that is, anti-imaging) filter is to smooth out the steps in the DAC output thereby removing the unwanted h.f. components. The roll-off requirements for the output filter depend on the effect the analog signal will have on subsequent analog stages. In general, the requirements of the anti-imaging filter are similar to those of the anti-aliasing filter.

2.9 Oversampling in D/A conversion

The motivation for oversampling DACs is similar to that for oversampling ADCs. In the case of oversampling DAC, the sample rate of the data to be converted into analog

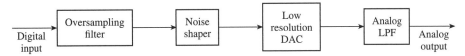

Figure 2.41   A block diagram illustrating the principles of oversampling D/A conversion.

is raised manyfold (e.g. 64 times) to produce analog signal samples with much finer spacing between them. Thus, only a relatively simple analog anti-imaging filter is required to smooth out or remove the out-of-band noise. Sampling at a very high rate creates a signal with a bandwidth that extends from 0 to half the sampling frequency. The quantization noise power is spread evenly over this wider band, making it possible to achieve high resolution D/A conversion with a low resolution DAC.

As in the case of the ADC, oversampling on its own is not adequate for achieving the desired DAC resolution and so noise shaping is necessary. Thus a practical, oversampling DAC typically consists of four main parts: an oversampling digital filter, a noise shaper (e.g. sigma delta modulator), a low resolution DAC (e.g. a single-bit DAC) and a simple analog anti-imaging filter: see Figure 2.41.

The oversampling filter is used to raise the sampling rate and to attenuate the image components. Reducing the wordlength (e.g. from 16 bits to 1 bit) generates a great deal of quantization noise, but this is spread over a much wider bandwidth because the rate has been raised substantially. In addition, the noise shaper pushes much of the quantization noise out of the signal band into higher frequencies. The out-of-band noise is smoothed out using a simple anti-imaging filter.

2.9.1 Oversampling D/A conversion in the CD player

We will illustrate the principles of oversampling D/A conversion further by considering how it is done in some compact disc players: see Figure 2.42. After decoding and error correction the digital signals read from the compact disc are in 16-bit words, representing audio information at 44.1 kHz rate. If the digital codes were converted directly into analog, image frequency bands centred at multiples of the sampling frequency of 44.1 kHz would be produced (see Figure 2.43(a)). Although the image frequencies would be inaudible as they are above 20 kHz, they could cause over-

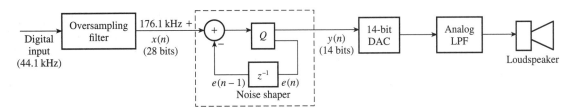

Figure 2.42   A simplified block diagram of audio signal reproduction in the CD player with four times oversampling and noise shaping.

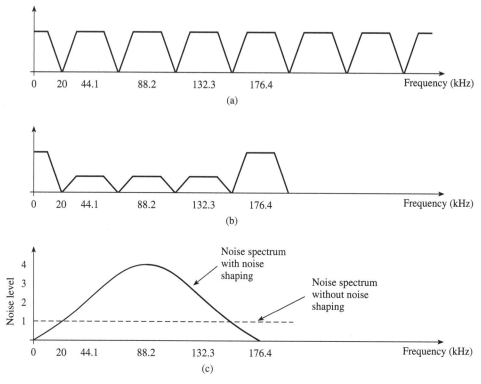

Figure 2.43 (a) Spectrum of audio signal samples at 44.1 kHz. (b) Spectrum of a four times oversampled audio signal showing much reduced image components. (c) The effect of the noise shaper on the noise spectrum. Note that in the audio band (0–20 kHz) the noise level is substantially less than at the high frequency and compared to the noise spectrum if there were no noise shaping (dashed line).

loading if passed on to the player's amplifier and loudspeaker, or they could set up intermodulation distortion. Thus, the frequency components above the baseband need to be attenuated by at least 50 dB. Analog filters that can provide this level of attenuation will have to meet very tight specifications, and would require trimming to ensure that the filters for the two stereo channels are matched.

To avoid such problems, an oversampling filter is used in the compact disc player. This is achieved by multiplying the sampling frequency of the data by four to 176.4 kHz (4 × 44.1 kHz) before D/A conversion. The frequency spectrum of the four times oversampled and digitally filtered signal is shown in Figure 2.43(b). It is seen that the image components above 20 kHz are substantially reduced after oversampling, making it easier to filter out. In the time domain, the effect of this is to produce a signal with much finer steps. In the frequency domain, the image frequencies are now pushed up to higher frequencies.

The output of the oversampling filter (28-bit words) is fed into a noise shaper and then quantized to 14-bit words, by rounding: see Figure 2.42. (The oversampling filter has a coefficient wordlength of 12 bits, but its input consists of 16-bit words. After

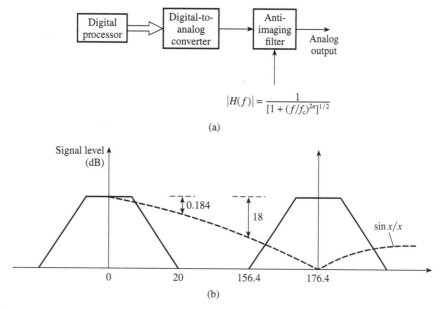

$$|H(f)| = \frac{1}{[1 + (f/f_c)^{2n}]^{1/2}}$$

(a)

(b)

Figure 2.44   (a) Back end of a real-time DSP system. (b) Spectrum at the DAC output.

filtering, the output consists of 28-bit words.) The quantization error is fed back and combined with the output of the oversampling filter. The noise shaper serves to push the quantization noise towards the high frequency end.

The combined effect of oversampling, filtering and noise shaping is to substantially reduce the image components and the quantization level in the signal band. This makes it possible to use a 14-bit DAC and still achieve a SNR performance that is equivalent to that of a 16-bit DAC. It can be shown that the four times oversampling filter and the noise shaper provide an improvement in the SNR of 6 dB and 7 dB, respectively.

The holding effect of the DAC produces a $\sin x / x$ effect in the spectrum which serves to further reduce the image components. A simple anti-imaging filter is then used to recover the audio signal.

Example 2.13

Figure 2.44(a) depicts the set-up used to recover an analog signal, after it has been digitally processed in a certain real-time digital audio system. The analog signal has a baseband that extends from dc to 20 kHz and the digital-to-analog converter is updated at a rate of 176.4 kHz. The image frequencies are to be suppressed by at least 50 dB and the signal components of interest are to be altered by a maximum of 0.5 dB. Determine the minimum values for the order and cutoff frequency for the anti-image filter, assuming that it has a Butterworth characteristic. State any reasonable assumptions made.

Solution   Assuming a zero-order hold, the spectrum at the DAC output is the product of the signal spectrum and the $\sin x/x$ response; see Figure 2.44(b). Attenuation of the signal due to the $\sin x/x$ spectrum at the two critical frequencies, 20 kHz and 156.4 kHz (the image frequencies closest to the baseband), is as follows:

$$\text{at 20 kHz:} \qquad \frac{\sin x}{x} = 0.9789 \text{ (with } x = \omega T/2) \rightarrow -0.184 \text{ dB}$$

$$\text{at 156.4 kHz:} \qquad \frac{\sin x}{x} = 0.125 \rightarrow -18 \text{ dB}$$

Thus, in the passband the output filter should not have more than $0.5 - 0.184 = 0.316$ dB deviation. In the stopband an additional attenuation of at least $50 - 18 = 32$ dB is required. Thus

$$20 \log [1 + (20/f_c)^{2n}]^{1/2} \leqslant 0.316 \text{ dB}$$

$$20 \log [1 + (156.4/f_c)^{2n}]^{1/2} \geqslant 32 \text{ dB}$$

Solving the simultaneous equations for n gives $n = 2.4 \simeq 3$ (integer) and $f_c = 30.76$ kHz.

2.10   Constraints of real-time signal processing with analog input/output signals

The main constraints and errors introduced by the analog-to-digital and digital-to-analog conversion processes in real-time DSP have already been discussed. We outline here these constraints and possible solutions.

- The use of a finite number of bits to represent data introduces an intrinsic error, the quantization error, which is propagated into subsequent signal processing. Two ways of dealing with this error are to increase the resolution of the ADC and to oversample the signal followed by a further DSP to improve the SNR (see Chapter 9 for more details).

- High resolution ADC and DAC are in general slow (except for the very expensive converters). Typically, an ADC takes a few microseconds to convert an analog sample, and a DAC takes a significant fraction of a microsecond to settle. Thus these delays impose a limit on the maximum sampling frequency achievable. In fact, with current technology, the ADCs/DACs are now a major bottleneck in most real-time DSP applications.

- The ADCs/DACs are subject to a variety of other errors which include temperature effects and nonlinearities. Thus a good real-time DSP system with analog inputs should have good quality analog input/output sections.

- The output of the sample and hold is wideband (because of the image frequencies) and will increase the noise at the ADC input.

- Aliasing error from signal energy outside the band of interest is always present. To reduce aliasing to an acceptable level, bandlimit the signal before sampling, and oversample if possible.

- The use of a zero-order DAC introduces the $\sin x/x$ effect which progressively reduces the high frequency components of a signal. This can be compensated for by using a digital filter with an $x/\sin x$ response.

- Errors are introduced by the anti-aliasing filters. Typically, these are amplitude and phase errors. The amplitude responses of these filters are not flat in the band of interest. Analog filters with reasonably good amplitude response have invariably poor phase response, which means that the harmonic relationships between signal components are distorted. In multichannel systems, the problem is compounded because the distortions introduced by the analog signal conditioners are different for each channel and may need to be compensated for.

- Sample and hold errors include acquisition time, aperture uncertainty, droop errors during the conversion interval, and feedthrough in the hold mode.

- The trend in modern DSP systems, especially digital audio systems such as the compact disc player, is to use single-bit ADCs and DACs. These newer devices exploit the advantages of multirate techniques (see Chapter 9 for details).

2.11 Application examples

Applications of the analog I/O interface techniques are pervasive as part of most real-time DSP systems. The newer application examples exploit the sampling theorem and factors associated with sampling, namely, the sampling rate, the fact that the spectrum of the sampled data repeats at multiples of the sampling frequency, the finite bandwidth of the valid signal (i.e. the actual signal components are constrained between say 0 and $F_s/2$), the effect of sampling on quantization noise, etc. This has led to the development of low cost, high resolution A/D and D/A converters (see Sections 2.5 and 2.9). These applications rely on multirate techniques which are covered in Chapter 9 and for this reason we will postpone the discussion of such applications to that chapter.

Bandpass sampling techniques are being exploited in communication systems to enhance receiver design. The issues involved in such applications require a good understanding of multirate and so we will also postpone this discussion.

2.12   Summary

A generalized view of a real-time DSP system is one which consists of an analog-to-digital conversion section, a digital processor and a digital-to-analog conversion section. The time required to convert between analog and digital in such a system limits the maximum signal bandwidth that can be handled, and the devices used in the conversion processes may introduce significant errors or signal degradation. Most of the errors can be minimized by a careful choice of devices (ADCs, DACs, and so on) and the system parameters (sampling frequency and so on). For example, aliasing can be reduced by sampling at a sufficiently high frequency and employing adequate bandlimiting filters.

Problems

Sampling and aliasing control

2.1   What do you understand by each of the following:

    (a)   Nyquist frequency?

    (b)   Nyquist rate?

    (c)   Sampling rate?

    (d)   Sampling frequency?

2.2   A signal has the spectrum depicted in Figure 2.45. Determine the minimum sampling frequency to avoid aliasing. Assume the signal is sampled at a rate of 16 kHz, and sketch the spectrum of the sampled signal in the interval ±16 kHz. Indicate the relevant frequencies including the foldover frequency in your sketch.

2.3   Explain why sampling theorem considerations alone are not sufficient for establishing the actual sampling frequencies used in practical DSP systems.

2.4   Explain clearly the role of the anti-aliasing filter and the anti-imaging filter in real-time DSP systems. Why are the requirements of the two filters often the same in DSP systems?

2.5   The requirements for the analog input section of a certain real-time DSP system are:

frequency band of interest	0–4 kHz
maximum permissible passband ripple	≤0.5 dB
stopband attenuation	≥50 dB

Determine the minimum order of an anti-aliasing filter with Butterworth characteristics and a suitable sampling frequency to satisfy the requirements.

2.6   The analog input to a real-time DSP system is digitized with a 16-bit ADC in the bipolar mode. The peak-to-peak amplitude of the input signal is in the range ±10 V and lies in the band 0–10 kHz. Estimate the minimum

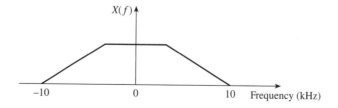

Figure 2.45

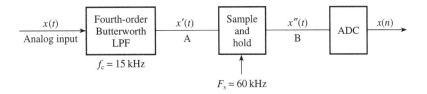

Figure 2.46

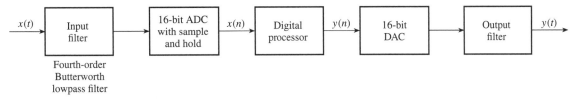

Figure 2.47 Real-time digital signal processing system.

(1) stopband attenuation, A_{min}, for the anti-aliasing filter, and

(2) sampling frequency, F_s, to keep the aliasing error in the passband to just below the quantization noise level (assume a sixth-order Butterworth filter is used for the anti-aliasing filtering).

2.7 An analog signal with a uniform power spectral density is bandlimited by a filter with the following amplitude response:

$$|H(f)| = \frac{1}{[1 + (f/f_c)^6]^{1/2}}$$

where $f_c = 3.4$ kHz. The signal is digitized using a linear 8-bit ADC. Determine the minimum sampling frequency so that the maximum aliasing error is less than the quantization error level in the passband.

2.8 The analog input signal to a real-time DSP system is bandlimited to 30 Hz with an analog filter having a third-order Butterworth characteristic before it is digitized. If the aliasing error due to sampling is to be less than 1% of the signal level in the passband, determine the minimum sampling frequency, F_s, required for the system.

If the signal, after digitization and processing, was converted back to analog what will be the average error introduced by the aperture effect at 30 Hz? Assume that the input signal was digitized using an ideal sampler and ADC, but was recovered using a zero-order hold DAC. A common sampling frequency of 256 Hz at the input and output may be assumed.

2.9 The front end of a real-time DSP system is depicted in Figure 2.46. Assume a wideband input signal.

(a) Sketch the spectrum of the signal before sampling (point A) and after sampling (point B) between the range $\pm F_s/2$.

(b) Determine the signal and aliasing error levels at 15 kHz and the Nyquist frequency (i.e. 30 kHz).

(c) Determine the minimum sampling frequency, F_s (min), to give a signal-to-aliasing error level of 10 : 1 at 15 kHz. State any assumptions made.

2.10 Figure 2.47 depicts a real-time DSP system. Assuming that the frequency band of interest extends from 0 to 100 Hz and that a 16-bit, *bipolar*, ADC is used, estimate:

(1) the minimum stopband attenuation, A_{min}, for the anti-aliasing filter,

(2) the minimum sampling frequency, F_s, and

(3) the level of the aliasing error relative to the signal level in the passband for the estimated A_{min} and F_s.

Sketch and label the spectrum of the signal at the output of the analog filter, assuming a wideband signal at the input, and that of the signal after sampling.

2.11 (a) Discuss, briefly, three main factors that determine the level of the aliasing error in practical DSP systems. Point out the relevance of each to aliasing control.

(b) An analog signal with uniform power spectrum density is bandlimited by an anti-aliasing filter with a magnitude–frequency response characterized by the following equation:

$$\frac{1}{\sqrt{1 + \left(\dfrac{f}{f_c}\right)^8}}$$

where $f_c = 40$ Hz. The signal is digitized with a linear 12-bit, bipolar, ADC. Determine

(1) the minimum sampling frequency to keep the maximum aliasing error in the passband to no greater than the quantization error level;

(2) the maximum passband signal level, in dB, relative to the ADC quantization noised floor.

State any reasonable assumptions made.

(c) Write down the equation for the bandpass sampling theorem. Explain why the theorem is of interest in digital communication systems.

(d) Starting with the equation in (c), derive an expression for the theoretical minimum sampling rate for a bandpass signal. Assume that the ratio of the upper bandedge frequency to the bandwidth of the signal is an integer. Comment on why the theoretical minimum sampling rate should not be used in practice.

Bandpass undersampling

2.12 The front end of the receiver for a multichannel communication system is depicted in Figure 2.48(a). The received signal has the spectrum shown in Figure 2.48(b), with the channel numbers indicated. A bandpass filter is used to isolate the signal in the desired channel before the signal is digitized at the lowest possible rate.

Assume an ideal bandpass filter with the following characteristics:

$$H(f) = 1 \quad \text{if } 10\text{ kHz} \leqslant f \leqslant 20\text{ kHz}$$
$$0 \quad \text{otherwise}$$

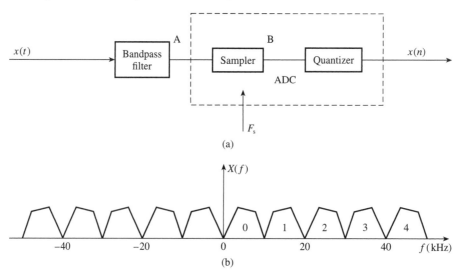

(a)

(b)

Figure 2.48   (a) Front end of system. (b) Spectrum of received signal.

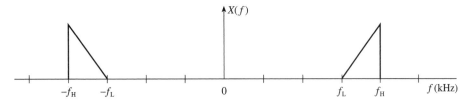

Figure 2.49

(a) (i) Determine the minimum theoretical sampling frequency.
   (ii) Sketch the spectrum of the signal before sampling (point A) and after sampling (point B).

(b) Repeat parts (i) and (ii) for a bandpass filter that passes channel no. 2.

2.13 The spectrum of a narrowband signal is depicted in Figure 2.49. Obtain and sketch the spectrum of the sample signal, in the range $\pm F_s/2$, for each of the following three cases:

(1) $\dfrac{f_H}{B} = 3$

(2) $\dfrac{f_H}{B} = 4$

(3) $\dfrac{f_H}{B} = 4.5$

Assume that the bandwidth of the signal, $B = 5$ kHz, and that the signal is sampled at the rate of $2B$ in each case.

2.14 (a) Determine the minimum theoretical sampling rate, F_s, to avoid aliasing for a bandpass signal with frequency components in the interval 20 MHz $< f < 30$ MHz. Justify your answer and explain why the minimum theoretical sampling rate should not be used in practice.

(b) Assume that the bandpass signal in (a) has a spectrum depicted in Figure 2.50. Determine the edge frequencies of all the frequency bands (including the image components) of the signal after sampling in the interval $\pm 2F_s$. Sketch and clearly label the spectrum of the sampled signal in that interval on graph paper.

(c) Calculate the allowable range of sampling rates to avoid aliasing if the analog bandpass signal is augmented by a 5 kHz guardband at either bandedge.

2.15 (a) Explain briefly the principles of bandpass undersampling technique. Comment on the benefits of the techniques in practice.

(b) A digital radio receiver uses an IF (intermediate frequency) of 50 kHz in the second stage.

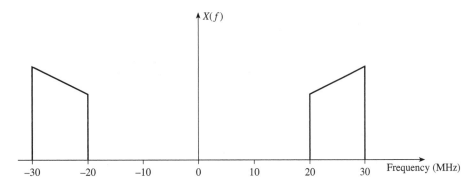

Figure 2.50

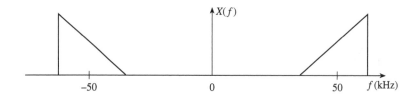

Figure 2.51

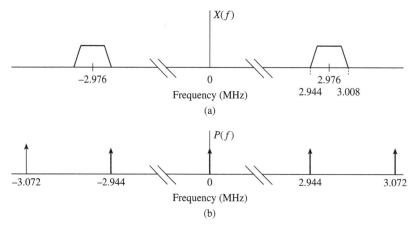

Figure 2.52   (a) Signal spectrum at the second IF stage. (b) Spectrum of sampling function.

(i)   Determine the minimum sampling frequency, F_s, for the system to avoid aliasing if the IF signal bandwidth is 6 kHz.

(ii) Sketch and label the spectrum of the sampled signal in the interval $\pm F_s$. Explain clearly how you obtained the spectrum of the sampled signal and comment on its shape.

Assume that the integer band sampling technique is used and that the signal spectrum at the second IF stage is as shown in Figure 2.51.

2.16 (a) The spectrum at the second IF stage of a digital receiver is depicted in Figure 2.52(a), where the IF frequency is 2.976 MHz. Show, with the aid of appropriate sketches, that the IF signal can be sampled at a rate of 128 kHz without aliasing.

(b) Show, with the aid of appropriate sketches, that if the IF frequency is 3 MHz there will

be an aliased output if the IF signal is the sampled signal at 128 kHz.

2.17 (a) Write down the equation for the bandpass sampling theorem. Explain why the bandpass sampling theorem is of interest in digital communication.

(b) Starting with the equation, derive an expression for the theoretical minimum sampling rate for a bandpass signal. Assume that the ratio of the upper bandedge frequency to the bandwidth of the signal is an integer. Comment on why the theoretical minimum sampling rate may be inappropriate in practice.

Quantization noise in A/D conversion

2.18 A real-time DSP system uses a linear 16-bit ADC in the bipolar mode, with an input range of ± 5 V.

What is the maximum quantization error? Calculate the theoretical maximum SQNR, in decibels, for the system.

2.19 A sinusoidal signal with peak-to-peak amplitude of 5 V is digitized with a 16-bit ADC. Assuming linear quantization, determine

(1) the quantization step size, and
(2) the rms signal-to-quantization noise ratio.

State any assumptions made.

2.20 The analog input to a DSP system is digitized at a rate of 100 kHz with uniform quantization. Assuming a sine wave input with a peak-to-peak amplitude of ±5 V, find the minimum number of bits for the ADC to achieve a SQNR of at least 90 dB. State any reasonable assumptions.

2.21 Show that the signal-to-quantization noise ratio of a linear ADC is given by

$$SQNR = 6.02B + 4.77 - 20 \log (A/\sigma_x) \text{ (dB)}$$

where B is the number of ADC bits, $\pm A$ is the input range of the ADC, and σ_x is the rms value of the input signal. Determine the SQNR if the resolution of the ADC is 16 bits and the input is

(1) a sine wave signal, and
(2) a signal with an rms value of $A/4$.

State any assumptions made.

2.22 The analog input signal to a B-bit ADC has an rms value of σ_x (V). The input range of the ADC is adjusted to the range $\pm 3\sigma_x$ (V). Find an expression for the SQNR, in decibels, for the converter. State any reasonable assumptions made.

Oversampling in A/D conversion – aliasing and quantization noise control

2.23 (a) Explain, with the aid of a suitable sketch, the principles of oversampling techniques and how they can be used to increase the effective resolution of Nyquist rate analog-to-digital converters.

(b) A digital audio system uses oversampling techniques and an 8-bit, bipolar, Nyquist rate converter to digitize an analog input

signal which has frequency components in the range 0–4 kHz. Estimate the effective resolution, in bits, of the converter if the sampling rate is 40 MHz. Show clearly how you obtained your answer. Comment on the practical problems associated with this approach.

2.24 (a) A requirement exists for a general purpose, multichannel (up to 64 channels) data acquisition system for collecting medical data. Each analog channel is to be individually configured, by the user, to have a passband edge frequency between 0.5 Hz and 200 Hz, and a selectable sampling frequency in the range of 1 Hz to 2 kHz. In the passband, the maximum permissible ripple is 0.5 dB and the image components must be at least 40 dB below the signal components.

Explain the strategy you would use to satisfy the above requirement. Your answer should address the following points:

(i) considerations for application-specific issues;
(ii) how oversampling techniques may be used in this application to satisfy the requirement in an efficient and economical way (in terms of cost/component count).

(b) Assume that identical anti-aliasing filters are used for all the channels in the system in (a), each with the following Butterworth characteristics:

$$A(f) = \frac{1}{\sqrt{1 + \left(\dfrac{f}{f_c}\right)^8}}$$

where $f_c = 3$ dB cutoff frequency of filter.

Determine, with the aid of sketches of the spectrum of the data before and after sampling:

(i) the cutoff frequency, f_c;
(ii) a suitable common sampling frequency, F_s.

Comment on your answer.

2.25 An audio system handles signals with a baseband that extends from 0 to 20 kHz. Determine the over-

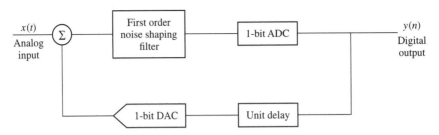

Figure 2.53   A first-order sigma delta modulator.

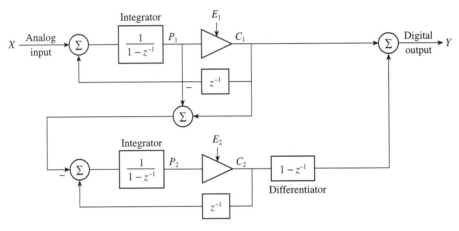

Figure 2.54   The output transform, $Y(z)$, of the second-order sigma delta modulator is given by
$$Y(z) = X(z) + E_2(z)(1 - z^{-1})^2.$$

sampling ratio and the minimum sampling frequency that will be necessary to achieve a performance that would be obtained with a 16-bit ADC using an 8-bit converter.

2.26 (a) In relation to 'single-bit' ADCs, write brief explanatory notes on each of the following techniques:

    (i)   oversampling;
    (ii) noise spectrum shaping.

    (b) Why are 'single-bit' ADCs preferred to conventional successive approximation ADCs in high fidelity DSP systems?

    (c) A digital signal processing system with analog audio signal input in the range 0–20 kHz uses oversampling techniques and the first-order sigma delta modulator

depicted in Figure 2.53 to digitize the analog signal. Assuming that the sampling frequency is 3.072 MHz, find the response of the noise shaping filter at 20 kHz. Estimate the effective resolution, in bits, of the digitizer.

2.27 A digital signal processing system, with analog audio signal input in the range 0–20 kHz, uses oversampling techniques and the second-order sigma delta modulator to convert the analog signal into a digital bit stream at a rate of 6.144 MHz. The z-plane model of the sigma delta modulator is depicted in Figure 2.54.

    Determine the overall improvement in signal-to-quantization noise ratio made possible by oversampling and noise shaping and hence estimate the effective resolution, in bits, of the digitizer.

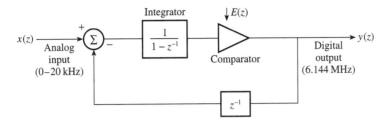

Figure 2.55   SDM *z*-plane model.

2.28   A digital signal processing system, with analog audio signal input in the range 0–20 kHz, uses oversampling techniques and the first-order sigma delta modulator to convert the analog signal into a digital bit stream at a rate of 6.144 MHz. The *z*-plane model of the sigma delta modulator is depicted in Figure 2.55.

  (i)   Explain how the digital bit stream may be converted into a digital multibit stream at 92 kHz rate.

  (ii)   Determine the overall improvement in signal-to-quantization noise ratio made possible by oversampling and noise shaping and hence estimate the effective resolution, in bits, of the digitizer.

D/A conversion and sin *x*/*x* effects

2.29   The block diagram of a real-time DSP system with analog output is depicted in Figure 2.56(a), and

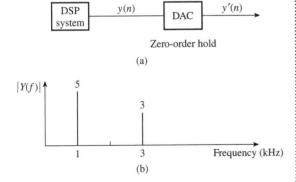

Figure 2.56   (a) Real-time DSP system with analog output. (b) Spectrum of signal applied to the DAC.

Figure 2.56(b) shows the baseband spectrum of the signal applied to the DAC. Sketch the spectrum of the signal at the output of the DAC in the interval 0–$2F_s$, where F_s is the sampling frequency. Determine the amplitudes of the signal components in your sketch. Assume a sampling frequency of 15 kHz.

2.30   A real-time DSP system uses a 16-bit processor, a 12-bit ADC with a conversion time of 15 μs, and a 12-bit DAC with a settling time of 500 ns. If the required DSP operation is the convolution summation given by

$$y(n) = \sum_{k=0}^{N-1} h(k)x(n-k)$$

where the variables have the usual meanings and the computation must be performed between samples, estimate the real-time capability of the system, stating any assumptions made.

2.31   The output of a digital-to-analog converter in response to a digital sequence is given by

$$y(t) = \sum_{n} y(n)h(t - nT)$$

where $h(t)$ is the impulse response of the DAC and $1/T$ is the rate at which data is fed to the DAC. Assume that the DAC is a zero-order hold and $h(t)$ a square pulse of duration T (s).

  Sketch the output of the DAC in response to the input sequence, $y(n)$, shown in Figure 2.57. Show that the spectral shaping effect of the DAC on the signal spectrum can be compensated for by a digital filter with a spectrum of the form

$$|H(\omega)| = \frac{\omega T}{2 \sin (\omega T/2)}$$

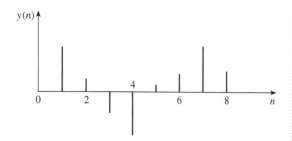

Figure 2.57

2.32 Critically examine the main constraints and errors introduced by analog/digital conversion processes in real-time digital signal processing, suggesting how each constraint or error may be reduced.

Oversampling in D/A conversion – imaging and quantization noise control

2.33 Figure 2.58 depicts the set-up used to recover an analog signal, after it has been digitally processed in a certain real-time digital audio system. The analog signal has a baseband that extends from 0 to 24 kHz and the sampling rate is 192 kHz.

The image frequencies are to be suppressed by at least 50 dB, without altering the audio signal components by more than 0.5 dB. Determine, with the aid of appropriate sketches of the signal spectrum, the minimum values for the order and cutoff frequency for the anti-image filter, assuming that it has a Butterworth characteristic. Stage any reasonable assumptions made.

2.34 Figure 2.59 depicts the set-up used to recover an analog signal, after it has been digitally processed in a certain real-time digital audio system. The analog signal has a baseband that extends from 0 to 20 kHz and the sampling rate is 176.4 kHz.

The noise shaper is characterized by the following equations:

$$y'(n) = x(n) - e(n - 1) \tag{a}$$

$$e(n) = y(n) - y'(n) \tag{b}$$

(a)   Derive an expression for the transfer function seen by the quantization noise and hence sketch the spectrum of the quantization noise after noise shaping.

(b)   Determine the improvement in signal-to-quantization-noise ratio made possible by oversampling and noise shaping and hence

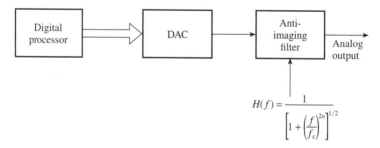

Figure 2.58   A back end of a real-time DSP system.

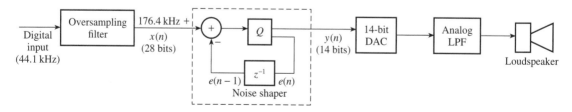

Figure 2.59   A simplified block diagram of audio signal reproduction in the CD player with four times oversampling and noise shaping.

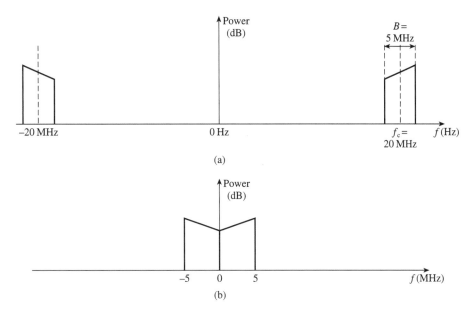

(a)

(b)

Figure 2.60

estimate the effective resolution, in bits, of the DAC. Assume a four times oversampling ratio.

2.35 A DSP system is preceded by a sample and hold with an aperture time of 10 ns and acquisition time of 1 μs followed by an 8-bit ADC. Determine the maximum ADC conversion time to support a sampling frequency of 100 kHz.

MATLAB problems

2.36 Figure 2.59 depicts the set-up used to recover an analog signal, after it has been digitally processed in a certain real-time digital audio system. The analog signal has a baseband that extends from 0 to 20 kHz and the sampling rate is 176.4 kHz.

(a) Use MATLAB to compute and plot the spectrum of a suitable four times oversampling FIR filter. List the coefficients of the filter.

(b) Represent the filter coefficients as 12-bit fixed-point numbers.

(c) Generate an audio signal (16 bits at 44.1 kHz).

(d) Simulate the D/A process in MATLAB using fixed-point arithmetic and plot the signals at the output of each block in the time and frequency domain.

2.37 The spectrum of a communication signal of bandwidth B and carrier frequency f_c is shown in Figure 2.60(a). The analog signal is to be passed through an anti-aliasing filter and sampled at a frequency of F_s. The desired spectrum of the sampled signal is given in Figure 2.60(b).

(1) Provide the specification of a suitable anti-aliasing filter and justify your solution.

(2) Provide a list of suitable sampling frequencies that would enable the spectrum in Figure 2.60(b) to be recovered without aliasing. Sketch the resulting spectrum for each case.

(3) In order to keep the sampling rate as low as possible, choose the lowest sampling frequency in (ii) and verify the spectrum of the sub-sampled signal using MATLAB. Provide your MATLAB code in your answer.

Hint Sampling is multiplication in the time domain, which is mathematically equivalent to convolution in the frequency domain. Look up the frequency spectrum of the sampling function and convolve it with that of Figure 2.60(a).

(4) Provide the specification of a digital filter to recover the signal of interest and discuss how adding guard-bands might help to simplify this filter.

2.38 (a) Design, at a block diagram level, a simple digital AM receiver that uses bandpass undersampling techniques and quadrature mixing to demodulate the received signal. Assume an IF bandwidth of 6 kHz. Your design should include:

∎ a specification (with justification) of a suitable IF centre frequency in the range 40 kHz and 60 kHz, an optimum sampling frequency to avoid aliasing, an appropriate quadrature oscillator frequency, and suitable digital filters;
∎ a sketch of the spectrum of the IF signal before and after sampling;
∎ a clear description of how your digital receiver works;
∎ a statement of any reasonable assumptions.

(b) Develop and test a simplified MATLAB model of your digital AM receiver.

References

CCITT (1989) Possible applications for 16 kbits/sec voice coding. Appendix 3 – Annex 1 to Question 21/XV, 13–22, March.

CCITT Recommendation G.726 (1990) 40, 32, 24 and 16 kbit/s Adaptive Differential Pulse Code Modulation (ADPCM), ITU Geneva, Switzerland.

CCITT Recommendation G.711 (1998) Pulse Code Modulation (PCM) of Voice Frequencies, ITU Geneva, Switzerland.

Del Re E. (1978) Bandpass signal filtering and reconstruction through minimum-sampling-rate digital processing. *Alta Frequenza*, **47**(9), September, 395E/675–398E/678.

Goedhart D., Van de Plassche R.J. and Stikvoort E.F. (1982) Digital-to-analog conversion in playing compact disc. *Philips Technical Rev.*, **40**(6), 174–9.

Vaughan R.G., Scott N.L. and White D.R. (1991) The theory of bandpass sampling. *IEEE Transactions on Signal Processing*, **39**(9), September, 1973–84.

Bibliography

Aziz P.M., Sorensen H.V. and Spiegel J.V.D. (1996) An overview of sigma-delta converters. *IEEE Signal Processing Magazine*, January, 61–84.

Bellamy J. (1982) *Digital Telephony*. New York: John Wiley & Sons.

Berkhout P.J. and Eggermont L.D.J. (1985) Digital audio systems. *IEEE ASSP Magazine*, October, 45–67.

Betts J.A. (1978) *Signal Processing, Modulation and Noise*. Unibooks, Hodder and Stoughton.

Blesser B.A. (1978) Digitization of audio: a comprehensive examination of theory, implementation, and current practice. *J. Audio Eng. Soc.*, **26**(10), 739–71.

Blesser B., Locanthi B. and Stockham Jr, T.G. (eds) (1982) *Digital Audio*. New York: Audio Engineering Society.

Candy J.C., Wooley B.A. and Benjamin O.J. (1981) A voice band codec with digital filtering. *IEEE Trans. Communications*, **COM-29**(6), June, 815–30.

Garret P.H. (1981) *Analog I/O Design*. Reston VA: Reston Publishing Co. Inc.

ITTCC (1986) Study Group XVIII – Report R26C, Recommendation G7221. 32 kbit/s Adaptive Differential Pulse-Code Modulation (ADPCM).

Jayant N.S. and Noll P. (1984) *Digital Coding of Waveforms*. Englewood Cliffs NJ: Prentice-Hall.

Macario R.C.V. (1991) *Signal Coding B: Speech Coding*. C. Xydeas, 82–99.

Mueller H.R., Schindler H.R. and Vettiger P. (1978) Signal-to-noise analysis of a PCM voice system based on analogue/digital filtering. *IEEE Trans. Communications*, **COM-26**(5), May, 653–9.

Natvig J.E. (1988) Speech coding in the pan-European digital mobile radio systems. *Speech Communication Magazine*, January.

Oliver B.M., Pierce J.R. and Shannon C.E. (1948) The philosophy of PCM. *Proc IRE*, November, 1324–31.

Oppenheim A. and Schaffer R.W. (1975) *Digital Signal Processing*. Englewood Cliffs NJ: Prentice-Hall.

Papamichalis P. (1987) *Practical Approaches to Speech Coding*. Englewood Cliffs NJ: Prentice-Hall.

Rabiner L.R. and Gold B. (1975) *Theory and Applications of Digital Signal Processing*. Englewood Cliffs NJ: Prentice-Hall.

Sheingold D.H. (ed.) (1986) *Analog-Digital Conversion Handbook*. Englewood Cliffs NJ: Prentice-Hall.

Steer Jr, R.W. (1989) Antialiasing filters reduce errors in A/D converters. *EDN*, March, 171–86.

Tiefenthaler C. (1987) Oversampling to increase signal to noise ratio of ADCs. *Electronic Product Design*, March, 59–62.

Van Doren A.H. (1982) *Data Acquisition Systems*. Reston VA: Reston Publishing Co. Inc.

Discrete transforms

3

This chapter includes introductory material on the usefulness of discrete transforms in digital signal processing, and the derivation of the widely used Fast Fourier Transform. The Discrete Cosine, Walsh, and Hadamard transforms are briefly described. An account is provided of the wavelet transform because of growing interest therein, together with explanations of its application for the denoising of signals based upon multiresolution analysis and also singularity detection.

3.1 Introduction

The transformation of discrete data between the time and frequency domains is described in this chapter. Voltage versus time representations become magnitude versus frequency and phase versus frequency representations, and vice versa. The two domains provide complementary information about the same data. Thus it may sometimes be more meaningful in an application to inspect the magnitude versus frequency plot for changes in the voltage amplitude at a particular frequency than to observe the voltage waveform in order, for example, to obtain an early indication of wear in a machine by fast Fourier transforming the output data. Another example might involve the use of a discrete Fourier transform analyzer and oscilloscope for checking the output of a modulator in a communication system to ensure correct functioning, in which case the test signals should produce amplitude components at certain known frequencies. Attention in these two cases of spectral analysis to a selected and restricted set of frequencies illustrates that transformation may produce the advantage of data reduction in which unimportant data is ignored, thus resulting in increased ease of interpretation. Discrete transforms, particularly the discrete cosine transform, are used in the data compression of speech and video signals to allow transmission with reduced bandwidth. They are also used in image processing to obtain a reduced set of features for pattern recognition purposes. The transforms are also useful as a mathematical tool for accelerating calculations in other signal processing applications such as correlation as used in, for example, sonars for range detection, or in convolution or deconvolution to determine the interrelationships between a system and its inputs and outputs. For these computations the transformation from the frequency to the time domain is as important as the converse. The entire subject is very mathematical, but it may be true to say that the use of discrete transforms in many applications is now standard so expert knowledge of the mathematics and theorems will rarely be required of the applications engineer. The spectral analysis of waveforms is the exception. Each problem here has to be treated on its own merits and it is important to understand the topic properly to avoid the numerous pitfalls associated with the need to acquire sufficient discrete regular data samples, and to avoid aliasing, picket fencing and spectral leakage. These topics will be discussed in detail in Chapter 11.

Of the available transforms, the discrete Fourier transform (DFT) and the algorithm for its fast computation, the fast Fourier transform (FFT), are the best known and probably the most important. Reasons for this are that they permit adequate representation in the frequency domain for all but the shortest of data lengths ($< 1\,$s), that the truncated Fourier frequency components give a more faithful representation of the data than any other exponential series, that the individual components are sinusoidal and are not distorted during transmission through linear systems, thereby constituting good test signals, and that the FFT can be computed so rapidly. Another reason is that Fourier analysis has been in existence since its publication in 1822 by Fourier and has therefore achieved a high degree of familiarity, respectability and development, as well as a wide range of applications.

Recently considerable effort has been devoted to the wavelet transform because of its ability to describe stochastic signals of time varying frequency content in terms of wavelet amplitudes. This subject is highly mathematical, but the basic principles are given here, together with two illustrative applications in the extraction of signals from noise.

Electrical and electronic engineering students are initially taught to analyze the electrical behaviour of circuits using the Laplace transform. This is because the Fourier transform can deal with neither non-zero initial conditions nor step inputs.

When they progress to the study of the frequency response and stability of discrete systems, such as finite impulse response filters, use of the z-transform is required. Thus, the applications of the Fourier transform lie primarily in fast signal processing computations using the FFT and in spectral analysis. However, the three transforms are related. The Laplace transform may be regarded as the more general since the other two may be derived from it. Thus, the Laplace variable is $s = \sigma + j\omega$ while the Fourier transform variable is $s = j\omega$ and the z-transform variable is given by $z = e^{sT}$, where T is the time between the discrete samples. Finally, the Fourier transform and z-transform variables are related by $z = e^{j\omega T}$ (see Chapter 4).

3.1.1   Fourier series

Any periodic waveform, $f(t)$, can be represented as the sum of an infinite number of sinusoidal and cosinusoidal terms, together with a constant term, this representation being the Fourier series given by

$$f(t) = a_0 + \sum_{n=1}^{\infty} a_n \cos(n\omega t) + \sum_{n=1}^{\infty} b_n \sin(n\omega t) \tag{3.1}$$

where t is an independent variable which often represents time but could, for example, represent distance or any other quantity, $f(t)$ is often a varying voltage versus time waveform, but could be any other waveform, $\omega = 2\pi/T_p$ is known as the first harmonic, or fundamental, angular frequency, related to the fundamental frequency, f, by $\omega = 2\pi f$, T_p is the repetition period of the waveform,

$$a_0 = \frac{1}{T_p} \int_{-T_p/2}^{T_p/2} f(t)\,dt$$

is a constant equal to the time average of $f(t)$ taken over one period which might represent, for example, a dc voltage level,

$$a_n = \frac{2}{T_p} \int_{-T_p/2}^{T_p/2} f(t)\cos(n\omega t)\,dt$$

and

$$b_n = \frac{2}{T_p} \int_{-T_p/2}^{T_p/2} f(t)\sin(n\omega t)\,dt$$

The frequencies $n\omega$ are known as the nth harmonics of ω. The infinite series 3.1 therefore includes cosinusoidal and sinusoidal frequency-dependent terms of different amplitudes a_n and b_n at the positive harmonic frequencies $n\omega$. This series may be written more compactly by using exponential notation and has the advantage in that form of being more easily manipulated mathematically. The series then becomes

$$f(t) = \sum_{n=-\infty}^{\infty} d_n \, e^{jn\omega t} \tag{3.2}$$

in which

$$d_n = \frac{1}{T_p} \int_{-T_p/2}^{T_p/2} f(t) \, e^{-jn\omega t} \, dt \tag{3.3}$$

is complex and $|d_n|$ has the unit of volts.

The summation includes negative values of n, so that half of the series consists of negative frequencies, $-n\omega$. These lack physical significance, being purely mathematical, but, as a result, the magnitudes $|d_n|$ of the complex amplitudes d_n are now halved numerically. This represents an equal sharing of the amplitudes between corresponding negative and positive frequencies. Thus the correct amplitude at frequency $n\omega$ is found by doubling the calculated value. The complex and trigonometric forms are related through

$$|d_n| = (a_n^2 + b_n^2)^{1/2} \tag{3.4}$$

and

$$\phi_n = -\tan^{-1}(b_n/a_n) \tag{3.5}$$

where ϕ_n is the phase angle of the nth harmonic component, also given by the arctangent of the ratio of the imaginary to the real parts of d_n. Thus each harmonic component of the waveform is characterized by both its phase angle and its amplitude.

Example 3.1

As an example consider the periodic unipolar pulse waveform shown in Figure 3.1(a). Deliberate choice of the time origin to be offset from the centre and edge of a pulse is intended to allow illustration of the phase feature of the Fourier series. Substituting appropriate values into Equation 3.3 gives

$$d_n = \frac{1}{T_p} \int_{-(\tau-x\tau)}^{x\tau} A \, e^{-jn\omega t} \, dt \tag{3.6}$$

$$= \frac{A}{T_p} \left[\frac{e^{-jn\omega t}}{-jn\omega} \right]_{-(\tau-x\tau)}^{x\tau}$$

$$= \frac{A}{n\omega T_p} \frac{e^{-jn\omega x\tau} - e^{jn\omega(\tau-x\tau)}}{-j}$$

$$= \frac{A}{n\omega T_p} e^{-jn\omega x\tau} \left[\frac{e^{jn\omega\tau} - 1}{j} \right]$$

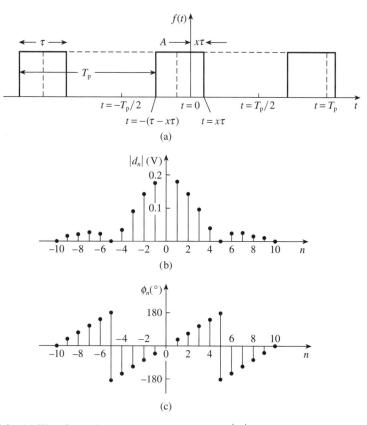

(a)

(b)

(c)

Figure 3.1 (a) Waveform, $f(t)$; (b) amplitude spectrum, $|d_A|$; and (c) phase spectrum, ϕ_n.

$$= \frac{2A}{n\omega T_p} e^{-jn\omega x\tau} \left[\frac{e^{jn\omega\tau/2} - e^{-jn\omega\tau/2}}{2j} \right] e^{jn\omega\tau/2}$$

$$= \frac{2A}{n\omega T_p} e^{jn\omega(\tau/2-x\tau)} \sin\left(\frac{n\omega\tau}{2} \right)$$

$$= \frac{2A}{n\omega T_p} \frac{n\omega\tau}{2} e^{jn\omega(\tau/2-x\tau)} \frac{\sin(n\omega\tau/2)}{n\omega\tau/2}$$

$$= \frac{A\tau}{T_p} \operatorname{Sa}\left(\frac{n\omega\tau}{2} \right) e^{jn\omega(0.5-x)\tau} \qquad (3.7)$$

where

$$\operatorname{Sa}\left(\frac{n\omega\tau}{2} \right) = \frac{\sin(n\omega\tau/2)}{n\omega\tau/2}$$

is known as the sampling function of the argument $n\omega\tau/2$. The modulus of d_n is

$$|d_n| = \frac{A\tau}{T_p}\left|\mathrm{Sa}\left(\frac{n\omega\tau}{2}\right)\right|$$

and is plotted in Figure 3.1(b). $n\omega(0.5 - x)\tau$ represents the phase angle, ϕ_n, in radians, associated with the nth harmonic component. In order to be able to plot this phase angle against harmonic number, n, a specific case is considered. Set $x = 0$, that is locate the time origin at the lagging edge of a pulse, and let $\tau = T_p/5$, when

$$\phi_n = \frac{n\omega\tau}{2} = n\frac{2\pi}{T_p}\frac{\tau}{2} = n\frac{2\pi}{T_p}\frac{T_p}{5}\frac{1}{2} = \frac{n}{5}\pi$$

ϕ_n is plotted in Figure 3.1(c), where the convention $-180° \leqslant \phi_n \leqslant 180°$ has been adopted. Selection of a different time origin would result in a different phase spectrum, ϕ_n versus n. The analysis is usually simplified by location of the time origin at a position of symmetry, for example at the midpoint of a pulse in a periodic pulse train. For the chosen case the amplitude spectrum (Figure 3.1(b)) is seen to be an even function ($|d_n| = |d_{-n}|$) while the phase spectrum (Figure 3.1(c)) is seen to be an odd function ($\phi_{-n} = -\phi_n$). The phase angles, ϕ_n, ϕ_{-n}, give the relative phase angles of the harmonic components to one another. At time t, the absolute phase angles are $\{n\omega(0.5 - x)\tau + n\omega t\}$, from Equation 3.2.

3.1.2 The Fourier transform

The Fourier series approach has to be modified when the waveform is not periodic. A most important example of this is the case of a single rectangular pulse such as might be obtained from the periodic waveform of Figure 3.1(a) by increasing the period T_p to be infinite. As T_p is increased the spacing between the harmonic components, $1/T_p = \omega/2\pi$, decreases to $d\omega/2\pi$, eventually becoming zero. This corresponds to a change from the discrete frequency variable $n\omega$ to the continuous variable ω, and the amplitude and phase spectra become continuous. Thus $d_n \to d(\omega)$ as $T_p \to \infty$. With these modifications, Equation 3.3 becomes

$$d(\omega) = \frac{d\omega}{2\pi}\int_{-\infty}^{\infty} f(t)\,\mathrm{e}^{-j\omega t}\,dt \tag{3.8}$$

It is conventional to normalize this formula by dividing by $d\omega/2\pi$ to obtain

$$\frac{d(\omega)}{d\omega/2\pi} = F(j\omega) = \int_{-\infty}^{\infty} f(t)\,\mathrm{e}^{-j\omega t}\,dt \tag{3.9}$$

$F(j\omega)$ is complex and is known as the Fourier integral or, more commonly, the Fourier transform. If we put

$$F(j\omega) = \text{Re}\,(j\omega) + j\,\text{Im}\,(j\omega) = |F(j\omega)|\,e^{j\phi(\omega)} \tag{3.10}$$

then

$$|F(j\omega)| = [\text{Re}^2\,(j\omega) + \text{Im}^2\,(j\omega)]^{1/2} \tag{3.11}$$

and has the units of volts per hertz rather than volts, $|F(j\omega)|$ is therefore an amplitude density called the amplitude spectral density. The associated phase angle, $\phi(\omega)$, is

$$\phi(\omega) = \tan^{-1}\,[\text{Im}\,(j\omega)/\text{Re}\,(j\omega)] \tag{3.12}$$

$|F(j\omega)|^2$ has the units of $V^2\,Hz^{-2}$. Since normalized electrical power, that is the power dissipated by a $1\,\Omega$ resistor, has the units of V^2 which are equivalent to $J\,s^{-1}$ or $J\,Hz$ (J denotes joules, the unit of energy) then $V^2\,Hz^{-2}$ is $J\,Hz \times Hz^{-2} = J\,Hz^{-1}$. Hence $|F(j\omega)|^2$ has the unit equivalent to energy Hz^{-1}, that is $|F(j\omega)|^2$ is an energy spectral density. The area under the $|F(j\omega)|$ versus f plot between frequencies $f_0 - df$ and $f_0 + df$ gives the mean voltage at frequency f_0, and similarly the area under the corresponding $|F(j\omega)|^2$ versus f plot will yield the mean energy at frequency f_0. It is quite usual in spectral analysis to plot the energy spectral density against frequency.

| **Example 3.2** | Returning to the consideration of a single pulse, let us calculate its amplitude spectral density using Equation 3.9 and Figure 3.1(a). The expression becomes |

$$F(j\omega) = \int_{-(\tau - x\tau)}^{x\tau} A\,e^{-j\omega t}\,dt \tag{3.13}$$

which differs from Equation 3.6 only by the constant $1/T_p$. It follows that

$$F(j\omega) = A\tau\,e^{j\omega(1/2 - x)\tau}\,\text{Sa}\,(\omega\tau/2) \tag{3.14}$$

which is a factor T_p larger than d_n corresponding to the fact that $|F(j\omega)|$ has units of volts multiplied by time or $V\,Hz^{-1}$. Incidentally, the result 3.14 may be obtained more simply than Equation 3.7 was obtained if certain Fourier transform properties are utilized. Thus a pulse of width τ and of unit height centred at $t = 0$ and denoted as rect (t/τ) has the Fourier transform $\tau\,\text{Sa}\,(\omega\tau/2)$. Since $Af(t)$ has the transform $AF[f(t)]$, where F denotes the Fourier transform, then the pulse of height A has the transform $A\tau\,\text{Sa}\,(\omega\tau/2)$. In the case of Figure 3.1(a), the pulse is shifted left by $\tau/2 - x\tau$ and so is actually the rectangular pulse, rect $\{[t + (\tau/2 - x\tau)]/\tau\}$. The delay property of the Fourier transform states that $f(t - t_0) = e^{-j\omega t_0}F[f(t)]$ for a pulse shifted right by t_0. Application of this property to $A\tau\,\text{Sa}\,(\omega\tau/2)$ gives the required form of the transform as

$$F(j\omega) = e^{+j\omega(+\tau/2 - x\tau)} A\tau \, Sa\left(\frac{\omega\tau}{2}\right)$$

$$= A\tau \, e^{j\omega(1/2 - x)\tau} \, Sa\left(\frac{\omega\tau}{2}\right)$$

which is the same result as Equation 3.14.

If the time origin is located at the centre of the pulse, that is $x = \frac{1}{2}$, then the Fourier transform of the pulse is given by

$$F(j\omega) = \frac{A\tau \, \sin\left(\omega\tau/2\right)}{\omega\tau/2} = A\tau \, Sa\left(\frac{\omega\tau}{2}\right) \tag{3.15}$$

and is real. $|F(j\omega)|$ is continuous and is plotted in Figure 3.2(a) for the values $A = 1$ V, $T_p = 10$ s and $\tau = 2$ s. This amplitude spectrum, shaped in proportion to the sampling function, is always associated with rectangular pulses, and also with any waveform of finite duration, τ. The latter may be regarded as an infinite waveform multiplied by rect $[(t \pm t_0)/\tau]$, that is by a unit pulse. Experimentally determined waveforms fall into this category, being of finite duration τ. The sampling function passes through zero whenever $\sin\left(\omega\tau/2\right) = 0$, that is whenever $\omega\tau/2 = m\pi$ ($m \neq 0$, m integer). Thus amplitude nulls occur at $f = 1/\tau$, $2/\tau$, $3/\tau$, ... When $\omega \to 0$, $\sin\left(\omega\tau/2\right) \to \omega\tau/2$ and $Sa\left(\omega\tau/2\right) = \sin\left(\omega\tau/2\right)/\left(\omega\tau/2\right) \to 1$, so $F(j\omega) = A\tau$ at $\omega = 0$, that is at $f = 0$. The energy spectral density of the pulse of amplitude 2 V is plotted in Figure 3.2(b) for comparison with the amplitude spectral density of Figure 3.2(a).

It is possible to transform from the frequency domain to the time domain by using the inverse Fourier transform. In this case

$$f(t) = \frac{1}{2\pi} \int_{-\infty}^{\infty} F(j\omega) \, e^{j\omega t} \, d\omega = \int_{-\infty}^{\infty} F(j\omega) \, e^{j\omega t} \, df \tag{3.16}$$

3.2 DFT and its inverse

In practice the Fourier components of data are obtained by digital computation rather than by analog processing. Because the analog waveform consists of an infinite number of contiguous points, the representation of all their values is a practical impossibility. Thus the analog values have to be sampled at regular intervals and the sample values are then converted to a digital binary representation. This is achieved using a sample-and-hold circuit followed by an analog-to-digital (AD) converter. Provided the number of samples recorded per second is high enough the waveform will be adequately represented. The theoretically necessary sampling rate is called the

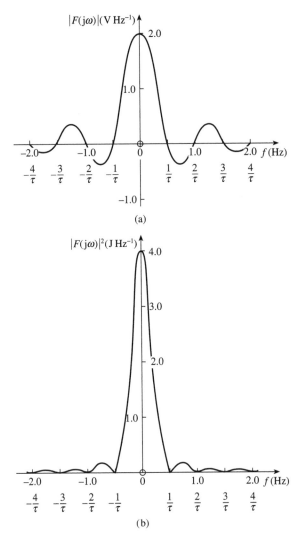

Figure 3.2 (a) Amplitude spectrum of a 2 V pulse; and (b) energy spectrum of a 2 V pulse.

Nyquist rate and is $2f_{max}$ where f_{max} is the frequency of the highest frequency sinusoidal component in the signal of significant amplitude. In this chapter, therefore, it is assumed that the digital values of the data are available for transformation, and aspects such as windowing, required for spectral analysis, will be dealt with in Chapter 11. Thus the data to be transformed is discrete and probably non-periodic. It is not possible to apply the Fourier transform because it is for continuous data. However, an analog transform for use with discrete data, known as the discrete Fourier transform (DFT), is available.

Assume that a waveform has been sampled at regular time intervals T to produce the sample sequence $\{x(nT)\} = x(0), x(T), \ldots, x[(N-1)T]$ of N sample values, where n is the sample number from $n = 0$ to $n = N - 1$. The data values $x(nT)$ will be real only when representing the values of a time series such as a voltage waveform. The DFT of $x(nT)$ is then defined as the sequence of complex values $\{X(k\Omega)\} = X(0)$, $X(\Omega), \ldots, X[(N - 1)\Omega]$ in the frequency domain, where Ω is the first harmonic frequency given by $\Omega = 2\pi/(N - 1)T \simeq 2\pi/NT$ for $N \gg 1$. Thus the $X(k\Omega)$ have real and imaginary components in general so that for the kth harmonic

$$X(k) = R(k) + jI(k) \tag{3.17}$$

and

$$|X(k)| = [R^2(k) + I^2(k)]^{1/2} \tag{3.18}$$

and $X(k)$ has the associated phase angle

$$\phi(k) = \tan^{-1}[I(k)/R(k)] \tag{3.19}$$

where $X(k)$ is understood to represent $X(k\Omega)$. These equations are therefore analogous to those for the Fourier transform: compare Equations 3.17–3.19 with 3.10–3.12.

Note that N real data values (in the time domain) transform to N complex DFT values (in the frequency domain). The DFT values, $X(k)$, are given by

$$X(k) = F_D[x(nT)] = \sum_{n=0}^{N-1} x(nT)\,e^{-jk\Omega nT}, \, k = 0, 1, \ldots, N - 1 \tag{3.20}$$

where F_D denotes the discrete Fourier transformation. In this equation, k represents the harmonic number of the transform component. The equation can be seen to be analogous to the Fourier transform of Equation 3.9 when $f(t) = 0$ for $T < 0$ and $t > (N - 1)T$ by putting $x(nT) = f(t)$, $k\Omega = \omega$, and $nT = t$, so the two transforms may be expected to have similar properties. The transforms are not, however, equal. Thus, making these substitutions in Equation 3.9, and putting $dt = T$, and replacing the integral by a summation gives for harmonic frequencies kf_s, where $f_s = 1/(N - 1)T = 2\pi/\Omega$,

$$\sum_{n=0}^{N-1} x(nT)\,e^{-jk\Omega nT}\, T = F(j\omega) \tag{3.21}$$

for $0 \leqslant t \leqslant (N - 1)T$. Then a comparison of Equation 3.20 with 3.21 reveals that

$$F(j\omega) = TX(k) \tag{3.22}$$

showing that the Fourier transform components are related to the DFT components by the sampling interval, and may be obtained by multiplying the DFT components by the sampling interval.

Note also that in practical applications $N \gg 1$ and the approximation $\Omega = 2\pi/NT$ is frequently used. This approximation is accordingly assumed for the illustrative calculations in this chapter, even for $N = 4$.

Example 3.3   It is appropriate at this point to illustrate the use of Equation 3.20 by evaluating a simple case. The DFT of the sequence $\{1, 0, 0, 1\}$ will be evaluated. It is worth noting here that if discontinuities in the actual data are present the average value of the data on each side of the discontinuity is taken for computational purposes to represent the value at the discontinuity. This will apply to the first and final data values as well as to other discontinuities. However, when obtaining the spectra of signals it is necessary to carry out a process known as windowing to avoid distortion of the spectra due to discontinuities at the beginning and end of the data. This topic is discussed in Chapter 11. In this section it is assumed that the sequence $\{1, 0, 0, 1\}$ has already been preprocessed. Assume that this data represents four consecutive voltages $x(0) = 1$, $x(T) = 0$, $x(2T) = 0$, $x(3T) = 1$, recorded at time intervals, T. Thus $N = 4$. It is required to find the complex values $X(k)$ for $k = 0$, $k = 1$, $k = 2$ and $k = 3$ (since $N - 1 = 3$). With $k = 0$, Equation 3.20 becomes

$$X(0) = \sum_{n=0}^{3} x(nT)\, e^{-j0} = \sum_{n=0}^{3} x(nT)$$

$$= x(0) + x(T) + x(2T) + x(3T)$$

$$= 1 + 0 + 0 + 1 = 2$$

so $X(0) = 2$ is entirely real, of magnitude 2 and phase angle $\phi(0) = 0$. With $k = 1$, Equation 3.20 becomes

$$X(1) = \sum_{n=0}^{3} x(nT)\, e^{-j\Omega nT}$$

T has not been given, but may be eliminated using $\Omega = 2\pi/NT$, giving

$$X(1) = \sum_{n=0}^{3} x(nT)\, e^{-j\Omega n 2\pi/N\Omega} = \sum_{n=0}^{3} x(nT)\, e^{-j2\pi n/N}$$

$$= 1 + 0 + 0 + 1e^{-j2\pi 3/4} = 1 + e^{-j3\pi/2}$$

$$= 1 + \cos\left(\frac{3\pi}{2}\right) - j\sin\left(\frac{3\pi}{2}\right) = 1 + j$$

Thus $X(1) = 1 + j$ and is complex with magnitude $\sqrt{2}$ and phase angle $\phi(\Omega) = \tan^{-1} 1 = 45°$. For $k = 2$, Equation 3.20 becomes

$$X(2) = \sum_{n=0}^{3} x(nT)\, e^{-j2\Omega nT} = \sum_{n=0}^{3} x(nT)\, e^{-j2n2\pi/N}$$

$$= \sum_{n=0}^{3} x(nT)\, e^{-j4\pi n/N}$$

$$= 1 + 0 + 0 + 1e^{-j4\pi 3/4} = 1 + 0 + 0 + e^{-j3\pi} = 1 - 1 = 0$$

Thus $X(2) = 0$, of magnitude zero and of indeterminate phase angle $\phi(2)$. Finally, for $k = 3$, Equation 3.20 becomes

$$X(3) = \sum_{n=0}^{3} x(nT) \, e^{-j3n2\pi/N}$$

$$= 1 + 0 + 0 + e^{-j9\pi/2} = 1 - j$$

Thus $X(3) = 1 - j$, of magnitude $\sqrt{2}$ and phase angle $\phi(3) = -45°$.

It has therefore been shown that the time series $\{1, 0, 0, 1\}$ has the DFT given by the complex sequence $\{2, 1 + j, 0, 1 - j\}$.

It is common practice to represent the DFT by the plots of $|X(k)|$ versus $k\Omega$ and of $\phi(k)$ versus $k\Omega$. This may be done in terms of harmonics of Ω, or in terms of frequency if Ω is known. To find Ω it is necessary to know the value of T, the sampling interval. If it is assumed that the above data sequence had been sampled at 8 kHz then $T = 1/(8 \times 10^3) = 125 \; \mu s$. Then $\Omega = 2\pi/NT = 2\pi/(4 \times 125 \times 10^{-6}) = 12.57 \times 10^3$ rad s^{-1}. Hence $2\Omega = 25.14 \times 10^3$ rad s^{-1} and $3\Omega = 37.71 \times 10^3$ rad s^{-1}. Figure 3.3(a) is a plot of $x(nT)$ versus t, Figure 3.3(b) is a plot of $|X(k)|$ versus $k\Omega$, and Figure 3.3(c) is a plot of $\phi(k)$ versus $k\Omega$. It is noteworthy that the 'amplitude' plot of Figure 3.3(b) is symmetrical about the second harmonic component, that is about harmonic number $N/2$, and that in Figure 3.3(c) the phase angles are an odd function centred round this component. These results are more generally true.

An important property of the DFT may be deduced if the kth component of the DFT, $X(k)$, is compared with the $(k + N)$th component, $X(k + N)$. Thus

$$X(k) = \sum_{n=0}^{N-1} x(nT) \, e^{-jk\Omega nT}$$

$$= \sum_{n=0}^{N-1} x(nT) \, e^{-jk2\pi n/N}$$

and

$$X(k + N) = \sum_{n=0}^{N-1} x(nT) \, e^{-jk2\pi n/N} \, e^{-jN2\pi n/N}$$

$$= \sum_{n=0}^{N-1} x(nT) \, e^{-jk2\pi n/N} \, e^{-j2\pi n}$$

$$= \sum_{n=0}^{N-1} x(nT) \, e^{-jk2\pi n/N} = X(k)$$

since n is integral so $e^{-j2\pi n} = 1$.

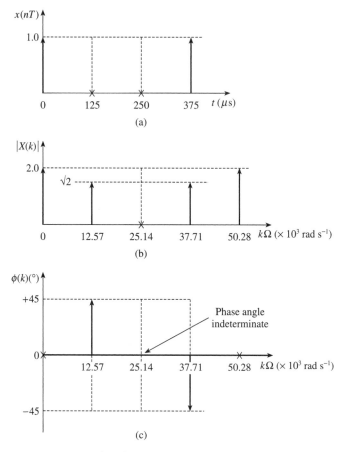

Figure 3.3 (a) $x(nT)$ versus t; (b) $|X(k)|$ versus k; and (c) $\phi(k)$ versus k.

The fact that $X(k + N) = X(k)$ shows that the DFT is periodic with period N. This is the cyclical property of the DFT. The values of the DFT components are repetitive. If $k = 0$, then $k + N = N$ and $X(0) = X(N)$. In the above example $X(0) = 2$ and therefore $X(4) = 2$ also. This is illustrated in Figure 3.3(b) where the fourth harmonic amplitude is drawn at 50.28 kHz. The symmetry of the amplitude distribution about the second harmonic is obvious. The general conclusion is that the amplitude spectrum of an N-point DFT is symmetrical about harmonic $N/2$ when both the zero and $(N + 1)$th harmonics are included in the plot. Similarly, the phase function being odd exhibits anti-symmetry about harmonic $N/2$. If $2f_{max}$ samples per second were taken of the signal for t seconds then $2f_{max}t = N$, so $1/t = 2f_{max}/N$ is the first harmonic frequency. The symmetry at harmonic $N/2$ therefore occurs at the frequency $(N/2)/(2f_{max}/N) = f_{max}$, the maximum frequency present in the signal. Thus all the signal components are fully represented in an amplitude spectrum plotted up to f_{max} or harmonic component $N/2$, and it is unnecessary to plot further points. In this context, f_{max} is known as the folding frequency, since the spectrum between harmonics $N/2$ and N may be folded about the axis of symmetry at f_{max} to superimpose exactly the low frequency half of the spectrum. It is now seen that N real data values transform to $N/2$ complex DFT values of practical significance. The latter consist of $N/2$ real values and $N/2$ imaginary values giving a total of N values derived from the initial N real data values. Finally the values of the Fourier transform components, $F(j\omega)$, of the data $\{1, 0, 0, 1\}$ in Example 3.3 may be obtained by multiplying the DFT components by $T = 125$ μs. Therefore $F(0) = 250$ μV Hz^{-1}, $F(12.57 \text{ kHz}) = (125 + j125)$ μV Hz^{-1}, $F(25.14 \text{ kHz}) = 0$ V Hz^{-1}, $F(37.71 \text{ kHz}) = (125 - j125)$ μV Hz^{-1}.

As explained in the introduction to the chapter, it is also necessary to be able to carry out discrete transformation from the frequency to the time domain. This may be achieved using the inverse discrete Fourier transform (IDFT), defined by

$$x(nT) = F_D^{-1}[X(k)] = \frac{1}{N} \sum_{k=0}^{N-1} X(k) \, e^{jk\Omega nT}, \, n = 0, 1, \ldots, N-1 \tag{3.23}$$

where F_D^{-1} denotes the inverse discrete Fourier transformation.

The analogy with Equation 3.16 for the inverse Fourier transform is obvious. This time it is quite simple to show that the inverse Fourier transform is obtainable from the IDFT by dividing the IDFT by T. The validity of Equation 3.23 can be demonstrated by substituting for $x(nT)$ in Equation 3.20.

| Example 3.4 |

It is useful to illustrate the inverse discrete Fourier transform by using it to derive the time series $\{1, 0, 0, 1\}$ from its DFT components $[2, 1 + j, 0, 1 - j]$.

With $n = 0$,

$$x(nT) = x(0) = \frac{1}{N} \sum_{k=0}^{N-1} X(k)$$

$$= \tfrac{1}{4}[X(0) + X(1) + X(2) + X(3)]$$

$$= \tfrac{1}{4}[2 + (1 + j) + 0 + (1 - j)] = 1$$

as expected. With $n = 1$,

$$x(nT) = x(T) = \frac{1}{N} \sum_{k=0}^{N-1} X(k) \, e^{jk\Omega T}$$

$$= \frac{1}{N} \sum_{k=0}^{N-1} X(k) \, e^{jk2\pi/N} = \frac{1}{4} \sum_{k=0}^{N-1} X(k) \, e^{jk\pi/2}$$

$$= \tfrac{1}{4}[2 + (1 + j) \, e^{j\pi/2} + 0 + (1 - j) \, e^{j3\pi/2}]$$

$$= \tfrac{1}{4}[2 + (1 + j)j + (1 - j)(- j)]$$

$$= \tfrac{1}{4}(2 + j - 1 - j - 1) = 0$$

as expected. With $n = 2$,

$$x(nT) = x(2T) = \frac{1}{N} \sum_{k=0}^{N-1} x(k) \, \mathrm{e}^{\mathrm{j}k\pi}$$

$$= \tfrac{1}{4} [2 + (1 + \mathrm{j}) \, \mathrm{e}^{\mathrm{j}\pi} + (1 - \mathrm{j}) \, \mathrm{e}^{\mathrm{j}3\pi}] = \tfrac{1}{4} [2 - (1 + \mathrm{j}) - (1 - \mathrm{j})]$$

$$= 0$$

again, as expected. Finally, with $n = 3$,

$$x(nT) = x(3T) = \frac{1}{N} \sum_{k=0}^{N-1} X(k) \, \mathrm{e}^{\mathrm{j}k3\pi/2}$$

$$= \tfrac{1}{4} [2 + (1 + \mathrm{j}) \, \mathrm{e}^{\mathrm{j}3\pi/2} + (1 - \mathrm{j}) \, \mathrm{e}^{\mathrm{j}9\pi/2}]$$

$$= \tfrac{1}{4} [2 + (1 + \mathrm{j})(- \mathrm{j}) + (1 - \mathrm{j})\mathrm{j}] = \tfrac{1}{4} (2 - \mathrm{j} + 1 + \mathrm{j} + 1) = 1$$

the correct final term of the series.

3.3 Properties of the DFT

The DFT has a number of mathematical properties which can be used to simplify problems or which lead to useful applications. Some of them are listed below. The data sequences $x(nT)$ are written $x(n)$.

(1) *Symmetry.*

$$\mathrm{Re} \, [X(N - k)] = \mathrm{Re} \, X(k) \tag{3.24}$$

(where Re denotes the real part) states the symmetry of the amplitude spectrum, discussed previously, and

$$\mathrm{Im} \, [X(N - k)] = -\mathrm{Im} \, [X(k)] \tag{3.25}$$

(where Im denotes the imaginary part) states the antisymmetrical property of the phase spectrum. This is a property to be aware of when considering component values.

(2) *Even functions.* If $x(n)$ is an even function $x_\mathrm{e}(n)$, that is $x_\mathrm{e}(n) = x_\mathrm{e}(-n)$, then

$$F_\mathrm{D}[x_\mathrm{e}(n)] = X_\mathrm{e}(k) = \sum_{n=0}^{N-1} x_\mathrm{e}(n) \cos (k\Omega nT) \tag{3.26}$$

(3) *Odd functions.* If $x(n)$ is an odd function $x_0(n)$, that is $x_0(n) = -x_0(-n)$, then

$$F_D[x_0(n)] = X_0(k) = -j \sum_{n=0}^{N-1} x_0(n) \sin(k\Omega nT) \tag{3.27}$$

(4) *Parseval's Theorem.* The normalized energy in the signal is given by either of the expressions

$$\sum_{n=0}^{N-1} x^2(n) = \frac{1}{N} \sum_{k=0}^{N-1} |X(k)|^2 \tag{3.28}$$

The right-hand side of Equation 3.28 is the mean square spectral amplitude, while the left-hand side is the sum of the squared magnitudes of the time series.

(5) *Delta function*:

$$F_D[\delta(nT)] = 1 \tag{3.29}$$

(6) The linear cross-correlation of two data sequences or series may be computed using DFTs. The linear cross-correlation of two finite length sequences $x_1(n)$ and $x_2(n)$, each of length N, is defined to be

$$r_{x_1x_2}(j) = \frac{1}{N} \sum_{n=-\infty}^{\infty} x_1(n)x_2(n+j), \quad -\infty \leqslant j \leqslant \infty \tag{3.30}$$

It is also necessary to define the circular correlation of finite length periodic sequences $x_{1p}(n)$, $x_{2p}(n)$ as

$$r_{cx_1x_2}(j) = \frac{1}{N} \sum_{n=0}^{N-1} x_{1p}(n)x_{2p}(n+j), \quad j = 0, \ldots, N-1 \tag{3.31}$$

because the circular correlation can be evaluated using DFTs. Thus

$$r_{cx_1x_2}(j) = F_D^{-1}[X_1^*(k)X_2(k)] \tag{3.32}$$

Equation 3.32 is known as the correlation theorem. The circular correlation given by Equation 3.32 can be converted into a linear correlation by using augmenting zeros. Now, if the sequences are $x_1(n)$ of length N_1, and $x_2(n)$ of length N_2, their linear correlation will be of length $N_1 + N_2 - 1$. To achieve this $x_1(n)$ is replaced by $x_{1a}(n)$ which consists of $x_1(n)$ with $N_2 - 1$ zeros added, and $x_2(n)$ is augmented by $N_1 - 1$ zeros to become $x_{2a}(n)$. The linear cross-correlation of $x_1(n)$ and $x_2(n)$ is then given by

$$r_{x_1x_2}(j) = F_D^{-1}[X_{1a}^*(k)X_{2a}(k)] \tag{3.33}$$

where

$$X_{1a}(k) = F_D[x_{1a}(n)] \text{ and } X_{2a}(k) = F_D[x_{2a}(n)]$$

This subject is treated more fully in Chapter 5.

(7) DFTS may also be used in the computation of circular convolutions, and, by using augmenting zeros, in linear convolutions. These may be either time or frequency domain convolutions. The time convolution theorem states that

$$x_{3p}(n) = x_{1p}(n) \circledast x_{2p}(n) = F_D^{-1}[X_1(k)X_2(k)] \qquad (3.34)$$

where $\circledast$ denotes circular convolution, and $x_{1p}(n)$, $x_{2p}(n)$ and $x_{3p}(n)$ are finite periodic sequences of equal length.

In an analogous manner to Equation 3.31, $x_{3p}(n)$ may also be written

$$x_{3p}(n) = \sum_{m=0}^{N-1} x_{1p}(m)x_{2p}(n-m) \qquad (3.35)$$

Furthermore,

$$X_3(k) = X_1(k)X_2(k) \qquad (3.36)$$

where $X_3(k) = F_D[x_3(n)]$.

The following equation is the statement of the frequency convolution theorem:

$$\frac{1}{N}X_1(k) \circledast X_2(k) = F_D[x_1(n)x_2(n)] \qquad (3.37)$$

where

$$X_1(k) \circledast X_2(k) = \sum_{m=0}^{N-1} X_1(m)X_2(k-m) \qquad (3.38)$$

Equation 3.34 gives rise to the statement that convolution in the time domain is equivalent to multiplication in the frequency domain, while Equation 3.37 has led to the observation that convolution in the frequency domain is equivalent to multiplication in the time domain. These statements may provide a means of remembering the relationships. Convolution will also be treated in more detail in Chapter 5.

3.4 Computational complexity of the DFT

A large number of multiplications and additions are required for the calculation of the DFT. For an 8-point DFT the expansion for $X(k)$ becomes (from Equation 3.20)

$$X(k) = \sum_{n=0}^{7} x(n) \, e^{-jk2\pi n/8}, \, k = 0, \ldots, 7 \qquad (3.39)$$

and letting $k2\pi/8 = K$ this expands to

$$X(k) = x(0) \, e^{-jK0} + x(1) \, e^{-jK1} + x(2) \, e^{-jK2} + x(3) \, e^{-jK3} + x(4) \, e^{-jK4}$$
$$+ x(5) \, e^{-jK5} + x(6) \, e^{-jK6} + x(7) \, e^{-jK7}, \, k = 0, \ldots, 7 \qquad (3.40)$$

Equation 3.40 contains eight terms on the right-hand side. Each term consists of a multiplication of an exponential term which is always complex by another term which is real or complex (for example, real for a voltage time series). Each of the product terms is added together. There are therefore eight complex multiplications and seven complex additions to be calculated. For an N-point DFT, there will be N and $N - 1$ of them respectively. There are also eight harmonic components to be evaluated ($k = 0, \ldots, 7$). This number becomes N for an N-point DFT. Therefore the calculation of the eight-point DFT requires $8^2 = 64$ complex multiplications and $8 \times 7 = 56$ complex additions. For an N-point DFT these become N^2 and $N(N - 1)$ respectively. If $N = 1024$, then approximately one million complex multiplications and one million complex additions are required. Clearly some means of reducing these numbers is desirable.

The amount of computation involved may be reduced if we note that there is a considerable amount of built-in redundancy in equations such as Equation 3.4. For example if $k = 1$ and $n = 2$, $e^{-jk2\pi n/8} = e^{-j\pi/2}$, and if $k = 2$ and $n = 1$, $e^{-jk2\pi n/8} = e^{-j\pi/2}$ also.

3.5 The decimation-in-time fast Fourier transform algorithm

In this section it will be shown how the computational redundancy inherent in the DFT is used to reduce the number of different calculations necessary, thereby speeding up the computation. For a 1024-point DFT the number of calculations required can be reduced by a factor of 204.8. The algorithms which can achieve this are given the title 'fast Fourier transform', or FFT in short. When applied in the time domain the algorithm is referred to as a decimation-in-time (DIT) FFT. The first DIT algorithm was due to Cooley and Tukey (1965), after whom it is often named. Decimation then refers to the significant reduction in the number of calculations performed on time domain data. It is noteworthy that the computational savings will be seen to increase as $N^2 - (N/2) \log_2 N$.

First the notation will be simplified and some mathematical relationships will be established. Thus Equation 3.20 will be re-written as

$$X_1(k) = \sum_{n=0}^{N-1} x_n \, e^{-j2\pi nk/N}, \, k = 0, \ldots, N - 1 \qquad (3.41)$$

Also, the factor $e^{-j2\pi/N}$ will be written as W_N, thus

$$W_N = e^{-j2\pi/N} \tag{3.42}$$

so that Equation 3.41 becomes

$$X_1(k) = \sum_{n=0}^{N-1} x_n W_N^{kn}, \; k = 0, \ldots, N-1 \tag{3.43}$$

It is worthwhile at this point to note some relationships involving W_N. First,

$$W_N^2 = (e^{-j2\pi/N})^2 = e^{-j2\pi2/N} = e^{-j2\pi/(N/2)} = W_{N/2} \tag{3.44}$$

Second,

$$W_N^{(k+N/2)} = W_N^k W_N^{N/2} = W_N^k \, e^{-j(2\pi/N)(N/2)} = W_N^k \, e^{-j\pi}$$
$$= -W_N^k \tag{3.45}$$

Summarizing the useful results concerning W_N for convenience, we have

$$W_N = e^{-j2\pi/N} \tag{3.46a}$$
$$W_N^2 = W_{N/2} \tag{3.46b}$$
$$W_N^{(k+N/2)} = -W_N^k \tag{3.46c}$$

In exploiting the computational redundancy expressed by Equations 3.46 the data sequence is divided into two equal sequences, one of even-numbered data, and one of odd-numbered data. For the sequences to be of equal length, they must all contain an even number of data. If the original sequence consists of an odd number of data, then an augmenting zero should be added to render the number of data even. This allows the DFT, $X_1(k)$, to be written in terms of two DFTs, $X_{11}(k)$ and $X_{12}(k)$, which are the DFTs of the even-valued data and of the odd-valued data respectively (see Table 3.1). Thus the N-point DFT is converted into two DFTs each of $N/2$ points. This process is then repeated until $X_1(k)$ is decomposed into $N/2$ DFTs, each of two points, both of which are initial data. Thus, in practice, the initial data is re-ordered and the $N/2$ two-point DFTs are calculated by taking the data in pairs. These DFT outputs are suitably combined in fours to provide $N/4$ four-point DFTs which are computed and appropriately combined to produce $N/8$ eight-point DFTs which are computed, and so on until the final N-point DFT, $X_1(k)$, is obtained. At each stage common factors which are powers of W_N are incorporated to reduce the number of complex calculations. This procedure is justified as follows.

The suffixes, n, in Equation 3.43 extend from $n = 0$ to $n = N - 1$, corresponding to the data values $x_0, x_1, x_2, x_3, \ldots, x_{N-1}$. The even-numbered sequence is $x_0, x_2, x_4, \ldots,$ x_{N-2}, and the odd-numbered sequence is $x_1, x_3, \ldots, x_{N-1}$. Both sequences contain $N/2$ points. The terms in the even sequence may be designated x_{2n} with $n = 0$ to $n = N/2 - 1$ while those in the odd sequence become x_{2n+1}. Then Equation 3.43 may be re-written

Table 3.1 Structure of an 8-point FFT.

Line number	Line content	Content							k ranges	N ranges
1	Data sequence A_0	A_0	$x_0\ x_1\ x_2\ x_3\ x_4\ x_5\ x_6\ x_7$								$0,\dots,7$
2	8-point DFT of A_0	$X_1(k) = X_{11}(k) + W_N^k X_{12}(k)$							$0,\dots,N-1$ $(0,\dots,7)$	$0,\dots,7$
3	Re-ordered A_0: two sequences, A_1 and A_2	A_1	$x_0\ x_2\ x_4\ x_6$			A_2	$x_1\ x_3\ x_5\ x_7$			$0,\dots,3$
4	4-point DFTs of A_1 and A_2	$X_{11}(k) = X_{21}(k) + W_{N/2}^k X_{22}(k)$				$X_{12}(k) = X_{23}(k) + W_{N/2}^k X_{24}(k)$			$0,\dots,N/2-1$ $(0,\dots,3)$	$0,\dots,3$
5	Re-ordered sequences A_1 and A_2: four sequences A_3, A_4, A_5, A_6	A_3	$x_0\ x_4$	A_4	$x_2\ x_6$	A_5	$x_1\ x_5$	$A_6\quad x_3\ x_7$		$0,1$
6	2-point DFTs of A_3, A_4, A_5, A_6	$X_{21}(k) = x_0 + W_{N/4}^k x_4$		$X_{22}(k) = x_2 + W_{N/4}^k x_6$		$X_{23}(k) = x_1 + W_{N/4}^k x_5$		$X_{24}(k) = x_3 + W_{N/4}^k x_7$	$0,\dots,N/4-1$ $(0,1)$	$0,1$

$$X_1(k) = \underbrace{\sum_{n=0}^{N/2-1} x_{2n} W_N^{2nk}}_{\text{even sequence}} + \underbrace{\sum_{n=0}^{N/2-1} x_{2n+1} W_N^{(2n+1)k}}_{\text{odd sequence}}$$

$$= \sum_{n=0}^{N/2-1} x_{2n} W_N^{2nk} + W_N^k \sum_{n=0}^{N/2-1} x_{2n+1} W_N^{2nk}, \quad k = 0, \ldots, N-1 \tag{3.47}$$

Using Equation 3.46b gives $W_N^{2nk} = W_{N/2}^{nk}$ so Equation 3.47 becomes

$$X(k) = \sum_{n=0}^{N/2-1} x_{2n} W_{N/2}^{nk} + W_N^k \sum_{n=0}^{n=N/2-1} x_{2n+1} W_{N/2}^{nk}, \quad k = 0, \ldots, N-1 \tag{3.48}$$

Equation 3.48 may be written

$$X_1(k) = X_{11}(k) + W_N^k X_{12}(k), \quad k = 0, \ldots, N-1 \tag{3.49}$$

On comparison of Equation 3.49 with Equation 3.43, it is seen that $X_{11}(k)$ is indeed the DFT of the even sequence, while $X_{12}(k)$ is that of the odd sequence. Therefore, as previously stated, the DFT, $X_1(k)$, can be expressed in terms of two DFTs: $X_{11}(k)$ and $X_{12}(k)$. The factor $W_{N/2}^k$ occurs in both $X_{11}(k)$ and $X_{12}(k)$ and needs calculation once only.

Table 3.1 illustrates the process for an 8-point DFT. Line 1 gives the data while line 2 gives an expression for the DFT of the data in terms of the DFTs of the even and odd sequences, $X_{11}(k)$ and $X_{12}(k)$ respectively. Line 3 shows the re-ordered data from which $X_{11}(k)$ and $X_{12}(k)$ are derived. Line 4 gives the DFTs of the data sequences of line 3 in terms of the DFTs of their even and odd sequences, $X_{21}(k)$, $X_{22}(k)$, $X_{23}(k)$ and $X_{24}(k)$. These sequences are shown in line 5, and are seen to be the ultimate 2-point sequences, whose DFTs are $X_{21}(k)$, $X_{22}(k)$, $X_{23}(k)$ and $X_{24}(k)$ and are expressed in terms of the data in line 6. Thus the single 8-point DFT has been decomposed into four 2-point DFTs, each one of which produces two values, for example $X_{21}(0)$ and $X_{21}(1)$ in the case of $X_{21}(k)$. This process involved two decompositions, and the weights W_N^k were squared at each step. Considering line 6 it is seen that

$$X_{21}(k) = x_0 + W_{N/4}^k x_4 \quad k = 0, \ldots, N/4 - 1, \quad \text{that is } k = 0, 1 \tag{3.50}$$

Thus

$$X_{21}(0) = x_0 + x_4$$

while

$$X_{21}(1) = x_0 + W_{N/4} x_4$$
$$= x_0 + W_2 x_4 = x_0 + e^{-j2\pi/2} x_4 = x_0 + e^{-j\pi} x_4 = x_0 - x_4$$

Similarly,

$$X_{22}(0) = x_2 + x_6, \qquad X_{22}(1) = x_2 - x_6$$

$$X_{23}(0) = x_1 + x_5, \qquad X_{23}(1) = x_1 - x_5$$

$$X_{24}(0) = x_3 + x_7, \qquad X_{24}(1) = x_3 - x_7,$$

from which we observe that the values with $k = 1$ differ only by a sign from those with $k = 0$. This point is emphasized if $X_{11}(k)$ ($k = 0, 1, 2, 3$) is considered. Now,

$$X_{11}(k) = X_{21}(k) + W_{N/2}^k X_{22}(k) \tag{3.51}$$

so,

$$X_{11}(0) = X_{21}(0) + W_{N/2}^0 X_{22}(0) = X_{21}(0) + X_{22}(0) \tag{3.52}$$

$$X_{11}(1) = X_{21}(1) + W_{N/2}^1 X_{22}(1) = X_{21}(1) + e^{-j\pi/2} X_{22}(1)$$

$$= X_{21}(1) - jX_{22}(1) \tag{3.53}$$

$$X_{11}(2) = X_{21}(2) + W_{N/2}^2 X_{22}(2) = X_{21}(2) + e^{-j(2\pi/8)2\times2} X_{22}(2)$$

$$= X_{21}(2) + e^{-j\pi} X_{22}(2) = X_{21}(2) - X_{22}(2) \tag{3.54}$$

Now

$$X_{21}(2) = x_0 + W_{N/4}^2 x_4 = x_0 + W_2^2 x_4 = x_0 + x_4 = X_{21}(0)$$

and

$$X_{22}(2) = x_2 + W_{N/4}^2 x_6 = x_2 + x_6 = X_{22}(0)$$

Hence Equation 3.54 is equivalent to

$$X_{11}(2) = X_{21}(0) - X_{22}(0) \tag{3.55}$$

$$X_{11}(3) = X_{21}(3) + W_{N/2}^3 X_{22}(3) \tag{3.56}$$

Now

$$X_{21}(3) = x_0 + W_{N/4}^3 x_4 = x_0 + e^{-j(2\pi/2)3} x_4$$

$$= x_0 + e^{-j3\pi} x_4 = x_0 - x_4 = X_{21}(1)$$

and

$$X_{22}(3) = x_2 - x_6 = X_{22}(1)$$

Hence Equation 3.56 is equivalent to

$$X_{11}(3) = X_{21}(1) + e^{-j(2\pi/4)3} X_{22}(1) = X_{21}(1) + e^{-j3\pi/2} X_{22}(1)$$

$$= X_{21}(1) + jX_{22}(1) \tag{3.57}$$

Drawing these results together gives

$$X_{11}(0) = X_{21}(0) + X_{22}(0) = X_{21}(0) + W_8^0 X_{22}(0) \tag{3.58a}$$

$$X_{11}(2) = X_{21}(0) - X_{22}(0) = X_{21}(0) - W_8^0 X_{22}(0) \tag{3.58b}$$

$$X_{11}(1) = X_{21}(1) - jX_{22}(1) = X_{21}(1) + W_8^2 X_{22}(1) \tag{3.58c}$$

$$X_{11}(3) = X_{21}(1) + jX_{22}(1) = X_{21}(1) - W_8^2 X_{22}(1) \tag{3.58d}$$

Inspection of these equations shows how the DFTs $X_{11}(k)$ are related to the DFTs of their even-numbered and odd-numbered data, and that $X_{11}(0)$ and $X_{11}(2)$ are given by expressions with common terms which differ only in one sign. The same is true of $X_{11}(1)$ and $X_{11}(3)$. Equations such as these are known as recomposition equations because starting from the data pairs and forming $X_{21}(k)$, $X_{22}(k)$, $X_{23}(k)$ and $X_{24}(k)$ allows $X_{11}(k)$ and $X_{12}(k)$ to be found, and hence $X_1(k)$. The number of complex additions and multiplications involved is reduced in this way because (i) the

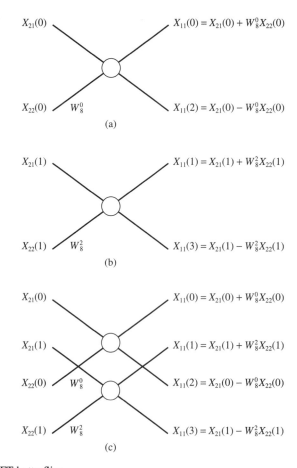

Figure 3.4   FFT butterflies.

recomposition equations are expressed in terms of powers of the recurring factor W_N, (ii) use is also made of relationships of the type $X_{21}(2) = X_{21}(0)$ and $X_{21}(3) = X_{21}(1)$, and (iii) the presence of only sign differences in the pairs of expressions is exploited. This algorithm is known as the Cooley–Tukey algorithm.

3.5.1 The butterfly

Equations 3.58 may be represented diagrammatically by exploiting the symmetry centred on the sign differences and taking the equations in pairs. Thus from Equations 3.58a and 3.58b the outputs of the recompositions are $X_{11}(0)$ and $X_{11}(2)$ formed from the inputs $X_{21}(0)$ and $X_{22}(0)$. This is illustrated in Figure 3.4(a). The inputs are at the left-hand side of the cross, the outputs to the right. Figure 3.4(b) shows how the outputs $X_{11}(1)$ and $X_{11}(3)$ are obtained diagrammatically. By overlapping Figures 3.4(a) and 3.4(b), a composite diagram is obtained in which the output DFTs are arranged in order of increasing k. This is shown in Figure 3.4(c). The structure of Figure 3.4(a) or 3.4(b) is referred to as a 'butterfly', being reminiscent of the symbolic representation of the insect. The entire 8-point FFT may be depicted in this manner as in Figure 3.5.

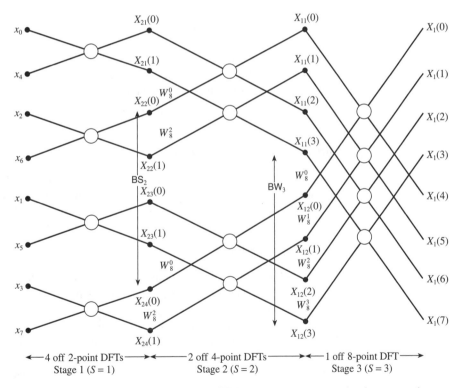

Figure 3.5 FFT butterflies for an 8-point DFT: BW_3, memory separation between points contributing to uppermost butterfly of stage 3; BS_2, memory separation between bottom points of butterflies in stage 2 with the same weighting factor.

Example 3.5   It will now be instructive to obtain the DFT of the sequence $\{1, 0, 0, 1\}$, previously evaluated in Section 3.2, by means of the decimation-in-time FFT algorithm. This is a four-point DFT with $x_0 = 1$, $x_1 = 0$, $x_2 = 0$, $x_3 = 1$, and $X_1(k) = X_{11}(k) + W_N^k X_{12}(k)$, $k = 0, 1, 2, 3$. The re-ordered sequence is x_0, x_2, x_1, x_3.

We can now utilize the top left-hand corner of Figure 3.5 to work out the DFT. Points x_0, x_4, x_2, x_6 are replaced by x_0, x_2, x_1, x_3, and the required DFT values are $X_{11}(0)$, $X_{11}(1)$, $X_{11}(2)$ and $X_{11}(3)$. Therefore,

$$X_{21}(0) = x_0 + x_2 = 1$$

$$X_{21}(1) = x_0 - x_2 = 1$$

$$X_{22}(0) = x_1 + x_3 = 1$$

$$X_{22}(1) = x_1 - x_3 = -1$$

$$X_{11}(0) = X_{21}(0) + W_8^0 X_{22}(0) = 1 + 1 = 2$$

$$X_{11}(1) = X_{21}(1) + W_8^2 X_{22}(1) = 1 + e^{-j\pi/2}(-1) = 1 + j$$

$$X_{11}(2) = X_{21}(0) - W_8^0 X_{22}(0) = 1 - 1 = 0$$

$$X_{11}(3) = X_{21}(1) - W_8^2 X_{22}(1) = 1 - j$$

These values are the same as those obtained in Section 3.2, but were more readily obtained using the FFT algorithm. This conclusion is general and the computational savings increase as the number of data increases.

3.5.2   Algorithmic development

Inspection of Figure 3.5 shows that in order to execute the FFT the program must re-order the input data and perform the butterfly computations. These will now be considered in turn. (See also Strum and Kirk, 1988.)

3.5.2.1   Re-ordering the input data

While it might at first appear that there is no obvious way to program the re-ordering of the input data, there is. The secret is to think in binary terms. Table 3.2 shows in the first column the required ordering of the data for input to the butterfly network as given by Figure 3.5. Each value is assumed to be stored in a binary memory address. These addresses are given in the second column. The third column shows these memory addresses bit reversed. If these bit-reversed addresses are taken to correspond to the binary addresses of the original data sequence, commencing with $x(0)$ at 000, then the corresponding data values are given in the fourth column, which is seen to contain the original data sequence. Thus the addresses of the re-ordered data are seen to be the bit-reversed addresses of the original data sequence. The program is

Table 3.2 Sequence re-ordering by bit reversal.

Required sequence for butterfly computation	Binary addresses of required sequence data	Bit-reversed addresses	Corresponding sequence = original data sequence
x_0	000	000	x_0
x_4	100	001	x_1
x_2	010	010	x_2
x_6	110	011	x_3
x_1	001	100	x_4
x_5	101	101	x_5
x_3	011	110	x_6
x_7	111	111	x_7

therefore required to convert the data point numbers (0 to $N-1$) to binary, to bit reverse these binary numbers and to convert them back to the denary numbers which are the addresses of the re-ordered data. Conversion to binary may be achieved by repetitively dividing by two when the remainders give the reverse-ordered digits of the corresponding binary number which is therefore the required binary address of the re-ordered sequence. These remainders may be obtained by using the MOD function obtainable in a high level language, for example MOD(K,2) gives the remainder on dividing the denary value K by 2. The integer part of $K/2$ is found by performing integer division. The remaining digits are found by repeating this process until $\log_2 N$ divisions have been made. This is because the data is to consist of $2^m = N$ data points so each address requires m digits and is divisible by two m times, where $m = \log_2 N$. The Ith bit of the new address is the binary coefficient of 2^{m-1-I} so the new address (NADDR) may be found using a DO loop for a particular value of K (K=IADDR) which runs from I=0 to I=m−1 and which takes the remainders (RMNDR) from the successive integers obtained after dividing K=IADDR by two. The pseudo-code required for this is

```
DO FOR I=0 TO m−1
    RMNDR:=MOD(IADDR,2)
    NADDR:=NADDR+RMNDR×2^m-1-I
    IADDR:=IADDR/2
END DO
```

This DO loop has to be nested inside another DO loop, the function of which is to extract the original data from a complex array DATA(K), K=0 to N−1 being the datum point number which is also the number of the complex element in the array and corresponds to the initial address of the datum, and to insert the re-ordered data into the array NEWDATA(NADDR). The data in NEWDATA is now in the correct sequence for the butterfly computations. The complete pseudo-code is

```
DO FOR K=0 TO N–1
    NADDR:=0
    IADDR:=K
    DO FOR I=0 TO m–1
        RMNDR:=MOD(IADDR,2)
        NADDR:=NADDR+RMNDR×2^{m-1-I}
        IADDR:=IADDR/2
    END DO
    NEWDATA(NADDR):=DATA(K)
END DO
```

3.5.2.2   Butterfly computations

Three stages of computation are involved:

(1)   calculate weighting factors, $W_N^R = e^{-j(2\pi/N)R}$;

(2)   evaluate butterflies in a stage (see Figure 3.5 for stage definitions);

(3)   compute over all stages.

An effective procedure is to calculate the weighting factors W_N^R involved in a stage, and for each one to evaluate all the butterflies in the stage with that factor. Having evaluated all the butterflies in that stage, the procedure is repeated for all stages. Thus, referring to stage 2 in Figure 3.5, W_8^0 and the two butterflies involving W_8^0 are calculated. Then W_8^2 and the two butterflies involving it are calculated. This procedure is carried out for each of the three stages in turn starting with stage 1 and the re-ordered data. In order to achieve the FFT computation using a relatively short program, a number of properties of the algorithm are required. Let the butterfly width BWIDTH represent the separation in memory of the two points which contribute to a butterfly. For the bottom butterfly of stage 3 in Figure 3.5 this is BW_3. It is a spacing of four points. Consideration of the butterflies in the other stages leads to the conclusion that in general

$$BWIDTH=2^{S-1} \tag{3.59}$$

where S is the stage number. (BWIDTH–1) is the number of steps over which the exponent of the weighting factor changes. Let BSEP, the butterfly separation, be the separation in memory of like points between the nearest butterflies in a stage which have the same weighting factor. For stage 2 in Figure 3.5, BS_2 represents BSEP for butterflies with the weighting factor W_8^0. Inspection of the figure shows that for the Sth stage

$$BSEP=2^S \tag{3.60}$$

Finally for an N-point FFT the exponents of the weighting factors change by

$$P = N/2^S \tag{3.61}$$

in the Sth stage. This can be seen in Figure 3.5. For example, in stage 2, $S = 2$, $P = 8/2^2 = 2$ and the weighting factors are W_8^0 and W_8^2.

Each butterfly may be calculated as

$$XNEW(TOP)=XOLD(TOP)+W_N^R\times XOLD(BOTTOM)$$
$$XNEW(BOTTOM)=XOLD(TOP)-W_N^R\times XOLD(BOTTOM) \qquad (3.62)$$

where the left-hand side refers to the butterfly inputs and the right-hand side to the butterfly outputs. However, memory may be conserved by re-writing these equations as

$$TEMP=W_N^R\times X(BOTTOM)$$

where X(BOTTOM) refers to the output XOLD(BOTTOM) of the butterfly in the previous stage,

$$X(BOTTOM)=X(TOP)-TEMP \qquad (3.63a)$$

where X(BOTTOM) now refers to the required butterfly output, X(TOP) refers to the previous value, and

$$X(TOP)=X(TOP)+TEMP \qquad (3.63b)$$

where the new left-hand side value of X(TOP) is the required butterfly output.

With all this knowledge available it is now possible to write the pseudo-code for the FFT as shown below:

```
10   PI:=3.141593
20   DO FOR S=1 TO m                              evaluate for m stages
30       BSEP:=2^S                                evaluating for the stage S
40       P:=N/BSEP                                (P = N/2^S), change in
                                                  exponent
50       BWIDTH:=BSEP/2                           (BWIDTH = 2^{S-1}),
                                                  separation between
                                                  butterfly inputs
60       DO FOR J=0 TO (BWIDTH-1)                 work out weighting factors
                                                  for a particular stage
70           R:=P.J                               calculate power of W_N
80           THETA:=2×PI×R/N                      calculate exponent of e^{-J}
90           WN:=CMPLX{cos(THETA),-sin(THETA)}
                                                  calculate W_N^R
100          DO FOR TOPVAL=J STEP BSEP UNTIL N/2
                                                  for all the Jth butterflies
                                                  in the stage
110              BOTVAL:=TOPVAL+BWIDTH
120              TEMP:=X(BOTVAL)×WN
130              X(BOTVAL):=X(TOPVAL)-TEMP
140              X(TOPVAL):=X(TOPVAL)+TEMP
150          END DO
160      END DO
170  END DO
```

Lines 100–150 work out all butterflies in a stage which have the same weighting factors. There are always $N/2$ butterflies in a stage.

Table 3.3   Savings in complex multiplications and additions when the FFT is used instead of the DFT.

	DFT		FFT		Ratio of DFT multiplications to FFT multiplications	Ratio of DFT additions to FFT additions
N	Number of complex multiplications	Number of complex additions	Number of complex multiplications	Number of complex additions		
2	4	2	1	2	4	1
4	16	12	4	8	4	1.5
8	64	56	12	24	5.3	2.3
16	256	240	32	64	8.0	3.75
32	1 024	992	80	160	12.8	6.2
64	4 096	4 032	192	384	21.3	10.5
128	16 384	16 256	448	896	36.6	18.1
256	65 536	65 280	1 024	2 048	64.0	31.9
512	262 144	261 632	2 304	4 608	113.8	56.8
1024	1 048 576	1 047 552	5 120	10 240	204.8	102.3
2048	4 194 304	4 192 256	11 264	22 528	372.4	186.1
4096	16 777 216	16 773 120	24 576	49 152	682.7	341.3
8192	67 108 864	67 100 672	53 248	106 496	1 260.3	630.0

3.5.3   Computational advantages of the FFT

The computational advantages of the FFT may be illustrated by considering first the FFT algorithm of Figure 3.5. This figure shows that an N-point FFT contains $N/2$ butterflies per stage and $\log_2 N$ stages, that is it contains a total of $(N/2) \log_2 N$ butterflies. Figure 3.4(a) shows that each butterfly involves one complex multiplication of the form $W_N^R X_{ij}(k)$. Hence the FFT involves $(N/2) \log_2 N$ complex multiplications compared with N^2 in the case of the DFT, as shown in Section 3.4. Thus the computational saving in complex multiplications is $N^2 - (N/2) \log_2 N$. Each butterfly contains two complex additions so the FFT requires $N \log_2 N$ complex additions compared with $N(N-1)$ for the DFT. Thus the saving in complex additions is $N(N-1) - N \log_2 N$. These savings are illustrated in Table 3.3. For a typical 1024-point DFT it is seen that the computation time will be reduced by two orders of magnitude if the FFT algorithm is employed.

3.6   Inverse fast Fourier transform

An FFT algorithm for determining the inverse fast Fourier transform (IFFT) is readily obtainable from the FFT algorithm. Its use lies in transforming spectra into their corresponding waveforms and in checking that the FFT has been correctly computed by using basically the same algorithm to obtain the original data. To see how the IFFT is derived make the following substitutions in Equation 3.20. Sum over the variable λ rather than n, let the variable k become μ, set $\Omega = 2\pi/NT$ so that the exponent of e becomes $-jk(2\pi/N)\lambda$. Again using the notation $x(\lambda T) = x(\lambda)$ and so on, Equation 3.20 then becomes

$$X(\mu) = \sum_{\lambda=0}^{N-1} x(\lambda)\, e^{-j\mu(2\pi/N)\lambda} \quad \mu = 0, 1, \ldots, N-1 \tag{3.64}$$

Now make similar substitutions in Equation 3.23, that is put $k = \lambda$ and $n = \mu$, so that Equation 3.23 becomes

$$x(\mu) = \frac{1}{N} \sum_{\lambda=0}^{N-1} X(\lambda)\, e^{j\lambda(2\pi/N)\mu} \quad \mu = 0, 1, \ldots, N-1 \tag{3.65}$$

In the last two equations $X(\mu)$, $X(\lambda)$, $x(\lambda)$ and $x(\mu)$ are all elements of the equidimensional arrays X and x, and so it can be seen that the IFFT, $x(\mu)$, differs from the FFT, $X(\mu)$, only in the factor $1/N$ and the sign of the exponent. Thus with small modifications the FFT may be used to calculate the IFFT. The two transforms may be included in the same algorithm by making the following modifications to the preceding pseudo-code:

line 5	K:=1 FOR FFT, K:=−1 FOR IFFT
line 80	THETA:=K×2×PI×R/N
line 145	IF K=−1 DO
line 146	X(BOTVAL):=X(BOTVAL)/N
line 147	X(TOPVAL):=X(TOPVAL)/N
line 148	END DO

3.7 Implementation of the FFT

Basically it should be clear now that in principle the FFT or IFFT is computed by providing a data array and operating on the data using the FFT or IFFT algorithms (including the bit-reversal algorithms). However, there remain some other considerations. So far the effects of discontinuities which occur at the two ends of the data have been ignored, together with the phenomena known as aliasing and picket fencing. In order to compute good approximations to the true data spectra it is necessary to take these effects into account using the techniques described in Chapter 11. Another aspect lies in the fact that so far in this chapter attention has been confined to the radix-2 decimation-in-time algorithm, but other algorithms, including decimation in frequency (DIF), exist. Some of these points will be discussed in what follows.

3.7.1 The decimation-in-frequency FFT

Section 3.5 described the decimation-in-time fast Fourier transform algorithm which was obtained by repeatedly dividing an initial discrete Fourier transform of the form of Equation 3.43 into two transforms, one consisting of the even terms and one of the odd terms, until the initial transform was reduced to two-point transforms of the initial data. An alternative approach is to separate the initial transform into two transforms, one containing the first half of the data and the other containing the second half of the data. The resulting decimation-in-frequency algorithm was first derived by Gentleman and Sande (1966) and is often known as the Sande–Tukey algorithm, unhappily for Gentleman. Overall there is little to choose between the two algorithms.

3.7.2 Comparison of DIT and DIF algorithms

For the DIF algorithm the order of the input data is unaltered but the output FFT sequence is bit reversed. Both the DIF and DIT algorithms are in-place algorithms. By re-drawing them it is possible to maintain the order of both the inputs and the outputs, but the resulting algorithms are no longer in-place algorithms and extra storage space is required (see Chapter 12). The number of complex multiplications required for both algorithms is the same. Overall, there is little to choose between the two.

3.7.3 Modifications for increased speed

Further increases in computational speed are possible for the DIT algorithm. For example, a radix-4 FFT can be used to reduce the number of complex multiplications by a factor approaching 2. The number of additions is also reduced. Another enhancement in speed may be obtained by removing unnecessary multiplications by the weighting factor W_N (often called the twiddle factor!) which occur when $W_N = \pm 1$, or $\pm j$. This also reduces the number of additions required. Implementation is by including a separate butterfly for the cases for which $W_N = \pm 1$ or $\pm j$. This, for example, would give a radix-2, two-butterfly, in-place, DIT algorithm. Further time saving is obtainable by first calculating the sine and cosine parts of W_N and storing them in a look-up table from which their values are obtained as required. Various other improvements are possible. It is therefore apparent that there exist, not just one FFT, but many. A full appreciation of the subject requires a considerable investment of study time and significant mathematical ability. It is advantageous to be adept with advanced matrix theory, multidimensional index mapping, and number theory. This more comprehensive approach to the subject is beyond the aim of this chapter, which is to explain the concept of discrete transforms in a way that most people involved in digital signal processing might understand. There are a number of other books, for example Burrus and Parks (1985) and Beauchamp (1987), which present the more specialized approach and these are recommended to the reader. As already stated, there are a number of algorithms from which to choose, and an expert might well write his or her own which is more suitable for the application. However, reducing the numbers of multiplications and additions required does not always result in a proportional increase in speed of computation. Decreasing the multiplications may result in more program code and more additions. If hardware signal processing chips are to be used these will impose their own limitations which may negate the improvements in the algorithm. Since increases in speed are not likely to exceed a factor of 2, the authors are inclined to recommend the basic radix-2 DIT FFT described above for general-purpose use where speed is not that critical. A large number of FORTRAN FFT programs are presented in Burrus and Parks (1985), together with discussion as to their advantages and disadvantages. Programs are also given for hardware implementation. A C language program for the radix-2 DIT FFT is given in Appendix 3A.

3.8 Other discrete transforms

A variety of other transforms are available. The Winograd Fourier transform (Winograd, 1978; see also Burrus and Parks, 1985; Beauchamp, 1987; Rader, 1968; Signal Processing Committee, 1979; McClellan and Rader, 1979) and the prime factor algorithm (Beauchamp, 1987; see also McClellan and Rader, 1979) provide ingenious but complicated methods of increasing the speed of computation of the FFT. The discrete cosine transform is particularly useful in data compression applications (see Section 3.8.1). The Walsh transform (Section 3.8.2) analyzes signals into rectangular waveforms rather than sinusoidal ones and is computed more rapidly than the FFT. The Hadamard transform (Section 3.8.3), constructed by re-ordering the Walsh ordered sequence, is even faster to compute. While displaying advantages for some purposes the Walsh and Hadamard transforms suffer from some disadvantages which limit their applicability; see Sections 3.8.2 and 3.8.3. Finally, the Haar transform is particularly useful for edge detection in image processing (Rosenfield and Thurston, 1971) and for similar applications. Beauchamp (1987) provides a good source of starting material for those wishing to know more about the above transforms and their applications. During the 1990s there has been growing interest in the wavelet transform (Chan, 1995; Daubechies, 1990, 1992; Burrus, Gopinath and Guo, 1998) and so it is introduced below in Section 3.8.4.

3.8.1 Discrete cosine transform

In addition to their uses in speeding correlation and convolution computations, and in spectral analysis, transform methods are also used to achieve data compression in, for example, speech and video transmissions and for recording biomedical signals such as EEGs and ECGs. They are also utilized for pattern recognition. In these applications only the more significant transform components are utilized. Hence the number of coding bits required is reduced. This allows faster transmission, the use of narrower bandwidth transmission lines, and easier pattern recognition (because of the data reduction). The three important features of a suitable transform are its compressional efficiency, which relates to concentrating the energy at low frequencies, ease of computation, and minimum mean square error. The ideal transform for achieving these is the Karhunen–Loève transform, but this cannot be represented algorithmically. However, the discrete cosine transform (DCT) has virtually the same properties and does possess an algorithm. It consists essentially of the real part of the DFT. This definition is reasonable since the Fourier series of a real and even function contains only the cosine terms, and in, for example, the case of sampled voltage values the data used is real and can be made symmetrical by doubling the data by adding its mirror image. Thus the DFT is given by (Equation 3.41)

$$X(k) = \sum_{n=0}^{N-1} x_n\, e^{-j2\pi nk/N}, \quad k = 0, 1, \ldots, N-1$$

Defining the DCT $X_c(k)$ as the real part of this gives

$$X_c(k) = \mathrm{Re}\,[X(k)] = \sum_{n=0}^{N-1} x_n \cos\left(\frac{k2\pi n}{N}\right), \quad k = 0, 1, \ldots, N-1 \qquad (3.66)$$

This is one of several forms of the DCT. A more common form is (Beauchamp, 1987; Yip and Ramamohan, 1987; Ahmed and Rao, 1975)

$$X_c(k) = \frac{1}{N}\sum_{n=0}^{N-1} x_n \cos\left(\frac{k2\pi n + k\pi}{2N}\right) = \frac{1}{N}\sum_{n=0}^{N-1} x_n \cos\left[\frac{k\pi(2n+1)}{2N}\right]$$
$$k = 0, 1, \ldots, N-1 \quad (3.67)$$

and other forms also exist (Yip and Ramamohan, 1987; Pennebaker and Mitchell, 1993; Pitas, 1993; Bailey and Birch, 1989).

Implementations of the DCT exist based on the FFT as might be expected (Narasinka and Petersen, 1978), and a fast DCT which is six times as fast as these has been developed (Chen *et al.*, 1977). Another version is the C-matrix transform which can be more simply constructed in hardware (Srinivassan and Rao, 1983).

3.8.2   Walsh transform

The transforms discussed so far have been based on cosine and sine functions. Transforms based on pulse-like waveforms which take only values of ±1 are much simpler and faster to compute. They are also more appropriate for the representation of waveforms which contain discontinuities, for example in images. Conversely, they are less appropriate for describing continuous waveforms and may not be phase invariant in which case the derived spectrum may be distorted. However, such waveforms are used in image processing (astronomy and spectroscopy), signal coding, and filtering.

Just as the DFT is based on a set of harmonically related cosine and sine waveforms, so is the discrete Walsh transform (DWT) based on a set of harmonically related rectangular waveforms, known as Walsh functions. However, frequency is not

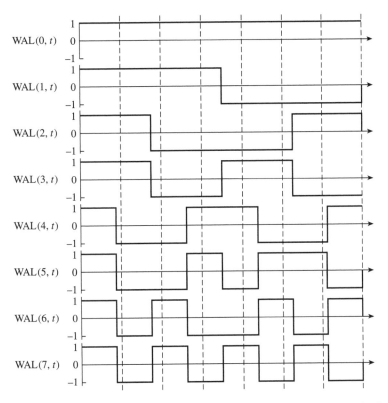

Figure 3.6 Sequency-ordered Walsh functions to $n = 7$ showing sampling times for 8×8 Walsh transform matrix.

defined for rectangular waveforms and so the analogous term sequency is used. Sequency is half the average number of zero crossings per unit time. Figure 3.6 shows the set of Walsh functions up to the order of $N = 8$ drawn in order of increasing sequency. They are said to be sequency, or Walsh, ordered. The Walsh function at time t and of sequency n is designated WAL(n, t). Inspection of Figure 3.6 shows that there are equal numbers of even and odd Walsh functions, just as there are corresponding cosinusoidal and sinusoidal Fourier series components. The even functions, WAL$(2k, t)$, are written CAL(k, t), and the odd functions, WAL$(2k + 1, t)$, are written SAL(k, t), where $k = 1, 2, \ldots, N/2 - 1$.

Any waveform, $f(t)$, may be written as a Walsh function series, analogous to a Fourier series, as

$$f(t) = a_0 \, \text{WAL}(0, t) + \sum_{i=1}^{N/2-1} \sum_{j=1}^{N/2-1} [a_i \, \text{SAL}(i, t) + b_j \, \text{CAL}(j, t)] \tag{3.68}$$

where the a_i and b_j are the series coefficients.

For any two Walsh functions,

$$\sum_{t=0}^{N-1} \text{WAL}(m, t)\, \text{WAL}(n, t) = \begin{cases} N & \text{for } n = m \\ 0 & \text{for } n \neq m \end{cases}$$

that is, Walsh functions are orthogonal.

The discrete Walsh transform pair is

$$X_k = \frac{1}{N} \sum_{i=0}^{N-1} x_i\, \text{WAL}(k, i) \quad k = 0, 1, \ldots, N-1 \tag{3.69}$$

and

$$x_i = \sum_{i=0}^{N-1} X_k\, \text{WAL}(k, i) \quad i = 0, 1, \ldots, N-1 \tag{3.70}$$

where we note that, apart from the factor of $1/N$, the inverse transform is the same as the transform, and that $\text{WAL}(k, i) = \pm 1$. The transform pair may, therefore, be calculated by matrix multiplication by digital means as mentioned above. However, the lack of phase invariance means that the DWT is unsuitable for fast correlations or convolutions.

Equation 3.69 shows that the kth DWT component is obtained by multiplying each waveform sample x_i by the Walsh function of sequency k and summing for $k = 0$, $1, \ldots, N-1$. This may be expressed for all k DWT components in matrix notation as

$$\mathbf{X}_K = \mathbf{x}_i \mathbf{W}_{ki} \tag{3.71}$$

where $\mathbf{x}_i = [x_0\ x_1\ x_2 \ldots x_{N-1}]$, the data sequence,

$$\mathbf{W}_{ki} = \begin{bmatrix} W_{01} & W_{02} & \cdots & W_{0,N-1} \\ W_{11} & & & \\ \vdots & & & \vdots \\ W_{N-1,1} & W_{N-1,2} & \cdots & W_{N-1,N-1} \end{bmatrix}$$

the Walsh transform matrix, and $\mathbf{X}_k = [X_0\ X_1 \ldots X_{N-1}]$, the $N-1$ components of the DWT. Note that $\mathbf{W}_{ki}$ is an $N \times N$ matrix where N is the number of data points, that is sampled waveform points. Thus, if there are N data points it is necessary to consider the first N sequency-ordered Walsh functions. Each one is sampled N times. The kth row of $\mathbf{W}_{ki}$ corresponds to the N sampled values of the kth sequency component.

Example 3.6

As an example, let us compute the DWT of the data sequence $(1, 2, 0, 3)$. This consists of $N = 4$ data points and so $\mathbf{W}_{ki}$ is a 4×4 matrix obtainable from the first four rows of Figure 3.6 as

$$\mathbf{W}_{ki} = \begin{bmatrix} 1 & 1 & 1 & 1 \\ 1 & 1 & -1 & -1 \\ 1 & -1 & -1 & 1 \\ 1 & -1 & 1 & -1 \end{bmatrix} \tag{3.72}$$

Therefore, from Equation 3.71, $\mathbf{X}_k$ is given by

$$\mathbf{X}_k = \tfrac{1}{4}[1 \quad 2 \quad 0 \quad 3]\begin{bmatrix} 1 & 1 & 1 & 1 \\ 1 & 1 & -1 & -1 \\ 1 & -1 & -1 & 1 \\ 1 & -1 & 1 & -1 \end{bmatrix} = \tfrac{1}{4}[6 \quad 0 \quad 2 \quad -4]$$

so that $X_0 = 1.5$, $X_1 = 0$, $X_2 = 0.5$ and $X_3 = -1$. This is considerably easier to calculate than the corresponding DFT! Needless to say, fast DWTs (FDWTs) exist.

The corresponding spectrum can be calculated with power components given by

$$P(k) = [|\text{CAL}(k, t)|^2 + |\text{SAL}(k, t)|^2]^{1/2}$$

where

$$P(0) = X_c^2(0)$$
$$P(k) = X_c^2(k, t) + X_s^2(k, t) \tag{3.73}$$
$$P\left(\frac{N}{2}\right) = X_s^2\left(\frac{N}{2}, t\right)$$

where $k = 1, 2, \ldots, N/2 - 1$, and with phase components

$$\phi(0) = 0, \pi$$
$$\phi(k) = \tan^{-1}\left[\frac{X_s(k)}{X_c(k)}\right], \quad k = 1, 2, \ldots, N/2 - 1 \tag{3.74}$$

and

$$\phi\left(\frac{N}{2}\right) = 2k\pi \pm \pi/2, \quad k = 0, 1, 2, \ldots$$

For the above DWT we have, therefore

$$P(0) = 1.5^2 = 2.25; \ \phi(0) = 0, \pi$$

$$P(1) = 0^2 + 0.5^2 = 0.25; \ \phi(1) = \tan^{-1}\left(\frac{0}{0.5}\right) = 0$$

$$P(2) = (-1)^2 = 1; \ \phi(2) = \frac{\pi}{2} + 2k\pi, \quad k = 0, 1, 2$$

3.8.3 Hadamard transform

The Hadamard transform, or Walsh–Hadamard transform, is basically the same as the Walsh transform, but with the Walsh functions and therefore the rows of the transform

$$
\begin{array}{c}
i \longrightarrow \\
\begin{array}{cccccccc}
0 & 1 & 2 & 3 & 4 & 5 & 6 & 7
\end{array}
\end{array}
$$

$$
\begin{array}{c}
k \\
\downarrow
\end{array}
\begin{array}{c}
0 \\ 1 \\ 2 \\ 3 \\ 4 \\ 5 \\ 6 \\ 7
\end{array}
\left[
\begin{array}{cccccccc}
1 & 1 & 1 & 1 & 1 & 1 & 1 & 1 \\
1 & -1 & 1 & -1 & 1 & -1 & 1 & -1 \\
1 & 1 & -1 & -1 & 1 & 1 & -1 & -1 \\
1 & -1 & -1 & 1 & 1 & -1 & -1 & 1 \\
1 & 1 & 1 & 1 & -1 & -1 & -1 & -1 \\
1 & -1 & 1 & -1 & -1 & 1 & -1 & 1 \\
1 & 1 & -1 & -1 & -1 & -1 & 1 & 1 \\
1 & -1 & -1 & 1 & -1 & 1 & 1 & -1
\end{array}
\right]
$$

Figure 3.7 8×8 Hadamard transform matrix.

matrix re-ordered. The resultant Hadamard matrix comprises subarrays of second-order matrices. The Hadamard matrix of order 8×8 is given as $^8\mathbf{H}$ in Figure 3.7, and it can be seen to consist of matrices

$$
^2\mathbf{H} = \begin{bmatrix} 1 & 1 \\ 1 & -1 \end{bmatrix} \quad \text{and} \quad -^2\mathbf{H} = \begin{bmatrix} -1 & -1 \\ -1 & 1 \end{bmatrix}
$$

Any Hadamard matrix of order $2N$ may be obtained recursively from $^2\mathbf{H}$ as

$$
^{2N}\mathbf{H} = \begin{bmatrix} {}^N\mathbf{H} & {}^N\mathbf{H} \\ {}^N\mathbf{H} & -{}^N\mathbf{H} \end{bmatrix} \tag{3.75}
$$

The value of this recursive property is that by Hadamard ordering the Walsh functions the resultant fast Walsh–Hadamard transform may be more quickly computed than the DWT. Hadamard-ordered (or naturally ordered) Walsh functions are shown in Figure 3.8. The Hadamard-ordering sequence is obtained from the Walsh-ordered sequence by

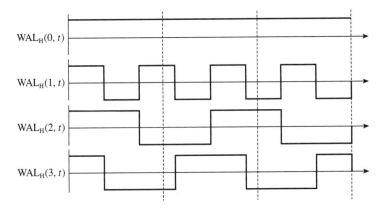

Figure 3.8   Hadamard-ordered Walsh functions to $n = 7$ showing sampling times for 4×4 Hadamard transform matrix.

(1) expressing the order of the Walsh-ordered function in binary;

(2) bit-reversing the binary values;

(3) converting the binary values to Gray code; and

(4) converting this value to decimal.

Example 3.7

By way of example the discrete Walsh–Hadamard transform of the sequence (1, 2, 0, 3) will now be calculated. The 4×4 Hadamard matrix $\mathbf{H}_{ki}$ is

$$\mathbf{H}_{ki} = \begin{bmatrix} 1 & 1 & 1 & 1 \\ 1 & 1 & -1 & -1 \\ 1 & -1 & 1 & -1 \\ 1 & -1 & -1 & 1 \end{bmatrix} = \begin{bmatrix} 1 & 1 & 1 & 1 \\ 1 & -1 & 1 & -1 \\ 1 & 1 & -1 & -1 \\ 1 & -1 & -1 & 1 \end{bmatrix} \qquad (3.76)$$

by the properties of a Hadamard matrix. The DWHT of (1, 2, 0, 3) is, therefore, given by

$$\mathbf{X}_k^{WH} = \tfrac{1}{4}[1 \quad 2 \quad 0 \quad 3] \begin{bmatrix} 1 & 1 & 1 & 1 \\ 1 & -1 & 1 & -1 \\ 1 & 1 & -1 & -1 \\ 1 & -1 & -1 & 1 \end{bmatrix} = \tfrac{1}{4}[6 \quad -4 \quad 0 \quad 2]$$

so that $X_0^{WH} = 1.5$, $X_1^{WH} = -1$, $X_2^{WH} = 0$, $X_3^{WH} = 0.5$. The magnitudes of these Walsh–Hadamard components are the same as the previously calculated Walsh components, but are re-ordered.

3.8.4 Wavelet transform

In physics the Heisenberg Uncertainty Principle states that it is impossible to simultaneously know both the position, x, and the momentum, p, of a particle with precision. In fact

$$xp \geqslant h = 6.626 \times 10^{-34} \, \text{J s} \qquad (3.77)$$

where h is Planck's constant. With the help of Einstein's equation, $E = mc^2$, this principle can be translated to the domain of signal processing to state that time and frequency cannot be resolved simultaneously to any precision. Thus,

$$\Delta f.T \geqslant 1 \qquad (3.78)$$

where Δf and T represent the resolutions in frequency and time. If T is highly resolved, the frequency will be less accurate, and vice versa. It therefore may become difficult to simultaneously measure to the required degree of accuracy the frequency of a signal component and the time at which the component occurs, or to resolve in time different frequency components. This could be the case for signals containing high frequency components of short duration which occur close together in time together with longer duration components closely spaced in frequency and occurring at different times. Such signals are non-periodic. The wavelet transform addresses this general problem of effecting a time-frequency analysis, and provides the means to analyze non-stationary signals. Wavelet transforms also have applications for filtering signals, for denoising signals, and for finding the locations and distributions of singularities.

Whereas in the Fourier transforms the signal values are weighted with an exponential of imaginary and harmonic frequency dependent argument, which is essentially a sinusoidal term, in the wavelet transform the signal values are weighted by wavelet functions.

All wavelets are derived from a basic (mother) wavelet. There are a number of possible mother wavelets, chosen to have the following properties. They should be oscillatory, there should be no DC component, they should be bandpass, they should decay rapidly towards zero with time, and they should be invertible. The last property ensures the wavelet transform of a signal is unique. We can write the basic wavelet as $\Psi(t)$. For example, the Morlet or modified Gaussian mother wavelet is

$$\Psi(t) = e^{j\omega_0 t}\, e^{-t^2/2} \tag{3.79}$$

and has the Fourier transform

$$H(\omega) = \sqrt{2\pi}\, e^{-(\omega-\omega_0)^2/2} \tag{3.80}$$

These two waveforms are sketched in Figure 3.9, and $\Psi(t)$ is seen to satisfy the above oscillatory and decay properties.

The remaining (baby) wavelets are obtained by scaling the mother wavelet to form a family of wavelets. Each baby wavelet may be written as

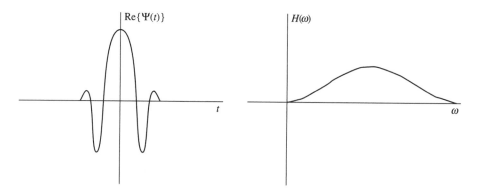

Figure 3.9   Modified Gaussian or Morlet mother wavelet, $\Psi(t)$, and its Fourier transform, $H(\omega)$.

$$\frac{1}{\sqrt{a}}\Psi\{(t-\tau)/a\}$$

where a is a variable scaling constant and τ is a constant of translation. If the scaling constant, a, is increased, both the magnitude and the argument of the wavelet are decreased. A decrease in argument for a given magnitude represents a decrease in frequency. Thus, an increase in scale, a, corresponds to a decrease in frequency and so the wavelet is dilated in time horizontally. Positive values of the translation cause a translation of the wavelet along the positive time axis. Hence wavelets of larger or smaller magnitudes, of higher or lower frequency content, may be generated and located at different positions in time by adjusting the scaling constant, a, and the translation constant, τ. By this means non-stationary signals with different frequency components extending over different intervals of time may be represented by a sum of different wavelets. This is achieved using the wavelet transform.

There are several versions of the wavelet transform as defined below. These form a natural progression in increasing discretization. In these definitions the signal, $s(t)$, is assumed to be square integrable, i.e.

$$\int s^2(t)dt < \infty \tag{3.81}$$

This assumption is true for all signals of finite magnitude and of short time duration. Sinusoidal and DC signals do not satisfy the condition, so are excluded in the following considerations.

The Continuous Wavelet transform, $CWT(a, \tau)$, may be defined as

$$CWT(a, \tau) = \left(1/\sqrt{a}\right)\int s(t)\Psi\{(t-\tau)/a\}dt \tag{3.82}$$

The parameters in this equation may be discretized to yield the Discrete Parameter Wavelet transform, $DPWT(m, n)$, defined as

$$DPWT(m, n) = a_0^{-m/2}\int s(t)\Psi\{(t-n\tau_0 a_0^m)/a_0^m\}dt \tag{3.83}$$

where the following substitutions have been made: $a = a_0^m$, $\tau = n\tau_0 a_0^m$. In these substitutions a_0 and τ_0 are the sampling intervals for a and τ, and m and n are integers. It is quite usual to choose $a_0 = 2$ and $\tau_0 = 1$. Then

$$DPWT(m, n) = 2^{-m/2}\int s(t)\Psi\{(t-n2^m)/2^m\}dt = 2^{-m/2}\int s(t)\Psi\{2^{-m}t - n\}dt \tag{3.84}$$

This dilates the time axis by the factor 2^{-m} and the wavelet is translated positively in time by $2^m n$.

Discretizing the time produces the Discrete Time Wavelet transform, $DTWT(m, n)$, defined as

$$DTWT(m, n) = a_0^{-m/2}\sum_k s(k)\Psi(a_0^{-m}k - n\tau_0) \tag{3.85}$$

Again, if $a_0 = 2$ and $\tau_0 = 1$, the $DTWT(m, n)$ becomes

$$DTWT(m, n) = 2^{-m/2}\sum_k s(k)\Psi(2^{-m}k - n) \tag{3.86}$$

which is known as the Discrete Wavelet transform. Thus the Discrete Wavelet transform is obtained from the Continuous Wavelet transform by discretizing the scale parameter, a, the translation parameter, τ, and the time, and setting $a_0 = 2$ and τ_0 to 1.

Besides using the wavelet transforms to study the time–frequency content of signals, they can be used to filter signals, e.g. to remove some of any noise present. First the signals are transformed into their components. Then the noise components are identified and removed. Finally the denoised signal is reconstructed from the component wavelets. For the Continuous Wavelet transform the reconstruction formula (inverse transform) is

$$s(t) = \frac{1}{C_\Psi} \int_{-\infty}^{\infty} \int_{a>0}^{\infty} CWT(a, \tau)\left\{\frac{1}{\sqrt{a}}\right\}\Psi\{(t - \tau)/a\}\left\{\frac{1}{a^2}\right\} da\,dt \tag{3.87}$$

where

$$C_\Psi = \int_0^{\infty} \{|H(\omega)|^2/\omega\}\,d\omega < \infty$$

and $H(\omega)$ is the Fourier transform of the basic wavelet, $\Psi(t)$.

Readers requiring further information on reconstruction formulae are referred to references such as Chan (1995), Daubechies (1990, 1992), Chui (1992), Mallat and Hwang (1992), and Burrus (1998).

Insight into the wavelet transforms can be gained by considering the possible interpretations of the Continuous Wavelet transform, Equation 3.82. It is seen that one interpretation is that $CWT(a, \tau)$ represents the cross-correlation of $s(t)$ with $\Psi(t/a)/\sqrt{a}$ at a lag of $-\tau/a$. Similarly it could be the cross-correlation of the scaled signal $s(at)$ with $\sqrt{a}\Psi(t)$ at a lag $-\tau/a$. The $CWT(a, \tau)$ may also represent the output of a bandpass filter of impulse response $CWT(a, \tau)$ at time τ/a to the input $s(t)$. Alternatively it may represent the output of a bandpass filter of impulse response $\sqrt{a}\Psi(-t)$ at time τ/a to an input $s(at)$. Thus, wavelet transformations may be viewed as effecting bandpass filtering or cross-correlation. By changing the scale, a, different frequency components are filtered.

3.8.5   Multiresolution analysis by the wavelet method

Multiresolution analysis (MRA) refers to the partitioning of the signal components into a number of frequency bands. It may be implemented using low and high pass filters and sub-sampling. The inverse procedure is also possible, which allows signal reconstruction. MRA may be performed with the aid of the discrete wavelet transform using sampled signals. It has been used in the analysis of the information content of images (Mallat, 1989) and for analyzing evoked potentials of the

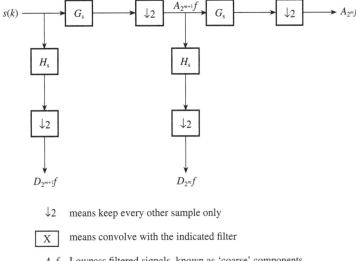

$\downarrow 2$ means keep every other sample only

$\boxed{X}$ means convolve with the indicated filter

$A_x f$ Lowpass filtered signals, known as 'coarse' components

$D_x f$ Highpass filtered signals, known as 'detail' signals. These
represent the difference signal between two adjacent lowpass
filtered signals, i.e. between two adjacent levels of resolution

Figure 3.10   MRA decomposition of a signal using the wavelet method.

electroencephalogram (Thakor *et al.*, 1993). Mallat (1989) presents the wavelet
version of MRA and explains how to apply it. Provided $a = 2^m$ and $\tau = 2^m n$ it is
possible to design a lowpass filter $G_s(\omega)$ and a highpass filter $H_s(\omega)$, which when used
repeatedly in a pyramidal structure, will resolve the signal into frequency bands of
increasingly high resolution. This pyramidal algorithm is illustrated in Figure 3.10.
$G_s(\omega)$ and $H_s(\omega)$ are symmetric to the lowpass filter $G(\omega)$ and the highpass filter $H(\omega)$
respectively. These are quadrature mirror filters which are related to a suitable wavelet
function. Signal reconstruction may be achieved using the pyramidal algorithm shown
in Figure 3.11.

MRA may be used to study signal components and also to filter the signals. After
signal decomposition unwanted components may be removed, and the filtered signal
may then be reconstructed.

In Figure 3.12 typical results of applying MRA to extract an event-related potential
(the CNV) within a noisy background electroencephalogram (EEG) from a single trial
recording in which the signal-to-noise ratio was -14 dB are shown. Since the purpose
is to obtain the ERP it was only possible to demonstrate the method by simulating a
number of trials by modelling some variable ERPs and adding them to EEGs. Both
adaptive MRA (Saatchi *et al.*, 1997) and a modified version of it were applied to the
single trial data. It is seen that the denoised CNV waveforms are quite unlike the true
CNV. Since the signal-to-noise ratios of relatively large ERPs are about -15 dB,
these results show that the methods are more suitable for denoising waveforms
with considerably larger signal-to-noise ratios.

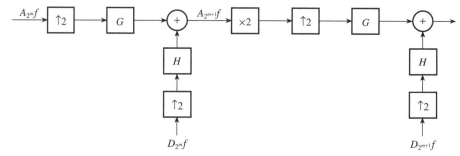

$\uparrow 2$ means insert a zero between each sample

$\boxed{\text{X}}$ means convolve with the indicated filter

$\boxed{\times 2}$ means multiply by 2

$A_x f$ Lowpass filtered signals, known as 'coarse' components

$D_x f$ Highpass filtered signals, known as 'detail' signals

Figure 3.11   MRA reconstruction of a signal using the wavelet method.

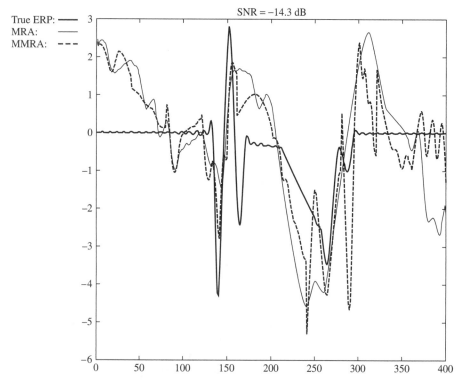

Figure 3.12   True and denoised ERPs extracted from a single trial by multiresolution analysis at a signal-to-noise ratio of −14.3 dB.

3.8.6 Signal representation by singularities: the wavelet transform method

It has been shown (Mallat and Hwang, 1992) that all signals and noise may be completely represented by their singularities. Singularities are defined in terms of their Lipschitz exponents. If some function $f(t)$ is continuously differentiable at t_0, it is non-singular and is said to be Lipschitz 1. Signals which are not Lipschitz 1 are singular. If $f(t)$ is n times differentiable at t_0, its nth derivative is singular and it is described as being Lipschitz α where $\alpha > n$. It is also possible to have negative Lipschitz exponents. The Lipschitz exponents characterize the singularities by magnitude and sign. By describing signals in terms of their singularities, and then removing unwanted singularities, filtered versions of the signals can be reconstructed from the remaining singularities. This technique has applications in, for example, edge detection in images (Mallat and Hwang, 1992) and signal denoising (Mallat and Hwang, 1992; Zhang and Zheng, 1997). Signal denoising refers to the removal of noise from signals to increase the signal-to-noise ratio.

Many signals have singularities with positive Lipschitz exponents while noise is characterized by negative Lipschitz exponents. This therefore offers the possibility of separating signals from noise if the singularities associated with the noise can be detected and removed. Mallat and Hwang (1992) have shown that singularities may be located and characterized by the moduli and signs of the wavelet transform maxima plotted against time at different scales as in Figure 3.13. Thus, each plotted maximum represents a singularity. By comparing the maxima between the different scales, it becomes possible to identify those associated with the noise and to determine the values of the Lipschitz exponents.

In what follows, it is assumed that the signals do not exhibit local oscillations. The considerations are more complicated for signals which do (Mallat and Hwang, 1992).

First consider signals whose wavelet transform modulus maxima lie in a cone $|(t - t_0)| \leq Ca$, where C is any constant. Such a cone is indicated in Figure 3.13. Now consider those maxima within the cone which lie on a connected curve between different scales. Such a curve is known as a maximum line. Then the wavelet transforms, $W(a, \tau)$, associated with the modulus maxima on such a maximum line vary with the scale, a, as

$$|W(a, \tau)| \leq Aa^{\alpha} \tag{3.88}$$

where A is a constant, $a > 0$ and $0 < \alpha < 1$.

Hence, as a increases the modulus of the wavelet transform increases. In other words the moduli increase on passing to lower frequency scales. Thus, their signal energy content increases at lower frequencies or equivalently decreases at higher frequencies. By taking logarithms of Equation 3.88, Equation 3.89 is obtained:

$$\log |W(a, \tau)| \leq \log A + \alpha \log a \tag{3.89}$$

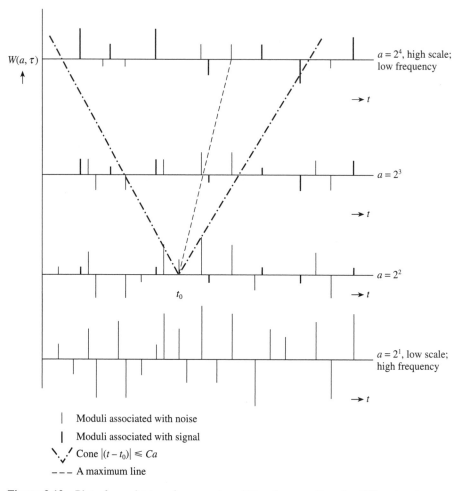

Figure 3.13   Plot of wavelet transform maxima, $W(a, \tau)$, versus time for different scales, a.

This means that α, the Lipschitz exponent, is the maximum slope of the logarithmically scaled straight lines of Equation 3.89 which remain above $\log|W(a, \tau)|$. This will soon be seen to be useful.

Now, in comparison with the above case, white noise is characterized by a negative Lipschitz exponent and the expectation of the squared modulus of the wavelet transform maximum varies as

$$E[|W(a, \tau)|^2] = \|\Psi\|^2 \sigma^2 / a \tag{3.90}$$

where σ^2 is the noise variance and Ψ is the wavelet function. Thus the modulus maxima decrease in proportion to $1/\sqrt{a}$ as the scale increases. Otherwise expressed, the modulus maxima increase as $\sqrt{a}$ with frequency. This means that maxima associated with white noise increase across scales with frequency. This is the opposite in sense to the changes in magnitude associated with signals with positive Lipschitz exponents. The same is true of other noise-like signals.

The modulus maxima of a noisy signal are illustrated purely schematically in Figure 3.13, which has not been drawn to scale. It is seen that the signal maxima at a given location increase in magnitude at higher scales (lower frequencies), while the noise maxima increase in magnitude at lower scales (higher frequencies) and also increase in number. Generally, as the scale is halved, the number of noise maxima is doubled. Thus, by observing the changes in maxima at the same positions from scale to scale it is possible to identify those maxima associated with noise. These may therefore be removed and not used in the reconstruction of a denoised signal.

This technique (Mallat and Hwang, 1992; Zhang and Zheng, 1997) was applied to the simulated ERPs in EEG data of Section 3.8.5. The denoised and the true ERPs for a single trial are shown in Figure 3.14 for a signal-to-noise ratio of −5.3 dB. This is a large signal-to-noise ratio compared with −15 dB, which is about the largest to be

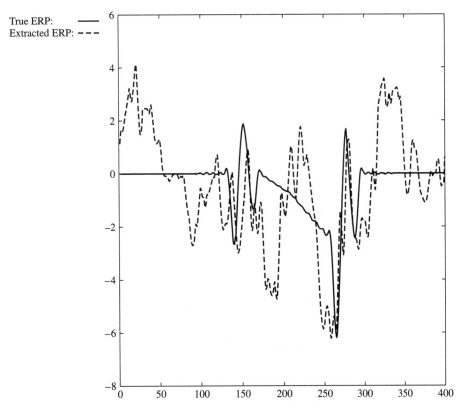

Figure 3.14 ERP denoised by singularity detection technique and true ERP, EEG, signal-to-noise ratio −5.3 dB.

True ERP: ——
Extracted ERP: ---

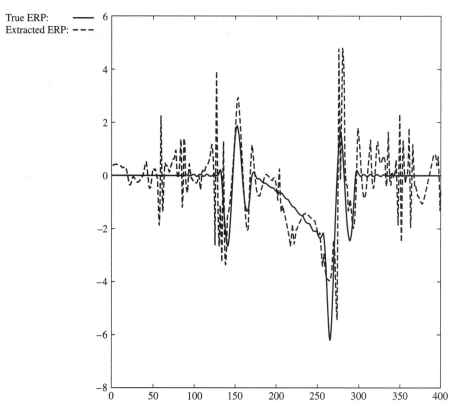

Figure 3.15 ERP denoised by singularity detection technique and true ERP, white noise, signal-to-noise ratio −5.67 dB.

expected, and the extracted CNV is nothing like the true CNV. Why is this? The singularity detection method is based upon the assumption of white noise and the EEG signal is not white. If the simulations are repeated using white noise instead of EEG, then results such as in Figure 3.15 are obtained. This time the extracted CNV is much more similar to the true one. It is therefore crucial that for the method to work satisfactorily the noise must be white.

A problem arises when the locations of signal and noise maxima coincide. If the maximum is not removed, then noise will be retained in the signal. If the maximum is removed, some of the signal will be removed, resulting in distortion of the reconstructed signal. A solution to this is to seek high scales where the signal is dominant (relatively large signal-to-noise ratio) and to apply Equation 3.89 within the cone, $|(t - t_0)| \leq Ca$, to determine α. Then the ratio of the modulus maxima at the same locations at adjacent lower scales within the cone is 2^α. Hence the true magnitudes of the maxima at these lower scales may be calculated and used in the signal reconstruction. Of course, a cone, $|(t - t_0)| \leq Ca$, may be drawn for any modulus maximum, t_0.

Another limitation of this denoising technique lies in the accuracy to which the modulus maxima of the noise may be identified and removed. This depends upon the precision to which α may be determined which will depend upon the chosen mother wavelet and upon the singularity (modulus maximum) removal algorithm, and upon the reconstruction algorithm. The reader is referred to the references for details of these.

3.9 An application of the DCT: image compression

As described in Section 3.8.1 the Discrete Cosine transform (DCT) is used to compress signal data, and this is particularly important for the storage and transmission of images, since each image involves a large amount of data. For example a 320 × 240 picture element (pixel) image represented by eight bits per pixel would occupy 76.8 kbytes, equivalent to about 25 pages of text. Hence the importance of image compression. After the image data has been compressed it has usually been encoded for transmission, which results in further compression. Different organizations had been developing their own approaches and standards until in 1986 a group of experts formed the Joint Photographic Experts Group (JPEG) to attempt the standardization of the compression and coding of still greyscale and colour images. The JPEG committee was a sub-committee of a group from the International Organization for Standardization (ISO), and also contained members drawn from the CCITT (International Telegraph and Telephone Consultative Committee) and the IEC (International Electrotechnical Commission). Standardization is necessary to allow the interchange of images between different applications, such as between PCs, LANs, CD ROMs, and digital cameras. The JPEG standard established an architecture for a set of image compression functions which allows a wide amount of variation in detail. There is a basic structure which all systems must implement, but beyond that there are extended and hierarchical structures. The success of the JPEG encouraged the formation of MPEG (Moving Pictures Expert Group) and of JBIG (Joint Bi-level Image Experts Group). In this section attention is confined to some basic introductory aspects of the JPEG standard. The JPEG standard is fully described in Pennebaker and Mitchell (1993). This text incorporates the requirements and guidelines published as the ISO DIS (Draft International Standard) 10918-1 as an appendix. Brief descriptions appear in Pitas (1993) and Bailey and Birch (1989).

Figure 3.16 is the basic block diagram of the JPEG compression system for transmission. An inverse system is needed for reception and decompression. A two-dimensional DCT of the image data is first computed. The DCT coefficients are then quantized and thresholded. The sequential zero-frequency or DC coefficients are then differentially pulse code modulated, and the resulting bit stream is either Huffman coded or arithmetically coded. The other frequency coefficients (AC coefficients) are Huffman coded or arithmetically coded. Long runs of zeros are run-length coded. Two streams of compressed data are produced consisting of the coded DC and AC coefficients.

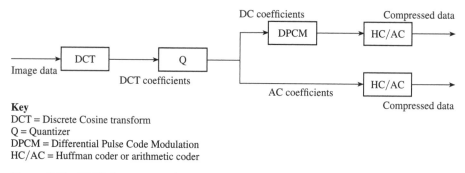

Key
DCT = Discrete Cosine transform
Q = Quantizer
DPCM = Differential Pulse Code Modulation
HC/AC = Huffman coder or arithmetic coder

Figure 3.16   JPEG data compression.

3.9.1   The Discrete Cosine transform

Any rectangular image is represented as an array of numerical values which indicate the image attributes of intensity, tone, and colour in some way. See, for example, Pennebaker and Mitchell (1993) for further details. Each of these values is known as a picture element or pixel, and as we have seen, there is likely to be a large number of them in an image. The image statistics in different regions may be quite different, and so it is preferable to divide the image up into a number of contiguous smaller blocks of more similar statistical characteristics for transformation. Smaller blocks also yield higher compression because of the higher correlation between adjacent pixels. Furthermore the transforms of smaller blocks are more easily computed. Thus the basic block in the JPEG standard consists of a square of 8×8 pixels, so that the image is sub-divided into immediate neighbours of 8×8 pixel blocks.

These 8×8 pixel blocks are two dimensional (2D), and so a 2D DCT is appropriate to transform the block. This is achieved by first calculating the DCT of each horizontal row of pixels, replacing the horizontal rows of pixels by the DCT components (the horizontal DCT), calculating the DCTs of the columns, and then replacing each column by its DCT (the vertical DCT). Since the component frequencies of the horizontal DCTs increase from left to right, and those of the vertical DCTs increase from top to bottom, the resultant 2D DCT contains the lower frequencies in its top-left part and the higher frequencies in its bottom-right part. Since the lower frequency components are often of larger amplitude than the higher frequency components, the top left of the map tends to contain relatively large values, and the bottom right part smaller values. In the JPEG standard the 2D DCT is given as (Pennebaker and Mitchell, 1993):

$$S(v, u) = \frac{1}{4} C(v)C(u) \sum_{x=0}^{7} s(y, x) \cos\{(2x + 1)u\pi/16\} \cos\{(2y + 1)v\pi/16\}$$

(3.91)

where the $S(v, u)$ are the 2D DCT coefficients,

$$C(v) = \begin{cases} 1/\sqrt{2} & \text{for } v = 0 \\ 1 & \text{for } v > 0 \end{cases}$$

$$C(u) = \begin{cases} 1/\sqrt{2} & \text{for } u = 0 \\ 1 & \text{for } u > 0 \end{cases}$$

and $s(y, x)$ is a pixel value in the 8×8 pixel block.

The inverse 2D DCT required for image reconstruction is

$$s(y, x) = \frac{1}{4} \sum_{u=0}^{7} \sum_{v=0}^{7} C(u)C(v)S(v, u) \cos\{(2x + 1)u\pi/16\}\cos\{(2y + 1)v\pi/16\}$$

(3.92)

As with the DCT, there are fast 2D DCT transforms (Pennebaker and Mitchell, 1993).

3.9.2 2D DCT coefficient quantization

Each of the 64 DCT coefficients, $S(v, u)$, is now separately quantized using a uniform quantizer (see Section 2.4.2). Each of the 64 quantizers has a different step size. Each of the coefficients is normalized to the quantization step size of its quantizer, and the result rounded to the nearest integer. This procedure creates an array of integers containing some zeros, particularly in the bottom right of the array.

3.9.3 Coding

The top-left coeffecient in the array corresponds to the DC term of the 2D DCT. It represents a mean signal level for the 8×8 block, and its value will not normally change rapidly between adjacent blocks. Therefore this coefficient is treated differently from the other AC coefficients (see Figure 3.16). It is differentially encoded by differential pulse code modulation (DPCM). Its value is set to the difference between its value and the value of the DC coefficient from the previous 8×8 block. This usually produces relatively small values.

The AC coefficients are arranged sequentially in the order of a zig-zag sequence taken through the 8×8 array. This zig-zag is illustrated in Figure 3.17. The sequence follows the numbers within the matrix. The purpose is to order the 2D coefficients from the largest to the smallest.

Both sets of coefficients are now coded (see Figure 3.16), which further compresses the data. Where longer runs of zeros occur these may be compressed using run-length coding. The code indicates how many sequential zeros there are. Where these occur at the end of the zig-zag an end of block codeword is used. The remaining coefficients are Huffman coded in the baseline sequential system or are arithmetically coded in extended DCT schemes. Both codes contain variable length codewords, in which the most commonly occurring codewords are the shortest. This reduces the number of bits to be transmitted, or in other words increases the degree of compression. These two

$$\begin{pmatrix} 0 & 1 & 5 & 6 & 14 & 15 & 27 & 28 \\ 2 & 4 & 7 & 13 & 16 & 26 & 29 & 42 \\ 3 & 8 & 12 & 17 & 25 & 30 & 41 & 43 \\ 9 & 11 & 18 & 24 & 31 & 40 & 44 & 53 \\ 10 & 19 & 23 & 32 & 39 & 45 & 52 & 54 \\ 20 & 22 & 33 & 38 & 46 & 51 & 55 & 60 \\ 21 & 34 & 37 & 47 & 50 & 56 & 59 & 61 \\ 35 & 36 & 48 & 49 & 57 & 58 & 62 & 63 \end{pmatrix}$$

Figure 3.17   The 2D DCT zig-zag sequence.

codes are the most efficient in that they transmit the most information with the least number of bits. The Huffman code uses an optimally selected set of codewords which contain an integer number of information bits. Arithmetic coding is used to obtain about 10% increased compression performance. It is a form of one-pass adaptive coding in which the book of codewords adapts dynamically to the data being coded. The topics of information theory and coding, however, are outside the scope of this book.

3.10   Worked examples

Example 3.8

The first four sampled voltage values of a 10 Hz bandwidth signal sampled at 125 Hz were (0.5, 1, 1, 0.5). Demonstrate how the discrete Fourier transform of this sequence may be obtained using the fast Fourier transform, and hence obtain the Fourier transform of the data.

Solution

The flowgraph for the FFT is given in Figure 3.18. We have

$$X(0) = G(0) + W_4^0 H(0) = G(0) + H(0)$$

$$G(0) = x(0) + W_2^0 x(2) = x(0) + x(2)$$

$$H(0) = x(1) + W_2^0 x(3) = x(1) + x(3)$$

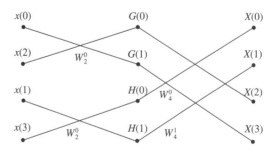

Figure 3.18   Flowgraph of the FFT in Example 3.8.

Since $W^0 = 1$, substituting values gives

$$X(0) = x(0) + x(2) + x(1) + x(3)$$
$$= 0.5 + 1 + 1 + 0.5 = 3$$
$$X(1) = G(1) + W_4^1 H(1)$$
$$G(1) = x(0) - W_2^0 x(2) = x(0) - x(2)$$
$$H(1) = x(1) - W_2^0 x(3) = x(1) - x(3)$$

Now, $W_N = e^{-j2\pi/N}$, and therefore $W_4^1 = e^{-j2\pi/4} = e^{-j\pi/2}$. Substituting gives

$$X(1) = x(0) - x(2) + e^{-j\pi/2}[x(1) - x(3)]$$
$$= 0.5 - 1 + \left[\cos\left(\frac{\pi}{2}\right) - j\sin\left(\frac{\pi}{2}\right)\right](1 - 0.5)$$
$$= 0.5 - 1 + (0 - j)0.5 = -0.5 - j0.5 = -0.5(1 + j)$$
$$X(2) = G(0) - W_4^0 H(0) = G(0) - H(0)$$
$$= x(0) + x(2) - [x(1) + x(3)]$$
$$= 0.5 + 1 - (1 + 0.5) = 0$$
$$X(3) = G(1) - W_4^1 H(1)$$
$$= x(0) - x(2) - e^{-j\pi/2}[x(1) - x(3)]$$
$$= 0.5 - 1 - \left[\cos\left(\frac{\pi}{2}\right) - j\sin\left(\frac{\pi}{2}\right)\right](1 - 0.5)$$
$$= -0.5 - (-j)0.5 = -0.5 + j0.5 = 0.5(-1 + j)$$

Therefore

$$X(\Omega) = \{3, -0.5(1 + j), 0, 0.5(-1 + j)\}$$

When T, the sampling interval, is small,

$$FT = T\,DFT$$

where FT is the Fourier transform. Here, $T = 1/125$ s $= 0.008$ s. The signal period is $1/10$ s $= 0.1$ s; therefore

$$\frac{T}{\text{period}} = \frac{0.008}{0.1} = 0.08 \ll 1$$

and it is a good approximation to put FT = T DFT, so that

$$FT = \{0.024, -0.004(1 + j), 0, 0.004(-1 + j)\}$$

Example 3.9

In a data compression system the data is first transformed and then the transform values are threshold limited to a threshold magnitude of 0.375. Two transforms are under consideration, the discrete cosine transform, $X_c(k)$, defined by

$$X_c(k) = \frac{1}{N} \sum_{n=0}^{N-1} x_n \cos\left(\frac{k2\pi n}{N}\right), \quad k = 0, 1, \ldots, N-1$$

and the Walsh transform, X_k, defined by

$$X_k = \frac{1}{N} \sum_{i=0}^{N-1} x_i \, \text{WAL}(k, i), \quad k = 0, 1, \ldots, N-1$$

Assuming representative results to be given by the data sequence {1, 2, 0, 3}, determine

(1) which is the more efficient of the two transforms for data compression in this case, and

(2) the percentage data compression achieved.

Inverse transform the compressed data obtained using the Walsh transform and compare with the original data sequence.

Solution

(1) For the DCT with $x_0 = 1$, $x_1 = 2$, $x_2 = 0$, $x_3 = 3$,

$$X_c(0) = \frac{1}{4}(x_0 \cos 0 + x_1 \cos 0 + x_2 \cos 0 + x_3 \cos 0)$$

$$= \frac{1}{4}(1 + 2 + 0 + 3) = \frac{6}{4} = 1.5$$

$$X_c(1) = \frac{1}{4}\sum_{n=0}^{3} x_n \cos\left(\frac{2\pi n}{4}\right) = \frac{1}{4}\sum_{n=0}^{3} x_n \cos\left(\frac{n\pi}{2}\right)$$

$$= \frac{1}{4}\left[x_0 + x_1 \cos\left(\frac{\pi}{2}\right) + x_2 \cos\left(\frac{2\pi}{2}\right) + x_3 \cos\left(\frac{3\pi}{2}\right)\right]$$

$$= \frac{1}{4}[1 + 2 \times 0 + 0 \times (-1) + 3 \times 0] = 0.25$$

$$X_c(2) = \frac{1}{4}\left[x_0 \cos\left(\frac{4\pi \times 0}{4}\right) + x_1 \cos\left(\frac{4\pi \times 1}{4}\right) + x_2 \cos\left(\frac{4\pi \times 2}{4}\right)\right.$$

$$\left. + x_3 \cos\left(\frac{4\pi \times 3}{4}\right)\right]$$

$$= \frac{1}{4}(x_0 - x_1 + x_2 - x_3) = \frac{1}{4}(1 - 2 + 0 - 3) = -1$$

$$X_c(3) = \frac{1}{4}\left[x_0 \, \cos\left(\frac{6\pi \times 0}{4}\right) + x_1 \, \cos\left(\frac{6\pi}{4}\right) + x_2 \, \cos\left(\frac{6\pi \times 2}{4}\right) \right.$$

$$\left. + x_3 \, \cos\left(\frac{6\pi \times 3}{4}\right) \right]$$

$$= \frac{1}{4}[1 + 2 \times 0 + 0 \times (-1) + 3 \times 0] = 0.25$$

Therefore

$$\text{DCT} = \{1.5, 0.25, -1, 0.25\}$$

The values remaining after thresholding ($|\text{values}| > 0.375$) are 1.5 and -1.

For the Walsh transform, evaluated in Section 3.8.2, $X_k = \{1.5, 0, 0.5, -1\}$, so the values remaining after thresholding in this case are 1.5, 0.5 and -1. Therefore the DCT provides more efficient data compression in this case.

(2)   Assuming that the data compression efficiency, η, is given by

$$\eta = \frac{(\text{no. of data in original seq.} - \text{no. of data in transformed seq.}) \times 100\%}{\text{no. of data in original seq.}}$$

then

$$\eta = \frac{4 - 2}{4} \times 100\% = 50\%$$

Finally, the Walsh transform compressed data is $\{1.5, 0, 0.5, -1\}$. The inverse transformation is given by Equation 3.70:

$$x_i = \sum_{i=0}^{N-1} X_k \, \text{WAL}(k, i), \quad i = 0, 1, \ldots, N-1$$

$$x_i = [1.5 \quad 0 \quad 0.5 \quad -1]\begin{bmatrix} 1 & 1 & 1 & 1 \\ 1 & 1 & -1 & -1 \\ 1 & -1 & -1 & 1 \\ 1 & -1 & 1 & -1 \end{bmatrix} = [1 \quad 2 \quad 0 \quad 3]$$

This is identical with the initial sequence. This is because $X_1 = 0 < 0.375$ and although it is not transmitted it is represented in the inverse transform as 0, its exact value. Normally the reconstituted sequence will be an approximation to the original one.

Problems

3.1   Determine the Fourier series representation of the periodic waveforms of Figure 3.19.

3.2   Obtain the complex Fourier series representation of the waveforms in Figure 3.19.

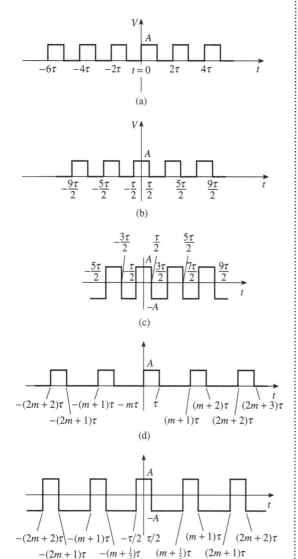

(a)

(b)

(c)

(d)

(e)

Figure 3.19   Periodic waveforms for Problems 3.1–3.4.

3.3   Show that the amplitudes found in Problem 3.1 agree with those found in Problem 3.2.

3.4   Plot the amplitude and phase spectra of the Fourier components of the waveforms in Figure 3.19.

3.5   Calculate the amplitude and energy spectral densities of the voltage waveform $v(t)$ given by

$$v(t) = \begin{cases} \dfrac{A}{\tau}t + A & -\tau \le t \le 0 \\[2mm] -\dfrac{A}{\tau}t + A & 0 \le t \le \tau \\[2mm] 0 & \tau \le t \le -\tau \end{cases}$$

where $A = 5$ V and $\tau = 20$ ms.

3.6   Calculate the energy spectral density and the phase spectrum of the waveform $v(t)$ given by

$$v = \begin{cases} 2\sin\left[\dfrac{2\pi}{T}\left(t + \dfrac{T}{4}\right)\right] & -\dfrac{T}{4} \le t \le \dfrac{T}{4} \\[3mm] 0 & \text{elsewhere} \end{cases}$$

where $T = 0.0167$ s.

3.7   Plot the energy spectral density of the function

$$w(t) = \begin{cases} \sin\left[\dfrac{2\pi}{T}\left(t + \dfrac{3T}{4}\right)\right] & -\dfrac{3T}{4} \le t \le -\dfrac{T}{2} \\[3mm] 1.0 & -\dfrac{T}{2} \le t \le \dfrac{T}{2} \\[3mm] \sin\left[\dfrac{2\pi}{T}\left(t - \dfrac{3T}{4}\right)\right] & \dfrac{T}{2} \le t \le \dfrac{3T}{4} \\[3mm] 0 & \dfrac{3T}{4} \le t \le -\dfrac{3T}{4} \end{cases}$$

where $T = 4$ s.

3.8   Calculate the DFT of the data sequence $\{0, 1, 1, 0\}$ and check the validity of your answer by calculating its IDFT.

3.9   Derive the dimensions of $X(k)$ and of $X^2(k)$. Hence calculate and plot the energy spectrum of the data sequence $\{0, 1, 1, 0\}$ whose DFT was calculated as the solution to Problem 3.8.

3.10 If the sequence {0, 1, 1, 0} of Problem 3.8 represented the digitized samples taken from a voltage waveform sampled at 125 Hz, determine the energy spectral density and phase spectrum of the Fourier transform of the data sequence.

3.11 Use the time-shifting property of the DFT and the solution to Problem 3.8 to obtain the amplitude and phase spectra of the time series {0, 0, 0, 0, 0, 1, 1, 0} for data sampled at the instants $t = 0, 1, 2, \ldots, 7$ ms.

3.12 Use the results of Problem 3.9 to verify Parseval's theorem for the data {0, 1, 1, 0}.

3.13 Use the correlation theorem to calculate the circular correlation of the data sequences {1, 1, 0, 1} and {1, 0, 0, 1}. Plot the correlation function against lag number, j.

3.14 Apply the correlation theorem to calculate the linear correlation of the data sequences {1, 1, 0, 1} and {1, 0, 0, 1}. Plot the correlation function against lag numbers and compare the result with the solution to Problem 3.13, explaining any differences.

3.15 Calculate the DFT of the data sequence {0, 1, 1, 0} using the decimation-in-time (Cooley–Tukey) FFT algorithm. Check the answer with that of Problem 3.8. Compare the numbers of complex additions and multiplications in the two methods.

3.16 Calculate the IFFT of the answer to Problem 3.15 to verify that the data sequence {0, 1, 1, 0} is obtained.

3.17 Calculate the FFT of the data sequence {0, 0, 1, 1, 1, 1, 0, 0} and plot the amplitude and phase spectra. Check the answer by calculating its IFFT to obtain the original sequence.

3.18 Write computer programs to compute the FFT and the IFFT. Check the FFT by computing the DFTs of the data sequences {0, 1, 1, 0} of Problem 3.8, and {1, 1, 0, 1} and {1, 0, 0, 1} of Problem 3.13. Check the IFFT by computing the IDFTs of the DFT sequences.

3.19 Use the FFT program to compute the 1024-point DFTs of the waveforms of Problems 3.5 and 3.7. Plot their energy and phase spectra and compare with the plots obtained in the solutions to Problems 3.5 and 3.7.

3.20 Using a 1024-point FFT compute and plot the energy spectrum of the rectangular pulse of amplitude 5 V and width $\tau = 6$ s. Compare the result with that of Problem 3.7.

3.21 (1) Use the convolution theorem (Equation 3.37) to obtain the convolution of the spectra of the two pairs of waveforms:

(a) $v_s = \sin(2\pi \times 100t)$ and the unit height pulse, v_w, centred at $t = 0$ and of width 2 s;

(b) $v_s = \sin(2\pi \times 100t)$

and

$v_w = \cos(2\pi \times 0.25t)$ for $1 \leqslant t \leqslant -1$ s

$\qquad 0 \qquad \qquad$ elsewhere

(2) When the Fourier components of a signal are obtained using the DFT of sampled data, a sample of the signal of length $(N-1)T$ is in effect used, where N is the number of data and T is the interval between the samples. The signal is said to have been windowed by a data window of length $(N-1)T$. The spectrum computed is then given by the convolution of the signal spectrum by the window spectrum. If in the case of the waveforms of part (1), v_s represents the signal and the v_w are the window data, comment on the relative suitabilities of the two data windows for defining the signal sample.

3.22 Calculate the discrete cosine, discrete Walsh, and discrete Hadamard transforms of the data sequence {0.1, −0.2, 0.3, −0.4, 0.5, 1.5, 2, 1.5, 0.5, −0.4, 0.3, −0.2, 0.1}. Assuming these transforms are being compared with respect to their data compression efficiencies with a preselected threshold value of 0.35, rank them in order of preference.

3.23 The sampled voltages obtained by scanning the intensity distribution of a photographic image are {3.2, 3.6, 3.3, 2.9, 1.7, 1.6, 1.8, 1.5}. Discuss the relative merits of transforming these data by the FFT and the DWT.

3.24 Extend the discussion of Problem 3.23 to include the amount of data compression available (including use of the DCT).

3.25 Draw up a table to show the advantages, disadvantages, and applications of the fast Fourier, discrete

Walsh, discrete cosine and discrete Hadamard transforms.

MATLAB problems

3.26 (a) Use an appropriate MATLAB function to find, by direct approach, the DFT coefficients of the following 8-point discrete-time sequence:

$$x(n) = \{4, 2, 1, 4, 6, 3, 5, 2\}$$

(b) Find, using an appropriate MATLAB function, the discrete-time sequence that corresponds to the following DFT coefficients:

$$27 + 0j$$
$$-4.12132 + 3.292893j$$
$$4 + j$$
$$0.12132 - 4.707107j$$
$$5 + 0j$$
$$0.12132 + 4.707107j$$
$$4 - j$$
$$-4.12132 - 3.292893j$$

3.27 (a) Compute, using MATLAB, the 32-point FFT of the discrete-time sequence given by:

$$x(n) = 1, n = 0, 1, \ldots, 15$$
$$0, n = 16, 17, \ldots, 31$$

(b) Compute, using MATLAB, the 64-point FFT of the data sequence in (a).

(c) Compare the results obtained in parts (a) and (b).

3.28 Explain clearly how a radix-2 FFT algorithm may be used to obtain an estimate of the frequency response of a discrete-time system with a z-transfer function in the form of a rational polynomial. Illustrate your answer by using the MATLAB FFT function to obtain the frequency response of a discrete-time filter with the following z-transfer function:

$$H(z) = \frac{1 - 1.618z^{-1} + z^{-2}}{1 - 1.516z^{-1} + 0.87z^{-2}}$$

Point out any practical issues that may be of concern.

References

Ahmed N. and Rao K.R. (1975) *Orthogonal Transforms for Digital Signal Processing.* Berlin: Springer.

Bailey D.J. and Birch N. (1989) Image compression using a discrete cosine transform image processor. *Electronic Engineering,* July, 9–44.

Beauchamp K.G. (1987) *Transforms for Engineers. A Guide to Signal Processing.* Oxford: Clarendon.

Burrus C.S. (1998) *Introduction to Wavelets and Wavelet Transforms: A Primer.* Englewood Cliffs NJ: Prentice-Hall.

Burrus C.S. and Parks T.W. (1985) *DFT/FFT and Convolution Algorithms. Theory and Implementation.* New York: Wiley.

Chan Y.T. (1995) *Wavelet Basics.* Boston MA: Kluwer Academic.

Chen W., Smith C.H. and Fialick S.C. (1977) A fast computational algorithm for the discrete cosine transform. *IEEE Trans. Communications,* **25**, 1004–9.

Chui C.K. (1992) *An Introduction to Wavelets.* Boston MA: Academic Press.

Cooley J.W. and Tukey J.W. (1965) An algorithm for the machine calculation of complex Fourier series. *Mathematics Computation,* **19**, 297–301.

Daubechies I. (1990) The wavelet transform, time-frequency localisation and signal analysis. *IEEE Trans. Information Theory,* **36**(5), 961–1005.

Daubechies I. (1992) *Ten Lectures on Wavelets.* Philadelphia: The Society for Industrial and Applied Mathematics.

Gentleman W.M. and Sande G. (1966) Fast Fourier transforms for fun and profit. In *Fall Joint Computing Conf., AFIPS Proc.,* **29**, 563–78.

Mallat S.G. (1989) A theory for multiresolution signal decomposition: the wavelet representation. *IEEE Trans. Pattern Analysis and Machine Intelligence*, **11**(7), 674–93.

Mallat S. and Hwang W.L. (1992) Singularity detection and processing with wavelets. *IEEE Trans. Information Theory*, **38**(2), 617–43.

McClellan J.H. and Rader C.M. (1979) *Number Theory in Digital Signal Processing*. Englewood Cliffs NJ: Prentice-Hall.

Narasinka M.J. and Petersen A.M. (1978) On the computation of the discrete cosine transform. *IEEE Trans. Communications*, **26**, 934–6.

Pennebaker W.B. and Mitchell J.L. (1993) *JPEG Still Image Data Compression Standard*. New York: Van Nostrand Reinhold.

Pitas I. (1993) *Digital Image Processing Algorithms*. New York: Prentice-Hall.

Rader C.M. (1968) Discrete Fourier transform when the number of data samples is prime. *IEEE Proc.*, **56**, 1107–8.

Rosenfield A. and Thurston M. (1971) Edge and curve detection for visual scene analysis. *IEEE Trans. Computing*, **20**, 562–9.

Saatchi M.R., Gibson C. and Rowe J.K.W. (1997) Adaptive multiresolution analysis based evoked potential filtering. *IEE Proc.-Sci. Meas. Technol.*, **144**(4), July, 149–55.

Signal Processing Committee (ed.) (1979) *Programs for Digital Signal Processing*. New York: IEEE.

Srinivassan R. and Rao K.R. (1983) An approximation to the discrete cosine transform. *Signal Processing*, **5**, 81–5.

Strum R.D. and Kirk D.E. (1988) *First Principles of Discrete Systems and Digital Signal Processing*. Reading MA: Addison-Wesley.

Thakor N.V., Xin-Rong G., Yi-Chun S. and Hanley D.F. (1993) Multiresolution wavelet analysis of evoked potentials. *IEEE Trans. Biomedical Engineering*, **40**(11), 1085–93.

Winograd S. (1978) On computing the discrete Fourier transform. *Mathematics Computation*, **32**, 175–99.

Yip P. and Ramamohan K. (1987) In *Handbook of Digital Signal Processing Engineering Applications* (Elliott D.E. (ed.)). New York: Academic Press.

Zhang J. and Zheng C. (1997) Extracting evoked potentials with the singularity detection technique. *IEEE Engineering in Medicine and Biology*. 155–61.

Appendices

3A C language program for direct DFT computation

The C language program given here evaluates, directly, the DFT or the IDFT of a discrete-time sequence, $x(n)$:

$$X(k) = \sum_{n=0}^{N-1} x(n)W^{nk}, \quad k = 0, 1, \ldots, N-1 \qquad \text{DFT} \qquad (3A.1a)$$

$$x(n) = \frac{1}{N} \sum_{k=0}^{N-1} X(k)W^{-nk} \qquad \text{IDFT} \qquad (3A.1b)$$

where $W = e^{j-2\pi/N}$ and N is the sequence length.

The input sequence, $x(n)$, must be in a complex form (real and imaginary). For a real data sequence, the imaginary parts of the data are set to zero. The main function, DFTD.c, is listed in Program 3A.1, and the function that computes the DFT or IDFT in Program 3A.2. Two functions, read__data() and save__data(), are required for reading the input data sequence and for saving the transformed data (Program 3A.3). The input data is held in the input file, coeff.dat, and the output is saved in the file dftout.dat.

Program 3A.1   Main function dftd.c, for derived computation of DFT.

```
/*-------------------------------------------------------------------------*/
/*                                                                         */
/*              Program to compute DFT coefficients directly               */
/*              3 other functions are used                                 */
/*                                                                         */
/*              E C Ifeachor. July, 1992                                   */
/*                                                                         */
/*-------------------------------------------------------------------------*/
#include     "dsp1.h"
#include     "dft.h"

main()
{
              extern   long npt;
              extern   int    inv;

              printf("select type of transform\n");
              printf("\n");
              printf("0      for forward DFT\n");
              printf("1      for inverse DFT\n");
              scanf("%d", &inv);
              read__data();
              dft();
              save__data();
              exit();
}
#include     "dft.c";
#include     "rdata.c";
#include     "sdata.c";
```

Program 3A.2   C language function for direct computation of the DFT of a discrete-time sequence. This function is held in a separate file.

```
/*-------------------------------------------------------------------------*/
/*                                                                         */
/*              Function to compute the DFT of a discrete-time             */
/*              sequence directly                                          */
/*                                                                         */
/*              E C Ifeachor. 31.10.91                                     */
/*                                                                         */
/*-------------------------------------------------------------------------*/
void          dft()
{
              extern int inv;
              extern long npt;
```

```
long      k, n;
double WN, wk, c, s, XR[size], XI[size];
extern complex x[size];

WN=2*pi/npt;
if(inv==1)
     WN=–WN;
for(k=0; k<npt; ++k){
     XR[k]=0.0; XI[k]=0.0;
     wk=k*WN;
     for(n=0; n<npt; ++n){
          c=cos(n*wk); s=sin(n*wk);
          XR[k]=XR[k]+x[n+1].real*c+x[n+1].imag*s;
          XI[k]=XI[k]–x[n+1].real*s+x[n+1].imag*c;
     }
     if(inv= =1){      /* divide by N for IDFT */
          XR[k]=XR[k]/npt;
          XI[k]=XI[k]/npt;
     }
}
for (k=1; k<=npt; ++k){       /* store transformed data in x */
     x[k].real=XR[k–1];
     x[k].imag=XI[k–1];
}
}
```

Program 3A.3   Function for reading the data, function for saving transformed data to disk file, header file containing constant structure definitions and header file containing common declarations and variables.

```
/* ------------------------------------------------------------------------------- */
/* *                                                                           */
/* *         Function to read data, in complex format, for the DFT or FFT      */
/* *                                                                           */
/* *         E C Ifeachor. Last modification: July, 1992.                      */
/* *                                                                           */
/* ------------------------------------------------------------------------------- */
void          read__data()
{
          extern    long      npt;
          int       n;
          extern    complex x[size];

          for(n=0; n<size; ++n){
               x[n].real=0;
               x[n].imag=0;
          }
          if((in=fopen("coeff.dat", "r"))= =NULL){
               printf("cannot open file coeff.dat\n");
               exit(1);
          }
          fscanf(in,"%ld", &npt);
```

```
                    for(n=1; n<=npt; n++){
                            fscanf(in,"%lf %lf ",&x[n].real,&x[n].imag);
                    }
                    fclose(in);
}


void            save__data()                                    /* file name sdata.c */
{
                long    k;
                int     k1;
                extern  long npt;
                extern  complex x[size];

                if((out=fopen("dftout.dat","w"))==NULL){
                        printf("cannot open file dftout.dat \n");
                        exit(1);
                }
                fprintf(out,"k \tXR(k) \t\tXI(k) \n");
                fprintf(out,"\n");
                for(k=1; k<=npt; ++k){
                        k1=k-1;
                        fprintf(out,"%d \t%f \t%f \n", k1, x[k].real, x[k].imag);
                }
                fclose(out);
}
/* This file contains common definitions and structures
        filename: dsp1.h
*/
#include         <stdio.h>
#include         <math.h>
#include         <dos.h>

#define size      600
#define pi        3.141592654
#define maxbits   30

typedef struct    {
        double    real;
        double    imag;
        double    modulus;
        double    angle;
        }complex;
/*
        filename: dft.h
*/
void            dft();
void            fft();
void            read__data();
void            save__data();
int             inv;
long            npt;
complex         x[size];
FILE            *in, *out, *fopen();
```

**Test Example
3A.1**

Use the direct DFT program to find the DFT coefficients of the following 8-point discrete-time sequence:

$$x(n) = \{4, 2, 1, 4, 6, 3, 5, 2\}$$

The input data file, created with PC edlin (most word processors may be used for this purpose) for the problem, has the following format:

```
8
4  0
2  0
1  0
4  0
6  0
3  0
5  0
2  0
```

The first line specifies the length of the data sequence.
    The DFT of the data, using the program, is given below:

k	XR(k)	XI(k)
0	27.000 000	0.000 000
1	−4.121 320	3.292 893
2	4.000 000	1.000 000
3	0.121 320	−4.707 107
4	5.000 000	−0.000 000
5	0.121 320	4.707 107
6	4.000 000	−1.000 000
7	−4.121 320	−3.292 893

**Test Example
3A.2**

Find, using the DFT program, the discrete-time sequence corresponding to the DFT coefficients above. The input data has the following format:

```
8
27.000 000      0.000 000
−4.121 320      3.292 893
 4.000 000      1.000 000
 0.121 320     −4.707 107
 5.000 000     −0.000 000
 0.121 320      4.707 107
 4.000 000     −1.000 000
−4.121 320     −3.292 893
```

The IDFT option was selected in response to the program prompts. The output of the program is the same as the discrete-time sequence in Test example 3A.1.

<table>
<tr><td>Test Example
3A.3</td><td>The third test example uses the complex data sequence (IEEE, 1979, Chapter 1):</td></tr>
</table>

$$x(n) = Q^n, \quad n = 0, 1, \ldots, 31$$

where $Q = 0.9 + j0.3$.

The input data sequence, $x(n)$, and its DFT, $X(k)$, using the direct DFT program are listed in Tables 3A.1 and 3A.2 respectively.

Table 3A.1   Complex input data sequence.

	$x(n)$	
n	*Real*	*Imaginary*
0	0.100000E01	0.
1	0.900000E00	0.300000E00
2	0.720000E00	0.540000E00
3	0.486000E00	0.702000E00
4	0.226800E00	0.777600E00
5	−0.291600E-01	0.767880E00
6	−0.256608E00	0.682344E00
7	−0.435650E00	0.537127E00
8	−0.553224E00	0.352719E00
9	−0.603717E00	0.151480E00
10	−0.588789E00	−0.447828E-01
11	−0.516476E00	−0.216941E00
12	−0.399746E00	−0.350190E00
13	−0.254714E00	−0.435095E00
14	−0.987144E-01	−0.467999E00
15	0.515569E-01	−0.450814E00
16	0.181645E00	−0.390265E00
17	0.280560E00	−0.296745E00
18	0.341528E00	−0.182903E00
19	0.362246E00	−0.621539E-01
20	0.344667E00	0.527352E-01
21	0.294380E00	0.150862E00
22	0.219684E00	0.224090E00
23	0.130488E00	0.267586E00
24	0.371637E-01	0.279974E00
25	−0.505440E-01	0.263125E00
26	−0.124428E00	0.221649E00
27	−0.178480E00	0.162156E00
28	−0.209279E00	0.923965E-01
29	−0.216070E00	0.203732E-01
30	−0.200575E00	−0.464851E-01
31	−0.166572E00	−0.102009E00

Table 3A.2   Transformed output for Test example 3A.3.

0.693972	3.499714
2.792268	8.050456
9.402964	−9.135013
1.866446	−3.833833
1.131822	−2.234158
0.904794	−1.534631
0.799557	−1.139607
0.739607	−0.882315
0.700858	−0.698566
0.673577	−0.558478
0.653112	−0.446244
0.636987	−0.352691
0.623790	−0.272085
0.612613	−0.200642
0.602885	−0.135703
0.594200	−0.075314
0.586276	−0.017948
0.578899	0.037651
0.571898	0.092607
0.565139	0.147983
0.558490	0.204882
0.551858	0.264523
0.545134	0.328363
0.538217	0.398257
0.531000	0.476679
0.523403	0.567133
0.515361	0.674850
0.506928	0.808100
0.498469	0.980906
0.491388	1.219210
0.490730	1.577083
0.517355	2.188832

3B   C program for radix-2 decimation-in-time FFT

The FFT program given here is a C language implementation of the radix-2 decimation-in-time FFT (Cooley and Tukey, 1965). The program evaluates the DFT or the IDFT of a discrete-time sequence as defined in Equations 3A.1. The program consists of a main function, dftf.c, and three functions: fft(), read__data(), and save__data(). As in the case of the direct DFT, all the functions are held in separate files and combined during compilation by include statements in the main function. The two functions read__data() and save__data() are used to read the data and to save the transformed data to file. These two files are identical to those used for the direct DFT. The main program, dftf.c, and the function fft() are listed in Programs 3B.1 and 3B.2 respectively.

Applying each of the test data described in Appendix 3A to the FFT program, in exactly the same format, yields identical results to those of the direct DFT. It is left as an exercise to the reader to confirm that this is the case.

Program 3B.1 Main function, dftf.c, for computing DFTs using decimation-in-time FFT.

```
/*-----------------------------------------------------------------------------------*/
/*                                                                                   */
/*           Program to compute DFT coefficients using DIT FFT                       */
/*           3 other functions are used                                              */
/*                                                                                   */
/*           E C Ifeachor. July, 1992                                                */
/*                                                                                   */
/*-----------------------------------------------------------------------------------*/
#include    "dsp1.h"
#include    "dft.h"
main()
{
            extern long npt;
            extern int     inv;

            printf("select type of transform \n");
            printf("\n");
            printf("0     for forward DFT\n");
            printf("1     for inverse DFT\n");
            scanf("%d", &inv);
            read__data();
            fft();
            save__data();
            exit();
}
#include    "fft.c";
#include    "rdata.c";
#include    "sdata.c";
```

Program 3B.2 C language implementation of radix-2, decimation-in-time FFT algorithm.

```
/*-----------------------------------------------------------------------------------*/
/*                                                                                   */
/*           file name: fft.c                                                        */
/*           E C Ifeachor. June, 1992                                                */
/*                                                                                   */
/*           Function computes the DFT of a sequence using radix2 FFT                */
/*                                                                                   */
/*                                                                                   */
/*-----------------------------------------------------------------------------------*/
void        fft()
{
            int         sign;
            long        m, irem, l, le, le1, k, ip,i,j;
            double      ur, ui, wr, wi, tr, ti, temp;
            extern      long npt;
            extern      int inv;
            extern      complex x[size];
```

```
/* in-place bit reverse shuffling of data */

j=1;
for(i=1; i<npt; ++i){
        if(i<j){
                tr=x[j].real; ti=x[j].imag;
                x[j].real= x[i].real;
                x[j].imag=x[i].imag;
                x[i].real=tr; x[i].imag=ti;
                k=npt/2;
                while(k<j){
                        j=j–k;
                        k=k/2;
                        }
                }
                else{
                        k=npt/2;
                        while(k<j){
                                j=j–k;
                                k=k/2;
                                }
                }
                j=j+k;
}
/* calculate the number of stages: m=log2(npt), and whether FFT or IFFT */
                m=0; irem=npt;
                while(irem>1){
                        irem=irem/2;
                        m=m+1;
                }
                if(inv==1)
                        sign=1;
                else
                        sign=–1;

/* perform the FFT computation for each stage */
                for(l=1; l<=m, l++){
                        le=pow(2, l);
                        le1= le/2;
                        ur=1.0; ui=0;
                        wr=cos(pi/le1);
                        wi=sign*sin(pi/le1);
                        for(j=1; j<=le1; ++j){
                                i=j;
                                while(i<=npt){
                                        ip=i+le1;
                                        tr=x[ip].real*ur–x[ip].imag*ui;
                                        ti=x[ip].imag*ur+x[ip].real*ui;
                                        x[ip].real=x[i].real–tr;
                                        x[ip].imag=x[i].imag–ti;
                                        x[i].real=x[i].real+tr;
                                        x[i].imag=x[i].imag+ti;
                                        i=i+le;
                                }
```

```
                        temp=ur*wr–ui*wi;
                        ui=ui*wr+ur*wi;
                        ur=temp;
                }
        }
        /* If inverse fft is desired divide each coefficient by npt */
        if(inv==1){
                for(i=1; i<=npt; ++i){
                        x[i].real=x[i].real/npt;
                        x[i].imag=x[i].imag/npt;
                }
        }
}
```

3C   DFT and FFT with MATLAB

The key functions in MATLAB and MATLAB Signal Processing Toolbox for performing one-dimensional DFT and FFT are **dftmtx**, **fft** and **ifft**. The Toolbox also contains functions for performing discrete cosine transform and two-dimensional FFT.

The **dftmtx** function may be used to compute the discrete Fourier transform of an *N*-point data sequence in the vector *x* using the following command:

$$X = x*dftmtx(N)$$

The **dftmtx** function computes and returns the twiddle factors as an $N \times N$ complex matrix. This is then multiplied by the data sequence, *x*, to yield its discrete Fourier transform.

The inverse DFT can be obtained by using the conj command:

$$x = X*conj(dftmtx(N))yN$$

The **fft** function computes the DFT of a one-dimensional data sequence using a radix-2 FFT algorithm (if the data length is a power of 2) and the **ifft** function is used to find the inverse DFT.

The MATLAB program (Program 3C.1) may be used to read data from a file and compute the forward or inverse DFT directly. Program 3C.2 may be used to compute the DFT or inverse DFT via the FFT.

Illustrative examples of the use of the programs can be found in the companion manual, *A Practical Guide for MATLAB and C Language Implementations of DSP Algorithms*, published by Pearson Education.

Program 3C.1   MATLAB program for direct computation of DFT

```
function DFTD

clear all;
% Program to compute DFT coefficients directly

direction=-1;  %1 – forward DFT, –1 – inverse DFT
in=fopen('datain.dat','r');
x=fscanf(in,'%g %g',[2,inf]);
```

```
fclose(in);
x=x(1,:)+x(2,:)*i; % form complex numbers

if direction==1
    y=x*dftmtx(length(x));          %compute DFT
else
    y=x*conj(dftmtx(length(x)))/length(x); %compute IDFT
end

% Save/Print the results
out=fopen('dataout.dat','w');
fprintf(out,'%g %g\n',[real(y); imag(y)]);
fclose(out);
subplot(2,1,1),plot(1:length(x),x); title('Input Signal');
subplot(2,1,2),plot(1:length(y),y); title('Output Signal');
```

Program 3C.2   MATLAB program for computation of DFT via FFT

```
function DFTF
% Program to compute DFT coefficients using DIT FFT
%

clear all;
direction=-1;   %1 – forward DFT, –1 – inverse DFT
in=fopen('dataout.dat','r');
x=fscanf(in,'%g %g',[2,inf]);
fclose(in);
x=x(1,:)+x(2,:)*i; % form complex numbers
if direction==1
    y=fft(x,length(x)) % compute FFT
else
    y=ifft(x,length(x)) % compute IFFT
end

% Save/Print the results
out=fopen('dataout.dat','w');
fprintf(out,'%g %g\n',[real(y); imag(y)]);
fclose(out);
subplot(2,1,1),plot(1:length(x),x); title('Input Signal');
subplot(2,1,2),plot(1:length(y),y); title('Output Signal');
```

References for Appendices

Cooley J.W. and Tukey J.W. (1965) An algorithm for the machine calculation of complex Fourier series. *Mathematics Computation*, **19**(90), April, 297–301.

IEEE (1979) *Programs for Digital Signal Processing*. New York: IEEE Press.

The *z*-transform and its applications in signal processing

4

The *z*-transform is a convenient yet invaluable tool for representing, analyzing and designing discrete-time signals and systems. It plays a similar role in discrete-time systems to that which the Laplace transform plays in continuous-time systems.

In this chapter, we present important aspects of the *z*-transform, especially those that will be used in subsequent chapters, and highlight its applications in discrete-time systems design. The applications include the use of the *z*-transform to describe discrete-time signals and systems so that we can readily infer their degree of stability and to visualize their frequency responses, the analysis of quantization errors in digital filters and the computation of the frequency response of discrete-time systems. Most of the applications are covered in more detail in subsequent chapters.

As in the rest of the book, the presentation in this chapter is practical in approach. Algorithms, C language programs and MATLAB programs, where necessary, are provided to enable readers to gain a deeper understanding of the subject. Much of the discussion in this chapter involves linear discrete-time signals and systems, and so we will begin by reviewing very briefly characteristic features of this class of signals and systems.

4.1   Discrete-time signals and systems

A discrete signal has values which are defined only at discrete values of time or some other appropriate variable, for example space. As discussed in Chapters 1 and 2, such a signal may be generated by sampling a continuous-time signal at regular time intervals nT, $n = 0, 1, \ldots$, where T is the sampling period. It may also be generated, artificially, via some algorithm in a computer. The amplitude of a discrete-time signal may have discrete values (discrete time, discrete amplitude), or it may be continuous.

By tradition, a discrete-time signal is represented as a sequence of numbers:

$$x(n), \qquad n = 0, 1, \ldots \tag{4.1a}$$

$$x(nT), \quad n = 0, 1, \ldots \tag{4.1b}$$

$$x_n, \qquad n = 0, 1, \ldots \tag{4.1c}$$

where the symbol $x(n)$, $x(nT)$ or x_n indicates the value of the signal at the discrete time n (or nT). For convenience we will use the symbol $x(n)$ to denote both the value of the sequence at the discrete time n and the sequence itself unless we wish to emphasize the difference. The meaning will be clear from the context. It is common practice in DSP to omit the symbol T since the sequence is not always a function of time (it may, for example, be a function of space). Sometimes T is omitted because the sampling frequency is assumed to be unity (that is, normalized) for convenience.

A discrete-time system is essentially a mathematical algorithm that takes an input sequence, $x(n)$, and produces an output sequence, $y(n)$. Examples of discrete-time systems are digital controllers, digital spectrum analyzers, and digital filters. A discrete-time system may be linear or nonlinear, time invariant or time varying. Linear time-invariant (LTI) systems form an important class of systems used in DSP. Examples are digital filters discussed in detail in Chapters 6 to 8.

A discrete-time system is said to be linear if it obeys the principles of superposition. That is, the response of a linear system to two or more inputs is equal to the sum of the responses of the system to each input acting separately in the absence of all the other inputs. For example, if an input $x_1(n)$ to the system gives rise to the output $y_1(n)$, and another input $x_2(n)$ produces the output $y_2(n)$, the response of the system to both inputs will be

$$a_1 x_1(n) + a_2 x_2(n) \rightarrow a_1 y_1(n) + a_2 y_2(n) \tag{4.2}$$

where a_1 and a_2 are arbitrary constants.

A discrete-time system is said to be time invariant (sometimes referred to as shift invariant) if its output is independent of the time the input is applied. For example, if the input $x(n)$ gives the output $y(n)$, then the input $x(n - k)$ will give the output $y(n - k)$:

$$x(n) \rightarrow y(n) \tag{4.3a}$$

$$x(n - k) \rightarrow y(n - k) \tag{4.3b}$$

that is, a delay in the input causes a delay by the same amount in the output signal.

The input–output relationship of an LTI system is given by the convolution sum

$$y(n) = \sum_{k=-\infty}^{\infty} h(k)x(n - k) \tag{4.4}$$

where $h(k)$ is the impulse response of the system. The values of $h(k)$ completely define the discrete-time system in the time domain. An LTI system is stable if its impulse response satisfies the condition

$$\sum_{k=-\infty}^{\infty} |h(k)| < \infty \tag{4.5}$$

This condition is satisfied if $h(k)$ is of finite duration or if $h(k)$ decays towards zero as k increases. Stability considerations are described in more detail in Section 4.5.7.

A causal system is one which produces an output only when there is an input. All physical systems are causal. In general, a causal discrete-time sequence, $x(n)$, or the impulse response, $h(k)$, of a discrete-time system is zero before time 0, that is $x(n) = 0$, $n < 0$, or $h(k) = 0$, $k < 0$. Much of the discussion in this book is about practical, that is causal, systems.

4.2 The z-transform

The z-transform of a sequence, $x(n)$, which is valid for all n, is defined as

$$X(z) = \sum_{n=-\infty}^{\infty} x(n)z^{-n} \tag{4.6}$$

where z is a complex variable.

In causal systems, $x(n)$ may be nonzero only in the interval $0 < n < \infty$ and Equation 4.6 reduces to the so-called one-sided z-transform:

$$X(z) = \sum_{n=0}^{\infty} x(n)z^{-n} \tag{4.7}$$

Clearly, the z-transform is a power series with an infinite number of terms and so may not converge for all values of z. The region where the z-transform converges is known as the region of convergence (ROC), and in this region the values of $X(z)$ are finite. Not surprisingly, the region of convergence is determined by the properties of $x(n)$ or equivalently by those of $X(z)$ as illustrated by the following set of examples.

Example 4.1

Find the z-transform and the region of convergence for each of the discrete-time sequences given in Figure 4.1.

(1)   The sequence of Figure 4.1(a) is noncausal, since $x(n)$ is not zero for $n < 0$, but it is of a finite duration. The sequence has values $x(-6) = 0$, $x(-5) = 1$, $x(-4) = 3$, $x(-3) = 5$, $x(-2) = 3$, $x(-1) = 1$ and $x(0) = 0$. From Equation 4.6, the z-transform is given by

$$X_1(z) = \sum_{n=-\infty}^{\infty} x(n)z^{-n}$$

$$= z^5 + 3z^4 + 5z^3 + 3z^2 + z$$

It is readily verified that the value of $X(z)$ becomes infinite when $z = \infty$. Thus the ROC is everywhere in the z-plane except at $z = \infty$.

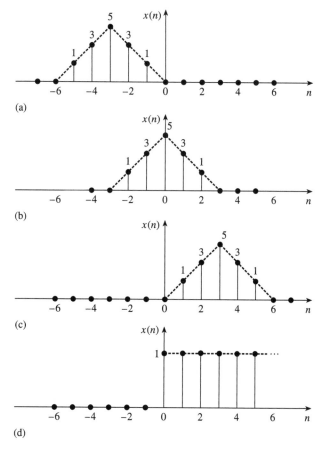

(a)

(b)

(c)

(d)

Figure 4.1   Causal and noncausal discrete-time sequences.

(2) Again, the sequence in Figure 4.1(b) is not causal. It is of a finite duration, and double sided. The values of the sequence are $x(3) = 0$, $x(-2) = 1$, $x(-1) = 3$, $x(0) = 5$, $x(1) = 3$, $x(2) = 1$ and $x(3) = 0$. From Equation 4.6, the z-transform is given by

$$X_2(z) = \sum_{n=-\infty}^{\infty} x(n)z^{-n}$$

$$= z^2 + 3z + 5 + 3z^{-1} + z^{-2}$$

It is evident that the value of $X(z)$ is infinite if $z = 0$ or if $z = \infty$. Therefore the region of convergence is everywhere except at $z = 0$ and $z = \infty$.

(3) Figure 4.1(c) represents a causal, finite duration sequence with values $x(0) = 0$, $x(1) = 1$, $x(2) = 3$, $x(3) = 5$, $x(4) = 3$, $x(5) = 1$ and $x(6) = 0$. The z-transform is given by

$$X_3(z) = \sum_{n=-\infty}^{\infty} x(n)z^{-n}$$

$$= z^{-1} + 3z^{-2} + 5z^{-3} + 3z^{-4} + z^{-5}$$

In this case, $X(z) = \infty$ for $z = 0$. Thus the region of convergence is everywhere except at $z = 0$.

(4) The discrete-time sequence in Figure 4.1(d) may be defined mathematically as

$$x(n) = 1 \quad 0 \leqslant n \leqslant \infty$$

$$= 0 \quad n < 0$$

Clearly it is a causal sequence of infinite duration. From Equation 4.6, the z-transform of the sequence is given by

$$X(z) = \sum_{n=-\infty}^{\infty} x(n)z^{-n}$$

$$= \sum_{n=0}^{\infty} z^{-n}$$

$$= 1 + z^{-1} + z^{-2} + \ldots$$

This is a geometric series with a common ratio of z^{-1}. The series converges if $|z^{-1}| < 1$ or equivalently if $|z| > 1$. Thus we may express $X(z)$ in closed form provided that $|z| > 1$:

$$X(z) = 1 + z^{-1} + z^{-2} + \ldots \tag{4.8}$$

$$= 1/(1 - z^{-1}) = z/(z - 1)$$

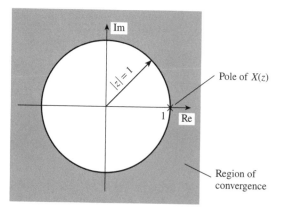

Figure 4.2   Region of convergence for Example 4.1, part (4).

In this case, the z-transform is valid everywhere outside a circle of unit radius whose centre is at the origin. The exterior of the circle is the region of convergence (Figure 4.2). We can readily verify that when $|z| > 1$ $X(z)$ converges whereas when $|z| < 1$ $X(z)$ diverges. For example, if we let $z = 2$ (outside the unit circle) we find that the series on the RHS of Equation 4.8 adds up to 2:

$$X(z) = 1 + 1/2 + (1/2)^2 + (1/2)^3 + \ldots = 2/(2 - 1) = 2$$

as it is clearly a geometric series with a common ration of $1/2$ and a first term of 1, giving the sum to infinity of $2/(2 - 1) = 2$. On the other hand if $z = 1/2$ (inside the unit circle) the series of Equation 4.8 becomes

$$X(z) = 1 + 1/0.5 + (1/0.5)^2 + (1/0.5)^3 + \ldots = 1 + 2 + 4 + 8 + \ldots$$

which is seen to be diverging. In Figure 4.2, the region of convergence (hatched) is seen to be bounded by the circle $|z| = 1$, the radius of the pole of $X(z)$. Values of z for which $X(z) = \infty$ are referred to as poles of $X(z)$. Values of z for which $X(z) = 0$ are referred to as the zeros of $X(z)$.

From the examples above we may infer that for causal sequences of finite duration the z-transform converges everywhere except at $z = 0$. For causal infinite duration sequences the z-transform converges everywhere outside a circle bounded by the radius of the pole with the largest radius. For stable causal systems the ROC always encloses the circle of unit radius which is important for the systems to have a frequency response.

The z-transforms of common sequences are available, in closed form, and are given in the form of tables, such as Table 4.1. Such tables are useful in finding the inverse z-transform.

Table 4.1   z-transforms of some common sequences.

Entry number	Discrete-time sequence x(n), n ⩾ 0	z-transform X(z)	Region of convergence of X(z)
1	$k\delta(n)$	k	Everywhere
2	k	$\dfrac{kz}{z-1}$	$\lvert z \rvert > 1$
3	kn	$\dfrac{kz}{(z-1)^2}$	$\lvert z \rvert > 1$
4	kn^2	$\dfrac{kz(z+1)}{(z-1)^3}$	$\lvert z \rvert > 1$
5	$ke^{-\alpha n}$	$\dfrac{kz}{z-e^{-\alpha}}$	$\lvert z \rvert > e^{-\alpha}$
6	$kne^{-\alpha n}$	$\dfrac{kze^{-\alpha}}{(z-e^{-\alpha})^2}$	$\lvert z \rvert > e^{-\alpha}$
7	$1 - e^{-\alpha n}$	$\dfrac{z(1-e^{-\alpha})}{z^2 - z(1+e^{-\alpha}) + e^{-\alpha}}$	$\lvert z \rvert > e^{-\alpha}$
8	$\cos(\alpha n)$	$\dfrac{z(z-\cos\alpha)}{z^2 - 2z\cos\alpha + 1}$	$\lvert z \rvert > 1$
9	$\sin(\alpha n)$	$\dfrac{z\sin\alpha}{z^2 - 2z\cos\alpha + 1}$	$\lvert z \rvert > 1$
10	$e^{-\alpha n}\sin(\alpha n)$	$\dfrac{ze^{-\alpha}\sin\alpha}{z^2 - 2e^{-\alpha}z\cos\alpha + e^{-2\alpha}}$	$\lvert z \rvert > e^{-\alpha}$
11	$e^{-\alpha n}\cos(\alpha n)$	$\dfrac{ze^{-\alpha}(ze^{\alpha} - \cos\alpha)}{z^2 - 2ze^{-\alpha}\cos\alpha + e^{-2\alpha}}$	$\lvert z \rvert > e^{-\alpha}$
12	$\cosh(\alpha n)$	$\dfrac{z^2 - z\cosh\alpha}{z^2 - 2z\cosh\alpha + 1}$	$\lvert z \rvert > \cosh\alpha$
13	$\sinh(\alpha n)$	$\dfrac{z\sinh\alpha}{z^2 - 2z\cosh\alpha + 1}$	$\lvert z \rvert > \sinh\alpha$
14	$k\alpha^n$	$\dfrac{kz}{z-\alpha}$	$\lvert z \rvert > \alpha$
15	$kn\alpha^n$	$\dfrac{k\alpha z}{(z-\alpha)^2}$	$\lvert z \rvert > \alpha$
16	$2\lvert c \rvert \lvert p \rvert^n \cos(n\angle p + \angle c)$	$\dfrac{cz}{z-p} + \dfrac{c^*z}{z-p^*}$	

k and α are constants; c is a complex number.

4.3   The inverse z-transform

The inverse z-transform (IZT) allows us to recover the discrete-time sequence $x(n)$, given its z-transform. The IZT is particularly useful in DSP work, for example in finding the impulse response of digital filters. Symbolically, the inverse z-transform may be defined as

$$x(n) = Z^{-1}[X(z)] \tag{4.9}$$

where $X(z)$ is the z-transform of $x(n)$ and Z^{-1} is the symbol for the inverse z-transform.

Assuming a causal sequence, the z-transform, $X(z)$, in Equation 4.7 can be expanded into a power series as

$$X(z) = \sum_{n=0}^{\infty} x(n)z^{-n}$$

$$= x(0) + x(1)z^{-1} + x(2)z^{-2} + x(3)z^{-3} + \ldots \tag{4.10}$$

It is seen that the values of $x(n)$ are the coefficients of z^{-n} ($n = 0, 1, \ldots$) and so can be obtained directly by inspection. In practice, $X(z)$ is often expressed as a ratio of two polynomials in z^{-1} or equivalently in z:

$$X(z) = \frac{b_0 + b_1 z^{-1} + b_2 z^{-2} + \ldots + b_N z^{-N}}{a_0 + a_1 z^{-1} + a_2 z^{-2} + \ldots + a_M z^{-M}} \tag{4.11}$$

In this form, the inverse z-transform, $x(n)$, may be obtained using one of several methods including the following three:

(1)   power series expansion method;

(2)   partial fraction expansion method;

(3)   residue method.

Each method has its own merits and demerits. In terms of mathematical rigour, the residue method is perhaps the most elegant. The power series method, however, lends itself most easily to computer implementation.

In the next few sections, we will describe each of the three methods in turn, using numerical examples to illustrate the principles involved. In the appendix, we describe a set of C language program listings for evaluating the inverse z-transform for methods (1) and (2). Several illustrative numerical examples using the programs are given.

4.3.1   Power series method

Given the z-transform, $X(z)$, of a causal sequence as in Equation 3.11, it can be expanded into an infinite series in z^{-1} or z by long division (sometimes called synthetic division):

$$X(z) = \frac{b_0 + b_1 z^{-1} + b_2 z^{-2} + \ldots + b_N z^{-N}}{a_0 + a_1 z^{-1} + a_2 z^{-2} + \ldots + a_M z^{-M}}$$

$$= x(0) + x(1)z^{-1} + x(2)z^{-2} + x(3)z^{-3} + \ldots \qquad (4.12)$$

In this method, the numerator and denominator of $X(z)$ are first expressed in either descending powers of z or ascending powers of z^{-1} and the quotient is then obtained by long division. We will illustrate the method by an example.

Example 4.2   Given the following z-transform of a causal LTI system, obtain its IZT by expanding it into a power series using long division:

$$X(z) = \frac{1 + 2z^{-1} + z^{-2}}{1 - z^{-1} + 0.3561z^{-2}}$$

Solution   First, we expand $X(z)$ into a power series with the numerator and denominator polynomials in ascending powers of z^{-1} and then perform the usual long division.

$$
\begin{array}{r}
1 + 3z^{-1} + 3.6439z^{-2} + 2.5756z^{-3} + \ldots \\
1 - z^{-1} + 0.3561z^{-2} \, \overline{\big)\, 1 + 2z^{-1} + z^{-2}} \\
\underline{1 - \ z^{-1} + 0.3561z^{-2}} \\
3z^{-1} + 0.6439z^{-2} \\
\underline{3z^{-1} - 3z^{-2} + 1.0683z^{-3}} \\
3.6439z^{-2} - 1.0683z^{-3} \\
\underline{3.6439z^{-2} - 3.6439z^{-3} + 1.297\,592\,7z^{-4}} \\
2.5756z^{-3} - 1.297\,592\,7z^{-4}
\end{array}
$$

Alternatively, we may express the numerator and denominator in positive powers of z, in descending order, and then perform the long division:

$$\frac{z^2 + 2z + 1}{z^2 - z + 0.3561}$$

$$
\begin{array}{r}
1 + 3z^{-1} + 3.6439z^{-2} + 2.5756z^{-3} + \ldots \\
z^2 - z + 0.3561 \, \overline{\big)\, z^2 + 2z + 1} \\
\underline{z^2 - \ z + 0.3561} \\
3z + 0.6439 \\
\underline{3z - 3 + 1.0683z^{-1}} \\
3.6439 - 3.643\,91z^{-1} + 1.297\,592\,7z^{-2} \\
2.5756z^{-1} - 1.297\,592\,7z^{-2}
\end{array}
$$

Either way, the z-transform is now expanded into the familiar power series, that is

$$X(z) = 1 + 3z^{-1} + 3.6439z^{-2} + 2.5756z^{-3} + \dots$$

The inverse z-transform can now be written down directly:

$$x(0) = 1; \ x(1) = 3; \ x(2) = 3.6439; \ x(3) = 2.5756; \dots$$

The long division approach can be reformulated (see Appendix 4A) so that the values of $x(n)$ are obtained recursively:

$$x(0) = b_0/a_0$$
$$x(1) = [b_1 - x(0)a_1]/a_0$$
$$x(2) = [b_2 - x(1)a_1 - x(0)a_2]/a_0$$
$$\vdots \qquad \vdots \quad \vdots \ \vdots \ \vdots \qquad \vdots$$

$$x(n) = \left[b_n - \sum_{i=1}^{n} x(n-i)a_i \right] \Big/ a_0, \quad n = 1, 2, \dots \tag{4.13a}$$

where

$$x(0) = b_0/a_0 \tag{4.13b}$$

We will repeat the previous example to illustrate the recursive approach.

Example 4.3

Find the first four terms of the inverse z-transform, $x(n)$, using the recursive approach. Assume that the z-transform, $X(z)$, is the same as in Example 4.2, that is

$$X(z) = \frac{1 + 2z^{-1} + z^{-2}}{1 - z^{-1} + 0.3561z^{-2}}$$

Solution

Comparing the coefficients of $X(z)$ above with those of the general transform in Equation 4.12 we have

$$a_0 = 1, \ a_1 = 2, \ a_2 = 1, \ b_0 = 1, \ b_1 = -1, \ b_2 = 0.3561; \ N = M = 2$$

From Equations 4.13 we have

$$x(0) = b_0/a_0 = 1$$
$$x(1) = [b_1 - x(0)a_1]/a_0 = [2 - 1 \times (-1)] = 3$$
$$x(2) = [b_2 - x(1)a_1 - x(0)a_2] = 1 - 3 \times (-1) - 1 \times 0.3561 = 3.6439$$
$$x(3) = [b_3 - x(2)a_1 - x(1)a_2 + x(0)a_3]$$
$$= 0 - x(2)a_1 - x(1)a_2 = 0 - 3.6439 \times (-1) - 3 \times 0.3561 = 2.5756$$

Thus the first four values of the inverse z-transform are

$$x(0) = 1, \; x(1) = 3, \; x(2) = 3.6439, \; x(3) = 2.5756$$

It is seen that both the recursive and direct, long division methods lead to identical solutions.

The recursion in Equation 4.13 can be readily implemented on a computer as shown in the partial C language code below:

```
x[0]=B[0]/A[0];
for(n=1;n<=npt;++n){
        sum=0;
        k=n;
        if(n>M)
                k=M;
        for(i=1;i<=k; ++i){
                sum=sum+x[n−i]*A[i];
        }
        x[n]=(B[n]−sum)/A[0];
}
```

In the code, M is the order of the denominator polynomial and npt is the number of data points for the IZT. The numerator and denominator polynomials are assumed to be in ascending powers of z^{-1}. A C language program based on the above code and MATLAB programs for evaluating the IZT are given in Appendices 4B and 4D respectively.

4.3.2   Partial fraction expansion method

In this method, the z-transform is first expanded into a sum of simple partial fractions. The inverse z-transform of each partial fraction is then obtained from tables, such as Table 4.1, and then summed to give the overall inverse z-transform. In many practical cases, the z-transform is given as a ratio of polynomials in z or z^{-1} and has the now familiar form

$$X(z) = \frac{b_0 + b_1 z^{-1} + b_2 z^{-2} + \ldots + b_N z^{-N}}{a_0 + a_1 z^{-1} + a_2 z^{-2} + \ldots + a_M z^{-M}} \tag{4.14}$$

If the poles of $X(z)$ are of first order and $N = M$, then $X(z)$ can be expanded as

$$X(z) = B_0 + \frac{C_1}{1 - p_1 z^{-1}} + \frac{C_2}{1 - p_2 z^{-1}} + \ldots + \frac{C_M}{1 - p_M z^{-1}}$$

$$= B_0 + \frac{C_1 z}{z - p_1} + \frac{C_2 z}{z - p_2} + \ldots + \frac{C_M z}{z - p_M} = B_0 + \sum_{k=1}^{M} \frac{C_k z}{z - p_k} \tag{4.15}$$

where p_k are the poles of $X(z)$ (assumed distinct), C_k are the partial fraction coefficients and

$$B_0 = b_N/a_N \qquad (4.16)$$

The C_k are also known as the residues of $X(z)$; see Section 4.3.3.

If the order of the numerator is less than that of the denominator in Equation 4.14, that is $N < M$, then B_0 will be zero. If $N > M$ then $X(z)$ must be reduced first, to make $N \leqslant M$, by long division with the numerator and denominator polynomials written in descending powers of z^{-1}. The remainder can then be expressed as in Equation 4.15.

The coefficient, C_k, associated with the pole p_k may be obtained by multiplying both sides of Equation 4.15 by $(z - p_k)/z$ and then letting $z = p_k$:

$$C_k = \left. \frac{X(z)}{z}(z - p_k) \right|_{z=p_k} \qquad (4.17)$$

If $X(z)$ contains one or more multiple-order poles (that is poles that are coincident) then extra terms are required in Equation 4.15 to take this into account. For example, if $X(z)$ contains an mth-order pole at $z = p_k$ the partial fraction expansion must include terms of the form

$$\sum_{i=1}^{m} \frac{D_i}{(z - p_k)^i} \qquad (4.18a)$$

The coefficients, D_i, may be obtained from the relationship

$$D_i = \frac{1}{(m-i)!} \frac{\mathrm{d}^{m-i}}{\mathrm{d}z^{m-i}} \left[(z - p_k)^m \frac{X(z)}{z} \right]_{z=p_k} \qquad (4.18b)$$

Evaluation of inverse z-transforms by the partial fraction expansion method is best illustrated by examples.

Example 4.4 *X(z) contains simple, first-order poles*   Find the inverse z-transform of the following:

$$X(z) = \frac{z^{-1}}{1 - 0.25z^{-1} - 0.375z^{-2}}$$

Solution   For simplicity, we first express the z-transform in positive powers of z by multiplying the numerator and denominator by z^2 (the highest power of z):

$$X(z) = \frac{z}{z^2 - 0.25z - 0.375} = \frac{z}{(z - 0.75)(z + 0.5)}$$

$X(z)$ contains first-order poles at $z = 0.75$ and at $z = -0.5$ (that is, only one pole occurs at each pole position). Since the order of the numerator is less than the order of the denominator ($N < M$), the partial fraction expansion has the form

$$X(z) = \frac{z}{(z - 0.75)(z + 0.5)} = \frac{C_1 z}{z - 0.75} + \frac{C_2 z}{z + 0.5} \qquad (4.19)$$

To make it easier to find the values of the C_k we divide both sides by z:

$$\frac{X(z)}{z} = \frac{z}{z(z - 0.75)(z + 0.5)} = \frac{C_1}{z - 0.75} + \frac{C_2}{z + 0.5} \qquad (4.20)$$

To obtain C_1, we simply multiply both sides of Equation 4.20 by $z - 0.75$ and let $z = 0.75$:

$$\frac{(z - 0.75)X(z)}{z} = \frac{(z - 0.75)}{(z - 0.75)(z + 0.5)} = C_1 + \frac{C_2(z - 0.75)}{z + 0.5}$$

$$C_1 = \frac{1}{z + 0.5}\bigg|_{z=0.75} = \frac{1}{0.75 + 0.5} = \frac{4}{5}$$

Similarly, C_2 is obtained as

$$C_2 = \frac{(z + 0.5)X(z)}{z}\bigg|_{z=-0.5}$$

$$= \frac{(z + 0.5)}{(z - 0.75)(z + 0.5)}\bigg|_{z=-0.5} = \frac{1}{-0.5 - 0.75} = -\frac{4}{5}$$

Using the values of C_1 and C_2 in Equation 4.19 we have

$$X(z) = \frac{(4/5)z}{z - 0.75} - \frac{(4/5)z}{z + 0.5} \qquad (4.21)$$

From the z-transform table, entry 14 in Table 4.1, the inverse z-transform of each term on the right-hand side of Equation 4.21 is given as

$$Z^{-1}\left[\frac{(4/5)z}{z - 0.75}\right] = \frac{4(0.75)^n}{5}$$

$$Z^{-1}\left[\frac{-(4/5)z}{z + 0.5}\right] = \frac{-4(-0.5)^n}{5}$$

The desired inverse z-transform, $x(n)$, is the sum of the two inverse z-transforms:

$$x(n) = \frac{4}{5}[(0.75)^n - (-0.5)^n], \quad n > 0$$

Example 4.5   *X(z) contains first-order, complex conjugate poles*   Find the discrete-time signal, *x(n)*, represented by the following z-transform using the partial fraction expansion method:

$$X(z) = \frac{1 + 2z^{-1} + z^{-2}}{1 - z^{-1} + 0.3561z^{-2}}$$

Solution   First, $X(z)$ is expressed in positive powers of z:

$$X(z) = \frac{N(z)}{D(z)} = \frac{z^2 + 2z + 1}{z^2 - z + 0.3561}$$

The poles of $X(z)$ are found by solving the quadratic $D(z) = z^2 - z + 0.3561 = 0$ using the formulae

$$p_1 = \frac{-b + (b^2 - 4ac)^{1/2}}{2a}$$

$$p_2 = \frac{-b - (b^2 - 4ac)^{1/2}}{2a}$$

(4.22)

where a and b are the coefficients of z^2 and z, respectively, and c is the constant term. With $a = 1$, $b = -1$, and $c = 0.3561$ the poles are

$$p_1 = \frac{-1 + (1 - 4 \times 0.3561)^{1/2}}{2}$$

$$= 0.5 + 0.3257j = re^{j\theta}$$

$$p_2 = p_1^* = 0.5 - 0.3257j = re^{-j\theta}$$

where $r = 0.5967$ and $\theta = 33.08°$. Thus we can express $X(z)$ in terms of its poles:

$$X(z) = \frac{z^2 + 2z + 1}{(z - p_1)(z - p_1^*)}$$

Since the numerator and denominator of $X(z)$ are of the same order, the partial fraction expansion has the form

$$\frac{X(z)}{z} = \frac{B_0}{z} + \frac{C_1}{z - p_1} + \frac{C_2}{z - p_1^*}$$

(4.23)

From Equation 4.16 $B_0 = 1/0.3561 = 2.8082$. To find C_1, we multiply both sides of Equation 4.23 by $z - p_1$ and then let $z = p_1$:

$$\frac{(z - p_1)X(z)}{z} = \frac{B_0(z - p_1)}{z} + C_1 + \frac{C_2(z - p_1)}{z - p_2}\bigg|_{z=p_1}$$

Thus

$$C_1 = \frac{(z - p_1)X(z)}{z} = \left. \frac{(z - p_1)(z^2 + 2z + 1)}{z(z - p_1)(z - p_2)} \right|_{z = p_1 = re^{j\theta}}$$

$$= \frac{(re^{j\theta})^2 + 2re^{j\theta} + 1}{re^{j\theta}(re^{j\theta} - re^{-j\theta})} \tag{4.24}$$

where $r = 0.5967$, $\theta = 33.08°$. After some manipulation and simplification we have

$$C_1 = \frac{2.1439 + 0.977\,19j}{-0.2122 + 0.3257j}$$

$$= -0.904\,099\,9 - 5.992\,847j$$

$$= 6.060\,66\angle{-98.58°}$$

Since p_1 and p_2 are complex conjugate pairs then

$$C_2 = C_1^* = -0.904\,099\,9 + 5.992\,847j = 6.060\,66\angle 98.58°$$

Thus the z-transform can be expressed as (from Equation 4.23)

$$X(z) = 2.8082 + \frac{C_1 z}{z - p_1} + \frac{C_2 z}{z - p_1^*} \tag{4.25}$$

where   .

$$p_1 = 0.5 + 0.3257j \qquad\qquad p_2 = 0.5 - 0.3257j$$

$$C_1 = -0.9041 - 5.599\,28j \qquad C_2 = -0.9041 + 5.599\,28j$$

From the z-transform table, entries 1 and 16 in Table 4.1, the inverse z-transform of the terms on the right-hand side of Equation 4.25 is

$$Z^{-1}(2.8082) = 2.8082u(n)$$

$$Z^{-1}\left[\frac{C_1 z}{z - p_1} + \frac{C_2 z}{z - p_1^*} \right] = 2 \times 6.060\,66(0.5967)^n \cos(33.08n - 98.58°)$$

$$= 12.1213(0.5967)^n \cos(33.08n - 98.58°)$$

Thus the discrete-time signal becomes

$$x(n) = 2.8082u(n) + 12.1213(0.5967)^n \cos(33.08n - 98.58°), \quad n \geq 0$$

A useful check for partial fraction results is to compute the values of $x(n)$ for $n = 0, 1, 2$ (say) and then to compare these with values obtained by the power series method. For example, from the expression for $x(n)$ we find that

$$x(0) = 2.8082 - 1.808\ 38 = 1; \ x(1) = 2.999\ 59 = 3; \ x(2) = 3.6436$$

which checks with the results obtained in Example 4.3 using the power series method.

Example 4.6   *$X(z)$ contains a second-order pole*   Find the discrete-time sequence, $x(n)$, with the following z-transform:

$$X(z) = \frac{z^2}{(z - 0.5)(z - 1)^2}$$

Solution $X(z)$ has a first-order pole at $z = 0.5$ and a second-order pole at $z = 1$. In this case, the partial fraction expansion has the form

$$X(z) = \frac{C}{z - 0.5} + \frac{D_1}{z - 1} + \frac{D_2}{(z - 1)^2} \tag{4.26}$$

To obtain C, we proceed as before and multiply both sides of Equation 4.26 by $z - 0.5$, set $z = 0.5$ and evaluate the expression

$$C = \frac{\cancel{(z - 0.5)}z^2}{z\cancel{(z - 0.5)}(z - 1)^2}\bigg|_{z=0.5}$$

$$= 0.5/(0.5 - 1)^2 = 2$$

To obtain D_1 we use Equation 4.18b, with $i = 1$ and $m = 2$. Thus

$$D_1 = \frac{d}{dz}\left[\frac{(z - 1)^2 X(z)}{z}\right]_{z=1} = \frac{d}{dz}\left[\frac{\cancel{(z-1)^2}z^{\cancel{2}}}{\cancel{z}(z - 0.5)\cancel{(z-1)^2}}\right]_{z=1}$$

$$= \frac{d}{dz}\left(\frac{z}{z - 0.5}\right)_{z=1} = \frac{z - 0.5 - z}{(z - 0.5)^2}\bigg|_{z=1} = -2$$

Similarly, D_2 is obtained from Equation 4.18b by letting $i = 2$ and $m = 2$:

$$D_2 = \frac{(z - 1)^2 X(z)}{z}\bigg|_{z=1} = \frac{\cancel{(z-1)^2}z^{\cancel{2}}}{\cancel{z}(z - 0.5)\cancel{(z-1)^2}}\bigg|_{z=1}$$

$$= 1/(1 - 0.5) = 2$$

Combining the results, $X(z)$ becomes

$$X(z) = \frac{2z}{z - 0.5} - \frac{2z}{z - 1} + \frac{2z}{(z - 1)^2}$$

The inverse z-transform of each term on the right-hand side is obtained from Table 4.1 and then summed to give $x(n)$:

$$x(n) = 2(0.5)^n - 2 + 2n = 2[(n - 1) + (0.5)^n], \quad n \geqslant 0 \tag{4.27}$$

The reader can verify that the result is correct by comparing the first few values of $x(n)$ with values calculated with the power series method.

You will agree that the partial fraction expansion method is very tedious, except for simple cases, and mistakes are likely. C language and MATLAB programs are described in the appendix for computing the inverse z-transform, using partial fraction expansions, for $X(z)$ with first-order poles.

4.3.3   Residue method

In this method the IZT is obtained by evaluating the contour integral

$$x(n) = \frac{1}{2\pi j} \oint_C z^{n-1} X(z)\, dz \tag{4.28}$$

where C is the path of integration enclosing all the poles of $X(z)$. For rational polynomials, the contour integral in Equation 4.28 is evaluated using a fundamental result in complex variable theory known as Cauchy's residue theorem (Mathews, 1982):

$$x(n) = \frac{1}{2\pi j} \oint_C z^{n-1} X(z)\, dz \tag{4.29}$$

$$= \text{sum of the residues of } z^{n-1}X(z) \text{ at all the poles inside } C.$$

In the last section, it was stated that the partial fraction coefficients, the C_k, are also referred to as residues of $X(z)$ and a way of obtaining their values was given. The key point to remember is that every residue, C_k, is associated with a pole, p_k. In the present method, the residue of $z^{n-1}X(z)$ at the pole p_k (not the residue of $X(z)$) is given by

$$\text{Res}\,[F(z), p_k] = \frac{1}{(m - 1)!} \frac{d^{m-1}}{dz^{m-1}}[(z - p_k)F(z)]_{z=p_k} \tag{4.30}$$

where $F(z) = z^{n-1}X(z)$, m is the order of the pole at p_k and Res $[F(z), p_k]$ is the residue of $F(z)$ at $z = p_k$. For a simple (distinct) pole, Equation 4.30 reduces to

$$\text{Res}\,[F(z), p_k] = (z - p_k)F(z) = (z - p_k)z^{n-1}X(z)\big|_{z=p_k} \tag{4.31}$$

Example 4.7   Find, using the residue method, the discrete-time signal corresponding to the following z-transform:

$$X(z) = \frac{z}{(z - 0.75)(z + 0.5)}$$

Assume C is the circle $|z| = 1$.

Solution   This problem is the same as Example 4.4. In factored form, $X(z)$ is given by

$$X(z) = \frac{z}{(z - 0.75)(z + 0.5)}$$

If we let $F(z) = z^{n-1}X(z)$ then

$$F(z) = \frac{z^{n-1}z}{(z - 0.75)(z + 0.5)}$$

$$= \frac{z^n}{(z - 0.75)(z + 0.5)}$$

$F(z)$ has poles at $z = 0.75$ and $z = -0.5$. A sketch of the contour with the positions of the poles indicated by crosses is given in Figure 4.3. Both poles lie inside the contour (unit circle). From Equation 4.29, the inverse z-transform is given by

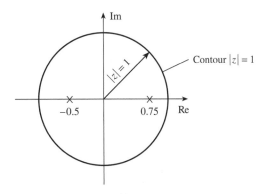

Figure 4.3   A sketch of the contour of integration showing the role of $X(z)$.

$$x(n) = \text{Res}\,[F(z),\,0.75] + \text{Res}\,[F(z),\,-0.5]$$

Since the poles are first order, Equation 4.31 will be used. Thus

$$\text{Res}[F(z),\,0.75] = (z - 0.75)F(z)\big|_{z=0.75}$$

$$= \frac{\cancel{(z - 0.75)}z^n}{\cancel{(z - 0.75)}(z + 0.5)}\bigg|_{z=0.75}$$

$$= \frac{(0.75)^n}{0.75 + 0.5}$$

$$= \frac{4}{5}(0.75)^n$$

$$\text{Res}[F(z),\,-0.5] = (z + 0.5)F(z)\big|_{z=-0.5}$$

$$= \frac{(z + 0.5)z^n}{(z - 0.75)(z + 0.5)}\bigg|_{z=-0.5}$$

$$= -\frac{4}{5}(-0.5)^n$$

The inverse z-transform is the sum of the residues at $z = 0.75$ and at $z = -0.5$:

$$x(n) = (4/5)[(0.75)^n - (-0.5)^n]$$

which is identical to the result obtained by the partial fraction expansion.

Example 4.8

The poles of X(z) are complex conjugate poles   Find the inverse z-transform, using the residue method, given the following z-transform:

$$X(z) = \frac{z^2 + 2z + 1}{z^2 - z + 0.3561}$$

Solution   In factored form $X(z)$ is given as:

$$X(z) = \frac{z^2 + 2z + 1}{(z - p_1)(z - p_2)}$$

where $p_1 = 0.5 + 0.3557j$ and $p_2 = 0.5 - 0.3557j$, that is $p_2 = p_1^*$. To find the inverse z-transform we evaluate the residues of $F(z)$, where in this case

$$F(z) = z^{n-1}X(z) = \frac{z^{n-1}(z^2 + 2z + 1)}{z^2 - z + 0.3561} = \frac{z^n(z^2 + 2z + 1)}{z(z^2 - z + 0.3561)}$$

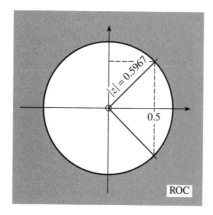

Figure 4.4   Contour for Example 4.8 showing the ROC.

$F(z)$ has the same poles as $X(z)$, that is at $z = p_1$ and $z = p_2$, plus a pole at $z = 0$ when $n = 0$. Figure 4.4 shows a sketch of the contour with the positions of the poles indicated. All the poles lie inside the contour. The pole at $z = 0$ does not exist for $n > 0$ and so we need to consider the two cases separately.

When $n = 0$, $F(z)$ reduces to

$$F(z) = \frac{z^2 + 2z + 1}{z(z^2 - z + 0.3561)}$$

and

$$x(0) = \text{Res}[F(z), 0] + \text{Res}[F(z), p_1] + \text{Res}[F(z), p_2]$$

Therefore

$$\text{Res}[F(z), 0] = zF(z)\big|_{z=0} = \frac{z(z^2 + 2z + 1)}{z(z^2 - z + 0.3561)}\bigg|_{z=0}$$

$$= 1/0.3561 = 2.8082$$

$$\text{Res}[F(z), p_1] = (z - p_1)F(z)\big|_{z=p_1}$$

$$= \frac{(z - p_1)(z^2 + 2z + 1)}{z(z - p_1)(z - p_2)}$$

$$= \frac{(re^{j\theta})^2 + 2re^{j\theta} + 1}{re^{j\theta}(re^{j\theta} - re^{-j\theta})}$$

where $r = 0.5967$ and $\theta = 33.08°$. Noting that this expression is identical to the Equation 4.24, we can write

$$\text{Res}[F(z), p_1] = -0.9041 - 5.9928j$$

Since p_1 and p_2 are complex conjugate pairs then

$$\text{Res}[F(z), p_2] = -0.9041 + 5.9928j$$

Thus

$$x(0) = \text{Res}[F(z), 0] + \text{Res}[F(z), p_1] + \text{Res}[F(z), p_2]$$

$$= 2.8082 - 0.9041 - 5.9928j - 0.9041 + 5.9928j$$

$$= 1$$

When $n > 0$, the pole at $z = 0$ vanishes and we have

$$F(z) = \frac{z^n(z^2 + 2z + 1)}{z(z^2 - z + 0.3561)}$$

$$x(n) = \text{Res}[F(z), p_1] + \text{Res}[F(z), p_2]$$

$$\text{Res}[F(z), p_1] = (z - p_1)F(z)\big|_{z=p_1}$$

$$= \frac{\cancel{(z - p_1)}z^n(z^2 + 2z + 1)}{z\cancel{(z - p_1)}(z - p_2)}\bigg|_{z=p_1} \tag{4.32}$$

$$= \frac{(re^{j\theta})^n[(re^{j\theta})^2 + 2re^{j\theta} + 1]}{re^{j\theta}(re^{j\theta} - re^{-j\theta})}$$

where $r = 0.5967$ and $\theta = 33.08°$. Noting that this expression is similar to Equation 4.24, we can write

$$\text{Res}[F(z), p_1] = (0.5967e^{j33.08})^n(6.060\,66e^{-j98.58})$$

$$= 6.060\,66(0.5967)^n[\cos(33.08n - 98.58)$$

$$+ j\sin(33.08n - 98.58)]$$

Since p_2 and p_1 are complex conjugate pole pairs we can write

$$\text{Res}[F(z), p_2] = 6.060\,66(0.5967)^n[\cos(33.08n - 98.58)$$

$$- j\sin(33.08n - 98.58)]$$

Thus

$$x(n) = \text{Res}[F(z), p_1] + \text{Res}[F(z), p_2]$$

$$= 12.1213(0.5967)^n\cos(33.08n - 98.58°), \quad n > 0$$

which checks with the results for the partial fraction expansion.

Example 4.9   *X(z) contains a second-order pole*   Find the discrete-time sequence, $x(n)$, with the following z-transform:

$$X(z) = \frac{z^2}{(z - 0.5)(z - 1)^2}$$

Solution   This example is the same as Example 4.6 under partial fraction expansion. According to the residue method the discrete-time sequence is given by

$$x(n) = \sum_{k=1}^{M} \text{Res}[F(z), p_k]$$

where

$$F(z) = z^{n-1}X(z) = \frac{z^{n+1}}{(z - 0.5)(z - 1)^2}$$

$F(z)$ has a simple pole at $z = 0.5$ and a second-order pole at $z = 1$; thus $x(n)$ is given by

$$x(n) = \text{Res}[F(z), p_1] + \text{Res}[F(z), p_2]$$

$$\text{Res}[F(z), 0.5] = \frac{(z - 0.5)z^{n+1}}{(z - 0.5)(z - 1)^2} = \frac{z^{n+1}}{(z - 1)^2}\Big|_{z=0.5}$$

$$= 0.5(0.5)^n/(0.5)^2 = 2(0.5)^n$$

$$\text{Res}[F(z), 1] = \frac{d}{dz}\left[\frac{(z - 1)^2 z^{n+1}}{(z - 0.5)(z - 1)^2}\right]$$

$$= \frac{(z - 0.5)(n + 1)z^n - z^{n+1}}{(z - 0.5)^2}\Big|_{z=1}$$

$$= [(0.5)(n + 1) - 1]/(0.5)^2 = 2(n - 1)$$

Combining the results, we have

$$x(n) = 2[(n - 1) + (0.5)^n]$$

which is the same result as for the partial fraction expansion method.

The reader may have noticed that the partial fraction and residue methods are related. Both methods require the evaluation of residues albeit performed in different ways. The partial fraction method requires the evaluation of the residues of $X(z)$, that is the C_k, while the residue method requires the evaluation of the residues of $z^{n-1}X(z)$. When $X(z)$ has first-order poles we have

$$\text{Res}\,[z^{n-1}X(z), p_k] = z^n\text{Res}\,[X(z), p_k] = z^n C_k \qquad (4.33)$$

Thus the C language program for the partial fraction expansion described in the appendix may be exploited to obtain results for the residue method.

4.3.4   Comparison of the inverse z-transform methods

We have discussed in some detail three methods of obtaining the inverse z-transform: the power series, partial fraction expansion and the residue methods. A limitation of the power series method is that it does not lead to a closed form solution (although this can be deduced in simple cases), but it is simple and lends itself to computer implementation. However, because of its recursive nature care should be taken to minimize possible build-up of numerical errors when the number of data points in the inverse z-transform is large, for example by using double precision.

Both the partial fraction expansion and the residue methods lead to closed form solutions. The main disadvantage with both methods is the need to factorize the denominator polynomial, that is finding the poles of $X(z)$. If the order of $X(z)$ is high finding the poles of $X(z)$, if $X(z)$ is not in factored form, is quite a difficult task. This topic is discussed further in Section 4.5.1. Both methods may also involve high-order differentiation if $X(z)$ contains multiple-order poles. Clearly, if closed form solution is required then the partial fraction or residue method is the most appropriate. The partial fraction method is particularly useful in generating the coefficients of parallel structures for digital filters (see Section 4.5.11). The residue method is widely used in the analysis of quantization errors in discrete-time systems. (See Chapter 13.)

The use of a suitable tool such as MATLAB or the C language programs in this book greatly simplifies z-transform and inverse z-transform operations. Several illustrative examples are provided in the appendices.

4.4   Properties of the z-transform

Some useful properties of the z-transform which have found practical use in DSP are described briefly below. The proofs for some of these properties are given as problems at the end of the chapter.

(1)   *Linearity.* If the sequences $x_1(n)$ and $x_2(n)$ have z-transforms $X_1(z)$ and $X_2(z)$, then the z-transform of their linear combination is

$$ax_1(n) + bx_2(n) \rightarrow aX_1(z) + bX_2(z) \qquad (4.34)$$

(2) *Delays or shifts.* If the z-transform of a sequence, $x(n)$, is $X(z)$ then the z-transform of the sequence delayed by m samples is $z^{-m}X(z)$. This property is widely used in converting the z transfer function of discrete-time systems into time domain difference equations and vice versa; see Section 4.5.8.

$$x(n) \to X(z)$$

$$x(n - m) \to z^{-m}X(z)$$

(3) *Convolution.* Given a discrete-time LTI system with input $x(n)$ and impulse response $h(k)$, the output of the system is given by

$$y(n) = \sum_{k=-\infty}^{\infty} h(k)x(n - k) \tag{4.35a}$$

In terms of the z-transform, the input and output are related as

$$Y(z) = H(z)X(z) \tag{4.35b}$$

where $X(z)$, $H(z)$ and $Y(z)$ are, respectively, the z-transform of $x(n)$, $h(k)$ and $y(n)$. Given $X(z)$ and $H(z)$, the output $y(n)$ can be obtained by inverse z-transforming $Y(z)$.

It is seen that the convolution operation in Equation 4.35a has become a multiplicative process in the z-domain. $H(z)$ is often referred to as the system transfer function.

(4) *Differentiation.* If $X(z)$ is the z-transform of $x(n)$, then the z-transform of $nx(n)$ can be obtained by differentiating $X(z)$:

$$x(n) \to X(z)$$

$$nx(n) \to -z\frac{\mathrm{d}X(z)}{\mathrm{d}z} \tag{4.36}$$

This property is useful in obtaining the inverse z-transform when $X(z)$ contains multiple order poles.

(5) *Relationship with the Laplace transform.* Continuous-time systems or signals are normally described using the Laplace transform. If we let $z = e^{sT}$, where s is the complex Laplace variable given by

$$s = d + j\omega$$

then

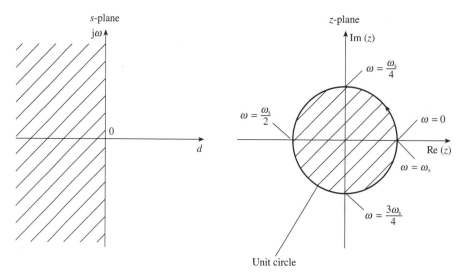

Figure 4.5   Mapping of *s*-plane to *z*-plane. The left-hand side of the *s*-plane maps to the interior of the *z*-plane, the right-hand side maps to the exterior and the jω-axis maps onto the unit circle.

$$z = e^{(d+j\omega)T} = e^{dT}e^{j\omega T} \tag{4.37}$$

Thus

$$|z| = e^{dT} \text{ and } \angle z = \omega T = 2\pi f/F_s = 2\pi\omega/\omega_s$$

where ω_s (rad s^{-1}) is the sampling frequency. As ω varies from $-\infty$ to ∞ the *s*-plane is mapped to the *z*-plane as shown in Figure 4.5. The entire jω axis in the *s*-plane is mapped onto the unit circle. The left-hand *s*-plane is mapped to the inside of the unit circle and the right-hand *s*-plane maps to the outside of the unit circle.

In terms of frequency response, the jω axis is the most important in the *s*-plane. In this case, $d = 0$, and frequency points in the *s*-plane are related to points on the *z*-plane unit circle by

$$z = e^{j\omega T} \tag{4.38}$$

Table 4.2 shows how some specific frequencies are mapped from the *s*-plane to the *z*-plane. It is clear that the mapping is not unique since, for example, the two frequencies $\omega = \omega_s$ and $\omega = 2\omega_s$ in the *s*-plane map to the same point on the unit circle.

Table 4.2 Mapping of frequencies from the s-plane to the z-plane.

s-plane: ω (rad s^{-1})	z-plane: ωT (rad)
0	0
$\omega_s/4$	$\pi/2$
$\omega_s/2$	π
$3\omega_s/4$	1.25π
ω_s	2π
$1.25\omega_s$	$\pi/2$
$1.5\omega_s$	π
$1.75\omega_s$	1.25π
$2\omega_s$	2π

4.5 Some applications of the z-transform in signal processing

Applications of the z-transform in DSP are many. Several of these are discussed in more detail in later chapters, especially in Chapter 8. The next few sections are intended to highlight some of these applications and to establish some fundamental issues common to them.

4.5.1 Pole–zero description of discrete-time systems

In most practical discrete-time systems the z-transform, that is the system transfer function $H(z)$, can be expressed in terms of its poles and zeros. Consider, for example, the following z-transform representing a general, Nth-order discrete-time filter (where $N = M$):

$$H(z) = \frac{N(z)}{D(z)} \tag{4.39}$$

where

$$N(z) = b_0 z^N + b_1 z^{N-1} + b_2 z^{N-2} + \ldots + b_N$$

$$D(z) = a_0 z^N + a_1 z^{N-1} + a_2 z^{N-2} + \ldots + a_N$$

The a_k and b_k are the coefficients of the filter.

If $H(z)$ has poles at $z = p_1, p_2, \ldots, p_N$ and zeros at $z = z_1, z_2, \ldots, z_N$, then $H(z)$ can be factored and represented as

$$H(z) = \frac{K(z - z_1)(z - z_2)\ldots(z - z_N)}{(z - p_1)(z - p_2)\ldots(z - p_N)} \tag{4.40}$$

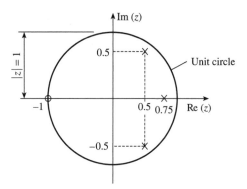

Figure 4.6   Description of a z-transform in the form of a pole–zero diagram: ✕, pole; ○, zero.

where z_i is the ith zero, p_i is the ith pole and K is the gain factor. You will recall that the poles of a z-transform such as $H(z)$ are the values of z for which $H(z)$ becomes infinity. The values of z for which $H(z)$ becomes zero are referred to as zeros. The poles and zeros of $H(z)$ may be real or complex. When they are complex, they occur in conjugate pairs, to ensure that the coefficients, a_k and b_k, are real. It should be clear from Equation 4.40 that if the locations of the poles and zeros of $H(z)$ are known, then $H(z)$ itself can be readily reconstructed to within a constant.

The information contained in the z-transform can be conveniently displayed as a pole–zero diagram; see for example Figure 4.6. In the diagram, ✕ marks the position of a pole and ○ denotes the position of a zero. For this example, the poles are located at $z = 0.5 \pm 0.5j$, and at $z = 0.75$. A single zero is at $z = -1$. An important feature of the pole–zero diagram is the unit circle, that is the circle defined by $|z| = 1$; see Figure 4.6. As will become clear, the unit circle plays an important role in the analysis and design of discrete-time systems.

The pole–zero diagram provides an insight into the properties of a given discrete-time system. For example, from the locations of the poles and zeros we can infer the frequency response of the system as well as its degree of stability. For a stable system, all the poles must lie inside the unit circle (or be coincident with zeros on the unit circle).

Often, the z-transform is not available in factored form but as a ratio of polynomials such as Equation 4.39. In these cases, describing the z-transform, $H(z)$, in terms of its poles and zeros will require finding the roots of the denominator polynomial, $D(z)$, and those of the numerator polynomial, $N(z)$.

For a second-order polynomial which has the form $ax^2 + bx + c$, the roots are given by

$$\frac{-b \pm (b^2 - 4ac)^{1/2}}{2a} \tag{4.41}$$

For higher-order polynomials, finding the roots of $N(z)$ or $D(z)$ is a difficult task. In practice, this is often achieved using numerical methods involving, for example, Newton's and/or Baistow's algorithms (see, for example, Atkinson and Harley (1983)). The need to find the poles and zeros often arises in connection with the design of discrete filters and in stability analysis. Fortunately, in the case of discrete-time filter design the poles and zeros are automatically generated by the filter design software, obviating the need to find the roots directly.

Example 4.10   (1) Express the following transfer function in terms of its poles and zeros and sketch the pole–zero diagram:

$$H(z) = \frac{1 - z^{-1} - 2z^{-2}}{1 - 1.75z^{-1} + 1.25z^{-2} - 0.375z^{-3}}$$

(2) Determine the transfer function, $H(z)$, of a discrete-time filter with the pole–zero diagram shown in Figure 4.7.

Solution   (1) First, we express $H(z)$ in positive powers of z and then factorize it so that its poles and zeros can be determined. If we multiply the denominator and numerator by z^3, the highest power of z, we obtain

$$H(z) = \frac{z^3 - z^2 - 2z}{z^3 - 1.75z^2 + 1.25z - 0.375}$$

Factoring, we have

$$H(z) = \frac{(z - 2)(z + 1)z}{(z - 0.5 + j0.5)(z - 0.5 - j0.5)(z - 0.75)}$$

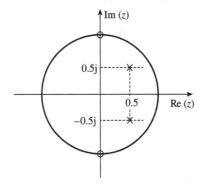

Figure 4.7   Pole–zero diagram for Example 4.10, part (2).

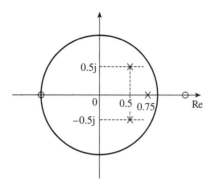

Figure 4.8   Pole–zero diagram for Example 4.10.

Thus, the pole locations are at $z = 0.5 \pm 0.5\mathrm{j}$ and at $z = 0.75$. The zeros are at $z = 2$, $z = -1$ and $z = 0$. The pole–zero diagram is depicted in Figure 4.8.

(2)   From the pole–zero diagram, the zeros of the transfer function are at $z = \pm\mathrm{j}$ and the poles are at $z = 0.5 \pm 0.5\mathrm{j}$. The transfer function can be written down directly:

$$H(z) = \frac{K(z - \mathrm{j})(z + \mathrm{j})}{(z - 0.5 - 0.5\mathrm{j})(z - 0.5 + 0.5\mathrm{j})}$$

$$= \frac{K(z^2 + 1)}{z^2 - z + 0.5}$$

$$= \frac{K(1 + z^{-2})}{1 - z^{-1} - 0.5z^{-2}}$$

4.5.2   Frequency response estimation

There are many instances when it is necessary to evaluate the frequency response of discrete-time systems. For example, in the design of discrete filters, it is often necessary to examine the spectrum of the filter to ensure that the desired specifications are satisfied. The frequency response of a system can be readily obtained from its *z*-transform.

For example, if we set $z = \mathrm{e}^{\mathrm{j}\omega T}$, that is evaluate the *z*-transform around the unit circle, we obtain the Fourier transform of the system:

$$H(z) = \sum_{n=-\infty}^{\infty} h(n)z^{-n} \Bigg|_{z=\mathrm{e}^{\mathrm{j}\omega T}} \tag{4.42a}$$

$$= H(\mathrm{e}^{\mathrm{j}\omega T}) = \sum_{n=-\infty}^{\infty} h(n)\mathrm{e}^{-\mathrm{j}n\omega T} \tag{4.42b}$$

$H(e^{j\omega T})$ is referred to as the frequency response of the system. We have used the symbol T to emphasize the dependence of the frequency response of discrete-time systems on the sampling frequency. In general, $H(e^{j\omega T})$ is complex. Its modulus gives the magnitude response and its phase the phase response of the system.

The frequency response may be obtained from the z-transform using several methods. We will describe three methods.

4.5.3   Geometric evaluation of frequency response

This is a simple but useful method of obtaining a rough idea of what the frequency response of a discrete-time system would look like, based on its pole–zero diagram. Recall that the z-transform of an LTI system may be expressed in terms of its poles and zeros:

$$H(z) = \frac{K(z - z_1)(z - z_2)\ldots(z - z_N)}{(z - p_1)(z - p_2)\ldots(z - p_N)} = \frac{\prod\limits_{i=1}^{N} K(z - z_i)}{\prod\limits_{i=1}^{N}(z - p_i)} \tag{4.43}$$

where we have assumed, for simplicity, that the orders of the numerator and denominator are equal. The frequency response is obtained by making the substitution $z = e^{j\omega T}$ in Equation 4.43 and evaluating $H(e^{j\omega T})$ in the interval $(0 \leqslant \omega \leqslant \omega_s/2)$.

$$H(e^{j\omega T}) = \frac{\prod\limits_{i=1}^{N} K(e^{j\omega T} - z_i)}{\prod\limits_{i=1}^{N}(e^{j\omega T} - p_i)} \tag{4.44}$$

A geometric interpretation of Equation 4.44 for a z-transform with only two zeros and two poles is shown in Figure 4.9. In this case, the frequency response is given by

$$H(e^{j\omega T}) = \frac{K(e^{j\omega T} - z_1)(e^{j\omega T} - z_2)}{(e^{j\omega T} - p_1)(e^{j\omega T} - p_2)}$$

$$= \frac{KU_1\angle\theta_1 U_2\angle\theta_2}{V_1\angle\phi_1 V_2\angle\phi_2} \tag{4.45}$$

where U_1 and U_2 represent the distances from the zeros to the point $z = e^{j\omega T}$, and V_1 and V_2 the distances of the poles to the same point as shown in Figure 4.9. Thus the magnitude and phase responses for the system, from Equation 4.45, are

$$|H(e^{j\omega T})| = \frac{U_1 U_2}{V_1 V_2}, \quad K = 1$$

$$\angle[H(e^{j\omega T})] = \theta_1 + \theta_2 - (\phi_1 + \phi_2)$$

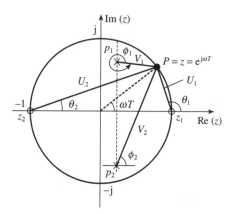

Figure 4.9 Geometric evaluation of the frequency response from the pole–zero diagram.

The complete frequency response is obtained by evaluating $H(e^{j\omega T})$ as the point P moves from $z = 0$ to $z = -1$. It is evident that, as the point P moves closer to the pole p_1, the length of the vector V_1 decreases and so the magnitude response increases. On the other hand, as the point moves closer to the zero z_1, the zero vector U_1 decreases and so the magnitude response, $|H(e^{j\omega T})|$, decreases. Thus at the pole the magnitude response exhibits a peak whereas, at the zero, the magnitude response falls to zero.

In general, in the geometric method, the frequency response at a given frequency ω (at an angle ωT) is determined by the ratio of the product of the zero vectors, $U_i\angle\theta_i$, $i = 1, 2, \ldots$, with the product of the pole vectors, $V_i\angle\phi_i$, $i = 1, 2, \ldots$.

Example 4.11

Determine, using the geometric method, the frequency response at dc, 1/8, 1/4, 3/8 and 1/2 the sampling frequency, of the causal discrete-time system with the following *z*-transform:

$$H(z) = \frac{z + 1}{z - 0.7071}$$

Sketch the amplitude frequency response in the interval $0 \leqslant \omega \leqslant \omega_s$, where ω_s (rad s^{-1}) is the sampling frequency.

Solution

In this example, $H(z)$ has a single pole and a single zero, as shown in the pole–zero diagram of Figure 4.10(a). From Equation 4.44, the response at ω is given by

$$H(e^{j\omega T}) = \frac{U\angle\theta}{V\angle\phi} = \frac{e^{j\omega T} + 1}{e^{j\omega T} - 0.7071} = \frac{1 + \cos(\omega T) + j\sin(\omega T)}{\cos(\omega T) - 0.7071 + j\sin(\omega T)} \tag{4.46}$$

At dc, $\omega T = 0$ and the zero and pole vectors to the point $z = 0$ are $2\angle 0°$ and $0.2929\angle 0°$. Thus the frequency response is given by

$$H(e^{j\omega T}) = 2/0.2929 = 6.828\angle 0°$$

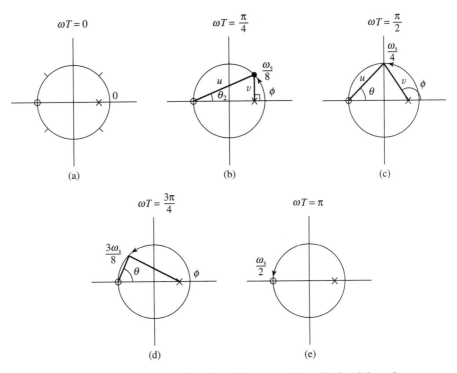

Figure 4.10   Frequency response estimation using geometric method and the pole–zero diagram.

At $\omega = \omega_s/8$, $\omega T = \omega_s/8F_s = \pi/4$. The pole and zero vectors in this case are shown in Figure 4.10(b). Rather than actually measuring the lengths and angles of the vectors we will use the explicit expression on the right-hand side of Equation 4.46. Thus

$$H(e^{j\omega T}) = \frac{1 + \cos(\pi/4) + j\sin(\pi/4)}{\cos(\pi/4) - 0.7071 + j\sin(\pi/4)}$$

$$= \frac{1.8477\angle 22.5°}{0.7071\angle 90°} = 2.6131\angle -67.5°$$

The responses at the remaining frequencies, obtained in a similar way, are summarized below, and the vectors are given in Figures 4.10(c)–4.10(e).

ω (rad s^{-1})	ωT (rad)	$\lvert H(e^{j\omega T})\rvert$	$\angle H(e^{j\omega T})$ (degrees)
0	0	6.828	0
$\omega_s/8$	$\pi/4$	2.6131	−67.5
$\omega_s/4$	$\pi/2$	1.1547	−80.26
$3\omega_s/8$	$3\pi/4$	0.4840	−85.93
$\omega_s/2$	π	0	0

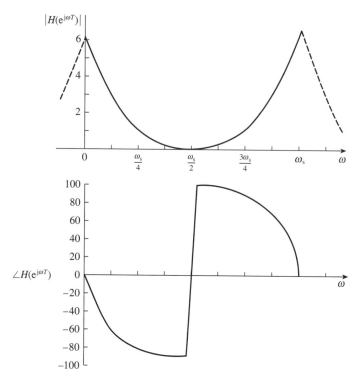

Figure 4.11   A sketch of the frequency response of the discrete-time system of Example 4.11.

A sketch of the magnitude and phase responses is shown in Figure 4.11. An important point to note is that the magnitude response, $|H(e^{j\omega T})|$, is symmetrical about half the sampling frequency (Nyquist frequency), and the phase response antisymmetrical about the same frequency. This is always the case when the coefficients, a_k and b_k, of a discrete-time system are real. Further, the frequency response of such systems is periodic with a period of ω_s (the sampling frequency), a behaviour that is consistent with the sampling theorem.

4.5.4   Direct computer evaluation of frequency response

Geometric evaluation of the frequency response gives one a feel for the frequency response, but it is clearly very tedious if the precise response is required at many frequencies. Although the process can be automated, the difficulty of finding the locations of the poles and zeros limits its usefulness. If the complete frequency response is required, it is common practice to make the substitution $z = e^{j\omega T}$ directly into the transfer function and then to evaluate the resulting expression:

$$H(e^{j\omega T}) = \frac{b_0 + b_1 z^{-1} + \ldots + b_N z^{-N}}{a_0 + a_1 z^{-1} + \ldots + a_M z^{-M}}\bigg|_{z = e^{j\omega T}} \quad (4.47)$$

$$= \frac{b_0 + b_1 e^{-j\omega T} + \ldots + b_N e^{-jN\omega T}}{a_0 + a_1 e^{-j\omega T} + \ldots + a_M e^{-jM\omega T}}$$

$$= \frac{b_0 + b_1[\cos(\omega T) - j\sin(\omega T)] + \ldots + b_N[\cos(N\omega T) - j\sin(N\omega T)]}{a_0 + a_1[\cos(\omega T) - j\sin(\omega T)] + \ldots + a_M[\cos(M\omega T) - j\sin(M\omega T)]}$$

$$(4.48)$$

A C language implementation of Equation 4.48 is discussed in Appendix 4C. The program evaluates $H(e^{j\omega T})$ in the interval $(0 \leqslant \omega \leqslant \omega_s/2)$. The use of MATLAB to obtain the frequency response is described in Appendix 4D and illustrated by an example.

4.5.5 Frequency response estimation via FFT

The FFT may also be used to evaluate the frequency response of discrete-time systems. A way of doing this, for IIR systems, is first to obtain the impulse response of the system using, for example, the power series method, and then to compute the FFT of the impulse response. This follows directly from Equation 4.42b which shows that the frequency response of a discrete-time system is simply the Fourier transform of its impulse response. To obtain a smooth frequency response, it is important to take a sufficient number of impulse response values and/or to zero-pad the impulse response values before the FFT is taken. C language and MATLAB implementations are discussed in the appendix.

An alternative technique is first to zero-pad the numerator and denominator coefficients, for example

$$\{b(n)\} = \{b_0, b_1, b_2, \ldots, b_M, 0, 0, \ldots, 0\} \quad (4.49)$$

$$\{a(n)\} = \{a_0, a_1, a_2, \ldots, a_N, 0, 0, \ldots, 0\}$$

and then to obtain the FFTs of $\{b(n)\}$ and $\{a(n)\}$, $A(k)$ and $B(k)$, respectively. The ratio of the two FFTs gives the frequency response:

$$H(e^{j\omega_k T}) = A(k)/B(k), \quad k = 0, 1, \ldots, N/2 \quad (4.50)$$

4.5.6 Frequency units used in discrete-time systems

Continuous-time systems or signals are normally described using the Laplace transform. Thus the frequency response of a continuous-time system is traditionally evaluated by letting $s = j\omega$ in the system transfer function, $H(s)$, where s is the

Table 4.3   Units of frequency used in discrete-time systems and their relationships to points on the unit circle.

f (Hz)	ω (rad s^{-1})	ωT (rad)	$z = e^{j\omega T}$
0	0	0	1
$\dfrac{F_s}{8}$	$\dfrac{\omega_s}{8}$	$\dfrac{\pi}{4}$	$\dfrac{\sqrt{2}}{2} + \dfrac{\sqrt{2}}{2}j$
$\dfrac{F_s}{4}$	$\dfrac{\omega_s}{4}$	$\dfrac{\pi}{2}$	j
$\dfrac{3F_s}{8}$	$\dfrac{3\omega_s}{8}$	$\dfrac{3\pi}{4}$	$-\dfrac{\sqrt{2}}{2} + \dfrac{\sqrt{2}}{2}j$
$\dfrac{F_s}{2}$	$\dfrac{\omega_s}{2}$	π	-1
$\dfrac{5F_s}{8}$	$\dfrac{5\omega_s}{8}$	$\dfrac{5\pi}{4}$	$-\dfrac{\sqrt{2}}{2} - \dfrac{\sqrt{2}}{2}j$
$\dfrac{3F_s}{4}$	$\dfrac{3\omega_s}{4}$	$\dfrac{3\pi}{2}$	$-j$
$\dfrac{7F_s}{8}$	$\dfrac{7\omega_s}{8}$	$\dfrac{7\pi}{4}$	$\dfrac{\sqrt{2}}{2} - \dfrac{\sqrt{2}}{2}j$
F_s	ω_s	2π	1

$F_s = 1/T$ is the sampling frequency in Hz; T is the sampling period; $\omega_s = 2\pi/T$ is the sampling frequency in rad s^{-1}.

complex Laplace variable. In DSP we deal with discrete-time systems and signals. In this case, the frequency response is found by letting $z = e^{j\omega T}$ and then evaluating the z-transfer function, $H(z)$, in the interval $0 \leqslant \omega \leqslant \omega_s/2$. The key point in discrete-time systems is the dependence of the useful frequency range on the sampling frequency, ω_s.

Table 4.3 shows how ωT and z change as ω varies from 0 to ω_s. It can be inferred that as the angle ωT goes from 0 to 2π the value of z varies from 1 through j and back to 1. This information is depicted in Figure 4.12. The figure also makes it evident that the frequency response of a discrete-time system is cyclic: as we go round the circle one or more revolutions the values of z simply repeat.

Two frequency units are normally used to describe the frequency response of discrete-time systems, namely, ω (rad s^{-1}) and f (Hz). When the frequency unit is rad s^{-1} the frequency response goes from $\omega = 0$ to $\omega = \omega_s/2$ or equivalently from $\omega = 0$ to $\omega = \pi/T$ (since $\omega_s = 2\pi F_s = 2\pi/T$). When the standard frequency unit of hertz is used the frequency range is from 0 to $F_s/2$ or equivalently 0 to $1/2T$. Both frequency units may also be expressed in normalized form, that is $T = 1$ or equivalently $F_s = 1$. Table 4.3 shows the relationship between the two frequency units. Thus

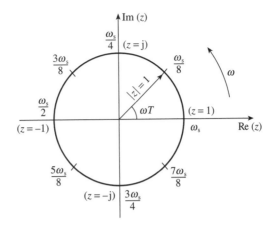

Figure 4.12   z-plane unit circle showing critical frequency points.

the frequency ranges of interest may be expressed in one of the following, equivalent, six ways:

$$
\left.\begin{array}{ll}
0 \leqslant \omega \leqslant \omega_s/2 & \text{rad s}^{-1} \\
0 \leqslant \omega \leqslant \pi/T & \text{rad s}^{-1} \\
0 \leqslant \omega \leqslant \pi & \text{(normalized)}
\end{array}\right\} \tag{4.51}
$$

$$
\left.\begin{array}{ll}
0 \leqslant f \leqslant F_s/2 & \text{Hz} \\
0 \leqslant f \leqslant 1/2T & \text{Hz} \\
0 \leqslant f \leqslant 1/2 & \text{(normalized)}
\end{array}\right\} \tag{4.52}
$$

The unit of Hz is more appealing (and less confusing) when we are examining frequency response plots or specifying discrete-time systems. However, when evaluating the numerous mathematical formulas in DSP, the unit of rad s^{-1} is more convenient.

Example 4.12

Given the frequency response specification for a bandpass discrete-time filter in Hz as

passband	6–10 kHz
stopbands	0–4 kHz and 12–16 kHz
sampling frequency	32 kHz

(1)   express the specifications in normalized frequency, f,

(2)   convert the specification from standard units of Hz to rad s^{-1}, and

(3)   convert the specifications from the units of rad s^{-1} in part (2) to normalized frequency, ω.

Solution   (1)   The bandedge frequencies, which are in units of Hz, can be expressed in normalized form by simply dividing each frequency by the sampling frequency. Thus, the specification in normalized form becomes

passband	0.1875–0.3125
stopbands	0–0.125 and 0.375–0.5
sampling frequency	1

(2)   Since $\omega = 2\pi f$, each bandedge frequency is simply multiplied by 2π to convert it into rad s^{-1}. The frequency response specifications now become

passband	12 000π–20 000π rad s^{-1}
stopbands	0–8000π and 24 000π–32 000π rad s^{-1}
sampling frequency	64 000π rad s^{-1}

(3)   The bandedge frequencies in (2) can be expressed in normalized form by dividing each frequency by 32 kHz (the sampling frequency), for example

$$12\,000\pi \rightarrow \frac{12\,000\pi}{32\,000} = \frac{3\pi}{8}$$

Thus the specifications become

passband	$3\pi/8$–$5\pi/8$
stopbands	0–$\pi/4$ and $3\pi/4$–π
sampling frequency	2π

4.5.7   Stability considerations

Stability analysis is often carried out as part of the design of discrete-time systems. A useful stability criterion for LTI systems is that all bounded inputs produce bounded outputs. This is the so-called BIBO (bounded input, bounded output) condition. An LTI system is said to be BIBO stable if and only if it satisfies the criterion

$$\sum_{k=0}^{\infty} |h(k)| < \infty \tag{4.53}$$

where $h(k)$ is the impulse response of the system. It is obvious that if the impulse response is of finite length the condition above is satisfied since the sum of the impulse response coefficients will be finite. Thus, stability considerations apply only to systems with impulse response of infinite duration.

	$h(n)$			
n	$\alpha = 0.5$	$\alpha = 0.99$	$\alpha = 1$	$\alpha = 1.5$
0	0.00000e+00	0.00000e+00	0.00000e+00	0.00000e+00
1	1.00000e+01	1.00000e+01	1.00000e+01	1.00000e+01
2	5.00000e+00	9.90000e+00	1.00000e+01	1.50000e+01
3	2.50000e+00	9.80100e+00	1.00000e+01	2.25000e+01
4	1.25000e+00	9.70299e+00	1.00000e+01	3.37500e+01
5	6.25000e−01	9.60596e+00	1.00000e+01	5.06250e+01
6	3.12500e−01	9.50990e+00	1.00000e+01	7.59375e+01
7	1.56250e−01	9.41480e+00	1.00000e+01	1.13906e+02
8	7.81250e−02	9.32065e+00	1.00000e+01	1.70859e+02
9	3.90625e−02	9.22745e+00	1.00000e+01	2.56289e+02

(a)

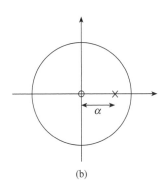

(b)

Figure 4.13 An illustration of the behaviour of the impulse response of a system for various degrees of stability: (a) impulse response; (b) *z*-plane pole–zero diagram. The system *z*-transform is $10z^{-1}/(1 - \alpha z^{-1})$. (i) For $\alpha = 0.5$ system is stable, (ii) for $\alpha = 0.99$ it is marginally stable, (iii) for $\alpha = 1$ it is potentially unstable, and (iv) for $\alpha = 1.5$ it is unstable. Notice, for example, that for $\alpha = 0.5$ the impulse response values decay rapidly as *n* increases, while for $\alpha = 1.5$ the impulse response values increase rapidly.

For the output to be bounded all the poles must lie inside the unit circle. When a pole lies outside the unit circle the system is unstable. In practice, a system with a pole on the unit circle is also regarded as unstable or potentially unstable, since a minor disturbance or error will invariably push the system into instability. An exception is when a pole is coincident with a zero on the unit circle so that its effects are nullified. For an unstable system, the impulse response will increase indefinitely with time.

In principle, testing for stability is simple: find the pole positions of the *z*-transform. If any pole is on or outside the unit circle (unless it is coincident with a zero on the unit circle) then the system is unstable. In practice, finding the positions of the poles may not be an easy task.

A simple test that may be used when the system *z*-transform, $H(z)$, is not available in factored form is to obtain and plot a sufficient number of values of the impulse response, by finding the inverse *z*-transform. If the impulse response increases indefinitely with time or fails to decay fast enough then the system is either unstable or marginally stable. Figure 4.13 shows examples of the behaviour of the impulse response of a simple discrete-time system for various degrees of stability. Other more sophisticated tests for stability are available in more advanced texts on *z*-transforms (for example Jury, 1964; Proakis and Manolakis, 1992). Further considerations of stability for second-order systems are given in Chapter 8.

4.5.8 Difference equations

The difference equation specifies the actual operations that must be performed by the discrete-time system on the input data, in the time domain, in order to generate the desired output. The difference equation, for most practical cases of interest, may be written as

$$y(n) = \sum_{k=0}^{N} a_k x(n-k) - \sum_{k=1}^{M} b_k y(n-k) \tag{4.54}$$

where $x(n)$ is the input sample, $y(n)$ is the output sample, $y(n-k)$ is the previous output and a_k, b_k are system coefficients. Equation 4.54 indicates that the current output, $y(n)$, is obtained from present and past input samples and previous outputs, $y(n-k)$.

The difference equations for discrete-time systems are readily obtained from their transfer functions and vice versa using the delay property of the *z*-transform:

$$a_k x(n) \leftrightarrow a_k X(z)$$

$$a_k x(n-k) \leftrightarrow a_k z^{-k} X(z)$$

Thus, Equation 4.54 may be written as

$$Y(z) = \sum_{k=0}^{N} a_k z^{-k} X(z) - \sum_{k=0}^{M} b_k z^{-k} Y(z) \tag{4.55}$$

Simplifying, we obtain the *z*-domain transfer function of the discrete system, $H(z)$:

$$H(z) = \frac{Y(z)}{X(z)} = \sum_{k=0}^{N} a_k z^{-k} \bigg/ \left(1 + \sum_{k=0}^{M} b_k z^{-k} \right) \tag{4.56}$$

If all the denominator coefficients, b_k, are zero Equations 4.54 and 4.55 reduce to

$$y(n) = \sum_{k=0}^{N} a_k x(n-k) \tag{4.57a}$$

$$H(z) = \frac{Y(z)}{X(z)} = \sum_{k=0}^{N} a_k z^{-k} \tag{4.57b}$$

The output of the system, $y(n)$, now depends only on the present and past input samples but not on previous outputs as in Equation 4.54. The coefficients, the a_k, in this case represent the impulse response of the system and are normally denoted by the symbol $h(k)$. This class of LTI system is referred to as finite impulse response (FIR) systems, since the length of $h(k)$ is clearly finite.

Systems characterized by Equations 4.54 and 4.56 where at least one of the denominator coefficients is nonzero are referred to as infinite impulse response (IIR) systems. In IIR systems at least one of the poles will be nonzero, but FIR systems normally contain no poles.

4.5.9   Impulse response estimation

In the design of discrete-time systems the need often arises to obtain values of impulse responses. For example, in FIR system design the impulse response is required to implement the system, and in IIR system design the values are required for stability analysis. The impulse response may also be used to evaluate the frequency response of the system.

The impulse response of a discrete-time system may be defined as the inverse z-transform of the system's transfer function, $H(z)$:

$$h(k) = Z^{-1}[H(z)], \quad k = 0, 1, \ldots$$

If the z-transform, $H(z)$, is available as a power series, that is

$$H(z) = \sum_{n=0}^{\infty} h(n)z^{-n}$$

$$= h(0) + h(1)z^{-1} + h(2)z^{-2} + \ldots \tag{4.58}$$

the coefficients of the z-transform give directly the impulse response. For IIR systems, $H(z)$ is often expressed as a ratio of polynomials such as Equation 4.47. In this case, the IZT methods described in Section 4.3 can be used to obtain the impulse response of the system. The C language and MATLAB programs described in the appendix may be used for this purpose.

The impulse response may also be viewed as the response of a discrete-time system to a unit impulse, $u(n)$, which has a value of 1 at $n = 0$ and a value of 0 at all other values of n. This view arises from the fact that if we make the input to the system equal to the unit impulse, $x(n) = u(n)$, the output of the system is in fact equal to $h(n)$, the system's impulse response (strictly speaking, the unit sample response):

$$y(n) = \sum_{k=0}^{\infty} h(k)x(n - k) = \sum_{k=0}^{\infty} h(k)u(n - k)$$

$$= h(0)u(n) + h(1)u(n - 1) + h(2)u(n - 2) + \ldots$$

$$= h(n), \quad n = 0, 1, \ldots \tag{4.59}$$

This provides a simple alternative method of computing $h(n)$ (indeed it provides another method of obtaining the inverse z-transform) as illustrated by the following example.

Example 4.13   Find the impulse response of the discrete-time filter characterized by the following z-transfer function (1) by using the power series method and (2) by applying a unit impulse to the system:

$$H(z) = \frac{1 - z^{-1}}{1 + 0.5z^{-1}}$$

Solution   (1)   Using the power series method, the values of the impulse response are obtained as

$$1 + 0.5z^{-1} \overline{\smash{\big)}\; 1 - z^{-1}} \quad \overset{\displaystyle 1 - 1.5z^{-1} + 0.75z^{-2} - 0.375z^{-3} \ldots}{}$$

$$\underline{1 + 0.5z^{-1}}$$

$$-1.5z^{-1}$$

$$\underline{-1.5z^{-1} - 0.75z^{-2}}$$

$$0.75z^{-2}$$

$$\underline{0.75z^{-2} + 0.375z^{-3}}$$

$$-0.375z^{-3}$$

From the quotients the impulse response values are

$$h(0) = 1,\; h(1) = -1.5,\; h(2) = 0.75,\; h(3) = -0.325$$

The impulse response values can of course be obtained with the aid of the C language program for the power series method given in the appendix.

(2)   First, we need to obtain the difference equation of the filter from the transfer function:

$$H(z) = \frac{Y(z)}{X(z)} = \frac{1 - z^{-1}}{1 + 0.5z^{-1}}$$

Cross-multiplying and using the delay property of the z-transform we obtain the difference equation:

$$Y(z) + 0.5Y(z)z^{-1} = X(z) - X(z)z^{-1}$$

$$y(n) + 0.5y(n - 1) = x(n) - x(n - 1)$$

Simplifying we have

$$y(n) = x(n) - x(n - 1) - 0.5y(n - 1)$$

The impulse response of the filter can now be obtained by letting $x(n) = u(n)$ where

$$u(n) = 1, \quad n = 0$$

$$= 0, \quad n \neq 0$$

and assuming an initial condition of $y(-1) = 0$:

$$y(0) = 1$$

$$y(1) = x(1) - x(0) - 0.5y(0) = 0 - 1 - 0.5 = -1.5$$

$$y(2) = x(2) - x(1) - 0.5y(1) = -0.5 \times -1.5 = 0.75$$

$$y(3) = x(3) - x(2) - 0.5y(2) = -0.5 \times 0.75 = -0.325$$

$$\vdots$$

From this the impulse response values are

$$h(0) = 1,\ h(1) = -1.5,\ h(2) = 0.75,\ h(3) = -0.325$$

It is seen that both methods lead to identical results.

4.5.10 Applications in digital filter design

One of the most important applications of the *z*-transform in DSP is in the design and analysis of errors in digital filters, especially IIR filters. It is used extensively to determine the coefficients of digital filters and to analyze the effects of various quantization errors on digital filter performance. For example, it is well known that quantization errors are inherent in discrete-time systems when they are implemented in hardware or software owing to finite register lengths of practical processors. The *z*-transform provides a convenient means of analyzing the effects of such errors on system performance. In particular, errors due to rounding or truncating the result of the multiplication operations indicated in the difference equations are often analyzed with the aid of the *z*-transform. Noise analysis in discrete-time filters is discussed in more detail in Chapter 13.

Another important application of the *z*-transform in discrete filter design is in the representation of digital filter structures. We will discuss this in more detail here because it requires the use of the partial fraction expansion program mentioned earlier.

4.5.11 Realization structures for digital filters

Discrete-time filters are often represented in the form of block or signal flow diagrams. The diagrams are a convenient way of representing the difference equations or equivalently the transfer functions. Consider, for example, a simple discrete filter with the following difference equation:

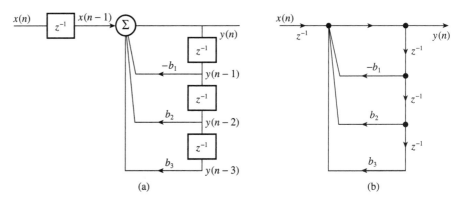

(a)   (b)

Figure 4.14   Realization diagrams for a difference equation: (a) in block diagram form;
(b) signal flow diagram.

$$y(n) = x(n - 1) - b_1 y(n - 1) + b_2 y(n - 2) + b_3 y(n - 3) \qquad (4.60)$$

The block diagram representation of this equation is shown in Figure 4.14(a). In the
figure, the symbol z^{-1} represents a delay of 1 unit of time as may be deduced from the
signals at various nodes, the arrows represent multipliers and the constants next to
them the multiplication factors. The relationship between the difference equation
and the block diagram should be evident. A signal flow diagram representation of
the same difference equation is shown in Figure 4.14(b). It is common to refer to
the block or flow diagram as a realization diagram.

When the order of $H(z)$ is high, the discrete-time filter is rarely realized directly as
Figure 4.14 implies because large errors will result if a small number of bits is used to
represent the coefficients and to execute the difference equations (see Chapters 8 and
13). The common practice is to decompose the transfer function into a cascade or
parallel combination of second- and/or first-order z-transforms. For cascade realization, the transfer function, $H(z)$, is factored as

$$H(z) = H_1(z)H_2(z) \ldots H_K(z) = \prod_{i=1}^{K} H_i(z) \qquad (4.61)$$

where $H_i(z)$ is either a second- or a first-order section:

$$H_i(z) = \frac{b_0 + b_{1i} z^{-1} + b_{2i} z^{-2}}{1 + a_{1i} z^{-1} + a_{2i} z^{-2}} \qquad \text{second order}$$

$$H_i(z) = \frac{b_0 + b_{1i} z^{-1}}{1 + a_{1i} z^{-1}} \qquad \text{first order}$$

and K is the integer part of $(M + 1)/2$. The overall z-transform is the product of the
individual z-transforms: see Figure 4.15.

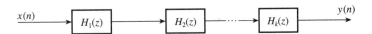

Figure 4.15 General structure for cascade realization.

For parallel realization, the transfer function is decomposed, using partial fractions, to give

$$H(z) = B_0 + \sum_{i=1}^{K} H_i(z) \tag{4.62}$$

where, as above, $H_i(z)$ is either a second-order or a first-order section but of the form

$$H_i(z) = \frac{a_{0i} + a_{1i}z^{-1}}{1 + b_{1i}z^{-1} + b_{2i}z^{-2}} \quad \text{second order}$$

$$H_i(z) = \frac{a_0}{1 + b_{1i}z^{-1}} \quad \text{first order}$$

where K is the integer part of $(M + 1)/2$ and

$$B_0 = a_N/b_M$$

Figure 4.16 shows the general structure for the parallel realization.

In the design of digital filters, software packages are normally used to obtain the coefficients, a_{ki} and b_{ki}, above. Unfortunately, most software packages produce coefficients only for the cascade structure. The coefficients for the parallel structure can be obtained from those of the cascade structure by using a partial fraction expansion. We will illustrate this by an example.

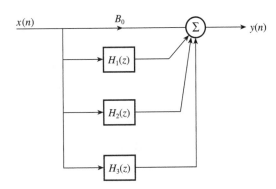

Figure 4.16 General structure for parallel realization.

Example 4.14   A discrete-time system is characterized by the following z-transfer function:

$$H(z) = \frac{1 - 2z^{-2} + z^{-4}}{1 - 0.414\,21z^{-1} + 0.085\,79z^{-2} + 0.292\,895z^{-3} + 0.5z^{-4}}$$

(1)   Express $H(z)$ in a form suitable for cascade realization using two second-order sections.

(2)   Repeat part (1) for parallel realization.

Solution   (1)   In factored form, $H(z)$ is given by

$$H(z) = H_1(z)H_2(z)$$

where

$$H_1(z) = \frac{1 - 2z^{-1} + z^{-2}}{1 - 1.414\,21z^{-1} + z^{-2}} \tag{4.63a}$$

$$H_2(z) = \frac{1 + 2z^{-1} + z^{-2}}{1 + z^{-1} + 0.5z^{-2}} \tag{4.63b}$$

(2)   To express $H(z)$ in a form suitable for parallel realization, we first expand it using partial fraction methods. Thus

$$H(z) = B_0 + \frac{C_1}{z - p_1} + \frac{C_2}{z - p_2} + \frac{C_3}{z - p_3} + \frac{C_4}{z - p_4} \tag{4.64}$$

Using the program for the PFE given in Appendix 4B the poles, p_1 to p_4, and coefficients, B_0 and C_1 to C_4, are obtained as

$$p_1 = 0.7071 + 0.7071\mathrm{j} = e^{\,\mathrm{j}0.785}; \qquad\qquad p_2 = p_1^*$$

$$p_3 = -0.5 + 0.5\mathrm{j} = 0.7071e^{\,\mathrm{j}2.356\,19}; \qquad\qquad p_4 = p_3^*$$

$$B_0 = 2$$

$$C_1 = 0.114\,383 + 0.666\,669\mathrm{j} = 0.676\,410\,4\angle 1.400\,877; \qquad C_2 = C_1^*$$

$$C_3 = -0.614\,382\,76 - 0.580\,880\,79\mathrm{j}$$

$$= 0.845\,510\,897\,6\angle 3.898\,969; \qquad\qquad C_4 = C_3^*$$

where the angles are in radians. Having found the poles and coefficients, C_k and B_0, the partial fractions in Equation 4.64 must be combined such that $H(z)$ is a sum of second-order sections of the form

$$H(z) = B_0 + \sum_{i=1}^{2} H_i(z) \tag{4.65a}$$

where

$$H_i(z) = \frac{a_0 + a_{1i} z^{-1}}{1 + b_{1i} z^{-1} + b_{2i} z^{-2}} \tag{4.65b}$$

To ensure that the coefficients a_{ki} and b_{ki} in Equation 4.65b are real, the partial fractions in Equation 4.64 involving C_1 and C_2 must be combined since they are complex conjugate pairs. Similarly the fractions with coefficients C_3 and C_4 must also be combined. Combining the fractions involving C_1 and C_2 we have

$$\frac{C_1 z}{z - p_1} + \frac{C_2 z}{z - p_2} = \frac{(C_1 + C_2)z^2 - (C_1 p_2 + C_2 p_1)z}{z^2 - (p_1 + p_2)z + p_1 p_2} \tag{4.66}$$

$$= \frac{C_1 + C_2 - (C_1 p_2 + C_2 p_1)z^{-1}}{1 - (p_1 + p_2)z^{-1} + p_1 p_2 z^{-2}} \tag{4.67}$$

Comparing Equations 4.65b and 4.67, with $i = 1$ in Equation 4.65b, we find that

$$a_{01} = C_1 + C_2, \qquad a_{11} = -(C_1 p_2 + C_2 p_1)$$
$$b_{11} = -(p_1 + p_2), \qquad b_{21} = p_1 p_2 \tag{4.68}$$

If we make use of the fact that $p_2 = p_1^*$, $C_2 = C_1^*$, and substitute the values of p_1 and C_1, then

$$a_{01} = C_1 + C_1^* = 2 \times 0.114\,383 = 0.2288$$

$$a_{11} = -(C_1 p_1^* + C_1^* p_1)$$

$$= -(|C_1| e^{j\theta_1} |p_1| e^{-j\phi_1} + |C_1| e^{-j\theta_1} |p_1| e^{j\phi_1})$$

$$= -|C_1||p_1|[e^{j(\theta_1 - \phi_1)} + e^{-j(\theta_1 - \phi_1)}]$$

$$= -2|C_1||p_1| \cos(\theta_1 - \phi_1)$$

$$= -2 \times 0.676\,410\,4 \times 1 \cos(1.400\,877 - 0.785\,400\,68)$$

$$= -1.1046 \tag{4.69}$$

(where $\theta_1 = \angle C_1$, $\phi_1 = \angle p_1$). Thus we have

$$H_1(z) = \frac{0.2288 - 1.1046 z^{-1}}{1 - 1.4142 z^{-1} + z^{-2}} \tag{4.70}$$

where the values of the denominator coefficients have been taken straight from Equation 4.63. Similarly, from the partial fractions involving C_3 and C_4 we have

$$H_2(z) = \frac{-1.2288 - 0.0335 z^{-1}}{1 + z^{-1} + 0.5 z^{-2}} \tag{4.71}$$

Combining the results,

$$H(z) = 2 + \frac{0.2288 - 1.1046z^{-1}}{1 - 1.4142z^{-1} + z^{-2}} + \frac{-1.2288 - 0.0335z^{-1}}{1 + z^{-1} + 0.5z^{-2}}$$

Although the above process is straightforward, it is very tedious and mistakes are likely, especially if the partial fraction coefficients are worked out by hand. We describe in Appendix 4B a C language routine that may be used to obtain the coefficients of the parallel structure, given the transfer function in cascade form. The program is really a simple extension of the partial fraction expansion routine described in the same appendix. In Chapter 8, we will discuss in some detail applications of cascade and parallel realization structures.

4.6 Summary

A knowledge of the *z*-transform is very important in DSP work, as it is an invaluable tool for representing, analyzing and designing discrete-time systems.

We have shown how to evaluate the *z*-transform of discrete-time sequences and how to recover the sequences from their *z*-transforms. Several C language and MATLAB programs are provided to enable readers to gain a practical understanding of the concepts and applications of the *z*-transform in signal processing. Try to use them whenever possible.

Problems

4.1 Find the *z*-transform of each of the following discrete-time sequences:

(1) $x(n) = \sin(n\omega T), n = 0, 1, \ldots$

(2) $x(n) = a^n, \quad n \geqslant 0$

$\qquad = 0, \quad n < 0$

(3) $x(n) = 1, \quad 0 \leqslant n \leqslant N - 1$

$\qquad = 0, \quad \text{elsewhere}$

4.2 An exponential sequence is defined as

$$x(n) = e^{-kn}, \quad n \geqslant 0$$

Find its *z*-transform, including the constraint on *z* for the *z*-transform to converge, for each of the following cases:

(1) *k* is real;

(2) *k* is complex.

4.3 Given the causal sequences $x(n)$ and $nx(n)$, with *z*-transforms $X(z)$ and $X'(z)$, show that

$$X'(z) = -z \frac{\mathrm{d}X(z)}{\mathrm{d}z}$$

4.4 The *z*-transform of a discrete-time sequence is given by

$$X(z) = \sum_{n=0}^{\infty} x(n)z^{-n}$$

Starting from the above equation show, stating any assumptions made, that the inverse *z*-transform is given by

$$x(n) = \frac{1}{2\pi j} \oint z^{n-1} X(z) \, dz, \quad n > 0$$

Discuss, briefly, the role of the residue theorem in evaluating the above integral.

4.5 (1) Find, using the power series method, the first five values of the causal discrete-time sequence corresponding to each of the following z-transforms:

(a) $X(z) = \dfrac{z - 1}{(z - 0.7071)^2}$

(b) $X(z) = \dfrac{1}{(z - 0.5)(z + 0.9)^3}$

(c) $X(z) = \dfrac{z^4 - 1}{z^4 + 1}$

(d) $X(z) = \dfrac{z^3 - z^2 + z - 1}{(z + 0.9)^3}$

(2) Repeat part (1) using the partial fraction expansion method.

(3) Repeat part (1) using the residue method.

4.6 (1) Given a z-transform, $X(z)$, in the form

$$X(z) = \frac{N(z)}{D(z)}$$

where $N(z)$ and $D(z)$ are polynomials, and assuming that $X(z)$ has a pole at $z = p_k$, show that

$$\mathrm{Res}\,[X(z), p_k] = \frac{N(p_k)}{D'(p_k)}$$

where

$$D'(p_k) = \frac{dD(z)}{dz}$$

(2) Use this result to find the inverse z-transform of the following:

$$X(z) = \frac{1}{z^4 - 1}$$

4.7 Given the following transfer function for a stable, causal system, find its impulse response, $h(n)$, in closed form using the residue method:

$$H(z) = \frac{1}{1 + b_1 z^{-1} + b_2 z^{-2}}$$

Assume the poles are distinct and complex.

4.8 The partial fraction expansion of an Nth-order discrete-time system is given by

$$X(z) = \frac{N(z)}{D(z)} = B_0 + \sum_{k=1}^{N} \frac{C_k z}{z - p_k z^k}$$

where

$$N(z) = a_0 + a_1 z^{-1} + a_2 z^{-2} + \ldots + a_N z^{-N}$$
$$D(z) = b_0 + b_1 z^{-1} + b_2 z^{-2} + \ldots + b_M z^{-M}$$

p_k are the poles of $X(z)$ (assumed distinct) and C_k are the partial fraction coefficients. Find a general expression for C_k, $k = 1, 2, \ldots$, in terms of the poles, the p_k, and $N(z)$. Assuming that $N = 3$ show, by long division, that B_0 is given by

$$B_0 = a_3/b_3$$

4.9 Given the difference equation

$$y(n) + B_1 y(n - 1) + B_2 y(n - 2) = A, \quad n \geq 0$$

where A, B_1 and B_2 are arbitrary constants, find an expression for the z-transform, $Y(z)$. Use an appropriate inverse z-transform technique to obtain a closed form expression for $y(n)$.

4.10 A second-order discrete-time system is characterized by the following z-transfer function:

$$H(z) = \frac{1}{(z - 0.9)^2}, \quad |z| > 0.9$$

Obtain the corresponding discrete-time sequence, $h(n)$, using the residue method.

4.11 For the discrete-time system shown in Figure 4.17, obtain the difference equation relating the output, $y(n)$, and the input, $x(n)$. Derive its transfer function, $H(z)$.

4.12 The transfer function of a discrete-time system has poles at $z = 0.5$, $z = 0.1 \pm j\,0.2$ and zeros at $z = -1$ and at $z = 1$.

(1) Sketch the pole–zero diagram for the system.

(2) Derive the system transfer function, $H(z)$, from the pole–zero diagram.

(3) Develop the difference equation.

(4) Draw the realization diagram in signal flowgraph form.

4.13 The signal flow diagram of a discrete-time system is shown in Figure 4.18. Obtain the two-step difference equation relating the output, $y(n)$, and

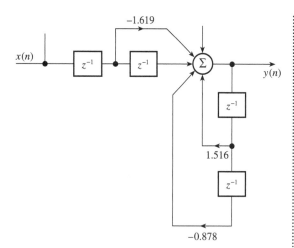

Figure 4.17 Block diagram for the discrete-time system of Problem 4.11.

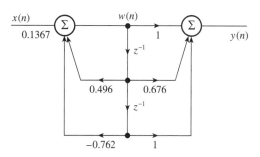

Figure 4.18 The signal flow diagram of the discrete-time system for Problem 4.13.

input, $x(n)$. Derive the transfer function, $H(z)$, from the difference equation.

4.14 The frequency response specification for a bandpass discrete-time filter in normalized form is as follows

passband	$0.4\pi-0.6\pi$
stopbands	$0-0.3\pi$ and $0.7\pi-\pi$
sampling interval	$T = 100\ \mu s$

(1) Express the specifications in rad s^{-1} (de-normalized).

(2) Convert the specification from rad s^{-1} to standard units of Hz.

(3) Convert the specifications from the units of hertz in part (2) to normalized form.

(4) Sketch the frequency responses for each of the three cases above in the interval from 0 to the sampling frequency.

4.15 An LTI system is characterized by the following *z*-transform:

$$\frac{1 + z^{-2}}{1 + 0.81z^{-2}}$$

Determine the frequency response at dc, 1/4 and 1/2 the sampling frequency. Sketch the frequency response in the interval $0 \leqslant \omega \leqslant \omega_s$, where ω_s is the sampling frequency in rad s^{-1}.

4.16 A requirement exists for a simple lowpass discrete-time filter with the following specifications:

| cutoff frequency | 1 kHz |
| sampling frequency | 10 kHz |

Specify and sketch a suitable pole–zero diagram for the filter.

Obtain the transfer function of the filter from the pole–zero diagram. Determine the amplitude and phase response at 1 kHz, 2.5 kHz and 5 kHz. Sketch the amplitude–frequency response.

4.17 The transfer function of a certain system is defined as

$$H(z) =$$
$$\frac{(1 - 1.094\,621z^{-1} - z^2)(1 - 0.350\,754z^{-1} + z^{-2})}{(1 - 1.340\,228z^{-1} + 0.796\,831z^{-2})(1 - 0.5z^{-1} - 0.5z^{-2})}$$

(1) Find the poles and zeros, and sketch the pole–zero diagram.

(2) State, with justification, whether the system is stable or not.

Computer-based problems using MATLAB and language tools

4.18 The transfer function of a discrete-time system is given by:

$$H(z) = \frac{z^2 - z}{z^2 - 0.9051z + 0.4096}$$

(a) Determine the locations of the poles and zeros with the aid of MATLAB **roots** command (zeros should be at $z = 0$, $z = \pm 1$; poles at a radius of 0.64 $\angle$ 45°).

(b) Repeat part (a) if the numerator and denominator polynomials are written in increasing negative powers of z as follows:

$$H(z) = \frac{1 - z^{-1}}{1 - 0.9051z^{-1} + 0.4096z^{-2}}$$

(c) Compare the results in (a) and (b) and comment on any difference. Suggest how to ensure to make the answers the same.

(d) Use the **zplane** command in MATLAB to plot the pole–zero diagram for $H(z)$ in (a) above for each of the following two cases:

(i) with the coefficients of the numerator and denominator polynomials, $b(z)$ and $a(z)$, as inputs.

(ii) with the locations of the poles and zeros of $H(z)$ as inputs.

4.19 The transfer function of a discrete-time notch filter is given by:

$$H(z) = \frac{1 - 2\cos\theta + z^{-2}}{1 - 2r\cos\theta z^{-1} + r^2 z^{-2}}$$

where r is the radius of the poles and θ the angle of the poles and zeros.

(a) Plot the magnitude frequency response of the filter, using MATLAB, and estimate the depth of the notch (relative to the magnitude response at 0 Hz) for each of the following cases:

(i) $r = 0.5$, $\theta = \pm 15°$
(ii) $r = 0.5$, $\theta = \pm 60°$
(iii) $r = 0.5$, $\theta = \pm 90°$
(iv) $r = 0.5$, $\theta = \pm 120°$

(b) Plot the magnitude frequency response of the filter and estimate the depth of the notch (relative to the magnitude response at 0 Hz) for each of the following cases:

(v) $r = 0.5$, $\theta = \pm 45°$
(vi) $r = 0.8$, $\theta = \pm 45°$
(vii) $r = 0.9$, $\theta = \pm 45°$
(viii) $r = 0.99$, $\theta = \pm 45°$

Comment on how the positions (radius and angles) of the poles and zeros affect the frequency response of the notch filter.

4.20 The z-transform of a discrete-time system is given by

$$H(z) = \sum_{k=0}^{8} a_k z^{-k} \bigg/ \sum_{k=0}^{8} b_k z^{-k}$$

where

$a_0 = 2.740\ 584 \times 10^{-2}$ $b_0 = 1$

$a_1 = 2.825\ 341 \times 10^{-3}$ $b_1 = 2.233\ 030 \times 10^{-1}$

$a_2 = -2.932\ 353 \times 10^{-2}$ $b_2 = 2.353\ 762$

$a_3 = 3.563\ 199 \times 10^{-4}$ $b_3 = 4.369\ 285 \times 10^{-1}$

$a_4 = 4.924\ 136 \times 10^{-2}$ $b_4 = 2.712\ 411$

$a_5 = 3.563\ 226 \times 10^{-4}$ $b_5 = 3.571\ 619 \times 10^{-1}$

$a_6 = -2.932\ 353 \times 10^{-2}$ $b_6 = 1.593\ 957$

$a_7 = 2.825\ 337 \times 10^{-3}$ $b_7 = 1.141\ 820 \times 10^{-1}$

$a_8 = 2.740\ 582 \times 10^{-2}$ $b_8 = 4.143\ 201 \times 10^{-1}$

Obtain and plot the impulse response of the system, with the aid of the inverse z-transform program (power series method) described in the appendix or a suitable MATLAB program. From your plot, state whether the system is stable, marginally stable or unstable.

4.21 The transfer function of a third-order IIR system, in factored form, is given by

$$H(z) = \frac{N_1(z)N_2(z)}{D_1(z)D_2(z)}$$

where

$N_1(z) = 1 - 0.971\ 426z^{-1} + z^{-2}$

$N_2(z) = 1 + z^{-1}$

$D_1(z) = 1 - 0.935\ 751z^{-1} + 0.726\ 879z^{-2}$

$D_2(z) = 1 + 0.183\ 11z^{-1}$

(1) Draw a realization diagram for the system, for a cascade structure, using a second-order and one first-order section.

(2) Express $H(z)$ as the sum of partial fractions,

$$H(z) = B_0 + \sum_{k=1}^{3} \frac{C_k}{z - p_k}$$

and use the partial fraction expansion program in Appendix 4C or a suitable MATLAB program to compute the coefficients, B_0 and C_k.

(3)   Combine the partial fractions such that the system can be realized in parallel using a second-order and one first-order section.

(4)   Draw a realization diagram for the parallel structure using the result from (3).

4.22   A digital notch filter is characterized by the following z-transform:

$$\frac{z^2 + 1}{z^2 + r^2}$$

Estimate the frequency response, using the FFT approach and a sampling frequency of 1 kHz, for each of the following cases: (i) $r = 0.8$, (ii) $r = 0.95$, (iii) $r = 1$. Explain your results.

4.23   A lowpass discrete-time filter has the following transfer function:

$$H(z) = \frac{a_0 + a_1 z^{-1} + a_2 z^{-2} + \ldots + a_4 z^{-4}}{b_0 + b_1 z^{-1} + b_2 z^{-2} + \ldots + b_4 z^{-4}}$$

where

$$a_0 = 0.193\,441 \quad b_0 = 1$$

$$a_1 = 0.378\,331 \quad b_1 = -2.516\,884$$

$$a_2 = 0.524\,14 \quad b_2 = 1.054\,118$$

$$a_3 = 0.378\,331 \quad b_3 = -0.240\,603$$

$$a_4 = 0.193\,441 \quad b_4 = 0.198\,586\,1$$

Estimate the frequency response of the filter using

(1)   the program for direct frequency response evaluation discussed in Appendix 4C, and

(2)   the power series method and FFT.

Compare the two results and comment on any differences.

4.24   The coefficients of a simple bandpass FIR system are listed in Table 4.4. Assuming a sampling frequency of 10 kHz, compute the magnitude–frequency response of the system using the program discussed in Appendix 4C.

4.25   Given the z-transform

$$H(z) =$$

$$\frac{1 + 3z^{-1} + z^{-2} + z^{-3}}{1 + (1 - k)z^{-1} + (k + 0.3561)z^{-2} + 0.3561k}$$

use the computer program for the power series expansion method in Appendix 4B to compute a

Table 4.4   Coefficients of the FIR bandpass filter for Problem 4.24.

H (1) = −0.67299600E−02	= H(35)
H (2) = 0.16799420E−01	= H(34)
H (3) = 0.17195700E−01	= H(33)
H (4) = −0.27849080E−01	= H(32)
H (5) = −0.17486810E−01	= H(31)
H (6) = 0.13515580E−01	= H(30)
H (7) = 0.45570510E−02	= H(29)
H (8) = 0.33293060E−01	= H(28)
H (9) = 0.95162150E−02	= H(27)
H(10) = −0.68548560E−01	= H(26)
H(11) = −0.68992230E−02	= H(25)
H(12) = 0.23802370E−01	= H(24)
H(13) = −0.11597510E−01	= H(23)
H(14) = 0.12073780E+00	= H(22)
H(15) = 0.23806900E−01	= H(21)
H(16) = −0.29095690E+00	= H(20)
H(17) = −0.12362380E−01	= H(19)
H(18) = 0.36717700E+00	= H(18)

sufficient number of values of the system's impulse response for each of the following cases:

(1) $k = -1$;

(2) $k = 1$;

(3) $k = 2$;

(4) $k = 0.9$.

For each case, plot the impulse response and from this state whether the system is stable, marginally stable or unstable.

4.26   Develop a C language program for checking the result of partial fraction expansion. The program should accept as inputs the values of the poles p_k ($k = 1, 2, \ldots, M$) and the associated coefficients C_k ($k = 1, 2, \ldots, M$), and produce as outputs the coefficients of the polynomials $A(z)$ and $B(z)$, where

$$X(z) = \frac{A(z)}{B(z)}$$

and

$$A(z) = a_0 + a_1 z^{-1} + a_2 z^{-2} + \ldots + a_N z^{-N}$$

$$B(z) = b_0 + b_1 z^{-1} + b_2 z^{-2} + \ldots + b_M z^{-M}$$

Extend your program to check the results of converting cascade structures to parallel.

4.27   Repeat Problem 4.26 using MATLAB.

References

Atkinson L.V. and Harley P.J. (1983) *An Introduction to Numerical Methods with Pascal*, Chapter 3. Wokingham: Addison-Wesley.

Jury E.I. (1964) *Theory and Applications of the z-transform Method*. New York: Wiley.

Mathews J.H. (1982) *Basic Complex Variables for Mathematics and Engineering*. Boston MA: Allyn and Bacon.

Proakis J.G. and Manolakis D.G. (1992) *Digital Signal Processing*, 2nd edn. New York: Macmillan.

Bibliography

Ahmed N. and Natarajan T. (1983) *Discrete-time Signals and Systems*. Reston VA: Reston Publishing Co. Inc.

Churchhill R.V., Brown J.W. and Verhey R.F. (1976) *Complex Variables and Applications*. New York: McGraw-Hill.

Jong M.T. (1982) *Methods of Discrete Signals and Systems Analysis*. New York: McGraw-Hill.

Oppenheim A.V. and Schafer R.W. (1975) *Digital Signal Processing*. Englewood Cliffs NJ: Prentice-Hall.

Rabiner L.R. and Gold B. (1975) *Theory and Application of Digital Signal Processing*. Englewood Cliffs NJ: Prentice-Hall.

Ragazzini J.R. and Zadeh L.A. (1952) Analysis of sampled data systems. *Trans. AIEE*, **71**(II), 225–34.

Steiglitz K. (1974) *An Introduction to Discrete Systems*. New York: Wiley.

Strum R.D. and Kirk D.E. (1988) *First Principles of Discrete Systems and Digital Signal Processing*. Reading MA: Addison-Wesley.

Appendices

4A   Recursive algorithm for the inverse z-transform

It was stated in the main text that the long division method can be recast in a recursive form. Specifically, we want to show here that given a z-transform, $X(z)$, such that

$$X(z) = \frac{b_0 + b_1 z^{-1} + b_2 z^{-2}}{a_0 + a_1 z^{-1} + a_2 z^{-2}}$$

the inverse z-transform, $x(n)$, can be obtained as (Jury, 1964)

$$x(n) = \frac{1}{a_0}\left[b_n - \sum_{i=1}^{n} x(n-i)a_i \right], \quad n = 1, 2, \ldots$$

$$x(0) = \frac{b_0}{a_0}$$

The result can be generalized. Using long division, we can express $X(z)$ as a power series as follows:

$$\frac{b_0}{a_0} + \left[\left(b_1 - \frac{b_0}{a_0}a_1\right)\bigg/a_0\right]z^{-1} + \frac{1}{a_0}\left[\left(b_2 - \frac{b_0}{a_0}a_2\right) - \frac{a_1}{a_0}\left(b_1 - \frac{b_0}{a_0}a_1\right)\right]z^{-2}$$

$$a_0 + a_1 z^{-1} + a_2 z^{-2} \,\Big|\, b_0 + b_1 z^{-1} + b_2 z^{-2}$$

$$b_0 + \left(\frac{b_0}{a_0}a_1\right)z^{-1} \quad + \left(\frac{b_0}{a_0}a_2\right)z^{-2}$$

$$\left(b_1 - \frac{b_0}{a_0}a_1\right)z^{-1} + \left(b_2 - \frac{b_0}{a_0}a_2\right)z^{-2}$$

$$\left(b_1 - \frac{b_0}{a_0}a_1\right)z^{-1} + \frac{a_1}{a_0}\left(b_1 - \frac{b_0}{a_0}a_1\right)z^{-2} + \frac{a_2}{a_0}\left(b_1 - \frac{b_0}{a_0}a_1\right)z^{-3}$$

$$\left[\left(b_2 - \frac{b_0}{a_0}a_2\right) - \frac{a_1}{a_0}\left(b_1 - \frac{b_0}{a_0}a_1\right)\right]z^{-2} - \frac{a_2}{a_0}\left(b_1 - \frac{b_0}{a_0}a_1\right)z^{-3}$$

$$\left[\left(b_2 - \frac{b_0}{a_0}a_2\right) - \frac{a_1}{a_0}\left(b_1 - \frac{b_0}{a_0}a_1\right)\right]z^{-2} + \frac{a_1}{a_0}\left[\left(b_2 - \frac{b_0}{a_0}a_2\right)\right.$$

$$\left. - \frac{a_1}{a_0}\left(b_1 - \frac{b_0}{a_0}a_1\right)\right]z^{-3} + \frac{a_2}{a_0}\left[\left(b_2 - \frac{b_0}{a_0}a_2\right) - \frac{a_1}{a_0}\left(b_1 - \frac{b_0}{a_0}a_1\right)\right]z^{-4}$$

$$\left\{\left[-\frac{a_2}{a_0}\left(b_1 - \frac{b_0}{a_0}a_1\right)\right] - \frac{a_1}{a_0}\left[\left(b_2 - \frac{b_0}{a_0}a_2\right)\right.\right.$$

$$\left.\left. - \frac{a_1}{a_0}\left(b_1 - \frac{b_0}{a_0}a_1\right)\right]\right\}z^{-3} - \frac{a_2}{a_0}\left[\left(b_2 - \frac{b_0}{a_0}a_2\right)\right.$$

$$\left. - \frac{a_1}{a_0}\left(b_1 - \frac{b_0}{a_0}a_1\right)\right]z^{-4}$$

$$\vdots$$

The quotient of the long division gives the coefficients of the power series:

$$X(z) = \frac{b_0}{a_0} + \left[\left(b_1 - \frac{b_0}{a_0}a_1\right)\bigg/a_0\right]z^{-1} + \frac{1}{a_0}\left[\left(b_2 - \frac{b_0}{a_0}a_2\right) - \frac{a_1}{a_0}\left(b_1 - \frac{b_0}{a_0}a_1\right)\right]z^{-2}$$

$$+ \frac{1}{a_0}\left\{\left[-\frac{a_2}{a_0}\left(b_1 - \frac{b_0}{a_0}a_1\right)\right] - \frac{a_1}{a_0}\left[\left(b_2 - \frac{b_0}{a_0}a_2\right) - \frac{a_1}{a_0}\left(b_1 - \frac{b_0}{a_0}a_1\right)\right]\right\}z^{-3} + \ldots$$

From the definition of the z-transform for causal systems, $X(z)$ is given by

$$X(z) = \sum_{n=0}^{\infty} x(n)z^{-n} = x(0) + x(1)z^{-1} + x(2)z^{-2} + \ldots$$

Thus, we can write

$$x(0) = \frac{b_0}{a_0}$$

$$x(1) = \frac{1}{a_0}\left(b_1 - \frac{b_0}{a_0}a_1\right) = \frac{1}{a_0}[b_1 - x(0)a_1]$$

$$x(2) = \frac{1}{a_0}\left[\left(b_2 - \frac{b_0}{a_0}a_2\right) - \frac{a_1}{a_0}\left(b_1 - \frac{b_0}{a_0}a_1\right)\right]$$

$$= \frac{1}{a_0}[b_2 - x(0)a_2 - a_1 x(1)]$$

$$x(3) = \frac{1}{a_0}\left[\left\{-\frac{a_2}{a_0}\left(b_1 - \frac{b_0}{a_0}a_1\right)\right\} - \frac{a_1}{a_0}\left\{\left(b_2 - \frac{b_0}{a_0}a_2\right) - \frac{a_1}{a_0}\left(b_1 - \frac{b_0}{a_0}a_1\right)\right\}\right]$$

$$= \frac{1}{a_0}[-a_2 x(1) - a_1 x(2)]$$

In general, we can write

$$x(n) = \frac{1}{a_0}\left[b_n - \sum_{i=1}^{n} x(n-i)a_i\right], \quad n = 1, 2, \ldots$$

$$x(0) = b_0/a_0$$

4B   C program for evaluating the inverse *z*-transform and for cascade-to-parallel structure conversion

A C language program has been developed for computing inverse *z*-transforms using the power series or partial fraction expansion method. The program can also be used to convert a discrete-time system transfer function, $H(z)$, from cascade to parallel structure. The program is fairly large and so for efficiency is organized into two program modules, izt.c and ltilib.c, held in separate files, which can be compiled separately and then linked:

izt.c   program for computing the inverse *z*-transform via power series or partial fraction expansion, and for converting a transfer function, $H(z)$, in cascade form to an equivalent transfer function in parallel form via partial fraction expansion.

ltilib.c   a library of functions including the power_series and partial_fraction functions.

The program and library are not listed here because of lack of space, but can be found on the CD in the companion handbook *A Practical Guide for MATLAB and C Language Implementations of DSP Algorithms* (see the Preface for details).

4B.1   Power series method

The inverse z-transform, $x(n)$, is computed recursively by the function power_series() (see the program) based on the following equation:

$$x(n) = \left[b_n - \sum_{i=1}^{n} x(n-i)a_i \right] \Big/ a_0, \quad n = 1, 2, \ldots \tag{4B.1a}$$

where

$$x(0) = b_0/a_0 \tag{4B.1b}$$

To use the program to calculate the inverse z-transform, via the power series method, the z-transform may be specified in either direct or cascade form:

$$X(z) = \frac{b_0 + b_1 z^{-1} + b_2 z^{-2} + \ldots + b_N z^{-N}}{a_0 + a_1 z^{-1} + a_2 z^{-2} + \ldots + a_M z^{-M}} \quad \text{direct form} \tag{4B.2a}$$

$$X(z) = \prod_{k=i}^{K} X_i(z) \qquad \text{cascade} \tag{4B.2b}$$

where $X_i(z)$ is a second-order section given by

$$X_i(z) = \frac{b_{0i} + b_{1i} z^{-1} + b_{2i} z^{-2}}{1 + a_{1i} z^{-1} + a_{2i} z^{-2}} \tag{4B.3}$$

An input data file, called coeff.dat, must be created. The file contains the number of stages, K (for the direct form, $K = 1$), and the denominator and numerator coefficients of the z-transform. The use of the input data file is convenient as it removes the task of typing coefficients and the possibility of mistakes. Further, it makes it easier to utilize the results of one investigation as input to another program. The following examples illustrate the use of the program to calculate the inverse z-transform via the power series method.

| Example 4B.1 | Find the first five values of the inverse z-transform of the discrete-time system characterized by the following z-transform using the power series method: |

$$X(z) = \frac{0.183\,301\,5 + 0.341\,956\,1 z^{-1} + 0.341\,956\,1 z^{-2} + 0.183\,301\,5 z^{-3}}{1 - 0.352\,518\,2 z^{-1} + 0.419\,402\,3 z^{-2} - 0.016\,369 z^{-3}}$$

Clearly, $X(z)$ is in direct form. The input data file, created with edlin that comes with all PCs, has the form:

```
1
1   -0.3525182   0.4194023   -0.016369
0.1833015   0.3419561   0.3419561   0.1833015
```

The output of the program is summarized below:

$h(0) = -0.016\,369;\ h(1) = 0.177\,531;\ h(2) = 0.411\,404;\ h(3) = 0.070\,570\,5;$

$h(4) = -0.147\,666\,6$

Example 4B.2

Obtain the first five values of the inverse z-transform of the following system using the power series method:

$$X(z) = \frac{N_1(z)N_2(z)N_3(z)}{D_1(z)D_2(z)D_3(z)}$$

where

$$N_1(z) = 1 - 1.122\ 346z^{-1} + z^{-2}$$

$$N_2(z) = 1 - 0.437\ 833z^{-1} + z^{-2}$$

$$N_3(z) = 1 + z^{-1}$$

$$D_1(z) = 1 - 1.433\ 509z^{-1} + 0.858\ 110z^{-2}$$

$$D_2(z) = 1 - 1.293\ 601z^{-1} + 0.556\ 929z^{-2}$$

$$D_3(z) = 1 - 0.612\ 159z^{-1}$$

Clearly, the transfer function consists of three stages: two second-order stages and one first-order stage. The first-order stage is entered as a second-order stage with a zero coefficient for the z^{-2} term. The input data file is given below.

3			/*number of stages; maximum 5*/
1	−1.433509	0.858110	/*coefficients of $D_1(z)$*/
1	−1.122346	1	/*coefficients of $N_1(z)$*/
1	−1.293601	0.556929	/*coefficients of $D_2(z)$*/
1	−0.437833	1	/*coefficients of $N_2(z)$*/
1	−0.6121593	0	/*coefficients of $D_3(z)$*/
1	1	0	/*coefficients of $N_3(z)$*/

The comments on the right-hand side are not part of the file; they are merely for explanation purposes. The output of the program is summarized below:

$$x(0) = 1;\ x(1) = 2.779\ 09;\ x(2) = 5.2725$$

$$x(3) = 8.7218;\ x(4) = 11.7438;\ x(5) = 13.4723$$

4B.2 Partial fraction expansion

Given an Nth-order z-transform, with distinct poles, that is

$$X(z) = \frac{N(z)}{D(z)} = \frac{b_0 z^N + b_1 z^{N-1} + \ldots + b_{N-1} z + b_N}{a_0 z^N + a_1 z^{N-1} + \ldots + a_{N-1} z + a_N}$$

then $X(z)$ can be expanded into partial fractions as

$$\frac{N(z)}{zD(z)} = \frac{N(z)}{z(z - p_1)(z - p_2)(z - p_3)\ldots(z - p_N)} = \frac{B_0}{z} + \sum_{k=1}^{M} \frac{C_k}{z - p_k} \tag{4B.4}$$

where

$$N(z) = b_0 z^N + b_1 z^{N-1} + \ldots + b_{N-1} z + b_N$$

$$D(z) = a_0 z^N + a_1 z^{N-1} + \ldots + a_{N-1} z + a_N$$

the p_k are the poles of $X(z)$ (assumed first order) and the C_k are the partial fraction coefficients. The constant, B_0, is given by

$$B_0 = b_N / a_N \tag{4B.5}$$

The partial fraction coefficient, C_k, associated with p_k is obtained by multiplying both sides of Equation 4B.4 by $z - p_k$ and then letting $z = p_k$:

$$C_k = \frac{N(z)(z - p_k)}{zD(z)} = \left. \frac{N(z)}{zD_k(z)} \right|_{z=p_k} \tag{4B.6}$$

where

$$D_k(z) = \prod_{\substack{i=1 \\ i \neq k}}^{M} (z - p_i)$$

For example, to find C_1, we multiply both sides of Equation 4B.4 by $z - p_1$ and then let $z = p_1$:

$$C_1 = \frac{N(z)(z - p_1)}{zD(z)} = \left. \frac{N(z)(z - p_1)}{z(z - p_1)(z - p_2)(z - p_3) \ldots (z - p_N)} \right|_{z=p_1} = \left. \frac{N(z)}{zD_1(z)} \right|_{z=p_1}$$

where

$$D_1(z) = (z - p_2)(z - p_3) \ldots (z - p_N)$$

With the poles expressed in polar coordinates, that is $p_k = r_k e^{j\theta k}$, the coefficients are given by

$$C_k = \frac{N(r_k e^{j\theta k})}{r_k e^{j\theta k} D_k(e^{j\theta k})} \quad k = 1, \ldots, N \tag{4B.7}$$

The partial fraction expansion function first finds the positions of the poles, p_k, $k = 1, 2, \ldots, N$, and then evaluates Equation 4B.7 for each pole.

When the values of B_0 and C_k have been obtained, the z-transform can be written as

$$X(z) = B_0 + \sum_{k=1}^{N} \frac{C_1 z}{z - p_k} \tag{4B.8}$$

For causal sequences, the inverse z-transform is the sum of the inverse z-transform of each term in Equation 4B.8:

$$x(n) = B_0 u(n) + C_1 (p_1)^n + C_2 (p_2)^n + \ldots + C_N (p_N)^n \tag{4B.9}$$

To use the program to calculate the partial fraction coefficients, the z-transform must be expressed in a cascade form using second-order factors. An example will make this clear.

Find, using the partial fraction expansion method, the inverse z-transform of the fifth-order transfer function given in Example 4B.2.

The partial fraction expansion for the transfer function has the form

$$X(z) = B_0 + \sum_{k=1}^{5} \frac{C_k z}{z - p_k} \tag{4B.10}$$

The input data file for the example, coeff.data, is identical to that of Example 4B.2. The output of the program is summarized below:

poles of the z-transform

pk	real	imag	mag	phase
1	0.716754	0.586833	0.926342	39.308436
2	0.716754	−0.586833	0.926342	−39.308436
3	0.646801	0.372261	0.746277	29.922232
4	0.646801	−0.372261	0.746277	−29.922232
5	0.612159	0.000000	0.612159	0.000000

partial fraction coeffs

B0 = −3.418163

Ck	real	imag	mag	phase
1	1.611473	5.209672	5.453212	72.811944
2	1.611473	−5.209672	5.453212	−72.811943
3	−19.580860	−9.681908	21.843751	−153.689550
4	−19.580861	9.681908	21.843751	153.689551
5	40.356939	0.000000	40.356939	0.000000

From these values, the inverse z-transform is given by

$$x(n) = B_0 u(n) + \sum_{k=1}^{5} C_k (p_k)^n, \quad n \geq 0$$

Table 4.1 may also be used to find the inverse z-transform, $x(n)$, in a form that combines the terms with complex conjugate poles. This is left as an exercise for the reader.

4B.3   Cascade-to-parallel structure conversion

The program can also be used for converting a z-transform from cascade to parallel, based on the principles described in Example 4.14.

The transfer function of a fourth-order discrete-time system, in cascade form, is given by

$$H(z) = \frac{N_1(z)N_2(z)}{D_1(z)D_2(z)}$$

where

$$D_1(z) = 1 + 0.052\ 921z^{-1} + 0.831\ 73z^{-2}$$

$$N_1(z) = 1 + 0.481\ 199z^{-1} + z^{-2}$$

$$D_2(z) = 1 - 0.304\ 609z^{-1} + 0.238\ 865z^{-2}$$

$$N_2(z) = 1 + 1.474\ 597z^{-1} + z^{-2}$$

Use the program to convert the transfer function from cascade into parallel structure. The input data file has the following form:

```
2
1   0.05292   0.83173
1   0.481199   1
1   -0.304609   0.238865
1   1.474597   1
```

This gives the following output from the program:

selected desired operation

0	for power series method of IZT
1	for partial fraction coeffs estimation
2	for cascade to parallel conversion
2	

poles of the *z*-transform

pk	real	imag	mag	phase
1	-0.026460	0.911413	0.911797	91.662967
2	-0.026460	-0.911413	0.911797	-91.662967
3	0.152305	0.464401	0.488738	71.842631
4	0.152305	-0.464401	0.488738	-71.842631

partial fraction coeffs

B0=5.035604

Ck	real	imag	mag	phase
1	-0.257338	0.421333	0.493705	121.415410
2	-0.257338	-0.421333	0.493705	-121.415409
3	-1.760464	-3.766287	4.157421	-115.052650
4	-1.760464	3.766287	4.157421	115.052650

press enter to continue

stage	Ni(z)	
0	-0.514677	-0.781635
1	-3.520927	4.034388
2	0.000000	-0.000000

stage	Di(z)		
0	1.000000	0.052921	0.831373
1	1.000000	-0.304609	0.238865
2	0.000000	0.000000	0.000000

4C C program for estimating frequency response

The program computes the frequency response using either the direct estimation method or via the FFT as described in Section 4.5.5. The z-transform of the system whose frequency response is to be estimated must be in either direct or cascade form. An example will make this clear.

Example 4C.1

Obtain the frequency response of the discrete-time system whose transfer function is given by

$$H(z) = \frac{1 - 1.6180z^{-1} + z^{-2}}{1 - 1.5161z^{-1} + 0.878z^{-2}}$$

using

(1) the direct estimation method and

(2) the FFT approach.

Assuming a sampling frequency of 500 Hz and a resolution of < 1 Hz.

Solution

To satisfy the desired resolution, the number of frequency points to use with the program, npt, is 512 for the FFT approach ($500/512 = 0.98$ Hz) and 256 for the direct estimation. With the input data file

```
1
1   −1.5161   0.878
1   −1.618    1
```

the frequency response was obtained for each method. In either case, the response is held in ASCII format in three files as follows:

magn.dat	contains the magnitude response in decibels
phase.dat	contains the phase response in radians
fresp.dat	contains the frequency response in rectangular form

The first 10 values of the magnitude and phase responses are listed in Table 4C.1 and the magnitude and phase response for the direct method are depicted in Figures 4C.1(a) and 4C.1(b) respectively.

The program for frequency response consists of five functions held in separate files as follows:

freqres1.c	main function
fixdata.c	computes the magnitude and phase angles
freqd.c	direct frequency response estimation
fft.c	radix-2 decimation in time FFT algorithm
ltilib.c	a collection of common DSP functions. This is identical to ltilib.c described in Appendix 4B except that it does not require a .h file

The functions are not listed here because of lack of space, but can be found on the CD in the companion handbook (see the Preface for details).

Table 4C.1 The first 10 values of the magnitude and phase responses for Example 4C.1, using direct estimation or FFT approach.

k	Direct estimation		FFT estimation	
	Magnitude (dB)	Phase (rad)	Magnitude (dB)	Phase (rad)
0	0.469 496	0.000 000	0.469 496	0.000 000
1	0.469 39	−0.004 155	0.469 391	−0.004 138
2	0.469 073	−0.008 318	0.469 076	−0.008 286
3	0.468 541	−0.012 500	0.468 549	−0.012 451
4	0.467 791	−0.016 710	0.467 805	−0.016 644
5	0.466 817	−0.020 956	0.466 839	−0.020 873
6	0.465 612	−0.025 249	0.465 643	−0.025 148
7	0.464 165	−0.029 599	0.464 208	−0.029 479
8	0.462 466	−0.034 016	0.462 523	−0.033 876
9	0.460 501	−0.038 511	0.460 574	−0.038 351

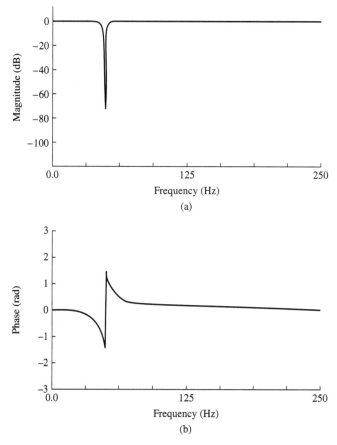

Figure 4C.1 (a) Magnitude–frequency response of the IIR system for Example 4C.1 using direct estimation. (b) Phase response of the IIR system for Example 4C.1 using direct estimation.

4D   z-transform operations with MATLAB

In the next few sections, we will illustrate the use of MATLAB functions for performing a variety of z-transform and inverse z-transform operations to complement the use of the C language programs described in previous sections. MATLAB and MATLAB Signal Processing Toolbox provide a fast and convenient means of performing a variety of z-transform and inverse z-transform operations for DSP systems design and analysis.

Further illustrative examples of z-transform operations using MATLAB can be found in the companion manual, *A Practical Guide for MATLAB and C Language Implementations of DSP Algorithms*, published by Pearson Education.

4D.1   Inverse z-transform

The key MATLAB functions for performing inverse z-transform operations are the **deconv** and **residuez**. The **deconv** function is used to perform the long division required in the power series method. The **residuez** function is used to find the partial fraction coefficients (residues) and poles of the z-transform.

4D.1.1   Power series expansion with MATLAB

In the power series method, the key operation is polynomial division. The MATLAB function **deconv** performs the deconvolution operation. In the power series method, we exploit the fact that the deconvolution operation is equivalent to polynomial division. Thus, given a z-transform, $X(z)$, of the form:

$$X(z) = \frac{b_0 + b_1 z^{-1} + \ldots + b_n z^{-n}}{a_0 + a_1 z^{-1} + \ldots + a_m z^{-m}} = \frac{b(z)}{a(z)}$$

the format of the command for deconvolution is

[q, r] = deconv(b,a)

where b and a are vectors representing the numerator and denominator polynomials, $b(z)$ and $a(z)$, respectively, in increasing negative powers of z. The quotient of the polynomial division is returned in the vector q and the remainder is contained in r. To implement the power series method, the long division operation is applied successively depending on the number of points required in the inverse operation.

Example 4D.1

Find the first five terms of the inverse z-transform, $x(n)$, using the power series (polynomial division) method and MATLAB. Assume that the z-transform, $X(z)$, has the following form:

$$X(z) = \frac{1 + 2z^{-1} + z^{-2}}{1 - z^{-1} + 0.3561z^{-2}}$$

Solution   The set of MATLAB commands and the answer are given below. First, the coefficient vectors for the numerator and denominator polynomials are formed, zeros are appended to the coefficient vector b to ensure the correct dimension for MATLAB, and then the **deconv** command is used to compute the inverse z-transform.

```
»
» b=[1 2 1];
» a=[1 −1 0.3561];
» n=5;
» b=[b zeros(1, n−1)];
» [x, r]=deconv(b,a);
» disp(x)
```

    1.0000   3.0000   3.6439   2.5756   1.2780

Thus, $x(0) = 1$, $x(1) = 3$, $x(2) = 3.6439$, $x(4) = 2.5756$, and $x(5) = 1.2780$.

Example 4D.2   Find the first five values of the inverse z-transform of the following using the power series (polynomial division) method and MATLAB:

$$X(z) = \frac{N_1(z)N_2(z)N_3(z)}{D_1(z)D_2(z)D_3(z)}$$

where

$$N_1(z) = 1 - 1.223\ 46z^{-1} + z^{-2}$$

$$N_2(z) = 1 - 0.437\ 833z^{-1} + z^{-2}$$

$$N_3(z) = 1 + z^{-1}$$

$$D_1(z) = 1 - 1.433\ 509z^{-1} + 0.858\ 11z^{-2}$$

$$D_2(z) = 1 - 1.293\ 601z^{-1} + 0.556\ 929z^{-2}$$

$$D_3(z) = 1 - 0.612\ 159z^{-1}$$

Solution   The z-transform has three pairs of numerator and denominator polynomials. In the MATLAB implementation (Program 4D.1), the vectors containing the polynomial coefficients are first formed. The MATLAB function **sos2tf** (second order sections-to-transfer function) is then used to convert the three pairs of polynomials into a transfer function with a pair of rational polynomials, $b(z)/a(z)$:

$$X(z) = \frac{b(z)}{a(z)} = \frac{b_0 + b_1z^{-1} + b_2z^{-2} + \ldots + b_mz^{-m}}{a_0 + a_1z^{-1} + a_2z^{-2} + \ldots + a_nz^{-n}}$$

The **deconv** function is used to generate the inverse z-transform coefficients. The first five values of the inverse z-transform are

$$x(0) = 1.0000, x(1) = 4.6915, x(2) = 11.4246, x(3) = 19.5863, x(4) = 27.0284$$

Program 4D.1

```
n = 5;   % number of power series points
N1 = [1 −1.122346 1]; D1 = [1 −1.433509 0.85811];
N2 = [1 1.474597 1]; D2 = [1 −1.293601 0.556929];
N3 = [1 1 0]; D3 = [1 −0.612159 0];
B = [N1; N2; N3]; A = [D1; D2; D3];
[b,a] = sos2tf([B A]);
b = [b zeros(1,n−1)];
[x,r] = deconv(b,a);   %perform long division
disp(x);
```

4D.1.2   Partial fraction expansion with MATLAB

The MATLAB function **residuez** may be used to perform partial fraction expansion of a z-transform, $X(z)$, expressed as a ratio of two polynomials. The syntax for the **residuez** command is

[r, p, k] = residuez(b, a)

where b and a are vectors representing the numerator and denominator polynomials, $b(z)$ and $a(z)$, respectively, in increasing negative powers of z as follows:

$$H(z) = \frac{b(z)}{a(z)} = \frac{b_0 + b_1 z^{-1} + b_2 z^{-2} + \ldots + b_m z^{-m}}{a_0 + a_1 z^{-1} + a_2 z^{-2} + \ldots + a_n z^{-n}}$$

If the poles of $H(z)$ are distinct, its partial fraction expansion has the form

$$\frac{b(z)}{a(z)} = \frac{r_1}{1 - p_1 z^{-1}} + \ldots + \frac{r_n}{1 - p_n z^{-1}} + k_1 + k_2 z^{-1} + \ldots + k_{m-n-1} z^{-(m-n)}$$

The **residuez** function returns the residues of the rational polynomial $b(z)/a(z)$ in the vector r, the pole positions in p, and the constant terms in k.

Example 4D.3

Find the partial fraction expansion of the following z-transform.

$$X(z) = \frac{1 + 2z^{-1} + z^{-2}}{1 - z^{-1} + 0.3561z^{-2}}$$

Solution

In the example, the polynomials are already in the correct format and so we can apply the command directly as

» [r, p, k] = residuez([1,2,1], [1, −1, 0.3561])

```
r =
  −0.9041 − 5.9928i
  −0.9041 + 5.9928i

p =
   0.5000 + 0.3257i
   0.5000 − 0.3257i

k = 2.8082
```

Thus, the z-transform, expressed as a partial fraction expansion, becomes

$$X(z) = 2.8082 + \frac{r_1}{1 - p_1 z^{-1}} + \frac{r_2}{1 - p_2 z^{-1}}$$

where

$$r_1 = -0.9041 - 5.9928j \qquad r_2 = -0.9041 + 5.9928j$$
$$p_1 = 0.5 + 0.3257j \qquad p_2 = 0.5 - 0.3257j$$

Example 4D.4   Find the partial fraction expansion of the following z-transform using MATLAB:

$$X(z) = \frac{N_1(z)N_2(z)N_3(z)}{D_1(z)D_2(z)D_3(z)}$$

where

$$N_1(z) = 1 - 1.223\ 46z^{-1} + z^{-2}$$
$$N_2(z) = 1 - 0.437\ 833z^{-1} + z^{-2}$$
$$N_3(z) = 1 + z^{-1}$$
$$D_1(z) = 1 - 1.433\ 509z^{-1} + 0.858\ 11z^{-2}$$
$$D_2(z) = 1 - 1.293\ 601z^{-1} + 0.556\ 929z^{-2}$$
$$D_3(z) = 1 - 0.612\ 159z^{-1}$$

The MATLAB function **sos2tf** is used to convert the numerator and denominator polynomials into a single pair of polynomials, $b(z)/a(z)$. The **residuez** function is then used to find the partial fraction expansion. The set of MATLAB commands for finding the partial fraction expansion of $X(z)$ is shown in Program 4D.2. Running the MATLAB program gives the partial fraction coefficients:

```
r =
   -1.9022 + 4.6797i
   -1.9022 - 4.6797i
   -9.0607 - 13.5515i
   -9.0607 + 13.5515i
   24.7049

p =
   0.7168 + 0.5868i
   0.7168 - 0.5868i
   0.6468 + 0.3723i
   0.6468 - 0.3723i
   0.6122

k = 1
```

Program 4D.2

```
N1 = [1 -1.122346 1];
N2 = [1 -0.437833 1];
N3 = [1 1 0];
D1 = [1 -1.433509 0.85811];
D2 = [1 -1.293601 0.556929];
D3 = [1 -0.612159 0];
sos = [N1 D1; N2 D2; N3 D3];
[b, a] = sos2tf(sos);
[r, p, k] = residuez(b, a)
```

4D.2 Conversion between structures – cascade-to-parallel conversion

MATLAB provides a set of functions that allow the conversion between different formats and structures that are used in DSP relatively easily. The ability to convert between the parallel and cascade structures is particularly useful.

Example 4D.5

Repeat Example 4B.4 using MATLAB.

The set of MATLAB commands for carrying out the conversion is given in Program 4D.3.

Program 4D.3

```
nstage=2;
N1 = [1 0.481199 1];
N2 = [1 1.474597 1];
D1 = [1 0.052921 0.83173];
D2 = [1 -0.304609 0.238865];
sos = [N1 D1; N2 D2];
[b, a] = sos2tf(sos);
[c, p, k] = residuez(b, a);
m = length(b);
b0 = b(m)/a(m);
j=1;
for i=1:nstage
        bk(j)=c(j)+c(j+1);
        bk(j+1)=-(c(j)*p(j+1)+c(j+1)*p(j));
        ak(j)=-(p(j)+p(j+1));
        ak(j+1)=p(j)*p(j+1);
        j=j+2;
end
b0
ak
bk
c
```

```
p
k
====
cprealization

b0 =
    5.0334
ck =
   -0.3766 - 0.2460i
   -0.3766 + 0.2460i
    1.4804 - 1.3903i
    1.4804 + 1.3903i

pk =
   -0.0265 + 0.9116i
   -0.0265 - 0.9116i
    0.1523 + 0.4644i
    0.1523 - 0.4644i

ks = 1

b0 = 1.2023

ak = 1.4746    1.0000   0.0529   0.8317

bk = -0.0000   -0.0000   0.4283   0.1683

c =
   -0.0000 + 0.0000i
   -0.0000 - 0.0000i
    0.2141 - 0.0861i
    0.2141 + 0.0861i
```

4D.3   Pole–zero diagram

The MATLAB function, **zplane**, allows the computation and display of the pole–zero diagram. The syntax for the command is

zplane(b, a)

where b and a are the coefficient vectors of the numerator and denominator polynomials, $b(z)/a(z)$. In this format, the command first finds the locations of the poles and zeros (i.e. the roots of $b(z)$ and $a(z)$, respectively) and then plots the *z*-plane diagram.

Example 4D.6   A discrete-time system is characterized by the following transfer function:

$$H(z) = \frac{1 - 1.6180z^{-1} + z^{-2}}{1 - 1.5161z^{-1} + 0.878z^{-2}}$$

Obtain and plot its pole–zero diagram. Use MATLAB in each case and assume a sampling frequency of 500 Hz and a resolution of < 1 Hz for the frequency response.

Solution   The MATLAB commands are

```
b = [1 −1.6180 1];          % form numerator and denominator polynomials
a = [1 −1.5161 0.878];
zplane(b,a)                  % compute and plot the pole–zero diagram
```

The pole–zero diagram is given in Figure 4D.1.

If the locations of the poles and zeros are known, these can be used as inputs to the **zplane** command. The syntax of the command in this case is **zplane(z, p)**, where z and p are the zeros and poles.

The locations of the poles and zeros can be found directly using the **roots** command. This is useful for converting between pole and zero and the transfer function representations.

For example, an IIR filter is represented by

$$H(z) = \frac{1 - 1.6180z^{-1} + z^{-2}}{1 - 1.5161z^{-1} + 0.878z^{-2}}$$

The poles and zeros of the filter can be found using the roots command as follows:

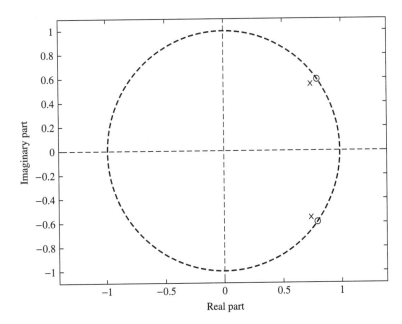

Figure 4D.1

```
b = [1 –1.618 1];
a = [1 –1.5161 0.878];
zk = roots(b);
pk=roots(a)
```

The numerator and denominator polynomials, $b(z)$ and $a(z)$, can be obtained using the **poly** function: **B = poly(zk); A=poly(pk)**.

4D.4   Frequency response estimation

The Signal Processing Toolbox contains many useful functions for computing and displaying the frequency response of discrete-time systems. The most widely used is the **freqz** function. Given the transfer function of a system in the following form:

$$X(z) = \frac{b_0 + b_1 z^{-1} + \ldots + b_n z^{-n}}{a_0 + a_1 z^{-1} + \ldots + a_m z^{-m}} = \frac{b(z)}{a(z)}$$

the **freqz** function uses an FFT-based approach to compute the frequency response. The function has a variety of formats. A useful format is **[h,f] = freqz(b, a, npt, Fs)**, where the variables b and a are the vectors of the numerator and denominator polynomials. Fs is the sampling frequency and npt the number of frequency points between 0 and $F_s/2$. In the MATLAB Toolbox, the Nyquist frequency (i.e. $F_s/2$) is the unit of normalized frequency. Using the **freqz** command without output arguments plots the magnitude and phase responses automatically. Example 4D.7 illustrates a way of using the **freqz** command to compute the frequency response of a discrete-time system.

| Example 4D.7 | A discrete-time system is characterized by the following transfer function: |

$$H(z) = \frac{1 - 1.6180z^{-1} + z^{-2}}{1 - 1.5161z^{-1} + 0.878z^{-2}}$$

Obtain and plot the frequency response of the system using MATLAB. Assume a sampling frequency of 500 Hz and a resolution of < 1 Hz.

Solution   The MATLAB commands are

```
b = [1 –1.6180 1];      % form numerator and denominator coefficient vectors
a = [1 –1.5161 0.878];
freqz(b,a,256,500)      % compute and plot the frequency response
```

The frequency response of the discrete-time system is shown in Figure 4D.2.

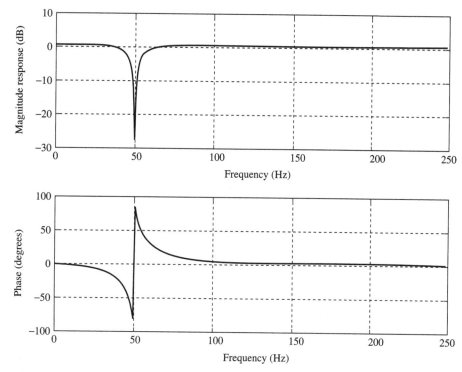

Figure 4D.2

References for Appendices

Jury E.I. (1964) *Theory and Applications of the z-transform Method*. New York: Wiley.
Signal Processing Toolbox User's Guide. The Math Works, 1998.

Correlation and convolution

5

The nature of the correlation process is first described in this chapter followed by an explanation using worked examples of the calculation of cross- and autocorrelations. The attenuating effects of correlation on the noise content of signals is described, as are a number of applications of correlation. The technique of fast correlation utilizing the FFT is then explained. The topic of convolution is covered in a similar manner to correlation. The treatment includes circular and linear convolution, fast linear convolution, and the sectioning methods (overlap–add, overlap–save) needed to handle large amounts of input data. Deconvolution is also included. The relationship between correlation and convolution is established. The chapter finishes with a section on implementation and some worked application examples.

5.1 Introduction

It is frequently necessary to be able to quantify the degree of interdependence of one process upon another, or to establish the similarity between one set of data and another. In other words, the correlation between the processes or data is sought. Correlation can be defined mathematically and can be quantified. The process of correlation occupies a significant place in signal processing. Applications are found in image processing for robotic vision or remote sensing by satellite in which data from different images is compared, in radar and sonar systems for range and position finding in which transmitted and reflected waveforms are compared, in the detection and identification of signals in noise, in control engineering for observing the effect of inputs on outputs, in the identification of binary codewords in pulse code modulation systems using correlation detectors, as an integral part of the ordinary least squares estimation technique, in the computation of the average power in waveforms, and in many other fields, such as, for example, climatology. Correlation is also an integral part of the process of convolution. The convolution process is essentially the correlation of two data sequences in which one of the sequences has been reversed. This means that the same algorithms may be used to compute correlations and convolutions simply by reversing one of the sequences. The process of convolution gives the output from a system which filters the input. The spectrum of a recorded signal consists of the convolution of the spectrum of the signal with the spectrum of its window function.

The determination of an unknown system impulse response is known as system identification. The determination of an unknown input from the system impulse response and the output signal is known as deconvolution. When the impulse response is unknown, the determination of the unknown input signal is known as blind deconvolution. Each of these important topics is described.

5.2 Correlation description

Consider how two data sequences, each consisting of simultaneously sampled values taken from the two corresponding waveforms, might be compared. If the two waveforms varied similarly point for point, then a measure of their correlation might be obtained by taking the sum of the products of the corresponding pairs of points. This proposal becomes more convincing when the case of two independent and random data sequences is considered. In this case the sum of the products will tend towards a vanishingly small random number as the number of pairs of points is increased. This is because all numbers, positive and negative, are equally likely to occur so that the product pairs tend to be self-cancelling on summation. By contrast, the existence of a finite sum will indicate a degree of correlation. A negative sum will indicate negative correlation, that is an increase in one variable is associated with a decrease in the other variable. The cross-correlation $r_{12}(n)$ between two data sequences $x_1(n)$ and $x_2(n)$ each containing N data might therefore be written as

$$r_{12} = \sum_{n=0}^{N-1} x_1(n)x_2(n)$$

This definition of cross-correlation, however, produces a result which depends on the number of sampling points taken. This is corrected for by normalizing the result to the number of points by dividing by N. Alternatively this may be regarded as averaging the sum of products. Thus, an improved definition is

$$r_{12} = \frac{1}{N} \sum_{n=0}^{N-1} x_1(n)x_2(n)$$

Example 5.1

The calculation of r_{12} is illustrated in the following example, in which the point numbers in the data sequences are the n, and the sequences are x_1 and x_2.

n	1	2	3	4	5	6	7	8	9
x_1	4	2	−1	3	−2	−6	−5	4	5
x_2	−4	1	3	7	4	−2	−8	−2	1

$$r_{12} = \frac{1}{9}(4 \times -4 + 2 \times 1 + -1 \times 3 + 3 \times 7 + -2 \times 4 + -6 \times -2 + -5 \times -8 +$$

$$4 \times -2 + 5 \times 1)$$

$$= 5$$

However, this definition needs modification to be useful. In some cases it may indicate zero correlation although the two waveforms are 100% correlated. This may occur, for example, when the two waveforms are out of phase, which will often be the case. The situation is illustrated by the waveforms of Figure 5.1. From this figure it is seen that each pair product in the correlation is zero, and hence the correlation is zero, because one of either x_1 or x_2 is always zero. However, the waveforms are clearly highly correlated, although they are out of phase. The phase difference could, for example, occur because x_1 is the reference signal while x_2 is the delayed output from a circuit. To overcome such phase differences it is necessary to shift, or lag, one of the waveforms with respect to the other. Typically x_2 is shifted to the left to align the

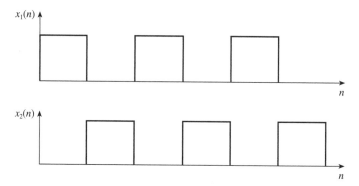

Figure 5.1   Out-of-phase 100% correlated waveforms with zero correlation at lag zero.

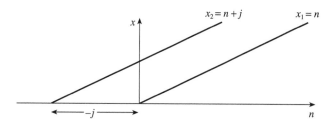

Figure 5.2 Waveform $x_2 = x_1 + j$ shifted j lags to the left of waveform x_1.

waveforms prior to correlation. As illustrated in Figure 5.2 this is equivalent to changing $x_2(n)$ to $x_2(n + j)$, where j represents the amount of lag which is the number of sampling points by which x_2 has been shifted to the left. An alternative, but equivalent, procedure is to shift x_1 to the right. The formula for the cross-correlation thus becomes

$$r_{12}(j) = \frac{1}{N} \sum_{n=0}^{N-1} x_1(n) x_2(n + j)$$

$$= r_{21}(-j) = \frac{1}{N} \sum_{n=0}^{N-1} x_2(n) x_1(n - j) \tag{5.1}$$

In practice when two waveforms are correlated their phase relationship will probably not be known and so the correlation will be computed for a number of different lags in order to establish the largest value of the correlation which is then taken to be the correct value.

| Example 5.2 | Consider the cross-correlation of the above two sequences $x_1(n)$ and $x_2(n)$ at a lag of $j = 3$, that is consider $r_{12}(3)$. The two sequences become |

n	1	2	3	4	5	6	7	8	9
x_1	4	2	-1	3	-2	-6	-5	4	5
x_2	7	4	-2	-8	-2	-1			

so

$$r_{12}(3) = \frac{1}{9}(4 \times 7 + 2 \times 4 + -1 \times -2 + 3 \times -8 + -2 \times -2 + -6 \times -1)$$

$$= 2.667$$

Of course, it is also possible to consider correlation in the continuous time domain, and some analog signal correlation is implemented this way. In the continuous domain $n \rightarrow t$ and $j \rightarrow \tau$ and

$$r_{12}(\tau) = \lim_{T \to \infty} \frac{1}{T} \int_{-T/2}^{T/2} x_1(t) x_2(t + \tau)\, dt \tag{5.2}$$

However, if $x_1(t)$ and $x_2(t)$ are periodic with period T_0 Equation 5.2 simplifies to

$$r_{12}(\tau) = \frac{1}{T_0} \int_{-T_0/2}^{T_0/2} x_1(t) x_2(t + \tau)\, dt \tag{5.3}$$

If the waveforms are finite energy waveforms, for example nonperiodic pulse-type waveforms, then the average evaluated over time T as $T \to \infty$ is not taken because then $1/T \to 0$ and $r_{12}(\tau)$ is always vanishingly small. For this case Equation 5.4 is used in principle:

$$r_{12}(\tau) = \int_{-\infty}^{\infty} x_1(t) x_2(t + \tau)\, dt \tag{5.4}$$

In practice, a finite record length will be processed and so Equation 5.5 or 5.1 will be applied:

$$r_{12}(\tau) = \frac{1}{T} \int_{0}^{T} x_1(t) x_2(t + \tau)\, dt \tag{5.5}$$

There is another difficulty associated with cross-correlating finite lengths of data. This can be seen in the above example in which $r_{12}(3) = 2.667$ was determined. As x_2 is shifted to the left the waveforms no longer overlap and data at the ends of the sequences no longer form pair products. This is known as the end effect. In the example the number of pairs has dropped from nine to six for a lag of three. The result is a linear decrease in $r_{12}(j)$ as j increases, leading to debatable values of $r_{12}(j)$. One possible solution is to make one of the sequences twice as long as the required length for correlation. This could be achieved by recording more data, or, if one of the sequences were periodic, by repeating the sequence (taking care to match the two ends). Another possibility is to add a correction to all computed values. Figure 5.3 shows how $r_{12}(j)$ decreases with j purely as a result of the end effect, that is actual variations in $r_{12}(j)$ are not included. At $j = 0$, $r_{12}(j) = r_{12}(0)$, which can be computed. At $j = N$, $r_{12}(N) = 0$, because the waveforms no longer overlap. In between, at some lag j, the true value of $r_{12}(j)$ is $r_{12}(j)_{\text{true}}$ while the actual value caused by the end effect is $r_{12}(j)$. Then, from the figure

$$\frac{r_{12}(j)_{\text{true}} - r_{12}(j)}{j} = \frac{r_{12}(0)}{N}$$

whence

$$r_{12}(j)_{\text{true}} = r_{12}(j) + \frac{j}{N} r_{12}(0) \tag{5.6}$$

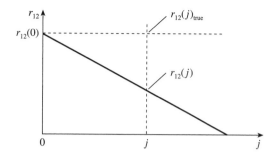

Figure 5.3 The effect of the end-effect on the cross-correlation $r_{12}(j)$.

Computed values of the cross-correlation are therefore easily corrected for end effects by adding $jr_{12}(0)/N$ to the values of $r_{12}(j)$.

The cross-correlation values computed according to the above formulae depend on the absolute values of the data. It is often necessary to measure cross-correlations according to the fixed scale between -1 and $+1$. This can be achieved by normalizing the values by an amount depending on the energy content of the data. For example, consider the two pairs of waveforms $x_1(n)$, $x_2(n)$, and $x_3(n)$, $x_4(n)$. The data values are given in the table below:

n	0	1	2	3	4	5	6	7	8
$x_1(n)$	0	3	5	5	5	2	0.5	0.25	0
$x_2(n)$	1	1	1	1	1	0	0	0	0
$x_3(n)$	0	9	15	15	15	6	1.5	0.75	0
$x_4(n)$	2	2	2	2	2	0	0	0	0

As may be seen from Figure 5.4, waveforms $x_1(n)$ and $x_3(n)$ are alike, differing only in magnitude. The same is true of the pair $x_2(n)$ and $x_4(n)$. The correlation between $x_1(n)$ and $x_2(n)$ is therefore the same as that between $x_3(n)$ and $x_4(n)$. However, the cross-correlations $r_{12}(1)$ and $r_{34}(1)$ are 1.47 and 8.83 respectively. They are different because they depend on the absolute values of the data. This situation can be rectified by normalizing the cross-correlation $r_{12}(j)$ by the factor

$$\left[\frac{1}{N} \sum_{n=0}^{N-1} x_1^2(n) \times \frac{1}{N} \sum_{n=0}^{N-1} x_2^2(n) \right]^{1/2} = \frac{1}{N} \left[\sum_{n=0}^{N-1} x_1^2(n) \sum_{n=0}^{N-1} x_2^2(n) \right]^{1/2} \qquad (5.7)$$

and similarly for $r_{34}(j)$. The normalized expression for $r_{12}(j)$ then becomes

$$\rho_{12}(j) = \frac{r_{12}(j)}{\dfrac{1}{N} \left[\displaystyle\sum_{n=0}^{N-1} x_1^2(n) \sum_{n=0}^{N-1} x_2^2(n) \right]^{1/2}} \qquad (5.8)$$

$\rho_{12}(j)$ is known as the cross-correlation coefficient. Its value always lies between -1 and $+1$. $+1$ means 100% correlation in the same sense, -1 means 100% correlation in

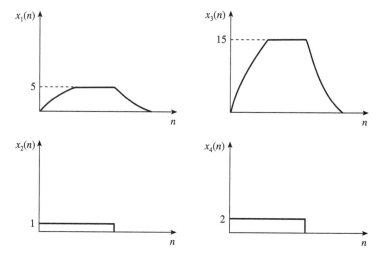

Figure 5.4   Pairs of waveforms $\{x_1(n), x_2(n)\}$, $\{x_3(n), x_4(n)\}$ of different magnitudes but equal cross-correlations.

the opposing sense, for example signals in antiphase. A value of 0 signifies zero correlation. This means the signals are completely independent. This would be the case, for example, if one of the waveforms were completely random. Small values of $\rho_{12}(j)$ indicate very low correlation. The normalizing factor for $r_{12}(j)$ in the above illustration is

$$\frac{1}{N}\left[\sum_{n=0}^{N-1} x_1^2(n) \sum_{n=0}^{N-1} x_2^2(n)\right]^{1/2} = \frac{1}{9}(88.31 \times 6)^{1/2} = 2.56$$

and for $r_{34}(j)$ it is

$$\frac{1}{N}\left[\sum_{n=0}^{N-1} x_3^2(n) \sum_{n=0}^{N-1} x_4^2(n)\right]^{1/2} = \frac{1}{9}(794.8 \times 24)^{1/2} = 15.35$$

Therefore

$$\rho_{12}(1) = \frac{r_{12}(1)}{2.56} = \frac{1.47}{2.56} = 0.57$$

and

$$\rho_{34}(1) = \frac{r_{34}(1)}{15.34} = \frac{8.83}{15.35} = 0.58$$

Now $\rho_{12}(1) = \rho_{34}(1)$ which demonstrates that this normalization process indeed allows a comparison of cross-correlations independently of the absolute data values.

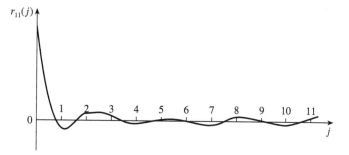

Figure 5.5   Autocorrelation function of a random waveform.

A special case occurs when $x_1(n) = x_2(n)$. The waveform is then cross-correlated with itself. This process is known as autocorrelation. The autocorrelation of a waveform is given by

$$r_{11}(j) = \frac{1}{N} \sum_{n=0}^{N-1} x_1(n)x_1(n+j)$$

The autocorrelation function has one very useful property in that

$$r_{11}(0) = \frac{1}{N} \sum_{n=0}^{N-1} x_1^2(n) = S$$

where S is the normalized energy of the waveform. This provides a method for calculating the energy of a signal. If the waveform is completely random, for example corresponding to that of white, gaussian noise in an electrical system, then the autocorrelation will have its peak value at zero lag and will reduce to a random fluctuation of small magnitude about zero for lags greater than about unity (see Figure 5.5). This constitutes a test for random waveforms. This topic will be more fully covered in Section 5.2.1. It is also true that

$$r_{11}(0) \geqslant r_{11}(j)$$

5.2.1 Cross- and autocorrelation

Care has to be exercised when cross-correlating two unequal length sequences when they are periodic. This is because the result of the correlation will be cyclic with the period of the shorter sequence. This result does not represent the full periodicity of the longer sequence and is, therefore, incorrect. This may be demonstrated by cross-correlating the sequences $a = \{4, 3, 1, 6\}$ and $b = \{5, 2, 3\}$ to obtain $r_{ab}(j)$. The sequence b is placed below sequence a, and b is shifted left by one lag on each of the subsequent rows, with the value of the cross-correlation appearing in the final column on the right.

Sequence				Lag	$r_{ab}(j)$
4	3	1	6		
3	5	2	3	0	47
5	2	3	5	1	59
2	3	5	2	2	34
3	5	2	3	3	47 $r_{ab}(j)$ repeats
5	2	3	5	4	59
etc.					

The result shows that $r_{ab}(j)$ is cyclic, repeating every third lag, that is $r_{ab}(j)$ has the same period as that of the shorter sequence, b. This procedure is known as cyclic correlation. To obtain the correct value in which each value in a is multiplied by each value in b, all the elements in b have to be shifted in turn below each value in a as shown below:

```
        4 3 1 6
            5 2 3
          5 2 3
        5 2 3
      5 2 3
    5 2 3
  5 2 3
5 2 3
```

This is seen to require 6 lags before the b sequence repeats. The sequence lengths are 4 and 3 and the number of lags necessary is $4 + 3 - 1 = 6$. This reveals the general rule for obtaining the linear cross-correlation of two periodic sequences of lengths N_1 and N_2: add augmenting zeros to each sequence to make the lengths of each sequence $N_1 + N_2 - 1$. This may be expressed as adding $N_2 - 1$ zeros to the sequence of length N_1 and adding $N_1 - 1$ zeros to the sequence of length N_2. This is now demonstrated for the given sequences a and b:

Sequence						Lag	$r_{ab}(j)$
4	3	1	6	0	0		
5	2	3	0	0	0	0	29
2	3	0	0	0	5	1	17
3	0	0	0	5	2	2	12
0	0	0	5	2	3	3	30
0	0	5	2	3	0	4	17
0	5	2	3	0	0	5	35
5	2	3	0	0	0	6	29 $r_{ab}(j)$ repeats
etc.							

Thus, the required linear cross-correlation of a and b is

$$r_{ab}(j) = \{29, 17, 12, 30, 17, 35\}$$

So far, the instances of cross-correlation taken have all assumed digitized data, but cross-correlation may also be performed analytically when analytical expressions can be written for the waveforms, including when this requires sectioning of the waveforms. In practice the analytical procedure has its equivalent in the use of analog circuits to effect the cross-correlation. An example of analytical cross-correlation follows.

Example 5.3

Obtain the cross-correlation $r_{12}(-\tau)$ between the waveforms $v_1(t)$ and $v_2(t)$ of Figure 5.6.

It is easy to express the waveforms analytically by dividing them into straight-line sections. It is only necessary to do this over one period, T, of the waveforms because $r_{12}(-\tau)$ will be periodic in τ with period T. Therefore, for $0 \leqslant t \leqslant T$, $v_1(t) = t/T$, and for $0 \leqslant t \leqslant T/2$, $v_2(t) = 1.0$, while for $T/2 \leqslant t \leqslant T$, $v_2(t) = -1.0$. The requirement is to obtain an expression for $r_{12}(-\tau)$, that is $v_2(t)$, the rectangular waveform, is to be shifted right with respect to $v_1(t)$. For $0 \leqslant \tau \leqslant T/2$, the situation is described by Figure 5.7

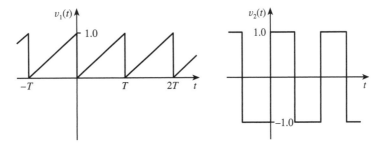

Figure 5.6   The waveform $v_1(t)$ and $v_2(t)$ for cross-correlation example.

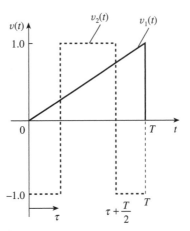

Figure 5.7   Sections of $v_2(t)$ for $\theta \leqslant \tau \leqslant T$.

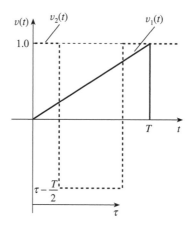

Figure 5.8   Sections of $v_2(t)$ for $T/2 \leqslant \tau \leqslant T$.

which shows that $v_1(t)$ has to be multiplied by three consecutive sections of $v_2(t)$ in which $v_2(t)$ has the consecutive values -1, 1, -1. For $T/2 \leqslant \tau \leqslant T$, Figure 5.8 applies in which the consecutive values of the set of $v_2(t)$ have changed to 1, -1, $+1$. This means there are two parts to the solution which must match at $\tau = T/2$.

Referring to Figure 5.7, the cross-correlation is split into the three sections with boundaries at $t = \tau$, $t = \tau + T/2$, and $t = T$. Hence

$$r_{12}(-\tau) = \frac{1}{T} \int_0^T v_1(t)v_2(t - \tau)\, \mathrm{d}t$$

$$= \frac{1}{T} \int_0^\tau \frac{t}{T}(-1)\, \mathrm{d}t + \frac{1}{T} \int_\tau^{\tau+T/2} \frac{t}{T}(1)\, \mathrm{d}t + \frac{1}{T} \int_{\tau+T/2}^T \frac{t}{T}(-1)\, \mathrm{d}t$$

$$= \frac{-1}{T^2}\left[\frac{t^2}{2}\right]_0^\tau + \frac{1}{T^2}\left[\frac{t^2}{2}\right]_\tau^{\tau+T/2} - \frac{1}{T^2}\left[\frac{t^2}{2}\right]_{\tau+T/2}^T$$

$$r_{12}(-\tau) = -\frac{1}{4} + \frac{\tau}{T} \quad \text{for } 0 \leqslant \tau \leqslant \frac{T}{2} \tag{5.9}$$

For $T/2 \leqslant \tau \leqslant T$, and referring to Figure 5.8, it is seen that

$$r_{12}(-\tau) = \frac{1}{T} \int_0^{\tau-T/2} \frac{t}{T}(1)\, \mathrm{d}t + \frac{1}{T} \int_{\tau-T/2}^\tau \frac{t}{T}(-1)\, \mathrm{d}t + \frac{1}{T} \int_\tau^T \frac{t}{T}(1)\, \mathrm{d}t$$

$$r_{12}(-\tau) = \frac{3}{4} - \frac{\tau}{T} \quad \text{for } \frac{T}{2} \leqslant \tau \leqslant T \tag{5.10}$$

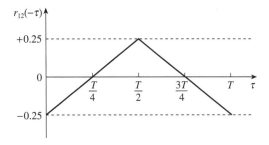

Figure 5.9 $r_{12}(-\tau)$ as a function of τ.

Substituting $\tau = T/2$ into Equations 5.9 and 5.10 gives $r_{12}(-\tau) = 1/4$ in both cases, confirming that the two functions match correctly. Figure 5.9 shows a plot of $r_{12}(-\tau)$ versus τ for $0 \leqslant \tau \leqslant T$.

It is of interest to give some consideration to the consequences of using finite lengths of data in the calculation of the correlation. In other words, what is the effect of using Equation 5.5, in which T is finite, instead of Equation 5.2?

This question can be answered by considering just one sinusoidal Fourier harmonic component of the signal. Equation 5.2 will give the correct autocorrelation, in which $T \gg T_p$, where T_p is the period of the sinusoid. Thus

$$r_{11}(\tau) = \lim_{T \to \infty} \frac{1}{2T} \int_{-T}^{T} A \sin(\omega t) A \sin(\omega t + \tau)\, \mathrm{d}t$$

$$= \lim_{T \to \infty} \frac{A^2}{2} \left[\cos(\omega \tau) - \frac{\cos(\omega T)}{2\omega T} \sin(\omega \tau) \right] \tag{5.11}$$

Inspection of this equation shows that the second term in the bracket $\to 0$ when $T \to \infty$, so when $T \neq \infty$ it represents an error. The $\cos(\omega T)$ term represents periodic error effects, while the term $1/2\omega T$ gives the trend in the error. Thus, as far as the correlation length, T, is concerned, the errors are greater the shorter the sequence, and are also largest for the lower frequency components of the waveform. The errors are also periodic in τ.

The $\cos(\omega T)$ term gives least errors when $\omega T = [(2n + 1)/2]\pi$. Since $\omega = 2\pi/T_p$ and large values of T are sought, this corresponds to

$$T \geqslant (2n + 1)\frac{T_p}{4} \tag{5.12}$$

The $\sin(\omega \tau)$ term is least when $\omega \tau = m\pi$, where m is integer. Hence,

$$\tau = \frac{m}{2} T_p \tag{5.13}$$

It is now necessary to make some reasonable assumptions. Assume the condition for large T is satisfied by $n \geqslant 10$. Then $T \geqslant nT_p/2$, or

$$T \geqslant 5T_p \tag{5.14}$$

From Equation 5.13, the largest allowable value of τ for the lowest frequency component ($m = 1$) satisfies

$$\tau < T_p \tag{5.15}$$

Combining Equations 5.14 and 5.15,

$$\tau \leqslant T/5$$

This means that when correlating waveforms the errors due to finite data lengths may be minimized by

(1)   ensuring that $T \geqslant 5T_p$, where T_p corresponds to the lowest frequency component of interest, and

(2)   overlapping the data by no more than 20% of their length.

Thus, for example, if telephone speech signals with a bandwidth of 300 Hz to 3.4 kHz and sampled at 40 kHz are to be correlated, $T_p = 1/300 = 3.3 \times 10^{-3}$ s. The least acceptable data length would be $5 \times 3.3 \times 10^{-3}$ s $= 16.7$ ms and the largest correlation shift would be 3.33 ms, or 133 data points.

Figure 5.10 shows the plot of $\rho_{11}(j)$, the autocorrelation coefficient of a purely random waveform, for example white noise. The expected value of $r_{11}(j)$ can be

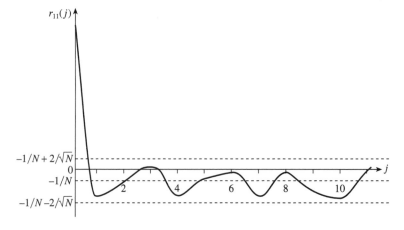

Figure 5.10   The autocorrelation coefficient of a random waveform.

shown to be $E[r_{11}(j)] \approx -1/N$ (Chatfield, 1980), where N is the number of data points, and its variance is $\text{var}[r_{11}(j)] \approx 1/N$. The expected value of $-1/N$ is shown on the figure as are the 95% confidence limits of $-1/N$, which are $\pm 2/N^{1/2}$. Values of $r_{11}(j)$ which fall outside these confidence limits may be significant, that is they may indicate that the waveform was not truly random. However, it should be noted that as many as one point in 20 may lie outside these limits even when the waveform is completely random. For a random waveform $r_{11}(j)$ should fall to within the 95% confidence limits within one or two lags. Experience and more sophistication is required to be sure that a waveform is random. For example, data pre-whitening may be advisable (Jenkins and Watts, 1968).

The autocorrelation function of a periodic waveform is itself a periodic waveform. This is easily proved as follows. The periodic waveform $x(t)$ of period T satisfies

$$x(t) = x(t + nT)$$

so,

$$r_{11}(\tau) = \lim_{T \to \infty} \frac{1}{T} \int_{-T/2}^{T/2} x(t)x(t + \tau)\, dt$$

$$= \lim_{T \to \infty} \frac{1}{T} \int_{-T/2}^{T/2} x(t)x(t + \tau + nT)\, dt$$

$$r_{11}(\tau) = r_{11}(\tau + nT) \tag{5.16}$$

Thus $r_{11}(\tau)$ is seen to be periodic in τ with period T. This is a useful property because it enables the detection of periodic signals in noise for small signal-to-noise ratios. Autocorrelating the waveform tends to reduce the noise while at the same time developing the periodic autocorrelation function of the signal. Once detected, further processing can be applied to determine its shape if this is required.

Equation 5.11 showed that the autocorrelation function of $A \sin(\omega t)$ is $(A^2/2) \cos(\omega \tau)$. In this case, as in others, the amplitude of the autocorrelation function is related simply to that of the signal, and may be used to estimate the signal amplitude. Another common example is that of the rectangular wave of amplitude A which the reader could show has a triangular autocorrelation function of amplitude A^2. Finally, it should be noted that autocorrelation functions are not unique. This means that a number of different waveforms may share the same autocorrelation function. Hence the shapes of waveforms should not be deduced from the detected autocorrelation functions.

Consider now the case in which the waveform, $v(t)$, is partially random. This represents the case of a noisy signal which may be written as the sum of a signal term, $s(t)$, and a noise term, $q(t)$. Thus

$$v(t) = s(t) + q(t) \tag{5.17}$$

$s(t)$ and $q(t)$ are assumed to be uncorrelated. The sampled autocorrelation function of $v(t)$ is $r_{vv}(j)$ given by

$$r_{vv}(j) = \frac{1}{N} \sum_{n=0}^{N-1} [s(n) + q(n)][s(n+j) + q(n+j)] \tag{5.18}$$

$$= \frac{1}{N} \sum_{n=0}^{N-1} s(n)s(n+j) + \frac{1}{N} \sum_{n=0}^{N-1} s(n)q(n+j) + \frac{1}{N} \sum_{n=0}^{N-1} q(n)s(n+j)$$

$$+ \frac{1}{N} \sum_{n=0}^{N-1} q(n)q(n+j) \tag{5.19}$$

$$= r_{ss}(j) + E[s(n)q(n+j)] + E[q(n)s(n+j)] + E[q(n)q(n+j)]$$

$$= r_{ss}(j) + E[s(n)]E[q(n+j)] + E[q(n)]E[s(n+j)] + E[q(n)]E[q(n+j)]$$

$$= r_{ss}(j) + \overline{s(n)}\,\overline{q(n)} + \overline{q(n)}\,\overline{s(n)} + \overline{q(n)}^2$$

$$= r_{ss}(j) + 2\bar{s}\bar{q} + \bar{q}^2 \tag{5.20}$$

Now, $\bar{q} \to 0$ for large N, for which

$$r_{vv}(j) \to r_{ss}(j) \tag{5.21}$$

For smaller N, the cross-correlation terms in Equation 5.19 and the autocorrelation of the noise tend towards zero with increasing lag j.

Thus it is seen that the autocorrelation function of a partially random, or noisy, waveform consists of the autocorrelation function of the signal component superimposed on a noisy decaying function which depends on both the random and signal components and which decays towards the value $2\bar{s}\bar{q} + \bar{q}^2$. Thus the plot of $r_{vv}(j)$ against j will display the periodicity of $s(t)$ provided $|r_{ss}(j)| > |(2\bar{s}\bar{q} + \bar{q}^2)|$: see Figure 5.11. This offers a method for identifying the period of a signal in noise (see Section 5.2.2).

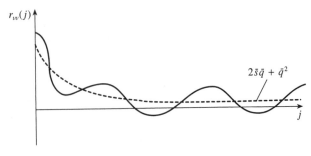

Figure 5.11   The autocorrelation function of a noisy signal.

Example 5.4   Derive the cross-correlation function of two noisy waveforms.

Let the two waveforms be $\{s_1(t) + q_1(t)\}$ and $\{s_2(t) + q_2(t)\}$. Their sampled cross-correlation, $r_{12}(j)$, is given by

$$r_{12}(j) = \frac{1}{N}\sum_{n=0}^{N-1}[\{s_1(n) + q_1(n)\}\{s_2(n + j) + q_2(n + j)\}] \tag{5.22}$$

$$= \frac{1}{N}\sum_{n=0}^{N-1}[s_1(n)s_2(n + j) + s_1(n)q_2(n + j) + q_1(n)s_2(n + j) + q_1(n)q_2(n + j)]$$

$$= \frac{1}{N}\sum_{n=0}^{N-1}s_1(n)s_2(n + j) + \frac{1}{N}\sum_{n=0}^{N-1}s_1(n)q_2(n + j) + \frac{1}{N}\sum_{n=0}^{N-1}q_1(n)s_2(n + j)$$

$$+ \frac{1}{N}\sum_{n=0}^{N-1}q_1(n)q_2(n + j)$$

$$= r_{s_1s_2}(j) + r_{s_1q_2}(j) + r_{q_1s_2}(j) + r_{q_1q_2}(j) \tag{5.23}$$

As in the previous case of autocorrelation the last three terms on the right-hand side of Equation 5.23 decay towards zero with increasing lag j. For large N, Equation 5.23 becomes

$$r_{12}(j) = r_{s_1s_2}(j) + \overline{s_1}\,\overline{q_2} + \overline{q_1}\,\overline{s_2} + \overline{q_1}\,\overline{q_2} \tag{5.24}$$

Thus as N increases $r_{12}(j) \rightarrow r_{s_1s_2}(j)$, the cross-correlation function of the two signals.

The above analyses illustrate that the cross- and autocorrelation processes emphasize signal properties by reducing the noise content.

5.2.2   Applications of correlation

5.2.2.1   Calculations of energy spectral density and energy content of waveforms

It can be shown that

$$F[r_{11}(\tau)] = G_E(f) \tag{5.25}$$

where $G_E(f)$ is the energy spectral density of the waveform, that is the energy spectral density and the autocorrelation function constitute a Fourier transform pair.

It can further be shown that

$$r_{11}(0) = E \tag{5.26}$$

where E is the total energy of the waveform.

Example 5.5   Obtain a relationship between the zero-lag correlation functions of two different waveforms and their total energy content.

Let the waveforms be $v_1(n)$ and $v_2(n)$, and let their summation be $V(n) = v_1(n) + v_2(n)$. The zero-lag autocorrelation function of $V(n)$ is

$$r_{vv}(0) = E_V = \frac{1}{N}\sum_{n=0}^{N-1} V^2(n) = \frac{1}{N}\sum_{n=0}^{N-1}[v_1(n) + v_2(n)]^2$$

where E_V is the energy of the waveform $V(n)$.

$$E_V = \frac{1}{N}\sum_{n=0}^{N-1}[v_1^2(n) + v_2^2(n) + 2v_1(n)v_2(n)]$$

$$= \frac{1}{N}\sum_{n=0}^{N-1} v_1^2(n) + \frac{1}{N}\sum_{n=0}^{N-1} v_2^2(n) + \frac{1}{N}\sum_{n=0}^{N-1} v_1(n)v_2(n)$$

so

$$E_V = r_{v_1}(0) + r_{v_2}(0) + 2r_{v_1v_2}(0) \tag{5.27}$$

Equation 5.27 is the first form of the required result. Alternatively it may be written as

$$E_V = E_{v_1} + E_{v_2} + 2r_{v_1v_2}(0) \tag{5.28}$$

Thus the energy of $V(n)$ equals the sum of the energies of its components plus $2r_{v_1v_2}(n)$, where $r_{v_1v_2}(n)$ is the zero-lag cross-correlation function of $v_1(n)$ and $v_2(n)$. If $v_1(n)$ and $v_2(n)$ are uncorrelated then the total energy is just the sum of the component energies.

If the signals $v_1(n)$ and $v_2(n)$ are noisy such that $v_1(n) = v_1'(n) + q_1(n)$ and $v_2(n) = v_2'(n) + q_2(n)$ then it is easy to show that

$$E_V = E_{v_1'} + E_{v_2'} + E_{q_1} + E_{q_2} + r_{v_1'v_2'}(0) \tag{5.29}$$

5.2.2.2   Detection and estimation of periodic signals in noise

The use of cross-correlation to detect and estimate periodic signals in noise will now be considered. The first proposal is that a signal buried in noise can be estimated by cross-correlating it with an adjustable 'template' signal. The template is adjusted by trial and error, guided by any foreknowledge, until the cross-correlation function has been maximized. This template is then the estimate of the signal. This proposal can be justified by referring to Equation 5.22 and assuming that for the template $q_2(n) = 0$. Equation 5.23 then becomes

$$r_{12}(j) = r_{s_1s_2}(j) + r_{q_1s_2}(j) \tag{5.30}$$

$$= r_{s_1s_2}(j) + \bar{q}_1\bar{s}_2 \tag{5.31}$$

Then, because $\bar{q}_1 \to 0$ as N increases,

$$r_{12}(j) \to r_{s_1 s_2}(j) \tag{5.32}$$

Clearly $r_{s_1 s_2}(j)$ will be a maximum when $s_2(n) = s_1(n)$ when $r_{s_1 s_2}(j)$ is the autocorrelation function of $s_1(n)$. Thus, changing the shape of the template $s_2(n)$ to maximize the cross-correlation function provides $s_2(n)$ as the estimate of $s_1(n)$.

The template method of signal estimation is convenient sometimes, for example when the shape of the signal is known approximately as for certain biomedically evoked potentials, but there is a more scientific approach that may be preferred. In this method the period of the signal is first estimated by autocorrelating the noisy waveform, and then the noisy waveform is cross-correlated with a periodic impulse train of period equal to that of the signal. The resulting cross-correlation function is the signal estimate.

Let the signal of period N_p points ($N_p < N$) be $s(n)$ and let the noise be $q(n)$ so that the noisy waveform is $S(n) = s(n) + q(n)$. Let $\delta(n - kN_p)$ be the periodic impulse train used for the cross-correlation and let N_δ be the number of impulses used in the cross-correlation. This is also equal to the number of periods of the signal over which the noise waveform is cross-correlated with the impulse train. Then

$$r_{S\delta}(-j) = \frac{1}{N_\delta} \sum_{n=0}^{N-1} [s(n) + q(n)]\delta(n - kN_p - j), \quad k = 0, 1, 2, \dots \tag{5.33}$$

For $j = 0$, and remembering that $\delta(n - kN_p) = 0$ for all $n \neq kN_p$,

$$r_{S\delta}(0) = \frac{1}{N_\delta}[s(0) + q(0) + s(N_p) + q(N_p) + s(2N_p) + q(2N_p) + \dots$$

$$+ s(N) + q(N)] \tag{5.34}$$

Now, because of the periodicity of the signal $s(n + kN_p) = s(n)$ and so Equation 5.34 becomes

$$r_{S\delta}(0) = \frac{1}{N_\delta}[Ns(0) + q(0) + q(N_p) + q(2N_p) + \dots + q(N)]$$

or

$$r_{S\delta}(0) = s(0) + \frac{1}{N_\delta} \sum_{k=0}^{N/N_p} q(kN_p) \tag{5.35}$$

As $N \to \infty$, $(1/N_\delta) \sum_{k=0}^{N/N_p} q(kN_p) \to 0$, and therefore $r_{S\delta}(0) \to s(0)$. Similarly, for other values of j,

$$r_{S\delta}(-j) = \frac{1}{N_\delta} \sum_{n=0}^{N-1} [s(n) + q(n)]\delta[(n - j) - kN_p], \quad k = 0, 1, 2, \dots$$

which also results in cancellation of the noise and yields values of $s(n)$ for $n = 1$, $2, \ldots$. Hence, from Equation 5.33

$$r_{s\delta}(-j) = s(0), s(1), \ldots, s(N-1), \quad j = 0, 1, 2, \ldots$$

which is the required signal. Thus, a signal lost in a noisy waveform may be estimated by

(1) autocorrelating the waveform to find the period of the signal, and

(2) cross-correlating the waveform with a periodic impulse train of the same period as the signal, during which procedure the impulse train is shifted to the right with respect to the signal waveform.

5.2.2.3   Correlation detection implementation of matched filter

Another application of correlation is in the correlation detection implementation of the matched filter. A matched filter is a filter which maximizes the S/N ratio at its output. A matched filter has the impulse response, $h(t)$, given by (Stremler, 1982)

$$h(t) = cs_i(T - t) \tag{5.36}$$

where c is an arbitrary constant, $s_i(t)$ is the input signal (noiseless) given by

$$s_i(t) = s_i(t) \quad \text{for } 0 \leqslant t \leqslant T$$
$$= 0 \quad \text{for } T < t < 0$$

and T is the time at which the filter output is sampled. The impulse response is seen to be obtained by reversing the signal in time and then advancing it T (s) along the time axis. As an example, Figure 5.12(a) shows a signal, which is actually an 8-bit PCM codeword, and Figure 5.12(b) shows the matched filter impulse response which will maximize the detection of this signal.

Matched filter detection will now be shown to be equivalent to correlation. The filter output, $y(t)$, is first expressed in terms of the convolution of its input, $s(t)$, with its impulse response (see Section 5.3 for convolution):

$$y(t) = \int_{-\infty}^{\infty} s(\tau)h(t - \tau)\,\mathrm{d}\tau \tag{5.37}$$

where

$$s(t) = s_1(t) + q(t) \tag{5.38}$$

τ is the lag, and $q(t)$ denotes the noise component as usual. Substituting Equation 5.38 into 5.37 gives

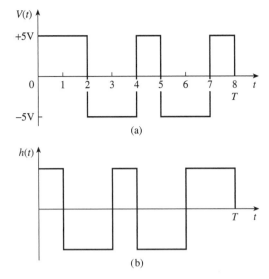

Figure 5.12 (a) A signal, which is an 8-bit PCM codeword. (b) The impulse response of the corresponding matched filter.

$$y(t) = \int_{-\infty}^{\infty} [s_1(\tau) + q(\tau)]h(t - \tau)\, d\tau$$

$$= \int_{-\infty}^{\infty} s_1(\tau)h(t - \tau)\, d\tau + \int_{-\infty}^{\infty} q(\tau)h(t - \tau)\, d\tau$$

The second term on the right-hand side tends to zero because $q(\tau)$ is random and is uncorrelated with $h(t - \tau)$. Therefore

$$y(t) \approx \int_{-\infty}^{\infty} s_1(\tau)h(t - \tau)\, d\tau \qquad (5.39)$$

Now, from Equation 5.36,

$$h(t - \tau) = cs_1(T - t + \tau) \qquad (5.40)$$

Combining Equations 5.39 and 5.40 gives

$$y(t) \approx \int_{-\infty}^{\infty} s_1(\tau)cs_1(T - t + \tau)\, d\tau \qquad (5.41)$$

If this output is sampled at time $t = T$,

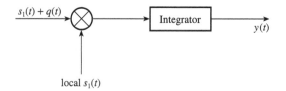

Figure 5.13   A schematic correlation detector.

$$y(T) \approx \int_{-\infty}^{\infty} s_1(\tau) c s_1(\tau)\, d\tau$$

$$\approx \int_{-\infty}^{\infty} s_1^2(\tau)\, d\tau = \int_{-\infty}^{\infty} s_1^2(t)\, dt = r_{11}(0) \tag{5.42}$$

if $c = 1$.

Thus $y(T)$ is the autocorrelation at lag zero of $s_1(t)$, and may be obtained by cross-correlating the noisy input with a locally produced noise-free signal. This constitutes the correlation detector. Figure 5.13 shows the schematic circuit of a correlation detector. For example, a PCM codeword detector would contain a correlation detector for each codeword as indicated in Figure 5.14.

In a digital m-bit codeword detector the codewords are stored and multiplied with m of the incoming bits. A peak value will occur whenever (i) the m incoming bits correspond exactly to the m-bit codeword or (ii) the m incoming bits correspond by chance with the m-bit codeword. The second outcome is highly undesirable. It might occur if two adjacent codewords happened to contain a bit sequence identical to the required m-bit codeword, or if a codeword had become corrupted. This possibility

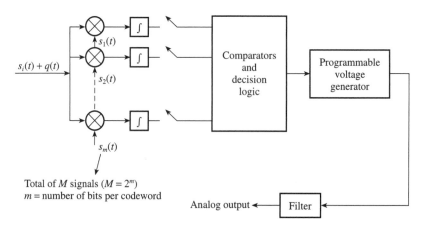

Figure 5.14   A PCM codeword detector based on the correlation detector.

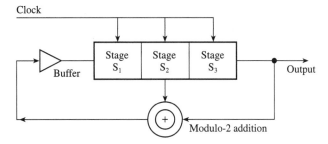

Figure 5.15   A three-stage pseudonoise sequence generator.

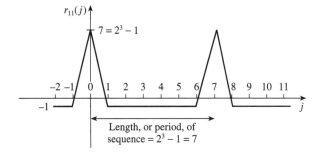

Figure 5.16   The autocorrelation function of a bipolar waveform three-stage pseudonoise generator.

makes it necessary to arrange for word synchronization as well as bit synchronization in the correlation receiver.

This leads to the requirement for synchronization codewords which have properties which include the following:

(1)   the correlation is small for sampling times $t \neq T$;
(2)   the correlation is large for samples taken at $t = T$.

A codeword with these properties will be one which has a large autocorrelation at zero lag, but a small autocorrelation at other lags. Thus detection of a large cross-correlation at the receiver will indicate alignment of the incoming codeword with the stored one. This synchronizes the receiver. Now, random waveforms have this autocorrelation property, and may be implemented in a digital receiver by a pseudonoise (PN) sequence, easily generated from a tapped shift register. A three-stage PN sequence generator is shown in Figure 5.15 as an example. The output sequence produced is 1, 1, 1, 0, 0, 1, 0 which then repeats. The autocorrelation function generated when the sequence is represented as a bipolar waveform is shown in Figure 5.16.

Some of the properties of the PN sequences are as follows.

(1)   An m-bit codeword produces a sequence of length $2^m - 1$.

(2)   The peak values are $2^m - 1$.

(3)   The autocorrelation function is equal to -1 other than at the peaks.

(4)   The output sequence contains 2^{m-1} ones and $2^{m-1} - 1$ zeros.

(5)   Their power density spectrum is uniform so they may be used as white noise sources.

The last of these properties offers another application for a PN sequence as a source of white noise.

5.2.2.4   The determination of the impulse response of electrical systems

A further application of correlation and of the PN sequence lies in the determination of the impulse response of electrical systems. There can be difficulties in impulse testing systems. For example, in the presence of noise, small impulses may be masked by the noise while larger impulses may cause system overload. It is also difficult to maintain a uniform energy spectral density over the bandwidth using a single impulse. The PN sequence, however, has a uniform energy spectrum, as explained above. Also, if the measurement time is a multiple of the sequence length the variance in the measurement will be zero. This leads to short measurement times and high accuracy.

The principle of the method is to apply a PN sequence to the input of the system. The impulse response is then given by the cross-correlation of the applied sequence with the output. This may be proved as follows.

Let $q(t)$ be the input PN sequence, and let $y(t)$ be the output of the system which has the impulse response $h(t)$. Then,

$$r_{qy}(\tau) = \lim_{T \to \infty} \frac{1}{T} \int_0^T q(t)y(t + \tau)\, dt \tag{5.43}$$

$$= \lim_{T \to \infty} \frac{1}{T} \int_0^T q(t)\, dt \int_{-\infty}^{\infty} h(v)q(t - v + \tau)\, dv \tag{5.44}$$

since $y(t)$ is given as the convolution of the input with the impulse response

$$y(t) = \int_{-\infty}^{\infty} h(v)q(t - v)\, dv \tag{5.45}$$

Changing the order of integration in Equation 5.44 gives

$$r_{qy}(\tau) = \int_{-\infty}^{\infty} h(v)\, dv \lim_{T \to \infty} \frac{1}{T} \int_0^T q(t)q(t - v + \tau)\, dt \tag{5.46}$$

$$= \int_{-\infty}^{\infty} h(v)r_{qq}(\tau - v)\, dv \tag{5.47}$$

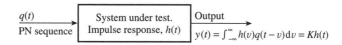

Figure 5.17   Determination of the impulse response of an electrical system.

Now, $r_{qq}(\tau - v)$ approximates to a delta function because it is the autocorrelation function of the PN sequence. Hence Equation 5.47 may be expressed as

$$r_{qy}(t) = K \int_{-\infty}^{\infty} h(v)\delta(\tau - v)\, dv = Kh(t) \tag{5.48}$$

where K is the area of the impulse function and is equal to the rms value of the noise (Beauchamp, 1973). Figure 5.17 illustrates the arrangement. The method is prone to some errors and precautions should be taken to avoid these.

5.2.2.5   Determination of the signal-to-noise ratio for a periodic noisy signal

By measuring the correlation coefficient of a noisy periodic signal its signal-to-noise ratio may be determined as well as the signal and noise powers (Main and Howell, 1993). The derivation of these expressions is given below. The period of noisy signals may be found as described in Section 5.2.2.2.

Let the signal be represented by $V_s(i)$, and the noise as $V_n(i)$, when the noisy signal, $V(i)$, is

$$V(i) = V_s(i) + V_n(i) \tag{5.49}$$

Then for the periodic signal of period n sampling intervals

$$V(i) = V(i + n) \tag{5.50}$$

The covariance of $V(i)$ is now defined as

$$\text{Cov}\,[V(i)] = \text{Cov}\,[V(i),\, V(i+n)] = \frac{1}{N}\sum_{i=1}^{N} [\{V(i) - \bar{V}(i)\}\{V(i+n) - \bar{V}(i+n)\}] \tag{5.51}$$

where

$$\bar{V}(i) = \frac{1}{N}\sum_{i=1}^{N} V(i)$$

the mean of $V(i)$. Thus Cov $[V(i)]$ is seen to be the zero-mean autocorrelation function of $V(i)$ at lag n. The autocorrelation coefficient of $V(i)$ is now defined in the usual way as

$$\rho[V(i)] = \frac{\mathrm{Cov}[V(i)]}{\sqrt{\left\{\dfrac{1}{N}\displaystyle\sum_{i=1}^{N}(V(i) - \bar{V}(i))^2\right\}\left\{\dfrac{1}{N}\displaystyle\sum_{i=1}^{N}(V(i+n) - \bar{V}(i+n))^2\right\}}}$$

$$= \frac{\mathrm{Cov}[V(i)]}{\sigma[V(i)]\sigma[V(i+n)]} \tag{5.52}$$

where

$$\sigma[V(i)] = \sqrt{\frac{1}{N}\sum_{i=1}^{N}(V(i) - \bar{V}(i))^2} \tag{5.53}$$

is the standard deviation of $V(i)$.

Note that normalization by the factor N in Equations 5.51 to 5.53 gives biased estimates. Unbiased estimates are obtained by replacing N by $N - 1$. Note also that terms such as

$$\sigma^2[V_x(i)] = \frac{1}{N}\sum_{i=1}^{N}\left(V_x(i) - \bar{V}_x(i)\right)^2 = \mathrm{Var}\,[V_x(i)]$$

are known as the variance of and represent the power associated with $V_x(i)$.

Now expand Equation 5.51 subject to the fact that correlations between noisy signals tend to zero with increasing numbers of samples. Thus,

$$\mathrm{Cov}\,[V(i), V(i+n)] = \mathrm{Cov}\,[(V_s(i) + V_n(i)), (V_s(i+n) + V_n(i+n))]$$

$$= \mathrm{Cov}\,[V_s(i), V_s(i+n)] = \mathrm{Var}\,[V_s(i)] \tag{5.54}$$

For sufficiently long data lengths $\sigma[V_n(i)] = \sigma[V_n(i+n)]$ and so

$$\sigma[V(i)]\sigma[V(i+n)] = \sigma^2[V(i)] = \mathrm{Var}\,[V(i)] = \mathrm{Var}\,[V_s(i) + V_n(i)]$$

$$= \mathrm{Var}\,[V_s(i)] + \mathrm{Var}\,[V_n(i)] + 2\mathrm{Cov}\,[V_s(i), V_n(i)] = \mathrm{Var}\,[V_s(i)] + \mathrm{Var}\,[V_n(i)] \tag{5.55}$$

since $\mathrm{Cov}\,[V_s(i), V_n(i)] = 0$.

An expression for the autocorrelation coefficient is now obtained by substituting Equations 5.54 and 5.55 into 5.52.

$$\rho[V(i)] = \frac{\mathrm{Var}\,[V_s(i)]}{\mathrm{Var}\,[V_s(i)] + \mathrm{Var}\,[V_n(i)]} = \frac{1}{1 + \dfrac{\mathrm{Var}\,[V_n(i)]}{\mathrm{Var}\,[V_s(i)]}} \tag{5.56}$$

The definition of the signal-to-noise ratio, S/N (dB), is $10 \log_{10}\{$(signal power)/(noise power)$\}$dB, which in the current notation becomes

$$\frac{S}{N}\,(\text{dB}) = 10\log_{10}\left|\frac{\text{Var}\,[V_s(i)]}{\text{Var}\,[V_n(i)]}\right|\text{dB} \tag{5.57}$$

Combining and transposing Equations 5.56 and 5.57 yields

$$\frac{S}{N}\,(\text{dB}) = 10\log_{10}\left|\frac{\rho[V(i)]}{1 - \rho[V(i)]}\right|\text{dB} \tag{5.58}$$

Thus, the signal-to-noise ratio of the noisy periodic waveform is easily obtained from its autocorrelation coefficient.

From Equation 5.55 we have

$$\text{Var}\,[V_s(i)] + \text{Var}\,[V_n(i)] \equiv S + N = \text{Var}\,[V(i)] \tag{5.59}$$

where S and N represent the signal and noise powers. Using Equations 5.56 and 5.59 expressions may be derived for S and N, namely

$$S = \rho[V(i)]\,\text{Var}\,[V(i)] \tag{5.60}$$

and

$$N = (1 - \rho[V(i)])\,\text{Var}\,[V(i)] \tag{5.61}$$

Equation 5.58 has been applied to the evaluation of the performance of magnetic recording channels in terms of their signal-to-noise ratio (Main and Howell, 1993).

5.2.3 Fast correlation

The correlation computation may be speeded up by exploiting the correlation theorem, usually stated as

$$r_{12}(j) = F_D^{-1}[X_1^*(k)X_2(k)] \tag{5.62}$$

but which is correctly written as

$$r_{12}(j) = \frac{1}{N}F_D^{-1}[X_1^*(k)X_2(k)] \tag{5.63}$$

where F_D^{-1} denotes the inverse discrete Fourier transform. This approach requires computation of two discrete Fourier transforms (DFTs) and one inverse DFT, each of which is most easily executed using an FFT algorithm (see Chapter 3). If the number of terms in the sequences is sufficiently large, it is quicker to use this FFT method than to calculate the cross-correlation directly.

Proof of the correlation theorem

Let $x_1(l)$, $x_2(r)$, and $x_3(n)$ be periodic sequences of length N, and let their DFTs be $X_1(k)$, $X_2(k)$, and $X_3(k)$ respectively. Furthermore, let

$$X_3(k) = X_1^*(k)X_2(k) \tag{5.64}$$

Now,

$$X_1^*(k) = \sum_{l=0}^{N-1} x_1(l)\, e^{j(2\pi/N)lk} \tag{5.65}$$

and

$$X_2(k) = \sum_{r=0}^{N-1} x_2(r)\, e^{j(2\pi/N)(-rk)} \tag{5.66}$$

Substituting Equations 5.65 and 5.66 into 5.64 gives

$$X_3(k) = \sum_{l=0}^{N-1} x_1(l)\, e^{j(2\pi/N)lk} \sum_{r=0}^{N-1} x_2(r)\, e^{j(2\pi/N)(-rk)} \tag{5.67}$$

$$= \sum_{l=0}^{N-1}\sum_{r=0}^{N-1} x_1(l)x_2(r)\, e^{j(2\pi/N)(lk-rk)} \tag{5.68}$$

Now,

$$x_3(n) = \frac{1}{N}\sum_{k=0}^{N-1} X_3(k)\, e^{j(2\pi/N)nk} \tag{5.69}$$

So, substituting Equation 5.68 into 5.69 yields

$$x_3(n) = \frac{1}{N}\sum_{k=0}^{N-1}\sum_{l=0}^{N-1}\sum_{r=0}^{N-1} x_1(l)x_2(r)\, e^{j(2\pi/N)(lk-rk+nk)}$$

$$= \frac{1}{N}\sum_{l=0}^{N-1} x_1(l)\sum_{r=0}^{N-1} x_2(r)\left[\sum_{k=0}^{N-1} e^{j(2\pi/N)(l-r+n)k}\right] \tag{5.70}$$

When $r = n + l$, the term in square brackets equals N. When $r \neq n + l$ it may be treated as a geometric series of the form

$$\sum ax^n$$

which sums over N terms to

$$\frac{a(1 - x^N)}{1 - x}$$

In this case the sum becomes

$$\frac{1[1 - e^{j(2\pi/N)(l-r+n)N}]}{1 - e^{j(2\pi/N)(l-r+n)}} \tag{5.71}$$

The exponent in the numerator is always an integral multiple of 2π, and so the exponential term is unity. Hence the summation equates to zero when $r \neq n + l$. Equation 5.70 can therefore be written

$$x_3(n) = \frac{1}{N} \sum_{l=0}^{N-1} x_1(l) \sum_{r=0}^{N-1} x_2(r) N \delta(l - r + n) \tag{5.72}$$

in which $\delta(l - r + n) = 1$ when $r = n + l$ and $\delta(l - r + n) = 0$ when $r \neq n + l$. Simplifying and putting $r = n + l$ gives

$$x_3(n) = \sum_{l=0}^{N-1} x_1(l) x_2(l + n) \tag{5.73}$$

or

$$\frac{1}{N} x_3(n) = \frac{1}{N} \sum_{l=0}^{N-1} x_1(l) x_2(l + n) \tag{5.74}$$

The right-hand side of this equation is equivalent to the cross-correlation of $x_1(n)$ and $x_2(n)$ and is seen to be equal to $(1/N)x_3(n)$. From Equation 5.69

$$x_3(n) = F_D^{-1}[X_3(k)] \tag{5.75}$$

Hence, by combining Equations 5.74, 5.75 and 5.63,

$$\frac{1}{N} F_D^{-1}[X_3(k)] = r_{12}(n) = \frac{1}{N} F_D^{-1}[X_1^*(k)X_2(k)] \tag{5.76}$$

Finally, replacing n by j gives

$$r_{12}(j) = \frac{1}{N} F_D^{-1}[X_1^*(k)X_2(k)] \tag{5.77}$$

Example 5.6   Work out the cross-correlation of the two sequences $x_1(n)$ and $x_2(n)$ below by applying the correlation theorem:

$$x_1(n) = \{1, 0, 0, 1\}$$
$$x_2(n) = \{0.5, 1, 1, 0.5\}$$

First use the correlation theorem, Equation 5.77. $X_1(k)$ was found in Section 3.5 to be

$$X_1(k) = 2, 1 + j, 0, 1 - j$$

and so

$$X_1^*(k) = 2, 1 - j, 0, 1 + j$$

$X_2(k)$ is most easily obtained using the FFT algorithm given in Section 3.5. Thus, with $x_0 = 0.5$, $x_2 = 1$, $x_1 = 1$, and $x_3 = 0.5$,

$$X_{21}(0) = x_0 + x_2 = 1.5$$
$$X_{21}(1) = x_0 - x_2 = -0.5$$
$$X_{22}(0) = x_1 + x_3 = 1.5$$
$$X_{22}(1) = x_1 - x_3 = 0.5$$
$$X_{11}(0) = X_{21}(0) + X_{22}(0) = 3$$
$$X_{11}(1) = X_{21}(1) + (-j)X_{22}(1) = -0.5 - j0.5$$
$$X_{11}(2) = X_{21}(0) - X_{22}(0) = 0$$
$$X_{11}(3) = X_{21}(1) - (-j)X_{22}(1) = -0.5 + j0.5$$

Bringing the values of the FFTs together gives

$$X_1^*(k) = 2, 1 - j, 0, 1 + j$$
$$X_2(k) = 3, -0.5 - j0.5, 0, -0.5 + j0.5$$

So,

$$X_1^*(0)X_2(0) = 2 \times 3 = 6$$
$$X_1^*(1)X_2(1) = (1 - j)(-0.5 - j0.5) = -1$$
$$X_1^*(2)X_2(2) = 0 \times 0 = 0$$
$$X_1^*(3)X_2(3) = 0.5(1 + j)(-1 + j) = -1$$

Thus

$$[X_1^*(k)X_2(k)] = 6, -1, 0, -1$$

It is now necessary to take the inverse DFT (IDFT) of this. As explained in Section 3.6, the IDFT is obtainable by changing the signs of the exponents (in the weighting factors W_N) in the above FFT algorithm and dividing the result by N. Hence, and without changing the notation in the algorithm,

$$X_{21}(0) = x_0 + x_2 = 6$$

$$X_{21}(1) = x_0 - x_2 = 6$$

$$X_{22}(0) = x_1 + x_3 = -2$$

$$X_{22}(1) = x_1 - x_3 = 0$$

$$X_{11}(0) = X_{21}(0) + X_{22}(0) = 4$$

$$X_{11}(1) = X_{21}(1) + jX_{22}(1) = 6$$

$$X_{11}(2) = X_{21}(0) - X_{22}(0) = 8$$

$$X_{11}(3) = X_{21}(1) - jX_{22}(1) = 6$$

The components of $F_D^{-1}[X_1^*(k)X_2(k)]$ are obtained by dividing the values of $X_{11}(0)$, $X_{11}(1)$, $X_{11}(2)$, and $X_{11}(3)$ by $N = 4$. Thus

$$F_D^{-1}[X_1^*(k)X_2(k)] = 1, 1.5, 2, 1.5$$

Now, from Equation 5.77,

$$r_{12}(j) = \frac{1}{4}F_D^{-1}[X_1^*(k)X_2(k)] = \{0.25, 0.375, 0.5, 0.375\} \tag{5.78}$$

This correlation will be circular because all the data is periodic with period N. The cross-correlation $r_{12}(j)$ may be worked out directly to be

$$r_{12}(0) = (1 \times 0.5 + 0 + 0 + 1 \times 0.5)/4 = 0.25$$

$$r_{12}(1) = (1 \times 1 + 0 + 0 + 1 \times 0.5)/4 = 0.375$$

$$r_{12}(2) = (1 \times 1 + 0 + 0 + 1 \times 1)/4 = 0.5$$

$$r_{12}(3) = (1 \times 0.5 + 0 + 0 + 1 \times 1)/4 = 0.375$$

The next value, $r_{12}(4)$, is 0.25, the same as $r_{12}(0)$, and the sequence repeats periodically. This is circular correlation, as discussed in Section 5.2.1, and this result agrees with that derived above using the correlation theorem. The correlation theorem can be used to obtain the linear correlation by adding augmenting zeros to the two sequences as explained in Section 5.2.1. Thus, if the sequence lengths are N_1 for $x_1(n)$ and N_2 for $x_2(n)$, then $N_2 - 1$ zeros are added to $x_1(n)$ and $N_1 - 1$ zeros are added to $x_2(n)$. The cross-correlation is then computed using these two augmented sequences. This method of evaluating cross-correlations by using the correlation theorem and FFTs is known as fast correlation.

Cross-correlation calculations may also be speeded up by implementing them recursively, and this will be illustrated for the case of zero lag. The cross-correlation at zero lag of two sampled waveforms $x_1(n)$ and $x_2(n)$ is

$$r_{12}(0) = \frac{1}{N} \sum_{n=0}^{N-1} x_1(n)x_2(n) \tag{5.79}$$

This involves the computation of N products, $N - 1$ sums and one division. This may occupy excessive time in an on-line application in which pairs of new data values are arriving at the sampling rate. The calculation has to be repeated when the next data pair is available. The new calculation will differ from the previous one only in that the product of the new data has to be added to the sum of product pairs, and the first product has to be subtracted. Thus, for each cross-correlation:

$$\text{new value} = \text{previous value} + \frac{1}{N}(\text{product of two new data})$$

$$- \frac{1}{N}(\text{product of first two data}) \tag{5.80}$$

This is the basis of the recursive algorithm. Each cross-correlation now only requires one multiplication, one subtraction, one addition, and one division, provided that the data pair products are saved. For an N-point correlation, the recursive approach gives correct values after the first $N - 1$ points have been computed.

In many applications it is required to set the means of the data to zero, for example to remove dc levels from electrical waveforms. This requires calculation of the average value of the waveforms and then subtraction of it from all the sampled values. This mean value calculation may also be done recursively since for each new data pair:

$$\text{new average} = \text{previous average} + \frac{1}{N}(\text{new datum} - \text{first datum}) \tag{5.81}$$

It is also possible to combine the mean level subtraction and cross-correlation calculations in one recursive algorithm. Consider

$$\bar{x}_1(k) = \frac{1}{N} \sum_{n=0}^{N-1} x_1(n) \tag{5.82}$$

and

$$\bar{x}_2(k) = \frac{1}{N} \sum_{n=0}^{N-1} x_2(n) \tag{5.83}$$

The value of the cross-correlation function of the kth set of N points is

$$r_{12}(k) = \frac{1}{N} \sum_{n=0}^{N-1} x_1(n)x_2(n) \tag{5.84}$$

When the means have been removed the cross-correlation function value becomes $r_{12}^0(k)$ where

$$r_{12}^0(k) = \frac{1}{N} \sum_{n=0}^{N-1} [x_1(n) - \bar{x}_1(k)][x_2(n) - \bar{x}_2(k)] \tag{5.85}$$

which, on expansion, simplifies to

$$r_{12}^0(k) = r_{12}(k) - \bar{x}_1(k)\bar{x}_2(k) \tag{5.86}$$

Combining Equations 5.80 and 5.83 gives

$$r_{12}(k) = r_{12}(k-1) + \frac{1}{N}[x_1(k)x_2(k) - x_1(k-N)x_2(k-N)] \tag{5.87}$$

From Equation 5.81,

$$\bar{x}_1(k) = \bar{x}_1(k-1) + \frac{1}{N}[x_1(k) - x_1(k-N)] \tag{5.88}$$

and

$$\bar{x}_2(k) = \bar{x}_2(k-1) + \frac{1}{N}[x_2(k) - x_2(k-N)] \tag{5.89}$$

Equations 5.86–5.89 constitute the recursive algorithm which combines subtraction of the mean from the data with the calculation of the cross-correlation. Each calculation requires only three multiplications, four subtractions, three additions, and four divisions. Practitioners should note that care has to be exercised in the choice of N when the mean values of the data are varying, or inaccurate results may be obtained.

5.3 Convolution description

The term convolution describes, amongst other things, how the input to a system interacts with the system to produce the output. Generally the system output will be a delayed and attenuated or amplified version of the input. It is particularly useful to consider the output from the system owing to an impulse input. This is because any input may be represented as a sequence of impulses of different strengths. The output

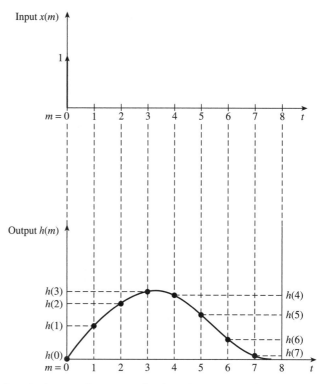

Figure 5.18   Impulse input and corresponding impulse response of a system.

of the system owing to the impulse input will not be a corresponding impulse, but will vary with time, passing through a maximum, as shown in Figure 5.18. This figure shows that at sampling instant m the output owing to the unit impulse applied at sampling instant 0 is $h(m)$. The characteristic is known as the impulse response $h(m)$ of the system.

Consider now the application of a sequence of impulses $x(m)$ to the system, applied at the sampling instants m. Referring to Figure 5.19, the output at instant 0 is $y(0)$ given by

$$y(0) = h(0)x(0)$$

At the sampling instant $m = 1$ the output will be given by $h(0)x(1)$, the effect of the current input $x(1)$, plus the delayed effect $h(1)x(0)$ of the input applied at sampling instant $m = 0$. Thus

$$y(1) = h(1)x(0) + h(0)x(1)$$

Similarly, subsequent outputs will be given by

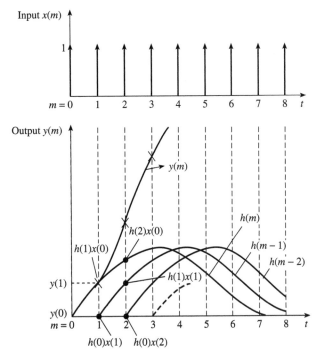

Figure 5.19   Applied impulse sequence and system response derived from the individual impulse responses.

$$y(2) = h(2)x(0) + h(1)x(1) + h(0)x(2)$$

$$y(3) = h(3)x(0) + h(2)x(1) + h(1)x(2) + h(0)x(3)$$

$$\vdots$$

$$y(n) = h(n)x(0) + h(n-1)x(1) + \ldots + h(0)x(n) \tag{5.90}$$

The output may only be written in this manner as a linear sum of the effects of previous inputs if the system is a linear one. Equation 5.90 describes the output of a first-order linear system.

Inspection of the above expressions reveals that the output is obtained by multiplying the input sequence by the corresponding points of the time-reversed impulse response function. Alternatively, since Equation 5.90 may equally well be written as

$$y(n) = h(0)x(n) + h(1)x(n-1) + \ldots + h(n)x(0) \tag{5.91}$$

the output may be regarded as the product of the corresponding pairs of points in the impulse response function by the time-reversed input sequence. Thus the convolution sum is equivalent to the cross-correlation of the one sequence by the time-reversed second sequence.

Equations 5.90 and 5.91 may be written compactly as

$$y(n) = \sum_{m=0}^{n} h(n-m)x(m) \tag{5.92}$$

and

$$y(n) = \sum_{m=0}^{n} h(m)x(n-m) \tag{5.93}$$

These latter two functions are referred to as the convolution sums of the inputs by the impulse response function, and the output is said to be given by the convolution of the input by the impulse response of the system.

Equations 5.92 and 5.93 may be extended to waveforms of infinite duration by writing them as

$$y(n) = \sum_{m=-\infty}^{\infty} x(m)h(n-m) = x(n) \circledast h(n) \tag{5.94}$$

and

$$y(n) = \sum_{m=-\infty}^{\infty} h(m)x(n-m) = h(n) \circledast x(n) \tag{5.95}$$

the generalized forms of the convolution sum. In these equations the symbol $\circledast$ signifies the operation of convolution.

If the input consists of a continuous sequence of impulses the above summations may be replaced by integrals, so, for example, Equation 5.94 becomes

$$y(t) = \int_{-\infty}^{\infty} x(\lambda)h(t-\lambda)\,\mathrm{d}\lambda \tag{5.96}$$

which is known as the convolution integral.

So far the term convolution has been taken to describe the result of convolving together the impulse response of a system and the input to the system. However, the idea can be extended to the convolution of any two sets of data and the term will henceforth be considered in this more general sense.

As an example, the two periodic time sequences (4, 3, 2, 1 $\{h(m)\}$) and (1, 2, 3, 4 $\{x(m)\}$) will now be convolved. Figure 5.20(a) shows the periodic sequence (4, 3, 2, 1 $\{h(m)\}$) and Figure 5.20(b) the time-reversed sequence $h(-m)$ which has become (1, 2, 3, 4). (Recall that the convolution sum requires that one sequence has to be multiplied point by point by the time-reversed second sequence, that is it corresponds to the cross-correlation of the one sequence by the time reversal of the other.) The figure also shows a window of width equal to one period over which the convolution is evaluated. Clearly the result obtained will be periodic

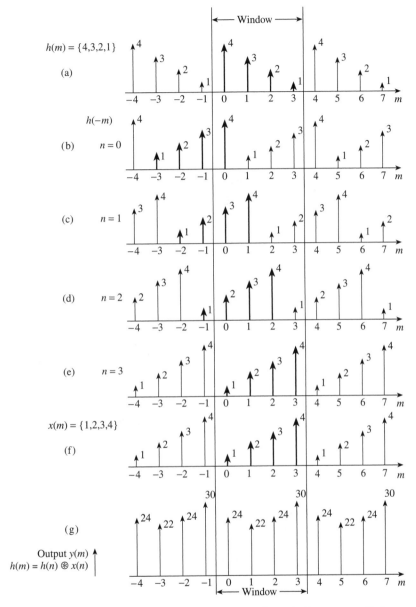

Figure 5.20 The convolution $y(m)$ of $h(n)$ and $x(n)$. (a) The periodic sequence $h(m)$. (b) The time-reversed sequence $h(-m)$. (c) to (e) Versions of $h(-m)$ shifted right by incremental lags. (f) The sequence $x(m)$. (g) The output sequence $y(m) = h(n) \circledast x(n)$.

as for the corresponding case of cyclic correlation (Section 5.2.1) and so it is only necessary to evaluate the convolution over the windowed interval. Figure 5.20(f) shows the second sequence $(1, 2, 3, 4 \{x(m)\})$ for reference.

Now, when $n = 0$, Equation 5.92 becomes

$$y(0) = \sum_{m=0}^{n} h(-m)x(m)$$

and is obtained by cross-correlating the windowed data in Figures 5.20(b) and 5.20(f):

$$y(0) = 4 \times 1 + 1 \times 2 + 2 \times 3 + 3 \times 4 = 24$$

When $n = 1$ Equation 5.92 becomes

$$y(1) = \sum_{m=0}^{n} h(1 - m)x(m)$$

and is obtained by cross-correlating the windowed data in Figures 5.20(c) and 5.20(f), giving

$$y(1) = 3 \times 1 + 4 \times 2 + 1 \times 3 + 2 \times 4 = 22$$

Similarly, it is found that

$$y(2) = 2 \times 1 + 3 \times 2 + 4 \times 3 + 1 \times 4 = 24$$

and

$$y(3) = 1 \times 1 + 2 \times 2 + 3 \times 3 + 4 \times 4 = 30$$

whereafter the output sequence repeats cyclically. This output sequence is shown in Figure 5.20(g).

When the waveforms are well defined mathematically the convolution may be performed analytically. By considering a similar example, and also illustrating the steps involved graphically, it is possible to obtain a better understanding of the convolution process.

Example 5.7   Convolve the waveforms $x(t)$ and $h(t)$ of Figure 5.21(a) analytically.

Let the convolution integral be

$$y(t) = x(t) \circledast h(t) = \int_{-\infty}^{\infty} x(\tau)h(t - \tau) \, d\tau \qquad (5.97)$$

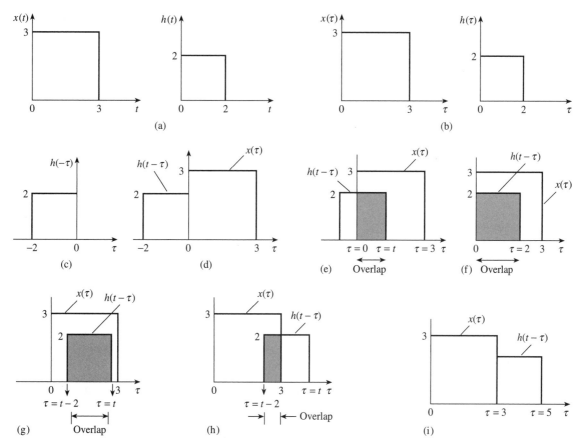

Figure 5.21   (a) The waveforms $x(t)$ and $h(t)$ which are to be convolved analytically. (b) $x(\tau)$ and $h(\tau)$ vs τ. (c) $h(-\tau)$ vs τ. (d) $h(t - \tau)$ and $x(\tau)$ vs τ. $t < 0$, $h(t - \tau)$ does not overlap $x(\tau)$. (e) $h(t - \tau)$ and $x(\tau)$ vs τ. $0 < t \leqslant 2$. First partial overlap occurs between $h(t - \tau)$ and $x(\tau)$. (f) $h(t - \tau)$ and $x(\tau)$ vs (τ). $t = 2$. End of first partial overlap. (g) $h(t - \tau)$ and $x(\tau)$ vs (τ). $2 \leqslant t \leqslant 3$. Complete overlap between $h(t - \tau)$ and $x(\tau)$ occurs. (h) $h(t - \tau)$ and $x(\tau)$ vs (τ). $3 \leqslant t \leqslant 5$. Second partial overlap occurs between $h(t - \tau)$ and $x(\tau)$. (i) $h(t - \tau)$ and $x(\tau)$ vs τ. $t > 5$. $h(t - \tau)$ does not overlap $x(\tau)$.

Equation 5.97 corresponds to Equation 5.96 in which the variable λ has been replaced by τ to indicate that time lags are being applied. The convolution integral depends on the variable τ and so Figure 5.21(a) has to be replaced by Figure 5.21(b).

It is now necessary to time reverse $h(\tau)$ as shown in Figure 5.21(c). $h(-\tau)$ is next shifted with respect to $x(\tau)$ in the direction of positive τ. The resulting waveform $h(t - \tau)$ then overlaps $x(\tau)$ in five separate geometrical stages as shown in Figures 5.21(d), 5.21(e), 5.21(g), 5.21(h), and 5.21(i). For each of these stages there is a corresponding convolution integral. Thus $x(t) \circledast h(t)$ exists as five separate contiguous regions.

- *Stage 1* $t < 0$ and $h(t - \tau)$ does not overlap $x(\tau)$ (Figure 5.21(d)). As the functions do not overlap $x(\tau)h(t - \tau) = 0$ for all t and there is no contribution to the convolution integral.

- *Stage 2* $0 < t \leq 2$ and partial overlap occurs between $h(t - \tau)$ and $x(\tau)$ (Figure 5.21(e)). Over this range

$$y(t) = \int_{\tau=0}^{\tau=t} x(\tau)h(t - \tau)\,d\tau = \int_{\tau=0}^{\tau=t} (3) \times (2)\,d\tau$$

$$y(t) = 6[\tau]_0^t = 6t, \quad 0 < t \leq 2 \tag{5.98}$$

This geometrical stage terminates when $t = 2$, as shown in Figure 5.21(f).

- *Stage 3* $2 \leq t \leq 3$ and there is complete overlap of $h(t - \tau)$ and $x(\tau)$ (Figure 5.21(g)). Over this range of t,

$$y(t) = \int_{\tau=t-2}^{\tau} (3) \times (2)\,d\tau = 6[\tau]_{t-2}^t$$

$$y(t) = 6(t - t + 2) = 12, \quad 2 \leq t \leq 3 \tag{5.99}$$

- *Stage 4* $3 \leq t \leq 5$. This is another type of overlap region shown in Figure 5.21(h):

$$y(t) = \int_{\tau=t-2}^{\tau=3} (3) \times (2)\,d\tau = 6[\tau]_{t-2}^3 = 6(5 - t) = 30 - 6t \tag{5.100}$$

- *Stage 5* $t > 5$. As seen in Figure 5.21(i) this is a second region of no overlap, and so there is no contribution to the convolution integral.

Stages 2 to 4 thus each make a contribution to the convolution integral, with the convolution integral having a different expression for each of the three regions corresponding to the three stages, as summarized below:

$$0 < t \leq 2 \quad y(t) = 6t$$

$$2 \leq t \leq 3 \quad y(t) = 12$$

$$3 \leq t \leq 5 \quad y(t) = 30 - 6t$$

From these expressions $y(t)$ may be plotted against t as shown in Figure 5.22.

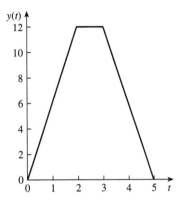

Figure 5.22   The convolution $y(t) = x(t) \circledast h(t)$ vs t.

It is now convenient to repeat Equations 5.94 and 5.96 here in order to discuss them.

$$y(n) = \sum_{m=-\infty}^{\infty} x(m)h(n-m) = x(n) \circledast h(n) \tag{5.94}$$

and

$$y(t) = \int_{-\infty}^{\infty} x(\lambda)h(t-\lambda)\, d\lambda \tag{5.96}$$

Inspection of these equations serves as a reminder that the convolution was carried out in time. This is known as convolution in the time domain. It is known that in the frequency domain the output component of a system at frequency f is $Y(f)$, given by

$$Y(f) = H(f)X(f) \tag{5.101}$$

where $H(f)$ is the frequency response function of the system at frequency f, and $X(f)$ is the Fourier transform of the input $x(t)$. It can also be demonstrated that $H(f)$ is the Fourier transform of $h(t)$. The inverse Fourier transform of Equation 5.101 is

$$F^{-1}[Y(f)] = y(t) = F^{-1}[H(f)X(f)] \tag{5.102}$$

Bringing together Equations 5.96 and 5.102 shows that

$$y(t) = \int_{-\infty}^{\infty} x(\lambda)h(t-\lambda)\, d\lambda = x(t) \circledast h(t) = F^{-1}[H(f)X(f)] \tag{5.103}$$

Thus, it is seen that the convolution of two waveforms in the time domain is equivalent to the inverse Fourier transform of the product of the Fourier transforms of the two waveforms. This useful fact is often stated in the abbreviated form that convolution in the time domain is equivalent to multiplication in the frequency domain.

The dual of this relationship exists, that is convolution in the frequency domain is equivalent to multiplication in the time domain. Thus it can be shown (McGillem and Cooper, 1974) that

$$Y(\omega) = \frac{1}{2\pi} \int_{-\infty}^{\infty} X(\omega - u)H(u)\, du = X(f) \circledast H(f)$$

$$= F[y(t)] = F[x(t)h(t)] \tag{5.104}$$

Thus, the Fourier transform of the product of two time sequences corresponds to the convolution of the Fourier transforms of the two sequences. This result is of practical use in explaining one of the effects of windowing data prior to spectral analysis (see Chapter 11). In this procedure the digitized data sequence is multiplied point by point by another sequence which consists of the sampled points of a window function. This is known as windowing and is carried out to reduce the errors on computing the energy spectrum of the data. The discrete Fourier transform of the windowed data is then computed and from this the energy spectrum is calculated. The purpose is to obtain the energy spectrum of the data sequence, but, from the above, what is actually obtained is the spectrum of the data sequence convolved with the spectrum of the window sequence.

5.3.1 Properties of convolution

(1) *Commutative law*

$$x_1(t) \circledast x_2(t) = x_2(t) \circledast x_1(t) \tag{5.105}$$

Note that this is identical to

$$\int_{-\infty}^{\infty} x_1(\tau)x_2(t - \tau)\, d\tau = \int_{-\infty}^{\infty} x_2(\tau)x_1(t - \tau)\, d\tau$$

(2) *Distributive law*

$$x_1(t) \circledast [x_2(t) + x_3(t)] = x_1(t) \circledast x_2(t) + x_1(t) \circledast x_3(t) \tag{5.106}$$

(3) *Associative law*

$$x_1(t) \circledast [x_2(t) \circledast x_3(t)] = [x_1(t) \circledast x_2(t)] \circledast x_3(t) \tag{5.107}$$

These properties may be proved either by manipulating the integrations involved or by considering the convolutions in terms of cross-correlating the one sequence by the time-reversed second sequence.

5.3.2   Circular convolution

Section 5.2.1 illustrated that the result of correlating two unequal length periodic sequences was a cyclic sequence of period equal to that of the shorter sequence, which was, therefore, an incorrect result. Because convolution is equivalent to the cross-correlation of one sequence by the reverse of a second sequence, the same will be true of convolution. Therefore, as with correlation, in convolution it is necessary that the two sequences be of the same length. Thus if the sequence lengths are N_1 and N_2, then $N_2 - 1$ augmenting zeros must be added to the sequence of length N_1, and $N_1 - 1$ augmenting zeros must be added to the sequence of length N_2. Both sequences will now be of identical length $N_1 + N_2 - 1$, and the correct linear convolution will be obtained, subject to taking other precautions as described for correlation.

5.3.3   System identification

Equation 5.95 gives the relation between the input to a system, $x(n)$, and its output, $y(n)$. The term system identification refers to the determination of $h(n)$ when it is unknown. If a test signal $x(n)$ is applied and $y(n)$ measured, $h(n)$ may be found as follows, apart from the method discussed in Section 5.2.2.4.

Equation 5.91 states that $y(n) = h(0)x(n) + h(1)x(n - 1) + \ldots + h(n)x(0)$. When $n = 0$, $y(0) = h(0)x(0)$, so

$$h(0) = \frac{y(0)}{x(0)} \tag{5.108}$$

Now, expanding and rearranging 5.93 gives

$$y(n) = h(n)x(0) + \sum_{m=0}^{n-1} h(m)x(n - m), \quad n \geq 1 \tag{5.109}$$

so

$$h(n) = \frac{y(n) - \displaystyle\sum_{m=0}^{n-1} h(m)x(n - m)}{x(0)} \quad n \geq 1, x(0) \neq 0 \tag{5.110}$$

Equations 5.108 and 5.110 allow $h(n)$ to be computed.

Example 5.8   A test signal, $x(n) = \{1, 1, 1\}$, is applied to a system with an unknown impulse response, $h(n)$. The observed system output is $y(n) = \{1, 4, 8, 10, 8, 4, 1\}$. Determine $h(n)$.

From 5.108

$$h(0) = \frac{y(0)}{x(0)} = \frac{1}{1} = 1$$

Using 5.110

$$h(n) = \frac{y(n) - \sum_{m=0}^{n-1} h(m)x(n-m)}{x(0)}$$

For $h(1)$:

$$h(1) = \frac{y(1) - \sum_{m=0}^{0} h(m)x(1-m)}{x(0)} = \frac{y(1) - h(0)x(1)}{x(0)} = \frac{4 - 1 \times 1}{1} = 3$$

For $h(2)$:

$$h(2) = \frac{y(2) - \sum_{m=0}^{1} h(m)x(2-m)}{x(0)} = \frac{y(2) - h(0)x(2) - h(1)x(1)}{x(0)}$$

$$= \frac{8 - 1 \times 1 - 3 \times 1}{1} = 4$$

For $h(3)$:

$$h(3) = \frac{y(3) - \sum_{m=0}^{2} h(m)x(3-m)}{x(0)} = \frac{y(3) - h(0)x(3) - h(1)x(2) - h(2)x(1)}{x(0)}$$

$$= \frac{10 - 1 \times 0 - 3 \times 1 - 4 \times 1}{1} = 3$$

For $h(4)$:

$$h(4) = \frac{y(4) - \sum_{m=0}^{3} h(m)x(4-m)}{x(0)}$$

$$= \frac{8 - h(0)x(4) - h(1)x(3) - h(2)x(2) - h(3)x(1)}{x(0)}$$

$$= \frac{8 - 1 \times 0 - 3 \times 0 - 4 \times 1 - 3 \times 1}{1} = 1$$

For $h(5)$:

$$h(5) = \frac{y(5) - \sum_{m=0}^{4} h(m)x(5 - m)}{x(0)}$$

$$= \frac{y(5) - h(0)x(5) - h(1)x(4) - h(2)x(3) - h(3)x(2) - h(4)x(1)}{x(0)}$$

$$= \frac{4 - 0 - 0 - 0 - 3 \times 1 - 1 \times 1}{1} = 0$$

In fact, $h(n) = 0$, $n \geqslant 5$. Hence $h(n) = \{1, 3, 4, 3, 1\}$.

5.3.4   Deconvolution

If the impulse response and the output of a system are known, then the procedure to obtain the unknown input is referred to as deconvolution. Deconvolution can be achieved by a similar procedure to that described in Section 5.3.3 on system identification. Expanding and rearranging Equation 5.93 in a different way this time gives

$$y(n) = h(0)x(n) + \sum_{m=1}^{n} h(m)x(n - m) \tag{5.111}$$

When $n = 0$, $y(0) = h(0)x(0)$.
   Hence

$$x(0) = \frac{y(0)}{h(0)} \tag{5.112}$$

Rearranging 5.111 gives

$$x(n) = \frac{y(n) - \sum_{m=1}^{n} h(m)x(n - m)}{h(0)} \tag{5.113}$$

Equations 5.112 and 5.113 are similar to Equations 5.108 and 5.110, and so the calculation of the $x(n)$ is analogous to that of $h(n)$.

Example 5.9   Assuming the same system as in Example 5.8, calculate the input, $x(n)$, given as in that example that $h(n) = \{1, 3, 4, 3, 1\}$ and $y(n) = \{1, 4, 8, 10, 8, 4, 1\}$.

From 5.112

$$x(0) = \frac{y(0)}{h(0)} = \frac{1}{1} = 1$$

From 5.113

$$x(1) = \frac{y(1) - h(1)x(0)}{h(0)} = \frac{4 - 3 \times 1}{1} = 1$$

$$x(2) = \frac{y(2) - h(1)x(1) - h(2)x(0)}{h(0)} = \frac{8 - 3 \times 1 - 4 \times 1}{1} = 1$$

$$x(3) = \frac{y(3) - h(1)x(2) - h(2)x(1) - h(3)x(0)}{h(0)} = \frac{10 - 3 \times 1 - 4 \times 1 - 3 \times 1}{1} = 0$$

and for $n \geqslant 3$, $x(n) = 0$.

Thus $x(n) = \{1, 1, 1\}$, in agreement with the value used in Example 5.8.

5.3.5   Blind deconvolution

The process of determining the input signal from the output signal when the impulse response of the system is unknown is known as blind deconvolution. A method of achieving this is described based upon the development of Bell and Sejnowski (1995). The problem and its solution are illustrated in Figure 5.23. In Figure 5.23(a) the required unknown source signal, $x(n)$, is passed through a system of impulse response, $h(n)$, to produce the measurable output signal, $f(n)$. $f(n)$ is the result of the convolution of $h(n)$ with $x(n)$, $h(n) \otimes x(n)$, and is therefore distorted by delayed versions of $x(n)$.

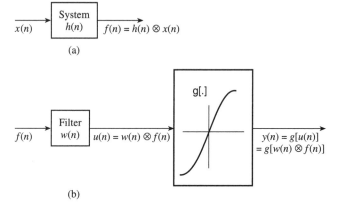

Figure 5.23   Blind deconvolution.

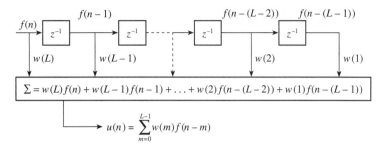

Figure 5.24   Transversal filter $w(n)$ for blind deconvolution.

The aim is to compute a signal, $u(n)$, which is a good approximation to $x(n)$. Thus, a causal filter, $w(n)$, which, when convolved with $f(n)$ produces $u(n)$ is required as shown in Figure 5.23(b). A simple filter would be the transversal filter as shown in Figure 5.24 (cf. Figure 1.4(a)). The output of this filter is

$$u(n) = \sum_{m=0}^{L-1} w(m) f(n - m)$$

and may alternatively be written in matrix notation as

$$\mathbf{U} = \mathbf{WF} \tag{5.114}$$

where $\mathbf{U} = \{u(0), u(1), \dots, u(N)\}^T$,

$$\mathbf{W} = \left\{ \begin{array}{cccccc} w(L) & 0 & \cdots & 0 & \cdots & 0 \\ w(L-1) & w(L) & \cdots & 0 & \cdots & 0 \\ \vdots & \cdots & \cdots & \vdots & \cdots & \vdots \\ w(1) & w(2) & \cdots & w(L) & \cdots & 0 \\ \vdots & \cdots & \cdots & \vdots & \cdots & \vdots \\ 0 & \cdots & \cdots & w(1) & \cdots & w(L) \end{array} \right\}$$

$\mathbf{F} = \{f(0), f(1), \dots, f(N)\}^T$, and N is the number of terms in the time series.

Bell and Sejnowski (1995) used the principle of information maximization to derive an algorithm for adaptively computing the weights in $\mathbf{W}$. Thus they are adjusted to reduce the statistical correlations between the points in $u(n)$. This is known as whitening $u(n)$, since the samples in a white noise sequence are statistically independent. To achieve this it is necessary to remove higher order statistical correlations. This is accomplished by applying $u(n)$ to a nonlinear transfer function, $g[u(n)]$, and maximizing the information in its output, $y(n) = g[u(n)]$. The formulae for updating the weights are

$$\Delta w(L) \propto \sum_{n=1}^{N} \left(\frac{1}{w(L)} - 2x(n)y(n) \right) \tag{5.115}$$

and

$$\Delta w(L - j) \propto \sum_{n=j}^{N} (-2x(n - 1)y(n)) \qquad (5.116)$$

The algorithm is computed until changes in $\Delta w(L)$ and $\Delta w(L - j)$ become small. The transversal filter is then implemented using the derived delay weights and the data deconvolved.

5.3.6   Fast linear convolution

In Section 5.2.3 it was shown that correlation computations could be speeded up using the correlation theorem. A similar theorem, the convolution theorem, exists in the case of convolution. Thus, in discrete terminology and for the time domain

$$x_1(l) \circledast x_2(r) = F_D^{-1}[X_1(k)X_2(k)] \qquad (5.117)$$

Equation 5.117 is the convolution theorem, in which F_D^{-1} denotes the inverse discrete Fourier transform, $X_1(k)$ is the discrete Fourier transform of $x_1(l)$, and $X_2(k)$ is the discrete Fourier transform of $x_2(r)$. As in Section 5.2.3, $x_1(l)$ and $x_2(r)$ are periodic sequences of length N.

Proof of convolution theorem

The proof of this theorem is almost identical to that of the correlation theorem as given in Section 5.2.3. In convolution one of the data sequences is reversed, and so, instead of Equation 5.65, its conjugate is employed, that is

$$X_1(k) = \sum_{l=0}^{N-1} x_1(l)\, e^{j(2\pi/N)(-lk)} \qquad (5.118)$$

while Equation 5.66 is used again:

$$X_2(k) = \sum_{r=0}^{N-1} x_2(r)\, e^{j(2\pi/N)(-rk)} \qquad (5.119)$$

Then, once more defining $x_3(n)$ as a periodic sequence of length N with DFT $X_3(k)$, $X_3(k)$ is written as

$$X_3(k) = X_1(k)X_2(k) \qquad (5.120)$$

The procedure of Section 5.2.3 is then followed, leading to the required result

$$x_1(l) \circledast x_2(r) = F_D^{-1}[X_1(k)X_2(k)] \qquad (5.121)$$

for time domain convolution. For convolution in the frequency domain the analogous equation below applies:

$$\frac{1}{N}[X_1(k) \circledast X_2(k)] = F_D[x_1(l)x_2(r)] \tag{5.122}$$

The last two equations represent periodic, or circular, convolutions which may be converted to linear by adding augmenting zeros as described in Section 5.3.2.

5.3.7 Computational advantages of fast linear convolution

The method of fast linear convolution offers the advantage of greater computational speed over the direct approach only if the number of values to be convolved is sufficiently large. The number of multiplications required to perform the convolution by the direct and fast methods is compared here as a measure of their relative computational efficiencies.

The necessary computations for the direct method were given in Equations 5.90. It can be seen from these equations that to obtain the linear convolution of the two N-point sequences $h(n-m)$ and $x(m)$ it is necessary to multiply each value of $h(n-m)$ by each value of $x(m)$. Thus N values of $h(n-m)$ are each to be multiplied by N values of $x(m)$, making $N \times N = N^2$ multiplications in all.

Now consider linear convolution of the same two N-point sequences by the fast method according to Equation 5.121. The addition of the necessary augmenting zeros means each sequence is of length $2N-1$ points. Assume $2N-1 \approx 2N$, for example $N \geqslant 8$, and that in order to use a radix-2 FFT N is given by an integer power of 2, that is $N = 2^d$ where d is an integer. The number of complex multiplications for an N-point FFT was shown to be $(N/2)\log_2 N$ (Section 3.5.3), so for the $2N$-point FFT $(2N/2)$ $\log_2 2N$ or $N\log_2 2N$ complex multiplications are necessary. Equation 5.121 requires two DFTs and one inverse DFT to be computed. The inverse will be computed using the modified DFT (Section 3.6). The computation therefore requires computation of three $2N$-point FFTs involving $3N\log_2 2N$ complex multiplications. Further, for each of the $2N$ values of Equation 5.121 it is necessary to evaluate the complex multiplications $X_1(k)X_2(k)$, thus increasing the number of complex multiplications necessary to $3N\log_2 2N + 2N$. Now, each complex multiplication of the form $(A + jB)(C + jD)$ requires four real multiplications: AC, AD, BC, and BD. Hence $12N\log_2 2N + 8N$ real multiplications are necessary.

It is therefore concluded that the direct method requires N^2 real multiplications while the method of fast convolution requires $12N\log_2 2N + 8N$ of them. Table 5.1 compares the numbers of real multiplications required for different values of N. The table demonstrates that fast convolution is faster than the direct method for sequences containing in excess of 128 data points, being approximately ten times faster for sequences containing 1024 points. The same conclusion also holds for direct and fast correlation.

Table 5.1　The numbers of real multiplications necessary for the convolution of two N-point sequences.

N	*Direct method*	*Fast convolution*	*Number ratio, fast: direct*
8	64	448	7
16	256	1 088	4.25
32	1 024	2 560	2.5
64	4 096	5 888	1.4375
128	16 384	13 312	0.8125
256	65 536	29 696	0.4531
512	262 144	65 536	0.250
1024	1 048 576	143 360	0.1367
2048	4 194 304	311 296	0.0742

5.3.8　Convolution and correlation by sectioning

So far it has been assumed that the two functions to be convolved (or correlated) are of finite duration. This, however, may not be the case. For example the input data may be considered to be of infinite duration, either because it is in fact continuous, or more likely because the available memory is not large enough to store all of it. In these cases it is necessary to perform the convolution (or correlation) in stages by dividing the input data into separate sections, performing the calculation for each of the input sections and then combining the results. The two methods used for this are known as the overlap–add and the overlap–save methods and will be described below but, first, these methods will be introduced by considering how to make the computations more efficient when the two functions do not start at the time origin.

Figure 5.25 shows two sampled waveforms $x(n)$ and $h(n)$ and their convolution $x(n) \circledast h(n) = y(n)$. $x(n)$ and $h(n)$ commence respectively at sample points a and b, so that if a and b are large compared with the number of data, N_1 and N_2 in $x(n)$ and

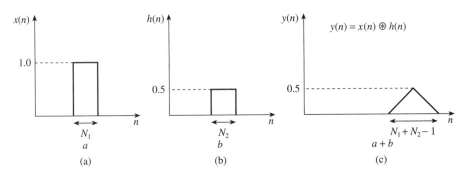

Figure 5.25　The convolution, $y(n) = x(n) \circledast h(n)$, of the two waveforms $x(n)$ and $h(n)$ which do not commence at the origin.

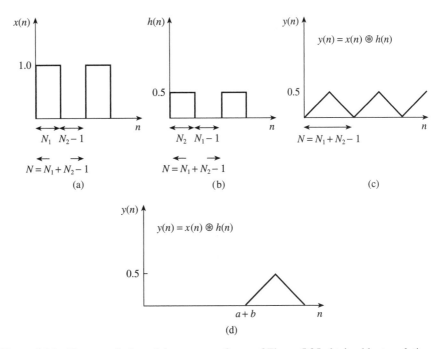

Figure 5.26   The convolution of the two waveforms of Figure 5.25 obtained by translating $x(n)$ and $h(n)$ to commence at the origin. (a) Addition of $N_2 - 1$ augmenting zeros to $x(n)$. (b) Addition of $N_1 - 1$ augmenting zeros to $h(n)$. (c) The convolution $y(n) = x(n) \circledast h(n)$. (d) The correct linear convolution obtained by displacing $y(n)$ along the n-axis to commence at $n = a + b$.

$h(n)$ respectively, then a considerable number of calculations involving zero data will be performed. The number of such calculations can be reduced by shifting the waveforms to commence at the origin as in Figure 5.26. Augmenting zeros must then be added to each waveform so that they both contain the same number of points $N = N_1 + N_2 - 1$ so that their periodic convolution corresponds to the linear convolution of the two waveforms. The convolution is performed by applying the convolution theorem, Equation 5.117, and an FFT algorithm. The correct result is then obtained by displacing the resulting convolution along the n-axis to commence at $n = a + b$ (Figure 5.26(d)). It is assumed in this figure that $N = 2^d$ where d is integral so that a radix-2 FFT may be used.

Figure 5.27 shows the analogous case for the correlation of $x(n)$ and $h(n)$, $r_{xh}(n)$. When these waveforms are transposed to the origin, augmenting zeros are added so that $N = 2^d \geqslant N_1 + N_2 - 1$, and the correlation is carried out using the correlation theorem, Equation 5.77, the resulting waveform is as shown in Figure 5.28. This is not a periodic version of Figure 5.27(c), although it has the correct basic waveshape. The desired periodic result may be obtained by commencing $x(n)$ at the point $n = N - N_1 + 1$ while continuing to commence $h(n)$ at $n = 0$, as in Figure 5.29 which shows that this produces the required periodic correlation function, Figure 5.29(c).

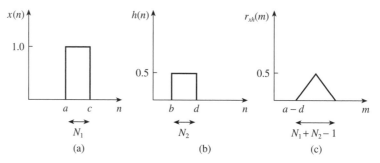

Figure 5.27   The cross-correlation, $r(m)$, of the two waveforms $x(n)$ and $h(n)$ which do not commence at the origin. (a) $x(n)$, (b) $h(n)$, (c) $r_{xh}(m)$.

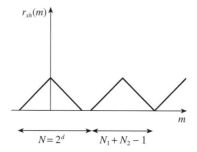

Figure 5.28   Incorrect cross-correlation obtained when $x(n)$ and $h(n)$ are translated to the origin for the cross-correlation.

The result has to be shifted by $a - d - N + N_1 + N_2$ data points to the right to commence at the correct value of $a - d$ (compare with Figure 5.29(c)).

It is now possible to extend these considerations to the case of an infinite sequence $x(n)$ to be convolved with the finite one $h(n)$.

5.3.9   Overlap–add method

Assume $x(n)$ is divided up into equal length sections of N_1 data points. Now assume that these are periodic and are convolved with the $N_2 h(n)$ data augmented with $N_1 - N_2$ zeros so that both sequences are periodic and of length N_1. The result of this convolution will be incorrect because to obtain a correct result each sequence has to be of length $N = N_1 + N_2 - 1$. However, each section of $x(n)$ is of length N_1 (and cannot be increased). The problem may be overcome by considering sections of $x(n)$ of length N and replacing the last $N_2 - 1$ data by zeros which augment the first $N - N_2 + 1 = N_1$ data (Figure 5.30). In this way a sequence of N_1 data of $x(n)$ with $N_2 - 1$ augmenting zeros is convolved with N_2 data of $h(n)$ with $N_1 - 1$ augmenting zeros. Both sequences contain $N = N_1 + N_2 - 1$ data and are correctly convolved (Figure 5.31). The same

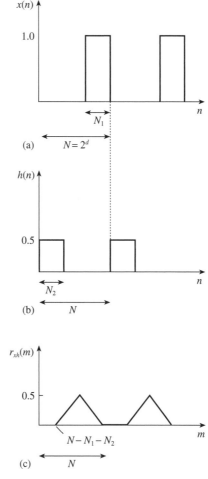

Figure 5.29   Method for obtaining the correct cross-correlation of the two sequences $x(n)$ and $h(n)$. (a) $x(n)$ commencing at $N - N_1 + 1$. (b) $h(n)$ commencing at the origin. (c) The resultant correct periodic cross-correlation coefficient, $r_{xh}(m)$.

procedure is carried out on the remaining sequences of $x(n)$ of length N. Because the final $N_2 - 1$ data of the $x(n)$ sections have been replaced by zeros, the resulting convolution functions are erroneous for the first and last $N_2 - 1$ points of each convolution, but these points sum to give the correct convolution when each convolved waveform is translated to its proper origin $(a + b)$ and the final $N_2 - 1$ points of the convolution derived from one section overlap those of the next. Figure 5.31 illustrates the process. Thus enough zeros are first added to eliminate end effects and then the convolution results are overlapped and added together exactly where the zeros were added to the sequence N_1. This is why it is called the overlap–add method.

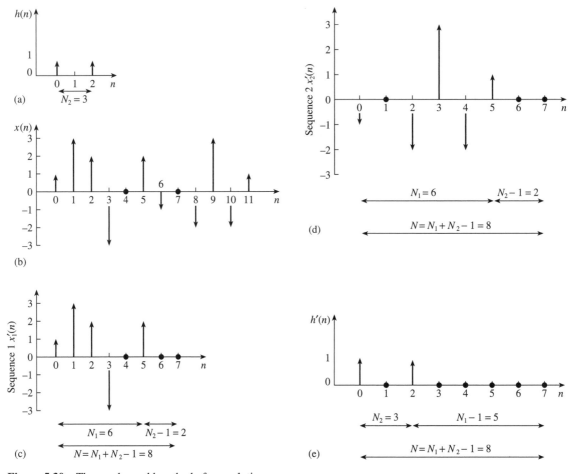

Figure 5.30 The overlap–add method of convolution.

<table>
<tr><td>Example 5.10</td><td>Use the overlap–add procedure to convolve the two sequences $h(n) = \{1, 0, 1\}$ and $x(n) = \{1, 3, 2, -3, 0, 2, -1, 0, -2, 3, -2, 1, \ldots\}$.</td></tr>
</table>

Solution   Let $x(n)$ be divided up into sections of length $N_1 = 6$ so that N, the number of points in the DFT, is $N_1 + N_2 - 1 = 6 + 3 - 1 = 8 = 2^d$ where $d = 3$, thus satisfying the requirements for linear convolution and for the use of a radix-2 FFT.

Adding zeros to $h(n)$ gives the augmented sequence $h'(n)$:

$$h'(n) = \{1, 0, 1, 0, 0, 0, 0, 0\}$$

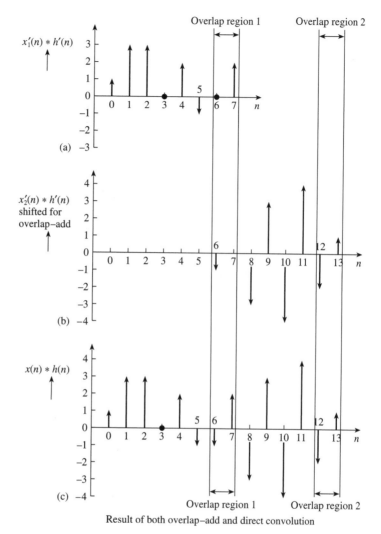

Figure 5.31 Equivalence of overlap–add method of convolution and of direct convolution.

The first two augmented sequences of $x(n)$ are

$$x_1'(n) = \{1, 3, 2, -3, 0, 2, 0, 0\}$$

and

$$x_2'(n) = \{-1, 0, -2, 3, -2, 1, 0, 0\}$$

The terms in the convolution sum $x_1'(n) \circledast h'(n)$ are

$$y_{10} = h'_0 x'_{10} = 1$$

$$y_{11} = h'_0 x'_{11} + h'_1 x'_{10} = 3 + 0 = 3$$

$$y_{12} = h'_0 x'_{12} + h'_1 x'_{11} + h'_2 x'_{10} = 2 + 0 + 1 = 3$$

$$y_{13} = h'_0 x'_{13} + h'_1 x'_{12} + h'_2 x'_{11} = -3 + 0 + 3 = 0$$

$$y_{14} = h'_0 x'_{14} + h'_1 x'_{13} + h'_2 x'_{12} = 0 + 0 + 2 = 2$$

$$y_{15} = h'_0 x'_{15} + h'_1 x'_{14} + h'_2 x'_{13} = 2 + 0 - 3 = -1$$

$$y_{16} = h'_0 x'_{16} + h'_1 x'_{15} + h'_2 x'_{14} = 0 + 0 + 0 = 0$$

$$y_{17} = h'_0 x'_{17} + h'_1 x'_{16} + h'_2 x'_{15} = 0 + 0 + 2 = 2$$

The terms in the convolution sum $x'_2(n) \circledast h'(n)$ are

$$y_{20} = h'_0 x'_{20} = -1$$

$$y_{21} = h'_0 x'_{21} + h'_1 x'_{20} = 0 + 0 = 0$$

$$y_{22} = h'_0 x'_{22} + h'_1 x'_{21} + h'_2 x'_{20} = -2 + 0 - 1 = -3$$

$$y_{23} = h'_0 x'_{23} + h'_1 x'_{22} + h'_2 x'_{21} = 3 + 0 + 0 = 3$$

$$y_{24} = h'_0 x'_{24} + h'_1 x'_{23} + h'_2 x'_{22} = -2 + 0 - 2 = -4$$

$$y_{25} = h'_0 x'_{25} + h'_1 x'_{24} + h'_2 x'_{23} = 1 + 0 + 3 = 4$$

$$y_{26} = h'_0 x'_{26} + h'_1 x'_{25} + h'_2 x'_{24} = 0 + 0 - 2 = -2$$

$$y_{27} = h'_0 x'_{27} + h'_1 x'_{26} + h'_2 x'_{25} = 0 + 0 + 1 = 1$$

The above two convolution sums are shown in Figures 5.31(a) and 5.31(b) respectively. If the first $N_2 - 1 = 2$ data of x'_2 are overlapped with the last $N_2 - 1$ data of x'_1 and the convolution sums added then the first 12 data of the resulting convolution waveform given by this overlap–add method are as shown in Figure 5.31(c).

The above result can be demonstrated to be identical to that obtained by carrying out the convolution directly as follows. The original sequence $x(n)$ contains 12 data and $h(n)$ contains 3 data. To obtain the linear convolution of the two, augmenting zeros must be added to both sequences so that they both contain $12 + 3 - 1 = 14$ data. Thus the sequences become

$$h'(n) = \{1, 0, 1, 0, 0, 0, 0, 0, 0, 0, 0, 0, 0, 0\}$$

and

$$x'(n) = \{1, 3, 2, -3, 0, 2, -1, 0, -2, 3, -2, 1, 0, 0\}$$

The first nine terms in the convolution sum are

$$y_0 = h'_0 x'_0 = 1$$

$$y_1 = h'_0 x'_1 + h'_1 x'_0 = 3$$

$$y_2 = h'_0 x'_2 + h'_1 x'_1 + h'_2 x'_0 = 2 + 0 + 1 = 3$$

$$y_3 = h'_0 x'_3 + h'_1 x'_2 + h'_2 x'_1 = -3 + 0 + 3 = 0$$

$$y_4 = h'_0 x'_4 + h'_1 x'_3 + h'_2 x'_2 = 0 + 0 + 2 = 2$$

$$y_5 = h'_0 x'_5 + h'_1 x'_4 + h'_2 x'_3 = 2 + 0 - 3 = -1$$

$$y_6 = h'_0 x'_6 + h'_1 x'_5 + h'_2 x'_4 = -1 + 0 + 0 = -1$$

$$y_7 = h'_0 x'_7 + h'_1 x'_6 + h'_2 x'_5 = 0 + 0 + 2 = 2$$

$$y_8 = h'_0 x'_8 + h'_1 x'_7 + h'_2 x'_6 = -2 + 0 - 1 = -3$$

The terms of this convolution sum are indeed identical to those obtained by the overlap–add method plotted in Figure 5.31(c).

The overlap–add procedure for fast convolution (or correlation) by sectioning is therefore as follows.

(1) Select the number of $x(n)$ data, N_1, to be of the order of the number of $h(n)$ data, N_2, with $N_1 > N_2$, and also select the number of DFT points to be $N = 2^d$ where d is an integer and $N \gg N_1 + N_2 - 1$. Satisfy these conditions by adding augmenting zeros to the data sequences as necessary.

(2) Shift the augmented sections of $x(n)$ data to the origin.

(3) For each section of augmented $x(n)$ data, $x'(n)$, perform the fast convolution $x'(n) \circledast h'(n)$, that is compute $X(k)H(k)$ and then its inverse transform.

(4) Sequentially overlap the resulting convolutions by their final and first $N_2 - 1$ values and then add them.

5.3.10 Overlap–save method

Consider again the convolution $x(n) \circledast h(n)$ illustrated in Figure 5.32 in which $N_2 - 1$ zeros have been added to $h(n)$ so that both sequences are of length N_1. The linear convolution sum of these may be obtained by successively sliding $h(n)$ to the right by

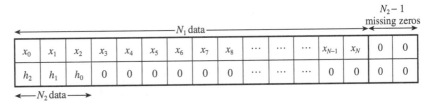

Figure 5.32 $x(n)$ and $h(n)$ with $N_2 - 1$ zeros added to $h(n)$.

one datum, cross-multiplying corresponding terms, and summing. However, because neither sequence is of length $N_1 + N_2 - 1$ the result will not be $x(n) \circledast h(n)$. There are in fact $N_2 - 1$ zeros missing from $x(n)$ of length N_1. This means that the first $N_2 - 1$ terms of the convolution sum will be incorrect and should be discarded. Therefore if the data $x(n)$ is divided up into contiguous sections of length N_1 the first $N_2 - 1$ values of each of the convolution sums must be discarded. The convolution of $x(n) \circledast h(n)$ will therefore contain a periodic sequence of gaps of length $N_2 - 1$. These gaps may be correctly filled by overlapping the final $N_2 - 1$ data of each $x(n)$ sequence of length N_1 with the first $N_2 - 1$ data of the following sequence, and then discarding these first $N_2 - 1$ data. This procedure is called the overlap–save method.

Example 5.11

Use the overlap–save procedure to convolve the same two sequences convolved in Section 5.3.9, that is

$$h(n) = \{1, 0, 1\}$$

and

$$x(n) = \{1, 3, 2, -3, 0, 2, -1, 0, -2, 3, -2, 1\}$$

Solution

Since $h(n)$ has $N_2 = 3$ the amount of overlap is $N_2 - 1 = 2$. The overlapping of the sections is as shown in Figure 5.33. The convolutions are evaluated for each section as below.

For section 1,

$$y_{10} = h_0 x_{10} = 1$$

$$y_{11} = h_0 x_{11} + h_1 x_{10} = 3 + 0 = 3$$

$$y_{12} = h_0 x_{12} + h_1 x_{11} + h_2 x_{10} = 2 + 0 + 1 = 3$$

$$y_{13} = h_1 x_{13} + h_1 x_{12} + h_2 x_{11} + h_3 x_{10} = -3 + 0 + 3 + 0 = 0$$

$h(n)$	1	0	1									
$x(n)$	1	3	2	-3	0	2	-1	0	-2	3	-2	1
Section 1	1	3	2	-3								
Section 2			2	-3	0	2						
Section 3					0	2	-1	0				
Section 4							-1	0	-2	3		
Section 5									-2	3	-2	1

Figure 5.33   Overlapping of sections for the overlap–save method of convolution.

Table 5.2 Results for Example 5.11.

Section		1	2	3	4	5	6	7	8	9	10	11	12
Section 1	y_0	~~1~~ 3		3	0								
Section 2	y_1			~~2~~ 3		2	−1						
Section 3	y_2					~~0~~ 2		−1	2				
Section 4	y_3							~~1~~ 0		−3	3		
Section 5	y_4									~~2~~ 3		−4	4
$x(n) \circledast h(n)$		~~1~~	3	3	0	2	−1	−1	2	−3	3	−4	4

Thus

$$y_1 = \{1, 3, 3, 0\}$$

For the remaining sections remember $h_1 = h_3 = 0$. We obtain, for section 2,

$$y_{20} = h_0 x_{20} = 2$$

$$y_{21} = h_0 x_{21} = -3$$

$$y_{22} = h_0 x_{22} + h_2 x_{20} = 2 + 0 = 2$$

$$y_{23} = h_0 x_{23} + h_2 x_{21} = 2 - 3 = -1$$

$$y_2 = \{2, -3, 2, -1\}$$

Similarly, for section 3,

$$y_3 = \{0, 2, -1, 2\}$$

For section 4,

$$y_4 = \{-1, 0, -3, 3\}$$

and finally, for section 5,

$$y_5 = \{-2, 3, -4, 4\}$$

These results are illustrated in Table 5.2 which shows that the first $N_2 - 1$ results of each sequence are discarded. Apart from the first $N_2 - 1$ points the last row in the table corresponds to the correct convolution.

The overlap–save procedure is therefore as follows.

(1)   Select the number of $x(n)$ data, $N_1 = 2^d$, to be convolved with $h(n)$ and add $N_2 - 1$ zeros to $h(n)$ so that both sequences are of length N_1.

(2)   Locate both sequences at the origin.

(3)   For each sequence compute the corresponding values of $X(k)$ and $H(k)$ using an FFT.

(4)   Compute $X(k)H(k)$ and its inverse, which is the convolution of each sequence with $h(n)$.

(5)   Adjust each of the convolutions to overlap the preceding one by $N_2 - 1$ data.

(6)   Discard the first $N_2 - 1$ data of each convolution and read out the remaining values which correspond to the correct convolution.

5.3.11   Computational advantages of fast convolution by sectioning

It was shown in Section 5.3.8 that unnecessary computational effort may be avoided by first setting every section of waveform to be convolved at the origin and it is assumed here that this has been done. It is further assumed that the computational requirements of the overlap–add and overlap–save methods are similar so that it is only necessary to consider the overlap–add method. It is assumed that the sequence $x(n)$ of length N is divided up into N/N_1 sections each of length N_1, that the sequence $h(n)$ is of length N_2, and that the lengths of the sequences for linear convolution are $N^1 = 2^d \geqslant N_1 + N_2 - 1$. It has also been shown, Section 5.3.7, that $12N^1 \log_2 2N^1 + 8N^1$ real multiplications are required to perform the fast convolution of two N^1-point sequences. Thus, in order to carry out the fast convolution of the N-point sequence $x(n)$ by the overlap–add method $(N/N_1)(12N^1 \log_2 2N^1 + 8N^1) = R_m(S)$ real multiplications will be required. This shows that the sequence lengths to be convolved, N^1, should be short while the lengths, N_1, of the $x(n)$ data sections should approach N^1. Ideally $N^1 = 2^d = N_1 + N_2 - 1$. The number of real multiplications for the original N-point sequence is $12N \log_2 2N + 8N = R_m(N)$. Table 5.3 shows that for the example of Section 5.3.9 the ratio $R_m(S)/R_m(N) \leqslant 1$ typically, with savings in computational time of the order of 50% being possible.

Table 5.3   Ratio $R_m(S)/R_m(N)$, number of real multiplications for sectioning method: number of real multiplications by straightforward method for fast convolution.

N	N^1	N_1	N/N_1	N_2	$R_m(S)/R_m(N)$	*Comment*
1020	8	6	170	3	0.54	Best result with N^1 short, $N_1 \approx N^1$ $N_1 \approx N^1$
1024	256	254	4	3	0.83	
1020	128	102	10	3	0.93	
1020	256	204	5	3	1.04	

5.3.12 The relationship between convolution and correlation

In convolution the value of the nth output is given by the convolution sum of Equation 5.93:

$$y(n) = \sum_{m=0}^{n} h(m)x(n-m) = h(0)x(n) + h(1)x(n-1) + \ldots + h(n)x(0) \quad (5.123)$$

The value of the cross-correlation function for the waveforms $h(n)$ and $x(n)$ for the jth lag is given by Equation 5.1 modified slightly to be

$$r_{hx}(j) = \frac{1}{N} \sum_{n=0}^{N-1} h(n)x(j+n)$$

$$= \frac{1}{N}[h(0)x(j) + h(1)x(j+1) + \ldots + h(N-1)x(j+N-1)] \quad (5.124)$$

It is easier to compare $y(n)$ and $r_{hx}(j)$ if the case $j = 0$, that is zero lag in the cross-correlation, is considered. Equation 5.124 then becomes

$$r_{hx}(0) = \frac{1}{N} \sum_{n=0}^{N-1} h(n)x(n)$$

$$= \frac{1}{N}[h(0)x(0) + h(1)x(1) + \ldots + h(N-1)x(N-1)] \quad (5.125)$$

Comparing Equations 5.123 and 5.125 reveals that they are of similar form except that the $x(n)$ sequence in the cross-correlation is in the reverse order to that in the convolution. Thus convolution equates to the cross-correlation of the two waveforms in which one of the original sequences has been time reversed, and the normalizing factor $1/N$ has been set to unity. This means that convolutions and correlations may be computed by the same computer programs simply by reversing one of the sequences.

5.4 Implementation of correlation and convolution

In considering the implementation of these operations it should be remembered that the two are intimately related. Two data sequences may be either correlated or convolved simply by reversing the order of one of the data sequences. Furthermore, for longer data sequences the operations can be speeded up using fast Fourier transform methods to achieve fast correlation or fast convolution. Where one of the data sequences is very long the overlap–add or overlap–save technique will be appropriate: see Sections 5.3.9 and 5.3.10 and Brigham (1974), Strum and Kirk (1988) and DeFatta *et al.* (1988).

Convolution or correlation can be achieved using, for example, an FIR filter, which may be implemented by FFTs. Correlation or convolution can also be achieved using a matched filter as shown in Section 5.2.2, Figure 5.13 which illustrates the correlation detector. For digital processing charge-coupled device (CCD) technology may be used to implement transversal filters. These give a linear phase response with data rates in excess of 100 MHz possible for basic delay line configurations (Grant *et al.*, 1989). Analog processing may be carried out by implementing tapped delay lines using surface acoustic wave (SAW) devices (Grant *et al.*, 1989). These operate over a range from about 2 MHz to 2 GHz. Other implementations include dedicated convolver and correlator chips, general-purpose digital signal processors, standard microprocessors, and transputers. An example of the latter would be a real-time system for the removal of ocular artefacts from all 16 channels of human EEG (Jervis *et al.*, 1990).

The computational time required for fast correlations and convolutions may be further halved as follows (Brigham, 1974). Consider the convolution of $x(n)$ with $h(n)$. When computing $X(k)$ fill the real part of the FFT with the even terms of $x(n)$ and the imaginary part with the odd terms, thereby halving the length of the FFT. The real part of $(1/N)F_D^{-1}[X(k)H(k)]$ then gives the even terms of the desired convolution and the imaginary part gives the odd terms.

Likewise, the convolutions of two data sequences $x_1(n)$ and $x_2(n)$ with $h(n)$ may be computed simultaneously. Fill the real part of an FFT with $x_1(n)$ and the imaginary part with $x_2(n)$ and transform to obtain $X^1(k)$. Then the real part of $(1/N)F_D^{-1}[X^1(k)H(k)]$ is $x_1(n) \circledast h(n)$ and the imaginary part is $x_2(n) \circledast h(n)$.

5.5 Application examples

5.5.1 Correlation

Example 5.12

This simplified example concerns the application of correlation theory to the control of the attitude of a spacecraft to ensure that the solar panel always faces the sun. Attitude errors are represented as multilevel pulses with level separations $a = 0.2$ mV and pulse widths $T_s = 1$ μs. An initial attempt is made to control the attitude error by transmitting a sequence of negative pulses of height a when there is a positive error. The control system is only considered satisfactory if the correlation coefficient between the error and control signals is better than -0.5. Figure 5.34(a) shows three error pulses while Figure 5.34(b) shows the corresponding control signal pulses. For this example it is assumed that these pulses are sufficient, and that it is unnecessary to consider a lag greater than T_s. The problem is to determine whether or not the system may be regarded as satisfactory.

The system can be proven by ascertaining that $|r_{12}(\tau)| > 0.5$ for $0 \leqslant \tau \leqslant T_s$. The cross-correlation may be achieved by shifting the control signal to the right while keeping the error signal fixed. This means that $r_{12}(-\tau)$ should be determined.

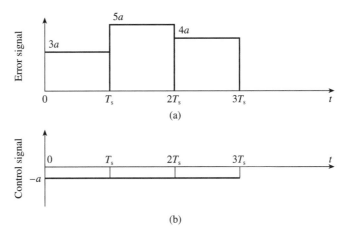

Figure 5.34   Spacecraft attitude control. (a) Error signal. (b) Control signal.

Now

$$r_{12}(-\tau) = \int_{-\infty}^{\infty} v_1(t)v_2(t - \tau)\, \mathrm{d}t$$

where $v_1(t)$ is the error signal and $v_2(t)$ is the control signal.

$$
\begin{aligned}
r_{12}(-\tau) &= \frac{1}{3T_s}\int_{\tau}^{T_s} 3a(-a)\, \mathrm{d}t + \frac{1}{3T_s}\int_{T_s}^{2T_s} 5a(-a)\, \mathrm{d}t + \frac{1}{3T_s}\int_{2T_s}^{3T_s} 4a(-a)\, \mathrm{d}t \\
&= \frac{a^2}{3T_s}\{[-3t]_{\tau}^{T_s} + [-5t]_{T_s}^{2T_s} + [-4t]_{2T_s}^{3T_s}\} \\
&= \frac{a^2}{3T_s}(-3T_s + 3\tau - 10T_s + 5T_s - 12T_s + 8T_s) \\
&= \frac{a^2}{3T_s}(-12T_s + 3\tau)
\end{aligned}
$$

$r_{12}(\tau)$ is now normalized to confine values to the range $-1 \ll r_{12}(\tau) \ll 1$ by dividing by the normalizing factor

$$\frac{1}{3T_s}\left[\int_{-\infty}^{\infty} v_1^2(t)\, \mathrm{d}t \int_{-\infty}^{\infty} v_2^2(t)\, \mathrm{d}t\right]^{1/2}$$

Now,

$$\int_{-\infty}^{\infty} v_1^2(t)\, dt = \int_0^{T_s} (3a)^2\, dt + \int_{T_s}^{2T_s} (5a)^2\, dt + \int_{2T_s}^{3T_s} (4a)^2\, dt$$

$$= a^2\{[9t]_0^{T_s} + [25t]_{T_s}^{2T_s} + [16t]_{2T_s}^{3T_s}\}$$

$$= a^2(9T_s + 25T_s + 16T_s) = 50a^2T_s$$

Also

$$\int_{-\infty}^{\infty} v_2^2(t)\, dt = \int_0^{3T_s} (-a)^2\, dt = a^2[t]_0^{3T_s} = 3a^2T_s$$

Hence the normalizing factor is

$$\frac{1}{3T_s}[(50a^2T_s)(3a^2T_s)]^{1/2} = \frac{1}{3T_s}150^{1/2}a^2T_s$$

and the normalized expression for $r_{12}(-\tau)$ is

$$r_{12}^N(-\tau) = \frac{3\tau - 12T_s}{150^{1/2}\,T_s} = \frac{3\tau}{12.25T_s} - \frac{12}{12.25}$$

$$r_{12}^N(-\tau) = 0.245 \times 10^6\tau - 0.98$$

When $\tau = 0$,

$$r_{12}^N(0) = -0.98$$

When $\tau = 1\ \mu s$ (the largest allowed value),

$$r_{12}^N(10^{-6}) = -0.735$$

Thus, over the range considered $|r_{12}^N(-\tau)| > |0.5|$ which satisfies the criterion for good control of the spacecraft's attitude.

Example 5.13

A sonar system is required for the determination of the distance of a sound source. The source is broadband and gaussian with zero mean. The system consists of two sonar transducers separated by a distance, d, and an associated signal processing system. The transducers, T_1 and T_2, receive the broadband noise signals $q_1(t)$ and $q_2(t) = Aq_1(t + \Delta t)$ respectively, Δt being the time lag due to the different path lengths to the source from the two transducers, and A the associated attenuation factor (assume $A = 1$ in this case). The signal processing system computes the correlation function of equal lengths of the two transducer outputs.

Draw a labelled block diagram of a simple system designed to achieve the correlation in the shortest possible time, and explain the principles upon which it is based.

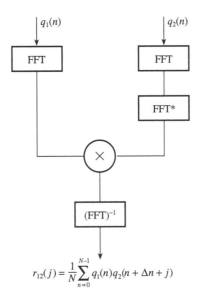

Figure 5.35   Sonar system block diagram.

Sketch the transducer output signals and their cross-correlation function, indicating noteworthy features.

If the peak value of the cross-correlation function is 10, and the receiver has a bandwidth of 1–10 Hz, calculate the received energy.

Solution   The system block diagram is shown in Figure 5.35. This system speeds up the correlation calculations in this design by applying the correlation theorem and computing the FFTs involved. This will be faster than a straightforward computation of correlation when the number of data points in the data sequences exceeds about 128. Thus the system computes $r_{12}(\tau)$ which expressed digitally is

$$r_{12}(j) = F_D^{-1}[F_1(k)F_2^*(k)]$$

The system output, $r_{12}(j)$, is

$$r_{12}(j) = \frac{1}{N}\sum_{n=0}^{N-1} q_1(n)Aq_1(n + \Delta n + j)$$

As $q_1(n)$ and $q_2(n)$ are random the system will only produce a significant output when the waveforms are relatively shifted to be in phase. This occurs when $j = -\Delta n$. The output then is

$$\frac{1}{N}\sum_{n=0}^{N-1} q_1^2(n) = P_{AV}, \text{ the average power}$$

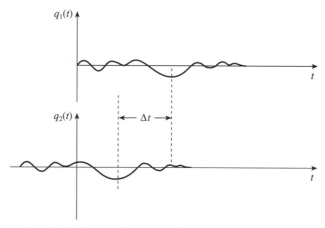

Δt = time lag between the two transducers

Figure 5.36   Broadband noise signals detected by sonar system.

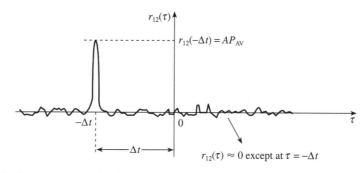

Figure 5.37   Cross-correlation function of signals detected by the sonar system.

Figures 5.36 and 5.37 show the waveforms and their cross-correlation function. The cross-correlation of the two waveforms is

$$r_{12}(\tau) = \frac{1}{T} \int_{-T/2}^{T/2} q_1(t) q_2(t + \tau)\, dt$$

On substituting for $q_2(t)$ this becomes

$$r_{12}(\tau) = \frac{1}{T} \int_{-T/2}^{T/2} q_1(t) A q_1(t + \Delta t + \tau)\, dt$$

which may be written

$$r_{12}(\tau) = \frac{A}{T} \int_{-T/2}^{T/2} q_1(t) q_1(t + \tau') \, dt, \quad \text{where } \tau' = \Delta t + \tau$$

The integrand is seen to be equivalent in magnitude to the autocorrelation of $q_1(t)$ at zero lag, and therefore represents the power of this signal, P_{AV}. Hence

$$r_{12}(\tau) = A P_{AV} \delta(t + \Delta t)$$

where δ represents the delta function. The magnitude of $r_{12}(\tau)$ is seen to be $A P_{AV}$. Therefore $A P_{AV} = 10$.

We may obtain the received energy over the required bandwidth by first applying the Wiener–Khintchine theorem to obtain the energy spectral density. The theorem is

$$G_E(f) = F_D[r_{12}(\tau)]$$
$$= F_D[A P_{AV} \delta(t + \Delta t)] = A P_{AV} \, e^{j\omega \Delta t}$$

Therefore $|G_E(f)| = A P_{AV} = 10 \text{ J Hz}^{-1}$. The signal bandwidth is $10 - 1 \text{ Hz} = 9 \text{ Hz}$. Therefore the received energy is $10 \times 9 = 90 \text{ J}$.

5.5.2   Convolution

5.5.2.1   FIR and IIR filters

The operation of transversal filters, both FIR and IIR, provides good application examples of convolution (Stremler, 1982; DeFatta *et al.*, 1988). They may be designed to convolve sequences or to perform more general digital filtering, for example two-dimensional filtering as employed in image processing (Grant *et al.*, 1989), for noise reduction, image enhancement, and pattern recognition.

Consider a linear time-independent (LTI) system describable by

$$y(n) = \sum_{k=1}^{N} a_k y(n-k) + \sum_{k=0}^{L} b_k x(n-k) \tag{5.126}$$

in which $y(n)$ represents the output sequence and $x(n)$ the input sequence. The output is seen to depend on the current input and also on the previous inputs and outputs. a_k and b_k are real constants and N is the order of the equation and represents the number of previous outputs that have to be considered.

Because the current output depends on previous outputs the system is recursive. If the system output depends only on previous inputs it is said to be nonrecursive and is describable by

$$y(n) = \sum_{k=0}^{L} b_k x(n-k) \tag{5.127}$$

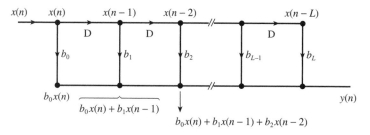

Figure 5.38   Diagrammatic representation of a nonrecursive filter.

and this equation is descriptive of a transversal filter (or tapped delay line).

Figure 5.38 shows a diagrammatic description of the system of Equation 5.127. The terms in the summation, which represents the system output, are obtained by summing the delayed and weighted values of the inputs. It will now be shown that these weights correspond to the impulse response of the system. Assume that the input, $x(n)$, is a unit impulse, $\delta(n)$, where

$$x(n) = \delta(n) = \begin{cases} 1, & n = 0, \quad \text{that is } x(0) = 1 \\ 0, & n \neq 0, \quad \text{that is } x(n \neq 0) = 0 \end{cases}$$

The corresponding output is the impulse response, $h(n)$. Substitution of consecutive input values into Equation 5.127 gives

$$y(0) = h(0) = b_0 x(0) + b_1 \times 0 = b_0 \times 1 = b_0$$

$$y(1) = h(1) = b_0 \times 0 + b_1 x(0) + b_2 \times 0 = b_1 \times 1 = b_1$$

$$\vdots$$

$$y(L) = h(L) = b_0 \times 0 + 0 + 0 + \ldots + 0 + b_L \times 1 = b_L$$

Therefore

$$h(n) = \{b_0, b_1, \ldots, b_L\} \tag{5.128}$$

showing that the weights on the system diagram correspond to the coefficients of its impulse response function. Such systems are known as finite impulse response (FIR) filters.

Now consider the output corresponding to the general input sequence $x(n)$. Substituting consecutive values into Equation 5.127 gives

$$y(n) = b_0 x(n) + b_1 x(n-1) + \ldots + b_n x(0)$$

$$\equiv h(0)x(n) + h(1)x(n-1) + \ldots + h(n)x(0) \tag{5.129}$$

which can be recognized as the convolution of the input with the output as would be expected. Thus FIR filters may also be regarded as convolvers in which the filter weights correspond to the coefficients of their impulse response.

A different, but similar, relationship applies in the case of infinite impulse response (IIR) filters. Consider the first-order recursive filter described by the equation

$$y(n) = a_1 y(n-1) + b_0 x(n) \tag{5.130}$$

It is easily demonstrable that for a unit impulse input

$$y(n) = h(n) = b_0 a_1^n \quad n \geqslant 0 \tag{5.131}$$

For the general input sequence $x(n)$, assuming $y(-1) = 0$,

$$y(0) = b_0 x(0)$$

$$y(1) = a_1 b_0 x(0) + b_0 x(1)$$

$$y(2) = a_1^2 b_0 x(0) + a_1 b_0 x(1) + b_0 x(2)$$

$$\vdots$$

$$y(n) = a_1^n b_0 x(0) + a_1^{n-1} b_0 x(1) + \ldots + a_1 b_0 x(n-1) + b_0 x(n)$$

Substituting the known values of the weights from Equation 5.131 gives

$$y(n) = h(n)x(0) + h(n-1)x(1) + \ldots + h(0)x(n) \tag{5.132}$$

Equations 5.131 and 5.132 show that the IIR filter corresponding to the first-order system is a convolver for which the impulse response coefficients are given by $h(n) = b_0 a_1^n$.

FIR filters are used in speech processing (Grant *et al.*, 1989) to achieve reduced bandwidth PCM, in sub-band coders, for parametric spectral analysis, and in linear predictive vocoders. FIR filters also find applications in radars, and in spread spectrum communications (Grant *et al.*, 1989).

5.5.2.2 Convolution coding

Convolutional codes allow burst error correction by distributing the parity check digits of the code over a long stream of symbols (Stremler, 1982; Taub and Schilling, 1986). The outputs of the flip–flops of a shift register provide delays and are tapped and appropriately combined using modulo-2 adders. This produces a number of outputs which are read consecutively each clock cycle (Figure 5.39). The system is essentially causal and nonrecursive and produces an output which depends on its previous inputs and convolves new input data with its impulse response.

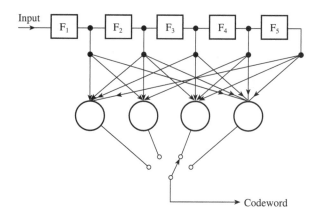

Figure 5.39 Convolution coder.

5.5.2.3 Deconvolution

The input to all systems is convolved with the impulse response of the system and may distort the output. This occurs, for example, in telecommunication systems and may necessitate the design of an equalizer, which is a linear filter which deconvolves the convolved output. Before a deconvolution filter can be designed the impulse response of the system must be measured (system identification). The subject of system identification and deconvolution is quite extensive (Proakis and Manolakis, 1988) and is not discussed here.

5.5.2.4 Speech

There is great interest in analyzing and coding speech for purposes such as human–machine interaction and data compression. Use is sometimes made of the fact that the speech waveform can be modelled as the convolution of a train of impulses representing pitch, an excitation pulse, and the impulse response of the vocal tract (Rabiner and Gold, 1975). The resulting triple convolution can easily be converted to a form suitable for processing by an LTI system. The use of FIR filters in speech processing was highlighted in Section 5.5.2.1.

5.6 Summary

The topics of correlation and convolution and their interrelationship have been thoroughly discussed in this chapter. Normalization procedures and the avoidance of end effects have been discussed for correlation. The effects of correlating noisy signals and the identification of signals in noise by correlation procedures have been described as well as other applications. The techniques of fast correlation and convolution based on the correlation and convolution theorems and using FFTs have been described, and it was shown how to obtain linear convolution. The fast overlap–add and overlap–save methods for obtaining the convolution of a long data sequence were obtained. A further means of speeding up the computations for real data by a factor of 2 by exploiting both the real and the imaginary parts of an FFT has also been described.

Problems

5.1 Two separate recordings of equal length are made of a periodic pulse train being transmitted down a noisy channel. Table 5.4 shows the recorded values of the sampled voltages.

(1) Determine the amount of lag between the two recordings and the period of the waveform.

(2) Derive the periodic waveform.

5.2 Evaluate the cross-correlation functions of recordings 1 and 2 of Table 5.4 with and without

correcting for the end effect. Estimate the errors introduced by the end effect.

5.3 What is the percentage correlation between recordings 1 and 2 of Table 5.4 evaluated at zero lag? Assume percentage correlation is defined as the correlation coefficient, ρ_{12}, multiplied by 100%.

5.4 The sampled voltages from a noisy waveform are given in Table 5.5. Use the technique of cross-correlation with a template waveform to discover the exact shape of the periodic waveform present. Check your conclusion using another method.

Table 5.4 Sampled voltages (volts) for two separate recordings from the same channel.

Recording 1	6.02	−5.98	7.92	−7.96	−0.78
Recording 2	8.93	−7.20	−0.82	3.23	1.44
	−8.34	9.22	−2.65	−3.7	9.51
	5.43	−9.88	−1.13	0.79	9.83
	5.53	3.50	−3.18	−8.85	8.21
	−8.73	4.64	−8.49	−4.66	−8.84
	1.69	−0.06	6.65	−8.00	−9.21
	5.55	−8.24	−0.37	2.71	4.63
	−0.78	7.27	−5.98	−3.97	9.11
	1.88	−0.92	−5.33	9.01	9.23
	4.23	2.99	−1.85	−5.27	3.81
	−3.7	5.08	−0.72	−5.08	−2.6
	6.62	−2.64	2.08	−5.91	−3.58
	9.67	−8.55	−3.08	4.18	8.11
	−1.65	3.64	−8.19	−3.50	4.84
	0.74	−3.87	−4.09	8.03	6.91
	7.25	2.93	−4.42	−8.21	3.61
	−9.87	−3.62	−8.29	−5.8	−7.04

Table 5.5   The sampled voltages from a noisy waveform.

−7.37,	−7.99,	3.31,	−8.59,	−1.68,	3.01,	12.21,	−2.38,	7.46,
−9.84,	1.48,	1.1,	−1.8,	5.48,	8.93,	0,	−9.36,	−10.11,
1.61,	3.36,	−4.86,	6.27					

Table 5.6   Digitized voltage values.

−0.92,	−3.71,	3.11,	−0.24,	4.65,	0.84,	−2.98,	−3.94,	−4.03,	−2.51,	0.17,
3.85,	2.58,	0.38,	4.58,	3.4,	−3.46					

5.5   Calculate the autocorrelation function of the periodic waveform of Problem 5.4 (a) numerically and (b) analytically. Compare these solutions with each other and with the autocorrelation function of the noisy waveform. Account for any discrepancies from the anticipated results.

5.6   A voltage waveform is sampled and digitized. The digitized voltage values are given in Table 5.6. Determine whether or not the waveform may be regarded as random. On the assumption that the sampling interval was 1 ms and that a periodic signal component was present with a period of 4 ms, obtain an estimate of the periodic waveform and plot it.

5.7   Compare the signal-to-noise ratios of

(1)   the noisy periodic waveform of recording 1 in Table 5.4,

(2)   the autocorrelation function of recording 1 in Table 5.4, and

(3)   the cross-correlation function of recordings 1 and 2 of Table 5.4.

5.8   Evaluate the theoretical signal-to-noise ratios for

(1)   the periodic waveform obtained from the data of recording 1, Table 5.4, by cross-correlation with the appropriate impulse train,

(2)   the autocorrelation function of recording 1 in Table 5.4, given that (i) the signal-to-noise ratio $(S/N)_{r0}$ of the autocorrelation function of a noisy sinusoidal signal is

$$(S/N)_{r0} = \cfrac{N}{1 + 8 \left/ \left(\dfrac{S_i}{N_i}\right)\right. + 2 \left/ \left(\dfrac{S_i}{N_i}\right)^2\right.}$$

where N is the number of data, S_i is the signal power, and N_i is the noise power, and (ii) the signal-to-noise ratio $(S/N)_\delta$ of the cross-correlation of a noisy-sinusoidal signal with a periodic impulse train of the same period as the signal is

$$(S/N)_\delta = \cfrac{N}{1 + 1 \left/ \left(\dfrac{S_i}{N_i}\right)\right.}$$

5.9   Compare the answers obtained to Problems 5.7 and 5.8.

5.10   A matched filter has the impulse response function $h(n) = \{1, -1, -1, 1, 1, -1, 1\}$ and is used to detect the arrival at a receiver of the corresponding signal transmitted down a noisy channel. Table 5.7 shows the sampled signal values, each of which represents the value of a pulse in a bipolar pulse train of amplitude ± 1.5 V and of pulse width 1 μs. Determine the time of arrival of the signal and the value of the matched filter constant.

5.11   Find the impulse response function of a system if its response to the PN sequence $\{1, 1, -1, 1, -1, -1, 1, -1\}$ is $y(n) = \{0, 0, 0.5, 1.5, 1.5, 1.5, 1, -1, -1, -1, -1.5, -0.5, -0.5, -0.5\}$.

5.12   A unit amplitude pulse of width 2 ms is applied to the circuit which has the impulse response shown

Table 5.7   Sampled voltages from a noisy bipolar signal.

$t\,(\mu s)$	0	1	2	3	4	5	6
Voltage	0.14	0.48	1.61	2.09	−2.40	0.40	2.35
$t\,(\mu s)$	7	8	9	10	11	12	13
Voltage	−0.59	−1.81	0.32	−0.47	1.81	−1.63	−2.28

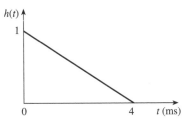

Figure 5.40   System impulse response for Problem 5.12.

in Figure 5.40. Determine the output waveform numerically. Sample the waveforms at 0.5 ms intervals.

5.13 (1) Figure 5.41 shows two functions $x_1(t)$ and $x_2(t)$. Evaluate

   (a) their convolution, $x_3(t)$, numerically, taking sampled values at $t = 0, 1, 2, 3, 4, 5$ s, and

   (b) $x_3(t)$ analytically.

   (2) Sketch the functions $x_3(t)$ and give reasons for any differences between them.

5.14 Determine the shape of the output pulse when a rectangular pulse of magnitude 5 V and width

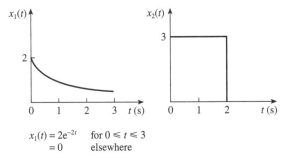

$x_1(t) = 2e^{-2t}$ for $0 \leqslant t \leqslant 3$
$= 0$ elsewhere

Figure 5.41   The functions $x_1(t)$ and $x_2(t)$ for Problem 5.13.

0.4 μs is applied at the input of a single-stage lowpass RC filter with a cutoff frequency of 6 MHz. Assume that the impulse response of the filter is given by

$$h(t) = \frac{1}{CR} e^{-t/CR} u(t)$$

5.15 A rectangular pulse of height 5 V and width 1.0 μs is applied to the input of a system with the response function, $h(t)$, given by

$$h(t) = 0.1[1 - e^{-t/(1.09\times10^{-6})}] \quad 0 \leqslant t \leqslant 10\,\mu s$$
$$= 0 \qquad 10\,\mu s < t < 0$$

Determine the system output

   (1) analytically, and

   (2) by sampling $h(t)$ every 1 μs and representing the pulse by an impulse function located at $t = 0$ s.

Critically compare your results.

5.16 Obtain the cross-correlation function between the two sets of data {1.5, 2.0, 1.5, 2.0, 2.5} and {0, 0.33, 0.67, 1.0}

   (1) by direct cross-correlation, and

   (2) by applying the correlation theorem.

5.17 Determine the output of an electrical system of impulse response function {0, 0.899, 0.990, 0.991, 1} when the input {0, 2.5, 5.0, 0} (volts) is applied

   (1) by direct convolution, and

   (2) by applying the convolution theorem.

5.18 Use the overlap–add method to calculate the output of the system with the impulse response function $h(n) = \{0, 0.899, 0.990, 0.999, 1\}$ for the input data given in Table 5.5 (but ignoring the last two data). Assume the data were sampled every 2.5 μs, and divide the input data into five equal length sections. Calculate the phase shift between the output and input to the system. Check your answer using the method of direct convolution.

5.19 Repeat Problem 5.18 using the overlap–add method in which the convolutions are obtained using the convolution theorem. Compare the result with the answer to Problem 5.18.

5.20 Find the output of the system of Problem 5.18 which has the impulse response function $h(n) = \{0, 0.899, 0.990, 0.999, 1\}$ for all but the last two data of Table 5.5 using the overlap–save method. Compare the result with the solution to Problem 5.18.

5.21 Repeat Problem 5.20 applying the convolution theorem to evaluate the convolutions. Compare the result with the answers to Problems 5.18–5.20.

5.22 Consider Problems 5.18–5.21 and your solutions to them and compare the numbers of calculations required for the different methods with each other and with the calculation by direct convolution.

5.23 Write a program to perform convolution by the overlap–add method. Use it to confirm the results of Problem 5.18 and then to investigate the output of various systems for inputs of your choice.

5.24 (1) Write a program to carry out fast correlation and use it to cross-correlate the data recordings 1 and 2 of Table 5.4.

 (2) Investigate the cross-correlations and autocorrelations of various waveforms such as square waves, rectangular waves, sine waves, random noise, and waveforms with various signal-to-noise ratios.

 (3) Compare the relative abilities of correlation and spectrum estimation methods to detect signals in noise.

MATLAB problems

5.25 The successive sample values of two discrete-time signals are as follows:

$$x = 4, 2, -1, 3, -2, -6, -5, 4, 5$$

$$y = -4, 1, 3, 7, 4, -2, -8, -2, 1$$

(a) Compute the estimates of the normalized and denormalized autocorrelation of each of the data sequences using MATLAB.

(b) Compute and plot estimates of the biased and unbiased autocorrelation of each of the data sequences using MATLAB.

(c) Compute and plot estimates of the normalized and denormalized cross-correlation function of the two data sequences using MATLAB.

(d) Compute and plot estimates of the biased and unbiased cross-correlation function of the two data sequences using MATLAB.

(e) Determine the estimates of the cross- or autocorrelation functions, as appropriate, at zero lag for parts (a) to (d) above.

(f) Determine the lengths of the cross- or autocorrelation functions, as appropriate, for parts (a) to (d) above.

(g) Compare the results from (a) to (e) above and comment on any differences.

5.26 Compare the normalized cross-correlation function of the data sequences obtained in Problem 5.25(a) with that obtained with the data sequences swapped (i.e. the cross-correlation function of x and y versus that for y and x).

5.27 (a) Generate a 1000 point random Gaussian white noise data sequence (using the **randn** function).

(b) Compute and plot the estimates of the autocorrelation function of the sequence in (a) for the first 30 lags.

5.28 A continuous time signal is characterized by the following equation:

$$x(t) = A \cos(2\pi f_1 t) + B \cos(2\pi f_2 t)$$

(a) Generate, with the aid of MATLAB, a discrete-time equivalent of the signal. Assume a sampling frequency of 1 kHz, $f_1 = 50$ Hz, $f_2 = 100$ Hz and that the ratio of the amplitudes of the frequency components, $A/B = 1.5$.

(b) Compute and plot the estimates of the autocorrelation function of the sequence in (a).

5.29 (a) Generate, with the aid of appropriate MATLAB functions, and plot each of the following waveforms:

(i) a sine wave – use sin(2*pi*t/100), with $t = 0 : 1 : 1000$.

(ii) a noise waveform – use the **randn** function.

(iii) a noisy sine wave – add the two waveforms in (i) and (ii).

(iv) a square waveform – use *square*(2*pi*t/100).

(b) Compute and plot the normalized autocorrelation of each of the waveforms in (a).

(c) Describe briefly the unique and shared properties of the autocorrelation functions computed in (b).

5.30 In this problem, the goal is to simulate the problem of estimating distance of objects by correlation. The cross-correlation of the transmitted signal and noisy reflected signal from the object should reveal a peak at a time lag that corresponds to twice the required distance.

(a) Generate, with the aid of appropriate MATLAB functions, each of the waveforms depicted in Figure 1.3(a) and (b) (where the top and bottom traces represent the transmitted and received waveforms, respectively).

(b) Compute the cross-correlation function between the two waveforms and hence estimate the distance of the object from the transmitter.

Assume a radio wave travelling at 3×10^8 m/s and a sampling frequency of 4 MHz.

5.31 Repeat Problem 5.25 using the **xcov** function. Comment on any differences between your results and those obtained in Problem 5.25.

References

Beauchamp K.G. (1973) *Signal Processing Using Analog and Digital Techniques*. London: Allen and Unwin.

Bell A.J. and Sejnowski T.J. (1995) An information-maximisation approach to blind separation and blind deconvolution. *Neural Computation*, **7**, 1129–59.

Brigham E.O. (1974) *The Fast Fourier Transform*, Sections 13.3 and 13.4. Englewood Cliffs NJ: Prentice-Hall.

Chatfield C. (1980) *The Analysis of Time Series*, p. 62. London: Chapman and Hall.

DeFatta D.J., Lucas J.G. and Hodgkiss W.S. (1988) *Digital Signal Processing: A System Design Approach*, Section 6.9, p. 306. New York: Wiley.

Grant P.M., Cowan C.F.N., Mulgrew B. and Dripps J.H. (1989) *Analogue and Digital Signal Processing and Coding*, Chapters 16, 17, 19 and 20. Bromley, UK: Chartwell-Bratt.

Jenkins G.M. and Watts D.G. (1968) *Spectral Analysis and its Applications*. San Francisco CA: Holden-Day.

Jervis B.W., Goude A., Thomlinson M., Mir S. and Miller G. (1990) Least squares artefact removal by transputer. In *IEE Colloquium on the Transputer and Signal Processing*, Savoy Place, London, 5 March 1990.

Main G. and Howell T.D. (1993) Determining a signal to noise ratio for an arbitrary data sequence by a time domain analysis. *IEEE Transactions on Magnetics*, **29**(6), November, 3999–4001.

McGillem C.D. and Cooper G.R. (1974) *Continuous and Discrete Signal and System Analysis*. New York: Holt, Rinehart, and Winston.

Proakis J.G. and Manolakis D.G. (1988) *Introduction to Digital Signal Processing*, p. 429. Basingstoke: Macmillan.

Rabiner L.R. and Gold B. (1975) *Theory and Application of Digital Signal Processing*, Chapters 12 and 13. Englewood Cliffs NJ: Prentice-Hall.

Stremler F.G. (1982) *Introduction to Communication Systems*, 2nd edn, Section 3.10 and p. 407. Reading MA: Addison-Wesley.

Strum R.D. and Kirk D.E. (1988) *First Principles of Discrete Systems and Digital Signal Processing*, Chapter 3. Reading MA: Addison-Wesley.

Taub H. and Schilling D.L. (1986) *Principles of Communication Systems*, 2nd edn, p. 562. New York: McGraw-Hill.

Appendix

5A   C language program for computing cross- and autocorrelation

The program, **correltn.c**, for computing the cross- or autocorrelation values of data sequences is available on the CD in the companion handbook (see Preface for details).

A framework for digital filter design

6

The purpose of this chapter is to provide a common framework for digital filter design. A simple step-by-step guide for designing digital filters, from specifications to implementation, is described. The options open to the designer at each step of the design process and factors that influence their choice are highlighted using several illustrative examples. Most DSP texts devote substantial space to the theory of digital filters, especially approximation methods, reflecting the considerable research effort that has gone into finding useful methods of calculating filter coefficients and the significant advances that have been made in filter design. However, such a coverage often overwhelms the inexperienced filter designer and leaves him or her not knowing how actually to go about designing a filter or how it all fits together. Thus the framework provided here, in our experience, is valuable to the designer who actually wants to design digital filters, as opposed to just learning about them from a purely theoretical point of view. This chapter sets the scene for Chapters 7 and 8 in which actual digital filter design is fully covered.

6.1   Introduction to digital filters

A filter is essentially a system or network that selectively changes the wave shape, amplitude–frequency and/or phase–frequency characteristics of a signal in a desired manner. Common filtering objectives are to improve the quality of a signal (for example, to remove or reduce noise), to extract information from signals or to separate two or more signals previously combined to make, for example, efficient use of an available communication channel.

A digital filter, as we shall see later, is a mathematical algorithm implemented in hardware and/or software that operates on a digital input signal to produce a digital output signal for the purpose of achieving a filtering objective. The term digital filter refers to the specific hardware or software routine that performs the filtering algorithm. Digital filters often operate on digitized analog signals or just numbers, representing some variable, stored in a computer memory.

A simplified block diagram of a real-time digital filter, with analog input and output signals, is given in Figure 6.1. The bandlimited analog signal is sampled periodically and converted into a series of digital samples, $x(n)$, $n = 0, 1, \ldots$. The digital processor implements the filtering operation, mapping the input sequence, $x(n)$, into the output sequence, $y(n)$, in accordance with a computational algorithm for the filter. The DAC converts the digitally filtered output into analog values which are then analog filtered to smooth and remove unwanted high frequency components.

Digital filters play very important roles in DSP. Compared with analog filters they are preferred in a number of applications (for example data compression, biomedical signal processing, speech processing, image processing, data transmission, digital audio, telephone echo cancellation) because of one or more of the following advantages.

∎ Digital filters can have characteristics which are not possible with analog filters, such as a truly linear phase response.

∎ Unlike analog filters, the performance of digital filters does not vary with environmental changes, for example thermal variations. This eliminates the need to calibrate periodically.

∎ The frequency response of a digital filter can be automatically adjusted if it is-implemented using a programmable processor, which is why they are widely used in adaptive filters.

∎ Several input signals or channels can be filtered by one digital filter without the need to replicate the hardware.

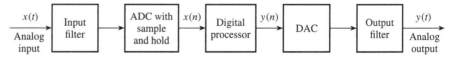

Figure 6.1   A simplified block diagram of a real-time digital filter with analog input and output signals.

- Both filtered and unfiltered data can be saved for further use.

- Advantage can be readily taken of the tremendous advancements in VLSI technology to fabricate digital filters and to make them small in size, to consume low power, and to keep the cost down.

- In practice, the precision achievable with analog filters is restricted; for example, typically a maximum of only about 60 to 70 dB stopband attenuation is possible with active filters designed with off-the-shelf components. With digital filters the precision is limited only by the wordlength used.

- The performance of digital filters is repeatable from unit to unit.

- Digital filters can be used at very low frequencies, found in many biomedical applications for example, where the use of analog filters is impractical. Also, digital filters can be made to work over a wide range of frequencies by a mere change to the sampling frequency.

The following are the main disadvantages of digital filters compared with analog filters:

- *Speed limitation*   The maximum bandwidth of signals that digital filters can handle, in real time, is much lower than for analog filters. In real-time situations, the analog–digital–analog conversion processes introduce a speed constraint on the digital filter performance. The conversion time of the ADC and the settling time of the DAC limit the highest frequency that can be processed. Further, the speed of operation of a digital filter depends on the speed of the digital processor used and on the number of arithmetic operations that must be performed for the filtering algorithm, which increases as the filter response is made tighter.

- *Finite wordlength effects*   Digital filters are subject to ADC noise resulting from quantizing a continuous signal, and to roundoff noise incurred during computation. With higher order recursive filters, the accumulation of roundoff noise could lead to instability.

- *Long design and development times*   The design and development times for digital filters, especially hardware development, can be much longer than for analog filters. However, once developed the hardware and/or software can be used for other filtering or DSP tasks with little or no modifications (several examples of this are given in subsequent chapters). Good computer aided design (CAD) support can make the design of digital filters an enjoyable task, but some expertise is required to make full and effective use of such design aids.

6.2   Types of digital filters: FIR and IIR filters

Digital filters are broadly divided into two classes, namely infinite impulse response (IIR) and finite impulse response (FIR) filters. Either type of filter, in its basic form,

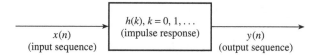

Figure 6.2   A conceptual representation of a digital filter.

can be represented by its impulse response sequence, $h(k)$ ($k = 0, 1, \ldots$), as in Figure 6.2. The input and output signals to the filter are related by the convolution sum, which is given in Equations 6.1 for the IIR and 6.2 for the FIR filter.

$$y(n) = \sum_{k=0}^{\infty} h(k)x(n-k) \tag{6.1}$$

$$y(n) = \sum_{k=0}^{N-1} h(k)x(n-k) \tag{6.2}$$

It is evident from these equations that, for IIR filters, the impulse response is of infinite duration whereas for FIR it is of finite duration, since $h(k)$ for the FIR has only N values. In practice, it is not feasible to compute the output of the IIR filter using Equation 6.1 because the length of its impulse response is too long (infinite in theory). Instead, the IIR filtering equation is expressed in a recursive form:

$$y(n) = \sum_{k=0}^{\infty} h(k)x(n-k) = \sum_{k=0}^{N} b_k x(n-k) - \sum_{k=1}^{M} a_k y(n-k) \tag{6.3}$$

where the a_k and b_k are the coefficients of the filter. Thus, Equations 6.2 and 6.3 are the difference equations for the FIR and IIR filters respectively. These equations, and in particular the values of $h(k)$, for FIR, or a_k and b_k, for IIR, are often very important objectives of most filter design problems. We note that, in Equation 6.3, the current output sample, $y(n)$, is a function of past outputs as well as present and past input samples, that is the IIR is a feedback system of some sort. This should be compared with the FIR equation in which the current output sample, $y(n)$, is a function only of past and present values of the input. Note, however, that when the b_k are set to zero, Equation 6.3 reduces to the FIR Equation 6.2.

Alternative representations for the FIR and IIR filters are given in Equations 6.4a and 6.4b respectively. These are the transfer functions for these filters and are very useful in evaluating their frequency responses (see Chapters 4, 7 and 8 for details).

As will become clear in the next few sections, factors that influence the choice of options open to the digital filter designer at each stage of the design process are strongly linked to whether the filter in question is IIR or FIR. Thus, it is very important to appreciate the differences between IIR and FIR, their peculiar characteristics, and more importantly, how to choose between them.

$$H(z) = \sum_{k=0}^{N-1} h(k) z^{-k} \tag{6.4a}$$

$$H(z) = \sum_{k=0}^{N} b_k z^{-k} \left/ \left(1 + \sum_{k=1}^{M} a_k z^{-k} \right) \right. \tag{6.4b}$$

6.3 Choosing between FIR and IIR filters

The choice between FIR and IIR filters depends largely on the relative advantages of the two filter types.

(1) FIR filters can have an exactly linear phase response. The implication of this is that no phase distortion is introduced into the signal by the filter. This is an important requirement in many applications, for example data transmission, biomedicine, digital audio and image processing. The phase responses of IIR filters are nonlinear, especially at the band edges.

(2) FIR filters realized nonrecursively, that is by direct evaluation of Equation 6.2, are always stable. The stability of IIR filters cannot always be guaranteed.

(3) The effects of using a limited number of bits to implement filters such as roundoff noise and coefficient quantization errors are much less severe in FIR than in IIR.

(4) FIR requires more coefficients for sharp cutoff filters than IIR. Thus for a given amplitude response specification, more processing time and storage will be required for FIR implementation. However, one can readily take advantage of the computational speed of the FFT and multirate techniques (see Chapter 9) to improve significantly the efficiency of FIR implementations.

(5) Analog filters can be readily transformed into equivalent IIR digital filters meeting similar specifications. This is not possible with FIR filters as they have no analog counterpart. However, with FIR it is easier to synthesize filters of arbitrary frequency responses.

(6) In general, FIR is algebraically more difficult to synthesize, if CAD support is not available.

From the above, a broad guideline on when to use FIR or IIR would be as follows.

■ Use IIR when the only important requirements are sharp cutoff filters and high throughput, as IIR filters, especially those using elliptic characteristics, will give fewer coefficients than FIR.

■ Use FIR if the number of filter coefficients is not too large and, in particular, if little or no phase distortion is desired. One might also add that newer DSP processors have architectures that are tailored to FIR filtering, and indeed some are designed specifically for FIRs (see Chapter 12).

Example 6.1   The following transfer functions represent two different filters meeting identical amplitude–frequency response specifications:

(1) $H(z) = \dfrac{b_0 + b_1 z^{-1} + b_2 z^{-2}}{1 + a_1 z^{-1} + a_2 z^{-2}}$

where

$$b_0 = 0.498\ 181\ 9$$
$$b_1 = 0.927\ 477\ 7$$
$$b_2 = 0.498\ 181\ 9$$
$$a_1 = -0.674\ 487\ 8$$
$$a_2 = -0.363\ 348\ 2$$

(2) $H(z) = \displaystyle\sum_{k=0}^{11} h(k) z^{-k}$

where

$$h(0) = \quad 0.546\ 032\ 80 \times 10^{-2} = h(11)$$
$$h(1) = -0.450\ 687\ 50 \times 10^{-1} = h(10)$$
$$h(2) = \quad 0.691\ 694\ 20 \times 10^{-1} = h(9)$$
$$h(3) = -0.553\ 843\ 70 \times 10^{-1} = h(8)$$
$$h(4) = -0.634\ 284\ 10 \times 10^{-1} = h(7)$$
$$h(5) = \quad 0.578\ 924\ 00 \times 10^{0} = h(6)$$

For each filter,

(a)   state whether it is an FIR or IIR filter,

(b)   represent the filtering operation in a block diagram form and write down the difference equation, and

(c)   determine and comment on the computational and storage requirements.

Solution   (a)   Filters (1) and (2) are IIR and FIR respectively.

(b)   The block diagram for filter (1) is given in Figure 6.3(a). The corresponding set of difference equations is

$$w(n) = x(n) - a_1 w(n-1) - a_2 w(n-2)$$
$$y(n) = b_0 w(n) + b_1 w(n-1) + b_2 w(n-2)$$

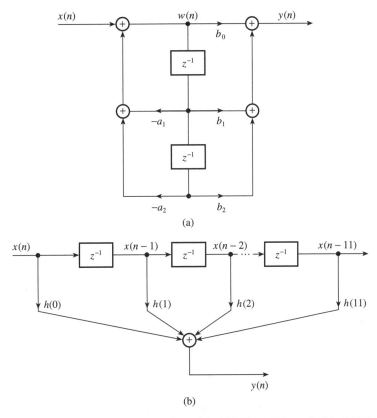

Figure 6.3 (a) Block diagram representation of the IIR filter of Example 6.1. (b) Block diagram representation of the FIR filter of Example 6.1.

The block diagram for filter (2) is given in Figure 6.3(b). The corresponding difference equation is

$$y(n) = \sum_{k=0}^{11} h(k)x(n-k)$$

(c) From examination of the two difference equations the computational and storage requirements for both filters are summarized below:

	FIR	IIR
Number of multiplications	12	5
Number of additions	11	4
Storage locations (coefficients and data)	24	8

It is evident that the IIR filter is more economical in both computational and storage requirements than the FIR filter. However, we could have exploited the symmetry in the FIR coefficients to make the FIR filter more efficient, although at the expense of its obvious implementational simplicity. A point worth making is that, for the same amplitude response specifications, the number of FIR filter coefficients (12 in this example) is typically six times the order (the highest power of z in the denominator) of the IIR transfer function (2 in this case).

6.4   Filter design steps

The design of a digital filter involves five steps:

(1)   Specification of the filter requirements.

(2)   Calculation of suitable filter coefficients.

(3)   Representation of the filter by a suitable structure (realization).

(4)   Analysis of the effects of finite wordlength on filter performance.

(5)   Implementation of filter in software and/or hardware.

The five steps are not necessarily independent; nor are they always performed in the order given. In fact techniques are now available that combine the second and aspects of the third and fourth steps. However, the approach discussed here gives a simple step-by-step guide that will ensure a successful design. To arrive at an efficient filter, it may be necessary to iterate a few times between the steps, especially if the problem specification is not watertight, as is often the case, or if the designer wants to explore other possible designs. Detailed discussions of these steps now follow.

6.4.1   Specification of the filter requirements

Requirement specifications include specifying (i) signal characteristics (types of signal source and sink, I/O interface, data rates and width, and highest frequency of interest), (ii) the characteristics of the filter (the desired amplitude and/or phase responses and their tolerances (if any), the speed of operation and modes of filtering (real time or batch)), (iii) the manner of implementation (for example, as a high level language routine in a computer or as a DSP processor-based system, choice of signal processor), and (iv) other design constraints (for example, the cost of the filter). The designer may not have enough information to specify the filter completely at the outset, but as many of the filter requirements as possible should be specified to simplify the design process.

Although the above requirements are application dependent it will be helpful to devote some time on some aspects of (ii). The characteristics of digital filters are often specified in the frequency domain. For frequency selective filters, such as lowpass and

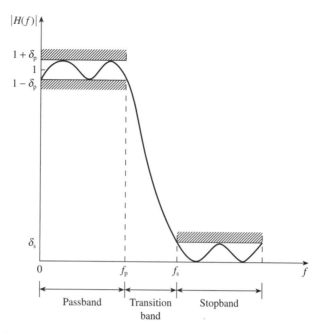

Figure 6.4   Tolerance scheme for a lowpass filter.

bandpass filters, the specifications are often in the form of tolerance schemes. Figure 6.4 depicts such a scheme for a lowpass filter. The shaded horizontal lines indicate the tolerance limits. In the passband, the magnitude response has a peak deviation of δ_p and, in the stopband, it has a maximum deviation of δ_s.

The width of the transition band determines how sharp the filter is. The magnitude response decreases monotonically from the passband to the stopband in this region. The following are the key parameters of interest:

δ_p passband deviation
δ_s stopband deviation
f_p passband edge frequency
f_s stopband edge frequency

The edge frequencies are often given in normalized form, that is as a fraction of the sampling frequency (f/F_s), but specifications using standard frequency units of hertz or kilohertz are valid and sometimes are more meaningful, especially to the inexperienced designer. Passband and stopband deviations may be expressed as ordinary numbers or in decibels when they specify the passband ripple and minimum stopband attenuation respectively. Thus the minimum stopband attenuation, A_s, and the peak passband ripple, A_p, in decibels are given as (for FIR filters)

$$A_s \text{ (stopband attenuation)} = -20\log_{10}\delta_s \tag{6.5a}$$

$$A_p \text{ (passband ripple)} \quad = 20\log_{10}(1+\delta_p) \tag{6.5b}$$

The phase response of digital filters is often not as meticulously specified as the amplitude response. In many cases it is sufficient to indicate that phase distortion is of concern or that linear phase response is desirable. However, in some applications where filters are used to equalize or compensate a system's phase response, for example, or as phase shifters, then the desired phase response will need to be specified.

Example 6.2   An FIR bandpass filter is to be designed to meet the following frequency response specifications:

passband	0.18–0.33	(normalized)
transition width	0.04	(normalized)
stopband deviation	0.001	
passband deviation	0.05	

(1)   Sketch the tolerance scheme for the filter.

(2)   Express the filter bandedge frequencies in the standard unit of kilohertz, assuming a sampling frequency of 10 kHz, and the stopband and passband deviations in decibels.

Solution   (1)   The tolerance scheme for the filter is given in Figure 6.5.

(2)   The bandedge frequencies, at a sampling frequency of 10 kHz, and the stopband and passband deviations are given below:

passband	1.8–3.3 kHz
stopbands	0–1.4 kHz and 3.7–5 kHz
stopband attenuation	$-20 \log_{10}(0.001) = 60$ dB
passband ripple	$20 \log_{10}(1 + 0.05) = 0.42$ dB

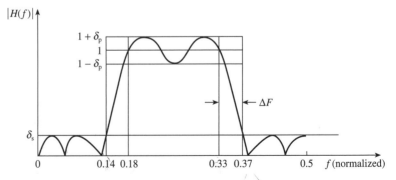

Figure 6.5   Tolerance scheme for the bandpass filter of Example 6.2.

6.4.2   Coefficient calculation

In this step, we select one of a number of approximation methods and calculate the values of the coefficients, $h(k)$, for FIR, or a_k and b_k, for IIR, such that the filter characteristics given in Section 6.4.1 are satisfied. The method used to calculate the filter coefficients depends on whether the filter is IIR or FIR type.

Calculations of IIR filter coefficients are traditionally based on the transformation of known analog filter characteristics into equivalent digital filters. The two basic methods used are the impulse invariant and the bilinear transformation methods. With the impulse invariant method, after digitizing the analog filter, the impulse response of the original analog filter is preserved, but not its magnitude–frequency response. Because of inherent aliasing, the method is inappropriate for highpass or bandstop filters. The bilinear method, on the other hand, yields very efficient filters and is well suited to the calculation of coefficients of frequency selective filters. It allows the design of digital filters with known classical characteristics such as Butterworth, Chebyshev and elliptic. Digital filters resulting from the bilinear transform method will, in general, preserve the magnitude response characteristics of the analog filter but not the time domain properties. Efficient computer programs now exist for calculating filter coefficients, using the bilinear method, by merely specifying filter parameters of interest (see Chapter 8). The impulse invariant method is good for simulating analog systems, but the bilinear method is best for frequency selective IIR filters.

The pole–zero placement method offers an alternative approach to calculating the coefficients of IIR filters. It is an easy way of calculating the coefficients of very simple filters. However, for filters with good amplitude response it is not recommended as it relies on 'trial and error' shuffling of the pole and zero positions.

As with IIR filters there are several methods of calculating the coefficients of FIR filters. The three methods discussed in this book in detail are the window, frequency sampling, and the optimal (Parks–McClellan algorithm). The window method offers a very simple and flexible way of computing FIR filter coefficients, but it does not allow the designer adequate control over the filter parameters. The main attraction of the frequency sampling method is that it allows a recursive realization of FIR filters which can be computationally very efficient. However, it lacks flexibility in specifying or controlling filter parameters. With the availability of an efficient and easy-to-use program, the optimal method is now widely used in industry and, for most applications, will yield the desired FIR filters. Thus, for FIR filters, the optimal method should be the method of first choice unless the particular application dictates otherwise or a CAD facility is unavailable.

In summary, there are several methods of calculating filter coefficients of which the following are the most widely used:

- impulse invariant   (IIR);
- bilinear transformation   (IIR);
- pole–zero placement   (IIR);

- window   (FIR);
- frequency sampling   (FIR);
- optimal   (FIR).

We choose the method that best suits our particular application. Our choice will be influenced by several factors, the most important of which are the critical requirements in the specifications. In general, the crucial choice is really between FIR and IIR. In most cases, if the FIR properties are vital then a good candidate is the optimal method, whereas, if IIR properties are desirable, then the bilinear method will in most cases suffice.

6.4.3   Representation of a filter by a suitable structure (realization)

Realization involves converting a given transfer function, $H(z)$, into a suitable filter structure. Block or flow diagrams are often used to depict filter structures and they show the computational procedure for implementing the digital filter. The structure used depends on whether the filter is an IIR or FIR filter.

For IIR filters, three structures commonly used are the direct form, cascade and parallel forms. The direct form is simply a straightforward representation of the IIR transfer function. In the cascade form, the transfer function of the IIR filter, Equation 6.4b, is factored and expressed as the product of second-order sections. In the parallel form, $H(z)$ is expanded, using partial fractions, as the sum of second-order sections. As an illustration and for comparison, Figure 6.6 shows the block diagrams for a fourth-order (that is, $N = 4$) IIR filter represented in the direct, cascade and parallel structures. The corresponding set of transfer functions and the difference equations describing the filter structures are also given in the figure.

The parallel and cascade structures are the most widely used for IIR because they lead to simpler filtering algorithms and are far less sensitive to the effects of implementing the filter using a finite number of bits than the direct structure. The direct structure suffers severe coefficient sensitivity problems, especially for high order filters, and should be avoided in these cases.

The most widely used structure for FIR is the direct form, Figure 6.7(a), because it is particularly simple to implement. In this form, the FIR is sometimes called a tapped delay line (because it resembles a tapped delay line) or transversal filter. Two other FIR structures that are also used are the frequency sampling structure and the fast convolution technique, Figures 6.7(b) and 6.7(c). Compared with the transversal structure, the frequency sampling structure can be computationally more efficient as it leads to fewer coefficients, but it may not be as simple to implement and would require more storage. The fast convolution uses the computational advantage of the fast Fourier transform (FFT) and is particularly attractive in situations where the power spectrum of the signal is also required.

There are many other practical structures for digital filters, but most of these are popular only in specific application areas. An example is the lattice structure which finds use in speech processing and linear prediction applications. The lattice structure

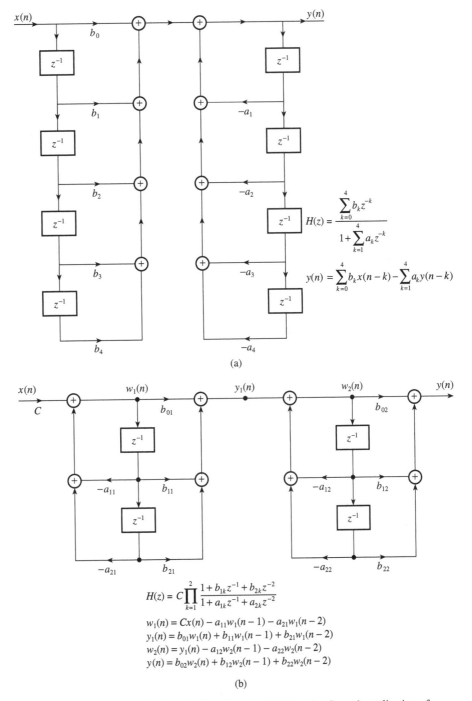

Figure 6.6 (a) Direct realization of a fourth-order IIR filter. (b) Cascade realization of a fourth-order IIR filter.

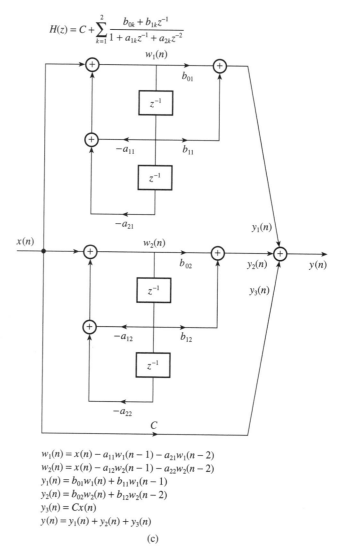

$$H(z) = C + \sum_{k=1}^{2} \frac{b_{0k} + b_{1k}z^{-1}}{1 + a_{1k}z^{-1} + a_{2k}z^{-2}}$$

$w_1(n) = x(n) - a_{11}w_1(n-1) - a_{21}w_1(n-2)$

$w_2(n) = x(n) - a_{12}w_2(n-1) - a_{22}w_2(n-2)$

$y_1(n) = b_{01}w_1(n) + b_{11}w_1(n-1)$

$y_2(n) = b_{02}w_2(n) + b_{12}w_2(n-2)$

$y_3(n) = Cx(n)$

$y(n) = y_1(n) + y_2(n) + y_3(n)$

(c)

Figure 6.6   (c) Parallel realization of a fourth-order IIR filter.

may be used to represent FIR as well as IIR filters. The basic lattice structure is characterized by a single input and a pair of outputs, as shown in Figure 6.8(a). A lattice structure, derived from the basic structure in Figure 6.8(a), for an N-point FIR filter is shown in Figure 6.8(b), and that for a second-order all-pole IIR filter (that is, with only denominator coefficients) is given in Figure 6.8(c). Further details of the lattice structure are given in Example 6.5.

In summary, the following are the commonly used realization structures for FIR and IIR filters:

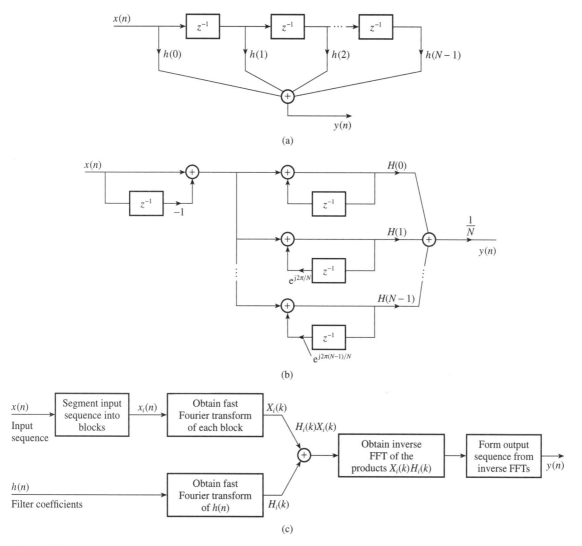

Figure 6.7 Realization structures for FIR filters: (a) transversal filter (or direct); (b) frequency sampling structure; (c) fast convolution.

- transversal (direct)   (FIR);
- frequency sampling   (FIR);
- fast convolution   (FIR);
- direct form   (IIR):
- cascade   (IIR);
- parallel   (IIR);
- lattice   (IIR or FIR).

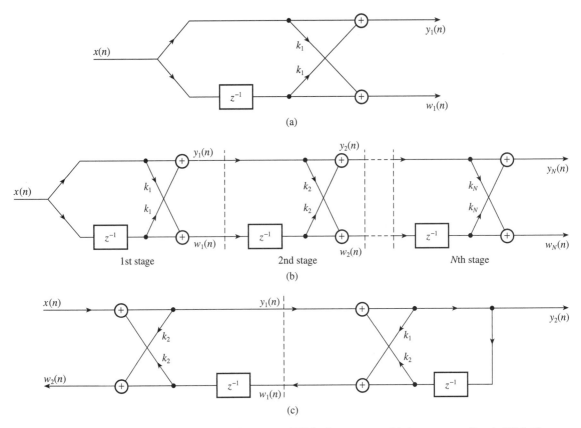

Figure 6.8 (a) The basic lattice structure. (b) An N-stage FIR lattice structure. (c) A two-stage all-pole IIR lattice structure.

For a given filter the choice between structures depends largely on (i) whether it is FIR or IIR, (ii) the ease of implementation and (iii) how sensitive the structure is to the effects of finite wordlength. Realization structures for FIR and IIR filters are discussed more fully in Chapters 7 and 8 respectively.

6.4.4 Analysis of finite wordlength effects

The approximation and realization steps assume infinite or very high precision. However, in actual implementation it is often necessary to represent the filter coefficients using a limited number of bits, typically 8 to 16 bits, and the arithmetic operations indicated in the difference equation are performed using finite precision arithmetic.

The effects of using a finite number of bits are to degrade the performance of the filter and in some cases to make it unusable. The designer must analyze these effects and choose suitable wordlengths (that is, numbers of bits) for the filter coefficients, filter variables, that is the input and output samples, and for the arithmetic operations within the filter.

The main sources of performance degradation in digital filters are as follows:

- *Input/output signal quantization*   In particular, the ADC noise due to quantizing of the input signal samples is significant (see Chapter 2 for details).

- *Coefficient quantization*   This leads to deviations in the frequency response of both FIR and IIR filters, and possibly to instability in IIR filters.

- *Arithmetic roundoff errors*   The use of finite precision arithmetic to perform filtering operations yields results that require additional bits to represent. When these are quantized to the permissible wordlength, often by rounding, roundoff noise is the result. This can cause undesirable effects such as instability in an IIR filter.

- *Overflow*   This occurs when the result of an addition exceeds permissible wordlength. It leads to wrong output samples and to possible instability in IIR filters.

The extent of filter degradation depends on (i) the wordlength and type of arithmetic used to perform the filtering operation, (ii) the method used to quantize filter coefficients and variables to the chosen wordlengths, and (iii) the filter structure. From a knowledge of these factors, the designer can assess the effects of finite wordlength on the filter performance and take remedial action if necessary.

Depending on how the filter is to be implemented, some of the effects may be insignificant. For example, when implemented as a high level language program on most large computers, coefficient quantization and roundoff errors are not important. For real-time processing, finite wordlengths (typically 8 bits, 12 bits, and 16 bits) are used to represent the input and output signals, filter coefficients and the results of arithmetic operations. In these cases, it is nearly always necessary to analyze the effects of quantization on the filter performance.

Detailed discussions of quantization and its effects on digital filter performance are given in Chapter 7 for FIR and Chapter 8 for IIR filters.

6.4.5   Implementation of a filter

Having calculated the filter coefficients, chosen a suitable structure, and verified that the filter degradation, after quantizing the coefficients and filter variables to the selected wordlengths, is acceptable, the difference equation must be implemented as a software routine or in hardware. Whatever the method of implementation, the output of the filter must be computed, for each sample, in accordance with the difference equation (assuming a time domain implementation).

As the examination of any difference equation will show (Equations 6.2 and 6.3), the computation of $y(n)$ (the filter output) involves only multiplications, additions/subtractions, and delays. Thus to implement a filter, we need the following basic building blocks:

- memory (for example ROM) for storing filter coefficients;
- memory (such as RAM) for storing the present and past inputs and outputs, that is $\{x(n), x(n-1), \dots\}$; and $\{y(n), y(n-1), \dots\}$;
- hardware or software multiplier(s);
- adder or arithmetic logic unit.

The designer provides these basic blocks and also ensures that they are suitably configured for the application. The manner in which the components are configured depends to a large extent on whether batch (that is, non-real-time) or real-time processing is required. In batch processing, the entire data is already available in some memory device. Such is the case in applications where, for example, experimental data is acquired from elsewhere for later analysis. In this case, the filter is often implemented in a high level language and runs in a general-purpose computer, such as a personal computer or a mainframe computer, where all the basic blocks are already configured. Thus, batch processing may be described as a purely software implementation (although the designer may wish to incorporate additional hardware to increase the speed of processing).

In real-time processing, the filter is required either (i) to operate on the present input sample, $x(n)$, to produce the current output sample, $y(n)$, before the next input sample arrives, that is within the intersample period, or (ii) to operate on an input block of data, using an FFT algorithm for example, to produce an output block of data within a period proportional to the block length. Real-time filtering may require fast and dedicated hardware if the sample rate is very high or if the filter is of a high order. For most audio frequency work, DSP processors such as the DSP56000 (by Motorola) and the TMS320C25 (by Texas Instruments) will be adequate and offer considerable flexibility. These processors have all the basic blocks on board, including built-in hardware multiplier(s). In some applications, standard 8-bit or 16-bit microprocessors such as the Motorola 6800 or 68000 families offer attractive alternative implementations. In addition to the signal processing hardware, the designer must also provide suitable input–output (for example, analog–digital conversion) interfaces to the digital hardware, depending on the type of data source and sink. Detailed discussions of filter implementations are given in Chapter 7 for FIR and Chapter 8 for IIR filters. DSP hardware is covered in Chapters 12, 13 and 14.

6.5 Illustrative examples

Example 6.3

Discuss the five main steps involved in the design of digital filters, using the following design problem to illustrate your answer.

A digital filter is required for real-time physiological noise reduction. The filter should meet the following amplitude response specifications:

passband	0–10 Hz
stopband	20–64 Hz
sampling frequency	128 Hz
maximum passband deviation	<0.036 dB
stopband attenuation	>30 dB

Other important requirements are that

(1) minimal distortion of the harmonic relationships between the components of the in-band signals is highly desirable,

(2) the time available for filtering is limited, the filter being part of a larger process, and

(3) the filter will be implemented using the Texas Instruments TMS32010 DSP processor with the analog input digitized to 12 bits.

Solution   This filter was designed and used in a certain biomedical signal processing project. We shall give here only an outline discussion of the design, postponing detailed discussion to Chapter 7 where FIR filter design methods are fully covered.

(1) *Requirement specification*   As discussed previously, the designer must give the exact role and performance requirements for the filter together with any important constraints. These have already been given for the example.

(2) *Calculation of suitable coefficients*   The requirements of minimal distortion and limited processing time are best achieved with a linear phase FIR filter, with coefficients obtained using the optimal method.

(3) *Selection of filter structure*   The transversal structure will lead to the most efficient implementation using the TMS32010 processor.

(4) *Analysis of finite wordlength effects*   Since the TMS32010 processor will be used, fixed point arithmetic should be used with each coefficient represented by 16 bits (after rounding) for efficiency. FIR filter degradation may result from input signal quantization, coefficient quantization, roundoff and overflow errors. A check should be made to ensure that the wordlengths are sufficiently long. Analysis of finite wordlength effects for this case showed that the input quantization noise and deviation in the frequency response due to coefficient quantization are both insignificant. The use of the TMS32010's 32-bit accumulator to sum the coefficient–data products, rounding only the final sum, would reduce roundoff errors to negligible levels. To avoid overflow, each coefficient should be divided by $\sum_{k=0}^{N-1}|h(k)|$ before quantizing to 16 bits.

(5) *Implementation*   Design and configure the TMS32010-based hardware (if it does not already exist) with the necessary input/output interfaces. Then write a TMS32010 program to handle the I/O protocols and calculate filter output, $y(n) = \sum_{k=0}^{N-1} h(k)x(n-k)$, for each new input, $x(n)$.

Example 6.4   An analog filter is to be converted into an equivalent digital filter that will operate at a sampling frequency of 256 Hz. The analog filter has the transfer function:

$$H(s) = \frac{1}{s^3 + 2s^2 + 2s + 1}$$

(1)   Obtain suitable coefficients for the digital filter.

(2)   Assuming that the digital filter is to be realized using the cascade structure, draw a suitable realization block diagram and develop the difference equations.

(3)   Repeat (2) for the parallel structure.

Solution   (1)   To preserve the amplitude response of the analog function, the bilinear method was used to obtain the filter coefficients. The application of the bilinear transformation approach to the analog transfer function (see Chapter 8 for details) yields the following transfer functions:

$$H(z) = \frac{0.1432(1 + 3z^{-1} + 3z^{-2} + z^{-3})}{1 - 0.1801z^{-1} + 0.3419z^{-2} - 0.0165z^{-3}}$$

(2)   For the cascade realization, $H(z)$ is factorized using partial fractions:

$$H(z) = 0.1432 \frac{1 + 2z^{-1} + z^{-2}}{1 - 0.1307z^{-1} + 0.3355z^{-2}} \frac{1 + z^{-1}}{1 - 0.0490z^{-1}}$$

The block diagram representation and the corresponding set of difference equations are given in Figure 6.9 and as follows:

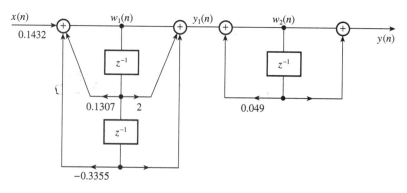

Figure 6.9

$$w_1(n) = 0.1432x(n) + 0.1307w_1(n-1) - 0.3355w_1(n-2)$$

$$y_1(n) = w_1(n) + 2w_1(n-1) + w_1(n-2)$$

$$w_2(n) = y_1(n) + 0.049w_2(n-1)$$

$$y_2(n) = w_2(n) + w_2(n-1)$$

(3) For the parallel realization, $H(z)$ is expressed using partial fractions (see Chapters 4 and 8 for details) as

$$H(z) = \frac{1.2916 - 0.084\,07z^{-1}}{1 - 0.131z^{-1} + 0.3355z^{-2}} + \frac{7.5268}{1 - 0.049z^{-1}} - 8.6753$$

The parallel realization diagram and its corresponding set of difference equations are given in Figure 6.10 and as follows:

$$w_1(n) = x(n) + 0.131w_1(n-1) - 0.3355w_1(n-2)$$

$$y_1(n) = 1.2916w_1(n) - 0.084\,07w_1(n-1)$$

$$w_2(n) = x(n) + 0.049w_2(n-1)$$

$$y_2(n) = 7.5268w_2(n)$$

$$y_3(n) = -8.6753x(n)$$

$$y(n) = y_1(n) + y_2(n\ ) + y_3(n)$$

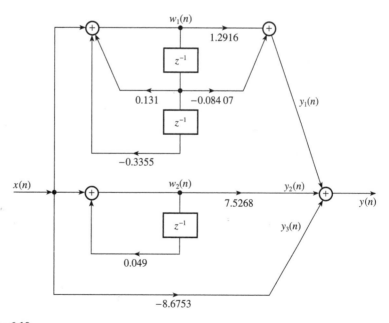

Figure 6.10

Example 6.5   The transfer function for an FIR filter is given by

$$H(z) = 1 - 1.3435z^{-1} + 0.9025z^{-2}$$

Draw the realization block diagram for each of the following cases:

(1)   transversal structure;
(2)   a two-stage lattice structure.

Calculate the values of the coefficients for the lattice structure.

Solution   (1)   From the transfer function, the diagram for the transversal structure is given in Figure 6.11. The input and output of the transversal structure are given by

$$y(n) = x(n) + h(1)x(n - 1) + h(2)x(n - 2) \tag{6.6}$$

(2)   A two-stage lattice structure for the filter is given in Figure 6.12. The outputs of the structure are related to the input as

$$y_2(n) = y_1(n) + k_2 w_1(n - 1)$$
$$= x(n) + k_1(1 + k_2)x(n - 1) + k_2 x(n - 2) \tag{6.7a}$$
$$w_2(n) = k_2 x(n) + k_1(1 + k_2)x(n - 1) + x(n - 2) \tag{6.7b}$$

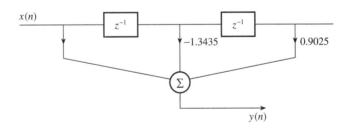

Figure 6.11

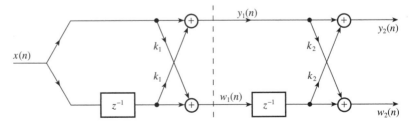

Figure 6.12

Comparing Equations 6.6 and 6.7a, and equating coefficients, we have

$$k_1 = \frac{h(1)}{1 + h(2)} \qquad k_2 = h(2)$$

from which $k_2 = 0.9025$ and $k_1 = -1.3435/(1 + 0.9025) = -0.7062$.

Notice that the coefficients of $y_2(n)$ and $w_2(n)$ (Equations 6.7a and 6.7b) are identical except that one is written in reverse order. This is a characteristic feature of the FIR lattice structure. Further details of the lattice structure, including recursive techniques for converting the coefficients of an FIR or IIR filter to those of equivalent lattice structures, are available in several books (see, for example, Proakis and Manolakis, 1992).

6.6   Summary

The term digital filter refers to a hardware or software implementation of a mathematical algorithm which accepts, as input, a digital signal, and produces as output another digital signal whose wave shape and/or amplitude and phase responses have been modified in a specified manner. In many applications, the use of digital filters is preferred to analog filters because they can meet much tighter magnitude and phase specifications, which eliminates the temperature and voltage drifts common with analog filters.

We have given in this chapter a common framework for designing FIR and IIR filters, from specification to implementation. A simple step-by-step procedure for designing these filters involves five key steps: (i) filter specification, (ii) calculation of suitable filter coefficients, (iii) realization of the filter using a suitable structure, (iv) quantization of filter coefficient and variables to suitable wordlengths and analysis of any resultant errors, and, finally, (v) implementation which is concerned with the hardware or software coding of the filter in a processor which will perform the actual filtering on input data.

Problems

6.1   Assume that the six methods of calculating filter coefficients given in Section 6.4.2 are all available. State and justify which of the methods you should use in each of the following applications:

(1)   phase (delay) equalization for a digital communication system;

(2)   simulation of analog systems;

(3)   a high throughput noise reduction system requiring a sharp magnitude frequency response filter;

(4)   image processing;

(5)   high quality digital audio processing;

(6) real-time biomedical signal processing with minimal distortion.

6.2 The following transfer functions represent two different filters meeting identical amplitude frequency response specifications:

(1) $H(z) =$
$$\frac{b_0 + b_1 z^{-1} + b_2 z^{-2}}{1 + a_1 z^{-1} + a_2 z^{-2}} \times \frac{b_3 + b_4 z^{-1} + b_5 z^{-2}}{1 + a_3 z^{-1} + a_4 z^{-2}}$$

where

$b_0 = 3.136\,362 \times 10^{-1}$

$b_1 = -5.456\,657 \times 10^{-2}$

$b_2 = 4.635\,728 \times 10^{-1}$

$b_3 = -5.456\,657 \times 10^{-2}$

$b_4 = 3.136\,362 \times 10^{-1}$

$b_5 = 4.635\,728 \times 10^{-1}$

$a_1 = -8.118\,702 \times 10^{-1}$

$a_2 = 3.339\,288 \times 10^{-1}$

$a_3 = -2.794\,577 \times 10^{-1}$

$a_4 = 3.030\,631 \times 10^{-1}$

(2) $H(z) = \sum_{k=0}^{22} h_k z^{-k}$

where

$h_0 = 0.398\,264\,80 \times 10^{-1} = h_{22}$

$h_1 = -0.168\,743\,80 \times 10^{-1} = h_{21}$

$h_2 = 0.347\,811\,30 \times 10^{-1} = h_{20}$

$h_3 = 0.120\,528\,90 \times 10^{-1} = h_{19}$

$h_4 = -0.447\,318\,60 \times 10^{-1} = h_{18}$

$h_5 = 0.278\,946\,10 \times 10^{-1} = h_{17}$

$h_6 = -0.875\,733\,60 \times 10^{-1} = h_{16}$

$h_7 = -0.909\,720\,60 \times 10^{-1} = h_{15}$

$h_8 = -0.156\,675\,50 \times 10^{-1} = h_{14}$

$h_9 = -0.284\,995\,60 \times 10^{0} = h_{13}$

$h_{10} = 0.740\,350\,30 \times 10^{-1} = h_{12}$

$h_{11} = 0.623\,495\,60 \times 10^{0}$

For each filter,

(a) state whether it is an FIR or IIR filter;

(b) represent the filtering operation in a block diagram form and write down the difference equation;

(c) determine and comment on the computational and storage requirements.

6.3 A digital filter is required to remove the mains component from fetal electrocardiogram (ECG) data stored in the memory of a mainframe computer. The data was digitized to 12-bit accuracy.
The specification for the filter includes:

attenuation at the mains frequency	>50 dB
passband ripple	<0.05 dB
passband edges	0–0.09 and 0.11–0.5 (normalized)
sampling frequency	500 Hz

The filter should be implemented as a high level language routine, callable from a main analysis program. Any signal distortion by the filter should be kept to a minimum as important ECG waves are easily distroyed.

Discuss fully the issues involved in the design of such a filter, pointing out the options open to the designer and making recommendations, with justifications.

6.4 A digital filter is required to pre-process raw fetal electrocardiogram (ECG) data, that is the electrical activity of the heart, to make it easier to detect fetal heartbeats in the presence of baseline wander, mains interference, uterine contractions, and movements of the baby or mother. Baseline sways and movement artefacts tend to occupy the frequency range 0–10 Hz while the mains interference is centred around about 50 or 60 Hz (depending on the country). Most of the energy in the ECG necessary for detecting heartbeats is thought to lie between about 5 and 50 Hz.

Assume that you have available fetal ECG data, analog bandpass filtered between 0.05 and 100 Hz and digitized at 500 samples s^{-1} at a resolution of 8 bits.

(1) Assume that an IIR filter is to be used, and develop a set of specifications for the digital filter. Justify your answer.

(2) Repeat (1) for an FIR filter.

Which of the two filter types is best for this application and why?

6.5   A digital filter is to be used to provide the bulk of the anti-aliasing filtering in a certain speech transmission system in preference to an existing active filter. Currently, the analog input signal to the system is sampled at 8 kHz after bandlimiting by the active filter with the following specifications:

passband	0–3.4 kHz
stopband edge frequency	8 kHz
attenuation in the stopband	30 dB
attenuation at 4 kHz	14 dB
passband ripple	<0.1 dB

Discuss, with the aid of a block diagram, how the specifications can be met by digital filtering. Specify, with reasons, the type of digital filter that will be used and its characteristics.

6.6   A requirement exists in a communication system to recover the clock frequency from noisy data received from a remote transmitter, using digital filtering techniques, to allow for the data to be extracted reliably. The clock frequency at the remote transmitter is known to be 2.048 MHz. Discuss the characteristics of a suitable digital filter for the task and specify its transfer function.

6.7   The coefficients of a lattice FIR filter are $k_1 = -0.266$ and $k_2 = 0.69$. Draw the realization diagram for the lattice filter. Compute the impulse response coefficients for the filter and draw the diagram for the equivalent transversal structure.

6.8   A second-order IIR digital filter is characterized by the following transfer function:

$$H(z) = \frac{1}{1 - 0.9z^{-1} + 0.81z^{-2}}$$

Draw the realization diagram for each of the following structures:

(1)   direct;

(2)   lattice.

Find the coefficients of the lattice structure from the given transfer function.

Reference

Proakis J.G. and Manolakis D.G. (1992) *Digital Signal Processing*, 2nd edn. New York: Macmillan.

Bibliography

DeFatta D.J., Lucas J.G. and Hodgkiss W.S. (1988) *Digital Signal Processing*. New York: Wiley.

Elliott D.F. (ed.) (1987) *Handbook of Digital Signal Processing*. London: Academic Press.

Oppenheim A.V. and Schafer R.W. (1975) *Digital Signal Processing*. Englewood Cliffs NJ: Prentice-Hall.

Parks T.W. and Burrus C.S. (1987) *Digital Filter Design*. New York: Wiley.

Rabiner L.R. and Gold B. (1975) *Theory and Application of Digital Signal Processing*. Englewood Cliffs NJ: Prentice-Hall.

Rabiner L.R., Cooley J.W., Helms H.D., Jackson L.B., Kaiser J.F., Rader C.M., Schafer R.W., Steiglitz K. and Weinstein C.J. (1972) Terminology in digital signal processing. *IEEE Trans. Audio Electroacoustics*, **20** (December), 322–37.

Taylor F.J. (1983) *Digital Filter Design Handbook*. New York: Dekker.

Finite impulse response (FIR) filter design

7

This chapter is concerned with the design of FIR filters from specifications, through coefficient calculation, to analysis of finite wordlength effects and implementations. Several fully worked examples are given throughout the chapter to illustrate the various design stages and to consolidate the important concepts. A complete filter design is included to show how all the stages fit together and to assist the readers who wish to design their own filters. PC-based MATLAB programs are available on the web (see Preface for details). In addition, the MATLAB and C-language programs are available on the CD in the companion handbook and may be used to replicate the results presented here or for designing user-specified filters.

7.1 Introduction

First, we will summarize important characteristics of FIR filters before devoting attention to their design.

7.1.1 Summary of key characteristic features of FIR filters

(1) The basic FIR filter is characterized by the following two equations:

$$y(n) = \sum_{k=0}^{N-1} h(k)x(n-k) \tag{7.1a}$$

$$H(z) = \sum_{k=0}^{N-1} h(k)z^{-k} \tag{7.1b}$$

where $h(k)$, $k = 0, 1, \ldots, N-1$, are the impulse response coefficients of the filter, $H(z)$ is the transfer function of the filter and N is the filter length, that is the number of filter coefficients. Equation 7.1a is the FIR difference equation. It is a time domain equation and describes the FIR filter in its nonrecursive form: the current output sample, $y(n)$, is a function only of past and present values of the input, $x(n)$. When FIR filters are implemented in this form, that is by direct evaluation of Equation 7.1a, they are always stable. Equation 7.1b is the transfer function of the filter. It provides a means of analyzing the filter, for example evaluating the frequency response.

(2) FIR filters can have an exactly linear phase response. The implications of this will be discussed in the next section.

(3) FIR filters are very simple to implement. All DSP processors available have architectures that are suited to FIR filtering. Nonrecursive FIR filters suffer less from the effects of finite wordlength than IIR filters. Recursive FIR filters also exist and may offer significant computational advantages (see Section 7.7 for details).

FIR filters should be used whenever we wish to exploit any of the advantages above, in particular the advantage of linear phase. Issues to consider when choosing between FIR and IIR filters are given in Section 6.3.

7.1.2 Linear phase response and its implications

The ability to have an exactly linear phase response is one of the most important properties of FIR filters. For this reason we shall look more closely at this property. When a signal passes through a filter, it is modified in amplitude and/or phase. The nature and extent of the modification of the signal is dependent on the amplitude and phase characteristics of the filter. The phase delay or group delay of the filter provides a useful measure of how the filter modifies the phase characteristics of the signal. If we consider a signal that consists of several frequency components (such as a speech waveform or a modulated signal) the phase delay of the filter is the amount of time delay each frequency component of the signal suffers in going through the filter. The group delay on the other hand is the average time delay the composite signal suffers at each frequency. Mathematically, the phase delay is the negative of the phase angle divided by frequency whereas the group delay is the negative of the derivative of the phase with respect to frequency:

$$T_p = -\theta(\omega)/\omega \tag{7.2a}$$

$$T_g = -\mathrm{d}\theta(\omega)/\mathrm{d}\omega \tag{7.2b}$$

A filter with a nonlinear phase characteristic will cause a phase distortion in the signal that passes through it. This is because the frequency components in the signal will each be delayed by an amount not proportional to frequency thereby altering their harmonic relationships. Such a distortion is undesirable in many applications, for example music, data transmission, video, and biomedicine, and can be avoided by using filters with linear phase characteristics over the frequency bands of interest.

A filter is said to have a linear phase response if its phase response satisfies one of the following relationships:

$$\theta(\omega) = -\alpha\omega \tag{7.3a}$$

$$\theta(\omega) = \beta - \alpha\omega \tag{7.3b}$$

where α and β are constant. If a filter satisfies the condition given in Equation 7.3a it will have both constant group and constant phase delay responses. It can be shown that for condition 7.3a to be satisfied the impulse response of the filter must have positive symmetry. The phase response in this case is simply a function of the filter length:

$$h(n) = h(N - n - 1), \quad \begin{cases} n = 0, 1, \ldots, (N-1)/2 & (N \text{ odd}) \\ n = 0, 1, \ldots, (N/2) - 1 & (N \text{ even}) \end{cases}$$

$$\alpha = (N - 1)/2$$

When the condition given in Equation 7.3b only is satisfied the filter will have a constant group delay only. In this case, the impulse response of the filter has negative symmetry:

$$h(n) = -h(N - n - 1)$$

$$\alpha = (N - 1)/2$$

$$\beta = \pi/2$$

Linear phase FIR filters form an important class of FIR filters. They possess a unique set of properties that influence how they are designed and implemented. We will explore some of these by way of an example.

Example 7.1   (1)   Discuss briefly the conditions necessary for a realizable digital filter to have a linear phase characteristic, and the advantages of filters with such a characteristic.

(2)   An FIR digital filter has impulse response, $h(n)$, defined over the interval $0 \leqslant n \leqslant N - 1$. Show that if $N = 7$ and $h(n)$ satisfies the symmetry condition

$$h(n) = h(N - n - 1)$$

the filter has a linear phase characteristic.

(3)   Repeat (2) if $N = 8$.

Solution   (1)   The necessary and sufficient condition for a filter to have a linear phase response is that its impulse response must be symmetrical (Rabiner and Gold, 1975):

$$h(n) = h(N - 1 - n) \text{ or } h(n) = -h(N - 1 - n)$$

For nonrecursive FIR filters, the storage space for coefficients and the number of arithmetic operations are reduced by nearly a factor of 2. For recursive FIR filters, the coefficients can be made to be simple integers, leading to increased speed of processing. In linear phase filters, all frequency components experience the same amount of delay through the filter, that is no phase distortion.

(2)   Using the symmetry condition we find that for $N = 7$:

$$h(0) = h(6); \ h(1) = h(5); \ h(2) = h(4)$$

The frequency response, $H(\omega)$, for the filter is obtained from Equation 7.1b by making the substitution $z = e^{j\omega T}$:

$$H(\omega) = H(e^{j\omega T})$$

$$= \sum_{k=0}^{6} h(k)e^{-jk\omega T}$$

$$= h(0) + h(1)e^{-j\omega T} + h(2)e^{-j2\omega T} + h(3)e^{-j3\omega T} + h(4)e^{-j4\omega T}$$

$$+ h(5)e^{-j5\omega T} + h(6)e^{-j6\omega T}$$

$$= e^{-j3\omega T}[h(0)e^{j3\omega T} + h(1)e^{j2\omega T} + h(2)e^{j\omega T} + h(3) + h(4)e^{-j\omega T}$$

$$+ h(5)e^{-j2\omega T} + h(6)e^{-j3\omega T}]$$

Using the symmetry condition we can group terms whose coefficients are numerically equal:

$$H(\omega) = e^{-j3\omega T}[h(0)(e^{j3\omega T} + e^{-j3\omega T}) + h(1)(e^{j2\omega T} + e^{-j2\omega T})$$
$$+ h(2)(e^{j\omega T} + e^{-j\omega T}) + h(3)]$$
$$= e^{-j3\omega T}[2h(0)\cos(3\omega T) + 2h(1)\cos(2\omega T)$$
$$+ 2h(2)\cos(\omega T) + h(3)]$$

If we let $a(0) = h(3)$ and $a(k) = 2h(3 - k)$, $k = 1, 2, 3$, then $H(\omega)$ can be written in a compact form:

$$H(\omega) = \sum_{k=0}^{3} a(k)\cos(n\omega T)e^{-j3\omega T} = \pm|H(\omega)|e^{j\theta(\omega)}$$

where

$$\pm|H(\omega)| = \sum_{k=0}^{3} a(k)\cos(n\omega T); \quad \theta(\omega) = -3\omega T$$

Clearly, the phase response is linear.

(3)   In this case, the symmetry conditions lead to

$$h(0) = h(7); \quad h(1) = h(6); \quad h(2) = h(5); \quad h(3) = h(4)$$

Following a similar approach to the above and using the symmetry condition we have

$$H(\omega) = e^{-j7\omega T/2}[h(0)(e^{j7\omega T/2} + e^{-j7\omega T/2}) + h(1)(e^{j5\omega T/2} + e^{-j5\omega T/2})$$
$$+ h(2)(e^{j3\omega T/2} + e^{-j3\omega T/2}) + h(3)(e^{j\omega T/2} + e^{-j\omega T/2})]$$
$$= e^{-j7\omega T/2}[2h(0)\cos(7\omega T/2) + 2h(1)\cos(5\omega T/2)$$
$$+ 2h(2)\cos(3\omega T/2) + 2h(3)\cos(\omega T/2)]$$
$$= \pm|H(\omega)|e^{j\theta(\omega)}$$

where

$$\pm|H(\omega)| = \sum_{k=1}^{4} b(k)\cos[\omega(k - 1/2)]; \quad \theta(\omega) = -(7/2)\omega T$$
$$b(k) = 2h(N/2 - k), \quad k = 1, 2, \ldots, N/2$$

The results above can be generalized for FIR filters: see Table 7.1.

Table 7.1   A summary of the key points about the four types of linear phase FIR filters.

Impulse response symmetry	Number of coefficients N	Frequency response $H(\omega)$	Type of linear phase
Positive symmetry, $h(n) = h(N-1-n)$	Odd	$e^{-j\omega(N-1)/2} \displaystyle\sum_{n=0}^{(N-1)/2} a(n) \cos(\omega n)$	1
	Even	$e^{-j\omega(N-1)/2} \displaystyle\sum_{n=1}^{N/2} b(n) \cos\left[\omega\left(n - \tfrac{1}{2}\right)\right]$	2
Negative symmetry, $h(n) = -h(N-1-n)$	Odd	$e^{-j[\omega(N-1)/2 - \pi/2]} \displaystyle\sum_{n=1}^{(N-1)/2} a(n) \sin(\omega n)$	3
	Even	$e^{-j[\omega(N-1)/2 - \pi/2]} \displaystyle\sum_{n=1}^{N/2} b(n) \sin\left[\omega\left(n - \tfrac{1}{2}\right)\right]$	4

$a(0) = h[(N-1)/2]; \; a(n) = 2h[(N-1)/2 - n]$
$b(n) = 2h(N/2 - n)$

7.1.3   Types of linear phase FIR filters

There are exactly four types of linear phase FIR filters, depending on whether N is even or odd and whether $h(n)$ has positive or negative symmetry. Two of the four types of linear phase filters were considered in the example above. Figure 7.1 illustrates how the impulse responses of the four types of linear phase FIR filters differ. Table 7.1 summarizes their key features.

The frequency response of a type 2 filter (positive symmetry and even length) is always zero at $f = 0.5$ (half the sampling frequency, as all frequencies are normalized to the sampling frequency); see Problem 7.1. Thus this type of filter is unsuitable as a highpass filter. Types 3 and 4 (both negative symmetry) each introduce a 90° phase shift. The frequency response is always zero at $f = 0$, making them unsuitable for lowpass filters. In addition, the type 3 response is always zero at $f = 0.5$, making it also unsuitable as a highpass filter. Type 1 is the most versatile of the four. Types 3 and 4 are often used to design differentiators and Hilbert transformers, because of the 90° phase shift that each can provide.

It is important to note that the phase delay (for type 1 and 2 filters) or group delay (for all four types of filters) is expressible in terms of the number of coefficients of the filter and so can be corrected to give a zero phase or group delay response. For example, for type 1 and 2 filters, the phase delay is given by

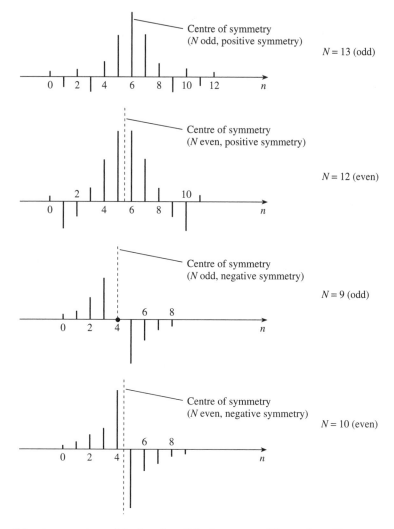

Figure 7.1 A comparison of the impulse of the four types of linear phase filters.

$$T_p = \left(\frac{N-1}{2}\right)T \qquad (7.4a)$$

and for types 3 and 4 the group delay is given by

$$T_p = \left(\frac{N-1-\pi}{2}\right)T \qquad (7.4b)$$

where T is the sampling period.

7.2 FIR filter design

As discussed in Chapter 6, the design of a digital filter involves five steps, viz:

(1) *Filter specification*   This may include stating the type of filter, for example lowpass filter, the desired amplitude and/or phase responses and the tolerances (if any) we are prepared to accept, the sampling frequency, and the wordlength of the input data.

(2) *Coefficient calculation*   At this step, we determine the coefficients of a transfer function, $H(z)$, which will satisfy the specifications given in (1). Our choice of coefficient calculation method will be influenced by several factors, the most important of which are the critical requirements in step (1).

(3) *Realization*   This involves converting the transfer function obtained in (2) into a suitable filter network or structure.

(4) *Analysis of finite wordlength effects*   Here, we analyze the effect of quantizing the filter coefficients and the input data as well as the effect of carrying out the filtering operation using fixed wordlengths on the filter performance.

(5) *Implementation*   This involves producing the software code and/or hardware and performing the actual filtering.

These five interrelated steps are summarized in Figure 7.2. We shall now go through the steps in detail for FIR filters, illustrating with examples as appropriate.

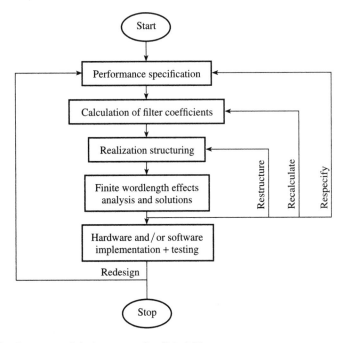

Figure 7.2   Summary of design stages for digital filters.

7.3 FIR filter specifications

Filter specifications were discussed in detail in Chapter 6. Here we shall deal with aspects of filter specification relating to the FIR filter. Several examples in this chapter will also illustrate various aspects of filter specification.

For the phase response, we need only state whether positive symmetry or negative symmetry is required (assuming linear phase). The amplitude–frequency response of an FIR filter is often specified in the form of a tolerance scheme. Figure 7.3 shows such a scheme for the lowpass filter. A similar scheme can be readily drawn for other frequency selective filters. Referring to the figure, the following parameters are of interest:

δ_p peak passband deviation (or ripples)
δ_s stopband deviation
f_p passband edge frequency
f_s stopband edge frequency
F_s sampling frequency

In practice it is often more convenient to express δ_p and δ_s in decibels as indicated in the figure. The difference between f_s and f_p gives the transition width of the filter. Another important parameter is the filter length, N, which defines the number of filter coefficients given. These parameters, in most cases, define completely the frequency response of the FIR filter.

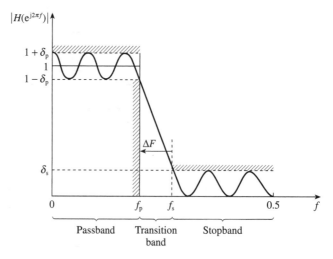

Figure 7.3 Magnitude–frequency response specification for a lowpass filter. The passband and stopband deviations are often expressed in decibels: passband deviation, $20 \log (1 + \delta_p)$ dB; stopband deviation, $-20 \log (\delta_s)$ dB.

Other specifications that may be of interest include the maximum number of filter coefficients we can accept (this may be forced on us by the particular application, such as the speed at which we wish to operate). We may not have any idea what constitutes a good choice for one or more of the parameters above and so may have to deduce them by trial and error.

Example 7.2   *Amplitude specification example*   A lowpass digital filter is required for physiological noise reduction. The filter should meet the following specifications:

passband edge frequency	10 Hz
stopband edge frequency	<20 Hz
stopband attenuation	>30 dB
passband ripple	<0.026 dB
sampling frequency	256 Hz

Important requirements in this application are (i) the filter should introduce as small a distortion as possible to the in-band signals and (ii) the length of filter should be as small as possible and should not exceed 37.

7.4   FIR coefficient calculation methods

Recall that an FIR filter is characterized by the following equations:

$$y(m) = \sum_{n=0}^{N-1} h(n)x(m-n)$$

$$H(z) = \sum_{n=0}^{N-1} h(n)z^{-n}$$

The sole objective of most FIR coefficient calculation (or approximation) methods is to obtain values of $h(n)$ such that the resulting filter meets the design specifications, such as amplitude–frequency response and throughput requirements. Several methods are available for obtaining $h(n)$. The window, optimal and frequency sampling methods, however, are the most commonly used. All three can lead to linear phase FIR filters.

7.5 Window method

In this method, use is made of the fact that the frequency response of a filter, $H_D(\omega)$, and the corresponding impulse response, $h_D(n)$, are related by the inverse Fourier transform:

$$h_D(n) = \frac{1}{2\pi} \int_{-\pi}^{\pi} H_D(\omega) e^{j\omega n} d\omega \tag{7.5}$$

The subscript D is used to distinguish between the ideal and practical impulse responses. The need for this distinction will soon become clear. If we know $H_D(\omega)$ we can obtain $h_D(n)$ by evaluating the inverse Fourier transform of Equation 7.5. As an illustration, suppose we wish to design a lowpass filter. We could start with the ideal lowpass response shown in Figure 7.4(a) where ω_c is the cutoff frequency and the frequency scale is normalized: $T = 1$. By letting the response go from $-\omega_c$ to ω_c we simplify the integration operation. Thus the impulse response is given by:

$$h_D(n) = \frac{1}{2\pi} \int_{-\pi}^{\pi} 1 \times e^{j\omega n} d\omega = \frac{1}{2\pi} \int_{-\omega_c}^{\omega_c} e^{j\omega n} d\omega$$

$$= \frac{2 f_c \sin(n\omega_c)}{n\omega_c}, \quad n \neq 0, -\infty \leqslant n \leqslant \infty \tag{7.6}$$

$$= 2 f_c, \qquad n = 0 \text{ (using L'Hôpital's rule)}$$

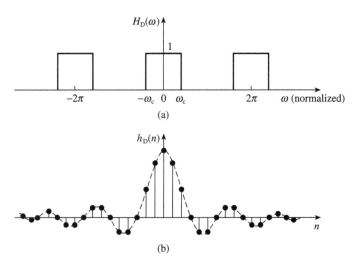

(a)

(b)

Figure 7.4   (a) Ideal frequency response of a lowpass filter. (b) Impulse response of the ideal lowpass filter.

Table 7.2   Summary of ideal impulse responses for standard frequency selective filters.

Filter type	Ideal impulse response, $h_D(n)$	
	$h_D(n), n \neq 0$	$h_D(0)$
Lowpass	$2f_c \dfrac{\sin(n\omega_c)}{n\omega_c}$	$2f_c$
Highpass	$-2f_c \dfrac{\sin(n\omega_c)}{n\omega_c}$	$1 - 2f_c$
Bandpass	$2f_2 \dfrac{\sin(n\omega_2)}{n\omega_2} - 2f_1 \dfrac{\sin(n\omega_1)}{n\omega_1}$	$2(f_2 - f_1)$
Bandstop	$2f_1 \dfrac{\sin(n\omega_1)}{n\omega_1} - 2f_2 \dfrac{\sin(n\omega_2)}{n\omega_2}$	$1 - 2(f_2 - f_1)$

f_c, f_1 and f_2 are the normalized passband or stopband edge frequencies; N is the length of filter.

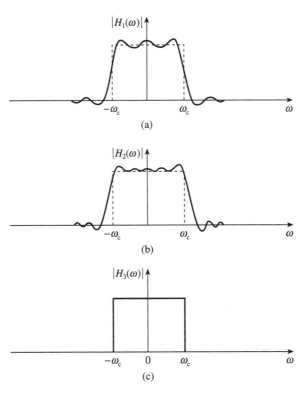

Figure 7.5   Effects on the frequency response of truncating the ideal impulse response to
(a) 13 coefficients, (b) 25 coefficients and (c) an infinite number of coefficients.

The impulse responses for the ideal highpass, bandpass and bandstop filters are obtained from the lowpass case of Equation 7.6 and are summarized in Table 7.2. The impulse response for the lowpass filter is plotted in Figure 7.4(b) from which we note that $h_D(n)$ is symmetrical about $n = 0$ (that is $h_D(n) = h_D(-n)$), so that the filter will have a linear (in this case zero) phase response. Several practical problems with this simple approach are apparent. The most important of these is that, although $h_D(n)$ decreases as we move away from $n = 0$, it nevertheless carries on, theoretically, to $n = \pm\infty$. Thus the resulting filter is not an FIR.

An obvious solution is to truncate the ideal impulse response by setting $h_D(n) = 0$ for n greater than M (say). However, this introduces undesirable ripples and overshoots – the so-called Gibb's phenomenon. Figure 7.5 illustrates the effects of discarding coefficients on the filter response. The more coefficients that are retained, the closer the filter spectrum is to the ideal response (Figures 7.5(b) and 7.5(c)). Direct truncation of $h_D(n)$ as described above is equivalent to multiplying the ideal impulse response by a rectangular window of the form

$$w(n) = 1, \quad |n| = 0, 1, \ldots, (M-1)/2$$
$$= 0, \quad \text{elsewhere}$$

In the frequency domain this is equivalent to convolving $H_D(\omega)$ and $W(\omega)$, where $W(\omega)$ is the Fourier transform of $w(n)$. As $W(\omega)$ has the classic $(\sin x)/x$ shape, truncation of $h_D(n)$ leads to the overshoots and ripples in the frequency response.

A practical approach is to multiply the ideal impulse response, $h_D(n)$, by a suitable window function, $w(n)$, whose duration is finite. This way the resulting impulse response decays smoothly towards zero. The process is illustrated in Figure 7.6. Figure 7.6(a) shows the ideal frequency response and the corresponding ideal impulse response. Figure 7.6(b) shows a finite duration window function and its spectrum. Figure 7.6(c) shows $h(n)$ which is obtained by multiplying $h_D(n)$ by $w(n)$. The corresponding frequency response shows that the ripples and overshoots, characteristic of direct truncation, are much reduced. However, the transition width is wider than for the rectangular case. The transition width of the filter is determined by the width of the main lobe of the window. The side lobes produce ripples in both passband and stopband.

7.5.1   Some common window functions

Several window functions have been proposed. One of the most widely used window functions is the Hamming window which is defined as

$$w(n) = 0.54 + 0.46 \cos{(2\pi n/N)} \begin{cases} -(N-2)/2 \leqslant n \leqslant (N-1)/2 & (N \text{ odd}) \\ -N/2 \leqslant n \leqslant N/2 & (N \text{ even}) \end{cases}$$
$$= 0 \qquad\qquad\qquad\qquad \text{elsewhere}$$

$$(7.7)$$

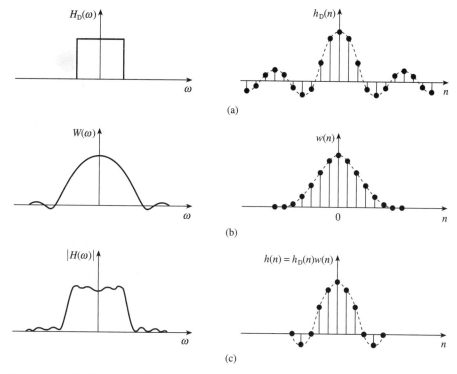

(a)

(b)

(c)

Figure 7.6 An illustration of how the filter coefficients, $h(n)$, are determined by the window method.

Figure 7.7 compares its time and frequency domain characteristics with those of the rectangular window. In the time domain, the Hamming window function decreases more gently towards zero on either side. In the frequency domain, the amplitude of the main lobe is wider (about twice) than that of the rectangular window, but its side lobes are smaller relative to the main lobe – about 40 dB down on the main lobes, compared with 14 dB for the rectangular window. The implication of this is that, compared with the rectangular window, the Hamming window will lead to a filter with wider transition width (because of the wider main lobe) but higher stopband attenuation (because of the smaller side lobe levels).

The appropriate relationship between the transition width (from passband to stopband) for a filter designed with the Hamming window and filter length is given by

$$\Delta f = 3.3/N \tag{7.8}$$

where N is the filter length and Δf the normalized transition width. The maximum stopband attenuation possible with the Hamming window is about 53 dB, and the minimum peak passband ripple is about 0.0194 dB.

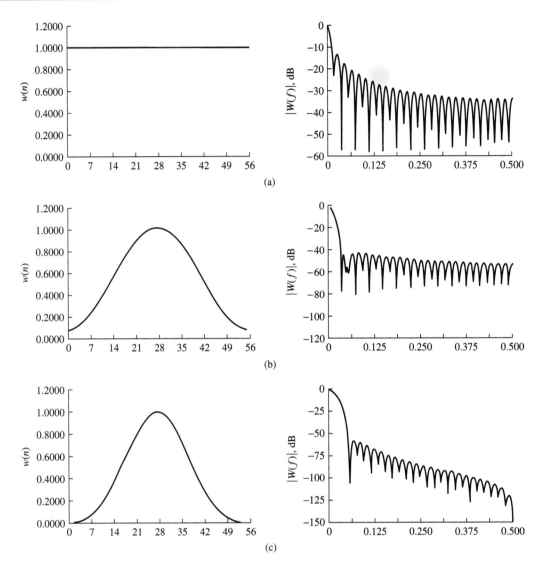

Figure 7.7   Comparison of the time and frequency domain characteristics of common window functions:
(a) rectangular; (b) Hamming; (c) Blackman.

The most relevant features of some of the most popular window functions are summarized in Table 7.3. We note that the first four window functions have fixed characteristics, such as transition width and stopband attenuation. Thus their use imposes a restriction on the filter designer. We also note that a filter designed by the window method has equal passband and stopband ripples, that is $\delta_p = \delta_s$ (in Figure 7.3). In practice, this restriction may lead to a filter whose passband ripple is unnecessarily small.

Table 7.3 Summary of important features of common window functions.

Name of window function	Transition width (Hz) (normalized)	Passband ripple (dB)	Main lobe relative to side lobe (dB)	Stopband attenuation (dB) (maximum)	Window function $w(n), \lvert n \rvert \leqslant (N-1)/2$
Rectangular	$0.9/N$	0.7416	13	21	1
Hanning	$3.1/N$	0.0546	31	44	$0.5 + 0.5\cos\left(\dfrac{2\pi n}{N}\right)$
Hamming	$3.3/N$	0.0194	41	53	$0.54 + 0.46\cos\left(\dfrac{2\pi n}{N}\right)$
Blackman	$5.5/N$	0.0017	57	75	$0.42 + 0.5\cos\left(\dfrac{2\pi n}{N-1}\right) + 0.08\cos\left(\dfrac{4\pi n}{N-1}\right)$
Kaiser	$2.93/N\ (\beta = 4.54)$	0.0274		50	$\dfrac{I_0(\beta\{1 - [2n/(N-1)]^2\}^{1/2})}{I_0(\beta)}$
	$4.32/N\ (\beta = 6.76)$	0.002\,75		70	
	$5.71/N\ (\beta = 8.96)$	0.000\,275		90	

The Kaiser window function goes some way in overcoming the above problems by incorporating a ripple control parameter, β, which allows the designer to trade-off the transition width against ripple. The Kaiser window is given by

$$w(n) = I_0 \left\{ \beta \left[1 - \left(\frac{2n}{N-1} \right)^2 \right]^{1/2} \right\} \Big/ I_0(\beta) \quad -(N-1)/2 \leqslant n \leqslant (N-1)/2$$

$$= 0 \qquad \qquad \text{elsewhere} \qquad (7.9)$$

where $I_0(x)$ is the zero-order modified Bessel function of the first kind. β controls the way the window function tapers at the edges in the time domain. $I_0(x)$ is normally evaluated using the following power series expansion (Rabiner and Gold, 1975):

$$I_0(x) = 1 + \sum_{k=1}^{L} \left[\frac{(x/2)^k}{k!} \right]^2$$

where typically $L < 25$. An algorithm due to Kaiser (Rabiner and Gold, 1975) gives an efficient implementation of this equation.

When $\beta = 0$, the Kaiser window corresponds to the rectangular window, and when it is 5.44, the resulting window is very similar, though not identical, to the Hamming window. The value of β is determined by the stopband attenuation requirements and may be estimated from one of the following empirical relationships:

$$\beta = 0 \qquad \qquad \text{if } A \leqslant 21\,\text{dB} \qquad (7.10\text{a})$$

$$\beta = 0.5842(A - 21)^{0.4} + 0.078\,86(A - 21) \qquad \text{if } 21\,\text{dB} < A < 50\,\text{dB} \qquad (7.10\text{b})$$

$$\beta = 0.1102(A - 8.7) \qquad \qquad \text{if } A \geqslant 50\,\text{dB} \qquad (7.10\text{c})$$

where $A = -20 \log_{10}(\delta)$ is the stopband attenuation, $\delta = \min(\delta_\text{p}, \delta_\text{s})$, since the passband and stopband ripples are nearly equal, δ_p is the desired passband ripple and δ_s is the desired stopband ripple. The number of filter coefficients, N, is given by

$$N \geqslant \frac{A - 7.95}{14.36\,\Delta f} \qquad (7.11)$$

where Δf is the normalized transition width. The values of β and N are used to compute the coefficients for the Kaiser window $w(n)$.

7.5.2   Summary of the window method of calculating FIR filter coefficients

- *Step 1*   Specify the 'ideal' or desired frequency response of filter, $H_\text{D}(\omega)$.
- *Step 2*   Obtain the impulse response, $h_\text{D}(n)$, of the desired filter by evaluating the inverse Fourier transform (Equation 7.6b). For the standard frequency selective filters the expressions for $h_\text{D}(n)$ are summarized in Table 7.2.

■ *Step 3*   Select a window function that satisfies the passband or attenuation specifications and then determine the number of filter coefficients using the appropriate relationship between the filter length and the transition width, Δf (expressed as a fraction of the sampling frequency).

■ *Step 4*   Obtain values of $w(n)$ for the chosen window function and the values of the actual FIR coefficients, $h(n)$, by multiplying $h_D(n)$ by $w(n)$:

$$h(n) = h_D(n)w(n) \tag{7.12}$$

It is clear that the window method is straightforward and involves a minimal amount of computational effort. Indeed you could obtain the coefficients with your pocket calculators. However, a PC-based program is available on the CD in the companion handbook (see the Preface for details) for calculating $h(n)$. It should be said that the resulting filter is not optimal, that is in many cases a filter with a smaller number of coefficients can be designed using other methods.

Example 7.3

Obtain the coefficients of an FIR lowpass filter to meet the specifications given below using the window method.

passband edge frequency	1.5 kHz
transition width	0.5 kHz
stopband attenuation	>50 dB
sampling frequency	8 kHz

Solution

From Table 7.2, we select $h_D(n)$ for lowpass filter which is given by

$$h_D(n) = 2f_c \frac{\sin(n\omega_c)}{n\omega_c} \quad n \neq 0$$

$$h_D(n) = 2f_c \quad n = 0$$

Table 7.3 indicates that the Hamming, Blackman or Kaiser window will satisfy the stopband attenuation requirements. We will use the Hamming window for simplicity. Now $\Delta f = 0.5/8 = 0.0625$. From $N = 3.3/\Delta f = 3.3/0.0625 = 52.8$, let $N = 53$. The filter coefficients are obtained from

$$h_D(n)w(n) \qquad\qquad -26 \leqslant n \leqslant 26$$

where

$$h_D(n) = \frac{2f_c \sin(n\omega_c)}{n\omega_c} \qquad\qquad n \neq 0$$

$$h_D(n) = 2f_c \qquad\qquad n = 0$$

$$w(n) = 0.54 + 0.46 \cos(2\pi n/53) \quad -26 \leqslant n \leqslant 26$$

Because of the smearing effect of the window on the filter response, the cutoff frequency of the resulting filter will be different from that given in the specifications. To account for this, we will use an f_c that is centred on the transition band:

$$f'_c = f_c + \Delta f/2 = (1.5 + 0.25) \text{ kHz} = 1.75 \text{ kHz} \rightarrow 1.75/8 = 0.218\,75$$

Noting that $h(n)$ is symmetrical, we need only compute values for $h(0)$, $h(1)$, ..., $h(26)$ and then use the symmetry property to obtain the other coefficients.

$n = 0$: $h_D(0) = 2f_c = 2 \times 0.218\,75 = 0.4375$

$\qquad\quad w(0) = 0.54 + 0.46 \cos(0) = 1$

$\qquad\quad h(0) = h_D(0)w(0) = 0.4375$

$n = 1$: $h_D(1) = \dfrac{2 \times 0.218\,75}{2\pi \times 0.218\,75} \sin(2\pi \times 0.218\,75)$

$\qquad\qquad\quad = \dfrac{\sin(360° \times 0.218\,75)}{\pi} = 0.312\,19$

$\qquad\quad w(1) = 0.54 + 0.46 \cos(2\pi/53)$

$\qquad\qquad\quad = 0.54 + 0.46 \cos(360°/53) = 0.996\,77$

$\qquad\quad h(1) = h(-1) = h_D(1)w(1) = 0.311\,18$

$n = 2$: $h_D(2) = \dfrac{2 \times 0.218\,75}{2 \times 2\pi \times 0.218\,75} \sin(2 \times 2\pi \times 0.218\,75)$

$\qquad\qquad\quad = \dfrac{\sin(157.5°)}{2\pi} = 0.060\,13$

$\qquad\quad w(2) = 0.54 + 0.46 \cos(2\pi \times 2/53)$

$\qquad\qquad\quad = 0.54 + 0.46 \cos(720°/53) = 0.987\,13$

$\qquad\quad h(2) = h(-2) = h_D(2)w(2) = 0.060\,12$

$\qquad\qquad \vdots \qquad \vdots \qquad\quad \vdots \qquad \vdots$

$n = 26$: $h_D(26) = \dfrac{2 \times 0.218\,75}{26 \times 2\pi \times 0.218\,75} \sin(26 \times 2\pi \times 0.218\,75)$

$\qquad\qquad\qquad = -0.011\,31$

$\qquad\quad w(26) = 0.54 + 0.46 \cos(2\pi \times 26/53)$

$\qquad\qquad\qquad = 0.54 + 0.46 \cos(9360°/53) = 0.080\,81$

$\qquad\quad h(26) = h(-26) = h_D(26)w(26) = -0.000\,914$

Table 7.4 FIR coefficients for Example 7.3 ($N = 53$, Hamming window, $f_c = 1750$ Hz).

h[0] =	−9.1399895e−04	= h[52]
h[1] =	2.1673690e−04	= h[51]
h[2] =	1.3270280e−03	= h[50]
h[3] =	3.2138355e−04	= h[49]
h[4] =	−1.9238177e−03	= h[48]
h[5] =	−1.4683633e−03	= h[47]
h[6] =	2.3627318e−03	= h[46]
h[7] =	3.4846558e−03	= h[45]
h[8] =	−1.9925839e−03	= h[44]
h[9] =	−6.2837232e−03	= h[43]
h[10] =	4.5320247e−09	= h[42]
h[11] =	9.2669460e−03	= h[41]
h[12] =	4.3430586e−03	= h[40]
h[13] =	−1.1271299e−02	= h[39]
h[14] =	−1.1402453e−02	= h[38]
h[15] =	1.0630714e−02	= h[37]
h[16] =	2.0964392e−02	= h[36]
h[17] =	−5.2583216e−03	= h[35]
h[18] =	−3.2156086e−02	= h[34]
h[19] =	−7.5449714e−03	= h[33]
h[20] =	4.3546153e−02	= h[32]
h[21] =	3.2593190e−02	= h[31]
h[22] =	−5.3413653e−02	= h[30]
h[23] =	−8.5682029e−02	= h[29]
h[24] =	6.0122145e−02	= h[28]
h[25] =	3.1118568e−01	= h[27]
h[26] =	4.3750000e−01	= h[26]

We note that the indices of the filter coefficients run from −26 to 26. To make the filter causal (necessary for implementation) we add 26 to each index so that the indices start at zero. The filter coefficients, with indices adjusted, are listed in Table 7.4. The spectrum of the filter (not shown) indicates that the specifications were satisfied.

Example 7.4 A requirement exists for an FIR digital filter to meet the following specifications:

passband	150–250 Hz
transition width	50 Hz
passband ripple	0.1 dB
stopband attenuation	60 dB
sampling frequency	1 kHz

Obtain the filter coefficients and spectrum using the window method.

Solution From the specification, the passband and stopband ripples are

$$20 \log (1 + \delta_p) = 0.1 \text{ dB, giving } \delta_p = 0.0115$$

$$-20 \log (\delta_s) = 60 \text{ dB, giving } \delta_s = 0.001$$

Thus

$$\delta = \min (\delta_p, \delta_s) = 0.001$$

The attenuation requirements can only be met by the Kaiser or the Blackman window. For the Kaiser window, the number of filter coefficients is

$$N \geqslant \frac{A - 7.95}{14.36 \, \Delta F} = \frac{60 - 7.95}{14.36(50/1000)} = 72.49$$

Let $N = 73$. The ripple parameter is given by

$$\beta = 0.1102(60 - 8.7) = 5.65$$

With $N = 73$, $\beta = 5.65$, the program window.c (see the Appendix) is used to compute the values of $w(n)$, the ideal impulse response $h_D(n)$ and the filter coefficients. To account for the smearing effects of the window functions, in computing the ideal impulse response cutoff frequencies of $f_{c1} - \Delta f/2$ and $f_{c2} + \Delta f/2$ were used, that is $f_{c1} = 125$ Hz and f_{c2} 275 Hz respectively. The filter coefficients are given in Table 7.5 and the filter spectrum in Figure 7.8.

For the Blackman window, an estimate of the number of filter coefficients is obtained as

$$N = 5.5/\Delta f = 5.5/(50/1000) \approx 110$$

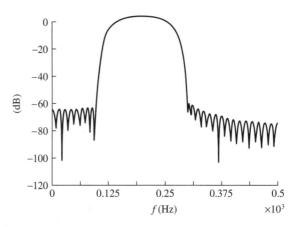

Figure 7.8 Spectrum of the filter (Example 7.4).

Table 7.5   Coefficients of the Kaiser filter (Example 7.4).

h[0] =	−1.0627330e−04	= h[72]
h[1] =	−3.9118142e−04	= h[71]
h[2] =	−7.5561629e−05	= h[70]
h[3] =	−1.3695577e−04	= h[69]
h[4] =	−6.8122013e−04	= h[68]
h[5] =	5.0929290e−04	= h[67]
h[6] =	2.3413494e−03	= h[66]
h[7] =	8.0280013e−04	= h[65]
h[8] =	−1.7031635e−04	= h[64]
h[9] =	−5.5034956e−04	= h[63]
h[10] =	−4.9912488e−04	= h[62]
h[11] =	−4.4036355e−03	= h[61]
h[12] =	−2.1639856e−03	= h[60]
h[13] =	6.9094151e−03	= h[59]
h[14] =	6.6067599e−03	= h[58]
h[15] =	−1.6445200e−03	= h[57]
h[16] =	4.5229777e−09	= h[56]
h[17] =	2.1890066e−03	= h[55]
h[18] =	−1.1720511e−02	= h[54]
h[19] =	−1.6377726e−02	= h[53]
h[20] =	6.8804519e−03	= h[52]
h[21] =	1.8882837e−02	= h[51]
h[22] =	2.9068601e−03	= h[50]
h[23] =	4.3925286e−03	= h[49]
h[24] =	1.8839744e−02	= h[48]
h[25] =	−1.2481155e−02	= h[47]
h[26] =	−5.2063428e−02	= h[46]
h[27] =	−1.6557375e−02	= h[45]
h[28] =	3.3298453e−02	= h[44]
h[29] =	1.0439025e−02	= h[43]
h[30] =	9.4320244e−03	= h[42]
h[31] =	8.5673629e−02	= h[41]
h[32] =	4.5314758e−02	= h[40]
h[33] =	−1.6657147e−01	= h[39]
h[34] =	−2.0669512e−01	= h[38]
h[35] =	8.9135544e−02	= h[37]
h[36] =	3.0000000e−01	= h[36]

The filter coefficients for the Blackman window are not given here owing to lack of space. It is evident that the Kaiser window is more efficient than the Blackman window in terms of the number of coefficients required to meet the same specifications. In general, the Kaiser window is more efficient compared to the other windows in this respect.

Example 7.5   Obtain the coefficients of a linear phase FIR filter using the Kaiser window to satisfy the following amplitude response specifications:

stopband attenuation	40 dB
passband ripple	0.01 dB
transition width	500 Hz
sampling frequency	10 kHz
ideal cutoff frequency	1200 Hz

Solution   From the specifications,

$$20 \log (1 + \delta_p) = 0.01 \text{ dB, giving } \delta_p = 0.001\,15$$

$$-20 \log (\delta_s) = 40 \text{ dB, giving } \delta_s = 0.01$$

Since both the passband and stopband ripples are equal (as they cannot be specified independently) in the window method, we use the smaller of the ripples:

$$\delta = \delta_s = \delta_p = 0.001\,15$$

This means that the stopband attenuation is more than actually required, in this case $-20 \log (0.001\,15) = 58.8$ dB.

From Equation 7.11, the number of filter coefficients required is

$$N = \frac{A - 7.95}{14.36 \, \Delta f} = \frac{58.8 - 7.95}{14.36(500/10\,000)} \approx 71$$

If the required attenuation specification of 40 dB was used N would have been 45. Thus the need for δ_p to be equal to δ_s in the window method has led to a higher than necessary number of filter coefficients.

The ripple parameter is obtained from Equation 7.10:

$$\beta = 0.5842(58.8 - 21)^{0.4} + 0.078\,86(58.8 - 21) = 5.48$$

The FIR coefficients are obtained from $h(n) = h_D(n)w(n)$ where, from Table 7.2,

$$h_D(n) = 2f_c \frac{\sin (n\omega_c)}{n\omega_c} \quad n \neq 0$$

$$h_D(n) = 2f_c \qquad n = 0$$

and $w(n)$ is given by Equation 7.9.

As explained before, the cutoff frequency, f_c, used in calculating $h(n)$ is different from that given in the specifications to account for the smearing effect of the window function. We select f_c that is in the middle of the transition band: $f'_c = 1200 + \Delta f/2 = 1450$ Hz.

Table 7.6   Filter coefficients using Kaiser window (Example 7.5).

h[0] =	9.8470163e−05	= h[70]
h[1] =	−1.3972411e−04	= h[69]
h[2] =	−4.5442489e−04	= h[68]
h[3] =	−4.8756977e−04	= h[67]
h[4] =	2.6173965e−05	= h[66]
h[5] =	8.6653647e−04	= h[65]
h[6] =	1.2967984e−03	= h[64]
h[7] =	6.1688894e−04	= h[63]
h[8] =	−1.0445340e−03	= h[62]
h[9] =	−2.4646644e−03	= h[61]
h[10] =	−2.1059775e−03	= h[60]
h[11] =	4.4371801e−04	= h[59]
h[12] =	3.5954580e−03	= h[58]
h[13] =	4.5526695e−03	= h[57]
h[14] =	1.5922295e−03	= h[56]
h[15] =	−3.8904820e−03	= h[55]
h[16] =	−7.6398162e−03	= h[54]
h[17] =	−5.6061945e−03	= h[53]
h[18] =	2.2010888e−03	= h[52]
h[19] =	1.0450148e−02	= h[51]
h[20] =	1.1760002e−02	= h[50]
h[21] =	2.8239875e−03	= h[49]
h[22] =	−1.1380549e−02	= h[48]
h[23] =	−1.9631856e−02	= h[47]
h[24] =	−1.2665935e−02	= h[46]
h[25] =	8.0061777e−03	= h[45]
h[26] =	2.8182781e−02	= h[44]
h[27] =	2.9474031e−02	= h[43]
h[28] =	3.8724896e−03	= h[42]
h[29] =	−3.5942288e−02	= h[41]
h[30] =	−5.9766794e−02	= h[40]
h[31] =	−3.7113570e−02	= h[39]
h[32] =	4.1378026e−02	= h[38]
h[33] =	1.5291289e−01	= h[37]
h[34] =	2.5100632e−01	= h[36]
h[35] =	2.9000000e−01	= h[35]

The following filter parameters are used with the computer program window.c (see the Appendix):

cutoff frequency	1450 Hz
ripple parameter, β	5.48
number of filter coefficients	71
sampling frequency	10 kHz

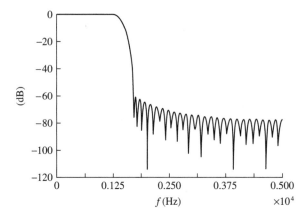

Figure 7.9   Filter spectrum using the Kaiser window (Example 7.5).

The resulting filter coefficients are given in Table 7.6 and the filter spectrum in Figure 7.9.

7.5.3   Advantages and disadvantages of the window method

- An important advantage of the window method is its simplicity: it is simple to apply and simple to understand. It involves a minimum amount of computational effort, even for the more complicated Kaiser window.
- The major disadvantage is its lack of flexibility. Both the peak passband and stopband ripples are approximately equal, so that the designer may end up with either too small a passband ripple or too large a stopband attenuation.
- Because of the effect of convolution of the spectrum of the window function and the desired response, the passband and stopband edge frequencies cannot be precisely specified.
- For a given window (except the Kaiser) the maximum ripple amplitude in the filter response is fixed regardless of how large we make N. Thus the stopband attenuation for a given window is fixed. Thus, for a given attenuation specification, the filter designer must find a suitable window.
- In some applications, the expression for the desired filter response, $H_D(\omega)$, will be too complicated for $h_D(n)$ to be obtained analytically from Equation 7.5. In these cases $h_D(n)$ may be obtained via the frequency sampling method before the window function is applied (see Section 7.7.1).

7.6 The optimal method

The optimal (in the Chebyshev sense) method of calculating FIR filter coefficients is very powerful, very flexible and, because of the existence of an excellent design program, very easy to apply. For these reasons and because the method yields excellent filters it has become the method of first choice in many FIR applications. The concept on which the method is based, the design program and its use will be discussed. Several design examples are presented to illustrate the method.

7.6.1 Basic concepts

It is evident in the window method that inherent in the process of calculating suitable filter coefficients is the problem of finding a suitable approximation to a desired or ideal frequency response. The peak ripple of filters designed by the window method occurs near the band edges, and decreases away from the band edges (Figure 7.10(a)). It turns out that if the ripples were distributed more evenly over the passband and stopband, for example as in Figure 7.10(b), a better approximation of the desired frequency response can be achieved.

The optimal method is based on the concept of equiripple passband and stopband. Consider the lowpass filter frequency response depicted in Figure 7.11. In the passband, the practical response oscillates between $1 - \delta_p$ and $1 + \delta_p$. In the stopband the filter response lies between 0 and δ_s. The difference between the ideal filter and the practical response can be viewed as an error function:

$$E(\omega) = W(\omega)[H_D(\omega) - H(\omega)] \tag{7.13}$$

where $H_D(\omega)$ is the ideal or desired response and $W(\omega)$ is a weighting function that allows the relative error of approximation between different bands to be defined. In the optimal method, the objective is to determine the filter coefficients, $h(n)$, such that the value of the maximum weighted error, $|E(\omega)|$, is minimized in the passband and stopband. Mathematically, this may be expressed as

$$\min[\max|E(\omega)|]$$

over the passbands and stopbands. It has been established (see, for example, Rabiner and Gold, 1975) that when $\max|E(\omega)|$ is minimized the resulting filter response will have equiripple passband and stopband, with the ripple alternating in sign between two equal amplitude levels (Figure 7.10(b)). The minima and maxima are known as extrema. For linear phase lowpass filters, for example, there are either $r + 1$ or $r + 2$ extrema, where $r = (N + 1)/2$ (for type 1 filters) or $r = N/2$ (for type 2 filters). The extremal frequencies are indicated as small circles in Figure 7.10(b).

For a given set of filter specifications, the locations of the extremal frequencies, apart from those at band edges (that is at $f = f_p$ and $f = F_s/2$), are not known *a priori*.

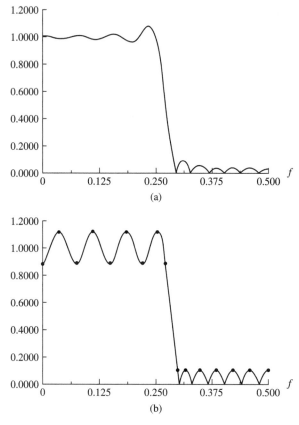

Figure 7.10   Comparison of the frequency response of (a) the window filter and (b) the optimal filter. In (a) the ripples are largest near the bandedge; in (b) the ripples have the same peaks (equiripple) in the passband or stopband.

Thus the main problem in the optimal method is to find the locations of the extremal frequencies. A powerful technique which employs the Remez exchange algorithm to find the extremal frequencies has been developed (Rabiner and Gold, 1975; McClellan *et al.*, 1973; Oppenheim and Schaffer, 1975). Knowing the locations of the extremal frequencies, it is a simple matter to work out the actual frequency response and hence the impulse response of the filter. For a given set of specifications (that is passband edge frequencies, N, and the ratio between the passband and stopband ripples) the optimal method involves the following key steps (see Figure 7.12):

■ use the Remez exchange algorithm to find the optimum set of extremal frequencies;

■ determine the frequency response using the extremal frequencies;

■ obtain the impulse response coefficients.

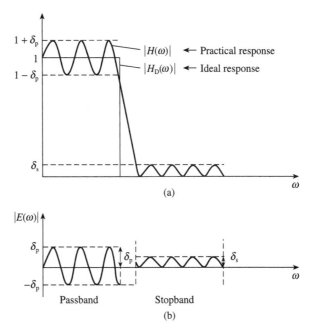

Figure 7.11   (a) Frequency response of an optimal lowpass filter. (b) Response of the error between the ideal and practical responses ($\delta_p = 2\delta_s$).

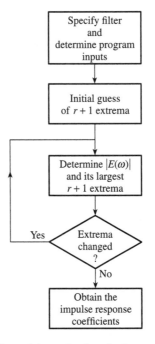

Figure 7.12   Simplified flowchart of the optimal method.

The heart of the optimal method is the first step where an iterative process is used to determine the extremal frequencies of a filter whose amplitude–frequency response satisfies the optimality condition. This step relies on the alternation theorem which specifies the number of extremal frequencies that can exist for a given value of N.

A FORTRAN program that implements the above process is available and is now widely used (for example McClellan *et al.*, 1973). An equivalent C language program is available on the CD in the companion handbook. The program allows the design of a variety of frequency selective filters including lowpass, highpass, bandpass and bandstop as well as differentiators and Hilbert transformers. It is also capable of computing the coefficients of a user-specified arbitrary frequency response. Further details of the optimal method can be found in the references given above. MATLAB-based implementation of the optimal method is discussed in the appendix.

7.6.2   Parameters required to use the optimal program

To use the design program, the user must provide a set of input parameters describing the filter. These consist of the following parameters:

N
: Number of filter coefficients, that is filter length. An estimate of this can be found from the relationships given in the next section.

Jtype
: This parameter specifies the type of filter. Three types of filter are possible: Jtype = 1 (multiple passband/stopband filters, including lowpass, highpass, bandpass, and bandstop filters), Jtype = 2 (specifies differentiator), Jtype = 3 (specifies Hilbert transformer).

$W(\omega)$
: The weighting function. This parameter specifies the relative importance of each band. In effect it allows a trade-off between the passband ripple and stopband attenuation. A weight is specified for each band.

Ngrid
: This parameter specifies the grid density. This is the number of frequency points at which, during the process of finding the extremal frequencies, the frequency response is checked to see whether the optimality condition has been satisfied (optimum in the sense that the maximum amplitude of the error, $|E(\omega)|$, is minimized in the passband(s) and stopband(s)). The default value for Ngrid is 16. In most designs, an Ngrid value of 16, 32 or 64 is adequate.

Edge
: This specifies the bandedge frequencies (that is, the lower and upper bandedge frequencies for the filter). All the frequencies must be entered in normalized form. The first edge is normally 0 and the last 0.5 (corresponding to half the sampling frequency). A maximum of 10 bands (passbands and/or stopbands) is supported.

The symbols used above are not necessarily those used in the more recent implementations. In fact some implementations now have a more user-friendly interface to the design program than the original implementation.

7.6.3   Relationships for estimating filter length, N

In practice, the number of filter coefficients is unknown. Its value may be estimated using the empirical relationships below.

7.6.3.1   Lowpass filter (Herrman *et al.*, 1973)

$$N \simeq \frac{D_\infty(\delta_p, \delta_s)}{\Delta F} - f(\delta_p, \delta_s)\Delta F + 1 \tag{7.14}$$

where ΔF is the width of the transition band normalized to the sampling frequency,

$$D_\infty(\delta_p, \delta_s) = \log_{10} \delta_s [a_1(\log_{10} \delta_p)^2 + a_2 \log_{10} \delta_p + a_3]$$
$$+ [a_4(\log_{10} \delta_p)^2 + a_5 \log_{10} \delta_p + a_6]$$
$$f(\delta_p, \delta_s) = 11.012\,17 + 0.512\,44[\log_{10} \delta_p - \log_{10} \delta_s]$$
$$a_1 = \;\;\;5.309 \times 10^{-3}; \qquad a_2 = \;\;\;7.114 \times 10^{-2}$$
$$a_3 = -4.761 \times 10^{-1}; \qquad a_4 = -2.66 \times 10^{-3}$$
$$a_5 = -5.941 \times 10^{-1}; \qquad a_6 = -4.278 \times 10^{-1}$$

δ_p is the passband ripple or deviation and δ_s is the stopband ripple or deviation.

7.6.3.2   Bandpass filter (Mintzer and Liu, 1979)

$$N \simeq \frac{C_\infty(\delta_p, \delta_s)}{\Delta F} + g(\delta_p, \delta_s)\Delta F + 1 \tag{7.15}$$

where

$$C_\infty(\delta_p, \delta_s) = \log_{10} \delta_s [b_1(\log_{10} \delta_p)^2 + b_2 \log_{10} \delta_p + b_3]$$
$$+ [b_4(\log_{10} \delta_p)^2 + b_5 \log_{10} \delta_p + b_6]$$
$$g(\delta_p, \delta_s) = -14.6 \log_{10}\left(\frac{\delta_p}{\delta_s}\right) - 16.9$$
$$b_1 = \;\;\;0.012\,01; \qquad b_2 = \;\;\;0.096\,64$$
$$b_3 = -0.513\,25; \qquad b_4 = \;\;\;0.002\,03$$
$$b_5 = -0.5705; \qquad a_6 = -0.443\,14$$

and ΔF is the transition width normalized to the sampling frequency.

A C language program is given in the appendix for computing the value of N using Equation 7.14 or 7.15.

7.6.4   Summary of procedure for calculating filter coefficients by the optimal method

- *Step 1*   Specify the bandedge frequencies (that is, passband and stopband frequencies), passband ripple and stopband attenuation (in decibels or ordinary units), and sampling frequency.

- *Step 2*   Normalize each bandedge frequency by dividing it by the sampling frequency, and determine the normalized transition width.

- *Step 3*   Use the passband ripple and stopband attenuation, expressed in ordinary units, and the normalized transition width (see note below) to estimate the filter length, N, from Equation 7.14 or 7.15. Typically, the value of N required to meet the specifications would be slightly higher (2 or 3) than the value determined from these equations.

- *Step 4*   Obtain the weights for each band from the ratio of the passband to stopband ripples (or stopband to passband ripples), expressed in ordinary units. It is convenient to express the weights for each band as an integer. For example, a lowpass filter with a passband and stopband ripples of 0.01 and 0.03 (passband ripple and stopband attenuation of 0.09 dB and 30.5 dB respectively) would have a weight of 3 for the passband and 1 for the stopband. Bandpass filter deviations (ripples) of 0.001 in the passband and 0.0105 for each of the stopbands would have weights of 21 for the passband and 2 for each of the stopbands.

- *Step 5*   Input the parameters to the optimal design program to obtain the coefficients: N, bandedge frequencies and weights for each band, together with a suitable grid density (typically 16 or 32).

- *Step 6*   Check the passband ripple and stopband attenuation produced by the program.

- *Step 7*   If the specifications are not satisfied, increase the value of N and repeat steps 5 and 6 until they are; then obtain and check the frequency response to ensure that it satisfies the specifications.

It should be noted that the optimal program considers only the passband and stopband during its approximation stage, treating the transition region as a 'don't care' region. To avoid failure or problems with convergence of the algorithm, it is best to set the transition regions equal to the width of the smallest transition region when designing bandpass or multiple-band filters. If unequal transition widths are used, the frequency response should always be checked to ensure that the specification is met. Local maxima and minima may occur in the transition bands, giving unexpected filter characteristics.

7.6.5   Illustrative examples

The following examples illustrate the use of the optimal program.

Example 7.6   A linear phase bandpass filter is required to meet the following specifications:

passband   900–1100 Hz
passband ripple   <0.87 dB
stopband attenuation >30 dB
sampling frequency   15 kHz
transition frequency 450 Hz

Use the optimal method to obtain suitable coefficients. Plot the filter spectrum.

Solution   From the specifications, the filter has three bands: a lower stopband (0 to 450 Hz), a passband (900 to 1100 Hz), and an upper stopband (1550 to 7500 Hz). To use the optimal design program the bandedge frequencies must be normalized, that is expressed as fractions of the sampling frequency:

$$450 \rightarrow 450/15\,000 = 0.03$$

$$900 \rightarrow 900/15\,000 = 0.06$$

$$1100 \rightarrow 1100/15\,000 = 0.0733$$

$$1550 \rightarrow 1550/15\,000 = 0.1033$$

$$7500 \rightarrow 7500/15\,000 = 0.5$$

Thus the three normalized bands are (0 to 0.03), (0.06 to 0.0733), (0.1033 to 0.5).

Next, we must choose weights for the bands. The weights are dependent on the passband and stopband deviations. The deviations, in ordinary units, can be obtained from the given passband ripple and stopband attenuation:

0.87 dB ripple: $20 \log (1 + \delta_p) \quad \rightarrow \delta_p = 0.105\,35$

30 dB attenuation: $-20 \log (\delta_s) \quad \rightarrow \delta_s = 0.031\,623$

The ratio of δ_p to δ_s is 3.33 = 10/3:

$$\frac{\delta_p}{\delta_s} = \frac{10}{3} = \frac{\text{stopband weight}}{\text{passband weight}}$$

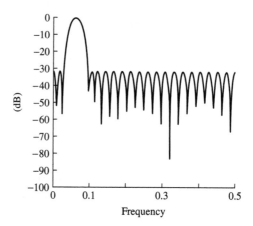

Figure 7.13   Frequency response of filter (normalized frequency scale).

Thus we could use weights of 3 for the passband and 10 for the stopband (notice that the weighting is applied in the opposite sense to the ratio of δ_p/δ_s). Equally valid are weights of 1 for the passband and 3.33 for the stopband. A grid density of 32 is used. Using the program which implements the relationships for N, the filter length was found to be 40. We will use $N = 41$.

The input to the optimal program can be summarized:

filter length, N	41
type of filter, Jtype	1
weights, $W(\omega)$	10, 3, 10
Ngrid	32
edge frequencies	0, 0.03, 0.06, 0.0733, 0.1033, 0.5

A printout of the output of the design program is given in Table 7.7 and the frequency spectrum is given in Figure 7.13. A few comments are in order.

■ The passband deviation is 3.33 times that of the stopband. This is because the errors in the passband and stopbands were given weights of 3 and 10, respectively. The higher the weight is for a band, the smaller is the resulting ripple or deviation.

■ The passband ripples and stopband attenuation (in decibels) are well within the specifications.

■ There are 22 extremal frequencies, that is $(N + 3)/2$ maxima and minima in the amplitude response. Notice that the bandedge frequencies are also extremal frequencies, as are $f = 0$ and $f = 0.5$ Hz. The bandedge frequencies are always extremal frequencies.

Table 7.7   Impulse response coefficients of the optimal filters (Example 7.6).

```
H( 1) = -0.15346380E-01 = H(41)
H( 2) = -0.57805500E-04 = H(40)
H( 3) =  0.50234820E-02 = H(39)
H( 4) =  0.12667060E-01 = H(38)
H( 5) =  0.21082060E-01 = H(37)
H( 6) =  0.27764180E-01 = H(36)
H( 7) =  0.30053620E-01 = H(35)
H( 8) =  0.25869350E-01 = H(34)
H( 9) =  0.14445660E-01 = H(33)
H(10) = -0.31893230E-02 = H(32)
H(11) = -0.24161370E-01 = H(31)
H(12) = -0.44207120E-01 = H(30)
H(13) = -0.58574530E-01 = H(29)
H(14) = -0.63185570E-01 = H(28)
H(15) = -0.55754610E-01 = H(27)
H(16) = -0.36546910E-01 = H(26)
H(17) = -0.85400990E-02 = H(25)
H(18) =  0.23083860E-01 = H(24)
H(19) =  0.52013800E-01 = H(23)
H(20) =  0.72248070E-01 = H(22)
H(21) =  0.79516810E-01 = H(21)
```

	BAND 1	BAND 2	BAND 3
LOWER BAND EDGE	0.000000000	0.060000000	0.103300000
UPPER BAND EDGE	0.030000000	0.073300000	0.500000000
DESIRED VALUE	0.000000000	1.000000000	0.000000000
WEIGHTING	10.000000000	3.000000000	10.000000000
DEVIATION	0.028891690	0.096305620	0.028891690
RIPPLE IN DB	-30.784510000	0.798631800	-30.784510000

EXTREMA FREQUENCIES

0.0000000	0.0208333	0.0300000	0.0600000	0.1033000
0.1122285	0.1308297	0.1538951	0.1777045	0.2015139
0.2260674	0.2506209	0.2759184	0.3004719	0.3257694
0.3503229	0.3756204	0.4001739	0.4254714	0.4500249
0.4753224	0.5000000			

■ The impulse response is symmetrical about the middle coefficient. The symmetry property is a necessary condition for a linear phase response. Note that, for type 1 filter, the middle coefficient has the maximum value.

For a given design, with the band edges fixed, the designer can adjust the passband ripple and stopband attenuation relative to each other as desired using the weights and N.

Example 7.7

A digital FIR notch filter satisfying the specifications given below is required:

notch frequency	1.875 kHz
attenuation at notch frequency	>60 dB
passband edge frequencies	1.575 and 2.175 kHz
passband ripple	<0.01 dB
sampling frequency	7.5 kHz
number of coefficients	61

Use the optimal method to obtain the filter coefficients of the FIR filter satisfying the specifications.

Solution

The filter has three bands. The normalized frequencies of the three bands and the deviations are

lower passband	0 to 0.21
notch frequency	0.25
upper passband	0.29 to 0.5
passband deviation	0.001 15 (from $20 \log_{10} (1 + \delta_p)$)
stopband deviation	0.001 (from $-20 \log_{10} (\delta_s)$)

The weights for the bands are 1, 1.1519, 1 (from δ_p/δ_s). The results are summarized in Table 7.8 and Figure 7.14. Notice that for a notch the stopband is effectively a single frequency.

Thus the bandedge frequencies entered into the design program are 0, 0.21, 0.25, 0.25, 0.29 and 0.5. By entering the notch frequency twice the stopband is effectively reduced to a single frequency, which is the desired result.

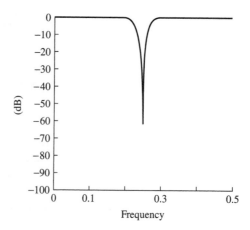

Figure 7.14   Filter response (normalized frequency scale).

Table 7.8   Impulse response coefficients for optimal filter (Example 7.7).

```
H(  1) =  0.12743640E–02 = H(61)
H(  2) =  0.26730640E–05 = H(60)
H(  3) = –0.23681110E–02 = H(59)
H(  4) = –0.17416350E–05 = H(58)
H(  5) =  0.43428480E–02 = H(57)
H(  6) =  0.53579250E–05 = H(56)
H(  7) = –0.71570240E–02 = H(55)
H(  8) = –0.49028620E–05 = H(54)
H(  9) =  0.10897540E–01 = H(53)
H(10) =  0.89629280E–05 = H(52)
H(11) = –0.15605960E–01 = H(51)
H(12) = –0.85508990E–05 = H(50)
H(13) =  0.21226410E–01 = H(49)
H(14) =  0.12250150E–04 = H(48)
H(15) = –0.27630130E–01 = H(47)
H(16) = –0.11091200E–04 = H(46)
H(17) =  0.34579770E–01 = H(45)
H(18) =  0.13800660E–04 = H(44)
H(19) = –0.41774130E–01 = H(43)
H(20) = –0.11560390E–04 = H(42)
H(21) =  0.48832790E–01 = H(41)
H(22) =  0.12787590E–04 = H(40)
H(23) = –0.55359840E–01 = H(39)
H(24) = –0.90065860E–05 = H(38)
H(25) =  0.60944450E–01 = H(37)
H(26) =  0.88997300E–05 = H(36)
H(27) = –0.65232190E–01 = H(35)
H(28) = –0.38167120E–05 = H(34)
H(29) =  0.67925720E–01 = H(33)
H(30) =  0.27041150E–05 = H(32)
H(31) =  0.93115220E+00 = H(31)
```

	BAND 1	BAND 2	BAND 3
LOWER BAND EDGE	0.000000000	0.250000000	0.290000000
UPPER BAND EDGE	0.210000000	0.250000000	0.500000000
DESIRED VALUE	1.000000000	0.000000000	1.000000000
WEIGHTING	1.000000000	1.151900000	1.000000000
DEVIATION	0.000978727	0.000849663	0.000978727
RIPPLE IN DB	0.008496785	–61.414990000	0.008496785

EXTREMA FREQUENCIES

0.0000000	0.0161290	0.0322580	0.0483871	0.0645161
0.0806451	0.0962701	0.1123991	0.1280241	0.1431450
0.1582660	0.1728829	0.1864918	0.1980845	0.2066530
0.2100000	0.2500000	0.2900000	0.2930243	0.3020971
0.3136902	0.3272994	0.3414128	0.3565342	0.3721596
0.3877850	0.4034105	0.4195400	0.4356695	0.4517990
0.4679285	0.4840580			

Example 7.8

It is important that the designer appreciates how the parameters interact so that appropriate trade-offs can be made when necessary. This example allows us to examine the effects of the parameters, δ_p, δ_s, W and the various possibilities.

A requirement exists for a linear phase FIR filter for reducing physiological noise (Hamer *et al.*, 1985). The filter is intended to be part of a large time-critical DSP system and so the number of coefficients should be kept as low as possible. The filter characteristics should meet the following specifications:

passband ripple	<0.026 dB
stopband	>30 dB
passband edge frequency	10 Hz
stopband edge	<20 Hz
sampling frequency	128 Hz

Solution

The normalized bandedge frequencies, passband and stopband deviations are

passband edge frequency	0.078
stopband edge frequency	<0.156 25
passband deviation	<0.003
stopband deviation	>0.0316

Since most of the filter specifications are variable, it is clear that there will be a range of possible solutions. The problem then is one of finding the best solution.

Now using the limiting values above in Equation 7.14, we find that $N > 25.6$ (this represents the smallest possible value of N). For each value of N in the range 25–37, the stopband frequency, f_s, that satisfies the specifications is computed using the following relationship:

$$f_s = f_p + \Delta f$$

where f_s and f_p are the stopband and passband edge frequencies, and Δf the transition width given by ($\Delta f_{max} = 20 - 10$ Hz = 10 Hz)

$$\Delta f = \frac{N-1}{2f(\delta_p, \delta_s)}\left[1 + \frac{4f(\delta_p, \delta_s)D_\infty(\delta_p, \delta_s) - 1}{(N-1)^2}\right]^{1/2}$$

Figure 7.15 gives the solution space (above the curve) which is bounded by stopband edge frequency of 20 Hz and $N = 26$ and 37. A value of 27 is chosen as a good solution. An odd value of N is preferred to avoid a noninteger number of sample delays through the filter. The following parameters were used: passband (0 to 0.078), stopband (0.152 388 5 to 0.5, that is 19 Hz to 64 Hz). Weights of 10.5 and 1 were used for the passband and stopband respectively. The resulting filter coefficients and parameters are given in Table 7.9. The filter parameters as well as its spectrum (not given) showed that the specifications were satisfied.

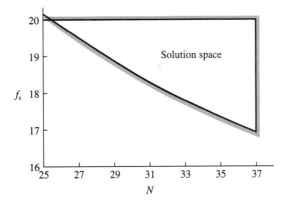

Figure 7.15   Stopband frequency versus filter length showing the range of possible solutions.

Table 7.9   Impulse response coefficients for the optimal filter (Example 7.8).

$$
\begin{aligned}
H(\ 1) &= -0.13614960E{-}01 = H(27)\\
H(\ 2) &= \ \ 0.34793330E{-}02 = H(26)\\
H(\ 3) &= \ \ 0.11140420E{-}01 = H(25)\\
H(\ 4) &= \ \ 0.16664540E{-}01 = H(24)\\
H(\ 5) &= \ \ 0.12807340E{-}01 = H(23)\\
H(\ 6) &= -0.33202110E{-}02 = H(22)\\
H(\ 7) &= -0.26167170E{-}01 = H(21)\\
H(\ 8) &= -0.42047790E{-}01 = H(20)\\
H(\ 9) &= -0.34767040E{-}01 = H(19)\\
H(10) &= \ \ 0.55338630E{-}02 = H(18)\\
H(11) &= \ \ 0.75072090E{-}01 = H(17)\\
H(12) &= \ \ 0.15527810E{+}00 = H(16)\\
H(13) &= \ \ 0.21933680E{+}00 = H(15)\\
H(14) &= \ \ 0.24378330E{+}00 = H(14)
\end{aligned}
$$

	BAND 1	BAND 2
LOWER BAND EDGE	0.000000000	0.152388500
UPPER BAND EDGE	0.078000000	0.500000000
DESIRED VALUE	1.000000000	0.000000000
WEIGHTING	10.500000000	1.000000000
DEVIATION	0.002604177	0.027343860
RIPPLE IN DB	0.022589730	-31.262770000

EXTREMA FREQUENCIES

0.0089286	0.0468750	0.0691964	0.0780000	0.1523885
0.1668974	0.1981473	0.2338614	0.2706916	0.3086379
0.3465842	0.3845305	0.4235928	0.4615391	0.5000000

7.7 Frequency sampling method

The frequency sampling method allows us to design nonrecursive FIR filters for both standard frequency selective filters (lowpass, highpass, bandpass filters) and filters with arbitrary frequency response. A unique attraction of the frequency sampling method is that it also allows recursive implementation of FIR filters, leading to computationally efficient filters. With some restrictions, recursive FIR filters whose coefficients are simple integers may be designed, which is attractive when only primitive arithmetic operations are possible, as in systems implemented with standard microprocessors.

7.7.1 Nonrecursive frequency sampling filters

Suppose we wish to obtain the FIR coefficients of the filter whose frequency response is depicted in Figure 7.16(a). We could start by taking N samples of the frequency response at intervals of kF_s/N, $k = 0, 1, \ldots, N-1$. The filter coefficients $h(n)$ can be obtained as the inverse DFT of the frequency samples:

$$h(n) = \frac{1}{N} \sum_{k=0}^{N-1} H(k)\, e^{j(2\pi/N)nk} \tag{7.16}$$

where $H(k)$, $k = 0, 1, \ldots, N-1$, are samples of the ideal or target frequency response.

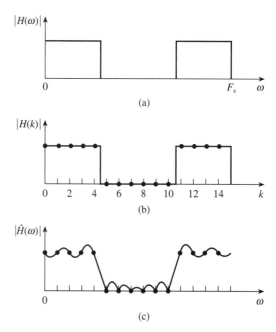

Figure 7.16 Concept of frequency sampling. (a) Frequency response of an ideal lowpass filter. (b) Samples of the ideal lowpass filter. (c) Frequency response of lowpass filter derived from the frequency samples of (b).

It can be shown (see Example 7.9) that for linear phase filters, with positive symmetrical impulse response, we can write (for N even),

$$h(n) = \frac{1}{N} \left[\sum_{k=1}^{N/2-1} 2|H(k)| \cos[2\pi k(n - \alpha)/N] + H(0) \right] \tag{7.17}$$

where $\alpha = (N - 1)/2$. For N odd, the upper limit in the summation is $(N - 1)/2$. The resulting filter will have a frequency response that is exactly the same as the original response at the sampling instants. However, between the sample instants, the response may be significantly different (Figure 7.16(c)). To obtain a good approximation to the desired frequency response, clearly we must take a sufficient number of frequency samples.

An alternative frequency sampling filter, known as type 2, results if we take frequency samples at intervals of

$$f_k = (k + 1/2)F_s/N, \quad k = 0, 1, \ldots, N - 1 \tag{7.18}$$

Figure 7.17 compares the sampling grids for both types of frequency sampling schemes. For a given filter specification, both methods will lead to somewhat different frequency responses. The designer needs to decide which of the two types best suits his or her needs.

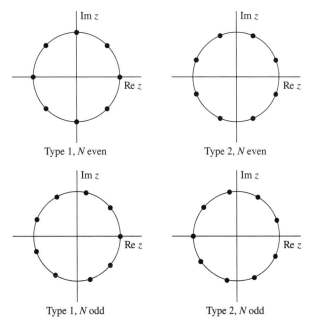

Figure 7.17 The four possible z-plane sampling grids for the two types of frequency sampling filters.

Example 7.9 (1) Show that the impulse response coefficients of a linear phase FIR filter with positive symmetry, for N even, can be expressed as

$$h(n) = \frac{1}{N}\left[\sum_{k=1}^{N/2-1} 2|H(k)|\cos[2\pi k(n-\alpha)/N] + H(0)\right]$$

where $\alpha = (N-1)/2$, and $H(k)$ are the samples of the frequency response of the filter taken at intervals of kF_s/N.

(2) A requirement exists for a lowpass FIR filter satisfying the following specifications:

passband	0–5 kHz
sampling frequency	18 kHz
filter length	9

Obtain the filter coefficients using the frequency sampling method.

Solution (1)

$$h(n) = \frac{1}{N}\sum_{k=0}^{N-1} H(k)\, e^{j(2\pi/N)nk} \tag{7.19}$$

$$= \frac{1}{N}\sum_{k=0}^{N-1} |H(k)|\, e^{-j2\pi\alpha k/N}\, e^{j2\pi kn/N}$$

$$= \frac{1}{N}\sum_{k=0}^{N-1} |H(k)|\, e^{j2\pi k(n-\alpha)/N}$$

$$= \frac{1}{N}\sum_{k=0}^{N-1} |H(k)|\, \cos[2\pi k(n-\alpha)/N] + j\sin[2\pi k(n-\alpha)/N]$$

$$= \frac{1}{N}\sum_{k=0}^{N-1} |H(k)|\, \cos[2\pi k(n-\alpha)/N] \tag{7.20}$$

since $h(n)$ is entirely real. For the important case of linear phase $h(n)$ will be symmetrical and so we can write

$$h(n) = \frac{1}{N}\left[\sum_{k=1}^{N/2-1} 2|H(k)|\cos[2\pi k(n-\alpha)/N] + H(0)\right] \tag{7.21}$$

For N odd, the upper limit in the summation is $(N-1)/2$.

(2) The ideal frequency response is depicted in Figure 7.18(a). The frequency samples are taken at intervals of kF_s/N, that is at intervals of $18/9 = 2$ kHz. Thus the frequency samples are given by

$$|H(k)| = 1 \quad k = 0, 1, 2$$
$$0 \quad k = 3, 4$$

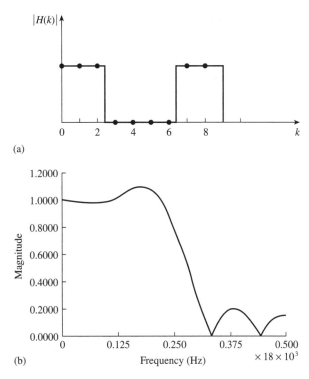

(a)

(b)

Figure 7.18   (a) Ideal frequency response showing sampling points. (b) Frequency response of frequency sampling filter.

Table 7.10   Nonrecursive coefficients for the FIR filter of Example 7.9.

h[0] =	7.2522627e−02	= h[8]
h[1] =	−1.1111111e−01	= h[7]
h[2] =	−5.9120987e−02	= h[6]
h[3] =	3.1993169e−01	= h[5]
h[4] =	5.5555556e−01	= h[4]

Using Equation 7.21 with a limit of $(N - 1)/2$ and the frequency samples we obtain the impulse response coefficients (see Table 7.10).

The CD in the companion handbook (see the Preface) contains a program to compute the FIR coefficients given the values of the frequency samples. The frequency response for the filter is shown in Figure 7.18(b). It is seen that the filter has a poor amplitude response, caused by the abrupt change from the passband (where $|H(k)| = 1$) to the stopband (where $|H(k)| = 0$).

7.7.1.1   Optimizing the amplitude response

The problem above is akin to that of the rectangular window. We recall that, in the case of the window method, we can trade off wider transition width for improved amplitude response. To improve the amplitude response of frequency sampling filters, at the expense of wider transition, we can introduce frequency samples in the transition band. Figure 7.19 illustrates a typical specification for a lowpass filter with three transition band frequency samples. For a lowpass filter, the stopband attenuation increases, approximately, by 20 dB for each transition band frequency sample (Rabiner *et al.*, 1970), with a corresponding increase in the transition width:

approximate stopband attenuation $(25 + 20M)$ dB
approximate transition width $(M + 1)F_s/N$

where M is the number of transition band frequency samples and N is the filter length.

The values of the transition band frequency samples that will give the optimum stopband attenuation are determined by an optimization process (Rabiner *et al.*, 1970). A useful optimization objective is to find values of the transition band frequency samples, $T_1, T_2, \ldots, T_M$ that minimize the peak stopband ripple (that is, they maximize the stopband attenuation). Mathematically, this may be stated as:

$$\underset{\{T_1, T_2, \ldots, T_M\}}{\text{minimize}} \left[\begin{array}{c} \max | W[H_D(\omega) - H(\omega)]| \\ \{\omega \text{ in the stopband}\} \end{array} \right] \tag{7.22}$$

where $H_D(\omega)$ and $H(\omega)$ are, respectively, the ideal and actual frequency responses of the filter; W is a weighting factor.

Rabiner *et al.* (1970) have provided a table of optimal (in the sense of Equation 7.22) values of transition band frequency samples which is widely used. A sample of the optimal values of transition band frequency samples is given in Table 7.11 for

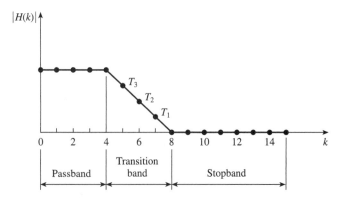

Figure 7.19   Lowpass filter frequency samples including three transition band samples. Note: because of the symmetry in the amplitude response only one half of the filter response is shown.

Table 7.11   Optimum transition band frequency samples for type 1 lowpass frequency sampling filters for $N = 15$ (adapted from Rabiner *et al.*, 1970).

BW	Stopband attenuation (dB)	T_1	T_2	T_3
One transition band frequency sample, $N = 15$				
1	42.309 322 83	0.433 782 96		
2	41.262 992 86	0.417 938 23		
3	41.253 337 86	0.410 473 63		
4	41.949 077 13	0.404 058 84		
5	44.371 245 38	0.392 681 89		
6	56.014 165 88	0.357 665 25		
Two transition band frequency samples, $N = 15$				
1	70.605 405 85	0.095 001 22	0.589 954 18	
2	69.261 681 56	0.103 198 24	0.593 571 18	
3	69.919 734 95	0.100 836 18	0.589 432 70	
4	75.511 722 56	0.084 074 93	0.557 153 12	
5	103.460 783 00	0.051 802 06	0.499 174 24	
Three transition band frequency samples, $N = 15$				
1	94.611 661 91	0.014 550 78	0.184 578 82	0.668 976 13
2	104.998 130 80	0.010 009 77	0.173 607 13	0.659 515 26
3	114.907 193 18	0.008 734 13	0.163 973 10	0.647 112 64
4	157.292 575 84	0.003 787 99	0.123 939 63	0.601 811 54

BW refers to the number of frequency samples in the passband.

$N = 15$. In the table, the bandwidth refers to the number of frequency samples in the passband of the filter.

In most cases, the values of the transition band frequency samples normally lie in the following ranges: for one transition frequency sample,

$$0.250 < T_1 < 0.450$$

for two transition frequency samples,

$$0.040 < T_1 < 0.150$$
$$0.450 < T_2 < 0.650$$

for three transition frequency samples,

$$0.003 < T_1 < 0.035$$
$$0.100 < T_2 < 0.300$$
$$0.550 < T_3 < 0.750$$

The lower values are for filters with wide bandwidth and lead to more stopband attenuation.

Example 7.10   (1)   A linear phase 15-point FIR filter is characterized by the following frequency samples:

$$|H(k)| = 1 \quad k = 0, 1, 2, 3$$
$$0 \quad k = 4, 5, 6, 7$$

Assuming a sampling frequency of 2 kHz, obtain its frequency response.

(2)   Compare the frequency response of the filter if (a) one transition band frequency sample is used, (b) two transition band frequency samples are used, and (c) three transition band frequency samples are used.

Solution   (1)   With the frequency samples as input to the design program fresamp.c (see the appendix), the coefficients of the filter are given in column 2 of Table 7.12. The corresponding frequency response is given in Figure 7.20(a).

(2)   For case (a) the value of the transition band frequency sample, from Table 7.11, is 0.4041. Thus, the frequency samples for the filter are:

$$|H(k)| = 1 \qquad k = 0, 1, 2, 3$$
$$0.404\,06 \quad k = 4$$
$$0 \qquad k = 5, 6, 7$$

With these frequency samples as input to the design program, the coefficients of the filter were computed and are summarized in Table 7.12. The corresponding frequency response is given in Figure 7.20(b).

Table 7.12   Nonrecursive filter coefficients for various transition bands frequency samples.

	No transition samples	*One transition sample*	*Two transition samples*	*Three transition samples*
h[0] =	−4.9815884e−02	−1.3766696e−02	−5.7195305e−03	−4.2282741e−03
h[1] =	4.1202267e−02	−2.3832554e−03	−7.6781827e−03	−7.6031627e−03
h[2] =	6.6666666e−02	3.9729333e−02	2.3920000e−02	1.8793332e−02
h[3] =	−3.6487877e−02	1.2729081e−02	2.5763613e−02	2.8145113e−02
h[4] =	−1.0786893e−01	−9.1220745e−02	−7.3701817e−02	−6.6396840e−02
h[5] =	3.4078020e−02	−1.8619356e−02	−4.4185450e−02	−5.2511978e−02
h[6] =	3.1889241e−01	3.1326097e−01	3.0552137e−01	3.0183514e−01
h[7] =	4.6666667e−01	5.2054133e−01	5.5216000e−01	5.6393334e−01

Because of symmetry, only the first half of the coefficients are listed here.

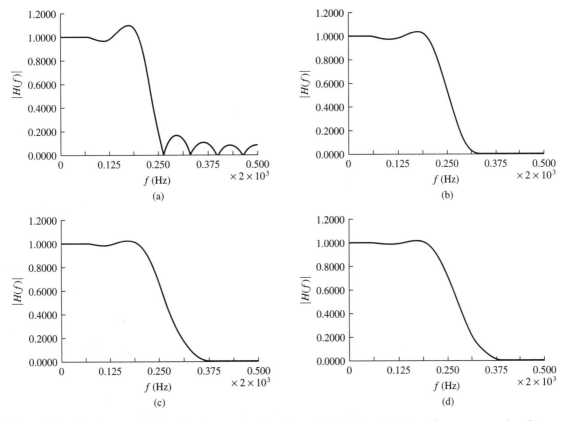

Figure 7.20   Frequency response of frequency sampling filter with (a) no transition band frequency samples; (b) one transition band frequency sample; (c) two transition band frequency samples and (d) three transition band frequency samples.

For cases (b) and (c), the frequency samples are defined respectively as

$$|H(k)| = 1 \qquad k = 0, 1, 2, 3$$
$$0.5571 \quad k = 4$$
$$0.0841 \quad k = 5$$
$$0 \qquad k = 6, 7$$
$$|H(k)| = 1 \qquad k = 0, 1, 2, 3$$
$$0.6018 \quad k = 4$$
$$0.1239 \quad k = 5$$
$$0.0038 \quad k = 6$$
$$0 \qquad k = 7$$

The coefficients for these cases are summarized in the fourth and fifth columns of Table 7.12. The corresponding frequency responses are given in Figures 7.20(c) and 7.20(d). It is seen that, as the number of transition band frequency samples increases, the amplitude response (in terms of the passband and stopband ripples) improves, but at the expense of increasing transition width or roll-off.

An alternative approach that may be used to improve the amplitude response is to obtain a large number of frequency samples, by sampling at closer intervals, to compute the impulse response using Equation 7.21, and then to apply one of the window functions discussed earlier to reduce the filter to the desired length.

7.7.1.2   Automatic design of frequency sampling filters

As stated before, tables of optimum values of the transition band frequency samples are available in the literature (Rabiner *et al.*, 1970) and are widely used for designing frequency sampling filters. If the designer wants a filter not tabulated, approximate values of the transition band frequency samples may be obtained by linear interpolation, but this is not always possible especially if the design involves a large number of transition band samples. Further, the information in the tables is not in a form filter designers are familiar with; for example, bandedge frequencies and passband ripples are not given. A general purpose computer program has recently been developed to automate many aspects of the design of nonrecursive and recursive frequency sampling filters (Ifeachor and Harris, 1993; Harris and Ifeachor, 1998). Essentially, in the program the values of the transition band samples are optimized by a hybrid genetic algorithm (GA) approach to give maximum attenuation in the stopband for a given set of filter specifications. The approach has been tested against tabulated results in the literature and was found to equal or improve on them in every case. It also allows filters not tabulated to be designed.

| Example 7.11 | Find the optimum transition band frequency samples and the corresponding filter coefficients for a lowpass filter meeting the following specifications: |

passband edge frequency	0.143 (normalized)
stopband edge frequency	0.245 (normalized)
number of filter coefficients	49

Solution   From the specifications, the number of frequency samples, $N = 49$. The sample numbers corresponding to the passband and stopband edge frequencies are 6 and 12, respectively. The number of transition band samples, $M = 5$. Thus the frequency samples for the ideal magnitude–frequency response are given by:

$$|H(k)| = 1, \qquad k = 0, 1, \ldots, 6$$
$$T_{k-6}, \qquad k = 7, \ldots, 11$$
$$0 \qquad k = 12, \ldots, 24$$

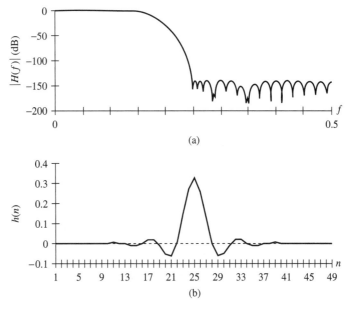

Figure 7.21   (a) The interpolated frequency response; (b) the filter coefficients. Passband ripple: 0.046 dB; stopband attenuation: 139.64 dB; passband width: 0.15; five transition samples; 49 filter coefficients. Transition sample values: 0.855 456, 0.485 507, 0.148 961, 0.019 693, 0.000 644.

The values of T_1 to T_5 are unspecified and are found by optimization using the hybrid GA program. The results of the optimization process are summarized in Figure 7.21.

Although the hybrid GA approach produces results that improve slightly on those in the literature, its main strength lies in the fact that it can quickly produce coefficients for filters that are not tabulated which are much more useful than those found by interpolation. It is also able to design filters with more transition samples.

7.7.2   Recursive frequency sampling filters

Recursive forms of the frequency sampling filter offer significant computational advantages over the nonrecursive forms if a large number of frequency samples are zero valued. It can be shown (see Example 7.12) that the transfer function of an FIR filter, $H(z)$, can be expressed in a recursive form:

$$H(z) = \frac{1 - z^{-N}}{N} \sum_{k=0}^{N-1} \frac{H(k)}{1 - e^{j2\pi k/N} z^{-1}} = H_1(z) H_2(z) \tag{7.23}$$

where

$$H_1(z) = \frac{1 - z^{-N}}{N}$$

$$H_2(z) = \sum_{k=0}^{N-1} \frac{H(k)}{1 - e^{j2\pi k/N} z^{-1}}$$

Thus we see that, in recursive form, $H(z)$ can be viewed as a cascade of two filters: a comb filter, $H_1(z)$, which has N zeros uniformly distributed around the unit circle, and a sum of N single all-pole filters, $H_2(z)$. The zeros of the comb filter and the poles of the single pole filters are coincident on the unit circle at points $z_k = e^{2\pi k/N}$. Thus the zeros cancel the poles, making $H(z)$ an FIR as it effectively has no poles.

In practice, finite wordlength effects cause the poles of $H_2(z)$ not to be located exactly on the unit circle so that they are not cancelled by the zeros, making $H(z)$ an IIR and potentially unstable. Stability problems can be avoided by sampling $H(z)$ at a radius, r, slightly less than unity. Thus the transfer function in this case becomes

$$H(z) = \frac{1 - r^N z^{-N}}{N} \sum_{k=0}^{N-1} \frac{H(k)}{1 - r e^{j2\pi k/N} z^{-1}} \qquad (7.24)$$

In general, the frequency samples, $H(k)$, are complex. Thus a direct implementation of Equation 7.23 or 7.24 would require complex arithmetic. To avoid this complication, we make use of the symmetry inherent in the frequency response of any FIR filter with real impulse response, $h(n)$. For the standard frequency selective linear phase filters (positive symmetrical impulse response), it can be shown (see Example 7.12) that Equation 7.24 can be expressed as

$$H(z) = \frac{1 - r^N z^{-N}}{N}$$

$$\times \left[\sum_{k=1}^{M} \frac{|H(k)| \{ 2\cos(2\pi k\alpha/N) - 2r\cos[2\pi k(1+\alpha)/N] z^{-1} \}}{1 - 2r\cos(2\pi k/N) z^{-1} + r^2 z^{-2}} + \frac{H(0)}{1 - z^{-1}} \right]$$

$$(7.25)$$

where $\alpha = (N-1)/2$. For N odd $M = (N-1)/2$ and for N even $M = N/2 - 1$. The realization diagram for Equation 7.25 is depicted in Figure 7.22.

7.7.3 Frequency sampling filters with simple coefficients

Recursive implementation of FIR filters greatly reduces the number of arithmetic operations in digital filters. If a filter in addition has coefficients that are simple

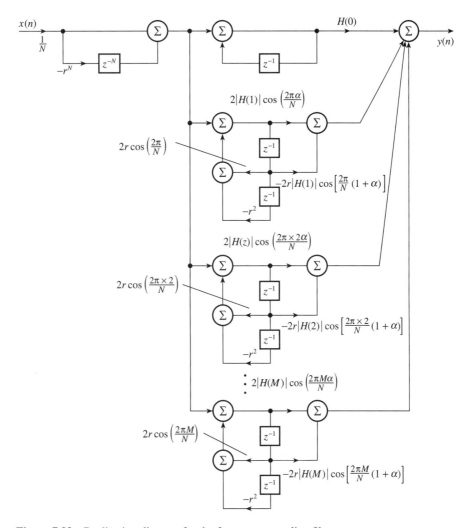

Figure 7.22   Realization diagram for the frequency sampling filter.

integers (or powers of 2) its computational efficiency is greatly improved, making it attractive in applications where processors with primitive arithmetic operations, such as ordinary microprocessors, are used. Lynn (1975) has developed a family of frequency sampling filters with small integer coefficients.

However, integer coefficients are only possible if some restrictions are placed on the locations of the poles of the transfer function (Equation 7.25). Equivalently, the passbands of the filters with integer coefficients can only be centred at restricted frequencies. Note that since the coefficients are integers we can place the poles on the unit circle and obtain a perfect cancellation. These filters are a special case of the frequency sampling filters.

Example 7.12   (1)   The transfer function of an FIR filter is defined as

$$H(z) = \sum_{n=0}^{N-1} h(n)z^{-n} \tag{7.26a}$$

Starting with the above equation, show that $H(z)$ for a linear phase FIR filter, with positive symmetrical impulse response, can be expressed in the following recursive form:

$$H(z) = \frac{1 - r^N z^{-N}}{N}$$

$$\times \left[\sum_{k=1}^{M} \frac{|H(k)|\{2\cos(2\pi k\alpha/N) - 2r\cos[2\pi k(1+\alpha)/N]z^{-1}\}}{1 - 2r\cos(2\pi k/N)z^{-1} + r^2 z^{-2}} \right.$$

$$\left. + \frac{H(0)}{1 - z^{-1}} \right]$$

where $\alpha = (N-1)/2$ and $H(k)$ are samples of the frequency response of the filter, taken at intervals of kF_s/N.

(2)   A requirement exists for a lowpass filter satisfying the following specifications:

passband	0–4 kHz
sampling frequency	18 kHz
filter length	9

Find the transfer function of the filter, in recursive form, using the frequency sampling method. Assume a radius $r = 1$. Draw the realization diagram and compare the computational complexities with direct form FIR.

Solution   (1)   The impulse response of the filter may be defined in terms of the frequency samples:

$$h(n) = \frac{1}{N} \sum_{k=0}^{N-1} H(k) r^n e^{j2\pi nk/N} \quad k = 0, 1, \ldots, N-1, r \leqslant 1 \tag{7.26b}$$

Using Equation 7.26b in Equation 7.26a the transfer function, $H(z)$, becomes

$$H(z) = \sum_{n=0}^{N-1} h(n)z^{-n} = \sum_{n=0}^{N-1} \left[\frac{1}{N} \sum_{k=0}^{N-1} H(k) r^n e^{j2\pi nk/N} \right] z^{-n}$$

Interchanging the order of the two summations we have

$$H(z) = \frac{1}{N} \sum_{k=0}^{N-1} H(k) \left\{ \sum_{n=0}^{N-1} [r e^{j(2\pi k/N)} z^{-1}]^n \right\} \tag{7.27}$$

Now the finite geometric series can be expressed as

$$S_N = \sum_{n=0}^{N-1} \delta^n = \frac{1 - \delta^N}{1 - \delta} \quad \delta \neq 1$$

In our case, with $\delta = r e^{j2\pi k/N} z^{-1}$, we can write

$$\sum_{n=0}^{N-1} [r e^{j(2\pi k/N)} z^{-1}]^n = \frac{1 - (r e^{j2\pi k/N} z^{-1})^N}{1 - r e^{j2\pi k/N} z^{-1}} = \frac{1 - r^N e^{j2\pi k} z^{-N}}{1 - r e^{j2\pi k/N} z^{-1}}$$

$$= \frac{1 - r^N z^{-N}}{1 - r e^{j2\pi k/N} z^{-1}}$$

since $e^{j2\pi k} = \cos(2\pi k) = 1$, $k = 0, 1, \ldots$. Thus we can write Equation 7.27 as

$$H(z) = \frac{1 - r^N z^{-N}}{N} \sum_{k=0}^{N-1} \frac{H(k)}{1 - r e^{j2\pi k/N} z^{-1}} = H_1(z) H_2(z) \qquad (7.28)$$

where

$$H_1(z) = \frac{1 - r^N z^{-N}}{N}$$

$$H_2(z) = \sum_{k=0}^{N-1} \frac{H(k)}{1 - r e^{j2\pi k/N} z^{-1}}$$

Now $H_2(z)$, on expansion, has the form:

$$H_2(z) = \frac{H(0)}{1 - r z^{-1}} + \frac{H(1)}{1 - r e^{j2\pi/N} z^{-1}} + \frac{H(2)}{1 - r e^{j2\pi 2/N} z^{-1}}$$

$$+ \ldots + \frac{H(N-2)}{1 - r e^{j2\pi(N-2)/N} z^{-1}} + \frac{H(N-1)}{1 - r e^{j2\pi(N-1)/N} z^{-1}}$$

For a filter with real coefficients, the following symmetry conditions hold:

$$H(k) = H^*(N - k), \quad e^{j2\pi(N-k)/N} = e^{-j2\pi k/N}$$

Thus we can write $H_2(z)$ as

$$H_2(z) = \frac{H(0)}{1 - r z^{-1}} + \frac{H(1)}{1 - r e^{j2\pi/N} z^{-1}} + \frac{H(2)}{1 - r e^{j2\pi 2/N} z^{-1}}$$

$$+ \ldots + \frac{H^*(2)}{1 - r e^{-j2\pi 2/N} z^{-1}} + \frac{H^*(1)}{1 - r e^{-j2\pi/N} z^{-1}}$$

Thus the poles occur in complex conjugate pairs (except the one at $k = 0$, for N odd, and the ones at $k = 0$ and $k = N/2$, for N even). For linear phase filters of even length, $H(N/2) = 0$. Combining the kth single-pole section and its conjugate we have

$$\frac{H(k)}{1 - r\,e^{j2\pi k/N}z^{-1}} + \frac{H^*(k)}{1 - r\,e^{-j2\pi k/N}z^{-1}}$$

$$= \frac{H(k)(1 - r\,e^{-j2\pi k/N}z^{-1}) + H^*(k)(1 - r\,e^{j2\pi k/N}z^{-1})}{(1 - r\,e^{j2\pi k/N}z^{-1})(1 - r\,e^{-j2\pi k/N}z^{-1})} \tag{7.29}$$

The denominator simplifies to

$$(1 - r\,e^{j2\pi k/N}z^{-1})(1 - r\,e^{-j2\pi k/N}z^{-1}) = 1 - 2r\cos(2\pi k/N)z^{-1} + r^2z^{-2} \tag{7.30}$$

For a linear phase filter, with positive symmetrical impulse response, $H(k)$ is given by

$$H(k) = |H(k)|\,e^{-j2\pi k\alpha/N}$$

where $\alpha = (N - 1)/2$. Thus the numerator may be simplified:

$$|H(k)|\,e^{-j2\pi k\alpha/N}(1 - r\,e^{-j2\pi k/N}z^{-1}) + |H(k)|\,e^{j2\pi k\alpha/N}(1 - r\,e^{j2\pi k/N}z^{-1})$$

$$= |H(k)|[e^{-j2\pi k\alpha/N}(1 - r\,e^{-j2\pi k/N}z^{-1}) + e^{j2\pi k\alpha/N}(1 - r\,e^{j2\pi k/N}z^{-1})]$$

$$= |H(k)|(e^{-j2\pi k\alpha/N} - r\,e^{-j2\pi k\alpha/N}e^{-j2\pi k/N}z^{-1} + e^{j2\pi k\alpha/N} - r\,e^{j2\pi k\alpha/N}e^{-j2\pi k/N}z^{-1})$$

$$= |H(k)|\{2\cos(2\pi k\alpha/N) - [r\,e^{-j2\pi k(1+\alpha)/N}z^{-1} + r\,e^{j2\pi k(1+\alpha)/N}z^{-1}]\}$$

$$= |H(k)|\{2\cos(2\pi k\alpha/N) - 2r\cos[2\pi k(1 + \alpha)/N]z^{-1}\} \tag{7.31}$$

Combining Equations 7.30 and 7.31 we can write $H(z)$ as:

$$H(z) =$$

$$\frac{1 - r^N z^{-N}}{N}\left[\sum_{k=1}^{M} \frac{|H(k)|\{2\cos(2\pi k\alpha/N) - 2r\cos[2\pi k(1 + \alpha)/N]z^{-1}\}}{1 - 2r\cos(2\pi k/N)z^{-1} + r^2z^{-2}}\right.$$

$$\left. + \frac{H(0)}{1 - rz^{-1}}\right] \tag{7.32a}$$

For N odd $M = (N - 1)/2$ and for N even $M = (N/2) - 1$.

(2) With $N = 9$, we will sample the response at intervals of $18/9 = 2$ kHz. Thus the frequency samples are defined as

$$|H(k)| = 1 \quad k = 0, 1, 2$$
$$0 \quad k = 3, 4$$

In this case, $\alpha = (N - 1)/2 = (9 - 1)/2 = 4$ and $r = 1$.

From Equation 7.32a and using the values of the frequency samples above $H(z)$ becomes

$$H(z) = \frac{1-z^{-9}}{9}\left\{ \frac{2|H(1)|[\cos(2\pi 4/9) - \cos(2\pi 5/9)z^{-1}]}{1 - 2\cos(2\pi/9)z^{-1} + z^{-2}} \right.$$

$$+ \frac{2|H(2)|[\cos(2\pi \times 2 \times 4/9) - \cos(2\pi \times 2 \times 5/9)z^{-1}]}{1 - 2z^{-1}\cos(4\pi/9) + z^{-2}}$$

$$\left. + \frac{1}{1-z^{-1}} \right\}$$

Now $\cos(8\pi/9) = -0.9397$, $\cos(10\pi/9) = -0.9397$, $\cos(2\pi/9) = 0.7660$, $\cos(16\pi/9) = 0.7660$, $\cos(20\pi/9) = 0.7660$ and $\cos(4\pi/9) = 0.1736$. Substituting these values into the equation above, we have

$$H(z) = \frac{1-z^{-9}}{9}\left[\frac{2(-0.9397 + 0.9397z^{-1})}{1 - 2\times 0.7660\,z^{-1} + z^{-2}} \right.$$

$$\left. + \frac{2(0.7660 - 0.7660\,z^{-1})}{1 - 2\times 0.1736\,z^{-1} + z^{-2}} + \frac{1}{1-z^{-1}} \right]$$

$$= \frac{1-z^{-9}}{9}\left[\frac{-1.8794(1 - z^{-1})}{1 - 1.5320\,z^{-1} + z^{-2}} + \frac{1.5320(1 - z^{-1})}{1 - 0.3472\,z^{-1} + z^{-2}} + \frac{1}{1-z^{-1}} \right]$$

The realization diagram is given in Figure 7.23. The computational complexities for both direct and frequency sampling filters are summarized below.

	Number of additions	Number of multiplications	Storage
direct	8	9	18
frequency sampling	10	7	25

With respect to Figure 7.23, the difference equations are

$$x'(n) = (1/9)[x(n) - x(n-9)]$$

$$y_1(n) = x'(n) + y(n-1)$$

$$w_2(n) = 1.5320w_2(n-1) - w_2(n-2) + x'(n)$$

$$y_2(n) = -1.8794w_2(n) + 1.8794w_2(n-1) \qquad (7.32b)$$

$$w_3(n) = 0.3472w_3(n-1) - w_3(n-2) + x'(n)$$

$$y_3(n) = 1.5320w_3(n) - 1.5320w_3(n-1)$$

$$y(n) = y_1(n) + y_2(n) + y_3(n)$$

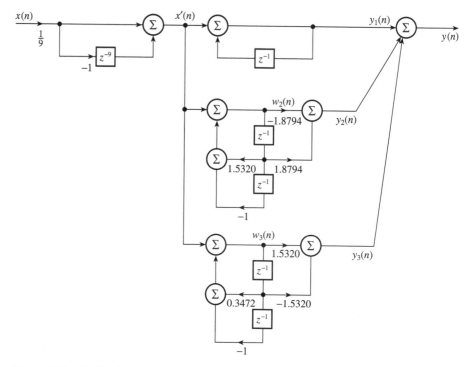

Figure 7.23 Realization diagram for the frequency sampling filter of Example 7.12.

Example 7.13

Obtain the transfer function and difference equation for

(1) a recursive FIR lowpass filter with simple integer coefficients meeting the following specifications:

> centre frequency   0 Hz
> sampling frequency   18 kHz

(2) a recursive FIR bandpass filter with simple integer coefficients meeting the following specifications:

> centre frequency   3 kHz
> sampling frequency   12 kHz

Solution

(1) If we set $N = 9$, the interval between frequency samples is $18/9 = 2$ kHz. The pole–zero diagram for this case is shown in Figure 7.24(a) and the corresponding magnitude response in Figure 7.24(b). From Figure 7.24(a), the transfer function is

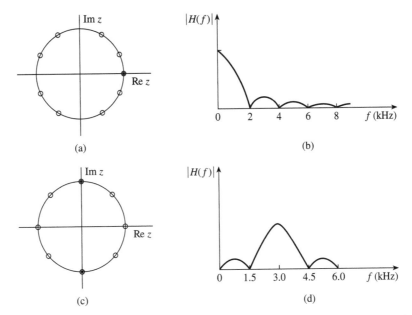

Figure 7.24 (a) Pole–zero diagram. (b) A sketch of the magnitude response of a recursive
FIR filter with integer coefficients. (c) Pole–zero diagram for a simple
bandpass filter with integer coefficients. (d) The corresponding magnitude
response.

$$H(z) = \frac{1 - z^{-9}}{9} \frac{1}{1 - z^{-1}}$$

The corresponding difference equation is

$$y(n) = y(n - 1) + (1/9)[x(n) - x(n - 9)]$$

(2) Since the passband is centred at 3 kHz, we must choose the sampling instants
carefully to ensure this is the case. Assuming $N = 8$, the z-plane diagram and
the corresponding magnitude response for one possible sampling grid are
shown in Figures 7.24(c) and 7.24(d) respectively.

   With $N = 8$, the sampling interval is $12/8 = 1.5$ kHz. The transfer function is

$$H(z) = \frac{1 - z^{-8}}{8} \frac{1}{1 + z^{-2}}$$

The corresponding difference equation is

$$y(n) = -y(n - 2) + (1/8)[x(n) - x(n - 8)]$$

It is apparent that the determination of the transfer function for frequency sampling with integer coefficients is a very simple process. However, the amplitude response of such filters is often poor and the designer is restricted as to where the passband can be located. To improve the attenuation and cutoff frequency characteristics of these filters, the transfer function can be raised to an integer value (Lynn, 1973, 1975).

7.7.4   Summary of the frequency sampling method

- *Step 1*   Specify the ideal or desired frequency response, the stopband attenuation and bandedge frequencies of the target filter.
- *Step 2*   From the specification select a type 1 frequency sampling filter, where frequency samples are taken at intervals of kF_s/N, or a type 2 frequency sampling filter, where frequency samples are taken at intervals of $(k + 1/2)F_s/N$.
- *Step 3*   Use the specification in step 1 and the design tables (Rabiner *et al.*, 1970) to determine N, the number of frequency samples of the ideal frequency response, M, the number of transition band frequency samples, BW, the number of frequency samples in the passband, and T_i, the values of the transition band frequency samples ($i = 1, 2, \ldots, M$).
- *Step 4*   Use the appropriate equation to calculate the filter coefficients.

Alternatively a computer-based program that uses genetic algorithms may be used to carry out Steps 2 to 4 (Harris and Ifeachor, 1998).

7.8   Comparison of the window, optimum and frequency sampling methods

The optimum method provides an easy and efficient way of computing FIR filter coefficients. Although the method provides total control of filter specifications, the availability of the optimal filter design software is mandatory. For most applications the optimal method will yield filters with good amplitude response characteristics for reasonable values of N. The method is particularly good for designing Hilbert transformers and differentiators. Other methods will yield larger approximation errors for differentiators and Hilbert transformers than the optimal method.

   In the absence of the optimal software or when the passband and stopband ripples are equal, the window method represents a good choice. It is a particularly simple method to apply and conceptually easy to understand. However, the optimal method will often give a more economic solution in terms of the number of filter coefficients. The window method does not allow the designer a precise control of the cutoff frequencies or ripples in the passband and stopband.

The frequency sampling approach is the only method that allows both nonrecursive and recursive implementation of FIR filters, and should be used when such implementations are envisaged as the recursive approach is computationally economical. The special form with integer coefficients should be considered only when primitive arithmetic and programming simplicity are vital (for example assembly language programming in a standard microprocessor), but a check should always be made to see whether its poor amplitude response is acceptable. Filters with arbitrary amplitude–phase response can be readily designed by the frequency sampling method. The frequency sampling method lacks precise control of the location of the bandedge frequencies or the passband ripples and relies on the availability of the design table of Rabiner *et al.* (1970) (although a PC-based program now exists for the design (Harris and Ifeachor, 1998)).

Example 7.14

Two linear phase FIR bandpass filters are required to satisfy the following specifications: for filter 1,

passband	8–12 kHz
stopband ripple	0.001
peak passband ripple	0.001
sampling frequency	44.14 kHz
transition width	3 kHz

and for filter 2,

passband	8–12 kHz
stopband ripple	0.001
peak passband ripple	0.01
sampling frequency	44.14 kHz
transition width	3 kHz

Obtain and compare the frequency response for each filter using

(1)   the window method,

(2)   the frequency sampling method, and

(3)   the optimal method.

Solution

(1)   *Window method*   For filter 1, from the specifications the passband ripple is $20 \log (1 + 0.001) = 0.008\ 68$ dB and the stopband attenuation is $-20 \log (0.001) = 60$ dB. From Equations 7.10 and 7.11 the parameters for the Kaiser window are

cutoff frequencies	6.5 kHz, 13.5 kHz
ripple parameter, β	5.653
number of filter coefficients	53
sampling frequency	44.14 kHz

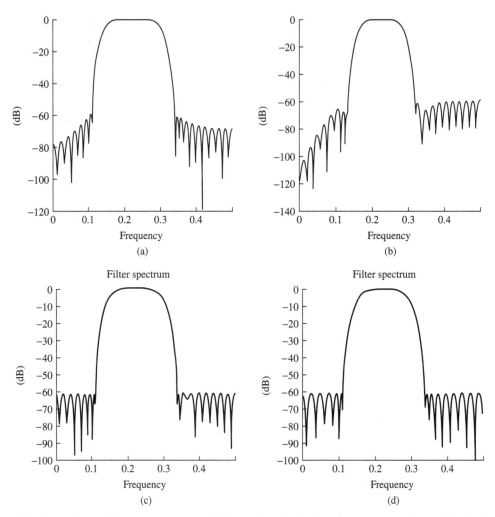

Figure 7.25 Comparison of frequency response of filters using the window, frequency sampling and optimal methods. (a) Filter response using Kaiser window (filter 1 and filter 2). (b) Filter response using frequency sampling method (filter 1 and filter 2). (c) Frequency response of optimal filter 1. (d) Frequency response of optimal filter 2.

For filter 2, the result is the same as for filter 1 since in the window method the passband and stopband ripples are always approximately equal.

The resulting filter spectra are given in Figure 7.25(a).

(2) *Frequency sampling method* For filter 1, we assume a type 1 sampling filter, and the filter length, N, is chosen as 53, the same as for the window method. From the design tables (Rabiner *et al.*, 1970), we find that we require two transition band frequency samples to achieve the desired stopband attenuation

of 60 dB, with $F_s = 44.14$ kHz, $M = 2$, $N = 53$. Sampling of the ideal frequency response, for $N = 53$, gives

$$|H(k)| = 0 \qquad k = 0, 1, \ldots, 7$$

$$0.106\,89 \quad k = 8$$

$$0.592\,53 \quad k = 9$$

$$1 \qquad k = 10\text{--}14$$

$$0.592\,53 \quad k = 15$$

$$0.106\,89 \quad k = 16$$

$$0 \qquad k = 17\text{--}26$$

Using the program fresamp.c (see the appendix), the filter was obtained and the corresponding frequency response is depicted in Figure 7.25(b).

Since the stopband attenuation is the same for both filters, filter 2 is the same as filter 1.

(3) *Optimal method*   For filter 1, from the specifications the normalized bandedge frequencies are 0, 5/44.14, 8/44.14, 12/44.14, 15/44.14 and 22.07/44.14, that is 0, 0.113 28, 0.181 24, 0.271 86, 0.339 83 and 0.5. Using the program in the appendix, we find $N = 49.6$. Since both the passband and stopband ripples are equal, the weights in the three bands are the same. The input parameters to the optimal design program are

number of filter coefficients	49
bandedge frequencies	0, 0.113 28, 0.181 24, 0.271 86, 0.339 83, 0.5
weights	5, 5, 5

The input parameters for filter 2 are

number of filter coefficients	39 (39.45)
bandedge frequencies	0, 0.113 28, 0.181 24, 0.271 86, 0.339 83, 0.5
weights	10, 1, 10

The resulting frequency responses for the optimal method are shown in Figures 7.25(c) and 7.25(d).

7.9   Special FIR filter design topics

7.9.1   Half-band FIR filters

Half-band filters are a special type of FIR filter. The main attractive feature of half-band filters is that nearly half the filter coefficients are zero which leads to a reduction in computational effort by a factor of 2. This feature makes half-band filters of interest in applications such as multirate processing where there is a need for efficient anti-aliasing filtering and/or anti-image filtering in order to alter the sampling rate of data (see Chapter 9 for details).

Causal half-band filters are characterized by the following features:

(1)   The passband and stopband ripples are equal, i.e.

$$\delta_p = \delta_s = \delta \tag{7.33}$$

(2)   The passband and the stopband edge frequencies are related in the following way:

$$f_s = \frac{F_s}{2} - f_p \tag{7.34}$$

(3)   The frequency response is symmetrical about a quarter of the sampling frequency. That is, at $f = F_s/4$

$$H\left(\frac{F_s}{4} + f\right) = 1 - H\left(\frac{F_s}{4} - f\right) \tag{7.35}$$

Also at this frequency, the normalized frequency response is down by a factor of 2, i.e.:

$$|H(f)| = 0.5 \left(\text{at } f = \frac{F_s}{4}\right)$$

(4)   In the unit impulse response, for N odd, every other coefficient is zero except $h(N-1)/2$:

$$h(2n) = 0, \qquad n = 0, 1, \ldots, (N-1)/4 \tag{7.36}$$

$$0.5, \qquad n = (N-1)/2$$

The coefficients of half-band filters can be obtained using the FIR methods described earlier, such as the window and optimal methods. In using the methods, the constraints given in Equations 7.33 and 7.34 must be imposed.

Example 7.15

Obtain the coefficients of an FIR lowpass filter to meet the specifications given below using the window method:

passband edge frequency	2 kHz
transition width	0.5 kHz
stopband attenuation	>50 dB
sampling frequency	8 kHz

Solution

The filter coefficients are listed in Table 7.13 and the filter spectrum is shown in Figure 7.26(a). It is evident from Table 7.13 that, starting from $h(0)$, every other coefficient is zero (ignoring errors due to numerical inaccuracy in computing $h(n)$). The implication of this is that during filtering every other input data sample can be ignored (effectively reducing the sampling frequency by a factor of 2).

Table 7.13   Coefficients of half-band lowpass filter (Hamming window, $N = 53$, $f_c = 2000$ Hz).

h[0] =	−1.1243421e−09	= h[52]
h[1] =	1.1109516e−03	= h[51]
h[2] =	1.3921496e−09	= h[50]
h[3] =	−1.6473646e−03	= h[49]
h[4] =	−2.0024685e−09	= h[48]
h[5] =	2.6429869e−03	= h[47]
h[6] =	2.9211490e−09	= h[46]
h[7] =	−4.1909615e−03	= h[45]
h[8] =	−4.0967870e−09	= h[44]
h[9] =	6.4068290e−03	= h[43]
h[10] =	5.4636006e−09	= h[42]
h[11] =	−9.4484947e−03	= h[41]
h[12] =	−6.9451110e−09	= h[40]
h[13] =	1.3555871e−02	= h[39]
h[14] =	8.4584215e−09	= h[38]
h[15] =	−1.9134767e−02	= h[37]
h[16] =	−9.9188559e−09	= h[36]
h[17] =	2.6953222e−02	= h[35]
h[18] =	1.1244697e−08	= h[34]
h[19] =	−3.8674295e−02	= h[33]
h[20] =	−1.2361758e−08	= h[32]
h[21] =	5.8666205e−02	= h[31]
h[22] =	1.3207536e−08	= h[30]
h[23] =	−1.0304890e−01	= h[29]
h[24] =	−1.3734705e−08	= h[28]
h[25] =	3.1728215e−01	= h[27]
h[26] =	5.0000000e−01	= h[26]

Note that every other coefficient is zero (numerical errors do not allow these coefficients to be exactly zero).

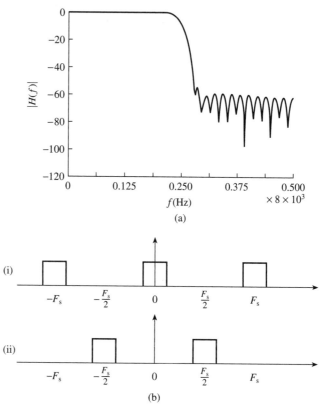

Figure 7.26 (a) Frequency response of a half-band lowpass filter. (b) Frequency response of (i) an ideal lowpass filter and (ii) an equivalent ideal highpass filter.

7.9.2 Frequency transformation

In some applications a need may arise to change, in real time, the characteristics of a filter from a lowpass to an equivalent highpass filter. A simple relationship exists between a lowpass and a highpass filter that permits such a change. The coefficients of an FIR highpass filter can be trivially obtained from those of an equivalent lowpass filter by changing the signs of the coefficients as follows:

$$h_{hp}(n) = (-1)^n h_{lp}(n) \tag{7.37}$$

This relationship is based on the knowledge that the frequency response of a highpass filter is the same as that of a lowpass filter but frequency translated by half the sampling frequency (see Figure 7.26(b)). Thus the frequency response of a highpass filter can be obtained from that of the lowpass by replacing f by $F_s/2 - f$:

$$H_{hp}(f) = H_{lp}\left(\frac{F_s}{2} - f\right) \tag{7.38}$$

Example 7.16   A lowpass filter is characterized by the following:

passband edge frequency	1.5 kHz
sampling frequency	10 kHz
number of coefficients	15

(1) Obtain the coefficients of the lowpass filter using the Hamming window.

(2) Write down the specifications for an equivalent highpass filter and use these to obtain its coefficients.

(3) Obtain the coefficients of the equivalent highpass filter by using the transformation above.

Solution   (1) Using the above parameters as inputs to the program window.c, the coefficients were obtained and are listed in Table 7.14.

(2) The specifications for the equivalent highpass filter are

passband edge frequency	$F_s/2 - f_c = 5000 - 1500$ kHz $= 3500$ kHz
sampling frequency	10 kHz
number of coefficients	15

Using these parameters as inputs to the design program window.c, the coefficients for the highpass filter were obtained and are listed in Table 7.14.

(3) Applying the simple transformation above, the coefficients for the highpass filter were obtained and are identical to those obtained in (2).

Table 7.14   Coefficients of a lowpass filter and those of an equivalent highpass filter.

	Lowpass	*Highpass*
$h(0)$	1.2654×10^{-3}	1.2654×10^{-3}
$h(1)$	-5.2341×10^{-3}	5.2341×10^{-3}
$h(2)$	-1.9735×10^{-2}	-1.9735×10^{-3}
$h(3)$	-2.3009×10^{-2}	2.3009×10^{-3}
$h(4)$	2.2366×10^{-2}	2.2366×10^{-2}
$h(5)$	1.2833×10^{-1}	-1.2833×10^{-1}
$h(6)$	2.4728×10^{-1}	2.4728×10^{-1}
$h(7)$	3.0000×10^{-1}	-3.0000×10^{-1}

7.9.3 Computationally efficient FIR filters

In some applications, the methods described in this chapter may not be appropriate. For example, in some applications the phase delay introduced by linear phase FIR filters may be unacceptably long (for example, the phase delay for type 1 FIR filters is $(N - 1)T/2$ which is large for large N). In a control system, for example, the use of such filters inside a feedback loop could cause instability. In such cases a minimum phase filter may be more appropriate (see Parks and Burrus, 1987).

The equiripple characteristics of the optimal method can lead to echoes in the impulse response of the filter which may be an undesirable effect. A smooth frequency response characteristic reduces the echoes in the impulse response tails.

In other applications, such as image processing, the number of arithmetic operations when standard FIR filters are used may be too large. Unfortunately, the integer coefficients filters are not suitable for such applications because of their poor amplitude response characteristics. FIR filters which require only very simple arithmetic operations, but whose amplitude responses are comparable with those of standard FIR filters, may be more appropriate (see, for example, Wade *et al.*, 1990 or Mitra and Kaiser, 1993).

The basis of the method is to cascade two or more elementary filter sections such as shown in Figure 7.27. Each elementary section involves hardly any multiplications.

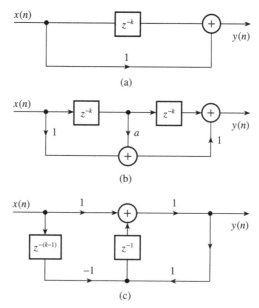

Figure 7.27 Examples of primitive FIR filter sections: (a) single adder section ($H(z) = 1 + z^{-k}$); (b) two-adder section ($H(z) = 1 + az^{-k} + z^{-2k}$; to avoid an explicit multiplication $a = \pm 2^n$, n integer; integer multiplications can be implemented as shifts); (c) recursive running sum lowpass section ($H(z) = (1 - z^{-k})/(1 - z^{-1}) = 1 + z^{-1} + z^{-2} + \ldots + z^{-(k-1)}$). A typical filter would consist of 4 to 7 such sections connected in cascade.

The main problems with this method include the difficulty of finding an efficient way of selecting the elementary filter sections to cascade and the fact that only filters of low order can be designed efficiently. Genetic algorithms have been employed to tackle the problems (Suckley, 1990).

7.10 Realization structures for FIR filters

The FIR filter is characterized by the transfer function, $H(z)$, given by

$$H(z) = \sum_{n=0}^{N-1} h(n) z^{-n}$$

Realization structures are essentially block (or flow) diagram representations of the different theoretically equivalent ways the transfer function can be arranged. In most cases, they consist of an interconnection of multipliers, adders/summers and delay elements. There are many FIR realization structures, but only those that are in common use are presented here.

7.10.1 Transversal structure

The transversal (or tapped delay) structure is depicted in Figure 7.28. The input, $x(n)$, and output, $y(n)$, of the filter for this structure are related simply by

$$y(n) = \sum_{m=0}^{N-1} h(m)x(n-m) \tag{7.39}$$

In the figure, the symbol z^{-1} represents a delay of one sample or unit of time. Thus $x(n-1)$ is $x(n)$ delayed by one sample. In digital implementations, the boxes labelled

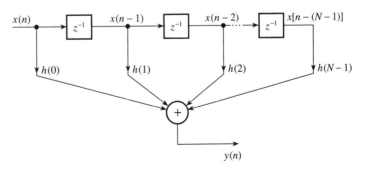

Figure 7.28 Transversal filter structure.

z^{-1} could represent shift registers or more commonly memory locations in a RAM. The transversal filter structure is the most popular FIR structure.

The output sample, $y(n)$, is a weighted sum of the present input, $x(n)$, and $N-1$ previous samples of the input, that is $x(n-1)$ to $x(n-N)$. For the transversal structure, the computation of each output sample, $y(n)$, requires

- $N-1$ memory locations to store the $N-1$ input samples,
- N memory locations to store the N coefficients,
- N multiplications, and
- $N-1$ additions.

7.10.2 Linear phase structure

A variation of the transversal structure is the linear phase structure which takes advantage of the symmetry in the impulse response coefficients for linear phase FIR filters to reduce the computational complexity of the filter implementation.

In a linear phase filter, the coefficients are symmetrical, that is $h(n) = \pm h(N-n-1)$. Thus the filter equation can be re-written to take account of this symmetry with a consequent reduction in both the number of multiplications and additions. For type 1 and 2 linear phase filters, the transfer function can be written as

$$H(z) = \sum_{n=0}^{(N-1)/2-1} h(n)[z^{-n} + z^{-(N-1-n)}] + h\left(\frac{N-1}{2}\right)z^{-(N-1)/2} \quad N \text{ odd} \qquad (7.40a)$$

$$H(z) = \sum_{n=0}^{N/2-1} h(n)[z^{-n} + z^{-(N-1-n)}] \qquad\qquad\qquad\qquad N \text{ even} \qquad (7.40b)$$

The corresponding difference equations are given by

$$y(n) = \sum_{k=0}^{(N-1)/2-1} h(k)\{x(n-k) + x[n-(N-1-k)]\}$$

$$\qquad\qquad + h[(N-1)/2]x[n-(N-1)/2] \qquad\qquad\qquad (7.41a)$$

$$y(n) = \sum_{k=0}^{(N-1)/2-1} h(k)\{x(n-k) + x[n-(N-1-k)]\} \qquad\qquad (7.41b)$$

A comparison of Equations 7.39 and 7.41 shows that the linear phase structure is computationally more efficient, requiring approximately half the number of multiplications and additions. However, in most DSP processors Equation 7.39 leads to a more efficient implementation, because the computational advantage in Equation 7.41 is lost in the more complex indexing of data implied.

Example 7.17   A linear phase FIR filter has seven coefficients which are listed below. Draw the realization diagrams for the filter using (a) direct (transversal) and (b) linear phase structures. Compare their computational complexities.

$$h(0) = h(6) = -0.032$$

$$h(1) = h(5) = \;\; 0.038$$

$$h(2) = h(4) = \;\; 0.048$$

$$h(3) \qquad\quad = -0.048$$

Solution   The realization diagrams are shown in Figure 7.29.

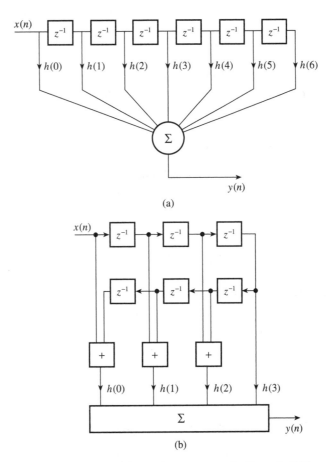

Figure 7.29   (a) Transversal and (b) linear phase structure for Example 7.17.

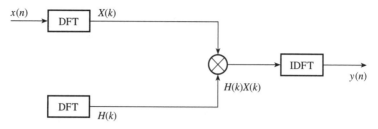

Figure 7.30   An illustration of fast convolution.

7.10.3   Other structures

7.10.3.1   Fast convolution

The fast convolution method involves performing the convolution operation of Equation 7.39 in the frequency domain. As was discussed in Chapter 5, convolution in the time domain is equivalent to multiplication in the frequency domain. In simple terms, filtering here is performed by first computing the DFTs of $x(n)$ and $h(n)$ (suitably zero padded), multiplying these together and then obtaining their inverse. The concept is depicted in Figure 7.30. In practice, techniques known as overlap–add and overlap–save are used in real-time filtering. These are discussed in Chapter 5.

7.10.3.2   Frequency sampling structure

In the frequency sampling structure, the filters are characterized by the samples of the desired frequency response, $H(k)$, instead of its impulse response coefficients. This case has already been discussed in detail. For narrowband filters, most of the frequency samples will be zero, and so the resulting frequency sampling filter will require a smaller number of coefficients and hence multiplications and additions than an equivalent transversal structure. A typical realization diagram is given in Figure 7.22.

7.10.3.3   Transpose and cascade structures

The transpose structure is similar to the direct structure, except that the partial sums feed into succeeding stages. This method is more susceptible to roundoff noise than the direct method. In the cascade realization, the transfer function, $H(z)$, is expressed as the product of second-order and first-order sections. The transpose and cascade structures are seldom used for FIR filters in current DSP implementations.

7.10.4   Choosing between structures

The choice between the structures depends on a number of factors and trade-offs which include ease of implementation, that is the implied hardware or software complexity, how difficult it is to obtain the impulse response or transfer function coefficients, and their relative sensitivity to coefficient quantization. In practice, the accuracy to which each coefficient is represented is limited by the wordlength of the processor used. The use of only a finite number of bits to represent each coefficient tends to move the zeros away from the desired locations, resulting in a deviation in the frequency response. The extent of deviation in the response depends on the number of bits and the structure used.

The direct structure is very easy to program and is efficiently implemented by most DSP chips as these have instructions tailored to transversal FIR filtering. It is the most common structure used to realize nonrecursive filters and its main attraction is its simplicity, requiring only a minimum of components and uncomplicated memory accesses for data. The cascade is less sensitive to coefficient errors and quantization noise, but the coefficients require more effort to obtain and the programming is not suited to DSP chips' architectures. The fast convolution structure offers significant computational advantages over the others, but requires the availability of the FFT.

The frequency sampling structure, for narrowband frequency selective filters, is computationally more efficient than an equivalent transversal structure. In such filters, only a relatively small number of frequency samples are nonzero for this structure, so that only very few multiplications per output sample are required. However, the frequency sampling structure may require more complicated programming, because of the more complex indexing inherent in its difference equation (compare for example Equations 7.39 and 7.32b). To avoid stability problems the poles and zeros of the frequency sampling structure should be located slightly inside the unit circle, for example at a radius $r = 0.99$. This structure is the natural choice when recursive implementation of FIR filters is mandatory. The structure is very modular and lends itself to parallel processing.

In general, the tranversal structure should be used unless the specification requirements dictate the use of the frequency sampling structure or there is a need to compute the spectrum of data as well when the fast convolution should be used.

7.11 Finite wordlength effects in FIR digital filters

In practice, FIR digital filters are often implemented using DSP processors (for example the Texas Instruments TMS320C50), algorithmic-specific DSP chips designed for FIR filtering (such as the INMOS A100) or, where high speed is desired, building blocks of multipliers, memory elements, adders and controllers (for example Plessey's PDSP1600 family). In these cases, the number of bits used to represent the input data to the filter and the filter coefficients and in performing arithmetic operations must be small for efficiency and to limit the cost of the digital filter. The problems caused by using a finite number of bits are referred to as finite wordlength effects, and in general lead to a lowering of the performance of the filter.

In this section, we will discuss the effects of finite wordlength on the performance of FIR digital filters and suggest ways of minimizing these effects. The discussion will centre on the direct form FIR structure as it is the most attractive FIR structure in modern signal processing, and rounding will be used, being the simplest and most widely used method of quantization.

There are four ways in which finite wordlength affects the performance of FIR digital filters.

(1) *ADC noise*   This is the familiar ADC quantization noise which results when the filter input is derived from analog signals. ADC noise limits the signal-to-noise ratio (SNR) obtainable. The effects can be reduced by using additional bits, consistent with inherent signal noise (see Chapter 13), and/or by using multirate techniques to enhance the signal to noise ratio (see Chapter 9).

(2) *Coefficient quantization errors*   These result from representing filter coefficients with a limited number of bits. This has the adverse effect of modifying the desired frequency response. In the stopband of a filter, for example, it limits the maximum attenuation possible, thus allowing additional signal transmission. A straightforward solution is to use enough bits to represent filter coefficients. However, optimization techniques allow efficient selection of coefficients to minimize coefficient wordlength.

(3) *Roundoff errors from quantizing results of arithmetic operations*   These can occur, for example, by discarding the lower-order bits before storing the results of a multiplication. This is normally forced on us by the wordlength of the processor used. This error reduces the SNR and may be reduced by rounding after double-length summing of products. The extent of the errors introduced depends on the type of arithmetic used and the filter structure.

(4) *Arithmetic overflow*   This occurs when partial sums or filter output exceeds the permissible wordlength of the system. Essentially, when an overflow occurs, the output sample will be wrong (normally the sign changes). A way to reduce or avoid an overflow is to scale the filter coefficients by dividing each coefficient by a factor such that the filter output sample never exceeds the permissible wordlength. This is clearly at the cost of reduced signal to noise ratio.

In the next few sections we will discuss items (2) to (4).

7.11.1   Coefficient quantization errors

Filter coefficients obtained by any of the approximation methods, for example window or optimal (Remez exchange), are usually very accurate to several places of decimals. To implement the filter, the coefficients must be represented by a fixed number of bits and very often this is determined by the wordlength of the processor used. For example, if we use one of the 16-bit DSP processors to implement the filter then the logical thing to do is to represent each filter coefficient with 16 bits. In doing so, however, we automatically introduce an error which causes the frequency response of the finite wordlength filter to deviate from the desired response. This deviation, in some cases, will mean that the initial specifications are no longer met.

Example 7.18   Determine the effects of quantizing, by rounding, the coefficients of the following filter to 8 bits:

stopband attenuation	>90 dB
passband ripple	<0.002 dB
passband edge frequency	3.375 kHz
stopband edge frequency	5.625 kHz
sampling frequency	20 kHz
number of coefficients	45

Solution   Use the design program optimal.c with the following input:

number of filter coefficients	45
bandedge frequencies	0, 0.168 75, 0.281 25, 0.5
weights	1, 7.28

The coefficients of the filter, before and after rounding to 8 bits, are listed in Table 7.15. The corresponding frequency responses are given in Figure 7.31. It is seen that, after quantization, the minimum stopband attenuation is 36 dB, a degradation of more than 58 dB. Clearly, more than 8 bits of resolution is required for the coefficients in this particular example.

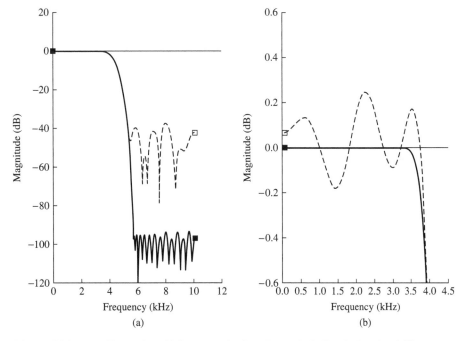

Figure 7.31   (a) Effects of coefficient quantization (Example 7.18). (b) Passband. ■, unquantized; □, quantized.

Table 7.15 Filter coefficients before and after quantization to 8 bits.

$h(n)$	$h_q(n)$
−1.05023e−04	0.00000e+00
−1.25856e−04	0.00000e+00
3.07141e−04	0.00000e+00
6.79484e−04	0.00000e+00
−2.89029e−04	0.00000e+00
−1.77474e−03	0.00000e+00
4.08318e−04	0.00000e+00
3.43482e−03	0.00000e+00
2.66515e−03	0.00000e+00
−5.00314e−03	−7.81250e−03
−7.30591e−03	−7.81250e−03
5.09712e−03	7.81250e−03
1.48422e−02	1.56250e−02
−1.40255e−03	0.00000e+00
−2.49785e−02	−2.34375e−02
−9.39383e−03	−7.81250e−03
3.64568e−02	3.90625e−02
3.28505e−02	3.12500e−02
−4.72008e−02	−4.68750e−02
−8.52427e−02	−8.59375e−02
5.48855e−02	5.46875e−02
3.10921e−01	3.12500e−01
4.42322e−01	4.45212e−01
3.10921e−01	3.12500e−01
5.48855e−02	5.46875e−02
−8.52427e−02	−8.59375e−02
−4.72008e−02	−4.68750e−02
3.28505e−02	3.12500e−02
3.64568e−02	3.90625e−02
−9.39383e−03	−7.81250e−03
−2.49785e−02	−2.34375e−02
−1.40255e−03	0.00000e+00
1.48422e−02	1.56250e−02
5.09712e−03	7.81250e−03
−7.30591e−03	−7.81250e−03
−5.00314e−03	−7.81250e−03
2.66515e−03	0.00000e+00
3.43482e−03	0.00000e+00
4.08318e−04	0.00000e+00
−1.77474e−03	0.00000e+00
−2.89029e−04	0.00000e+00
6.79484e−04	0.00000e+00
3.07141e−04	0.00000e+00
−1.25856e−04	0.00000e+00
−1.05023e−04	0.00000e+00

The effect of coefficient errors is to cause the frequency response to deviate from the desired response. This deviation in the extreme case will mean that the specifications are no longer met. For a particular filter design problem, a suitable coefficient wordlength can be determined by obtaining the frequency response for the filter for different coefficient wordlengths. From these, the minimum number of bits required to meet the desired specifications can be determined. However, valuable insight into the design of finite wordlength filters can be gained by analyzing the errors introduced by coefficient quantization.

Now, the quantized and unquantized coefficients, $h(n)$ and $h_q(n)$, respectively, are related as

$$h_q(n) = h(n) + e(n), \quad n = 0, 1, \ldots, N - 1 \tag{7.42}$$

where $e(n)$ is the error between the quantized and unquantized coefficients. In the frequency domain, Equation 7.42 can be written as

$$H_q(\omega) = H(\omega) + E(\omega) \tag{7.43}$$

where $E(\omega)$, the error in the desired frequency response, is given by

$$E(\omega) = \sum_{m=0}^{N-1} e(m) \exp(-j\omega m)$$

and $H_q(\omega)$ and $H(\omega)$ are the frequency responses of the filters with quantized and unquantized coefficients, respectively. Figures 7.32(a) and 7.32(b) give diagrammatic representations of Equations 7.42 and 7.43, respectively. We see that physically $e(n)$ can be viewed as the impulse response of another filter in parallel with the impulse response of the desired filter (Rabiner and Gold, 1975). The effect of coefficient error in the frequency domain is represented as a stray transfer function in parallel with that of the very accurate filter. An objective of the designer is to limit

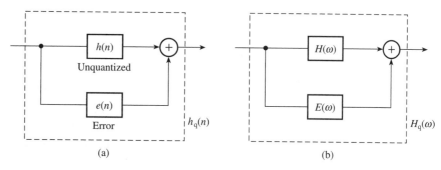

Figure 7.32   Illustration of the effects of coefficient quantizations. (a) Effects on the impulse response coefficients. (b) Effects on the frequency response of the filter.

the amplitude of $E(\omega)$ so that the frequency response of the actual filter meets the specification.

For frequency selective filters (lowpass, bandpass, bandstop filters), several researchers have developed bounds on the errors in the frequency response. These bounds could serve as useful guides in determining a suitable coefficient wordlength for a given filter. The bounds are useful in estimating coefficient wordlength requirements for adaptive FIR filters as the exact characteristics of these filters are not known *a priori* (see Chapter 10).

For direct form FIR structure, assuming rounding, the following are the most widely used bounds:

$$|E(\omega)| = N2^{-B} \tag{7.44a}$$

$$|E(\omega)| = 2^{-B}(N/3)^{1/2} \tag{7.44b}$$

$$|E(\omega)| = 2^{-B}[(N\log_e N)/3]^{1/2} \tag{7.44c}$$

where B is the number of bits used to represent each coefficient and N is the filter length. Bound 7.44a is an absolute upper bound, derived under worst-case assumptions (see Example 7.19) and so it is overly pessimistic. Bounds 7.44b and 7.44c are statistical bounds and could give a more accurate estimate of the errors in the frequency response and coefficient wordlengths to use. The statistical bounds assume that the quantization errors, $e(n)$, are uniformly distributed and have zero means.

Example 7.19

(1) Show, stating any assumptions, that the maximum stopband attenuation possible, $A_{\max}$, for a direct form lowpass FIR filter, with coefficients quantized by rounding, is bounded by

$$A_{\max} \leq 20\log_{10}(2^{-B}N) \tag{7.45}$$

(2) A lowpass FIR filter has the following specifications:

passband deviation	0.05 dB
sampling frequency	10 kHz
passband edge	1.8 kHz
transition width	500 Hz
number of coefficients	65

(a) Estimate the number of bits required to represent each coefficient for the filter to have an attenuation of at least 60 dB in the stopband.

(b) If the coefficient wordlength in (a) is used, estimate the expected increase in passband ripple and the reduction in stopband attenuation in decibels.

(c)   Compare the actual stopband attenuation and passband ripple of the filter using the coefficient wordlength obtained in (a).

Solution   (1)   Define the response, $E(\omega)$, due to the coefficient quantization error, $e(m)$, as

$$E(\omega) = \sum_{m=0}^{N-1} e(m) \exp(-j\omega m)$$

where N is the filter length. For rounding, the worst-case quantization error is $|e(m)| = 2^{-(B-1)}/2 = 2^{-B}$ where B is the coefficient wordlength (assuming two's complement representation). If we assume the worst-case error for all the coefficients, then we have

$$|E(\omega)| = \sum_{m=0}^{N-1} |e(m)| \exp(-j\omega m) = \sum_{m=0}^{N-1} 2^{-B} \exp(-j\omega m)$$

$$= 2^{-B} \sum_{m=0}^{N-1} \exp(-j\omega m) = 2^{-B} N$$

If $e(m)$ is viewed as the impulse response of another filter in parallel with the desired filter, then the limiting deviation in the passband or stopband is $2^{-B}N$, and so

$$A_{max} < 20 \log_{10}(2^{-B}N) \text{ dB}$$

Clearly, this bound is overly conservative. Fewer bits than it suggests will often suffice. However, the bound serves as a guide that can be applied simply.

(2)   (a)   From the bound above, and setting $A_{max} = 60$ dB, $N = 65$, we find that $B = 15.988$ bits. The required coefficient wordlength is therefore $B = 16$ bits.

(b)   After quantization, the worst case peak ripple in the passband, R_{max}, and the stopband attenuation, A_{max}, may be expressed as

$$R_{max} = 20 \log(1 + \delta_p + |E(\omega)|) = 20 \log(1 + 0.005\ 773 + 0.001)$$

$$= 0.0586 \text{ dB}$$

that is an increase of 0.0086 dB, and

$$A_{max} = -20 \log(\delta_s + |E(\omega)|) = -20 \log(0.001 + 0.001) = 54 \text{ dB}$$

that is a reduction of 6 dB (δ_p and δ_s are the passband and stopband deviations for the unquantized filter).

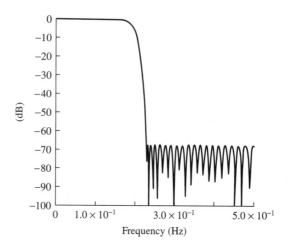

Figure 7.33   Filter spectrum before quantization for Example 7.19.

(c)   Use the optimal design program and the following parameters:

number of coefficients	65
bandedge frequencies	0, 0.18, 0.23, 0.5
passband–stopband weights	1, 5.773

The filter spectrum before quantization is shown in Figure 7.33. There was no significant difference between the quantized (16-bit wordlength) and unquantized frequency responses. The passband ripple and stopband attenuation after quantization were 0.0227 dB and 64.15 dB compared with 0.0224 dB and 66.96 dB before quantization.

It is clear that the main effects of coefficient quantization are a possible increase in the peak passband ripple and a reduction in the maximum attenuation in the stopband. Practical procedures are available for taking these effects into account when computing the coefficients of filters. Essentially, this involves mapping the unquantized filter specifications into a new set of specifications which are then used to obtain the coefficients. This mapping is such that after coefficient quantization the original specifications would be satisfied.

The resulting filter may not be optimal. This has led to the development of optimization techniques, such as mixed integer programming algorithms, for obtaining the coefficients of finite wordlength FIR filters (for example, Lawrence and Salazar, 1980). The new approaches lead to a significant reduction in coefficient wordlengths compared with straightforward rounding, but finding suitable coefficients often involves high computational overheads for even moderately large N. A pragmatic approach is to use one of the bounds in Equations 7.44 to estimate the number of bits required to represent the coefficients. The required coefficient wordlength will often be 1 to 4 bits above or below this value and can be determined by studying the frequency response corresponding to wordlengths in this range.

7.11.2   Roundoff errors

Recall that the difference equation of the FIR filter is given by

$$y(n) = \sum_{m=0}^{N-1} h(m)x(n-m) \tag{7.46}$$

where each variable is represented by a fixed number of bits. Typically, the input and output samples, $x(n-m)$ and $y(n)$, are each represented by 12 bits and the coefficients by 16 bits in 2's complement format.

It is seen from Equation 7.46 that the output of the filter is obtained as the sum of products of $h(m)$ and $x(n-m)$. After each multiplication, the product contains more bits than either $h(m)$ or $x(n-m)$. For example, if a 12-bit input is multiplied by a 16-bit coefficient the result is 28 bits long and will need to be quantized back to 16 bits (say) before it can be stored in a memory or to 12 bits before it can be output to the DAC (say). This quantization leads to errors whose effects are similar to those of the ADC noise, but could be more severe. The common way to quantize the result of an arithmetic operation is either (a) to truncate the results, that is to retain the most significant higher-order bits and to discard the lower-order bits, or (b) to round the results, that is to choose the higher-order bits closest to the unrounded result. This is achieved by adding half an LSB to the result.

Roundoff errors can be minimized by representing all products exactly, with double-length registers, and then rounding the results after obtaining the final sum, that is after obtaining $y(n)$. This approach introduces a smaller error than the alternative approach of rounding each product separately before summing.

7.11.3   Overflow errors

Overflow occurs when the sum of two numbers, usually two large numbers of the same sign, exceeds the permissible wordlength. Thus in Equation 7.46 overflow could occur when we add the two products: $h(0)x(n)$ and $h(1)x(n-1)$.

Provided that the final output, $y(n)$, is within the permissible wordlength overflow in partial sums is unimportant. This is a desirable property of 2's complement arithmetic. However, if the output, $y(n)$, exceeds the permissible limit then clearly the value of the output sample to the DAC, for example, will be wrong and steps should be taken to prevent this. An approach is to detect and correct for an overflow, but this may be an expensive overhead. Another alternative is to avoid or allow limited overflow by scaling the coefficients and/or the input data. The coefficients may be scaled in one of the following ways:

$$h(m) = \frac{h(m)}{\displaystyle\sum_{k=0}^{N-1} |h(k)|} \tag{7.47a}$$

$$h(m) = \frac{h(m)}{\left[\displaystyle\sum_{k=0}^{N-1} h^2(k)\right]^{1/2}} \tag{7.47b}$$

For the method given in Equation 7.47a overflow will never occur, but this form of scaling is often unnecessary because it is based on the worst-case conditions for overflow which are unlikely in practice. It will also introduce more coefficient quantization noise than the method given in Equation 7.47b which allows for occasional overflow.

The input data may be scaled in a similar manner to the coefficients. Often this leads to a better SNR. A third approach is to scale both the input and the output in such a manner that the best possible SNR is achieved. An efficient scaling would use a scale factor for the input that is a power of 2.

7.12   FIR implementation techniques

The difference equation for an FIR digital filter is given by

$$y(n) = \sum_{k=0}^{N-1} h(k)x(n-k) \tag{7.48}$$

The coefficients $h(k)$ will have been obtained at the approximation stage, a suitable structure chosen, and an analysis carried out to verify that the number of bits to be used to represent variables and in carrying out arithmetic operations is adequate. The final stage is to implement the filter, and the key issue here is essentially to produce software code and/or hardware realization of the chosen filter structure. The discussions here will be based on the transversal structure which is characterized by Equation 7.48 as it is the most popular.

As examination of the equation will show, the computation of $y(n)$ involves only multiplications, additions/subtractions, and delays. Thus, to implement a filter, we need the following basic components:

∎ memory (RAM) to store the present and past input samples, $x(n)$ and $x(n-k)$;

∎ memory (RAM or ROM) for storing the filter coefficients, the $h(k)$;

∎ a multiplier (software or hardware);

∎ adders or arithmetic logic unit (ALU).

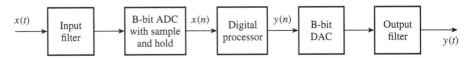

Figure 7.34   A simplified block diagram of a real-time digital filter with analog input/
output signals.

These components together with a means of controlling them constitute the digital
filter. If the source of the input data is analog, then we need an ADC as well.
Similarly, if the output destination is analog we need a DAC. Thus the structure of a
real-time filter has the form depicted in Figure 7.34. Filter implementation is by
tradition divided into two parts: hardware and software. This division, however, is
somewhat artificial in modern DSP because there is hardly any truly hardware
solution these days as most devices used in filtering are programmable. In this book,
we will regard any implementation on large systems, such as mainframe computers
and personal computers, as a software implementation. In such cases a high level
language would be used to code the filter equation and the operation would be carried
out offline. Implementations using DSP devices and special purpose hardware,
including standard microprocessors, would be regarded as hardware. In these cases,
the filtering equation may be firmware or assembly language code for the particular
device.

In most applications, real-time operation is often the main goal. In these cases
hardware implementation is the best option. Hardware implementation offers the
greatest speed, but is less flexible. Three approaches are now common in hardware
implementations: standard microprocessors (such as the Motorola 68000) and
DSP processors (such as the Texas Instruments TMS320), building block, and
algorithmic specific. In the building block approach, dedicated pieces of hardware
are used. In both the algorithm-specific and DSP processors the various devices
required for filtering – multipliers, adders, and so on – are implemented in hardware
and incorporated in a single IC using VLSI technology. Algorithm-specific processors, however, are already configured to perform FIR filtering. The designer merely
supplies the filter coefficients and the necessary glue logic to interface the processor
to the outside world. Examples are the Motorola DSP56200 and the INMOS A100.
DSP processors have architectures and instruction sets optimized for FIR filtering
operations. They are more flexible than algorithm-specific processors, but they are
slower.

The design of systems making use of software or hardware approaches is covered in
Chapters 12 and 13.

Figure 7.35 depicts a flowchart for the general FIR filtering operations from which
we see that, at each sampling instant, we must first shift the data by one place, read
and save the latest input sample, $x(n)$, and compute the current output sample using
the difference equation.

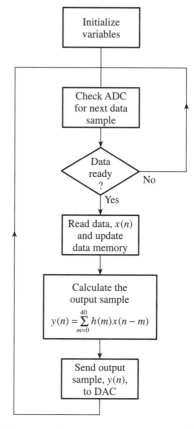

Figure 7.35   A simplified flowchart for a real-time, transversal, FIR filter.

7.13   Design example

Example 7.20   Design and implement a linear phase bandpass filter meeting the following specifications:

passband	900–1100 Hz
passband ripple	0.87 dB
stopband attenuation	>30 dB
sampling frequency	15 kHz
number of coefficients	41

The TMS32010 target board (see Chapter 13) is to be used to implement the filter.

Table 7.16 Unquantized, $h(m)$, and quantized, $h_q(m)$, coefficients for the design example.

m	Unquantized coefficients, $h(m)$	Quantized coefficients, $h_q(m)$
0	−1.534638e−02	−503
1	−5.780550e−05	−2
2	5.023483e−03	165
3	1.266706e−02	415
4	2.108206e−02	691
5	2.776418e−02	910
6	3.005362e−02	985
7	2.586935e−02	848
8	1.444566e−02	473
9	−3.189323e−03	−105
10	−2.416137e−02	−792
11	−4.420712e−02	−1449
12	−5.857453e−02	−1919
13	−6.318557e−02	−2070
14	−5.575461e−02	−1827
15	−3.654699e−02	−1198
16	−8.540099e−03	−280
17	2.308386e−02	756
18	5.201380e−02	1704
19	7.224807e−02	2367
20	7.951681e−02	2606
21	7.224807e−02	2367
22	5.201380e−02	1704
23	2.308386e−02	756
24	−8.540099e−03	−280
25	−3.654699e−02	−1198
26	−5.575461e−02	−1827
27	−6.318557e−02	−2070
28	−5.857453e−02	−1919
29	−4.420712e−02	−1449
30	−2.416137e−02	−792
31	−3.189323e−03	−105
32	1.444566e−02	473
33	2.586935e−02	848
34	3.005362e−02	985
35	2.776418e−02	910
36	2.108206e−02	691
37	1.266706e−02	415
38	5.023482e−03	165
39	−5.780550e−05	−2
40	−1.534638e−02	−503

Solution   As discussed earlier, there are five steps involved in the design of an FIR filter.

- ▪ *Step 1: specifications*   The specifications are already given.

- ▪ *Step 2: coefficient calculation*   We will use the optimal method to calculate the filter coefficients because it would yield the lowest number of filter coefficients (for nonrecursive FIR), and because it is available. We have already computed the coefficients for this filter in a previous example; see Table 7.7. The corresponding frequency response is given in Figure 7.13.

- ▪ *Step 3: realization*   The transversal structure is selected (Figure 7.29(a)), as it leads to the most efficient implementation using the TMS32010 processor. The difference equation for this structure is

$$y(n) = \sum_{m=0}^{40} h(m)x(n - m)$$

- ▪ *Step 4: quantization and analysis of errors*   Since the TMS32010 is to be used, each coefficient should be quantized to 16 bits for efficient operation. To do this, we multiply each coefficient by 2^{15} and then round up to the nearest integer. For example, the first two coefficients are quantized as follows:

$$h(0) = -0.015\ 346\ 38 \times 2^{15} = -502.87 \approx -503$$

$$h(1) = -0.000\ 057\ 805\ 5 \times 2^{15} = -1.89 \approx -2$$

The quantized and unquantized coefficients are listed in Table 7.16. The frequency response of the quantized filter should be checked to verify that the specifications are still met, particularly in the stopband. We found that, after quantization to 16 bits, there was little difference between the response of the quantized and unquantized filters.

With the TMS32010, partial sums implied in the difference equation will be carried out in a 32-bit accumulator. A fairly wide product register (32 bits) is used. Thus the effects of roundoff errors for $N = 41$ will be small. In this example, overflow is ignored. If it was a concern, we could overcome it by simply dividing each coefficient obtained in step 2 by a suitable scale factor, SF, for example

$$SF = \sum_{m=0}^{40} |h(m)|$$

The target board has only an 8-bit ADC. This would restrict the dynamic range of the signal that can be handled to only about 48 dB. In a high quality audio system, for example, the level of quantization noise would have been unacceptable, and in such cases the ADC resolution must be increased.

■ *Step 5: Implementation*   A flowchart for the FIR filtering operation is given in Figure 7.35. The flowchart is next translated into a TMS32010 assembly code and stored in the program memory (see Chapter 12 for the development and coding of FIR filtering operations).

7.14 Summary

The design of digital filters can be divided into five interdependent stages: filter specifications, coefficient calculation, realization, analysis of errors and filter implementation.

Filter specification is application dependent, but should include a specification of the amplitude and/or phase characteristics.

Coefficient calculation essentially involves finding values of $h(m)$ that will satisfy the desired specifications. The three most common methods of calculating FIR filter coefficients are (1) the window, (2) the frequency sampling, and (3) the optimal methods. The window method is the easiest, but lacks flexibility especially when the passband and stopband ripples are different. The frequency sampling method is well suited to recursive implementation of FIR filters and when filters other than the standard frequency selective filters (lowpass, highpass, bandpass and bandstop) are required. The optimal method is the most powerful and flexible. All three methods were covered in detail in this chapter.

The three most common FIR filter structures are the transversal, which involves a direct convolution using the filter coefficients, the frequency sampling structure, which is directly linked to the frequency sampling method of coefficient calculation, and the fast convolution. The choice between the structures is influenced by the intended application.

The performance of FIR filters of long lengths or high stopband attenuation may be affected by finite wordlength effects. For example, their frequency responses could be altered after coefficient quantization. Thus the characteristics of such filters should be checked to ensure that adequate wordlengths have been allowed, especially when wordlengths of less than about 12 bits are contemplated.

Implementation is normally embarked on when the first four steps are satisfactory and involves software coding or hardware realization of the chosen structure.

7.15 Application examples of FIR filters

There are many areas where FIR filters have been employed, including multirate processing (Crochiere and Rabiner, 1981), noise reduction (Hamer *et al.*, 1985), matched filtering (see Chapter 13), and image processing (Wade *et al.*, 1990).

In multirate processing, for example, FIR filters have been successfully used for efficient digital anti-aliasing and anti-imaging filtering for multirate systems such as high quality data acquisition and the compact disc player (see Chapter 9).

Problems

Concepts of FIR filters

7.1   The frequency response, $H(\omega)$, of a type 2, linear phase FIR filter may be expressed as (see Table 7.1)

$$H(\omega) = e^{-j\omega(N-1)/2} \sum_{n=1}^{N/2} b(n) \cos[\omega(n - \tfrac{1}{2})]$$

where $b(n)$ is related to the filter coefficients. Explain why filters with the response above are unsuitable as highpass filters. Use a simple case (such as $N = 4$) to illustrate your answer.

7.2   An FIR filter has an impulse response, $h(n)$, which is defined over the interval $0 \leqslant n \leqslant N - 1$. Show that if N is even and $h(n)$ satisfies the positive symmetry condition, that is $h(n) = h(N - n - 1)$, the filter has a linear phase response. Obtain expressions for the amplitude and phase responses of the filter.

Window method

7.3   Show that the impulse response for an ideal bandpass filter (see Table 7.2) is given by

$$h_{\mathrm{D}}(n) = 2 f_2 \frac{\sin n\omega_2}{n\omega_2} - 2 f_1 \frac{\sin n\omega_1}{n\omega_1} \quad n \neq 0$$

$$= 2(f_2 - f_1) \quad\quad\quad\quad n = 0$$

where f_1 is the lower passband frequency and f_2 is the upper passband frequency.

7.4   (1)   Obtain the coefficients of an FIR lowpass digital filter to meet the following specifications using the window method:

stopband attenuation	50 dB
passband edge frequency	3.4 kHz
transition width	0.6 kHz
sampling frequency	8 kHz

Include in your answer the type of window used and the reason for your choice.

(2)   Assuming that the filter coefficients are stored in contiguous memory locations in a microcomputer, list the values of the coefficients in the order in which they are stored.

(3)   Draw and briefly describe a flowchart of the direct software implementation of the filter in real time, and suggest two ways of improving the efficiency of the software implementation.

Note: you may use the information given in Table 7.2 in your design.

Optimal (Parks–McClellan) method

7.5   (1)   A linear phase FIR filter has an impulse response that satisfies the following symmetry condition:

$$h(n) = h(N - n - 1),$$
$$n = 0, 1, \ldots, (N - 1)/2$$

where N is the number of filter coefficients. Assuming that N is odd, determine the magnitude and phase responses of the filter and show that the filter has both constant phase and group delays. Comment on the practical significance of a linear phase response in a digital filter.

(2)   A linear phase bandpass digital filter is required for feature extraction in a certain signal analyzer. The filter is required to meet the following specification:

passband	12–16 kHz
transition width	3 kHz
sampling frequency	96 kHz
passband ripple	0.01 dB
stopband attenuation	80 dB

Assume that the coefficients of the filter are to be calculated using the optimal (Remez exchange) method. Determine the following parameters for the filter:

(a)   the number of filter coefficients, N;
(b)   suitable weights for the filter bands;
(c)   bandedge frequencies, in a form suitable for the optimal method.

Explain briefly the roles of the weights and grid frequencies in the optimal method.

Table 7.17 Relationship for estimating the length, N, for a bandpass filter.

$$N \approx \frac{C_\infty(\delta_p, \delta_s)}{\Delta F} + g(\delta_p, \delta_s)\Delta F + 1$$

where

$$C_\infty(\delta_p, \delta_s) = [\log_{10} \delta_s][b_1(\log_{10} \delta_p)^2 + b_2 \log_{10} \delta_p + b_3]$$
$$+ [b_4(\log_{10} \delta_p)^2 + b_5 \log_{10} \delta_p + b_6]$$

$$g(\delta_p, \delta_s) = -14.6 \log_{10}\left(\frac{\delta_p}{\delta_s}\right) - 16.9$$

$b_1 = 0.012\ 02 \qquad b_2 = 0.096\ 64$

$b_3 = -0.513\ 25 \qquad b_4 = 0.002\ 03$

$b_5 = -0.570\ 5 \qquad b_6 = -0.443\ 14$

ΔF, transition width normalized to the sampling frequency

δ_p, passband ripple or deviation

δ_s, stopband ripple or deviation

Suggest a suitable grid density for the above problem.

You may use the information given in Table 7.17.

7.6 An FIR lowpass digital filter is required to meet the following specifications:

stopband attenuation	>40 dB
passband edge frequency	100 Hz
passband ripple	<0.05 dB
transition width	10 Hz
sampling frequency	1024 Hz

(1) Calculate and list the coefficients of the filter, indicating clearly the method you used and why you chose it.

(2) The filter is to be implemented for real-time operation using the fast convolution method. Outline how you would implement the filter with the FT using the overlap–save technique. Indicate clearly parameters such as the number of samples by which the input sections overlap, the length of sections, the size of the transforms used and how the output samples are extracted from the transforms.

7.7 A linear phase, 41-point FIR differentiator is to be designed to meet the following specifications:

passband edge frequency	1 kHz
stopband edge frequency	1.5 kHz
sampling frequency	10 kHz
passband deviation	0.01
stopband deviation	0.01

Calculate the coefficients of the differentiator using the optimal method (Parks–McClellan/Remez exchange algorithm). Plot its magnitude–frequency response.

7.8 A linear phase, 43-point FIR Hilbert transform filter is to be designed to meet the following specifications:

lower bandedge frequency	1 kHz
upper bandedge frequency	4.5 kHz
sampling frequency	10 kHz
passband deviation	0.01

Calculate the coefficients of the filter using the optimal method. Plot its magnitude–frequency response in dB.

Frequency sampling filters

7.9 A 4-point, linear phase, FIR filter is characterized by the following frequency samples:

$$|H(k)| = 1, \quad k = 0$$
$$\tfrac{1}{2}, \quad k = 1, 3$$
$$0, \quad k = 2$$

(a) Starting from the general expression for the transfer function given in Equation 7.24, show that the transfer function of the above filter contains four zeros and three poles.

(b) Sketch the pole–zero diagram of the filter.

(c) Sketch the frequency response of the filter.

(d) Develop and sketch the realization diagram for the filter, with the complex conjugate poles combined, using the frequency sampling structure.

(e) Determine the four coefficients of the filter. The coefficients must be real.

7.10 A 4-point, linear phase, frequency sampling, FIR filter is characterized by the following frequency samples:

$$|H(k)| = 1, \quad k = 0$$
$$0, \quad k = 1, 2, 3$$

(a) Starting from the general expression for the transfer function given in Equation 7.24, determine the number of zeros and poles in the transfer function of the filter.

(b) Sketch the pole–zero diagram of the filter.

(c) Develop and sketch the realization of the filter, with the complex conjugate poles combined, using the frequency sampling structure.

(d) Determine the four coefficients of the filter. The coefficients must be real.

7.11 Frequency sampling filters have certain features in common with FIR filters and others in common with IIR filters. In this problem we will consider some of these features.

(a) What is the main advantage of recursive frequency sampling filters over non-recursive equivalents?

(b) Comment on the problem of finite wordlength effects associated with recursive frequency sampling filters, and suggest how it may be overcome in practice.

(c) The pole–zero diagram of a bandpass frequency sampling filter is depicted in Figure 7.36.

   (i) Write down, by inspection of the pole–zero diagram, the values of the frequency samples, $H(k)$, of the bandpass filter at the frequencies

   $$\omega_k = \frac{2\pi k}{N}, \quad k = 0, 1, \ldots, 7$$

   State any reasonable assumptions made.

   (ii) Sketch the magnitude–frequency response of the filter, with the sampling instants clearly labelled.

   (iii) Calculate the coefficients of the filter and hence determine its transfer function, $H(z)$, in recursive form.

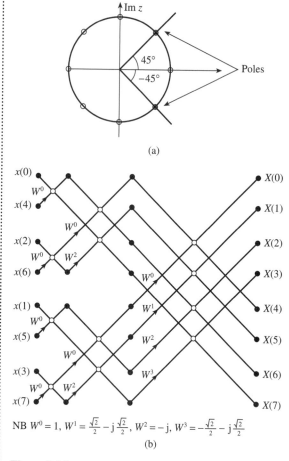

Figure 7.36

(d) Explain how the radix-2 FFT flow graph of Figure 7.36(b) may be used to calculate the impulse response of the filter from the frequency samples obtained above.

7.12 (a) Explain, with the aid of sketches, the basic concepts of the frequency sampling filter design method.

(b) A requirement exists for a lowpass digital filter satisfying the following requirements:

passband	0–20 Hz
sampling frequency	300 Hz
stopband attenuation	>50 dB
filter length	15

Table 7.18 Optimum transition band frequency samples for type 1 lowpass frequency sampling filters for $N = 15$ (adapted from Rabiner *et al.*, 1970).

BW	Stopband attenuation (dB)	T_1	T_2	T_3
One transition band frequency sample, N = 15				
1	42.309 322 83	0.433 782 96		
2	41.262 992 86	0.417 938 23		
3	41.253 337 86	0.410 473 63		
4	41.949 077 13	0.404 058 84		
5	44.371 245 38	0.392 681 89		
6	56.014 165 88	0.357 665 25		
Two transition band frequency samples, N = 15				
1	70.605 405 85	0.095 001 22	0.589 954 18	
2	69.261 681 56	0.103 198 24	0.593 571 18	
3	69.919 734 95	0.100 836 18	0.589 432 70	
4	75.511 722 56	0.084 074 93	0.557 153 12	
5	103.460 783 00	0.051 802 06	0.499 174 24	
Three transition band frequency samples, N = 15				
1	94.611 661 91	0.014 550 78	0.184 578 82	0.668 976 13
2	104.998 130 80	0.010 009 77	0.173 607 13	0.659 515 26
3	114.907 193 18	0.008 734 13	0.163 973 10	0.647 112 64
4	157.292 575 84	0.003 787 99	0.123 939 63	0.601 811 54

BW refers to the number of frequency samples in the passband.

(c) Find the coefficients of the transfer function of the digital filter, in recursive form, using the frequency sampling method and the information in Table 7.18.

(i) Develop and draw the realization diagram for the filter and compare the storage and computational requirements of the recursive implementation with direct form FIR.

(ii) Explain why the filter represented by the transfer function above is still an FIR filter even though its transfer function indicates that it is a recursive filter. Comment on the difficulties that may be encountered in practice with recursive frequency sampling filters, and indicate how these may be overcome.

7.13 The pole–zero diagram of a simple, frequency sampling bandpass filter is shown in Figure 7.37.

(a) Sketch the magnitude–frequency response of the filter and hence write down the values of the magnitude–frequency response at the sampling points.

(b) Obtain the transfer function of the filter, starting with the general transfer function of Equation 7.24 for a frequency sampling filter. Comment on your answer.

(c) Sketch the realization diagram of the filter and write down the difference equation.

(d) Compare the frequency sampling realization and the direct form realization in terms of their computational and storage requirements.

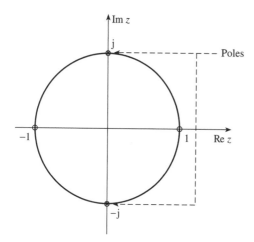

Figure 7.37

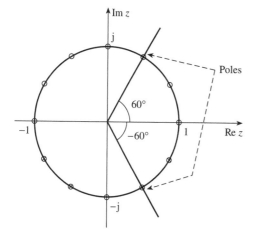

Figure 7.38

7.14 The pole–zero diagram of a simple, frequency sampling bandpass filter is shown in Figure 7.38.

(a) Sketch the magnitude–frequency response of the comb filter section (involves only zeros).

(b) Sketch the magnitude–frequency response of the filter and hence write down the values of the magnitude–frequency response at the sampling points.

(c) Write down the transfer function of the filter in recursive form.

(d) Obtain the transfer function of the filter, starting with the general transfer function of Equation 7.24 for a frequency sampling filter. Comment on your answer.

(e) Sketch the realization diagram of the filter and write down the difference equation.

7.15 (1) Discuss briefly the conditions necessary for a realizable digital filter to have a linear phase characteristic, and the advantages of filters with such a characteristic.

(2) In a certain signal processing application, the input signal, with significant frequency components in the range $0 \leqslant f \leqslant 10$ Hz, is contaminated by a 50 Hz mains interference. It is decided to remove the interference using a linear phase digital filter after digitizing the composite signal at a rate of 500 samples s^{-1}. As a first step in the design of the filter, the pole–zero diagram given in Figure 7.39 was developed. Obtain the transfer function, $H(z)$, of the filter and its difference equation.

(3) The filter obtained in part (2) is to be implemented in a microcomputer with simple arithmetic limited to only additions/subtractions and shifts. Redesign the filter so that its coefficients are integers. There should be no increase in the number of filter coefficients or sampling rate.

(4) Show that the phase response, $\theta(\omega)$, of the filter of part (3) is given by

$$\theta(\omega) = -\omega T$$

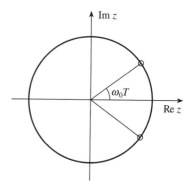

Figure 7.39 Pole–zero diagram for Problem 7.15: $\omega_0 T = \pi/5$ rad.

7.16 (1) A requirement exists for a real-time, narrowband, linear phase digital filter for a certain biomedical system. Justify the use of a frequency sampling filter for the system.

Assume that the transfer function of an N-point frequency sampling filter is given by

$$H(z) = \frac{1 - r^N z^{-N}}{N}$$

$$\left(\sum_{k=1}^{M} \frac{|H(k)|\,[2\cos(2\pi k\alpha/N) - 2r\cos[2\pi k(1 + \alpha)/N]}{1 - 2r\cos(2\pi k/N)z^{-1} + r^2 z^{-2}} \right.$$

$$\left. + \frac{H(0)}{1 - z^{-1}} \right)$$

where the $H(k)$ are the samples of the desired frequency response taken at intervals of F_s/N, $\alpha = (N-1)/2$.

(2) The desired filter is characterized by the following specifications:

passband	48–52 Hz
transition width	2 Hz
sampling frequency	500 Hz
stopband attenuation	>60 dB

Specify suitable frequency samples, $|H(k)|$. Develop and draw the realization diagram for the filter. How does the filter compare with an equivalent transversal structure in terms of storage and computational requirements?

(3) Comment on the $H(z)$ above and the difficulties that may be encountered in practice with recursive frequency sampling filters, and indicate how these may be overcome.

Explain why $H(z)$ describes a recursive filter and yet its unit impulse response, $h(n)$, is of a finite duration.

7.17 A requirement exists for an N-point FIR filter with the frequency response

$$H(e^{j\omega}) = |H(e^{j\omega})|\,e^{-j\omega\alpha}$$

where $\alpha = (N-1)/2$. Assume that N samples of $H(e^{j\omega})$ are taken at intervals of $f_k = (k + 1/2)F_s/N$, $k = 0, 1, \ldots, N-1$.

(1) Show that, for N even, the impulse response is given by

$$h(n) = \frac{1}{N}\left\{ \sum_{k=0}^{N/2-1} 2|H(k)|\cos[2\pi(n - \alpha)(k + 1/2)/N] \right\}$$

(2) Show that, for N odd, the impulse response is given by

$$h(n) = \frac{1}{N}\left\{ \sum_{k=0}^{(N-3)/2} 2|H(k)|\cos[2\pi(n - \alpha)(k + 1/2)/N] \right.$$

$$\left. + H[(N - 1)/2]\cos[\pi(n - \alpha)] \right\}$$

(3) Obtain an expression for the transfer function, $H(z)$, in recursive form for each of parts (1) and (2).

Special FIR filters

7.18 A highpass FIR filter is characterized by the following impulse response coefficients:

$$\{h(n)\} = \{0.127, -0.026, -0.237, 0.017, 0.434\}$$

Write down the coefficients of an equivalent lowpass filter with the aid of the frequency transformation given in Section 7.9.2.

7.19 Calculate the coefficients of an FIR half-band filter using the Kaiser window function. The half-band filter should meet the following specifications:

passband ripple	0.5 dB
stopband attenuation	45 dB
passband edge frequency	2 kHz
sampling frequency	10 kHz

7.20 Repeat Problem 7.19 using the optimal method.

FIR filter implementation

7.21 An analog signal is contaminated by a 50 Hz component and its harmonics at 100 Hz, 150 Hz, 200 Hz, 250 Hz and 300 Hz. Assume that the contaminated signal is sampled and digitized at 1 kHz.

Find the transfer function of a simple digital filter to remove the interference and its harmonics.

Draw a realization diagram for the digital filter. Compare and contrast the effects of finite wordlength on the performance of digital filters and those of component tolerances on the performance of analog filters. Use a notch filter to illustrate your answer.

7.22 (1)   Assess the effects of finite wordlength constraints on the performance of real-time FIR digital filter implementations, and suggest how these may be minimized.

(2)   In a certain real-time digital signal processing system, each coefficient of an N-point FIR filter is represented as an n-bit 2's complement number. Show that the maximum stopband attenuation, A_{max}, is bounded by

$$A_{max} < 20 \log_{10} N 2^{-B}$$

where B is the coefficient wordlength and N is the filter length.
State any assumptions made and comment on the bound given above.

(3)   The coefficients of a 7-point FIR filter are listed below. Draw a realization diagram for the filter such that a minimum number of multiplications is required for each output computation.

$$h(0) = -0.3$$
$$h(1) = 0.4$$
$$h(2) = 0.2$$
$$h(3) = 0.5$$
$$h(4) = 0.2$$
$$h(5) = 0.4$$
$$h(6) = -0.3$$

7.23 A fixed point FIR digital filter implementation uses 2's complement fractional arithmetic with the coefficients represented by three bits (including the sign).

(1)   Calculate and list all the possible decimal numbers that can be represented. State the largest and smallest representable decimal numbers.

(2)   The unquantized coefficients of the FIR filter are listed below. Assume that the

coefficients are quantized to three bits after truncation (sign included). List the quantized coefficients together with their quantization errors.

n	$h(n)$
0	−0.149 75
1	0.256 872
2	0.699 40
3	0.256 872
4	−0.149 725

(3)   Repeat part (2) if the coefficients are rounded.

7.24 The coefficients of an FIR filter are $\{h(n)\} = \{-1, 0.5, 0.75\}$.

(1)   Draw the structure for the filter assuming transversal realization.

(2)   Assuming that the coefficients as well as the input data samples are represented by three bits (including the sign bit) after truncation, determine and tabulate the values of the quantized coefficients in both binary and decimal.

(3)   Show that if the data $\{x(n)\} = \{0.5, -1, -0.5\}$ is applied to the filter the output, $y(n)$, will still be correct despite an overflow in the intermediate result (assume a double-length accumulator).

(4)   Show that the input $\{x(n)\} = \{-1, -0.75, 0.5\}$ will lead to wrong output values owing to overflow. How can the overflow be prevented?

7.25 Design a real-time lowpass digital filter for physiological noise reduction. The filter is intended to be part of a larger DSP system and so the number of arithmetic operations in the filter should be kept as low as possible.
The filter should meet the following amplitude specifications:

passband	8–12 Hz
passband ripple	0.1 dB
transition width	2 Hz
stopband attenuation	30 dB
sampling frequency	100 Hz

Other requirements are that

(1) minimal distortion of the harmonic relationships between the components of the in-band signals is highly desirable, and

(2) the filter will be implemented using the TMS320C25 DSP processor with analog input digitized to 12 bits.

7.26 Design a multiband FIR digital filter to meet the following specifications:

band 1	0–0.5 kHz	
	stopband attenuation	49 dB
band 2	1–1.5 kHz	
	passband ripple	0.3 dB
band 3	1.8–2.5 kHz	
	stopband attenuation	38 dB
band 4	3–3.6 kHz	
	passband ripple	0.3 dB
band 5	4.1–5 kHz	
	stopband attenuation	55 dB

The filter is to be implemented using a system with the TMS320C25 processor, a 12-bit ADC and 12-bit DAC at a sampling frequency of 10 kHz.

7.27 Discuss the five main steps involved in the design of digital filters, using the following design problem to illustrate your answer.

A digital filter is required for real-time physiological noise reduction. The filter should meet the following specifications:

passband	0–10 Hz
stopband	20–64 Hz
sampling frequency	256 Hz
maximum passband ripple	0.026 dB
stopband attenuation	30 dB

Other important requirements are that

(1) the filter should have a linear phase response so as to introduce as small a distortion as possible to the in-band signal components,

(2) the time available for filtering is limited, the filter being part of a larger process, and

(3) the filter will be implemented using a TMS32010 processor with the input digitized to 12 bits.

MATLAB problems

7.28 Use MATLAB to compute the coefficients, plot the magnitude–frequency response in dB, and determine the locations of the zeros of each of the following window-based filters (assume a sampling frequency of 2 kHz and a Hamming window function):

(1) A 7-point, bandpass FIR filter with pass- and stopband edge frequencies of 200 Hz and 500 Hz.

(2) An 8-point, bandpass FIR filter with pass- and stopband edge frequencies of 200 Hz and 500 Hz.

(3) A 7-point, FIR differentiator with pass- and stopband edge frequencies of 200 Hz and 500 Hz.

(4) An 8-point, FIR Hilbert transformer with bandedges of 200 Hz and 500 Hz.

Comment on the differences and/or similarities in the positions of the zeros.

7.29 A 41-point bandpass FIR filter is to be designed to approximate the following ideal magnitude response characteristics using the window method:

$$H(f) = 1 \quad 2\,\text{kHz} \leqslant f \leqslant 4\,\text{kHz}$$
$$0 \quad \text{otherwise}$$

Determine the impulse response coefficients of the filter and plot its magnitude and phase frequency responses with the aid of MATLAB for each of the following cases:

(1) Using a rectangular window.

(2) Using a Hamming window.

7.30 A requirement exists for a linear phase, low-pass, optimal FIR filter meeting the following specifications:

filter length	21
passband edge frequency	2 kHz
stopband edge frequency	3 kHz
sampling frequency	10 kHz

(1) Calculate the coefficients of the filter and plot its magnitude response in dB and phase response in degrees with the aid of MATLAB.

(2) Calculate and plot the phase and group delay responses of the filter.

(3) From examination of the magnitude and phase responses, determine the locations of the zeros.

(4) Explain why the phase response has discontinuities. How can the jumps in the phase response be corrected?

7.31 A requirement exists for an FIR digital filter to meet the following specifications:

passband	150–250 Hz
transition width	50 Hz
passband ripple	0.1 dB
stopband attenuation	60 dB
sampling frequency	1 kHz

Use the Hamming window and MATLAB to calculate the coefficients of the filter.

7.32 A linear phase FIR bandpass filter is required to satisfy the following specifications:

passband	8–12 kHz
stopband ripple	0.001
passband ripple	0.01
sampling frequency	48 kHz
transition width	3 kHz

Obtain the magnitude–frequency response of the filter with the aid of MATLAB for each of the following cases:

(1) Using the Hamming window.

(2) Using the Kaiser window.

(3) Using the optimal method.

(4) Using the frequency sampling method.

Compare the four cases.

7.33 A linear phase bandpass digital filter is required for feature extaction in a certain signal analyzer. The filter is required to meet the following specification:

passband	12–16 kHz
transition width	3 kHz
sampling frequency	96 kHz
passband ripple	0.01 dB
stopband attenuation	80 dB

The coefficients of the filter are to be calculated using the optimal method. Determine, with the aid of MATLAB, the following:

(1) The number of filter coefficients, N.

(2) Coefficients of the filter.

Plot the magnitude–frequency response.

7.34 A multi-band FIR digital filter is required to meet the following specifications:

band 1: 0–0.5 kHz, stopband attenuation ⩾49 dB
band 2: 1–1.5 kHz, passband ripple, 0.3 dB
band 3: 1.8–2.5 kHz, stopband attenuation, 38 dB
band 4: 3–3.6 kHz, passband ripple, 0.3 dB
band 5: 4.1–5 kHz, stopband attenuation, 55 dB

Use the optimal method and MATLAB to calculate the filter coefficients and plot the magnitude–frequency response. Assume a sampling frequency of 10 kHz and a transition width of 100 Hz.

7.35 An optimal lowpass filter is to be designed to meet the following specifications:

passband	0–6 kHz
transition width	1 kHz
passband ripple	0.1 dB
stopband attenuation	50 dB
sampling frequency	16 kHz

Determine the filter length and coefficients with the aid of the MATLAB commands **remezord** and **remez**.

Plot the magnitude–frequency response of the filter.

7.36 Calculate the coefficients of an FIR half-band filter using the Kaiser window function and MATLAB. The half-band filter should meet the following specifications:

passband ripple	0.5 dB
stopband attenuation	45 dB
passband edge frequency	2 kHz
sampling frequency	10 kHz

Repeat Problem 7.19 using the optimal method and the MATLAB commands **remezord** and **remez**.

7.37 A linear phase, 41-point FIR differentiator is to be designed to meet the following specifications:

passband edge frequency	1 kHz
stopband edge frequency	1.5 kHz
sampling frequency	10 kHz
passband deviation	0.01
stopband deviation	0.01

Calculate the coefficients of the differentiator using the optimal method (Parks–McClellan algorithm) and MATLAB. Plot its magnitude–frequency response.

7.38 A linear phase, 43-point FIR Hilbert transform filter is to be designed to meet the following specifications:

lower bandedge frequency	1 kHz
upper bandedge frequency	4.5 kHz
sampling frequency	10 kHz
passband deviation	0.01

Calculate the coefficients of the Hilbert transformer using the optimal method (Parks–McClellan algorithm) and MATLAB. Plot its magnitude–frequency response in dB.

References

Crochiere R.E. and Rabiner L.R. (1981) Interpolation and decimation of digital signals – a tutorial review. *Proc. IEEE*, **69**(3), 300–31.

Hamer C.F., Ifeachor E.C. and Jervis B.W. (1985) Digital filtering of physiological signals with minimal distortion. *Medical and Biol. Eng. and Computing*, **23**, 274–8.

Harris S.P. and Ifeachor E.C. (1998) Automatic design of frequency sampling filters by hybrid Genetic Algorithm Techniques. *IEEE Transactions on Signal Processing*, **46**(12), December, 3304–14.

Herrman O., Rabiner R.L. and Chan D.S.K. (1973) Practical design rules for optimum finite impulse response digital filters. *Bell System Technical J.*, **52**, 769–99.

Ifeachor E.C. and Harris S.P. (1993) A new approach to frequency sampling filter design, in *Proc. IEE/IEEE Workshop Natural Algorithms in Signal Processing*, 5/1–8.

Lawrence V.B. and Salazar A.C. (1980) Finite precision design of linear-phase FIR filters. *Bell System Technical J.*, **59**(9), 1575–98.

Lynn P.A. (1973) Recursive digital filters with linear phase characteristics. *Computer J.*, **15**, 337.

Lynn P.A. (1975) Frequency sampling filters with integer multipliers. In *Introduction to Digital Filtering*, Bogner R.E. and Constantinides A.G. (eds). New York: Wiley.

McClellan J.H., Parks T.W. and Rabiner L.R. (1973) A computer program for designing optimum FIR linear phase digital filters. *IEEE Trans. Audio Electroacoustics*, **21**, 506–26.

Mintzer F. and Liu B. (1979) Practical design rules for optimum FIR bandpass digital filters. *IEEE Trans. Acoustics, Speech Signal Processing*, **27**(2), 204–6.

Mitra S.K. and Kaiser J.F. (1993) *Handbook for Digital Signal Processing*. New York: Wiley.

Oppenheim A.V. and Schaffer R.W. (1975) *Digital Signal Processing*. Englewood Cliffs NJ: Prentice-Hall.

Parks T.W. and Burrus C.S. (1987) *Digital Filter Design*. New York: Wiley.

Rabiner L.R. and Gold B. (1975) *Theory and Applications of Digital Signal Processing*. Englewood Cliffs NJ: Prentice-Hall.

Rabiner L.R., Gold B. and McGonegal C.A. (1970) An approach to the approximation problem for nonrecursive digital filters. *IEEE Trans. Audio Electroacoustics*, **18**, 83–106.

Suckley D. (1990) Genetic algorithm in the design of FIR filters. *IEE Proc. Part G*, **138**(2), 234–8.

Wade G., van Eetvelt P. and Darwen H. (1990) Synthesis of efficient low-order FIR filters from primitive sections. *IEE Proc. Part G*, **137**(5), 367–72.

Bibliography

Bateman A. and Yates W. (1988) *Digital Signal Processing Design*. London: Pitman.

Chan D.S.K. and Rabiner L.R. (1973) Analysis of quantization errors in the direct form for finite impulse response digital filters. *IEEE Trans. Audio Electroacoustics*, **21**(4), 354–66.

Chan D.S.K. and Rabiner L.R. (1973) An algorithm for minimizing roundoff noise in cascade realizations of finite impulse response digital filters. *Bell System Technical J.*, **52**(3), 347–85.

DeFatta D.J., Lucas J.G. and Hodgkiss W.S. (1988) *Digital Signal Processing: A System Design Approach*. New York: Wiley.

Gersho A., Gopinath B. and Odlyzko A.M. (1979) Coefficient inaccuracy in transversal filtering. *Bell System Technical J.*, **58**(10), 2401–2416.

Gold B. and Jordan K.L., Jr (1968) A note on digital filter synthesis. *Proc. IEEE (Lett.)*, **56**, 1717–18.

Gold B. and Jordan K.L., Jr (1969) A direct search procedure for designing finite duration impulse response filters. *IEEE Trans. Audio Electroacoustics*, **17**, 33–6.

Gold B. and Rader C.M. (1969) *Digital Processing of Signals*. New York: McGraw-Hill.

Gore A.E. (1986) Cascadable digital signal processor. *New Electronics*, **19**, October, 39–41.

Heute U. (1977) Comments on Rabiner L.R. A simplified computational algorithm for implementing FIR digital filters. *IEEE Trans. Acoustics, Speech Signal Processing*, **25**, June, 266–7.

Hillman G.D. (1987) DSP56200: an algorithm-specific digital signal processor peripheral. *Proc. IEEE*, **75**, September, 1185–91.

Knowles J.B. and Olcayto E.M. (1968) Coefficient accuracy and digital filter response. *IEEE Trans. Circuit Theory*, **15**, 31–41.

Lin K., Frantz G.A. and Simar R. (1987) The TMS320 family of digital signal processors. *Proc. IEEE*, **75**, 1143–59.

Lynn P.A. (1970) Economic linear-phase recursive digital filters. *Electronics Lett.*, **6**, 143–5.

Lynn P.A. and Fuerst W. (1989) *Introductory Digital Signal Processing with Computer Applications*. New York: Wiley.

Mintzer F. (1982) On half-band, third-band and Nth-band FIR filters and their design. *IEEE Trans. Acoustics, Speech Signal Processing*, **30**, 734–8.

Mitra S.K. and Sherwood R.J. (1972) Canonic realizations of digital filters using the continued fraction expansion. *IEEE Trans. Audio Electroacoustics*, **20**, 185–94.

Proakis J.G. and Manolakis D.G. (1992) *Introduction to Digital Signal Processing*. New York: Macmillan.

Rabiner L.R. (1971) Techniques for designing finite-duration impulse response digital filters. *IEEE Trans. Communication Technology*, **19**, 188–95.

Rabiner L.R. (1973) Approximate design relationships for lowpass FIR digital filters. *IEEE Trans. Audio Electroacoustics*, **21**, 456–60.

Rabiner L.R. (1977) A simplified computational algorithm for implementing FIR digital filters. *IEEE Trans. Acoustics, Speech Signal Processing*, **25**, June, 259–61.

Rabiner L.R. and Schafer R.W. (1971) Recursive and nonrecursive realizations of digital filters designed by frequency sampling techniques. *IEEE Trans. Audio Electroacoustics*, **19**, 200–7.

Rabiner L.R. and Schafer R.W. (1972) Correction to 'Recursive and nonrecursive realizations of digital filters designed by frequency sampling techniques'. *IEEE Trans. Audio Electroacoustics (Corresp.)*, **20**, 104–5.

Rabiner L.R., Kaiser J.F. and Schafer R.W. (1974) Some considerations in the design of multiband finite impulse response digital filters. *IEEE Trans. Acoustics, Speech Signal Processing*, **22**(6), 462–72.

Rabiner L.R., McClellan J.H. and Parks T.W. (1975) FIR digital filter design techniques using weighted Chebyshev approximation. *Proc. IEEE*, **63**(4), 595–610.

Appendices

7A | C programs for FIR filter design

The following C language programs for designing FIR filters are available on the CD in the companion handbook (see Preface for details):

- fresamp.c, a program for computing filter coefficients via the frequency sampling approach;

- optimal.c, a program for computing filter coefficients via the optimal method;

- window.c, a program for computing filter coefficients via the window method;

- firfilt.c, a program for FIR filtering on data;

- ncoeff.c, a program for estimating the number of filter coefficients for optimal lowpass or bandpass filter.

To limit the size of the book, only the last program, ncoeff.c, is listed here (Program 7A.1). This program is a direct implementation of the equations given in Section 7.6.3. To illustrate the use of the program we will use it to estimate the length of a bandpass filter with the following specifications:

passband	1800–3300 Hz
stopbands	0–1400, 3700–5000 Hz
sampling frequency	10 kHz
passband ripple	1 dB
stopband attenuation	40 dB

Program 7A.1

```
* -------------------------------------------------------------------------------- *
*                                                                                  *
*       program for estimating the number of coefficients of                       *
*       optimal FIR lowpass or bandpass filter                                      *
*                                                                                  *
*       program name: ncoeff.c                                                      *
*                                                                                  *
*       Manny Ifeachor, 17.10.91                                                    *
*                                                                                  *
* -------------------------------------------------------------------------------- *
*/
#include    <stdio.h>
#include    <math.h>
#include    <dos.h>

int         filter__spec();
double      lpfcoeff();
```

```
double      bpfcoeff();
float       dp, ds, df;
int         ftype;


main()
{
        double N;
        ftype=filter__spec();                                    /* obtain filter specifications */
        switch(ftype){
                case 1:
                        N=lpfcoeff(); break;
                case 2:
                        N=bpfcoeff(); break;
                default:
                        printf("illegal filter type selected \n");
                        break;
        }
        printf("Number of coefficients      \t%f\n",N);
        printf("passband ripple in dB       \t%f\n",dp);
        printf("stopband attenuation in dB  \t%f\n",ds);
        printf("\n");
        printf("press enter to continue \n");
        getch();
        exit(0);
}
/* ---------------------------------------------------------------------------------------------------- */
int         filter__spec()
{
        int   itype;
        printf("program to estimate optimal filter length\n");
        printf("\n");
        printf("select filter type\n");
        printf("1    for optimal lowpass filter\n");
        printf("2    for optimal bandpass filter\n");
        scanf ("%d", &itype);
        printf("\n");
        printf("enter passband and stopband deviations in ordinary units\n");
        printf("deviations must be between 0 and 1\n");
        scanf("%f%f",&dp,&ds);
        switch(itype){
                case 1:
                        printf("enter normalized transition width \n");
                        scanf("%f", &df);
                        break;
                case 2:
                        printf("enter normalized transition width – the smaller width\n");
                        scanf("%f", &df);
                        break;
        }
                return(itype);
}
/* ---------------------------------------------------------------------------------------------------- */
```

```
double     lpfcoeff()
{
        float      ddp, dds, a1, a2, a3, a4, a5, a6, b1, b2;
        double     dinf, ff, t1, t2, t3, t4, Nl;

        /*    constants   */
        a1=0.005309; a2=0.07114; a3=-0.4761; a4=-0.00266;
        a5=-0.5941; a6=-0.4278;
        b1=11.01217; b2=0.5124401;

        ddp=log10(dp);
        dds=log10(ds);
        t1=a1*ddp*ddp;
        t2=a2*ddp;
        t3=a4*ddp*ddp;
        t4=a5*ddp;
        dinf=((t1+t2+a3)*dds) +(t3+t4+a6);
        ff=b1+b2*(ddp-dds);
        Nl=((dinf/df)-(ff*df)+1);
        dp=20*log10(1+dp); ds=-20*log10(ds);
        return(Nl);
}
/* ------------------------------------------------------------------------------------------------------------------ */
double     bpfcoeff()
{
        float      a1, a2, a3, a4, a5, a6, ddp, dds;
        double     t1, t2, t3, t4, cinf, ginf, Nb;

        a1=0.01201, a2=0.09664, a3=-0.51325; a4=0.00203;
        a5=-0.57054; a6=-0.44314;

        ddp=log10(dp);
        dds=log10(ds);
        t1=a1*ddp*ddp;
        t2=a2*ddp;
        t3=a4*ddp*ddp;
        t4=a5*ddp;
        cinf=dds*(t1+t2+a3)+t3+t4+a6;
        ginf=-14.6*log10(dp/ds)-16.9;
        Nb=(cinf/df) + ginf*df+1;
        dp=20*log10 (1+dp); ds=-20*log10(ds);
        return(Nb);
}
```

From the specifications, the normalized transition width is 0.04 (450/10 000), the passband deviation is 0.122, from 20 log (1 + 1), and the stopband deviation is 0.01, from −20 log (40). The prompts, responses and output of the program for the above example are given in Table 7A.1. The number of filter coefficients, 31 in this case, is only an estimate. In most practical cases, a higher value of filter length (that is, the number of filter coefficients) than given by the program is necessary to meet the specifications. In the above example, the actual filter length required to meet the specifications was 35. The designer should bear this in mind when using the program.

Table 7A.1 Prompts, responses and output of ncoeff.c.

program to estimate optimal filter length

select filter type
1   for optimal lowpass filter
2   for optimal bandpass filter
2

enter passband and stopband deviations in ordinary units deviations must be between 0 and 1
0.122 0.01
enter normalized transition width – the smaller width
0.04

Number of coefficients	31.261084
passband ripple in dB	0.999857
stopband attenuation in dB	40.000000

press enter to continue

7B   FIR filter design with MATLAB

The MATLAB Signal Processing Toolbox contains an excellent set of programs and functions for the design and analysis of different types of FIR digital filters. The programs and commands are readily accessible via high level commands and make the Toolbox a valuable tool for gaining an insight into the design and analysis of FIR filters without getting bogged down with extensive programming.

In this section, we will illustrate how to use some of the MATLAB functions and programs to design linear phase FIR filters. In particular, we will illustrate how to calculate the coefficients of linear phase FIR filters using window, optimal (Parks–McClellan) and frequency sampling methods and MATLAB to complement the use of C language programs discussed in the previous section.

7B.1   Window method

The steps involved in the calculation of coefficients of standard, frequency selective, linear phase FIR filters using the window method may be summarized as follows (see the main text for details):

1. Specify the desired frequency response.
2. Select a window function and estimate the number of filter coefficients, N.
3. Obtain the ideal impulse response, $h_D(n)$ (truncated to N values).
4. Obtain N coefficients of the window function, $w(n)$.
5. Obtain the FIR filter coefficients by applying the window, $h(n) = h_D(n) \times w(n)$.

For standard, frequency selective linear phase window-based FIR filter design (lowpass, highpass, bandpass and bandstop filters), the key high-level command in the Toolbox is the **fir1** command. The syntax for the basic **fir1** command is:

b = fir1(N–1, Fc)

The basic command computes and returns N-point impulse response coefficients of an FIR filter with a cutoff frequency F_c. The command returns the N-point coefficients in the vector b, arranged in ascending negative powers of z:

$$b(z) = b(0) + b(1)z^{-1} + b(2)z^{-2} + \ldots + b(N-1)z^{-(N-1)}$$

The parameter, $N - 1$, in the command specifies the order of the filter (normally one less than the number of FIR filter coefficients). The cutoff frequency, F_c, is normalized with respect to the Nyquist frequency (i.e., half the sampling frequency) and lies between 0 and 1 (where 1 corresponds to the Nyquist frequency).

By default, the basic **fir1** command applies a Hamming window and assumes a lowpass filter (or a bandpass filter if F_c specifies more than one cutoff frequency). The basic command can be extended by specifying the type of filter and/or the window function. The syntax in these cases is:

b = fir1(N–1,Fc,'filter-type')
b = fir1(N–1,Fc, window)
b = fir1(N–1,Fc,'filter-type', window)

For a highpass filter, the word 'high' specifies the filter type, and for bandstop the word 'stop' is used. For both bandpass and bandstop filters, the variable F_c is a vector that specifies the cutoff frequencies. For highpass and bandstop filters, the filter length must be an odd integer (even integers are unsuitable for highpass and bandstop filters because they lead to a zero magnitude response at the Nyquist frequency as described previously).

MATLAB supports the use of a variety of window functions including Hamming, Hanning, boxcar (rectangular), Kaiser and Chebyshev windows. The syntax for generating window coefficients is:

w = boxcar(N)
w = blackman(N)
w = hamming(N)
w = hanning(N)
w = kaiser(N, beta)

In practice, the window command is often embedded into the **fir1** command (see examples later).

It should be pointed out that there may be differences in the results obtained in using MATLAB to design window-based FIR filters compared to other programs because of differences in implementation. For example, in MATLAB, after windowing the impulse response coefficients may be scaled to give a magnitude–frequency response of unity in the middle of the passband. The word 'noscale' may be added to override this, e.g.: b = fir1(N–1, Fc, 'noscale'). Further, MATLAB implementation of most window functions may be slightly different from previous implementation and this may lead to slight differences in results. The designer should be aware of such differences and make an appropriate allowance to compensate for them, if necessary.

| Example 7B.1 | Determine the coefficients of a linear phase, FIR lowpass filter with a passband and stopband edge frequency of 1 kHz and 4.3 kHz, respectively. Use the Hamming window and assume a sampling frequency of 10 kHz. |

From Table 7.3, the approximate relationship between the transition width, Δf, and filter length, N, for a Hamming window-based filter is:

$$N \approx \frac{3.3}{\Delta f}$$

Now, Δf is 0.33 (from $(4.3 - 1)/10$) and so the filter length $N = 10$. Following the approach in the main text, the actual cutoff frequency (allowing for the smearing effect) is taken to lie half-way between the specified passband and the stopband edge frequencies, i.e. 2.65 kHz. In MATLAB, the cutoff frequency should be normalized to half the sampling frequency. Thus, f_c(normalized) $= 2.65/5 = 0.53$.

The MATLAB commands are given in Program 7B.1. The values of the truncated ideal impulse response, the window and the windowed FIR coefficients are given in Table 7B.1.

Program 7B.1   MATLAB m-file for computing FIR filter coefficients for Example 7B.1.

```
fc=0.53;                    %   Cutoff frequency (normalized to Fs/2)
N=10;                       %   Filter length (number of taps)
hd=fir1(N–1,fc,boxcar(N));  %   Truncated ideal impulse response
wn=hamming(N);              %   Calculate Hamming window coefficients
hn=fir1(N–1,fc,wn);         %   Obtain windowed coefficients
```

Table 7B.1   Filter parameters for Example 7B.1.

n	Truncated ideal impulse response, $hD(n)$	Window coeffs, $w(n)$	Windowed filter coeffs, $h(n)$
0	0.0641	0.0800	0.0053
1	−0.0388	0.1876	−0.0075
2	−0.1052	0.4601	−0.0496
3	0.1235	0.7700	0.0974
4	0.4564	0.9723	0.4544
5	0.4564	0.9723	0.4544
6	0.1235	0.7700	0.0974
7	−0.1052	0.4601	−0.0496
8	−0.0388	0.1876	−0.0075
9	0.0641	0.0800	0.0053

Example 7B.2   *Illustrating FIR coefficient calculation using the Kaiser window*   Determine the coefficients and plot the magnitude–frequency response of a bandpass FIR filter, using the Kaiser window and MATLAB, that meets the following specifications:

passband	150–250 Hz
transition width	50 Hz
passband ripple	0.1 dB
stopband attenuation	60 dB
sampling frequency	1 kHz

Solution   This problem is identical to Example 7.4 in the main text. Here, we will use MATLAB to solve the problem.

Example 7.4 gives the filter length, $N = 73$, and the ripple parameter, $\beta = 5.65$. The MATLAB program is listed in Program 7B.2. The filter coefficients and the magnitude spectrum are shown in Table 7B.2 and Figure 7B.1.

It should be noted that in this example, we have included the window type in the fir1 command, instead of computing the window coefficients separately.

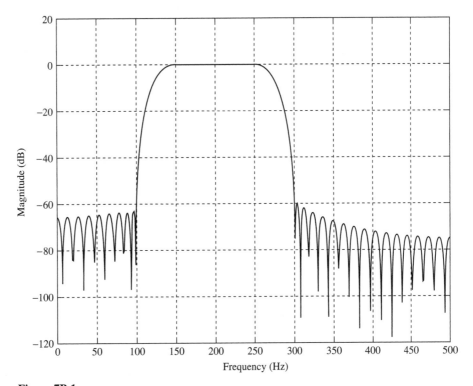

Figure 7B.1

Program 7B.2 MATLAB m-file for calculating FIR filter coefficients for Example 7B.2.

```
FS=1000;                              %  Sampling frequency
FN=FS/2;                              %  Nyquist frequency
N=73;                                 %  Filter length
beta=5.65;                            %  Kaiser window Ripple parameter
fc1=125/FN;                           %  Normalized cut off frequencies
fc2=275/FN;
FC=[fc1 fc2];                         %  Band edge frequency vector
hn=fir1(N–1, FC, kaiser(N, beta));    %  Obtain windowed filter coeffs
[H, f]=freqz(hn, 1, 512, FS);         %  Compute frequency response
mag=20*log10(abs(H));
plot (f, mag), grid on
xlabel ('Frequency (Hz)')
ylabel ('Magnitude Response (dB)')
```

Table 7B.2 Filter coefficients for Example 7B.2.

n	h(n)
0	−0.0001
1	−0.0004
2	−0.0001
3	−0.0001
4	−0.0007
5	0.0005
6	0.0023
7	0.0008
8	−0.0017
9	−0.0005
10	−0.0005
11	−0.0044
12	−0.0022
13	0.0069
14	0.0066
15	−0.0016
16	0.0000
17	0.0022
18	−0.0117
19	−0.0164
20	0.0069
21	0.0189
22	0.0029
23	0.0044
24	0.0188
25	−0.0125
26	−0.0520
27	−0.0165
28	0.0333
29	0.0104
30	0.0094
31	0.0856
32	0.0453
33	−0.1665
34	−0.2066
35	0.0891
36	0.2998

7B.2   The optimal method

The Signal Processing Toolbox in MATLAB contains a number of design programs and functions for designing optimal FIR filters based on Park–McClellan and Remez algorithms. The **remez** command is the key command for calculating FIR coefficients via the optimal method. The command may be used to design multiband linear phase FIR filters. In its basic form the command has the following syntax:

    b = remez(N–1, F, M)

where N is the length of the filter, F is a vector of the normalized bandedge frequencies, and M is a vector of the desired magnitude response of the filter at the specified bandedge frequencies. The bandedge frequencies are normalized to half the sampling frequency and lie in the range 0 to 1 (the Nyquist frequency corresponds to 1).

The basic command can be extended by specifying, for example, the relative weights of the ripples in the pass- and stopbands and/or the type of filter. In the case where the relative weights are specified and the deviations in the pass- and stopbands are required, the syntax of the command is:

    b = remez(N–1, F, M, WT)

where WT is a vector of the relative weights between the ripples in the bands.

A flag, 'ftype', may also be added to specify the type of filter required. There are four possible types of filter, depending on whether N is even or odd and the type of symmetry of the filter coefficients. Type 1 filters result when the filter length is odd (i.e. when the filter order which is $N - 1$ is even), and type 2 is when the filter length is even. There is no restriction on the use of type 1 filters for the standard frequency selective filters. Type 2 filters have a zero at the Nyquist frequency and so cannot be used to design highpass and bandstop filters. Type 3 (N is odd) filters lead to Hilbert transformers and type 4 (N is even) differentiators. For types 1 and 2 filters specifying the filter length is sufficient to indicate the choice, but for types 3 and 4 it is necessary to include the flag 'hilbert' or 'differentiator' to indicate the filter type.

Example 7B.3

Compute, using the optimal method, the coefficients and plot the frequency response of a bandpass, linear phase FIR filter with the following characteristics:

passband	1000–1500 Hz
transition band	500 Hz
filter length	41
sampling frequency	10 000 Hz

Solution

The frequency bands of the filter are 0 to 500 Hz (lower stopband), 1000–1500 Hz (passband), and 2000–5000 Hz (upper stopband). The bandedge frequencies must be normalized to half the sampling frequency:

$$500/5000 = 0.1$$
$$1000/5000 = 0.2$$
$$1500/5000 = 0.3$$
$$2000/5000 = 0.4$$
$$5000/5000 = 1$$

Thus the vector of normalized bandedge frequencies, F, becomes:

    F   [0, 0.1, 0.2, 0.3, 0.4 1]

The desired magnitude response is 1 in the passband and zero in the stopband, giving a vector of desired magnitude response of:

M = [0 0 1 1 0 0]

The MATLAB commands to calculate the coefficients and plot the magnitude response of the filter are given in Program 7B.3.

The filter coefficients and the magnitude response of the filter are shown in Table 7B.3 and Figure 7B.2.

Program 7B.3   MATLAB m-file for calculating optimal FIR filter coefficients and plotting frequency response.

```
%
%
Fs=10000;                        %   Sampling frequency
N=41;                            %   Filter length
M=[0 0 1 1 0 0];                 %   Desired magnitude response
F=[0, 0.1, 0.2, 0.3, 0.4 1];     %   Band edge frequencies
b = remez(N–1, F, M);            %   Compute the filter coefficients
[H, f] = freqz(b, 1, 512, Fs);   %   Compute the frequency response
mag = 20*log10(abs(H));          %   of filter and plot it
plot(f, mag)
xlabel('Frequency (Hz)')
ylabel('Magnitude (dB)')
```

Table 7B.3   Filter coefficients for Example 7B.2.

n	h(n)
0	−0.0001
1	−0.0004
2	−0.0001
3	−0.0001
4	−0.0007
5	0.0005
6	0.0023
7	0.0008
8	−0.0017
9	−0.0005
10	−0.0005
11	−0.0044
12	−0.0022
13	0.0069
14	0.0066
15	−0.0016
16	0.0000
17	0.0022
18	−0.0117
19	−0.0164
20	0.0069
21	0.0189

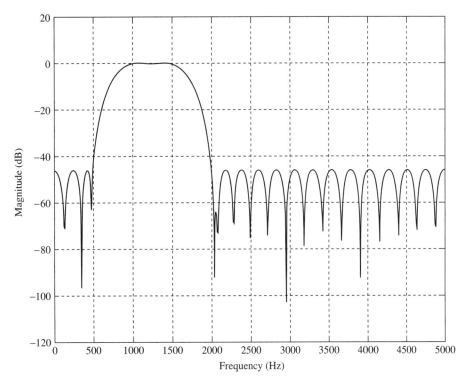

Figure 7B.2

<table>
<tr><td>Example 7B.4</td><td>A linear phase, bandpass filter is required to meet the following specifications:</td></tr>
</table>

passband	12–16 kHz
transition width	2 kHz
passband ripple	1 dB
stopband attenuation	45 dB
sampling frequency	50 kHz

Estimate the filter length, N, and use the optimal method to determine the filter coefficients and hence plot the magnitude–frequency response. Compare the pass- and stopband ripples of the filter with the specified values.

Solution

As before, the bandedge frequencies need to be determined and then normalized relative to the Nyquist frequency:

$$10/25 = 0.4$$
$$12/25 = 0.48$$
$$16/25 = 0.64$$
$$18/25 = 0.72$$

Using these values the vector of bandedge frequencies is:

F = [0 0.4 0.48 0.64 0.72 1]
M = [0 0 1 1 0 0]

An estimate of the filter length can be determined using the **remezord** command. This requires the pass- and stopband ripples to be in standard, linear units and so we must convert these from decibels:

$$\delta_p = \frac{10^{\frac{A_p}{20}} - 1}{10^{\frac{A_p}{20}} + 1}, \quad \delta_s = 10^{\frac{-A_s}{20}}$$

where A_p and A_s are the passband and stopband ripples, respectively, in dB.

The bandedge frequencies, desired magnitude response and values of the ripples and sampling frequencies are used to obtain an estimate of the filter order $(N-1)$ and hence the filter length (see Program 7B.4a, Figure 7B.3 and Table 7B.4).

Estimates of the filter parameters are: $N = 40$, weights $10.22 : 1 : 10.22$, and maximum deviation 0.0774. The value of N may be increased to achieve greater stopband attenuation and/or lower passband ripple (see Program 7B.4b).

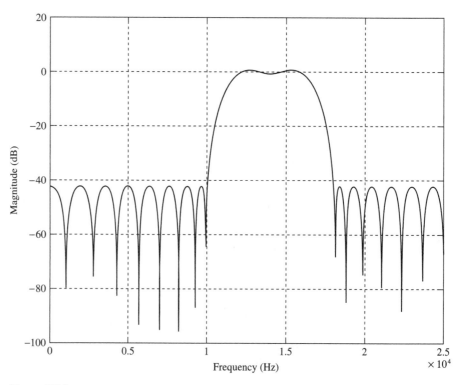

Figure 7B.3

Table 7B.4 Coefficients of the optimal FIR
filter of Example 7B.4.

n	h(n)
0	0.0005
1	−0.0017
2	−0.0088
3	0.0139
4	0.0136
5	−0.0273
6	−0.0060
7	0.0363
8	−0.0059
9	−0.0225
10	0.0054
11	−0.0080
12	0.0305
13	0.0293
14	−0.0988
15	−0.0085
16	0.1654
17	−0.0595
18	−0.1854
19	0.1411

Program 7B.4a MATLAB m-file for calculating the optimal FIR filter coefficients and
plotting frequency response for Example 7B.4.

```
%
%
Fs=50000;                                    %  Sampling frequency
Ap=1;                                        %  Pass band ripple in dB
As=45;                                       %  Stop band attenuation in dB
M=[0 1 0];                                   %  Desired magnitude response
F=[10000, 12000, 16000, 18000];             %  Band edge frequencies
dp=(10^(Ap/20)−1)/(10^(Ap/20)+1);           %  Pass and stop band ripples
ds=10^(−As/20);
dev=[ds dp ds];
[N1, F0, M0, W] = remezord(F, M, dev, Fs)    %  Determine filter order
[b delta] = remez(N1, F0, M0, W);            %  Compute the filter coefficients
[H, f] = freqz(b, 1, 1024, Fs);              %  Compute the frequency response
mag = 20*log10(abs(H));                      %  of filter and plot it
plot(f, mag), grid on
xlabel('Frequency (Hz)')
ylabel('Magnitude (dB)')
```

Program 7B.4b   An alternative MATLAB m-file for calculating the optimal FIR filter coefficients and plotting frequency response for Example 7B.4.

```
%
%
N=44
Fs=50000;                               %  Sampling frequency
Ap=1;                                   %  Pass band ripple in dB
As=45;                                  %  Stop band attenuation in dB
M=[0 0 1 1 0 0];                        %  Desired magnitude response
F=[0, 0.4, 0.48, 0.64, 0.72 1] ;        %  Band edge frequencies
dp=(10^(Ap/20)–1)/(10^(Ap/20)+1);
ds=10^(–As/20);
W=[dp/ds, 1, dp/ds];
dev=[ds ds dp dp ds ds];
[b delta] = remez(N–1, F, M, W);        %  Compute the filter coefficients
[H, f] = freqz(b, 1, 1024, Fs);         %  Compute the frequency response
mag = 20*log10(abs(H));                 %  of filter and plot it
plot(f, mag), grid on
xlabel('Frequency (Hz)')
ylabel('Magnitude (dB)')
```

7B.3   Frequency sampling method

The **fir2** command is used to design FIR filters with arbitrary frequency response characteristics such as those encountered in the frequency sampling method. The syntax for the basic command is

> b = fir2(N–1, F, H)

The **fir2** command calculates the coefficients of an N-length FIR filter. The vector F specifies the normalized frequency points in the range 0 to 1 (where the frequency points are normalized to half the sampling frequency as before). The vector H specifies the desired magnitude response at the frequency points specified by F. Both vectors are of the same length.

We will give examples to illustrate FIR filter design using the **fir2** command.

Example 7B.5

Frequency sampling filter design example   A linear phase, frequency sampling FIR filter has two transition band frequency samples. Assume that the filter has 15 taps and is characterized by the following frequency samples:

$$|H(k)| = 1 \qquad k = 0, 1, 2, 3$$
$$0.5571 \qquad k = 4$$
$$0.0841 \qquad k = 5$$
$$0 \qquad k = 6, 7$$

Determine the coefficients of the filter if the sampling frequency is 2 kHz.

Solution   The frequency samples above have been specified for the frequency range 0 to half the sampling frequency. Thus, the frequency points, normalized to half the sampling frequency, are: 0, 1/7, 2/7, 3/7, 4/7, 5/7, 6/7, 1.

The MATLAB program that uses the frequency samples to determine the FIR filter coefficients is given in Program 7B.5 and the magnitude–frequency response is depicted in Figure 7B.4. Filter coefficients are listed in Table 7B.5.

Program 7B.5   MATLAB m-file for computing the coefficients of an FIR frequency sampling filter.

```
N=15;
fd=[0 1/7 2/7 3/7 4/7 5/7 6/7 1];
Hd=[1 1 1 1 0.5571 0.0841 0 0];
hn=fir2(N-1, fd, Hd);
[H, f] = freqz(hn, 1, 512, Fs);
plot(f, abs(H)), grid on
xlabel('Frequency (Hz)')
ylabel('Magnitude')
```

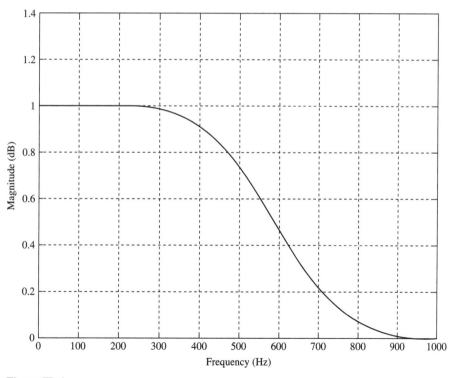

Figure 7B.4

Table 7B.5   Filter coefficients for Example 7B.5.

n	$h(n)$
0	−0.0001
1	−0.0006
2	0.0017
3	0.0128
4	−0.0299
5	−0.0571
6	0.2777
7	0.5910
8	0.2777
9	−0.0571
10	−0.0299
11	0.0128
12	0.0017
13	−0.0006
14	−0.0001

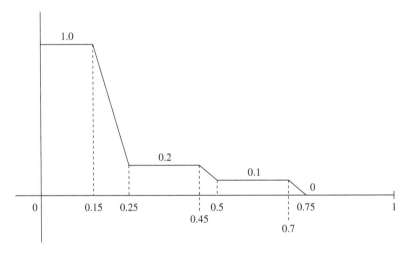

Figure 7B.5

Example 7B.6   *Design of a filter with arbitrary magnitude response*   A requirement exists for an FIR filter that approximates the magnitude–frequency response characteristics depicted in Figure 7B.5.

Determine the coefficients of a suitable FIR filter and plot its magnitude–frequency response. Assume a sampling frequency of 2 kHz and a filter length of 110.

Solution   The desired magnitude response has a level of 1 between the normalized frequencies of 0 and 0.15, a level of 0.3 between 0.25 and 0.45, a level of 0.1 between 0.5 and 0.75, and a level of 0

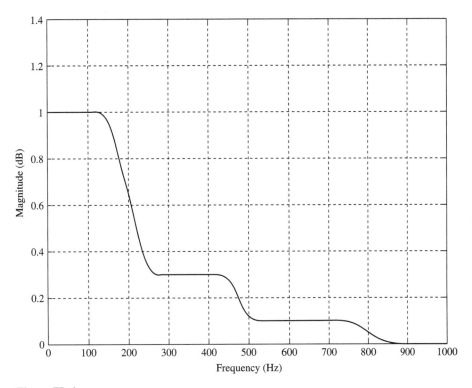

Figure 7B.6

Program 7B.6   MATLAB m-file for calculating the coefficients of an FIR filter with arbitrary magnitude response.

```
Fs=2000;                              %   Sampling frequency
N=110;                                %   Filter length
fd=[0 0.15 0.25 0.45 0.5 0.75 0.85 1]; %   Frequency sampling points
Hd=[1 1 0.3 0.3 0.1 0.1 0 0];         %   Frequency samples
hn=fir2(N−1, fd, Hd);                 %   Compute the impulse response
[H, f] = freqz(hn, 1, 512, Fs);
plot(f, abs(H)), grid on
xlabel('Frequency (Hz)')
ylabel('Magnitude')
```

between 0.85 and 1. These levels and the associated normalized frequencies must be specified in the MATLAB program.

The MATLAB program, with the frequency samples specified, is listed in Program 7B.6. The filter coefficients are calculated using the **fir2** command and the magnitude response by the **freqz** command. The coefficients are not listed here because of lack of space. The magnitude–frequency response of the FIR filter is depicted in Figure 7B.6.

Design of infinite impulse response (IIR) digital filters

8

This chapter presents practical design methods for digital infinite impulse response (IIR) filters, including popular methods which permit analog filters to be converted into equivalent digital filters. A simple but general step-by-step guide for designing digital IIR filters, from specifications to implementation, is described. Several fully worked examples are given to illustrate various aspects of digital IIR filter design, including the analysis of the effects of finite precision arithmetic on filter performance and real-time implementation.

A number of MATLAB and C language programs are provided to enable users to calculate filter coefficients and to carry out finite wordlength analysis. The reader is referred to Chapter 6 for a description of a general framework for filter design, comparison between IIR and FIR, and between digital and analog filters. In this chapter we will concentrate on IIR filter design and applications.

8.1 Introduction: summary of the basic features of IIR filters

Realizable IIR digital filters are characterized by the following recursive equation:

$$y(n) = \sum_{k=0}^{\infty} h(k)x(n-k) = \sum_{k=0}^{N} b_k x(n-k) - \sum_{k=1}^{M} a_k y(n-k) \tag{8.1}$$

where $h(k)$ is the impulse response of the filter which is theoretically infinite in duration, b_k and a_k are the coefficients of the filter, and $x(n)$ and $y(n)$ are the input and output to the filter. The transfer function for the IIR filter is given by

$$H(z) = \frac{b_0 + b_1 z^{-1} + \ldots + b_N z^{-N}}{1 + a_1 z^{-1} + \ldots + a_M z^{-M}} = \frac{\displaystyle\sum_{k=0}^{N} b_k z^{-k}}{1 + \displaystyle\sum_{k=1}^{M} a_k z^{-k}} \tag{8.2}$$

An important part of the IIR filter design process is to find suitable values for the coefficients b_k and a_k such that some aspect of the filter characteristic, such as frequency response, behaves in a desired manner. Equations 8.1 and 8.2 are the characteristic equations for IIR filters.

Note that, in Equation 8.1, the current output sample, $y(n)$, is a function of past outputs, $y(n-k)$, as well as present and past input samples, $x(n-k)$, that is the IIR filter is a feedback system of some sort. The strength of IIR filters comes from the flexibility the feedback arrangement provides. For example, an IIR filter normally requires fewer coefficients than an FIR filter for the same set of specifications, which is why IIR filters are used when sharp cutoff and high throughput are the important requirements. The price for this is that the IIR filter can become unstable or its performance significantly degraded if adequate care is not taken in its design.

The transfer function of the IIR filter, $H(z)$, given in Equation 8.2 can be factored as

$$H(z) = \frac{K(z - z_1)(z - z_2) \ldots (z - z_N)}{(z - p_1)(z - p_2) \ldots (z - p_M)} \tag{8.3}$$

where z_1, z_2, $\ldots$ are the zeros of $H(z)$, that is those values of z for which $H(z)$ becomes zero, and p_1, p_2, $\ldots$ are the poles of $H(z)$, that is values of z for which $H(z)$ is infinite.

A plot of the poles and zeros of the transfer function is known as the pole–zero diagram and provides a very useful way of representing and analyzing the filter in the complex z-plane; see Chapter 3 for details. For the filter to be stable, all its poles must lie inside the unit circle (or be coincident with zeros on the unit circle). There is no restriction on the zero locations.

8.2 Design stages for digital IIR filters

The design of IIR filters can be conveniently broken down into five main stages.

(1) Filter specification, at which stage the designer gives the function of the filter (for example, lowpass) and the desired performance.

(2) Approximation or coefficient calculation, where we select one of a number of methods and calculate the values of the coefficients, b_k and a_k, in the transfer function, $H(z)$, such that the specifications given in stage 1 are satisfied.

(3) Realization, which is simply converting the transfer function into a suitable filter structure. Typical structures for IIR filters are parallel and cascade of second and/or first-order filter sections.

(4) Analysis of errors that would arise from representing the filter coefficients and carrying out the arithmetic operations involved in filtering with only a finite number of bits.

(5) Implementation, which involves building the hardware and/or writing the software codes, and carrying out the actual filtering operation.

These stages are summarized in Figure 8.1. As indicated in the figure, the five stages are not independent and they are not always performed in the order given. In fact, techniques are now available that combine the second, third and fourth steps. However, the approach discussed here will ensure a successful design. To arrive at an efficient filter, it may be necessary to iterate a few times within and/or between the stages.

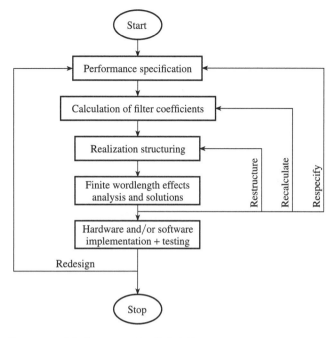

Figure 8.1   Summary of design stages for digital filters.

8.3   Performance specification

As with most other engineering problems, the design of digital IIR filters starts with an explicit specification of the performance requirements. These should include (i) signal characteristics (types of signal sources and sinks, I/O interface, data rates and wordlengths, and frequencies of interest), (ii) the frequency response characteristics of the filter (the desired amplitude and/or phase responses and their tolerances (if any), the speed of operation), (iii) the manner of implementation (for example, as a high level language routine in a computer or as a DSP processor-based system, choice of signal processor, modes of filtering (real-time or batch)), and (iv) other design constraints (such as costs and permissible signal degradation through the filter). In general, most of the above requirements are application dependent. The designer may not have enough information to specify the filter completely at the outset, but as many of the filter requirements as possible should be specified to simplify the design process.

For frequency selective filters, such as lowpass and bandpass filters, the frequency response specifications are often in the form of a tolerance scheme. Figure 8.2 depicts such a scheme for a bandpass IIR filter. The shaded horizontal lines indicate the tolerance limits. The following parameters are normally used to specify the frequency response.

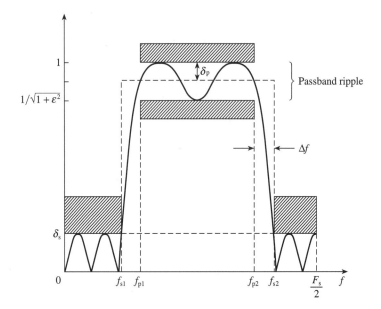

Figure 8.2   Tolerance scheme for an IIR bandpass filter.

ε^2	passband ripple parameter
δ_p	passband deviation
δ_s	stopband deviation
f_{p1} and f_{p2}	passband edge frequencies
f_{s1} and f_{s2}	stopband edge frequencies

The bandedge frequencies are sometimes given in normalized form, that is a fraction of the sampling frequency (f/F_s), but we shall specify them in standard frequency units of hertz or kilohertz as these are less confusing, especially to the inexperienced designer. Passband and stopband deviations may be expressed as ordinary numbers or in decibels: the passband ripple in decibels is

$$A_p = 10 \log_{10} (1 + \varepsilon^2) = -20 \log_{10} (1 - \delta_p) \tag{8.4a}$$

and the stopband attenuation in decibels is

$$A_s = -20 \log_{10} (\delta_s) \tag{8.4b}$$

As discussed in Chapter 6 and is evident in Figure 8.2, for IIR filters the passband ripple is the difference between the minimum and maximum deviation in the passband. For FIR filters, the passband ripple is the difference between the ideal response and the maximum (or minimum) deviation in the passband. Thus, for IIR, when we say passband ripple, what is meant is the peak-to-peak passband ripple.

8.4 Coefficient calculation methods for IIR filters

The task at this stage is to select one of a number of approximation methods and to use it to calculate the values of the coefficients, a_k and b_k, in Equation 8.2, such that the frequency response specifications given in the first design stage are satisfied.

A simple way to obtain the IIR filter coefficients is to place poles and zeros judiciously in the z-plane such that the resulting filter has the desired frequency response. This approach, known as the pole–zero placement method, is only useful for very simple filters, for example notch filtering, where the filter parameters (such as passband ripple) need not be specified precisely. A more efficient approach is first to design an analog filter satisfying the desired specifications and then to convert it into an equivalent digital filter. Most IIR digital filters are designed this way. The rationale behind this approach is that there already exists a wealth of information on analog filters in the literature which can be utilized. Three of the most common methods of converting analog filters into equivalent digital filters are the impulse invariant, the matched z-transform and the bilinear z-transform methods.

In the next few sections we will cover the following methods of calculating the coefficients of IIR filters:

- pole–zero placement;
- impulse invariant;
- matched z-transform;
- bilinear z-transform.

8.5 Pole–zero placement method of coefficient calculation

8.5.1 Basic concepts and illustrative design examples

When a zero is placed at a given point on the z-plane, the frequency response will be zero at the corresponding point. A pole on the other hand produces a peak at the corresponding frequency point; see Figure 8.3. Poles that are close to the unit circle give rise to large peaks, whereas zeros close to or on the circle produce troughs or minima. Thus, by strategically placing poles and zeros on the z-plane, we can obtain simple lowpass, or other frequency selective, filters. Lynn and Fuerst (1989) provide a more detailed discussion of filters of this type.

An important point to bear in mind is that for the coefficients of the filter to be real, the poles and zeros must either be real (that is lie on the positive or negative real axes) or occur in complex conjugate pairs. We will illustrate the method with examples.

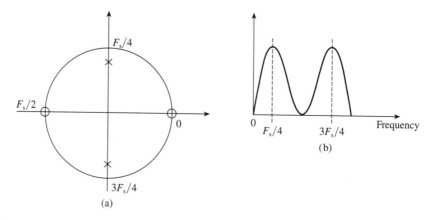

Figure 8.3 (a) Pole–zero diagram of a simple filter, and (b) a sketch of its frequency response.

Example 8.1

Illustrating the simple pole–zero method of calculating filter coefficients   A bandpass digital filter is required to meet the following specifications:

(1)   complete signal rejection at dc and 250 Hz;

(2)   a narrow passband centred at 125 Hz;

(3)   a 3 dB bandwidth of 10 Hz.

Assuming a sampling frequency of 500 Hz, obtain the transfer function of the filter, by suitably placing z-plane poles and zeros, and its difference equations.

Solution

First, we must determine where to place the poles and zeros on the z-plane. Since a complete rejection is required at 0 and 250 Hz, we need to place zeros at corresponding points on the z-plane. These are at angles of $0°$ and $360° \times 250/500 = 180°$ on the unit circle. To have the passband centred at 125 Hz requires us to place poles at $\pm 360° \times 125/500 = \pm 90°$. To ensure that the coefficients are real, it is necessary to have a complex conjugate pole pair.

The radius, r, of the poles is determined by the desired bandwidth. An approximate relationship between r, for $r > 0.9$, and bandwidth, bw, is given by

$$r \simeq 1 - (\text{bw}/F_s)\pi \tag{8.5}$$

For the problem, bw = 10 Hz and $F_s = 500$ Hz, giving $r = 1 - (10/500)\pi = 0.937$. The pole–zero diagram is given in Figure 8.4(a). From the pole–zero diagram, the transfer function can be written down by inspection:

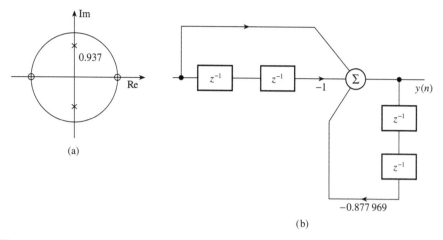

(a)

(b)

Figure 8.4   (a) Pole–zero diagram. (b) Block diagram representation of filter.

$$H(z) = \frac{(z-1)(z+1)}{(z - re^{j\pi/2})(z - re^{-j\pi/2})}$$

$$= \frac{z^2 - 1}{z^2 + 0.877\,969} = \frac{1 - z^{-2}}{1 + 0.877\,969 z^{-2}}$$

The difference equation is

$$y(n) = -0.877\,969\,y(n-2) + x(n) - x(n-2)$$

Comparing the transfer function, $H(z)$, with the general IIR equation (Equation 8.2), we find that the filter is a second-order section with the following coefficients:

$b_0 = 1$ $\quad a_1 = 0$

$b_1 = 0$ $\quad a_2 = 0.877\,969$

$b_2 = -1$

Example 8.2

Using the pole–zero placement method to calculate coefficients of a notch filter
Obtain, by the pole–zero placement method, the transfer function and the difference equation of a simple digital notch filter that meets the following specifications:

notch frequency	50 Hz
3 dB width of notch	±5 Hz
sampling frequency	500 Hz

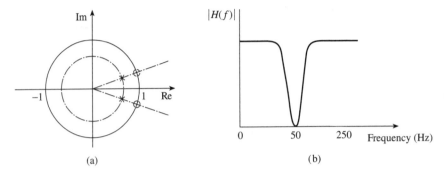

Figure 8.5   (a) Pole–zero diagram for Example 8.2 and (b) the corresponding frequency response.

Solution   ▪ To reject the component at 50 Hz, we place a pair of complex zeros at points on the unit circle corresponding to 50 Hz, that is at angles of $360° × 50/500$ $= ±36°$.

▪ To achieve a sharp notch filter and improved amplitude response on either side of the notch frequency, a pair of complex conjugate poles are placed at a radius $r < 1$. The width of the notch is determined by the locations of the poles. The relationship between the bandwidth and the radius of Example 8.1 is applicable. Thus the radius of the poles is 0.9372.

▪ The pole–zero diagram is given in Figure 8.5(a). From the figure, the transfer function of the filter is given by

$$H(z) = \frac{[z - \exp(-j36°)][z - \exp(j36°)]}{[z - 0.937\exp(-36°)][z - 0.9372\exp(36°)]}$$

$$= \frac{z^2 - 1.6180z + 1}{z^2 - 1.5164z + 0.8783} = \frac{1 - 1.6180z^{-1} + z^{-2}}{1 - 1.5164z^{-1} + 0.8783z^{-2}}$$

The difference equation is

$$y(n) = x(n) - 1.6180x(n-1) + x(n-2) + 1.5164y(n-1) - 0.8783y(n-2)$$

Comparing $H(z)$ with Equation 8.2 shows that the coefficients for the notch filter are

$b_0 = 1$ $a_1 = -1.5164$

$b_1 = -1.6180$ $a_2 = 0.8783$

$b_2 = 1$

8.6 Impulse invariant method of coefficient calculation

8.6.1 Basic concepts and illustrative design examples

In this method, starting with a suitable analog transfer function, $H(s)$, the impulse response, $h(t)$, is obtained using the Laplace transform. The $h(t)$ so obtained is suitably sampled to produce $h(nT)$, and the desired transfer function, $H(z)$, is then obtained by z-transforming $h(nT)$, where T is the sampling interval. We will illustrate the method by examples.

Example 8.3 *Illustrating the impulse invariant method* Digitize, using the impulse variant method, the simple analog filter with the transfer function given by

$$H(s) = \frac{C}{s - p} \tag{8.6}$$

Solution The impulse response, $h(t)$, is given by the inverse Laplace transform:

$$h(t) = L^{-1}[H(s)] = L^{-1}\left(\frac{C}{s - p}\right) = Ce^{pt}$$

where L^{-1} symbolizes the inverse Laplace transform. According to the impulse invariant method, the impulse response of the equivalent digital filter, $h(nT)$, is equal to $h(t)$ at the discrete times $t = nT$, $n = 0, 1, 2, \ldots$, that is

$$h(nT) = h(t)|_{t=nT} = Ce^{pnT}$$

The transfer function of $H(z)$ is obtained by z-transforming $h(nT)$:

$$H(z) = \sum_{n=0}^{\infty} h(nT)z^{-n} = \sum_{n=0}^{\infty} Ce^{pnT}z^{-1}$$

$$= \frac{C}{1 - e^{pT}z^{-1}}$$

Thus, from the result above, we can write

$$\frac{C}{s - p} \rightarrow \frac{C}{1 - e^{pT}z^{-1}} \tag{8.7}$$

To apply the impulse invariant method to a high-order (for example, Mth-order) IIR filter with simple poles, the transfer function, $H(s)$, is first expanded using partial fractions as the sum of single-pole filters:

$$H(s) = \frac{C_1}{s - p_1} + \frac{C_2}{s - p_2} + \ldots + \frac{C_M}{s - p_M}$$

$$= \sum_{K=1}^{M} \frac{C_K}{s - p_K} \tag{8.8}$$

where the p_K are the poles of $H(s)$. Each term on the right-hand side of Equation 8.8 has the same form as Equation 8.6 and so the transformation given in Equation 8.8 is applicable. Thus:

$$\sum_{K=1}^{M} \frac{C_K}{s - p_K} \rightarrow \sum_{K=1}^{M} \frac{C_K}{1 - e^{p_K T} z^{-1}} \tag{8.9}$$

High-order IIR filters are normally realized as cascades or parallel combinations of standard second-order filter sections. Thus the case when $M = 2$ is of particular interest. In this case the transform of Equation 8.9 becomes

$$\frac{C_1}{s - p_1} + \frac{C_2}{s - p_2} \rightarrow \frac{C_1}{1 - e^{p_1 T} z^{-1}} + \frac{C_2}{1 - e^{p_2 T} z^{-1}}$$

$$= \frac{C_1 + C_2 - (C_1 e^{p_2 T} + C_2 e^{p_1 T}) z^{-1}}{1 - (e^{p_1 T} + e^{p_2 T}) z^{-1} + e^{(p_1 + p_2) T} z^{-2}} \tag{8.10}$$

If the poles, p_1 and p_2, are complex conjugates, then C_1 and C_2 will also be complex conjugates and Equation 8.10 reduces to

$$\frac{C_1}{1 - e^{p_1 T} z^{-1}} + \frac{C_1^*}{1 - e^{p_1^* T} z^{-1}} = \frac{2C_r - [C_r \cos(p_i T) + C_i \sin(p_i T)] 2 e^{p_r T} z^{-1}}{1 - 2 e^{p_r T} \cos(p_i T) z^{-1} + e^{2 p_r T} z^{-2}} \tag{8.11}$$

where C_r and C_i are the real and imaginary parts of C_1, p_r and p_i are the real and imaginary parts of p_1, and * symbolizes a complex conjugate.

For most practical impulse invariant IIR filters, the transformations given in Equations 8.7, 8.10 and/or 8.11 are the only transformations required to obtain the coefficients of the transfer function. A C language program for computing the coefficients of impulse invariant filters is given in the appendix. We will illustrate the use of the transformations by an example.

Example 8.4 *Applying the impulse invariant method to filter design* It is required to design a digital filter to approximate the following normalized analog transfer function:

$$H(s) = \frac{1}{s^2 + \sqrt{2}s + 1}$$

Using the impulse invariant method obtain the transfer function, $H(z)$, of the digital filter, assuming a 3 dB cutoff frequency of 150 Hz and a sampling frequency of 1.28 kHz.

Solution
Before applying the impulse invariant method, we need to frequency scale the normalized transfer function. This is achieved by replacing s by s/α, where $\alpha = 2\pi \times 150 = 942.4778$, to ensure that the resulting filter has the desired response. Thus

$$H'(s) = H(s)|_{s=s/\alpha} = \frac{\alpha^2}{s^2 + \sqrt{2}\,\alpha s + \alpha^2} = \frac{C_1}{s - p_1} + \frac{C_2}{s - p_2}$$

where

$$p_1 = \frac{-\sqrt{2}\,\alpha(1 - \mathrm{j})}{2} = -666.4324(1 - \mathrm{j}),\ p_2 = p_1^*$$

$$C_1 = -\frac{\alpha}{\sqrt{2}}\mathrm{j} = -666.4324\mathrm{j};\ C_2 = C_1^*$$

Since the poles are complex conjugates, the transformation in Equation 8.11 is used to obtain the discrete-time transfer function, $H(z)$. For the problem, $C_r = 0$, $C_i = -666.4324$, $p_iT = 0.5207$, $p_rT = -0.5207$, $\mathrm{e}^{p_rT} = 0.5941$, $\sin(p_iT) = 0.4974$, $\cos(p_iT) = 0.8675$, and $\mathrm{e}^{p_rT} = 0.3530$. Substituting these values into Equation 8.11, we obtain $H(z)$:

$$H(z) = \frac{393.9264z^{-1}}{1 - 1.0308z^{-1} + 0.3530z^{-2}}$$

If we substitute $z = \mathrm{e}^{\mathrm{j}\omega T}$ in the equation above, the value of $H(z)$ at $\omega = 0$ is 1223, approximately equal to the sampling frequency. Such a large gain is characteristic of impulse invariant filters. In general, the gain of the transfer function obtained by this method is equal to the sampling frequency, that is $1/T$, and results from sampling the impulse response. To keep the gain down and to avoid overflows when the filter is implemented, it is common practice to multiply $H(z)$ by T (or equivalently to divide it by the sampling frequency). Thus, for the problem, the transfer function becomes

$$H(z) = \frac{0.3078z^{-1}}{1 - 1.0308z^{-1} + 0.3530z^{-2}}$$

Thus we have

$$b_0 = 0 \qquad a_1 = -1.0308$$
$$b_1 = 0.3078 \qquad a_2 = 0.3530$$

An alternative method of removing the effect of the sampling frequency on the filter gain is to work with normalized frequencies. Thus in the last example we would use $T = 1$ and $\alpha = 2\pi \times 150/1280 = 0.7363$. Using these values in Equation 8.11 leads

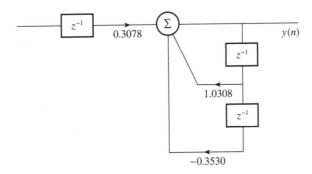

Figure 8.6 A block diagram representation of the filter in Example 8.4.

directly to the desired transfer function above. An important advantage of working with normalized frequencies is that the numbers involved are much simpler. It also means that the results can be generalized. The filter is represented in the form of a block diagram in Figure 8.6.

8.6.2 Summary of the impulse invariant method

(1) Determine a normalized analog filter, $H(s)$, that satisfies the specifications for the desired digital filter.

(2) If necessary, expand $H(s)$ using partial fractions to simplify the next step.

(3) Obtain the z-transform of each partial fraction to obtain Equation 8.9.

(4) Obtain $H(z)$ by combining the z-transforms of the partial fractions into second-order terms and possibly one first-order term. If the actual sampling frequency is used then multiply $H(z)$ by T.

8.6.3 Remarks on the impulse invariant method

(1) The impulse response of the discrete filter, $h(nT)$, is identical to that of the analog filter, $h(t)$, at the discrete time instants $t = nT$, $n = 0, 1, \ldots$; see Figure 8.7 for example. It is for this reason that the method is called the impulse invariant method.

(2) The sampling frequency affects the frequency response of the impulse invariant discrete filter. A sufficiently high sampling frequency is necessary for the frequency response to be close to that of the equivalent analog filter.

(3) As is the case with sampled data systems, the spectrum of the impulse invariant filter corresponding to $H(z)$ would be the same as that of the original analog filter, $H(s)$, but repeats at multiples of the sampling frequency as shown in Figure 8.8, leading to aliasing. However, if the roll-off of the original analog filter is sufficiently steep or if the analog filter is bandlimited before the

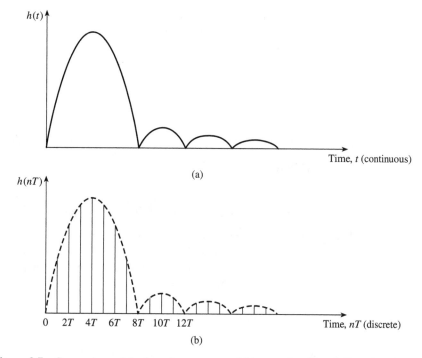

Figure 8.7   Comparison of the impulse response of (a) an analog filter, $h(t)$, and (b) its digital filter equivalent, $h(nT)$. In the impulse invariant method, the two impulse responses are identical at the sampling instants.

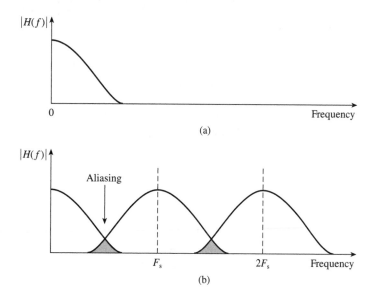

Figure 8.8   (a) Spectrum of an analog filter and (b) spectrum of an equivalent impulse invariant digital filter showing effects of aliasing.

impulse invariant method is applied, the aliasing will be low. Low aliasing can also be achieved by making the sampling frequency high. We conclude that the method may be used for very sharp cutoff lowpass filters with little aliasing, provided that the sampling frequency is reasonably high, but it is unsuitable for highpass or bandstop filters unless an anti-aliasing filter is used.

8.7   Matched z-transform (MZT) method of coefficient calculation

8.7.1   Basic concepts and illustrative design examples

The matched z-transform (MZT) method provides a simple way to convert an analog filter into an equivalent digital filter. In the MZT method, each of the poles and zeros of the analog filter is mapped directly from the s-plane to the z-plane using the following equation:

$$(s - a) \rightarrow (1 - z^{-1}e^{aT}) \tag{8.12}$$

where T is the sampling period. Equation 8.12 maps a pole (or zero) at the location $s = a$ in the s-plane onto a pole (or zero) in the z-plane at $z = e^{aT}$.

For higher order analog filters, the transfer function has several poles and/or zeros and these need to be mapped from the s-plane onto the z-plane. For a higher order analog filter with distinct poles and zeros, the transfer function may be written in the form:

$$H(s) = \frac{(s - z_1)(s - z_2) \ldots (s - z_M)}{(s - p_1)(s - p_2) \ldots (s - p_N)} \tag{8.13}$$

where z_k and p_k are the zeros and poles of $H(s)$, respectively.

The MZT may then be applied to each factor separately:

$$(s - z_k) \rightarrow (1 - z^{-1}e^{z_k T})$$

$$(s - p_k) \rightarrow (1 - z^{-1}e^{p_k T})$$

In higher order IIR filters, the second-order filter section is the basic building block. Thus, the case where $M = N = 2$ in Equation 8.13 is of particular interest. In this case, the analog transfer function reduces to

$$H(s) = \frac{(s - z_1)(s - z_2)}{(s - p_1)(s - p_2)} \tag{8.14}$$

The application of the MZT to this gives

$$\frac{(s - z_1)(s - z_2)}{(s - p_1)(s - p_2)} \rightarrow \frac{1 - (e^{z_1 T} + e^{z_2 T})z^{-1} + e^{(z_1 + z_2)T}z^{-2}}{1 - (e^{p_1 T} + e^{p_2 T})z^{-1} + e^{(p_1 + p_2)T}z^{-2}} \tag{8.15}$$

If the poles and zeros of the second-order section occur in complex conjugate pairs, then $p_2 = p_1^*$ and $z_2 = z_1^*$, and the right-hand side of Equation 8.15 simplifies to

$$\frac{1 - 2e^{z_r T}\cos(z_i T)z^{-1} + e^{z_r T}z^{-2}}{1 - 2e^{p_r T}\cos(p_i T)z^{-1} + e^{p_r T}z^{-2}} \tag{8.16}$$

where z_r and z_i, p_r and p_i, are the real and imaginary parts of z_1 and p_1, respectively.

In practice, it is more convenient to express the second-order analog filter section in the familiar rational polynomial format:

$$H(s) = \frac{(s - z_1)(s - z_2)}{(s - p_1)(s - p_2)} = \frac{A_0 + A_1 s + A_2 s^2}{B_0 + B_1 s + B_2 s^2}$$

The poles and zeros of $H(s)$ are then given by

$$p_{1,2} = -\frac{B_1}{2B_2} \pm \left[\left(\frac{B_1}{2B_2}\right)^2 - \frac{B_0}{B_2}\right]^{\frac{1}{2}} \tag{8.17a}$$

$$z_{1,2} = -\frac{A_1}{2A_2} \pm \left[\left(\frac{A_1}{2A_2}\right)^2 - \frac{A_0}{A_2}\right]^{\frac{1}{2}} \tag{8.17b}$$

In practice, Equations 8.17a and b allow us to determine in a straightforward way the locations of the poles and zeros (and hence their real and imaginary parts), given the transfer function of the analog filter. Once we know the real and imaginary parts of the zeros and poles of $H(s)$, we can determine the transfer function, $H(z)$, of the equivalent discrete filter using Equation 8.15 or 8.16.

Example 8.5 The normalized transfer function of an analog filter is given by

$$H(s) = \frac{1}{s^2 + \sqrt{2}s + 1}$$

Obtain the transfer function, $H(z)$, of an equivalent digital filter using the matched *z*-transform method. Assume a 3 dB cutoff frequency of 150 Hz and a sampling frequency of 1.28 kHz.

Solution   The cutoff frequency may be expressed as $\omega_c = 2\pi \times 150 = 942.4778$ rad/s. The transfer function of the denormalized analog filter is obtained by replacing s by s/ω_c:

$$H'(s) = H(s)\Big|_{s=\frac{s}{\omega_c}}$$

$$= \frac{\omega_c^2}{s^2 + \sqrt{2}\omega_c s + \omega_c^2}$$

The poles of the filter are located at

$$P_{1,2} = -\frac{\sqrt{2}\omega_c}{2} \pm \left[\left(\frac{\sqrt{2}\omega_c}{2}\right)^2 - \omega_c^2\right]^{\frac{1}{2}}$$

$$= -\frac{\sqrt{2}\omega_c}{2} \pm \omega_c\left[\left(\frac{\sqrt{2}}{2}\right)^2 - 1\right]^{\frac{1}{2}}$$

$$= -\frac{\sqrt{2}\omega_c}{2}(1 \mp j)$$

For the problem, the real and imaginary parts of the poles are

$$p_r = -\frac{\sqrt{2}\omega_c}{2} = -666.4324, \quad p_i = \frac{\sqrt{2}\omega_c}{2}j = 666.4324j$$

Thus, $p_r T = -0.520\,650\,3$, $p_i T = 0.520\,650\,3$, $\cos(p_i T) = 0.867\,496$ and $e^{p_r T} = 0.594\,134$. The resulting transfer function becomes

$$H(z) = \frac{8.8876 \times 10^5}{1 - 1.030\,818z^{-1} + 0.594\,134z^{-2}}$$

8.7.2   Summary of the matched z-transform method

(1) Determine a suitable analog transfer function, $H(s)$, that meets the specifications of the desired digital filter.

(2) Find the locations of the poles and zeros of $H(s)$. This may require factorizing the analog transfer function, $H(s)$.

(3) Map the poles and zeros from the s-plane to the z-plane using Equation 8.12. Equations 8.15 and 8.16 may be used for a second-order section.

(4) Combine the z-plane equations as appropriate to obtain the transfer function, $H(z)$.

8.7.3   Remarks on the matched *z*-transform method

(1)   The MZT method requires a knowledge of the locations of the poles and zeros of the analog filter. This can be obtained by factorizing the analog transfer function, $H(s)$. Thereafter, the MZT is relatively easy to apply.

(2)   The MZT and the impulse invariant methods lead to discrete filters with identical denominators. Compare, for example, the denominators of the MZT of Equation 8.15 and the result of the impulse invariant transformation given in Equation 8.10. This is also evident if we compare the denominators of the transfer functions obtained in Examples 8.4 and 8.5.

(3)   In digital filters, the useful frequency band extends from zero to the Nyquist frequency (half the sampling frequency) whereas in an analog filter it is from zero to infinity. Thus, the MZT mapping as with other mappings compresses the infinite analog frequency band into a finite frequency band. This causes a distortion in the frequency response of the equivalent digital filters compared to the analog filter. For the MZT, the resulting filters tend to provide less attenuation compared to the analog filters. In Section 8.12, we will discuss this further and show how this feature may be exploited.

(4)   If the analog filter has poles at frequencies close to or zeros beyond the Nyquist frequency (that is half the sampling frequency), the frequency response of the resulting MZT-based digital filter will be distorted because of aliasing (see later). In such cases, the response of the analog filter beyond the Nyquist frequency will still be significant. This is folded back into the desired band by the implicit sampling process.

(5)   The MZT is also unsuitable for digitizing an all-pole analog filter because of the absence of zeros above the Nyquist frequency. The problem in this case may be alleviated by adding zeros at $z = -1$ (i.e. at the Nyquist frequency).

The effects of the MZT on the frequency response will be discussed further in Section 8.12.

8.8   Bilinear *z*-transform (BZT) method of coefficient calculation

8.8.1   Basic concepts and illustrative design examples

This is by far the most important method of obtaining IIR filter coefficients. In the BZT method, the basic operation required to convert an analog filter $H(s)$ into an equivalent digital filter is to replace s as follows:

$$s = k\frac{z-1}{z+1}, \quad k = 1 \text{ or } \frac{2}{T} \tag{8.18a}$$

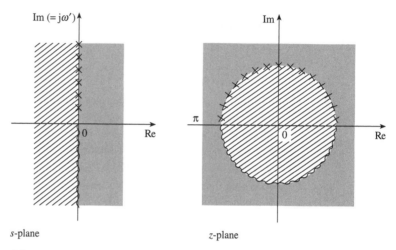

Figure 8.9   An illustration of the s-plane to z-plane mapping using the bilinear z-transformation. Note that the positive $j\omega'$ axis in the s-plane (that is, $s = 0$ to $s = j\infty$) maps to the upper half of the unit circle, and the negative $j\omega'$ axis maps to the lower half.

The above transformation maps the analog transfer function, $H(s)$, from the s-plane into the discrete transfer function, $H(z)$, in the z-plane as shown in Figure 8.9. Notice that in the figure the entire $j\omega$ axis in the s-plane is mapped onto the unit circle, the left-half s-plane is mapped inside the unit circle, and the right-half s-plane is mapped outside the z-plane unit circle. Thus, a stable analog filter, with poles on the left half of the s-plane, will lead to a digital filter with poles inside the unit circle.

Unfortunately, direct replacement of s in $H(s)$ as indicated in Equation 8.18a may lead to a digital filter with an undesirable response. This is readily shown by making the substitution $z = e^{j\omega T}$ and $s = j\omega'$ in Equation 8.18a. Simplifying, we find that the analog frequency ω' and the digital frequency ω are related as

$$\omega' = k \tan\left(\frac{\omega T}{2}\right), \quad k = 1 \text{ or } \frac{2}{T} \tag{8.18b}$$

Equation 8.18b is sketched in Figure 8.10. It is seen that the relationship between the analog frequency ω' and the digital frequency ω is almost linear for small values of ω, but becomes nonlinear for large values of ω, leading to a distortion (or warping) of the digital frequency response. Note, for example, that the passbands for the analog filter on the left-hand side are of constant width and centred at regular intervals, whereas the passbands for the digital equivalent are somewhat squashed up. This effect is normally compensated for by prewarping the analog filter before applying the bilinear transformation.

To compensate for the effect, we prewarp one or more critical frequencies before applying the BZT. For example, for a lowpass filter, we often prewarp the cutoff or bandedge frequency as follows:

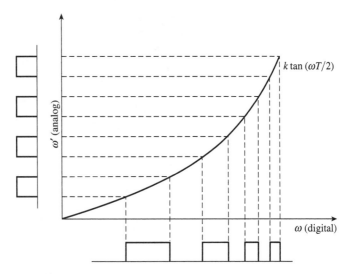

Figure 8.10   Relationship between analog and digital frequencies showing the warping effect. Notice that the equally spaced analog passbands are pushed together at the high frequency and, after transformation, in the digital domain.

$$\omega'_p = k \tan\left(\frac{\omega_p T}{2}\right) \tag{8.19}$$

where

ω_p = specified cutoff frequency
ω'_p = prewarped cutoff frequency
$k = 1$ or $\dfrac{T}{2}$
T = sampling period

8.8.2   Summary of the BZT method of coefficient calculation

For standard, frequency-selective IIR filters, the steps for using the BZT may be summarized as follows:

(1)   Use the digital filter specifications to find a suitable normalized, prototype, analog lowpass filter, $H(s)$.

(2)   Determine and prewarp the bandedge or critical frequencies of the desired filter. For lowpass or highpass filters there is just one bandedge or cutoff frequency (ω_p, say). For bandpass and bandstop filters, we have the lower and upper passband edge frequencies, ω_{p1} and ω_{p2}, each of which needs to be prewarped (the stopband edge frequencies may also be specified):

$$\omega_p' = \tan\left(\frac{\omega_p T}{2}\right) \tag{8.20a}$$

$$\omega_{p1}' = \tan\left(\frac{\omega_{p1} T}{2}\right); \quad \omega_{p2}' = \tan\left(\frac{\omega_{p2} T}{2}\right) \tag{8.20b}$$

(3) Denormalize the analog prototype filter by replacing s in the transfer function, $H(s)$, using one of the following transformations, depending on the type of filter required:

$$s = \frac{s}{\omega_p'} \qquad \text{lowpass to lowpass} \tag{8.21a}$$

$$s = \frac{\omega_p'}{s} \qquad \text{lowpass to highpass} \tag{8.21b}$$

$$s = \frac{s^2 + \omega_0^2}{Ws} \qquad \text{lowpass to bandpass} \tag{8.21c}$$

$$s = \frac{Ws}{s^2 + \omega_0^2} \qquad \text{lowpass to bandstop} \tag{8.21d}$$

where

$$\omega_0^2 = \omega_{p2}' \omega_{p1}', \quad W = \omega_{p2}' - \omega_{p1}'$$

(4) Apply the BZT to obtain the desired digital filter transfer function, $H(z)$, by replacing s in the frequency-scaled (i.e. denormalized) transfer function, $H'(s)$, as follows:

$$s = \frac{z - 1}{z + 1}$$

Example 8.6   *Lowpass filter*   It is required to design a digital lowpass filter to approximate the following transfer function:

$$H(s) = \frac{1}{s^2 + \sqrt{2}s + 1}$$

Using the BZT method obtain the transfer function, $H(z)$, of the digital filter, assuming a 3 dB cutoff frequency of 150 Hz and a sampling frequency of 1.28 kHz.

Solution   The critical frequency, $\omega_p = 2\pi \times 150$ rad/s, and $F_s = 1/T = 1.28$ kHz, giving a prewarped critical frequency of

$$\omega'_p = \tan{(\omega_p T/2)} = 0.3857$$

The frequency-scaled analog filter is given by

$$H'(s) = H(s)|_{s=s/\omega'_p} = \frac{1}{(s/\omega'_p)^2 + \sqrt{2}s/\omega'_p + 1}$$

$$= \frac{\omega'^2_p}{s^2 + \sqrt{2}\omega'_p s + \omega'_p s} = \frac{0.1488}{s^2 + 0.5455s + 0.1488}$$

Applying the BZT gives

$$H(z) = H'(s)\Big|_{s=\frac{z-1}{z+1}} = \frac{0.0878z^2 + 0.1756z + 0.0878}{z^2 - 1.0048z + 0.3561}$$

$$= \frac{0.0878(1 - 2z^{-1} + z^{-2})}{1 - 1.0048z^{-1} + 0.3561z^{-2}}$$

Example 8.7

Highpass filter   The normalized transfer function of a simple, analog lowpass, resistance–capacitance (RC) filter is given by

$$H(s) = \frac{1}{s + 1}$$

Starting from the s-plane equation, determine, using the BZT method, the transfer function of an equivalent discrete-time highpass filter. Assume a sampling frequency of 150 Hz and a cutoff frequency of 30 Hz.

Solution

The critical frequency for the digital filter is $\omega_p = 2\pi \times 30$ rad/s. The cutoff frequency, after prewarping, is $\omega'_p = \tan{(\omega_p T/2)}$. With $T = 1/150$ Hz, $\omega'_p = \tan{(\pi/5)} = 0.7265$.

Using the lowpass-to-highpass transformation of Equation 8.21a, the denormalized analog transfer function is obtained as

$$H'(s) = H(s)|_{s=\omega'_p/s} = \frac{1}{\omega'_p/s + 1} = \frac{s}{s + 0.7265}$$

The z-plane transfer function is obtained by applying the BZT:

$$H(z) = H'(s)|_{s=(z-1)/(z+1)} = \frac{(z-1)/(z+1)}{(z-1)/(z+1) + 0.7265}$$

Simplifying, we have

$$H(z) = 0.5792 \frac{1 - z^{-1}}{1 + 0.1584 z^{-1}}$$

The coefficients of the discrete-time filter are

$$b_0 = 0.5792, \quad a_1 = 0.1584$$
$$b_1 = -0.5792$$

Example 8.8   *Bandpass filter*   A discrete-time bandpass filter with Butterworth characteristics meeting the specifications given below is required. Obtain the coefficients of the filter using the BZT method:

passband	200–300 Hz
sampling frequency	2 kHz
filter order, N	2

Solution   A first order, normalized analog lowpass filter is required (since the frequency band transformation for bandpass filters – Equation 8.21c – will double the filter order). Thus

$$H(s) = \frac{1}{s + 1}$$

The prewarped critical frequencies are

$$\omega'_{p1} = \tan\left(\frac{\omega_{p1} T}{2}\right) = \tan\left(\frac{2\pi \times 200}{2 \times 2000}\right) = 0.3249$$

$$\omega'_{p2} = \tan\left(\frac{\omega_{p2} T}{2}\right) = \tan\left(\frac{2\pi \times 300}{2 \times 2000}\right) = 0.5095$$

$$\omega_0^2 = \omega'_{p1} \omega'_{p2} = 0.1655$$
$$W = \omega'_{p2} - \omega'_{p1} = 0.1846$$

Using the lowpass-to-bandpass transformation, Equation 8.21c, we have

$$H'(s) = H(s)\Big|_{s = \frac{s^2 + \omega_0^2}{Ws}} = \frac{1}{\dfrac{s^2 + \omega_0^2}{Ws} + 1}$$

$$= \frac{Ws}{s^2 + Ws + \omega_0^2}$$

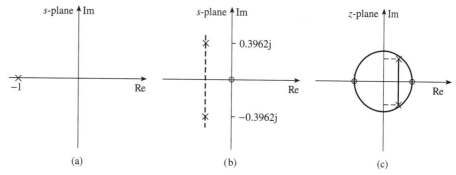

Figure 8.11 Pole–zero diagrams for (a) a prototype lowpass filter and those of
(b) intermediate analog bandpass and (c) discrete bandpass filters obtained
by band transformation.

Applying the BZT gives

$$H(z) = H'(s)\Big|_{s=\frac{z-1}{z+1}} = \frac{W\left(\dfrac{z-1}{z+1}\right)}{\left(\dfrac{z-1}{z+1}\right)^2 + W\left(\dfrac{z-1}{z+1}\right) + \omega_0^2}$$

Substituting the values of ω_0^2 and W, and simplifying, we have

$$H(z) = 0.1367\frac{1 - z^{-2}}{1 - 1.2362z^{-1} + 0.7265z^{-2}}$$

The pole–zero diagrams of the normalized prototype lowpass filter (LPF), the analog
bandpass filter and the discrete-time bandpass filter are depicted in Figure 8.11. Note
that the lowpass-to-bandpass transformation has introduced a single zero at the
origin of the s-plane and at infinity. The BZT then maps the zeros to $z = \pm 1$. Thus,
the zeros of the discrete bandpass filter are at $z = 1$ and $z = -1$. Its poles are at
$z = 0.6040 \pm 0.6015$j. The analog bandpass zeros are at $s = 0$ and infinity (not shown)
and the poles at $s = -0.0923 \pm 0.3962$j.

In practice, higher order IIR filters (for example, $N > 3$) are normally realized
as cascades or parallel combinations of second- and/or first-order filter sections to
reduce the effects of finite wordlength on filter performance (see later). Thus, after
converting an analog filter into discrete form the resulting z-transfer function $H(z)$, if it
is of a high order, will need to be expressed in factored form (for cascade realization)
or as the sum of second- and/or first-order terms (for parallel realization). To simplify
this task we can expressed $H(s)$, at the outset, in factored form and then transform
each factor separately. The resulting factors can then be combined or rearranged into
a suitable format for the desired realization. This is basically the approach used in
Example 8.15.

Comments on the bilinear transformation method

Essentially, the BZT method involves two separate transformations. First, the normalized analog transfer function is frequency scaled by replacing s as follows:

$$s = \frac{s}{\omega_p'} \tag{8.22a}$$

where

$$\omega_p' = k \tan\left(\frac{\omega_p T}{2}\right), \quad k = 1 \text{ or } \frac{2}{T}$$

Second, the BZT is applied by replacing s in the new transfer function as

$$s = k\frac{z-1}{z+1} \tag{8.22b}$$

(1) It is common practice in many texts (for example, Rabiner and Gold, 1975) to use the factor $k = 2/T$ in the two operations above. It should be mentioned that both $k = 1$ and $k = 2/T$ lead to the same results because k is cancelled out anyway. To illustrate, consider the following simple filter:

$$H(s) = \frac{1}{s+1}$$

Assuming that the digital filter is to have a cutoff frequency of ω_p, then we must frequency scale $H(s)$ with the following frequency:

$$\omega_p' = k \tan\left(\frac{\omega_p T}{2}\right)$$

Thus the transfer function is

$$H'(s) = H(s)\big|_{s=s/\omega_p'} = \frac{1}{s/k \tan\left(\omega_p T/2\right) + 1}$$

Next, we replace s by $k(z-1)/(z+1)$:

$$H(z) = H'(s)\big|_{s=k(z-1)/(z+1)} = \frac{1}{[\cancel{k}(z-1)/(z+1)]/\cancel{k}\tan\left(\omega_p T/2\right) + 1}$$

From the above we see that the factor k is cancelled out, and it would not have mattered whether k was 1 or $2/T$.

(2)   For computational efficiency, the two transformations can be combined into one:

$$s = \cot\left(\frac{\omega_p T}{2}\right)\frac{z-1}{z+1} \tag{8.23}$$

Example 8.15 illustrates this approach.

(3)   For lowpass and highpass filters, the order of $H(z)$ is the same as the order of $H(s)$. For example, if $H(z)$ is derived from a second-order analog filter $H(s)$ then $H(z)$ will also be a second-order system. For bandpass and bandstop filters, the order of $H(z)$ is twice the order of $H(s)$. This relationship is sometimes exploited to cut down on the algebraic manipulations in the BZT method (see for example Stanley *et al.*, 1984).

(4)   In practice, we sometimes encounter a situation where it is required to convert the *s*-plane transfer function, $H(s)$, of an existing analog filter, into an equivalent bandpass discrete-time filter. Such is the case, for example, in digital audio where an analog filter which has been successfully used for equalization may need to be converted into an equivalent digital filter (Clark *et al.*, 2000).

In such cases, the analog transfer function of the actual filter is already available and so the BZT can be applied directly, after prewarping and frequency scaling using a direct lowpass-to-lowpass frequency scaling. The following examples illustrate the design issues involved.

Example 8.9

An illustration of the BZT method when the analog transfer function of the actual filter is available   Determine, using the BZT technique, the coefficients of a discrete-time bell filter for audio signal processing in a digital mixer when the control settings correspond to a *Q*-factor of 2, a boost (peak) of 6.02 dB at 5 kHz. Assume a sampling frequency of 48 kHz, and that the equivalent analog filter has an *s*-plane transfer function of the form

$$H(s) = \frac{s^2 + (3+k)\dfrac{\omega_0}{Q}s + \omega_0^2}{s^2 + (3-k)\dfrac{\omega_0}{Q}s + \omega_0^2}$$

where

$$k = 3\left(\frac{G-1}{G+1}\right)$$

ω_0 = boost frequency

G = gain factor

Q = Q-factor

Solution   The gain or boost of 6.02 dB corresponds to

$$G = 10^{\frac{6.02}{20}} = 1.9999; \quad k = 0.9999$$

The s-plane transfer function becomes

$$H(s) = \frac{s^2 + 4\dfrac{\omega_0}{Q}s + \omega_0^2}{s^2 + 2\dfrac{\omega_0}{Q}s + \omega_0^2}$$

The analog frequency response may be obtained by making the substitution $s = j\omega$ in this equation.

Now the prewarped boost frequency is

$$\omega_0' = \tan\left(\frac{\omega_0 T}{2}\right) = 0.339\,454$$

and the frequency scaled factor (Clark *et al.*, 2000) is given by

$$p = \frac{\omega_0}{\tan\left(\omega_0 T / 2\right)} = \frac{\omega_0}{0.339\,454}$$

Thus, the prewarped analog transfer function is

$$H'(s) = H(s)\big|_{s=ps} = \frac{p^2 s^2 + 4\dfrac{\omega_0}{Q}ps + \omega_0^2}{p^2 s^2 + 2\dfrac{\omega_0}{Q}ps + \omega_0^2}$$

Applying the BZT,

$$H(z) = H'(s)\big|_{s=\frac{z-1}{z+1}}$$

$$= \frac{\left(\dfrac{z-1}{z+1}\right)^2 + 0.678\,908\,5\left(\dfrac{z-1}{z+1}\right) + 0.115\,229}{\left(\dfrac{z-1}{z+1}\right)^2 + 0.339\,454\left(\dfrac{z-1}{z+1}\right) + 0.115\,229}$$

$$= \frac{1.233\,352 - 1.216\,444z^{-1} + 0.299\,94z^{-2}}{1 - 1.216\,444z^{-1} + 0.533\,294\,6z^{-2}}$$

Example 8.10   A simple LRC notch filter has the following normalized, s-plane transfer function:

$$H(s) = \frac{s^2 + 1}{s^2 + s + 1}$$

Determine the transfer function of an equivalent discrete-time filter using the BZT method. Assume a notch frequency of 50 Hz and a sampling frequency of 500 Hz.

Solution   Since we already have the s-plane transfer function, it is inappropriate to apply a lowpass to bandstop transformation as this would amount to a double transformation. The critical frequency is

$$\omega_p' = \tan\left(\frac{\omega_p T}{2}\right) = \tan\left(\frac{2\pi \times 50}{500 \times 2}\right) = 0.324\,919\,6$$

The frequency-scaled s-plane transfer function is

$$H'(s) = H(s)\Big|_{s \to \frac{s}{\omega_p}}$$

$$= \frac{\left(\dfrac{s}{\omega_p'}\right)^2 + 1}{\left(\dfrac{s}{\omega_p'}\right)^2 + \dfrac{s}{\omega_p'} + 1}$$

$$= \frac{s^2 + 0.105\,572}{s^2 + 0.324\,919\,6s + 0.105\,572}$$

Apply the BZT:

$$H(z) = H'(s)\Big|_{s \to \frac{z-1}{z+1}}$$

$$= \frac{\left(\dfrac{z-1}{z+1}\right)^2 + 0.105\,572}{\left(\dfrac{z-1}{z+1}\right)^2 + 0.324\,919\,6\left(\dfrac{z-1}{z+1}\right) + 0.105\,572}$$

8.9 Use of BZT and classical analog filters to design IIR filters

In many practical cases, the analog transfer function, $H(s)$, from which $H(z)$ is obtained, may not be available and will have to be determined from the specifications of the desired digital filters. For standard frequency selective digital filtering tasks (i.e. involving lowpass, highpass, bandpass and bandstop filters), $H(s)$ can be derived from the classical filters with Butterworth, Chebyshev or elliptic characteristics (see Figure 8.12). In this section, we will present methods of designing IIR filters from such classical analog filters. We will consider the basic concepts in detail and illustrate the design issues with worked examples. First, we will review briefly the relevant features of three classical analog filters which are needed to design IIR filters. Only lowpass prototype filters are considered, since other filter types are normally derived from normalized lowpass filters as will be shown later.

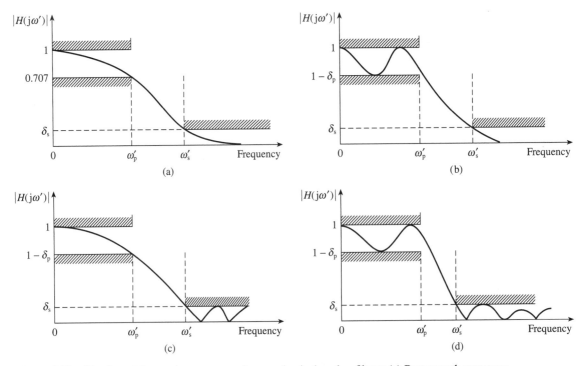

Figure 8.12 Sketches of frequency responses of some classical analog filters: (a) Butterworth response; (b) Chebyshev type I; (c) Chebyshev type II; (d) elliptic.

8.9.1 Characteristic features of classical analog filters

8.9.1.1 Butterworth filter

The lowpass Butterworth filter is characterized by the following magnitude-squared–frequency response:

$$|H(\omega)|^2 = \frac{1}{1 + \left(\dfrac{\omega}{\omega_p^p}\right)^{2N}} \tag{8.24}$$

where N is the filter order and ω_p^p is the 3 dB cutoff frequency of the lowpass filter (for the normalized prototype filter, $\omega_p^p = 1$). The magnitude–frequency response of a typical Butterworth lowpass filter is depicted in Figure 8.12(a), and is seen to be monotonic in both the passband and stopband. The response is said to be maximally flat because of its initial flatness (with a slope of zero at dc).

The filter order, N, is given by

$$N \geqslant \frac{\log\left(\dfrac{10^{\frac{A_s}{10}} - 1}{10^{\frac{A_p}{10}} - 1}\right)}{2 \log\left(\dfrac{\omega_s^p}{\omega_p^p}\right)} \tag{8.25}$$

where A_p and A_s are, respectively, the passband ripple and stopband attenuation in dB, and ω_s^p is the stopband edge frequency above.

The transfer function of the normalized analog Butterworth filter, $H(s)$, contains zeros at infinity and poles which are uniformly spaced on a circle of radius 1 in the s-plane at the following locations (Stearns and Hush, 1990; Jong, 1982):

$$s_k = e^{j\pi(2k+N-1)/2N} = \cos\left[\frac{(2k + N - 1)\pi}{2N}\right] + j\sin\left[\frac{(2k + N - 1)\pi}{2N}\right],$$
$$k = 1, 2, \ldots, N \tag{8.26}$$

The poles occur in complex conjugate pairs and lie on the left-hand side of the s-plane.

8.9.1.2 Chebyshev filter

The Chebyshev characteristic provides an alternative way of obtaining a suitable analog transfer function, $H(s)$. There are two types of Chebyshev filters, types I and II, with the following features (Figures 8.12(b) and 8.12(c)):

- Type I, with equal ripple in the passband, monotonic in the stopband;
- Type II, with equal ripple in the stopband, monotonic in the passband.

Type I Chebyshev filters, for example, are characterized by the magnitude-squared response

$$|H(\omega')|^2 = \frac{K}{1 + \varepsilon^2 C_N^2(\omega'/\omega_p')} \tag{8.27a}$$

where $C_N(\omega'/\omega_p')$ is a Chebyshev polynomial which exhibits equal ripple in the passband, N is the order of the polynomial as well as that of the filter, and ε determines the passband ripple, which in decibels is given by

$$\text{passband ripple} \leqslant 10 \log_{10}(1 + \varepsilon^2) = -20 \log_{10}(1 - \delta_p) \tag{8.27b}$$

A typical amplitude response of a type I Chebyshev characteristic is shown in Figure 8.12(b). The transfer function, $H(s)$, for the Chebyshev response depends on the desired passband ripple and the filter order, N. The filter order N is given by

$$N \geqslant \frac{\cosh^{-1}\left(\dfrac{10^{\frac{A_s}{10}} - 1}{10^{\frac{A_p}{10}} - 1}\right)}{\cosh^{-1}\left(\dfrac{\omega_s^p}{\omega_p^p}\right)} \tag{8.28}$$

where A_p and A_s are, respectively, the passband ripple and stopband attenuation in dB, and ω_s^p is the stopband edge frequency.

The poles of the normalized Chebyshev LPF lie on an ellipse in the s-plane and have coordinates given by (Stearns and Hush, 1990)

$$s_k = \sinh(\alpha)\cos(\beta_k) + j\cosh(\alpha)\sin(\beta_k) \tag{8.29}$$

where

$$\alpha = \frac{1}{N}\sinh^{-1}\left(\frac{1}{\varepsilon}\right); \ \beta_k = \frac{(2k + N - 1)\pi}{2N}, \quad k = 1, 2, \ldots, N$$

8.9.1.3   Elliptic filter

The elliptic filter exhibits equiripple behaviour in both the passband and the stopband; see Figure 8.12(d). It is characterized by the following magnitude-squared response:

$$|H(\omega')|^2 = \frac{K}{1 + \varepsilon^2 G_N^2(\omega')} \tag{8.30}$$

where $G_N(\omega')$ is a Chebyshev rational function. Unlike the Butterworth and Chebyshev filters, there is no simple expression for the poles of the elliptic filter. A procedure is available for computing locations of the poles (for example, in Antoniou, 1979; Jong, 1982; DeFatta *et al.*, 1988). The zeros of the elliptic lowpass filter are entirely imaginary.

The elliptic characteristic provides the most efficient filters in terms of amplitude response. It yields the smallest filter order for a given set of specifications and should be the method of first choice in IIR filter design, except where the phase response is of concern when the Butterworth response may be preferred.

Tables of the polynomials of $H(s)$ for the Butterworth, Chebyshev and elliptic characteristics are available in most analog design books in normalized form and can be used in the bilinear transformation. In practice, however, the computation of $H(z)$ from $H(s)$ is via a software package as we shall see later.

8.9.2 The BZT methodology using classical analog filters

In cases where the prototype lowpass filter does not exist, the stages of the BZT method are:

(1) Prewarp the bandedge or critical frequencies of the digital filter as described previously.

(2) Find a suitable lowpass prototype analog filter, based on the digital filter specifications and one of the classical filter characteristics. This involves using one of the frequency transformation equations (depending on the type of digital filter – LP, HP, BP or BS) in reverse to determine the specifications of the prototype LP filter. From this the order of the prototype filter and hence its transfer function, $H(s)$, are found.

(3) Denormalize the prototype analog LP filter, $H(s)$, by frequency transformation and scaling to obtain a new transfer function, $H'(s)$, as described previously.

(4) Apply the BZT to obtain the desired digital filter transfer function, $H(z)$, by replacing s in the frequency scaled transfer function, $H'(s)$, as described previously.

We will now look at the basic concepts for each of the filter types (LP, HP, BP and BS).

8.9.2.1 Lowpass filter – basic concepts

The lowpass-to-lowpass transformation is given by (see Equation 8.21a):

$$s = \frac{s}{\omega'_p}$$

If we replace s by $j\omega$ in the equation and denote frequencies for the prototype filter by ω^p and those for the lowpass filter to be designed by ω_{lp}, to distinguish between them, then the above equation becomes

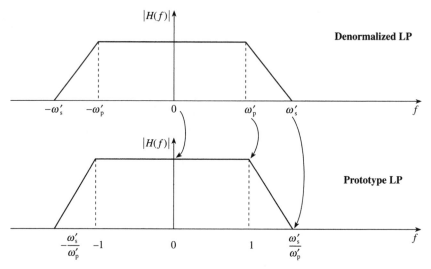

Figure 8.13   Relationships between frequencies in the denormalized LP and prototype LP filters.

$$j\omega^p = j\frac{\omega_{lp}}{\omega'_p}, \text{ i.e. } \omega^p = \frac{\omega_{lp}}{\omega'_p} \qquad (8.31)$$

Equation 8.31 defines the relationship between the frequencies in the prototype filter response and those of the denormalized lowpass filter that we wish to design. Given the critical frequencies for a denormalized lowpass filter, we use Equation 8.31 to determine the critical frequencies for the prototype filter and hence its specifications.

The three key critical frequencies for the prototype filter are: 0, passband edge frequency; ω^p_p (this is, in fact, always 1); and the stopband edge frequency, ω^p_s:

(1)   when $\omega_{lp} = 0$, $\omega^p = 0$ (from Equation 8.31)

(2)   when $\omega_{lp} = \omega'_p$ (i.e. the passband edge frequency), $\omega^p = \omega'_p/\omega'_p = 1 = \omega^p_p$

(3)   when $\omega_{lp} = \omega'_s$, $\omega^p = \omega'_s/\omega'_p = \omega^p_s$.

Thus, the critical frequencies for the prototype filter are: 0, 1, ω'_s/ω'_p.

The relationships between the frequencies of the denormalized lowpass filter and those of the prototype filter are shown in Figure 8.13.

8.9.2.2   Highpass filter – basic concepts

From the lowpass-to-highpass transformation, $s = \omega'_p/s$, and denoting the frequencies of the denormalized highpass filter by ω_{hp} and those of the prototype LP filter by ω^p (as before), the following relationship between the frequencies of the prototype LP filter and the desired highpass filter is obtained:

$$\omega^{p} = -\frac{\omega'_p}{\omega_{hp}} \tag{8.32}$$

Using Equation 8.32 we can now specify the critical frequencies of the prototype LP filter in terms of those of the desired highpass filter:

(1)   when $\omega_{hp} = 0$, $\omega^{p} = \infty$ (using Equation 8.32)

(2)   when $\omega_{hp} = \omega'_p$ (i.e. the passband edge frequency), $\omega^{p} = -1$

(3)   when $\omega_{hp} = \omega'_s$, $\omega^{p} = -\dfrac{\omega'_p}{\omega'_s}$

(4)   when $\omega_{hp} = -\omega'_p$, $\omega^{p} = 1$

(5)   when $\omega_{hp} = -\omega'_s$, $\omega^{p} = \dfrac{\omega'_p}{\omega'_s}$.

Thus, the three key critical frequencies for the prototype lowpass filter for designing the highpass filter are 0, 1 and ω'_p/ω'_s.

The critical frequencies for the prototype LP filter and their relationships with the frequencies of the denormalized highpass filter are depicted in Figure 8.14. We note that the lowpass-to-highpass transformation maps frequencies in the denormalized highpass filter as follows – it maps zero frequency to infinity, ω_p to unity and infinity to zero.

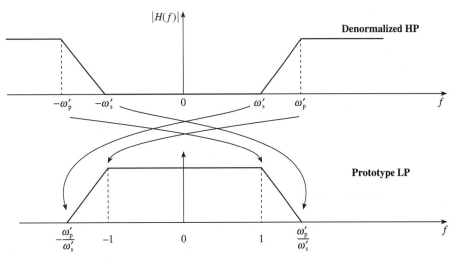

Figure 8.14   Relationships between frequencies in the denormalized HP and prototype LP filters.

8.9.2.3 Bandpass filters – basic concepts

The lowpass-to-bandpass transformation is given by

$$s = \frac{s^2 + \omega_0^2}{Ws}$$

From the lowpass-to-bandpass transformation, the frequencies of the bandpass filter, ω_{bp}, and those of the prototype LPF, ω^p, are related as

$$j\omega^p = \frac{(j\omega_{bp})^2 + \omega_0^2}{jW\omega_{bp}}$$

i.e.

$$\omega^p = \frac{\omega_{bp}^2 - \omega_0^2}{W\omega_{bp}} \tag{8.33}$$

Now, a bandpass filter has four critical or bandedge frequencies and a centre frequency:

ω_{p1}', ω_{p2}' = lower and upper passband edge frequencies

ω_{s1}', ω_{s2}' = lower and upper stopband edge frequencies

ω_0 = centre frequency ($\omega_0^2 = \omega_{p1}'\omega_{p2}'$)

Using the relationship given in Equation 8.33, the bandedge frequencies for the prototype LP filter can be found in terms of the bandedge frequencies for the bandpass filter:

(1)   when $\omega_{bp} = \omega_{s1}'$, $\omega^p = \omega_{s1}'^p = \dfrac{\omega_{s1}'^2 - \omega_0^2}{W\omega_{s1}'}$

(2)   when $\omega_{bp} = \omega_{p1}'$, $\omega^p = \dfrac{\omega_{p1}'^2 - \omega_0^2}{W\omega_{p1}'} = \dfrac{\omega_{p1}'^2 - \omega_{p1}'\omega_{p2}'}{(\omega_{p2}' - \omega_{p1}')\omega_{p1}'} = -1$

(3)   when $\omega_{bp} = \omega_{p2}'$, $\omega^p = \dfrac{\omega_{p2}'^2 - \omega_0^2}{W\omega_{p2}'} = \dfrac{\omega_{p2}'^2 - \omega_{p1}'\omega_{p2}'}{(\omega_{p2}' - \omega_{p1}')\omega_{p2}'} = 1$

(4)   when $\omega_{bp} = \omega_{s2}'$, $\omega^p = \omega_{s2}'^p = \dfrac{\omega_{s2}'^2 - \omega_0^2}{W\omega_{s2}'}$

(5) $\omega_{bp} = \omega_0$, $\omega^p = \dfrac{\omega_0'^2 - \omega_0^2}{W\omega_0^2} = 0$

(6) $\omega_s^p = \min(\omega_{s1}'^p, \omega_{s2}'^p)$.

Thus, the critical frequencies of interest for the prototype LP filter are

$$0, 1, \min(\omega_{s1}^p, |\omega_{s2}^p|)$$

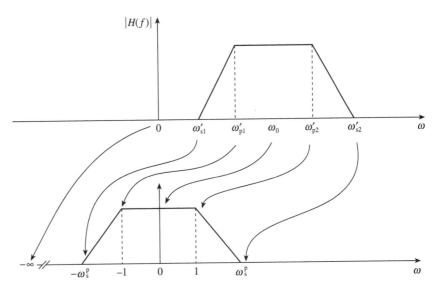

Figure 8.15 Mapping of the prototype LP to BPF.

The mappings of the frequencies between the bandpass filter and the prototype LP filter are depicted in Figure 8.15. We note, for example, that the centre frequency of the bandpass filter is mapped to zero in the prototype filter, and the upper passband and stopband edge frequencies, ω'_{p2} and ω'_{s2}, respectively, are mapped to the positive passband and stopband edge frequencies in the prototype filter. On the other hand, the lower passband and stopband edge frequencies, ω'_{p1} and ω'_{s1}, respectively, are mapped to the negative passband and stopband edge frequencies in the prototype filter. In practice, we set the stopband edge frequency for the prototype filter equal to the smaller of the two stopband frequencies ω^p_{s1} and ω^p_{s2} as indicated above. The passband ripples and stopband attenuation of the prototype LP filter are set equal to those of the digital bandpass filter.

From the specifications for the prototype LP filter, we can determine the order, N, and transfer function of the filter using Equation 8.25 or 8.28, for eaxmple. The order of the bandpass filter is twice that of the prototype LP, i.e. $2N$. The poles of the prototype filter are determined from Equations 8.26 and 8.29. For the Butterworth and Chebyshev (type I) filters, the zeros for the prototype LP are at infinity, but for the elliptic they are entirely imaginary. From the poles and zero locations (or standard tables for the classical filters), the transfer function of the prototype filter can be obtained.

8.9.2.4 Bandstop filters – basic concepts

The lowpass-to-bandstop transformation is given by

$$s = \frac{Ws}{s^2 + \omega_0^2}$$

The bandstop frequency, ω_{bs}, and those of the prototype LPF, ω^p, are related as

$$j\omega^p = \frac{jW\omega_{bs}}{(j\omega_{bs})^2 + \omega_0^2}$$

i.e.

$$\omega^p = \frac{W\omega_{bs}}{\omega_0^2 - \omega_{bs}^2} \tag{8.34}$$

From the relationship in Equation 8.34, we can determine the bandedge frequencies for the prototype LP filter from those of the desired digital bandstop filter. We recall that a bandstop filter has four bandedge frequencies – ω'_{p1}, ω'_{p2} (lower and upper passband edge frequencies), ω'_{s1}, ω'_{s2} (lower and upper stopband edge frequencies), and a centre frequency, ω_0 ($\omega_0^2 = \omega'_{p1}\omega'_{p2}$):

(1)   when $\omega_{bs} = \omega'_{p1}$, $\omega^p = \dfrac{W\omega'_{p1}}{\omega_0'^2 - \omega_{p1}'^2} = \dfrac{(\omega'_{p2} - \omega'_{p1})\omega'_{p1}}{\omega'_{p1}\omega'_{p2} - \omega_{p1}'^2} = 1$

(2)   when $\omega_{bs} = \omega'_{s1}$, $\omega^p = \omega_s^{p(1)} = \dfrac{W\omega'_{s1}}{\omega_0^2 - \omega_{s1}^2}$

(3)   when $\omega_{bs} = \omega'_{s2}$, $\omega^p = \omega_s^{p(2)} = \dfrac{W\omega'_{s2}}{\omega_0^2 - \omega_{s2}^2}$

(4)   when $\omega_{bs} = \omega_0$, $\omega^p = \dfrac{W\omega_0}{\omega_0^2 - \omega_0^2} = \infty$

(5)   when $\omega_{bs} = \omega'_{p2}$, $\omega^p = \dfrac{W\omega'_{p2}}{\omega_0^2 - \omega_{p2}'^2} = \dfrac{(\omega'_{p2} - \omega'_{p1})\omega'_{p2}}{\omega'_{p1}\omega'_{p2} - \omega_{p2}'^2} = -1.$

Thus, the stopband edge frequency for the prototype LP filter, $\omega_s^p = \min(\omega_s^{p(1)}, \omega_s^{p(2)})$ and its passband edge frequency is 1. The passband ripple and stopband attenuation are, respectively, A_p and A_s. The mappings between the frequencies of the bandstop filter and those of the prototype lowpass filter are shown in Figure 8.16. We see that the upper bandstop and bandpass edge frequencies in the bandstop filter are mapped to the negative frequencies in the prototype filter whereas the lower passband and stopband edge frequencies are mapped to the positive frequencies in the prototype filter.

The critical frequencies of interest for the prototype LP filter are

0, 1, ω_s^p (where $\omega_s^p = \min(\omega_s^{p(1)}, \omega_s^{p(2)})$).

From the specifications for the prototype LP filter, we can determine the order, N, and transfer function of the filter using Equation 8.25 or 8.28, for example. The order of the bandpass filter is twice that of the prototype LP, i.e. $2N$.

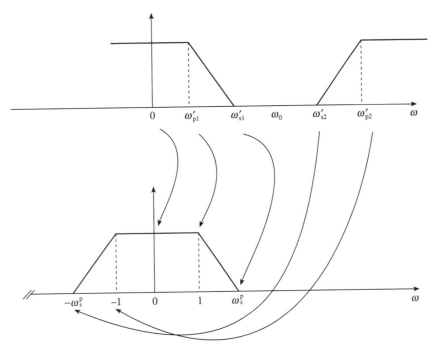

Figure 8.16   Relationships between the frequencies in the denormalized BS and prototype LP filters.

8.9.3   Illustrative design examples (lowpass, highpass, bandpass and bandstop filters)

Example 8.11   *Lowpass filter*   A lowpass digital filter meeting the following specifications is required:

passband	0–500 Hz
stopband	2–4 kHz
passband ripple	3 dB
stopband attenuation	20 dB
sampling frequency	8 kHz

Determine the following:

(1)   pass- and stopband edge frequencies for a suitable analog prototype lowpass filter;

(2)   order, N, of the prototype lowpass filter;

(3)   coefficients, and hence the transfer function, of the discrete-time filter using the bilinear z-transform.

Assume a Butterworth characteristic for the filter.

Solution   (1)   From the specifications, the prewarped frequencies are

$$\omega_p' = \tan\left(\frac{2\pi \times 500}{2 \times 8000}\right) = 0.198\,912$$

$$\omega_s' = \tan\left(\frac{2\pi \times 2000}{2 \times 8000}\right) = 1$$

$$\omega_s^p = \frac{\omega_s'}{\omega_p'} = 1/0.198\,912 = 5.0273$$

Thus, the prewarped pass- and stopband edge frequencies for the prototype LP filter are: 0, 1, 5.0273.

(2)   Next, we use Equation 8.25 and the values of the parameters above to determine the order of the filter.

Now

$$10^{A_s/10} - 1 = 10^{20/10} - 1 = 99; \quad 10^{A_p/10} - 1 = 10^{3/10} - 1 = 0.995\,262\,3;$$

$$\log\left(\frac{99}{0.995\,262\,3}\right) = 1.997\,697$$

For the prototype LPF

$$\omega_p^p = 1; \quad \omega_s^p = 5.0273; \quad \log\left(\frac{\omega_s^p}{\omega_p^p}\right) = 2\log(5.0273) = 1.402\,66.$$

$$N \geqslant \frac{1.997\,697}{1.402\,66} = 1.424. \text{ Let } N = 2$$

(3)   The poles of the prototype filter are (from Equation 8.26)

$$s_{p,1} = \cos\left[\frac{(2+2-1)\pi}{4}\right] + j\sin\left[\frac{(2+2-1)\pi}{4}\right] = -\frac{\sqrt{2}}{2} + j\frac{\sqrt{2}}{2}$$

$$s_{p,2} = -\frac{\sqrt{2}}{2} - j\frac{\sqrt{2}}{2}$$

The s-plane transfer function, $H(s)$, is

$$H(s) = \frac{1}{(s - s_{p,1})(s - s_{p,2})} = \frac{1}{s^2 + \sqrt{2}s + 1}$$

The frequency scaled s-plane transfer function is

$$H'(s) = H(s)\Big|_{\frac{s}{\omega'_p}} = \frac{1}{\left(\dfrac{s}{\omega'_p}\right)^2 + \sqrt{2}\,\dfrac{s}{\omega'_p} + 1}$$

$$= \frac{\omega'^2_p}{s^2 + \sqrt{2}s\omega'_p + \omega'^2_p}$$

Applying the BZT:

$$H(z) = H'(s)\Big|_{s=\frac{z-1}{z+1}} = \frac{\omega'^2_p}{\left(\dfrac{z-1}{z+1}\right)^2 + \sqrt{2}\omega'_p\left(\dfrac{z-1}{z+1}\right) + \omega'^2_p}$$

$$= \frac{\omega'^2_p(z+1)^2}{(z-1)^2 + \sqrt{2}\omega'_p(z-1)(z+1) + \omega'^2_p(z+1)^2}$$

After simplification and dividing top and bottom by z^2, we have

$$H(z) = \frac{\omega'^2_p}{1 + \sqrt{2}\omega'_p + \omega'^2_p} \times \frac{1 + 2z^{-1} + z^{-2}}{1 + \dfrac{2(\omega'^2_p - 1)z^{-1}}{1 + \sqrt{2}\omega'_p + \omega'^2_p} + \dfrac{(1 - \sqrt{2}\omega'_p + \omega'^2_p)z^{-2}}{1 + \sqrt{2}\omega'_p + \omega'^2_p}}$$

Using the values of the parameters

$$1 + \sqrt{2}\omega'_p + \omega'^2_p = 1.320\,87; \quad \omega'^2_p - 1 = -0.960\,43$$
$$1 - \sqrt{2}\omega'_p + \omega'^2_p = 0.758\,285\,8; \quad \omega'^2_p = 0.039\,565\,9$$

and substituting in the equation above and simplifying, we have

$$H(z) = \frac{0.029\,95(1 + 2z^{-1} + z^{-2})}{1 - 1.4542z^{-1} + 0.574\,08z^{-2}}$$

Example 8.12

Highpass filter   A highpass digital filter meeting the following specifications is required:

passband	2–4 kHz
stopband	0–500 Hz
passband ripple	3 dB
stopband attenuation	20 dB
sampling frequency	8 kHz

Determine the following:

(1)   pass- and stopband edge frequencies for a suitable analog prototype lowpass filter;

(2)   order, N, of the prototype lowpass filter;

(3)   coefficients, and hence the transfer function, of the discrete-time filter using the bilinear z-transform.

Assume a Butterworth characteristic for the filter.

Solution   (1)   From the specifications, the prewarped frequencies are

$$\omega'_s = \tan\left(\frac{2\pi \times 500}{2 \times 8000}\right) = 0.198\,912$$

$$\omega'_p = \tan\left(\frac{2\pi \times 2000}{2 \times 8000}\right) = 1$$

$$\omega^p_s = \frac{\omega'_p}{\omega'_s} = 1/0.198\,912 = 5.0273$$

Thus, the pass- and stopband edge frequencies for the prototype LP filter are: 0, 1, 5.0273.

(2)   Next, we use Equation 8.25 and the values of the parameters above to determine the order of the filter.
Now

$$10^{A_s/10} - 1 = 10^{20/10} - 1 = 99; \quad 10^{A_p/10} - 1 = 10^{3/10} - 1 = 0.995\,262\,3$$

$$\log\left(\frac{99}{0.995\,262\,3}\right) = 1.997\,697$$

For the prototype LPF

$$\omega^p_p = 1; \quad \omega^p_s = 5.0273; \quad \log\left(\frac{\omega^p_s}{\omega^p_p}\right) = 2\log(5.0273) = 1.402\,66$$

$$N \geqslant \frac{1.997\,697}{1.402\,66} = 1.424$$

Let $N = 2$.
The poles of the prototype filter are (from Equation 8.26)

$$s_{p,1} = \cos\left[\frac{(2+2-1)\pi}{4}\right] + j\sin\left[\frac{(2+2-1)\pi}{4}\right] = -\frac{\sqrt{2}}{2} + j\frac{\sqrt{2}}{2}$$

$$s_{p,2} = -\frac{\sqrt{2}}{2} - j\frac{\sqrt{2}}{2}$$

The s-plane transfer function, $H(s)$, is

$$H(s) = \frac{1}{(s - s_{p,1})(s - s_{p,2})} = \frac{1}{s^2 + \sqrt{2}s + 1}$$

The frequency scaled s-plane transfer function is

$$H'(s) = H(s)\Big|_{\frac{\omega'_p}{s}} = \frac{1}{\left(\dfrac{\omega'_p}{s}\right)^2 + \sqrt{2}\dfrac{\omega'_p}{s} + 1}$$

$$= \frac{s^2}{s^2 + \sqrt{2}s\omega'_p + \omega'^2_p}$$

Applying the BZT:

$$H(z) = H'(s)\Big|_{s=\frac{z-1}{z+1}} = \frac{\left(\dfrac{z-1}{z+1}\right)^2}{\left(\dfrac{z-1}{z+1}\right)^2 + \sqrt{2}\omega'_p\left(\dfrac{z-1}{z+1}\right) + \omega'^2_p}$$

$$= \frac{(z-1)^2}{(z-1)^2 + \sqrt{2}\omega'_p(z-1)(z+1) + \omega'^2_p(z+1)^2}$$

After simplification and dividing top and bottom by z^2, we have

$$H(z) = \frac{1}{1 + \sqrt{2}\omega'_p + \omega'^2_p} \times \frac{1 - 2z^{-1} + z^{-2}}{1 + \dfrac{2(\omega'^2_p - 1)z^{-1}}{1 + \sqrt{2}\omega'_p + \omega'^2_p} + \dfrac{(1 - \sqrt{2}\omega'_p + \omega'^2_p)z^{-2}}{1 + \sqrt{2}\omega'_p + \omega'^2_p}}$$

Using the values of the parameters

$$1 + \sqrt{2}\omega'_p + \omega'^2_p = 3.414\,21;\ \omega'^2_p - 1 = 0$$

$$1 - \sqrt{2}\omega'_p + \omega'^2_p = 0.585\,786;\ \omega'^2_p = 1$$

and substituting in the equation above and simplifying, we have

$$H(z) = \frac{0.292\,89(1 - 2z^{-1} + z^{-2})}{1 + 0.171\,57z^{-2}}$$

Example 8.13   *Bandpass filter*   A requirement exists for a bandpass digital filter, with a Butterworth magnitude–frequency response, that satisfies the following specifications:

lower passband edge frequency	200 Hz
upper passband edge frequency	300 Hz
lower stopband edge frequency	50 Hz
upper stopband edge frequency	450 Hz
passband ripple	3 dB
stopband attenuation	20 dB
sampling frequency	1 kHz

Determine the following:

(1) pass- and stopband edge frequencies of a suitable prototype lowpass filter;
(2) order, N, of the prototype lowpass filter;
(3) coefficients, and hence the transfer function, of the discrete-time filter using the BZT method.

Solution   The prewarped critical frequencies for the bandpass filter are

$$\omega_0 = 1; \qquad W = 0.6498$$

$$\omega'_{p1} = 0.7265; \qquad \omega'_{p2} = 1.376\,38$$

$$\omega'_{s1} = 0.1584; \qquad \omega'_{s2} = 6.3138$$

Thus, the bandedge frequencies for the prototype LP filter are (using the relationships above)

$$\omega_p^p = 1; \qquad \omega_s^p = 9.4721$$

Thus, we require a prototype LPF with $\omega_p^p = 1$; $\omega_s^p = 9.4721$, $A_p = 3$ dB; $A_s = 20$ dB. From Equation 8.25, the order of the prototype LPF is obtained as

$$10^{A_s/10} - 1 = 10^{20/10} - 1 = 99; \quad 10^{A_p/10} - 1 = 10^{3/10} - 1 = 0.9952623;$$

$$\log\left(\frac{99}{0.995\,262\,3}\right) = 1.997\,697\,6; \quad \frac{\omega_s^p}{\omega_p^p} = 9.4721; \quad 2\log(9.4721) = 1.952\,89$$

$$N = \frac{1.997\,697\,6}{1.952\,89} = 1.0229$$

N must be an integer, and for simplicity we will use $N = 1$. The s-plane transfer function for a first-order prototype LP filter is given by

$$H(s) = \frac{1}{s+1}$$

Using the lowpass-to-bandpass transformation from the table, we obtain

$$H'(s) = H(s)\Big|_{s = \frac{s^2 + \omega_0^2}{Ws}} = \frac{1}{\left(\dfrac{s^2 + \omega_0^2}{Ws}\right) + 1}$$

$$= \frac{Ws}{s^2 + Ws + \omega_0^2}$$

Applying the BZT gives

$$H(z) = H'(s)\Big|_{s = \frac{z-1}{z+1}}$$

$$= \frac{W\left(\dfrac{z-1}{z+1}\right)}{\left(\dfrac{z-1}{z+1}\right)^2 + W\left(\dfrac{z-1}{z+1}\right) + \omega_0^2}$$

Substituting the values for ω_0^2 and W, and simplifying, we have

$$H(z) = \frac{0.2452(1 - z^{-2})}{1 + 0.5095\, z^{-2}}$$

Example 8.14

Bandstop filter   A requirement exists for a bandstop digital IIR filter, with a Butterworth magnitude–frequency response, that meets the following specifications:

lower passband	0–50 Hz
upper passband	450–500 Hz
stopband	200–300 Hz
passband ripple	3 dB
stopband attenuation	20 dB
sampling frequency	1 kHz

Determine the following:

(1)   pass- and stopband edge frequencies of a suitable prototype lowpass filter;

(2)   order, N, of the prototype lowpass filter;

(3)   coefficients, and hence the transfer function, of the discrete-time filter using the BZT method.

Solution   The prewarped critical frequencies for the bandpass filter are

$$\omega_0 = 1; \qquad W = 6.1554$$

$$\omega'_{p1} = 0.1584; \qquad \omega'_{p2} = 6.3138$$

$$\omega'_{s1} = 0.7265; \qquad \omega'_{s2} = 1.376\,38$$

Thus, the bandedge frequencies for the prototype filter are (using the relationships above)

$$\omega_p^p = 1; \qquad \omega_s^p = 9.4721$$

We require a prototype LPF with $\omega_p^p = 1$, $\omega_s^p = 9.4721$, $A_p = 3$ dB, $A_s = 20$ dB. From Equation 8.25, the order of the prototype LPF is obtained as

$$10^{A_s/10} - 1 = 10^{20/10} - 1 = 99; \quad 10^{A_p/10} - 1 = 10^{3/10} - 1 = 0.995\,262\,3$$

$$\log\left(\frac{99}{0.995\,262\,3}\right) = 1.997\,697\,6; \quad \frac{\omega_s^p}{\omega_p^p} = 9.4721; \quad 2\log(9.4721) = 1.952\,89$$

$$N = \frac{1.997\,697\,6}{1.952\,89} = 1.0229$$

N must be an integer, and for simplicity we will use $N = 1$. The s-plane transfer function for a first-order prototype LP filter is given by

$$H(s) = \frac{1}{s + 1}$$

Using the lowpass-to-bandstop transformation from the table, we obtain

$$H'(s) = H(s)\Big|_{s = \frac{Ws}{s^2 + \omega_0^2}} = \frac{1}{\left(\dfrac{Ws}{s^2 + \omega_0^2}\right) + 1}$$

$$= \frac{s^2 + \omega_0^2}{s^2 + Ws + \omega_0^2}$$

Applying the BZT gives

$$H(z) = H'(s)\Big|_{s = \frac{z-1}{z+1}}$$

$$= \frac{\left(\dfrac{z-1}{z+1}\right)^2 + \omega_0^2}{\left(\dfrac{z-1}{z+1}\right)^2 + W\left(\dfrac{z-1}{z+1}\right) + \omega_0^2}$$

Substituting the values for ω_0^2 and W, and simplifying, we have

$$H(z) = \frac{0.2452(1 + z^{-2})}{1 - 0.5095\, z^{-2}}$$

Example 8.15   Obtain the transfer function of a lowpass digital filter meeting the following specifications:

passband	0–60 Hz
stopband	>85 Hz
stopband attenuation	>15 dB

Assume a sampling frequency of 256 Hz and a Butterworth characteristic.

Solution   This example illustrates how to combine steps 4 and 5 in the BZT process into one, as suggested by Equation 8.23.

(1) The critical frequencies for the digital filter are

$$\omega_1 T = \frac{2\pi f_1}{F_s} = \frac{2\pi 60}{256} = 2\pi \times 0.2344$$

$$\omega_2 T = \frac{2\pi f_2}{F_s} = \frac{2\pi 85}{256} = 2\pi \times 0.3320$$

(2) The prewarped equivalent analog frequencies are

$$\omega_1' = \tan\left(\frac{\omega_1 T}{2}\right) = 0.906\,347; \quad \omega_2' = \tan\left(\frac{\omega_2 T}{2}\right) = 1.715\,80$$

(3) Next we need to obtain $H(s)$ with Butterworth characteristics, a 3 dB cutoff frequency of 0.906 347, and a response at 85 Hz that is down by 15 dB. For an attenuation of 15 dB and a passband ripple of 3 dB, from Equation 8.25 $N = 2.68$. We use $N = 3$, since it must be an integer. A normalized third-order filter is given by

$$H(s) = \frac{1}{(s + 1)(s^2 + s + 1)} = \frac{1}{s + 1}\,\frac{1}{s^2 + s + 1}$$

$$= H_1(s)\,H_2(s)$$

$$\cot\left(\frac{\omega_1 T}{2}\right) = \cot\left(\frac{2\pi \times 0.2344}{2}\right) = 1.103\,155$$

Performing the transform in two stages, one for each of the factors of $H(s)$ above, we obtain

$$H_2(z) = H_2(s)\big|_{s=\cot(\omega_1 T/2)[(z-1)/(z+1)]}$$

$$= 0.3012 \, \frac{1 + 2z^{-1} + z^{-2}}{1 - 0.1307z^{-1} + 0.3355z^{-2}}$$

which we have arrived at after considerable manipulation. Similarly, we obtain $H_1(z)$ as

$$H_1(z) = 0.4754 \, \frac{1 + z^{-1}}{1 - 0.0490z^{-1}}$$

$H_1(z)$ and $H_2(z)$ may then be combined to give the desired transfer function, $H(z)$:

$$H(z) = H_1(z)H_2(z) = 0.1432 \, \frac{1 + 3z^{-1} + 3z^{-2} + z^{-3}}{1 - 0.1801z^{-1} + 0.3419z^{-2} - 0.0165z^{-3}}$$

8.10 Calculating IIR filter coefficients by mapping s-plane poles and zeros

8.10.1 Basic concepts

An alternative, and perhaps more powerful and flexible, method of computing the coefficients of $H(z)$ for practical IIR filters is to map the individual poles and zeros of a suitable analog filter from the s-plane into the z-plane and then to derive the digital filter coefficients from the z-plane poles and zeros. This approach is exploited by a number of commercial software implementations and is attractive when the filter is of a high order.

The procedure for calculating IIR coefficients by mapping s-plane poles and zeros into the z-plane is summarized below.

8.10.1.1 Step 1

As before, the designer starts with a normalized, Nth-order analog lowpass filter of Butterworth, Chebyshev or elliptic type, depending on the design requirements. The poles of the normalized LPF are then obtained using Equation 8.26 for a Butterworth or Equation 8.29 for a Chebyshev filter. For an elliptic filter each pole is complex and, in general, has the form

$$s_{1,k} = \alpha_{p,k} + j\beta_{p,k} \tag{8.35}$$

For Butterworth and Chebyshev (type I) filters, the zeros for the prototype LPF are at infinity, but for elliptic filters they are entirely imaginary. In general, the locations of the zeros of the normalized LPF are easier to determine than those of the poles.

8.10.1.2   Step 2

Next, the poles and zeros of the normalized analog LPF are converted into those of an LP, HP, BP or BSF using an appropriate transformation in Equations 8.21a–8.21d.

Lowpass and highpass filters

For a lowpass or highpass digital filter, the N poles of the normalized LP are transformed as follows (see Equations 8.21a and 8.21b):

$$s_{1,k} = s_{1,k}/\omega_p' \quad k = 1, 2, \ldots, N \text{ lowpass-to-lowpass} \tag{8.36a}$$

$$s_{h,k} = \omega_p'/s_{1,k} \quad k = 1, 2, \ldots, N \text{ lowpass-to-highpass} \tag{8.36b}$$

where ω_p' is the desired passband edge frequency, $s_{1,k}$ are the poles of the analog lowpass filter and $s_{h,k}$ are the poles of the analog highpass filter.

The similarity between Equations 8.36a and 8.36b is evident and is due to the duality between the lowpass and highpass characteristics. For N even, there will be $N/2$ complex pole pairs. For N odd there will be $(N-1)/2$ complex pole pairs and a single real pole.

For the classic filters – Butterworth, Chebyshev and elliptic – the transformations of Equations 8.36a and 8.36b map the zeros of the prototype LPF onto the imaginary axis in the *s*-plane. In the case of the Butterworth or Chebyshev filters, the zeros of the prototype filter are at infinity. In either case, the transformations map the zeros from infinity to infinity (for lowpass filters) or from infinity to the origin (for highpass filters), as shown in Figures 8.17(a)(ii) and 8.17(b)(ii).

Bandpass and bandstop filters

For bandpass digital filters, the analog BPF poles are obtained from those of the normalized prototype LPF using the transformation

$$s_{1,k} = \frac{s_{b,k}^2 + \omega_0^2}{Ws_{b,k}} \tag{8.37}$$

where $s_{1,k}$ are the poles of the prototype analog LPF, $s_{b,k}$ are the pole pairs of the intermediate analog BPF, $W = \omega_2' - \omega_1'$ is the width of the filter passband and $\omega_0^2 = \omega_1'\omega_2'$ gives the centre frequency of the passband. Equation 8.37 yields the following quadratic equation in $s_{b,k}$:

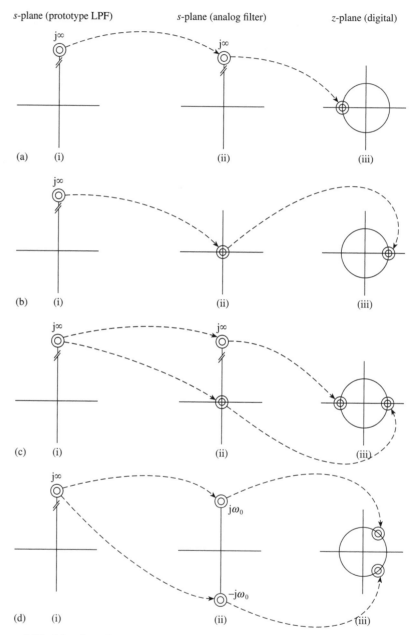

Figure 8.17   Mapping of zeros from a second-order prototype lowpass filter into (a) lowpass, (b) highpass, (c) bandpass and (d) bandstop.

$$s_{b,k}^2 - Ws_{1,k}s_{b,k} + \omega_0^2 = 0 \tag{8.38}$$

Solving for $s_{b,k}$ leads to the following expression for the bandpass analog poles:

$$s_{b,k} = \frac{W}{2}\left[s_{1,k} \pm \left(s_{1,k}^2 - \frac{4\omega_0^2}{W^2} \right)^{1/2} \right] \tag{8.39}$$

It is seen from Equation 8.39 that each analog LP pole, $s_{1,k}$ leads to a pair of analog BPF poles as a result of the s^2 term in the transformation. In general, the analog lowpass filter pole $s_{1,k}$ is complex and so the square-root term is complex. The evaluation of a complex square root using real arithmetic requires some care if the correct solution is to be obtained (see Appendix 8C).

For the bandstop digital filters, the following lowpass-to-bandstop transformation may be used:

$$s_{1,k} = \frac{Ws_{r,k}}{s_{r,k}^2 + \omega_0^2} \tag{8.40}$$

where $s_{1,k}$ are the poles of the analog prototype LPF, $s_{r,k}$ are the poles of the intermediate analog bandstop filter, W is the width of the stopband and ω_0^2 is the centre frequency in the stopband. Equation 8.40 leads to the following expression for obtaining analog bandstop poles from the prototype LPF:

$$s_{r,k} = \frac{W}{2}\left[s_{1,k}^{-1} \pm \left(s_{1,k}^{-2} - \frac{4\omega_0^2}{W^2} \right)^{1/2} \right] \tag{8.41}$$

Not surprisingly, Equation 8.41 has the same format as Equation 8.39 except for the inversion of the LPF poles due to the duality of the BPF and BSF.

As in the case of the lowpass and highpass filters, the transformations of Equations 8.39 and 8.41 map the zeros onto the imaginary axis. For Butterworth and Chebyshev filters, the lowpass-to-bandpass transformation maps the N zeros of the prototype lowpass filter from infinity to infinity and to the origin in the s-plane (see Figure 8.17(c)). The lowpass-to-bandstop transformation on the other hand maps the N zeros of the prototype lowpass filter from infinity to $\pm j\omega_0$ in the s-plane; see Figure 8.17(d). For elliptic bandpass and bandstop filters, the transformation maps the zeros of the prototype LPF which are entirely imaginary to other points on the $j\omega$ axis.

8.10.1.3   Step 3

The BZT is then used to map the poles and zeros from the s-plane into the digital z-plane. Each s-plane pole, $s_{p,k}$, is mapped as follows:

$$z_{p,k} = \frac{1 + s_{p,k}}{1 - s_{p,k}} \tag{8.42}$$

Similarly, each s-plane zero, $s_{z,k}$ of the transformed analog filters is mapped onto the z-plane as follows:

$$z_{z,k} = \frac{1 + s_{z,k}}{1 - s_{z,k}} \qquad (8.43)$$

Figures 8.17(a)–8.17(d) illustrate the way the zeros are mapped from the s-plane to the z-plane via the BZT for the Butterworth and Chebyshev filters. Note, for example, that for lowpass filters, Figure 8.17(a)(ii), the zeros are at infinity in the s-plane and the BZT of Equation 8.43 maps these to the point $z = -1$ in the z-plane, Figure 8.17(a)(iii), whereas for the bandpass filter the s-plane zeros, Figure 8.17(c)(ii), are mapped from the origin to the point $z = 1$, and from infinity to $z = -1$.

For the elliptic lowpass, highpass, bandpass and bandstop filters, $s_{z,k}$ are imaginary and the BZT maps these onto the unit circle in the z-plane. In general, the z-plane zeros for all the classical filters (Butterworth, Chebyshev and elliptic) obtained via the BZT lie on the unit circle, regardless of the filter type. As a result, the numerator coefficients of $H(z)$ of classical filters are always integers (0, ±1, ±2).

8.10.1.4 Step 4

The final step is to determine the numerator and denominator coefficients of the second- and/or first-order filter sections. This is achieved by combining complex conjugate pole and zero pairs as follows:

$$H_i(z) = \frac{(z - z_{z,k})(z - z_{z,k}^*)}{(z - z_{p,k})(z - z_{p,k}^*)} \qquad (8.44a)$$

$$= \frac{1 + b_{1i}z^{-1} + b_{2i}z^{-2}}{1 + a_{1i}z^{-1} + a_{2i}z^{-2}} \qquad (8.44b)$$

In general, a complex pole or zero pair located at $\alpha \pm j\beta$ leads to a quadratic in z of the form

$$[z - (\alpha + j\beta)][z - (\alpha - j\beta)] = z^2 - 2\alpha z + \alpha^2 + \beta^2$$

$$= 1 - 2\alpha z^{-1} + (\alpha^2 + \beta^2)z^{-2} \qquad (8.45a)$$

A single pole or zero on the real axis (that is, at $z = \pm\alpha$) leads to a first-order factor of the form

$$z \pm \alpha = 1 \pm \alpha z^{-1} \qquad (8.45b)$$

Often we find that an Nth-order filter has N real zeros on the real axis in the z-plane. In this case, we may pair the z-plane zeros so that the numerator of each filter section is a quadratic of the form

$$1 \pm 2z^{-1} + z^{-2} \tag{8.45c}$$

The overall transfer function, $H(z)$, is given by

$$H(z) = KH_1(z)H_2(z) \ldots H_M(z)$$

where K is a gain factor which is used to adjust the magnitude response in the passband to a desired level. In most cases, K is set to a value to make the maximum response in the passband equal to unity.

8.10.2 Illustrative examples

Example 8.16   A second-order digital bandpass filter (BPF), with Butterworth characteristics and a passband between 200 and 300 Hz at a sampling frequency of 2 kHz, is required for a certain DSP application. Determine the transfer function of the digital filter by mapping the s-plane poles and/or zeros of a suitable prototype analog lowpass filter into the z-plane.

Sketch and label the pole–zero diagrams of the prototype LPF, intermediate analog BPF, and the digital BPF.

Solution   A first-order normalized LPF is required, since the order of the filter is doubled by the bandpass transformation. Thus

$$H(s) = \frac{1}{s+1}$$

This transfer function has a single pole at $s_{1,1} = -1$. Now, $F_s = 2$ kHz $= 1/T$. Thus the prewarped bandedge frequencies are

$$\omega_1' = \tan\left(\frac{\omega_1 T}{2}\right) = \tan\left(\frac{2\pi \times 200}{2 \times 2000}\right) = 0.3249$$

$$\omega_2' = \tan\left(\frac{\omega_2 T}{2}\right) = \tan\left(\frac{2\pi \times 300}{2 \times 2000}\right) = 0.5095$$

Thus ω_0 and W are given by

$$\omega_0^2 = \omega_1'\omega_2' = 0.1655, \quad W = \omega_2' - \omega_1' = 0.1846$$

The single pole for the LPF is transformed into two poles for the BPF, using Equation 8.39, as follows:

$$s_{b,1} = \frac{0.1846}{2}\left\{-1 + \left[(-1)^2 - \frac{4 \times 0.1655}{(0.1846)^2}\right]^{1/2}\right\}$$

$$= -0.0923 + 0.4172j$$

$$s_{b,2} = \frac{0.1846}{2}\left\{-1 - \left[(-1)^2 - \frac{4 \times 0.1655}{(0.1846)^2}\right]^{1/2}\right\}$$

$$= -0.0923 - 0.4172j = s_{b,1}^*$$

From the BZT transformation, Equation 8.42, we have

$$z_{p,1} = \frac{1 - 0.0923 + 0.4172j}{1 + 0.0923 - 0.4172j} = 0.5979 + 0.6103j$$

$$z_{p,2} = z_{p,1}^*$$

The prototype LPF has a single zero at infinity. This is mapped by the lowpass-to-bandpass transformation to the origin and to infinity in the bandpass s-plane. That is, $s_{z,1} = 0$, $s_{z,2} = \infty$. The BZT maps these zeros to the points $z = 1$ and $z = -1$ in the z-plane.

$$s_{z,1} \to z_{z,1} = 1; \quad s_{z,2} \to z_{z,2} = -1$$

We can now determine the discrete transfer function, $H(z)$, from the poles and zeros:

$$H(z) = \frac{(z-1)(z+1)}{(z - z_{p,1})(z - z_{p,2})}$$

$$= \frac{z^2 - 1}{z^2 - 1.1958z + 0.7995} = \frac{1 - z^{-2}}{1 - 1.1958z^{-1} + 0.7995z^{-2}}$$

The transfer function is identical to that of Example 8.9 to within a constant factor. The pole–zero diagrams are also identical to those given in Figure 8.11.

Example 8.17   Starting with a suitable analog LPF, find the transfer function of a Chebyshev digital HPF in factored form to meet the following specifications:

passband edge frequency	15 kHz
attenuation at 18 kHz	>30 dB
passband ripple	1 dB
sampling frequency	48 kHz

Solution   The prewarped critical frequencies are

$$\omega_p' = \tan\left(\frac{15\,000\pi}{48\,000}\right) = 1.4966$$

$$\omega_s' = \tan\left(\frac{18\,000\pi}{48\,000}\right) = 2.4142$$

From the passband specifications, $\varepsilon = 0.3493$. The order of a suitable LP Chebyshev filter is 5 (the nearest integer).

With $\alpha = 1/N \sinh^{-1}(1/\varepsilon) = 0.3548$, $\sinh(\alpha) = 0.3623$, and $\cosh(\alpha) = 1.0636$, the left-hand poles of the normalized lowpass Chebyshev filter are located at ($\omega_p' = 1$)

$$s_{1,1} = 0.3623 \cos\left[\frac{(2+5-1)\pi}{10}\right] + 1.0636 \sin\left[\frac{(2+5-1)\pi}{10}\right]j$$

$$= -0.111\,96 + 1.0115j$$

$$s_{1,2} = 0.3623 \cos\left[\frac{(4+5-1)\pi}{10}\right] + 1.0636 \sin\left[\frac{(4+5-1)\pi}{10}\right]j$$

$$= -0.2931 + 0.6252j$$

$$s_{1,3} = 0.3623 \cos\left[\frac{(6+5-1)\pi}{10}\right] + 1.0636 \sin\left[\frac{(6+5-1)\pi}{10}\right]j = -0.3623$$

$$s_{1,4} = 0.3623 \cos\left[\frac{(8+5-1)\pi}{10}\right] + 1.0636 \sin\left[\frac{(8+5-1)\pi}{10}\right]j$$

$$= -0.2931 - 0.6252j$$

$$s_{1,5} = 0.3623 \cos\left[\frac{(10+5-1)\pi}{10}\right] + 1.0636 \sin\left[\frac{(10+5-1)\pi}{10}\right]j$$

$$= -0.111\,96 - 1.0115j$$

The reader should note the symmetry in the pole distribution and that $s_{1,1}$ and $s_{1,5}$ form a complex conjugate pair as do $s_{1,2}$ and $s_{1,4}$. Note also that each pole lies on the left-hand side of the *s*-plane, a necessary condition for stability.

Each prototype lowpass pole is transformed into the desired HP pole using Equation 8.36b:

$$s_{h,1} = -0.1618 - 1.4616j$$

$$s_{h,2} = -0.920\,13 - 1.9625j$$

$$s_{h,3} = -4.1306$$

$$s_{h,4} = s_{h,2}^* = -0.920\,13 + 1.9625j$$

$$s_{h,5} = s_{h,1}^* = -0.1618 + 1.4616j$$

Next, the poles are mapped from the s-plane into the z-plane using the BZT. From Equation 8.42 the z-plane poles after the BZT are given by (only the poles above the real axis are considered):

$$z_{h,1} = -0.3335 + 0.8386j$$

$$z_{h,2} = -0.4906 + 0.5207j$$

$$z_{h,3} = -0.6102$$

All the zeros, $z_{z,k}$, are located at $z = 1$. The coefficients of the second- and first-order filter sections can then be obtained from the poles and zeros as (Equations 8.45a and 8.45b)

$b_{11} = -2$	$b_{12} = -2$	$b_{13} = -1$
$b_{21} = 1$	$b_{22} = 1$	$b_{23} = 0$
$a_{11} = 0.6670$	$a_{12} = 0.9812$	$a_{13} = 0.6102$
$a_{21} = 0.8145$	$a_{22} = 0.5118$	$a_{23} = 0$

Finally, the transfer function is given by

$$H(z) = KH_1(z)H_2(z)H_3(z)$$

where

$$H_1(z) = \frac{1 - 2z^{-1} + z^{-2}}{1 + 0.6670z^{-1} + 0.8145z^{-2}}$$

$$H_2(z) = \frac{1 - 2z^{-1} + z^{-2}}{1 + 0.9812z^{-1} + 0.5118z^{-2}}$$

$$H_3(z) = \frac{1 - z^{-1}}{1 + 0.6102z^{-1}}$$

8.11 Using IIR filter design programs

Whatever the method we adopt, it is evident that the bilinear transform method involves considerable algebraic manipulation, with plenty of opportunity to make mistakes. Efficient computer programs now exist in the literature or commercially for calculating filter coefficients using the bilinear method, by merely specifying filter parameters of interest (IEEE, 1979; Gray and Markel, 1976; Parks and Burrus, 1987;

Jong, 1982; DeFatta *et al.*, 1988). Most of the computer programs in the literature are written in FORTRAN. There is a move away from such languages to more modern languages such as C or BASIC. A C language program for computing the filter coefficients using the BZT is on the CD in the companion handbook (see the Preface for details). The use of the program is illustrated in the following example. In Appendix 8B we illustrate how to use MATLAB to design a variety of IIR digital filters.

Example 8.18   *Illustrating the use of the IIR design program*   Obtain the coefficients of an audio digital filter with Chebyshev characteristics meeting the following specifications:

passband	0–2.5 kHz
stopband edge	2820 kHz
passband ripple	0.47 dB
sampling frequency	10 kHz
filter order	4

Solution   Using the program, the listing given below was obtained.

k	A_k	B_k
0	$1.000\,000 \times 10^{0}$	$1.934\,410 \times 10^{-1}$
1	$-2.516\,884 \times 10^{-1}$	$3.783\,311 \times 10^{-1}$
2	$1.054\,118 \times 10^{0}$	$5.241\,429 \times 10^{-1}$
3	$-2.406\,030 \times 10^{-1}$	$3.783\,311 \times 10^{-1}$
4	$1.985\,861 \times 10^{-1}$	$1.934\,410 \times 10^{-1}$

8.12 Choice of coefficient calculation methods for IIR filters

With the impulse invariant method, after digitizing the analog filter, the impulse response of the original analog filter is preserved, but not its magnitude–frequency response. Because of inherent aliasing, the method is inappropriate for highpass or bandstop filters. The bilinear z-transform method, on the other hand, yields very efficient filters and is well suited to the calculation of coefficients of frequency selective filters. It allows the design of digital filters with known classical characteristics such as Butterworth, Chebyshev and elliptic filters. Digital filters resulting from the bilinear transform method will, in general, preserve specific features of the magnitude response characteristics of the analog filter (e.g. bandedge frequencies, passband ripple and stopband attenuation) but not necessarily the time-domain properties. The impulse invariant method is good for simulating analog systems with lowpass characteristics, but the bilinear method is best for frequency selective IIR filters. The matched z-transform shares most of the inherent problems of the impulse invariant method. For simple filtering applications, the pole–zero placement method provides a simple, but effective, way of obtaining the filter coefficients.

8.12.1   Nyquist effect

The three methods for converting analog filters into equivalent discrete-time filters (namely, the matched z-transform, the impulse invariant and the bilinear z-transform methods) can have a significant effect on the filter characteristics (e.g. magnitude, phase and group delay responses), in certain cases. As stated earlier, the available frequency band for analog filters extends from zero to infinity, whereas for digital filters it is from zero to the Nyquist frequency (i.e. half the sampling frequency). Thus, the magnitude–frequency response of digital filters designed using either of the three methods may be significantly different from those of the analog filter because the entire analog frequency band (zero to infinity) is now compressed into a narrow band (zero to Nyquist frequency). This difference represents a distortion which is sometimes referred to as the Nyquist effect.

In many applications, the Nyquist effect is not harmful besides the provision of greater attenuation than specified. However, in applications where it is desirable to retain the analog filter response, such as in professional and semi-professional audio work (Clark *et al.*, 1996; 2000), the effect represents an undesirable distortion. In such applications, the extent of the distortion would be a factor in the choice of a transform for converting an analog filter into an equivalent discrete-time filter. The impact on other filter characteristics, such as group delay and impulse responses, may also be a factor in the choice of a method (Rabiner and Gold, 1975).

In this section, we will consider the consequences of the Nyquist effect briefly. A number of problems are provided at the end of the chapter to allow the reader to explore the relative advantages of the methods of converting analog filters into equivalent discrete filters.

Example 8.19   A lowpass, discrete-time filter, with Butterworth characteristics, is required to meet the following specifications:

cutoff frequency	300 Hz
filter order	5
sampling frequency	1000 Hz

(1)   With the aid of MATLAB, obtain and plot
   (a)   the magnitude–frequency and group delay responses of the filter using the impulse invariant method;
   (b)   the magnitude–frequency and group delay responses of the filter using the bilinear z-transform method.

(2)   Compare the two methods (bilinear z-transform and impulse invariant methods) in terms of the magnitude response distortions due to the Nyquist effect.

Solution   (1)   (a)   We will start by designing an analog filter to serve as a benchmark. The MATLAB m-file for the analog filter is listed in Program 8.1. The magnitude–frequency response of the filter, obtained with the aid of the program, is shown in Figure 8.18.

The magnitude–frequency and group delay responses for the equivalent discrete filter, designed via the impulse invariant method, are shown in Figures 8.19(a) and (b). The MATLAB m-file for the filter is shown in Program 8.2. The group delay response was obtained using the **grpdelay** command in the MATLAB Signal Processing Toolbox.

Program 8.1   MATLAB m-file for the analog filter design.

```
%
%  Program name: EX8-1.m
%
FN=1000/2;
fc=300;                        %  Cutoff frequency
N=5;                           %  Filter order
[z, p, k]=buttap(N);           %  Create an analog filter
w=linspace(0, FN/fc, 1000);    %  Plot the response of filter
h=freqs(k*poly(z), poly(p), w);
f=fc*w;
plot(f, 20*log10(abs(h))), grid
ylabel('Magnitude (dB)')
xlabel('Frequency (Hz)')
```

Program 8.2   MATLAB m-file for the design of the impulse invariant filter.

```
%
%  Program name: EX8-2.m
%
Fs=1000;                       %  Sampling frequency
fc=300;                        %  Cutoff frequency
WC=2*pi*fc;                    %  Cutoff frequency in radian
N=5;
[b,a]=butter(N, WC, 's');      %  Create an analog filter
[z,p,k]=butter(N, WC, 's');
[bz, az]=impinvar(b, a, Fs);   %  Determine coeffs of IIR filter
[h, f]=freqz(bz, az, 512, Fs);
plot(f, 20*log10(abs(h))), grid
xlabel('Frequency (Hz)')
ylabel('Magnitude Response (dB)')
```

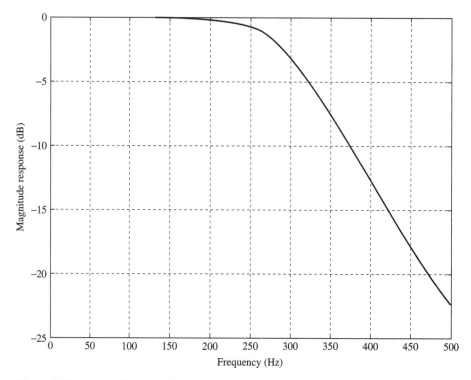

Figure 8.18   Analog filter magnitude response.

Program 8.3   MATLAB m-file for the design of the BZT filter.

```
%
%   Program name: EX8-3.m
%
Fs=1000;                            %   Sampling frequency
FN=Fs/2;
fc=300;                             %   Cutoff frequency
N=5;
[z, p, k]=butter(N, fc/FN);
[h, f]=freqz(k*poly(z), poly(p), 512, Fs);
plot(f, 20*log10(abs(h))), grid
ylabel('Magnitude (dB)')
xlabel('Frequency (Hz)')
```

    (b)   The magnitude–frequency and group delay responses for the equivalent discrete filter, designed via the BZT method, are shown in Figures 8.20(a) and (b). The MATLAB m-file for the filter is shown in Program 8.3.

(2) If we compare the magnitude–frequency responses of the impulse invariant filter and the bilinear z-transform filter to that of the analog filter, it is evident

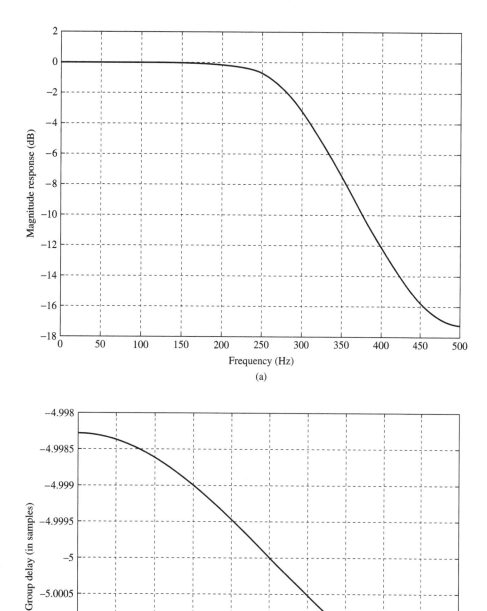

Figure 8.19 (a) Impulse invariant method. (b) Group delay response for the impulse invariant method.

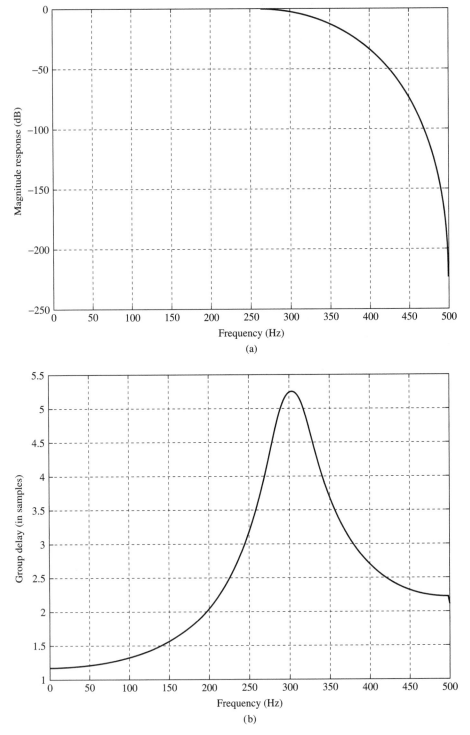

Figure 8.20 (a) Magnitude–frequency response for the BZT. (b) Group delay response for the BZT example.

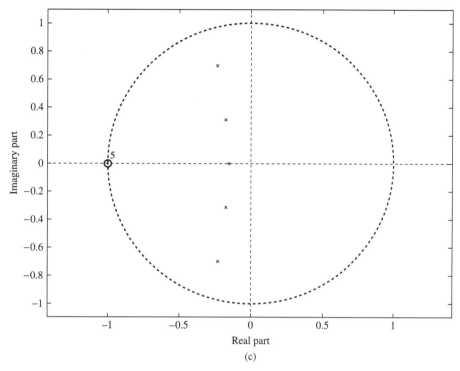

Figure 8.20 (c) BZT pole–zero diagram.

that the impulse invariant filter provides less and the BZT filter more attenuation nearer the Nyquist frequency. For example, at 500 Hz, the impulse invariant filter provides an attenuation of about 17 dB whereas the BZT filter has an attenuation in excess of 200 dB.

An analysis of the BZT filter shows that the transfer function has five zeros at the Nyquist frequency (see Figure 8.20(c)) which are responsible for the rapid drop in the magnitude response. This is often the case in BZT-based discrete filters and is the reason why it provides greater attenuation than the original analog filter. In contrast, the impulse invariant-based discrete filter has a zero at the origin and outside the unit circle. The differences between the magnitude responses of the two discrete filters and the analog filter represent a distortion.

In many cases, the distortions in the magnitude responses for the MZT (or impulse invariant) and the BZT have opposite effects. BZT-based filters tend to provide more attenuation whilst MZT and impulse invariant-based filters provide less. At low and mid-frequency bands (relative to the Nyquist frequency), the magnitude responses of lowpass and bandpass filters designed by the MZT and impulse invariant methods are reasonably close to those of the original analog filters, but deteriorate near the Nyquist frequency. The same is true of filters designed by the BZT method, except that the

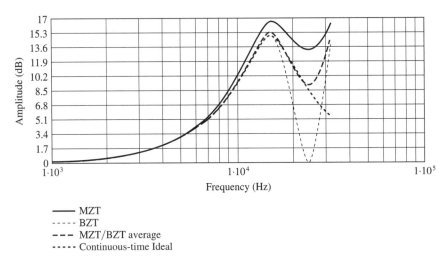

Figure 8.21 Magnitude response of the MZT, BZT, MZT/BZT average and the continuous-time ideal, Bell filter, 15 kHz, $Q = 2$, 15 dB boost.

deterioration in the response close to the Nyquist frequency is in the opposite sense. A simple, but effective way to combat response distortion for lowpass and bandpass filters is to combine the BZT with either MZT or the impulse invariant method by averaging their coefficients (Clark *et al.*, 1996; 2000). In this approach, the lowpass or bandpass filter is first designed separately by the BZT and MZT. The coefficients are then averaged to obtain a new set of coefficients as follows:

$$b_k' = [b_k(BZT) + b_k(MZT)]/2, \; k = 0, 1, 2$$

$$a_k' = [a_k(BZT) + a_k(MZT)]/2, \; k = 0, 1, 2$$

Figure 8.21 compares the responses of two discrete filters, one designed by the MZT method and the other by the BZT method. The filters were intended for an audio signal processing application where there is a need for efficient means of generating filter coefficients online at low response distortion. As before, the BZT-based filter gives a sharper magnitude–frequency response beyond 10 kHz (the Nyquist frequency is 24 kHz) whereas the MZT gives less attenuation. Both deviations in the magnitude response represent a distortion in this application. The averaged response (Figure 8.21) on the other hand has a reduced response distortion. The pole–zero diagram of Figure 8.22 compares the locations of the poles and zeros for the MZT and the averaged MZT/BZT filters. As would be expected, the averaging process has altered the positions of the poles and zeros of the MZT filter. In particular, it has introduced a 'soft' zero at the Nyquist frequency. The zero is said to be soft because it is located well inside the unit circle and so does not pull the magnitude response of the filter rapidly towards zero as in the BZT.

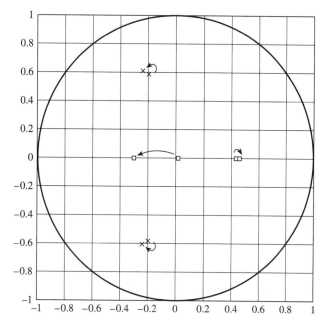

Figure 8.22 A comparison of the pole–zero plot of the MZT and the MZT/BZT average. The arrows indicate the pole–zero movement from the MZT to the MZT/BZT average.

8.13 Realization structures for IIR digital filters

Realization involves converting a given transfer function, $H(z)$, into a suitable filter structure. Flow or block diagrams are normally used to depict filter structures and they show the computational procedure for implementing the digital filter. The basic elements of realization structures are multipliers, adders, and delay elements; see Figure 8.23.

Recall that the IIR filter is characterized by the following equations:

$$H(z) = \sum_{k=0}^{N} b_k z^{-k} \bigg/ \left(1 + \sum_{k=1}^{M} a_k z^{-k}\right) \quad M \geqslant N \tag{8.46a}$$

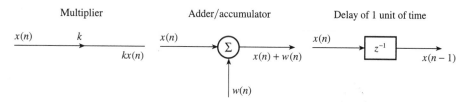

Figure 8.23 The basic elements of filter structures.

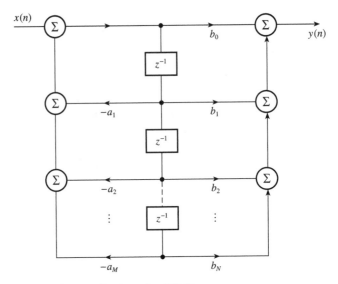

Figure 8.24   A direct form realization of an IIR filter.

$$y(n) = \sum_{k=0}^{N-1} b_k x(n-k) - \sum_{k=1}^{M} a_k y(n-k) \tag{8.46b}$$

A direct form realization of Equation 7.46 is shown in Figure 8.24, where $N = M$ for simplicity. Note that the coefficients used in the diagram are the same as in the transfer function, but with the signs reversed for the denominator coefficients. When the filter order is high, for example $M > 3$, direct realization of the filter as in Figure 8.24 is very sensitive to finite wordlength effects and should be avoided in these cases. In practice, $H(z)$ is normally broken down into smaller sections, typically second- and/or first-order blocks, which are then connected up in cascade or in parallel (see later).

8.13.1   Practical building blocks for IIR filters

Examples of practical second-order building blocks used in realizing higher-order IIR filters are depicted in Figure 8.25. The first (Figure 8.25(a)) is often called a canonic section (or direct form 2) because it has the minimum number of delay elements. This biquadratic section is characterized by the following equations:

$$w(n) = x(n) - \sum_{k=1}^{2} a_k w(n-k) \tag{8.47a}$$

$$y(n) = \sum_{k=0}^{2} b_k w(n-k) \tag{8.47b}$$

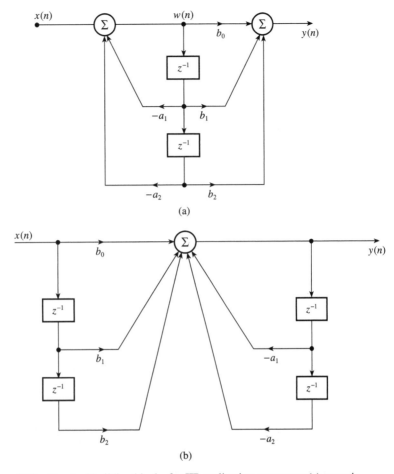

Figure 8.25   Practical building blocks for IIR realization structures: (a) canonic second-order section; (b) direct form second-order section.

$$H(z) = \frac{b_0 + b_1 z^{-1} + b_2 z^{-2}}{1 + a_1 z^{-1} + a_2 z^{-2}} \qquad (8.47c)$$

The second filter section (Figure 7.18(b)) is a direct realization of the second-order IIR equation. It is characterized by the following equations:

$$y(n) = \sum_{k=0}^{2} b_k x(n-k) - \sum_{k=1}^{2} a_k y(n-k) \qquad (8.48a)$$

$$H(z) = \frac{b_0 + b_1 z^{-1} + b_2 z^{-2}}{1 + a_1 z^{-1} + a_2 z^{-2}} \qquad (8.48b)$$

The canonic section (Figure 8.25(a)) is the most popular because it has a good roundoff noise property and requires a minimum number of storage elements, but it is susceptible to internal overflows. To avoid internal overflow it is necessary to scale the input to the filter section. Scaling is not mandatory for the direct form (Figure 8.25(b)) because it has only one adder, and may be preferred where scaling is not desired, for example high fidelity digital audio (Dattorro, 1988). Under certain conditions the direct form is superior to the canonic section, in terms of noise performance.

The coupled form has some desirable finite wordlength properties (Gold and Rader, 1969), but it requires more computational effort and cannot be readily used to implement transfer functions with second-order numerator coefficient.

The filter blocks given in Figure 8.25 are general, second-order, sections. Several other filter blocks can be derived from them. For example, if the numerator coefficients a_1 and a_2 in Figure 8.25(a) are both zero we have a purely recursive structure. On the other hand, if the filter coefficients were obtained using elliptical functions the coefficient a_2 is unity. A first-order filter block is readily obtained by setting $a_2 = b_2 = 0$ in any of the structures above.

Figure 8.26 shows the transposes of the second-order canonic and the direct form filter sections. These are obtained from Figures 8.26(a) and 8.26(b) respectively, by interchanging all the adders and branch nodes, and reversing the directions of the arrows. Although the transfer functions of the sections in Figure 8.26 are identical to their transposes, their finite wordlength properties are quite different. Other structures that are less sensitive to finite wordlength effects are available, but these are usually more complicated. Examples include minimum-noise, state-variable, and lattice structure.

8.13.2 Cascade and parallel realization structures for higher-order IIR filters

In practice higher-order transfer functions are realized as cascades or parallel combinations of second- and/or first-order building blocks described above. Typically, in cascade realization the transfer function is factored into $N/2$ second-order factors:

$$H(z) = \prod_{k=1}^{N/2} \left[\frac{b_{0k} + b_{1k}z^{-1} + b_{2k}z^{-2}}{1 + a_{1k}z^{-1} + a_{2k}z^{-2}} \right]$$

$$= \prod_{k=1}^{N/2} \frac{N_k(z)}{D_k(z)} \tag{8.49a}$$

where

$$N_k(z) = b_{0k} + b_{1k}z^{-1} + b_{2k}z^{-2}$$
$$D_k(z) = 1 + a_{1k}z^{-1} + a_{2k}z^{-2} \tag{8.49b}$$

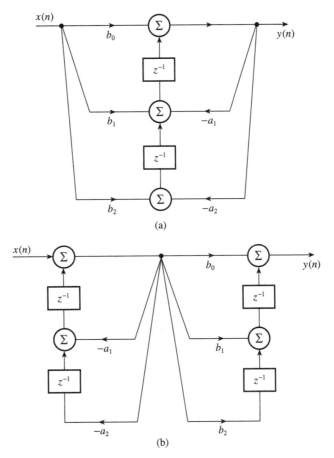

Figure 8.26 (a) Transpose canonic second-order section; (b) transpose direct form second-order section.

and N, the filter order, is assumed to be even. If N is odd, then one of the $H_k(z)$ will be a first-order section.

Each second-order factor, $H_k(z)$, can be realized using one of the building blocks and the resulting blocks connected in cascade; see Figure 8.27. Three difficulties arise with the cascade realization: (1) how to pair the numerator factors with the denominator factors; (2) the order in which the individual sections should be connected; (3) the need to scale the signal levels at various points within the filter to avoid the levels becoming too large or too small.

The numerator and denominator factors can be ordered in a variety of ways. For example, a fourth-order filter can be factored into two second-order sections, and then paired and ordered in one of four different ways:

$$(1) \quad H(z) = \frac{N_1(z)}{D_1(z)} \frac{N_2(z)}{D_2(z)}$$

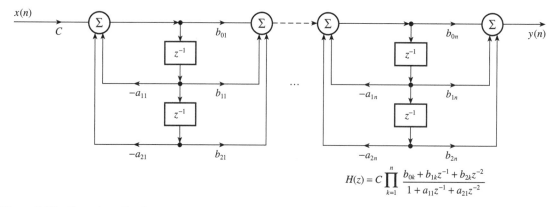

$$H(z) = C \prod_{k=1}^{n} \frac{b_{0k} + b_{1k}z^{-1} + b_{2k}z^{-2}}{1 + a_{11}z^{-1} + a_{21}z^{-2}}$$

Figure 8.27   Cascade realization.

(2) $$H(z) = \frac{N_2(z)}{D_2(z)} \frac{N_1(z)}{D_1(z)}$$

(3) $$H(z) = \frac{N_1(z)}{D_2(z)} \frac{N_2(z)}{D_1(z)}$$

(4) $$H(z) = \frac{N_2(z)}{D_1(z)} \frac{N_1(z)}{D_2(z)}$$

where each $N_k(z)$ and $D_k(z)$ is a second-order polynomial defined in Equation 8.49b. In the first case, the first filter section consists of the numerator and denominator pair $N_1(z)$ and $D_1(z)$, while the second filter section consists of the numerator and denominator pair $N_2(z)$ and $D_2(z)$. It is obvious that the number of possible pairings and orderings of the denominator and numerator is quite large. Typically, for an Nth-order filter, the number of different pairings and orderings possible is

$$\left(\frac{N}{2}! \right)^2 \tag{8.50}$$

A rule of thumb is to pair $N_i(z)$ with $D_k(z)$ if the zeros of $N_i(z)$ are closest to the poles of $D_k(z)$ to avoid having a large amplitude response at the frequency corresponding to the pole, and to place the second-order section with poles closest to the unit circle last in the cascade (Jackson, 1986). A number of efficient schemes have been developed for pairing and ordering the filter sections, based on which ordering gives the best signal-to-noise ratio, a topic intimately linked with pairing and ordering (see Chapter 13).

In parallel realization, an Nth-order transfer function $H(z)$ is expanded using partial fractions as

$$H(z) = C + \sum_{k=1}^{N/2} H_k(z) \tag{8.51}$$

where

$$C = \frac{b_N}{a_N}, \quad H_k(z) = \frac{b_{0k} + b_{1k}z^{-1}}{1 + a_{1k}z^{-1} + a_{2k}z^{-2}}$$

Again, each second-order section can be realized using the building blocks described previously as shown in Figure 8.28. It is worth noting that, in the parallel realization, the numerator coefficient for z^{-2} is zero. In the parallel structure, the order in which the sections is connected is not important. Furthermore, scaling is easier and can be carried out for each block independently (see later), and the SNRs are comparable with those of the best cascade realization (Jackson, 1986). However, the zeros of parallel structures are more sensitive to coefficient quantization errors. It should be mentioned that the sensitivity of the zeros of the parallel structure to coefficient quantization appears to be most serious when coefficient wordlength is down to as little as 5 bits or less. It appears that for coefficient wordlengths of 12 or more bits the differences between the parallel and cascade structures are small for most filters. However, an important advantage of the cascade method is that between 25% and 50% of the filter coefficients are simple integers (0, ±1 or ±2) when derived from classical analog filters via the BZT. This is attractive in systems with only primitive

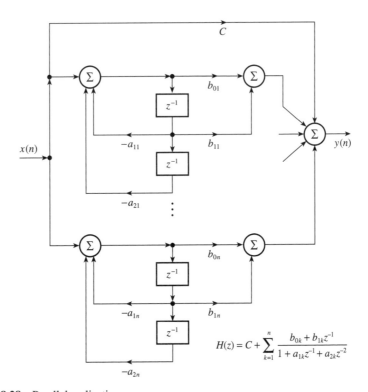

$$H(z) = C + \sum_{k=1}^{n} \frac{b_{0k} + b_{1k}z^{-1}}{1 + a_{1k}z^{-1} + a_{2k}z^{-2}}$$

Figure 8.28   Parallel realization.

arithmetic capability where the number of multiplications must be kept low. Further, most available software packages produce coefficients for cascade realization but not for the parallel structure. It is for these reasons that the cascade method has become very popular.

Example 8.20

Develop the transfer functions for (1) the cascade and (2) the parallel realization structures for the filter characterized by the following transfer function using second- and first-order sections:

$$H(z) = \frac{0.1432(1 + 3z^{-1} + 3z^{-2} + z^{-3})}{1 - 0.1801z^{-1} + 0.3419z^{-2} - 0.0165z^{-3}}$$

Solution

(1)   For cascade realization, $H(z)$ is expressed in factored form:

$$H(z) = 0.1432 \frac{1 + 2z^{-1} + z^{-2}}{1 - 0.1307z^{-1} + 0.3355z^{-2}} \frac{1 + z^{-1}}{1 - 0.0490z^{-1}}$$

(2)   For parallel realization $H(z)$ is expressed as the sum of second- and first-order sections using partial fractions:

$$H(z) = \frac{1.2916 - 0.08407z^{-1}}{1 - 0.131z^{-1} + 0.3355z^{-2}} + \frac{10.1764}{1 - 0.049z^{-1}} - 8.7107$$

The realization diagrams for the cascade and parallel realizations are given in Figures 8.29(a) and 8.29(b) respectively. The coefficients for the parallel realization were obtained using the C language program presented in Chapter 4.

8.14   Finite wordlength effects in IIR filters

The coefficients, a_k and b_k, obtained earlier (see Sections 8.4–8.10) are of infinite or very high precision, typically six to seven decimal places. When an IIR digital filter is implemented in a small system, such as an 8-bit microcomputer, errors arise in representing the filter coefficients and in performing the arithmetic operations indicated by the difference equation. These errors degrade the performance of the filter and in extreme cases lead to instability.

Before implementing an IIR filter, it is important to ascertain the extent to which its performance will be degraded by finite wordlength effects and to find a remedy if the degradation is not acceptable. In general, the effects of these errors can be reduced to acceptable levels by using more bits but this may be at the expense of increased cost.

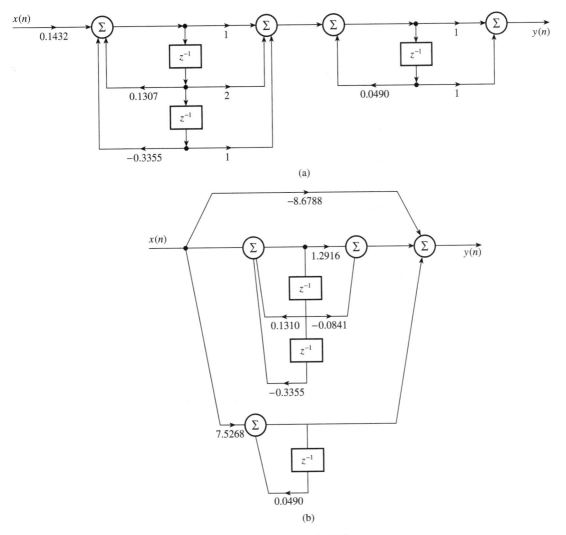

Figure 8.29 (a) Cascade and (b) parallel realizations for Example 8.20.

The main errors in digital IIR filters are as follows:

■ ADC quantization noise, which results from representing the samples of the input data, $x(n)$, by only a small number of bits;

■ coefficient quantization errors, caused by representing the IIR filter coefficients by a finite number of bits;

■ overflow errors, which result from the additions or accumulation of partial results in a limited register length;

■ product roundoff errors, caused when the output, $y(n)$, and results of internal arithmetic operations are rounded (or truncated) to the permissible wordlength.

The extent of filter degradation depends on (i) the wordlength and type of arithmetic used to perform the filtering operation, (ii) the method used to quantize filter coefficients and variables, and (iii) the filter structure. From a knowledge of these factors, the designer can assess the effects of finite wordlength on the filter performance and take remedial action if necessary. Depending on how the filter is to be implemented some of the effects may be insignificant. For example, when implemented as a high level language program on most large computers, coefficient quantization and roundoff errors are not important. For real-time processing, finite wordlengths (typically 8 bits, 12 bits, and 16 bits) are used to represent the input and output signals, filter coefficients and the results of arithmetic operations. In these cases, it is nearly always necessary to analyze the effects of quantization on the filter performance.

The effects of finite wordlength on performance are more difficult to analyze in IIR filters than in FIR filters because of their feedback arrangements. However, the use of MATLAB (see Appendix 8B) allows practical solutions to be obtained for specific filters. The effects of each of the four sources of errors listed above will be discussed, in turn, in the next few sections.

A more detailed analysis of the effects of finite wordlength on the performance of IIR filters and other DSP systems is given in Chapter 13.

8.14.1 Coefficient quantization errors

Recall that the IIR filter is characterized by the following equation:

$$H(z) = \frac{\sum\limits_{k=0}^{N} b_k z^{-k}}{1 + \sum\limits_{k=1}^{M} a_k z^{-k}}$$

When the coefficients are quantized to a finite number of bits, for example 8 or 16 bits, the quantized transfer function may be written as

$$[H(z)]_q = \frac{\sum\limits_{k=0}^{N} [b_k]_q z^{-k}}{1 + \sum\limits_{k=1}^{M} [a_k]_q z^{-k}} \tag{8.52}$$

where

$$[b_k]_q = b_k + \Delta b_k; \qquad [a_k]_q = a_k + \Delta a_k$$

$\Delta b_k, \Delta a_k$ are changes in the coefficients b_k and a_k, respectively
q denotes a quantized coefficient

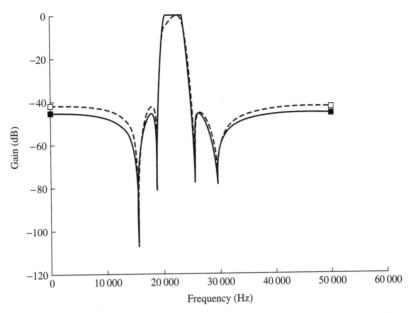

Figure 8.30   Practical effects of coefficient quantization on the frequency response: ■, unquantized; □, quantized, 5 bits.

The primary effect of quantizing filter coefficients using a finite number of bits is to alter the positions of the poles and zeros of $H(z)$ in the z-plane. This could lead to:

- instability or potential instability for high-order filters, with sharp transition widths and poles close to the unit circle;
- a change in the desired frequency response, as illustrated in Figure 8.30.

The quantized filter should be analyzed to ensure that its wordlength is sufficient for both stability and satisfactory frequency response. We will illustrate the effects of coefficient quantization on the frequency response of an IIR filter in the next example. A more detailed analysis of the effects of coefficient quantization on filter performance is given in Chapter 13.

Example 8.21   A bandpass digital filter is to be used for digital clock recovery at 4.8 kbaud and a sampling frequency of 153.6 kHz. The filter is characterized by the following transfer function:

$$H(z) = \frac{1}{1 + a_1 z^{-1} + a_2 z^{-2}}$$

where

$$a_1 = -1.957\,558 \text{ and } a_2 = 0.995\,913$$

Assess the effect of quantizing the coefficients to 8 bits on the pole positions and hence on the centre frequency.

Solution   First, we find the positions of the poles of the unquantized filter. The radius, r, and angle, θ, of each pole are:

$$r = \sqrt{a_2}, \quad \theta = \cos^{-1}\left(-\frac{a_1}{2r}\right)$$

Thus

$$r = \sqrt{0.995\,913} = 0.997\,95 \text{ and } \theta = \cos^{-1}\left(\frac{1.957\,558}{2 \times 0.997\,95}\right) = 11.25°$$

This corresponds to a centre frequency of

$$\left(\frac{11.25}{360}\right) \times 153.6 \times 10^3 = 4.7999 \text{ kHz}$$

As one of the coefficients is greater than unity, we will assign 1 bit as the sign bit, 1 bit for the integer and 6 bits for the fractional part of the coefficients. Thus, after quantizing to 8 bits the coefficient values become:

$$a_1 = -1.957\,558 \times 2^6 = -125 \; (\equiv 10000100)$$
$$a_2 = 0.995\,913 \times 2^6 = 63 \text{ (maximum positive fraction) } (\equiv 00111111)$$

In fractional notation, the quantized coefficient values are:

$$a_1 = -\frac{125}{64} = -1.953\,125; \quad a_2 = \frac{63}{64} = 0.984\,375.$$

The new pole position becomes: $r = 0.992\,156$, $\theta = 10.17°$; and the centre frequency now becomes

$$f_o = \left(\frac{10.17}{360}\right) \times 153.6 \times 10^3 = 4.34 \text{ kHz}$$

8.15 Implementation of IIR filters

In the IIR filter, the output, $y(n)$, is computed for each input sample, $x(n)$. Assuming cascade realization using the second-order direct form, the key filtering equation is

$$y(n) = \sum_{k=0}^{2} b_k x(n-k) - \sum_{k=1}^{2} a_k y(n-k)$$

This equation clearly shows that to implement the filter we need the following components:

- memory (for example ROM) for storing filter coefficients;
- memory (for example RAM) for storing the present and past inputs and outputs $\{x(n), x(n-1), \ldots\}$ and $\{y(n), y(n-1), \ldots\}$;
- hardware or software multiplier(s);
- adder or arithmetic logic unit.

In modern real-time DSP, the filtering operations are efficiently performed with a DSP processor such as the TMS320C50. These processors have all the basic blocks on-board, including in-built hardware multiplier(s). In some applications, standard 8-bit or 16-bit microprocessors such as the Motorola 6800 or 68000 families offer attractive alternative implementations. In addition to the signal processing hardware, the designer must also provide suitable input–output (such as analog–digital–analog conversion) interfaces to the digital hardware, depending on the type of data source and sink. This approach may be described as hardware implementation.

In batch or off-line processing a suitable high level language is used to implement the filter. In this case, the filter is often implemented in a high level language (for example C or FORTRAN) and runs in a general purpose computer, such as a personal computer or a mainframe computer, where all the basic blocks are already configured. Thus, batch processing may be described as a purely software implementation.

Computational requirements

The designer must analyze the impact of the computational requirements of a digital filter on the processor that will be used. The primary requirements for digital filters are multiplication, additions, accumulation and delays or shifts. For example, a filter consisting of a second-order section would require typically four multiplications, four additions, and some shifts and storage. If the filtering is performed in real time, for example at 44.1 kHz (for digital audio), the arithmetic operations must be performed once every $1/(44.1 \text{ kHz})$. Allowance must also be made for other overheads such as fetching the input data or saving or outputting the filtered data samples as well as other housekeeping operations.

8.16 A detailed design example of an IIR digital filter

This example will be used to illustrate some of the many concepts presented in this chapter. In particular, we shall see how the five-stage design procedure is applied.

Stage 1: filter specifications

Design and implement a lowpass IIR digital filter using a software package and the TMS320C50-based target board to meet the following specifications:

sampling frequency	15 kHz
passband	0–3 kHz
transition width	450 Hz
passband ripple	0.5 dB
stopband attenuation	45 dB

Note: The target board has a 12-bit ADC and 12-bit DAC.

Stage 2: coefficient calculation

Using the software design program (on the CD in the companion handbook) for IIR filters, it was found that a fourth-order elliptic filter, via the bilinear transform method, would be required to satisfy the specifications. The output listing of the design program is summarized below.

	Denominator A_k	Numerator B_k
1	1.000000E+00	5.846399E–02
2	–1.325263E+00	1.359507E–01
3	1.480202E+00	1.820297E–01
4	–7.841098E–01	1.359506E–01
5	2.339270E–01	5.846398E–02

Poles		Coefficients	
Real	Imaginary	z^{-1}	z^{-2}
0.247967	0.836885	–0.495935	0.761864
0.414664	0.367559	–0.829328	0.307046

Zeros		Coefficients	
Real	Imaginary	z^{-1}	z^{-2}
–0.337859	0.941197	0.675718	1.000000
–0.824828	0.565383	1.649656	1.000000

From the listing, the transfer function of the filter, in direct form, is given by

$$H(z) = $$

$$\frac{0.058\,463\,99 + 0.135\,950\,7z^{-1} + 0.182\,097\,9z^{-2} + 0.135\,950\,6z^{-3} + 0.058\,463\,98z^{-4}}{1 - 1.325\,263z^{-1} + 1.480\,202z^{-2} - 0.784\,109\,8z^{-3} + 0.233\,927z^{-4}}$$

Stage 3: realization

As explained earlier, direct form realization of $H(z)$ is very sensitive to many adverse effects of finite wordlength such as coefficient quantization errors, so it is important to break $H(z)$ down into smaller sections and then connect these up, for example in cascade or parallel structure. Assuming cascade structure, $H(z)$ is broken down into two second-order sections $H_1(z)$ and $H_2(z)$:

$$H(z) = H_1(z)H_2(z)$$

where

$$H_1(z) = \frac{b_{01} + b_{11}z^{-1} + b_{12}z^{-2}}{1 + a_{11}z^{-1} + a_{21}z^{-2}}$$

$$H_2(z) = \frac{b_{02} + b_{12}z^{-1} + b_{22}z^{-2}}{1 + a_{12}z^{-1} + a_{22}z^{-2}}$$

The realization diagram is depicted in Figure 8.31, with each filter section realized using a standard biquad structure. The corresponding sets of difference equations, which define how the filtering operation will be carried out, are as follows.

Filter section 1

$$w_1(n) = (1/s_1)x(n) - a_{11}w_1(n-1) - a_{21}w_1(n-2)$$

$$y_1(n) = b_{01}w_1(n)s_1/s_2 + b_{11}w_1(n-1)s_1/s_2 + b_{21}w_1(n-2)s_1/s_2$$

Filter section 2

$$w_2(n) = y_1(n) - a_{12}w_2(n-1) - a_{22}w_2(n-2)$$

$$y_2(n) = b_{02}w_2(n)s_2 + b_{12}w_2(n-1)s_2 + b_{22}w_2(n-2)s_2$$

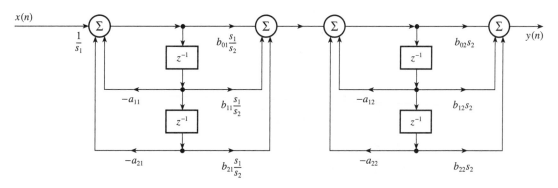

Figure 8.31 Realization diagram for the detailed design example.

The exact values of the coefficients, a_{ij} and b_{ij}, are dependent on how we pair the numerator and denominator polynomials of $H(z)$ and how the second-order filter sections used to realize the polynomials are ordered. The best pairing and ordering can only be determined from finite wordlength analysis.

Stage 4: analysis of finite wordlength effects

For the problem, based on the specifications given, we will assume that fixed point two's complement arithmetic will be used and that each coefficient will be quantized, by rounding, to a 16-bit wordlength.

   The main objective here is to assess the effects of the various quantization errors on the filter performance and to determine the best filter configuration to implement, in terms of signal-to-noise ratios. The sources of errors of concern are (see Chapter 13 for details):

- overflow errors,
- roundoff errors, and
- coefficient quantization errors.

To avoid overflow at the output of the adders in Figure 8.31, suitable scale factors are introduced before the adders as shown in the figure.

   Since $H(z)$ is fourth order to be realized as two second-order sections, its numerator and denominator factors can be paired and ordered in four possible ways:

$$H_A(z) = \frac{N_1(z)}{D_1(z)} \frac{N_2(z)}{D_2(z)}$$

$$H_B(z) = \frac{N_2(z)}{D_2(z)} \frac{N_1(z)}{D_1(z)}$$

$$H_C(z) = \frac{N_1(z)}{D_2(z)} \frac{N_2(z)}{D_1(z)}$$

$$H_D(z) = \frac{N_2(z)}{D_1(z)} \frac{N_1(z)}{D_2(z)}$$

where

$$N_1(z) = 1 + 0.675\ 718z^{-1} + z^{-2}$$

$$N_2(z) = 1 + 1.649\ 656z^{-1} + z^{-2}$$

$$D_1(z) = 1 - 0.495\ 935z^{-1} + 0.761\ 864z^{-2}$$

$$D_2(z) = 1 - 0.829\ 328z^{-1} + 0.307\ 046z^{-2}$$

Table 8.1   Scale factors for the four filter configurations.

Filter	Scale factor	L_1	L_2	L_∞
A	s_1	5.524 844	1.608 890	4.379 544
	s_2	11.821 571	3.677 381	7.262 393
B	s_1	2.479 158	1.359 467	2.175 539
	s_2	18.908 47	10.880 490	12.548 114
C	s_1	2.479 158	1.359 467	2.175 539
	s_2	11.821 571	10.880 490	7.262 393
D	s_1	5.524 844	1.608 890	4.379 544
	s_2	18.908 47	5.727 459	12.548 114

Each of the four possible filter configurations will have different scale factors as well as different signal to roundoff noise performance. An objective at this stage is to determine the best pairing and ordering, in terms of signal-to-noise performance, for the filter. Overflow and roundoff errors are intimately linked, and so scaling and roundoff analysis should be performed simultaneously.

Using the finite wordlength analysis program, scale factors based on the L_1 norm, L_2 norm and L_∞ norm for the four possible filters above were obtained and are summarized in Table 8.1. In this example, we have used the L_1 norm. For a fourth-order filter realized as a cascade of two second-order canonic sections, the roundoff noise at the output, after scaling, is given by

$$\sigma_o^2 = \frac{q^2}{12}[3s_1^2\|H_1(z)H_2(z)\|_2^2 + 5s_2^2\|H(z)\|_2^2 + 3]$$

where q is the quantization stepsize or rounding, $\|\cdot\|_2^2$ symbolizes the L_2 norm squared, $H_1(z)$ is the transfer function for the first filter stage, $H_2(z)$ is the transfer function for the second filter stage, s_1 is the scale factor for the first filter stage and s_2 is the scale factor for the second filter stage.

The noise performances for each of the four possible filter configurations, with each coefficient quantized to 16 bits (after scaling), are shown in Table 8.2. It is

Table 8.2   Comparison of roundoff noise performance of the four filter configurations.

Filter	Noise power
A	$703q^2$
B	$326.378q^2$
C	$382.32q^2$
D	$570.453q^2$

evident that filter B has the best roundoff noise performance. The scaled filter transfer function is given by

$$H(z) = H'_\mathrm{B}(z) = \frac{s_1}{s_2}\frac{N_2(z)}{D_2(z)}\frac{N_1(z)}{D_1(z)}s_2$$

$$= 0.131\,1136\,\frac{1 + 1.649\,656z^{-1} + z^{-2}}{1 - 0.829\,328z^{-1} + 0.307\,046z^{-2}}$$

$$\times\frac{1 + 0.675\,718z^{-1} + z^{-2}}{1 - 0.495\,935z^{-1} + 0.761\,864z^{-2}}10.880\,490$$

Next, we analyze the effects of coefficient quantization errors. Specifically, we check that the specified coefficient wordlength is adequate for stability and to meet the frequency response specifications. As the poles are not very close to the unit circle a 16-bit coefficient wordlength is adequate for stability. For example, for the first filter section the FWA program shows that as few as three bits are adequate for stability, and quantizing the coefficients to 16 bits altered the poles radius by a mere 0.000 48%. The FWA program also showed that only 12 bits are required to maintain the frequency response within the tolerance limits. With 16-bit coefficient wordlength, the response of the filter is essentially the same as the unquantized filter response. The frequency response and pole–zero diagram for the unquantized filter are depicted in Figure 8.32.

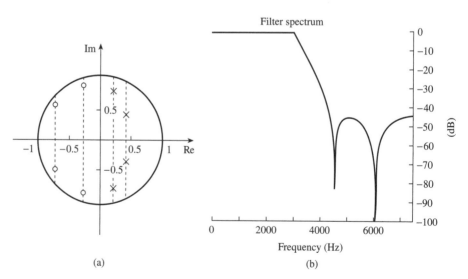

(a)   (b)

Figure 8.32   (a) The pole–zero diagram, and (b) the frequency response for the detailed design example.

Double-length accumulation of sums of products and post-accumulation quantization are used to keep the effects of roundoff noise to a minimum.

Stage 5

The quantized coefficients (obtained by multiplying the scaled coefficients by 2^{15}) are entered into the TMS320C25 IIR filter program listed in the appendix. Chapter 12 gives a more detailed discussion of the IIR filtering routines and their development.

8.17 Summary

The design of IIR filters can be divided into five interrelated stages (see Figure 8.1). Filter specifications are often application dependent, but should include details such as bandedges, tolerance limits for the amplitude response, sampling rate and I/O requirements. For filters with standard characteristics, the coefficients required to meet the amplitude response specifications can be efficiently obtained via the BZT. This approach as well as other useful coefficient calculation methods are described in the text and illustrated with many examples. Higher-order IIR filters are realized as cascades or parallel combinations of second- and first-order sections to keep changes in the positions of the poles and zeros due to the effects of finite wordlength small. Scaling the input to each section to prevent overflow in the internal nodes is necessary.

The performance of an IIR digital filter is limited by the number of bits used in its implementation. The four common sources of errors are those due to (1) input quantization, (2) coefficient quantization, (3) product roundoff and (4) addition overflow. Techniques for analyzing their effects on filter performance and, where possible, for eliminating or minimizing them have been presented. Coefficient wordlength must be adequate to minimize the effects of coefficient quantization on the frequency response and to prevent the possibility of instability. The stability of an IIR filter is always of concern. An IIR filter that is otherwise stable when implemented with infinite precision may become unstable if implemented with finite precision. In high fidelity audio work, for example, 24-bit coefficients are said to be necessary for processing low frequency audio signals. In most other cases, representing the coefficients with 16 or more bits and carrying out the arithmetic operations with double-length accumulators are sufficient to minimize the effects of finite wordlength.

Truncation or roundoff errors due to finite precision arithmetic operations create a nonlinear effect in the filter, such as limit cycles, whereby the filter output oscillates even in the absence of an input (or a constant input). The effects of roundoff error on filter performance can be quantified in terms of SNRs at the filter output. Reduction in the SNR due to roundoff error can be offset by the use of the error spectral shaping (ESS) scheme (see Chapter 13 for details). The primary effect of such schemes is to nullify the 'amplifying' effect of the poles of the filter on the roundoff errors. The price paid for this is an increase in the number of multiplications and additions, although first-order ESS with integer coefficients is computationally efficient.

Design programs are provided on the CD in the companion handbook to enable designers to calculate the filter coefficients and to analyze some of the effects of finite wordlength on filter performance (see the Preface for details).

8.18 Application examples in digital audio and instrumentation

This section provides an overview of several applications where the IIR filter either has been used or is appropriate.

8.18.1 Digital audio

Digital filters have found use in many areas of digital audio, especially in systems with high quality digital sources such as the CD player and DAT, where it makes sense to carry out digitally as many of the signal processing operations as possible. DSP also makes it possible to generate acoustic properties of locations such as concert halls, jazz clubs and discos. Applications in digital audio where an IIR filter has been used include graphic equalization, tone control, channel equalization, noise shaping in ADC/DAC, and band splitting.

In digital graphic equalizers, for example, IIR filters are used to split the entire audio frequency range into bands enabling a comprehensive tone adjustment of the reproduced sound to personal taste instead of the bass and treble controls. A typical five-band graphic equalizer would split the audio frequency range into five bands with centre frequencies of 100 Hz, 330 Hz, 3.3 kHz, 10 kHz and 16 kHz and allow adjustable signal level in each band in the range ±10 dB.

A simple filtering arrangement for graphic equalization is shown in Figure 8.33.

8.18.2 Digital control

With the increased awareness of the benefits of DSP and the availability of low cost processors, controllers are now being implemented digitally to achieve better accuracy and flexibility. Figure 8.34 shows the principles of digital control of an analog plant, $H(s)$, which could be a car or motor, for example. In general, the digital controller has an IIR characteristic.

8.18.3 Digital frequency oscillators

IIR filters have been used to generate accurate waveforms instead of the traditional look-up table approach. The approach exploits the fact that an IIR filter with poles on

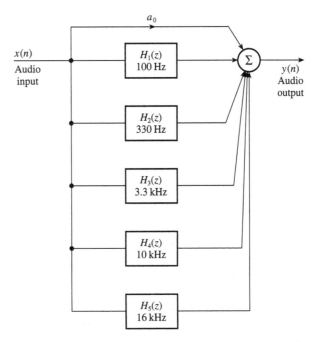

Figure 8.33   Simplified diagram of an all-digital graphic equalizer. The main component is a bank of parallel IIR filters with different centre frequencies. The gain of each filter is individually adjustable, for example with a sliding potentiometer, in the range ±10 dB, say.

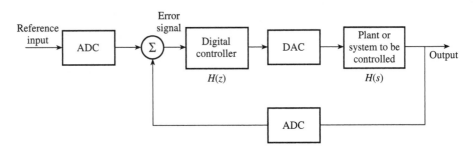

Figure 8.34   Principles of digital control of an analog plant.

the unit circle is essentially unstable. A simple sine wave oscillator is depicted in Figure 8.35(a). The poles of the IIR filter are located at $e^{\pm j\theta}$ and the frequency of oscillation is given by

$$\theta' = w_0 T_B$$

where T is the sampling period. The filter coefficient, $2 \cos \theta'$, is constrained to be an integer by taking the integer part of $2^B \times 2 \cos \theta'$ (B is the number of bits).

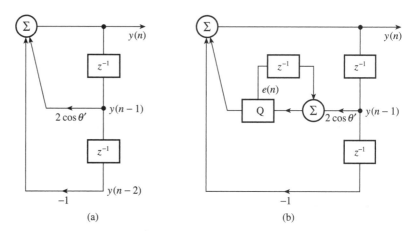

Figure 8.35 (a) A simple digital oscillator and (b) a simple digital oscillator with a first-order error spectral shaping.

A major issue in digital waveform generation using IIR filters concerns finite wordlength effects. For example, coefficient quantization would lead to unequally spaced frequencies, whereas product quantization leads to a buildup of roundoff errors which soon renders the waveform generator useless. However, using ESS techniques these errors can be reduced to a minimum. Figure 8.35(b) shows an oscillator utilizing the ESS technique (Abu-el-Haija and Al-Ibrahim, 1986), which offers a significant reduction in the roundoff noise effects.

8.19 Application examples in telecommunication

IIR filters are widely used in digital communication because of their characteristic features. In digital telephony (Feeney *et al.*, 1971), PCM permits the simultaneous transmission of many voice channels. Each channel is sampled at 8 kHz after band-limiting and is encoded using either the A-law or the μ-law. At the receiving end, the PCM data is converted back to analog and anti-image filtered. Digital IIR filters can be used to provide the necessary filtering at the transmit and receive ends (Figure 8.36). In this case, the filtering is performed at a higher sampling rate, for example 32 kHz, and then converted from linear to standard PCM codes.

In the next two sections, we will discuss in more detail two specific applications of IIR filters in digital communication – one for digital pushbutton telephones and the other for clock recovery in data communication.

8.19.1 Touch-tone generation and reception for digital telephones

An excellent application of IIR filter is the all digital dual-tone multifrequency touch-tone receiver (Jackson *et al.*, 1968; Mock, 1985).

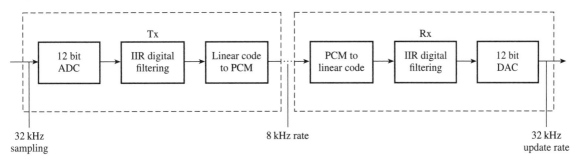

Figure 8.36   A PCM channel showing possible use of IIR filters for the main anti-aliasing filter (TX end) and anti-imaging filtering (RX end).

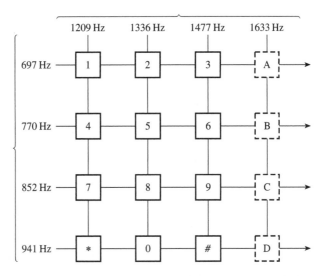

Figure 8.37   A simplified diagram of a 4×4 keypad for a touch-tone telephone. The buttons in broken lines are not available. Pressing a button generates a pair of tones, one from the low frequency group and the other from the high frequency group. For example, pressing 9 generates 852 Hz and 1477 Hz tones (after Mock, 1985).

In modern telephone systems, the information required to establish communication, and for maintenance and charging, is normally provided by a multifrequency code. Typically, the telephone set generates two tones, one low frequency tone and another high frequency tone (see Figure 8.37).

The tone generator can be implemented using a pair of programmable second-order IIR oscillators (Figure 8.38). When a button is pressed the code for the dialled digit is used to select the appropriate filter coefficients and initializing conditions from ROMs to produce a pair of tones (one high frequency tone and one low frequency tone). The tones are added to produce the touch-tone signal. As with the digital sine wave

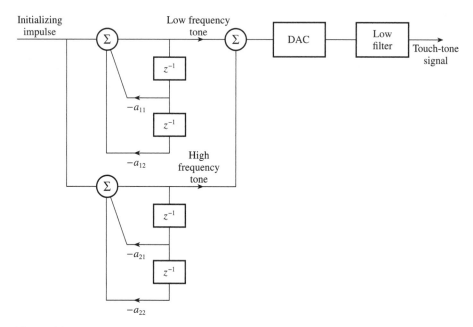

Figure 8.38　Touch-tone generator (after Mock, 1985). The code for the digit dialled is used to select filter coefficients and initial conditions and hence the oscillator frequency.

generator, the performance of the touch-tone generator can be improved by using an error feedback scheme.

At the receiving end, the information is digitized at a rate of 8 kHz and then separated into a low and a high frequency band by the front-end bandpass IIR filters. To detect the presence of a tone, level detection is carried out. This is performed by combined bandpass filtering and full wave rectification followed by lowpass filtering. To determine which of the low frequency tones is present the low frequency band is split into four bands by two sets of four BPFs. The same is true of the high frequency band. The resulting eight levels are passed to decision logic to determine the received code.

8.19.2　Digital telephony: dual tone multifrequency (DTMF) detection using the Goertzel algorithm

The Goertzel algorithm may be used as an alternative to standard IIR filters for the detection of DTMF tones (Mock, 1985; Marven, 1990; Chen, 1996; Texas Instruments, 1997). The Goertzel algorithm is a special IIR filter implementation of the discrete Fourier transform (DFT). The block diagram of a DTMF detection scheme based on the Goertzel algorithm is depicted in Figure 8.39. It consists of a parallel bank of eight pairs of Goertzel filters. Each filter pair detects a DTMF tone and its

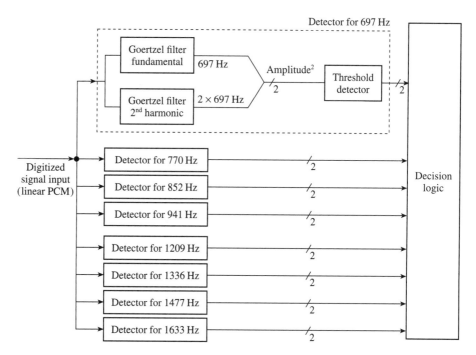

Figure 8.39 Principles of DTMF decoding using Goertzel filters (after Mock, 1985).

second harmonic. The second harmonic is required to distinguish between speech and DTMF tones. Speech has significant even-order harmonics whereas the DTMF signals do not. Each filter output is squared to obtain a measure of signal strength at each of the eight DTMF frequencies and their second harmonics. The strongest signal pair from the high and low frequency groups are used to determine the received digit.

Each Goertzel filter is a high Q, narrowband, second-order, bandpass IIR filter characterized by the following transfer function (see Figure 8.40):

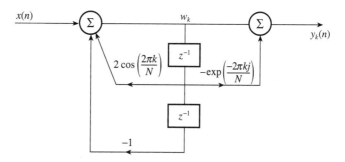

Figure 8.40 Structure of a second-order Goertzel filter.

$$H_k(z) = \frac{1 - W_N^k z^{-1}}{1 - 2\cos\left(\dfrac{2\pi k}{N}\right)z^{-1} + z^{-2}}$$ (8.53a)

where

$$W_N^k = \exp\left(-\frac{2\pi k j}{N}\right)$$

The difference equations for the filter are

$$v_k(n) = 2\cos\left(\frac{2\pi k}{N}\right)v_k(n-1) - v_k(n-2) + x(n), \; n = 0, 1, \ldots, N$$ (8.53b)

$$y_k(n) = v_k(n) - W_N^k v_k(n-1)$$ (8.53c)

where

$$w_k(-1) = w_k(-2) = 0$$

For DTMF tone detection, only the magnitude of the DTMF signal is required (the phase information is ignored) and so the GA is modified to produce magnitude squared outputs only:

$$|y_k(N)|^2 = v_k^2(N) + v_k^2(N-1) - 2\cos\left(\frac{2\pi k}{N}\right)v_k(N)v_k(N-1)$$ (8.53d)

The modified DTMF algorithm (Equations 8.53b and 8.53d) requires only one real coefficient, i.e.

$$2\cos\left(\frac{2\pi k}{N}\right)$$

to calculate the magnitude of each DTMF signal, and avoids complex arithmetic. The filter output, Equation 8.53d, is computed only once for each DTMF tone for $n = N$, i.e. at the end of the iterations for the feedback path of the filter (Equation 8.53b).

Advantages of the Goertzel algorithm include that it needs only one real coefficient per DTMF frequency to determine the magnitude of the signal, takes up little memory space and executes very fast. Unlike the FFT, it does not wait for the complete data set before processing, but processes each sample as it arrives. With FFTs, the value of N would normally be limited (typically to a power of 2) for efficiency. In the Goertzel algorithm, any integer value of N may be used, although the choice of N is a compromise between the frequency resolution and computation time.

The DFT size, N, and frequency bin number, k, determine the value of the coefficient of the filter and hence the frequency detected.

Table 8.3 Parameters for a DTMF decoding scheme (after Mock, 1985).
Sampling rate: 8 kHz
DFT size: 205 (first harmonic), 210 (second harmonic).

DTMF frequency (Hz)	Discrete frequency bin, k (first harmonic)	Discrete frequency bin, k (second harmonic)
697	18	35
770	20	39
852	22	43
941	24	47
1209	31	61
1336	34	67
1477	38	74
1633	42	82

The DFT size, N, discrete frequency bin, k, sampling interval, T, and DTMF tone frequency, f_k, are related as

$$f_k = \frac{k}{NT}$$

The sampling frequency and tone frequencies are set in accordance with international standards. The DFT size, N, may be varied. Table 8.3 lists the parameters for a possible decoding scheme (Mock, 1985).

It should be pointed out that the Goertzel filter has its poles on the unit circle and is sensitive to finite wordlength effects and these should not be ignored. Further, if the number of frequency points to be determined is relatively large the FFT may be more suitable.

Example 8.22   (a)   The discrete Fourier transform (DFT) of a data sequence, $x(n)$, $n = 0, 1, \ldots, N - 1$, may be defined as

$$X(k) = \sum_{m=0}^{N-1} x(m) W_N^{km}, \quad k = 0, 1, \ldots, N - 1$$

where W_N^{km} is the twiddle factor.

(i)   Starting with the above equation, show that the z-plane transfer function, $H_k(z)$, for a Goertzel filter for DTMF tone detection can be expressed in the following recursive form:

$$H_k(z) = \frac{1 - W_N^k z^{-1}}{1 - 2\cos\left(\dfrac{2\pi k}{N}\right) z^{-1} + z^{-2}}$$

(ii) Deduce an expression for the magnitude squared output of the Goertzel filter, $|y_k(n)|^2$, at the discrete time, $n = N$, and hence show that complex arithmetic is not required in the modified Goertzel algorithm.

(b) A DTMF tone detection scheme for a pushbutton telephone system is based on the specifications summarized in Table 8.3 and uses a second-order Goertzel filter.

Calculate the coefficients for the Goertzel filter to decode the digits at the receiver if the number dialled is '99'.

Solution   (a) (i) Now

$$X(k) = \sum_{m=0}^{N-1} x(m) W_N^{km}$$

Exploiting the periodicity of the twiddle factor, we can write the DFT equation as follows (since $W_N^{-kN} = 1$):

$$X(k) = W_N^{-kN} \sum_{m=0}^{N-1} x(m) W_N^{km}$$

$$= \sum_{m=0}^{N-1} x(m) W_N^{-k(N-m)} \tag{8.54}$$

Equation 8.54 has the same form as the convolution equation. Thus, if we define a data sequence, $y_k(n)$, as follows:

$$y_k(n) = \sum_{m=0}^{N-1} x(m) W_N^{-k(n-m)} \tag{8.55}$$

it is seen that $y_k(n)$ can be viewed as the output of an FIR filter which has as input the data sequence, $x(m)$, and N coefficients, $h_k(n)$, given by

$$h_k(n) = W_N^{-kn} \tag{8.56}$$

Comparing Equations 8.54 and 8.55, it is seen that the filter output when $n = N$ gives the DFT, $X(k)$, at the frequency bin, k:

$$X(k) = y_k(n)|_{n=N}$$

The z-plane transfer function of the filter is given by

$$H_k(z) = \sum_{n=0}^{\infty} W_N^{-kn} z^{-n}$$

$$= \frac{1}{1 - W_N^{-k} z^{-1}} \tag{8.57}$$

This is a first-order filter with a single complex pole (as the twiddle factors are complex) on the unit circle at $z = W_N^{-k}$. To avoid complex arithmetic, the single poles are combined into complex conjugate pole pairs to give a second-order filter section. This may be achieved by multiplying Equation 8.57 by

$$\left(\frac{1 - W_N^k z^{-1}}{1 - W_N^k z^{-1}} \right)$$

to give

$$H_k(z) = \frac{1 - W_N^k z^{-1}}{1 - 2\cos\left(\dfrac{2\pi k}{N}\right) z^{-1} + z^{-2}}$$

(ii) From Figure 8.40, the two-step difference equation for the Goertzel filter is

$$v_k(n) = 2\cos\left(\frac{2\pi k}{N}\right) v_k(n-1) - v_k(n-2) + x(n)$$

$$y_k(n) = v_k(n) - W_N^k v_k(n-1)$$

At $n = N$

$$y_k(N) = v_k(N) - W_N^k v_k(N-1)$$

$$= v_k(N) - v_k(N-1)\left[\cos\left(\frac{2\pi k}{N}\right) - j\sin\left(\frac{2\pi k}{N}\right)\right]$$

$$= v_k(N) - v_k(N-1)\cos\left(\frac{2\pi k}{N}\right) + j v_k(N-1)\sin\left(\frac{2\pi k}{N}\right)$$

$$|y_k(N)|^2 = (real\ part)^2 + (imaginary\ part)^2$$

$$= \left[v_k(N) - v_k(N-1)\cos\left(\frac{2\pi k}{N}\right)\right]^2 + \left[v_k(N-1)\sin\left(\frac{2\pi k}{N}\right)\right]^2$$

$$= v_k^2(N) - 2v_k(N)v_k(N-1)\cos\left(\frac{2\pi k}{N}\right)$$

$$+ v_k^2(N-1)\cos^2\left(\frac{2\pi k}{N}\right) + v_k^2(N-1)\sin^2\left(\frac{2\pi k}{N}\right)$$

$$= v_k^2(N) - 2v_k(N)v_k(N-1)\cos\left(\frac{2\pi k}{N}\right)$$

$$+ v_k^2(N-1)\left[\cos^2\left(\frac{2\pi k}{N}\right) + \sin^2\left(\frac{2\pi k}{N}\right)\right]$$

$$= v_k^2(N) + v_k^2(N-1) - 2\cos\left(\frac{2\pi k}{N}\right)v_k(N)v_k(N-1) \qquad (8.58)$$

(b)   The DTMF tones for the digit '9' are 1477 Hz and 852 Hz. The detection of each tone requires a pair of Goertzel IIR filters. For the 1477 tone, the frequency bins are 38 and 74:

$$a_1 = 2\cos\left(\frac{2\pi \times 38}{205}\right) = 0.79; \quad a_2 = -1$$

$$a_1' = 2\cos\left(\frac{2\pi \times 74}{210}\right) = -1.1996; \quad a_2' = -1$$

For the 852 Hz tone, the coefficients are:

$$a_1 = 1.5623, \, a_2 = -1; \qquad a_1' = 0.5, \, a_2' = -1$$

8.19.3   Clock recovery for data communication

A fundamental problem in most digital data communication over long distances is that of generating a clock at the receiving end at the correct frequency and phase so that the data may be correctly decoded. The clock is normally derived from the received data.

Traditionally, analog circuits, for example using phase lock loops, are used to recover the clock, but these are susceptible to drift with age and temperature. Further, such circuits are unsuitable in applications that involve burst transmissions because of their slow response or where more than one data rate is involved (Smithson, 1992).

The input data stream is normally scrambled at the transmitting end (to provide clock information during idle periods) and then encoded, with each code representing a symbol. The codes are then transmitted at the so-called symbol rate. The problem at the receiving end is to recover the symbol clock.

The principles of symbol clock recovery using DSP are shown in Figure 8.41. The data stream is added modulo-2, that is exclusive ORed, with a version of itself delayed by one-half of a clock period to produce an output which contains level changes at the symbol rate (point C). The data is then applied to a marginally stable bandpass IIR filter. The impulse response of such a filter decays very slowly with time, producing a 'damped oscillation' at a frequency w_0, the centre frequency of the filter. The use of a marginally stable filter ensures that there is an output even when there are no transitions in the input data stream for reasonably long periods. The sampling frequency of the filter is chosen to be a multiple of the symbol rate. The desired symbol clock is derived from the output of the filter by zero-crossing detection (point E in Figure 8.41). For a 2's complement representation, this is trivially implemented by examining the signs of the data samples at the output of the digital filter.

A simple all-pole, IIR filter of the form shown in Figure 8.42 may be used for the symbol clock recovery. The filter is characterized by the following transfer function:

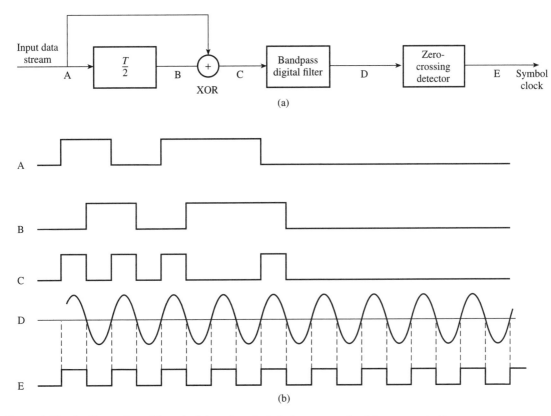

Figure 8.41   An illustration of the principles of symbol clock recovery for data communication.

$$H(z) = \frac{1}{[z - r\exp(-jw_0 T)][z - r\exp(jw_0 T)]}$$

$$= \frac{1}{z^2 - 2r\cos(w_0 T)z - r^2}$$

where w_0 is the centre frequency of the bandpass filter, r is the radius of the pole and T is the reciprocal of the sampling frequency. w_0 is normally chosen to be identical or very close to the symbol clock frequency to be recovered, and the sampling frequency is a multiple of the centre frequency. The bandwidth of the filter is determined by the radius of the pole (see Equation 8.4). To ensure that the impulse response decays slowly with time the pole is normally located very close to the unit circle, typically in the range $0.99 < r < 1$. As discussed in Section 8.5.1 (Equation 8.5), the pole radius, r, and the bandwidth of the filter, bw, are related as

$$r \approx 1 - (\text{bw}/F_s)\pi$$

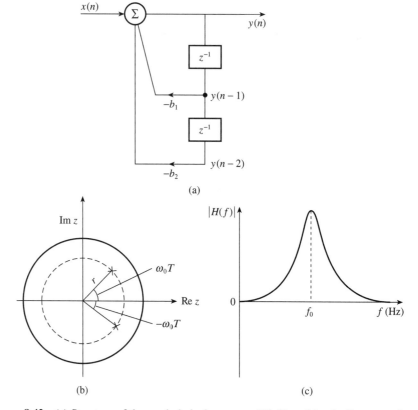

Figure 8.42 (a) Structure of the symbol clock recovery IIR filter, (b) pole diagram and (c) spectrum of filter.

where $F_s = 1/T$ is the sampling frequency of the filter.

For example, to recover the symbol clock for a hypothetical 4800 baud modem, suitable filter parameters are

data rate	4.8 kbaud
filter centre frequency, f_0	4.8 kHz
sampling frequency	153.6 kHz
bandwidth, bw	100 Hz

In this case, the pole radius (from the above equation), $r = 0.997\,954\,69$, and the pole angle $w_0 T = 2\pi f_0 T = (2\pi \times 4.8 \times 10^3 / 153.6 \times 10^3) = 0.196\,35$ rad $\approx 11.25°$. The resulting transfer function becomes

$$H(z) = \frac{1}{z^2 - 1.957\,558z + 0.995\,913}$$

As discussed earlier in the chapter, finite wordlength effects must be considered if the filter is to operate as desired. In particular, the input to the filter needs to be scaled to avoid the possibility of self-sustaining oscillations at its output due to overflow, and the use of a simple roundoff noise shaping scheme may help to produce a 'clean' clock. In a practical clock recovery system, a second filter stage will be necessary to improve the system performance when the input data is a sequence of 1s or 0s (Smithson, 1992).

Problems

8.1 A lowpass filter has poles and a zero at the following locations:

zero, -0.5; poles, 0.370, $0.6 \pm 0.5j$

(1) Plot the pole–zero diagram.

(2) Obtain the transfer function, $H(z)$.

8.2 Digitize, using the impulse invariant method, the analog filter with the transfer function

$$H(s) = \frac{\alpha}{s(s + \alpha)}, \quad \alpha = 0.5$$

Assume a sampling frequency of 1 (normalized).

8.3 A requirement exists to simulate in a digital computer an analog system with the following normalized characteristic:

$$H(s) = \frac{1}{s^2 + \sqrt{2}s + 1}$$

Obtain a suitable transfer function using

(1) the impulse invariant method, and

(2) the bilinear transform method.

Assume a sampling frequency of 5 kHz and a 3 dB cutoff frequency of 1 kHz.

8.4 Determine, using the BZT method, the transfer function and difference equation for the digital equivalent of the resistance–capacitance (RC) filter shown in Figure 8.43. Assume a sampling frequency of 150 Hz and a cutoff frequency of 30 Hz.

8.5 An IIR digital bandpass filter meeting the specifications given below is required. Starting with a suitable normalized analog LPF, (i) obtain

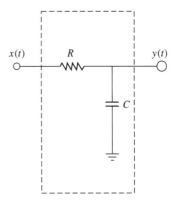

Figure 8.43

the coefficients of $H(z)$, in factored form, using an appropriate band transformation and the BZT, and (ii) sketch the pole–zero diagrams for the analog filters and the resulting digital bandpass filter.

passband	8–10 kHz
sampling frequency	32 kHz
bandpass filter order	4
filter type	Butterworth

8.6 (a) Comment on the practical implications of the 'warping effect' associated with the bilinear z-transform (BZT) design method in applications where the response of the analog filter to be digitized is close to the Nyquist frequency. Give an example of a specific application.

(b) A requirement exists to design an IIR highpass filter, with a Butterworth characteristic, meeting the following specifications:

passband	2–4 kHz
stopband	0–500 Hz
passband ripple	3 dB
stopband attenuation	20 dB
sampling frequency	8 kHz

Determine:

(i) the pass and stopband edge frequencies for a suitable analog prototype lowpass filter. Your answer must include details of the basic concepts of how the prototype lowpass filter is obtained from the specifications;

(ii) the order, N, of the prototype lowpass filter;

(iii) the transfer function, and hence the coefficients of the IIR filter using the BZT method.

8.7 A requirement exists for a digital bandpass filter with Butterworth characteristics and meeting the following specifications:

passband	200–300 Hz
sampling frequency	2000 Hz
filter order	2

(a) (i) Starting with a suitable prototype analog lowpass filter, determine the transfer function of the digital filter using the bilinear z-transform method.

(ii) Explain, with the aid of sketches of pole–zero diagrams, how the poles and zeros of the prototype filter have been successively mapped from the s-plane to the z-plane.

(b) Assume the filter in (a) is to be implemented in fixed-point arithmetic. Assess the effects of quantizing the coefficients to 8 bits on the pole positions and hence the centre frequency.

Note. The lowpass-to-bandpass transformation is given by

$$s \rightarrow \frac{s^2 + \omega_0^2}{Bs}$$

where

$$\omega_0^2 = \omega_1'\omega_2'; \qquad B = \omega_2' - \omega_1'$$

ω_1', ω_2' and ω_0 are the lower and upper band-edge frequencies, and the centre frequency respectively.

8.8 Obtain suitable coefficients for an IIR digital filter using the bilinear transform method and an elliptic characteristic to meet the following specifications:

passband	4–12 kHz
stopbands	0–3.4 kHz,
	12.6–16 kHz
passband ripples	<0.1 dB
stopband attenuation	>30 dB
sampling frequency	32 kHz

Determine a suitable coefficient wordlength to ensure that the filter is stable and the frequency response lies within the specified limits.

8.9 Design and implement in software a digital lowpass filter to meet the following specifications:

passband edge	2.5 kHz
stopband edge	3 kHz
passband deviation	<0.1 dB
stopband attenuation	>60 dB
sampling frequency	15 kHz

8.10 Obtain, via the bilinear transform, the coefficients of a digital filter that is maximally flat in the passband, 0 to 4 kHz, and has an attenuation of at least 25 dB at frequencies over 10 kHz. Assume a sampling frequency of 32 kHz.

8.11 The requirements for a certain lowpass filter are given below:

passband	0–30 Hz
stopband edge	50 Hz
stopband attenuation	>40 dB,
	at $f > 50$ Hz
sampling frequency	256 Hz

Assuming that the filter has a Butterworth response, determine via the bilinear transformation the transfer function, $H(z)$, of the filter. Obtain the realization diagram for the filter in cascade form using second- and/or first-order sections.

8.12 A bandpass digital filter with a Butterworth response is required for a certain real-time digital signal processing system. The filter is to satisfy the following requirements:

passband	0.3–3.4 kHz
stopband	0–0.2 kHz and 4–8 kHz
stopband attenuation	25 dB
sampling frequency	32 kHz

Obtain a suitable transfer function for the filter using the bilinear transform method.

8.13 A digital filter is required to remove baseline wander and artefacts due to body movement in a certain biomedical application. The filter is required to meet the following requirements:

passband	1–30 Hz
stopband	0–0.5 Hz and 40–128 Hz
passband ripple	<0.1 dB
stopband attenuation	>30 dB
sampling frequency	256 Hz

Determine the order of a suitable IIR filter and its transfer function $H(z)$.

8.14 A narrowband reject digital filter is required to remove an interfering signal. The filter should satisfy the following specifications:

passband edges	45 Hz and 55 Hz
passband ripple	< 0.1 dB
stopband attenuation	> 50 dB
sampling frequency	500 Hz

Obtain the coefficients of the filter.

8.15 Determine, using the impulse invariant method, the transfer function and difference equation for the digital equivalent of a single-pole RC lowpass filter. Assume a sampling frequency of 150 Hz and a 3 dB cutoff frequency of 30 Hz.

8.16 A standard second-order analog filter section, with simple poles, can be expressed as

$$\frac{A_0 + A_1 s}{B_0 + B_1 s + B_2 s^2} = \frac{C_1}{s + p_1} + \frac{C_2}{s + p_2}$$

where C_1 and C_2 are partial fraction coefficients and p_1 and p_2 are s-plane poles. Assume the impulse invariant transformation for the second-order section is given by

$$\frac{A_0 + A_1 s}{B_0 + B_1 s + B_2 s^2}$$

$$\rightarrow \frac{c_1 + c_2 - (c_1 e^{-p_2 T} + c_2 p^{-p_1 T}) z^{-1}}{1 - (e^{-p_1 T} + e^{-p_2 T}) z^{-1} + e^{-(p_1 + p_2) T} z^{-2}}$$

$$= \frac{a_0 - a_1 z^{-1}}{1 + b_1 z^{-1} + b_2 z^{-2}}$$

where T is the sampling interval.

(1) Find expressions for p_1, p_2, C_1 and C_2 in terms of $A_0, A_1, B_0, B_1,$ and B_2.

(2) Obtain expressions for the coefficients a_0, a_1, b_1 and b_2 for the case when the poles are complex conjugates.

(3) Repeat part (2) when the poles are real and unequal.

(4) Given the following normalized analog transfer function, use your results to obtain the coefficients of an equivalent discrete-time filter. Assume a sampling frequency of 10 kHz and a cutoff frequency of 2 kHz.

$$H(s) = \frac{1}{s^2 + \sqrt{2}s + 1}$$

8.17 Starting with a suitable analog Chebyshev LPF, obtain the transfer function of a digital bandstop filter to meet the following specifications:

stopband	10–15 kHz
sampling frequency	50 kHz
passband ripple	0.5 dB
filter order	6

8.18 Starting with a suitable analog Chebyshev LPF and the biquadratic transformation given below, determine, using the s-plane to z-plane pole–zero mapping approach, the transfer function of a digital bandpass filter to meet the following specifications:

passband	10–15 kHz
sampling frequency	50 kHz
passband ripple	0.5 dB
bandpass filter order	6

Alternatives to the BZT for the bandpass and bandstop filters are the following biquadratic transformations (Gold and Rader, 1969; Gray and Markel, 1976):

$$s = \cot \left[\frac{(\omega_2 - \omega_1) T}{2} \right] \left[\frac{z^2 - 2z \cos \gamma + 1}{z^2 - 1} \right]$$

lowpass to bandpass

$$s = \tan\left[\frac{(\omega_2 - \omega_1)T}{2}\right]\left[\frac{z^2 - 1}{z^2\ 2z\cos\gamma + 1}\right]$$

lowpass to bandstop

where

$$\cos\gamma = \cos\left[\frac{(\omega_2 + \omega_1)T}{2}\right]\Bigg/\cos\left[\frac{(\omega_2 - \omega_1)T}{2}\right]$$

ω_1 and ω_2 are the lower and upper bandedge frequencies (passband edge frequencies for BPFs, stopband edge frequencies for BSFs) and γ is the centre frequency.

8.19 An analog lowpass filter is characterized by a pair of s-plane poles at

$$p_{1,2} = -1.4 \pm 1.2936j$$

It is desired to convert the filter into a digital bandpass filter with passband edges of 3 kHz and 5 kHz at a sampling frequency of 15 kHz. Given the digital lowpass to bandpass transformation and the BZT in Equation 8.40, determine

(1) the poles and zeros of the digital bandpass filter, and

(2) its transfer function in factored form.

8.20 A dual-tone multifrequency (DTMF) detection scheme for a digital telephone uses a series of second-order Goertzel filters to extract the DTMF tones and their second harmonics. Assume that the DTMF tones for digit '0' are 941 Hz and 1336 Hz. Determine the values of the coefficients in the feedback path of the Goertzel filters for the low frequency tone, 941 Hz, if the sequence lengths $N = 205$ and 210 are used for the fundamental and second harmonic, respectively, with corresponding discrete frequency bins of 24 and 47.

8.21 (a) Explain, with the aid of a diagram, the principles of a dual-tone multifrequency (DTMF) detection scheme for push-button telephones using the Goertzel algorithm.

(i) The discrete Fourier transform (DFT) of a data sequence, $x(n)$, $n = 0, 1, \ldots,$ $N - 1$, may be defined as

$$X(k) = \sum_{m=0}^{N-1} x(m)W_N^{km},$$

$$k = 0, 1, \ldots, N-1$$

where W_N^{km} is the twiddle factor.

(ii) Starting with the above equation, show that the z-plane transfer function, $H_k(z)$, for a Goertzel filter for DTMF tone detection can be expressed in the following recursive form:

$$H_k(z) = \frac{1 - W_N^k z^{-1}}{1 - 2\cos\left(\dfrac{2\pi k}{N}\right)z^{-1} + z^{-2}}$$

(iii) Deduce an expression for the magnitude squared output of the Goertzel filter, $|y_k(n)|^2$, at the discrete time, $n = N$, and hence show that complex arithmetic is not required in the modified Goertzel algorithm.

(b) A DTMF tone detection scheme for a push-button telephone system is based on the specifications summarized in Table 8.3 and uses a second-order Goertzel filter.

Calculate the coefficients for the Goertzel filter to decode the digits at the receiver if the number dialled is '00'.

Compare the computational complexities of the Goertzel algorithm with the radix-2 FFT algorithm for DTMF tone detection, and hence justify the use of the Goertzel algorithm, instead of an FFT algorithm, for this application. State any reasonable assumptions made.

MATLAB problems

8.22 A requirement exists for a bandpass digital filter with a Chebyshev characteristic to meet the following specifications:

passband	1200–1800 Hz
stopband attenuation	>30 dB
passband ripple	<0.5 dB
transition width	400 Hz
sampling frequency	7.5 kHz

Using a suitable software program, obtain the filter coefficients.

8.23 An analog filter is to be converted into an equivalent digital filter that will operate at a sampling frequency of 256 Hz. Assume that the analog filter has the following transfer function:

$$H(s) = \frac{1}{s^3 + 2s^2 + 2s + 1}$$

(1) Obtain suitable coefficients for the digital filter.

(2) Assuming that the digital filter is to be realized using the cascade structure, draw a suitable realization block diagram and develop the difference equations.

(3) Repeat part (2) for the parallel structure.

8.24 An IIR filter has the following transfer function:

$$H(z) = \frac{0.1436 + 0.2872z^{-1} + 0.1436z^{-2}}{1 - 1.8353z^{-1} + 0.9748z^{-2}}$$

(1) Determine the positions of the poles and zeros and sketch the pole–zero diagram.

(2) Determine the radial distance of the pole from the origin.

(3) Estimate the number of bits required to represent each coefficient

 (a) to maintain stability, and

 (b) so that the amplitude response in the passband does not change by more than 1%.

8.25 Compare the matched z-transform, the impulse invariant and the bilinear z-transform methods in terms of

(a) the Nyquist effect on the magnitude frequency, phase and group delay responses;

(b) the distribution of the pole–zero diagrams.

Use MATLAB and the following filters as vehicles in your study.

(1) Lowpass filter
An elliptic, lowpass digital filter with the following specifications:

passband	0–1 kHz
stopband	3–5 kHz
passband ripple	1 dB
stopband attenuation	60 dB
sampling frequency	10 kHz

(2) Highpass filter
An elliptic, highpass digital filter with the following specifications:

stopband	0–1 kHz
passband	3–5 kHz
passband ripple	1 dB
stopband attenuation	60 dB
sampling frequency	10 kHz

(3) Bandpass filter

 (a) A Butterworth, bandpass digital filter with the following specifications:

passband	200–300 Hz
sampling frequency	2000 Hz
filter order	8

 (b) A Butterworth, bandpass digital filter with the following specifications:

passband	800–900 Hz
sampling frequency	2000 Hz
filter order	8

(4) Bandstop filter
An elliptic, bandstop digital filter with the following specifications:

passbands	0–15 kHz, 30–50 kHz
stopband	20–25 kHz
passband ripple	0.2 dB
stopband attenuation	40 dB
sampling frequency	100 kHz

8.26 Compare the features of digital Butterworth, Chebyshev type I, Chebyshev type II and elliptic filters in terms of:

(a) the distribution of their poles and zeros;

(b) filter order;

(c) transition widths, pass and stopband ripples of the magnitude–frequency responses;

(d) phase and group delay responses.

Use the following filters in your study.

(1) Lowpass filter

passband	0–500 Hz
stopband	2–4 kHz
passband ripple	3 dB
stopband attenuation	20 dB
sampling frequency	8 kHz

(2) Highpass filter

stopband	0–500 Hz
passband	2–4 kHz
passband ripple	3 dB
stopband attenuation	20 dB
sampling frequency	8 kHz

(3) Bandpass filters

(a)
lower passband edge frequency	250 Hz
upper passband edge frequency	300 Hz
lower stopband edge frequency	50 Hz
upper stopband edge frequency	450 Hz
passband ripple	3 dB
stopband attenuation	20 dB
sampling frequency	1 kHz

(b)
passbands	0–15 kHz, 30–50 kHz
stopband	20–25 kHz
passband ripple	0.2 dB
stopband attenuation	40 dB
sampling frequency	100 kHz

(4) Bandstop filters

(a)
passbands	0–50 Hz, 450–500 Hz
stopband	250–300 Hz
passband ripple	3 dB
stopband attenuation	20 dB
sampling frequency	1 kHz

(b)
passbands	0–15 kHz, 30–50 kHz
stopband	20–25 kHz
passband ripple	0.2 dB
stopband attenuation	40 dB
sampling frequency	100 kHz

8.27 (a) (i) Determine, using the BZT method and MATLAB, the transfer function of a discrete bell filter for audio signal processing in a digital mixer when the control settings correspond to a Q-factor of 2, a boots (peak) of 6.02 dB at 10 kHz. Assume a sampling frequency of 48 kHz, and that the equivalent analog filter has an s-plane transfer function given below.

Plot the magnitude, phase and group delay responses and the pole–zero diagram of the discrete filter.

(ii) Repeat the above problem at a sampling frequency of 96 kHz.

(b) Repeat (a) (i) and (ii) using the MZT method.

(c) Repeat (a) (i) and (ii) using the impulse invariant method.

(d) Compare the results in (a) to (c).

$$H(s) = \frac{s^2 + (3 + k)\dfrac{\omega_0}{Q}s + \omega_0^2}{s^2 + (3 - k)\dfrac{\omega_0}{Q}s + \omega_0^2}$$

where

$$k = 3\left(\frac{G-1}{G+1}\right); \quad \omega_0 = \text{boost frequency}$$

G = gain factor; Q = Q-factor.

8.28 (a) Develop a C-language pseudo-code for the detection of DTMF tones and their second-order harmonics. State any assumptions made.

(b) Repeat part (a) using MATLAB.

(c) Generate suitable frequency tones, using MATLAB, and use these to test your MATLAB program for detecting DTMF tones.

References

Abu-el-Haija A. and Al-Ibrahim M.M. (1986) Improving performance of digital sinusoidal oscillators by means of error feedback circuits. *IEEE Trans. Circuits and Systems*, **33**(4), 373–80.

Antoniou A. (1979) *Digital Filters Analysis and Design*. New York: McGraw-Hill.

Chen C.J. (1996) *Modified Goertzel Algorithm in DTMF Detection using the TMS320C80*, Texas Instruments, Application Report SPRA066, June.

Clark R.J., Ifeachor E.C. and Rogers G.M. (1996) Real-time equaliser coefficient realisation with minimised computational load and distortion. Preprint 4360, 101st Audio Engineering Society Convention.

Clark R.J., Ifeachor E.C., Rogers G.M. and Van Eetvelt P.W.J. (2000) Techniques for generating digital equaliser coefficients. *J. Audio Engineering Society*, **48**(4), 281–98.

Dattorro J. (1988) The implementation of recursive digital filters for high-fidelity audio. *J. Audio Engineering Society*, **36**(11), 851–78.

DeFatta D.J., Lucas J.G. and Hodgkiss W.S. (1988) *Digital Signal Processing*. New York: Wiley.

Feeney S.L., Kieburtz R.B., Mina K.V. and Tewksbury S.K. (1971) Design of digital filters for an all digital frequency division multiplex–time division multiplex translator. *IEEE Trans. Circuit Theory*, **18**, 702–11.

Gold B. and Rader C.M. (1969) *Digital Processing of Signals*. New York: McGraw-Hill.

Gray A.H. and Markel J.D. (1976) A computer program for designing elliptic filters. *IEEE Trans. Acoustics, Speech and Signal Processing*, **24**(6), 529–38.

IEEE (1979) *Programs for Digital Signal Processing*. New York: IEEE Press.

Jackson L.B. (1986) *Digital Filters and Signal Processing*. Boston MA: Kluwer.

Jackson L.B., Kaiser J.F. and McDonald H.S. (1968) An approach to the implementation of digital filters. *IEEE Trans. Audio and Electroacoustics*, **16**(3), 413–21.

Jong M.T. (1982) *Methods of Discrete Signal System Analysis*. New York: McGraw-Hill.

Lynn P.A. and Fuerst W. (1989) *Introductory Digital Signal Processing with Computer Applications*. Chichester: Wiley.

Marven C. (1990) General-purpose tone decoding and DTMF detection, in *Theory, Algorithms, and Implementations, Digital Signal Processing Applications with the TMS320 Family*, Vol. 2, literature number SPRA016, Texas Instruments.

Mock P. (1985) Add DTMF generation and decoding to DSP-μp designs. *EDN*, **30**.

Parks T.W. and Burrus C.S. (1987) *Digital Filter Design*. New York: Wiley.

Rabiner L.R. and Gold B. (1975) *Theory and Applications of Digital Signal Processing*. Englewood Cliffs NJ: Prentice-Hall.

Rader C.M. and Gold B. (1967) Effects of parameter quantization on the poles of a digital filter. *Proc. IEEE*, **55**, 688–9.

Smithson P. (1992) Clock recovery for a satellite data modem, University of Plymouth (personal communication).

Stanley W.D., Dougherty G.R. and Dougherty R. (1984) *Digital Signal Processing*, 2nd edn. Reston VA: Reston Publishing, Inc.

Stearns S.D. and Hush D.R. (1990) *Digital Signal Analysis*, 2nd edn. Englewood Cliffs NJ: Prentice-Hall.

Bibliography

Abu-el-Haija A.I. and Peterson A.M. (1979) An approach to eliminate roundoff errors in digital filters. *IEEE Trans. Acoustics, Speech and Signal Processing*, **27**, 195–8.

Ahmed N. and Natarajan T. (1983) *Discrete-time Signals and Systems*. Reston VA: Reston Publishing, Inc.

Allen J. (1975) Computer architecture for signal processing. *Proc. IEEE*, **63**(4), 624–48.

Arjmand M. and Roberts R.A. (1981) On comparing hardware implementations of fixed point digital filters. *IEEE Circuits Systems Mag.*, **3**(2), 2–8.

Avenhaus E. (1972) Filters with coefficients of limited wordlength. *IEEE Trans. Audio Electroacoustics*, **20**, 206–12.

Barnes C.W., Tran B.N. and Leung S.H. (1985) On the statistics of fixed-point roundoff error. *IEEE Trans. Acoustics, Speech and Signal Processing*, **33**, 595–606.

Bellanger M. (1984) *Digital Processing of Signals. Theory and Practice*. New York: Wiley.

Chang T.L. (1978) A low roundoff noise digital filter structure. In *Proc. IEEE Int. Symp. on Circuits and Systems*, May 1978, pp. 1004–8.

Chang T.L. (1979) Error-feedback digital filters. *Electronics Lett.*, **15**, 348–9.

Chang T.L. (1980) Comments on 'An approach to eliminate roundoff errors in digital filters'. *IEEE Trans. Acoustics, Speech and Signal Processing*, **28**(2), 244–5.

Chang T.L. (1981) Suppression of limit cycles in digital filters designed with one magnitude-truncation quantizer. *IEEE Trans. Circuits and Systems*, **28**(2), 107–11.

Chang T.L. (1981) On low-roundoff noise and low-sensitivity digital filter structures. *IEEE Trans. Acoustics, Speech and Signal Processing*, **29**(5), 1077–80.

Chang T.L. and White S.A. (1981) An error cancellation digital-filter structure and its distributed-arithmetic implementation. *IEEE Trans. Circuits and Systems*, **28**(4), 339–42.

Charalambous C. and Best M.J. (1974) Optimization of recursive digital filters with finite wordlengths. *IEEE Trans. Acoustics, Speech and Signal Processing*, **22**(6), 424–31.

Chassaing R. and Horning D.W. (1990) *Digital Signal Processing with the TMS320C25*. New York: Wiley.

Claasen T.A.C.M. and Kristiansson L.O.G. (1975) Necessary and sufficient conditions for the absence of overflow phenomena in a second order recursive digital filter. *IEEE Trans. Acoustics, Speech and Signal Processing*, **23**(6), 509–15.

Claasen T.A.C.M., Mecklenbrauker W.F.G. and Peek J.B.H. (1973) Second-order digital filter with only one magnitude-truncation quantiser and having practically no limit cycles. *Electronics Lett.*, **9**, 531–2.

Claasen T.A.C.M., Mecklenbrauker W.F.G. and Peek J.B.H. (1973) Some remarks on the classification of limit cycles in digital filters. *Philips Research Rep.*, **28**, 297–305.

Claasen T., Mecklenbrauker W.F.G. and Peek J.B.H. (1975) Frequency domain criteria for the absence of zero-input limit cycles in nonlinear discrete-time systems, with applications to digital filters. *IEEE Trans. Circuits and Systems*, **22**, 232–9.

Claasen T.A.C.M., Mecklenbrauker W.F.G. and Peek J.B.H. (1976) Effects of quantization and overflow in recursive digital filters. *IEEE Trans. Acoustics, Speech and Signal Processing*, **24**(6), 517–28.

Clark R.J., Ifeachor E.C., Rogers G.M. and Van Eetvelt P.W.J. (2000) Techniques for generating digital equaliser coefficients. *Journal of Audio Engineering Society*, **48**(4), 281–98.

Crochiere R.E. (1975) A new statistical approach to the coefficient wordlength problem for digital filters. *IEEE Trans. Circuits and Systems*, **22**, 190–6.

Crochiere R.E. and Oppenheim A.V. (1975) Analysis of linear digital networks. *Proc. IEEE*, **63**(4), 581–94.

Diniz P.S.R. and Antoniou A. (1985) Low-sensitivity digital filter structures which are amenable to error-spectrum shaping. *IEEE Trans. Circuits and Systems*, **32**(10), 1000–7.

Elliot D.F. (ed.) (1987) *Handbook of Digital Signal Processing*. London: Academic Press.

IEEE (1978) Digital Signal Processing II. Institute of Electrical and Electronics Engineers.

Jackson L.B. (1970) On the interaction of roundoff noise and dynamic range in digital filters. *BSTJ*, **49**(2), 159–84.

Jackson L.B. (1976) Roundoff noise bounds derived from coefficient sensitivities for digital filters. *IEEE Trans. Circuits and Systems*, **23**(8), 481–5.

Knowles J.B. and Olcayto E.M. (1968) Coefficient accuracy and digital filter response. *IEEE Trans. Circuit Theory*, **15**, 31–41.

Liu B. (1971) Effect of finite wordlength on the accuracy of digital filters – a review. *IEEE Trans. Circuit Theory*, **18**, 670–7.

Liu B. and Kaneko T. (1969) Error analysis of digital filters realized with floating-point arithmetic. *Proc. IEEE*, **57**(10), 1735–47.

Markel J.D. and Gray A.H. (1975) Fixed-point implementation algorithms for a class of orthogonal polynomial filter structures. *IEEE Trans. Acoustics, Speech and Signal Processing*, **23**(5), 486–94.

Markel J.D. and Gray A.H. (1975) Roundoff noise characteristics of a class of orthogonal polynomial structures. *IEEE Trans. Acoustics, Speech and Signal Processing*, **23**(5), 473–86.

Motorola (1988) *Digital Stereo 10-band Graphic Equalizer Using the DSP56001*. Motorola Application Note.

Mullis C.T. and Roberts R.A. (1976) Round-off noise in digital filters: frequency transformations and invariants. *IEEE Trans. Acoustics, Speech and Signal Processing*, **24**(6), 538–50.

Munson D.C. and Liu B. (1980) Low-noise realization for narrow-band recursive digital filters. *IEEE Trans. Acoustics, Speech and Signal Processing*, **28**, 41–54.

Nagle H.T. and Nelson V.P. (1981) Digital filter implementation on 16 bit microcomputers. *IEEE Micro*, **1**, 23–41.

Oppenheim A.V. and Schafer R.W. (1975) *Digital Signal Processing*. Englewood Cliffs NJ: Prentice-Hall.

Oppenheim A.V. and Weinstein, C.J. (1972) Effects of finite register length in digital filtering and the fast Fourier transform. *Proc. IEEE*, **60**, 957–76.

Peled A., Liu B. and Steiglitz K. (1974) A new hardware realization of digital filters. *IEEE Trans. Acoustics, Speech and Signal Processing*, **22**, 456–62.

Peled A., Liu B. and Steiglitz K. (1975) A note on implementation of digital filters. *IEEE Trans. Acoustics, Speech and Signal Processing*, **23**, 387–9.

Rabiner L.R., Cooley J.W., Helms H.D., Jackson L.B., Kaiser J.F., Rader C.M., Schafer R.W., Steiglitz K. and Weinstein C.J. (1972) Terminology in digital signal processing. *IEEE Trans. Audio and Electroacoustics*, **20**, 322–37.

Sandberg I.W. and Kaiser J.F. (1972) A bound on limit cycles in fixed-point implementations of digital filters. *IEEE Trans. Audio and Electroacoustics*, **20**, 110–12.

Schmalzel J.L., Heine D.N. and Ahmed N. (1980) Some pedagogical considerations of digital filter hardware implementation. *IEEE Circuits and Systems Mag.*, **2**(1), 4–13.

Sim P.K. and Pang K.K. (1985) Effects of input-scaling on the asymptotic overflow-stability properties of second recursive digital filters. *IEEE Trans. Circuits and Systems*, **32**(10), 1008–15.

Steiglitz K. (1971) Designing short-word recursive digital. *Proc. 9th Ann. Allerton Conf. on Circuit and System Theory*, 6–8 October, 1971, pp. 778–88.

Steiglitz K., Bede L. and Liu B. (1976) An improved algorithm for ordering poles and zeros of fixed-point recursive digital filters. *IEEE Trans. Acoustics, Speech and Signal Processing*, **24**, 341–3.

Taylor F.J. (1983) *Digital Filter Design Handbook*. New York: Marcel Dekker.

Thong T. (1976) Finite wordlength effects in the ROM digital filter. *IEEE Trans. Acoustics, Speech and Signal Processing*, **24**, 436–7.

Thong T. and Liu B. (1977) Error spectrum shaping in narrowband recursive digital filters. *IEEE Trans. Acoustics, Speech and Signal Processing*, **25**, 200–3.

Williamson D. and Sridharan S. (1985) An approach to coefficient wordlength reduction in digital filters. *IEEE Trans. Circuits and Systems*, **32**(9), 893–903.

Appendices

8A C programs for IIR digital filter design

Several C language programs for designing IIR digital filters have been developed for the book. Because of lack of space, only the program for the impulse invariant method of coefficient calculation is listed here. The following programs are not listed in the book, but are available on the CD in the companion handbook. MATLAB m-files that may be used to perform similar tasks to some of the C programs are available on the web (see the Preface for details).

- bilinear transformation;
- coefficient calculation for classical IIR filters (Butterworth, Chebyshev and elliptic) via the bilinear transform method;
- finite wordlength analysis for IIR filters.

8A.1   C program for the impulse invariant method

Please note that in the C programs the coefficients A_k and B_k for the analog transfer functions and a_k and b_k for the discrete-time filters are the numerator and denominator coefficient vectors, respectively. Unlike the programs, we have modified the use of these coefficients in Equations 8A.1 to 8A.8 to be consistent with those of MATLAB.

The program for the impulse invariant method is listed in Program 8A.1. First, we will summarize the concepts on which the program is based and then illustrate its use by an example. The discussion will be based on the second-order filter section, since this is the basic building block for digital IIR filters.

Consider a general second-order s-plane transfer function which is to be converted into a discrete transfer function:

$$H(s) = \frac{B_0 + B_1 s}{A_0 + A_1 s + A_2 s^2} \tag{8A.1}$$

The equivalent z-transfer function is also a second-order section of the form

$$H(z) = \frac{b_0 + b_1 z^{-1}}{1 + a_1 z^{-1} + a_2 z^{-2}} \tag{8A.2}$$

Program 8A.1   Impulse invariant method (*Note*: In the program the coefficients A's and B's, and a's and b's, are reversed compared to the equations above.).

```
/*---------------------------------------------------------------------------- *
 *       Impulse invariant method                                              *
 *                                                                             *
 *       The analog transfer function must be frequency-scaled                 *
 *       (normalized frequency) before using program                           *
 *       30.10.92                                                              *
 *---------------------------------------------------------------------------- *
 */
#include    <stdio.h>
#include    <math.h>
#include    <dos.h>

void        dfilter();
double      T;
double      a0, a1, a2, b0, b1, b2;
double      p1, p2, pr, pi;
double      c1, c2, cr, ci;
float       A0, A1, B0, B1, B2, temp;

main()
{
                                                    /* initialize coeffs */
            A0=0; A1=0; B0=1; B1=0; B2=0;
            a0=0; a1=0; b0=1; b1=0; b2=0;
            c1=0; c2=0; p1=0; p2=0; a2=0;
                                                    /* read s-plane coefficients */
```

```
                    printf("impulse invariant discrete filters \n");
                    printf("\n");
                    printf("enter s-plane coefficients \n");
                    printf("enter denominator coeffs: B0, B1, B2 \n");
                    scanf("%f %f %f", &B0, &B1, &B2);
                    printf("enter numerator coeffs: A0, A1 \n");
                    scanf("%f %f", &A0, &A1);
                    T=1;
                    dfilter();
                    printf("\n");
                    printf("press enter to continue\n");
                    getch();
                    exit(0);
}
/*-------------------------------------------------------------------------------------------------------- */
void        dfilter()
{
/* Find the s-plane pole positions */
                    temp = B1*B1 – 4*B0*B2;

                    if(B2= =0){                                           /* a single pole */
                            p1=–B0/B1;
                            a0=A0/B1;
                            b1=–exp(p1*T);
                    }
                    if(temp>0){                                          /* real and unequal poles */
                            pr=–B1/(2*B2);
                            pi=(pr*pr)–B0/B2;
                            pi=sqrt(pi);
                            p1=pr+pi;
                            p2=pr–pi;
                            c1=(A0+A1*p1)/((p1–p2)*B2);
                            c2=A1/B2–c1;
                            a0=c1+c2;
                            a1=–(c1*exp(p2*T) + c2*exp(p1*T));
                            b1=–exp(p1*T)–exp(p2*T);
                            b2=exp((p1+p2)*T);

                    }
                    if(temp<0){                                          /* complex conjugate poles */
                            pr=–B1/(2*B2);
                            pi=(pr*pr)–B0/B2;
                            pi=sqrt(–pi);
                            cr=A1/(B2*2);
                            ci=–(A0+A1*pr)/(2*pi*B2);
                            a0=2*cr;
                            a1=–(cr*cos(pi*T)+ci*sin(pi*T))*2*exp(pr*T);
                            b1=–2*exp(pr*T)*cos(pi*T);
                            b2=exp(2*pr*T);

                    }
                    printf("discrete filter coeffs: \n");
                    printf("a0 a1 a2: \t%f %f %f \n", a0, a1, a2);
                    printf("b0 b1 b2: \t%f %f %f \n", b0, b1, b2);

}
```

Given the values of the coefficients of the analog transfer function, $H(s)$, the C program listed in Program 8A.1 computes the coefficients of the equivalent z-transfer function, $H(z)$. To see how the program works, we will establish the relationship between the coefficients of $H(s)$ and those of $H(z)$.

Using a partial fraction expansion, the s-plane transfer function of Equation 8A.1 can be expressed as

$$\frac{B_0/A_2 + (B_1/A_2)s}{A_0/A_2 + (A_1/A_2)s + s^2} = \frac{c_1}{s - p_1} + \frac{c_2}{s - p_2} \tag{8A.3}$$

where p_1 and p_2 are the s-plane poles of $H(s)$ given by

$$p_{1,2} = \frac{-A_1}{2A_2} \pm \left[\left(\frac{A_1}{2A_2} \right)^2 - \frac{A_0}{A_2} \right]^{1/2} \tag{8A.4}$$

Multiplying both sides of Equation 8A.3 by $(s - p_1)(s - p_2)$ and equating coefficients of s and the constant terms we have

$$\frac{B_0}{A_2} = -(c_1 p_2 + c_2 p_1) \tag{8A.5a}$$

$$\frac{B_1}{A_2} = c_1 + c_2 \tag{8A.5b}$$

Solving for c_1 and c_2 we have

$$c_1 = \frac{B_0 + B_1 p_1}{(p_1 - p_2)A_2} \tag{8A.6a}$$

$$c_2 = \frac{B_1}{A_2} - c_1 \tag{8A.6b}$$

Applying the impulse invariant transformation to Equation 8A.3 gives the discrete transfer function, $H(z)$:

$$H(z) = \frac{c_1 + c_2 - (c_1 e^{p_2 T} + c_2 e^{p_1 T})z^{-1}}{1 - (e^{p_1 T} + e^{p_2 T})z^{-1} + e^{(p_1 + p_2)T} z^{-2}} \tag{8A.7}$$

$$= \frac{b_0 + b_1 z^{-1}}{1 + a_1 z^{-1} + a_2 z^{-2}}$$

where

$$b_0 = c_1 + c_2, \qquad b_1 = -(c_1 e^{p_2 T} + c_2 e^{p_1 T})$$
$$a_1 = -(e^{p_1 T} + e^{p_2 T}), \qquad a_2 = e^{(p_1 + p_2)T}$$

p_1 and p_2 are defined in Equation 8A.4, and c_1 and c_2 are defined in Equation 8A.6.

Thus, given the s-plane coefficients for a second-order filter (that is A_0, A_1, A_2, B_0 and B_1), the coefficients of the equivalent discrete filter can be obtained directly using the relationships above. Evaluation of the coefficients of $H(z)$ in Equation 8A.7 depends on the type of the s-plane poles, p_1 and p_2. In practice, three cases arise: these are when the two poles are (i) real and unequal, (ii) a complex conjugate pair, or (iii) real and equal (that is, coincident poles). Only the first two cases are considered as the third does not occur often and is more involved.

In the first case, Equation 8A.7 can be used directly to obtain the coefficients of $H(z)$. In the second case, a simpler form of Equation 8A.7 is used to avoid complex arithmetic. Exploiting the properties of the poles, Equation 8A.7 (for the second case) becomes

$$H(z) = \frac{(c_1 + c_1^*) - (c_1 e^{p_1^* T} + c_1^* e^{p_1 T}) z^{-1}}{1 - (e^{p_1 T} + e^{p_1^* T}) z^{-1} + e^{(p_1 + p_1^*) T} z^{-2}}$$

$$= \frac{2c_r - [c_r \cos(p_i T) + c_i \sin(p_i T)] 2 e^{p_r T} z^{-1}}{1 - 2 e^{p_r T} \cos(p_i T) z^{-1} + e^{2 p_r T} z^{-2}} \tag{8A.8}$$

where p_r is the real part of p_1, p_i the imaginary part of p_1, c_r the real part of c_1 and c_i the imaginary part of c_1. From Equation 8A.4, the real and imaginary parts of p_1 are given by

$$p_r = -\frac{A_1}{2A_2}, \quad p_i = \left\{ -\left[\left(\frac{A_1}{2A_2} \right)^2 - \frac{A_0}{A_2} \right] \right\}^{1/2}$$

and, from Equation 8A.6, the partial fraction coefficient, c_1, is given by

$$c_1 = \frac{B_1}{2A_2} - \frac{B_0 + B_1 p_r}{2 p_i A_2} j = c_r + c_i j$$

Thus, in standard form, the coefficients of the second-order z-transfer function for the case where the poles of $H(s)$ are complex conjugates are

$$b_0 = 2c_r, \qquad\qquad b_1 = -[c_r \cos(p_i T) + c_i \sin(p_i T)] 2 e^{p_r T}$$
$$a_1 = -2 e^{p_r T} \cos(p_i T), \quad a_2 = e^{2 p_r T}$$

Example 8A.1

We will use the C program in Program 8A.1 to compute the coefficients for the discrete filter of Example 8.4.

The program expects the frequencies to be normalized. With a sampling frequency of 1280 Hz and a cutoff frequency of 150 Hz, the normalized cutoff frequency is $150/1280$. The transfer function is first frequency scaled by replacing s by s/α where $\alpha = 2\pi \times 150/1280 = 0.736\,31$:

$$H'(s) = \frac{\alpha^2}{\alpha^2 + \sqrt{2}\alpha s + s^2} = \frac{1}{1 + (\sqrt{2}/\alpha)s + (1/\alpha^2)s^2} = \frac{1}{1 + 1.920\,675 s + 1.844\,96 s^2}$$

The prompts and outputs of the program are given below. The discrete coefficients are identical to the values obtained in Example 8.4.

```
impulse invariant discrete filters

enter s-plane coefficients
enter denominator coeffs: A0, A1, A2
1   1.920675   1.84496
enter numerator coeffs: B0, B1
1   0
discrete filter coeffs:
b0  b1  b2:   0.000000   0.307718   0.000000
a0  a1  a2:   1.000000  -1.030953   0.353088
```

From the above listings, the z-transfer function can be written down directly:

$$H(z) = \frac{0.307\,718z^{-1}}{1 - 1.030\,953z^{-1} + 0.353\,088z^{-2}}$$

The coefficients are identical to the values obtained in Example 8.4.

8B   IIR filter design with MATLAB

The MATLAB Signal Processing Toolbox provides many useful functions for the design and analysis of classical digital IIR filters (e.g. Butterworth, Chebyshev types I and II, and elliptic filters) for a given set of specifications (e.g. pass- and stopband edge frequencies, passband ripples and stopband attenuation).

In particular, the Toolbox provides functions for converting classical analog filters into equivalent discrete-time filters as discussed in the main text.

As the reader may recall, a crucial stage in the design of digital IIR filters is coefficient calculation. For classical digital IIR filters, the steps involved at this stage may be summarized as follows:

(1)   Specify the desired filter.

(2)   Determine a suitable analog prototype lowpass filter with, e.g. Butterworth, Chebyshev type I, Chebyshev type II or elliptic characteristics.

(3)   Transform the prototype analog filter to create a lowpass, highpass, bandpass or bandstop filter.

(4)   Convert the transformed filter into an equivalent discrete-time filter (for example, using the impulse invariant or the bilinear z-transform method).

The MATLAB Signal Processing Toolbox provides a number of high-level functions for carrying out steps 2 to 4 simultaneously or separately. For example, the syntax for the MATLAB command to create a lowpass, highpass, bandpass or bandstop filter with a Butterworth characteristic is (steps 1 and 2):

```
[b, a] = butter (N, Wc, options)
[z, p, k] = butter (N, Wc, options)
```

The first command calculates the numerator and denominator coefficients of an Nth order discrete-time Butterworth filter with a 3 dB cutoff frequency (or bandedge frequency), W_c, normalized to the Nyquist frequency. The numerator and denominator coefficients of the filter are returned in the vectors b and a, respectively, in ascending negative powers of z.

If the word 'options' is left out, the command defaults to a lowpass filter (unless W_c is a vector of frequencies, in which case it defaults to a bandpass filter). For highpass and bandstop filters the words 'high' and 'stop' are used as the options. For bandpass and bandstop filters, W_c is a two-element vector that specifies the cutoff (or bandedge) frequencies:

$$W_c = [\omega_{c1},\, \omega_{c2}]$$

where $\omega_{c1} < \omega < \omega_{c2}$ is the passband (bandpass filters) or stopband (bandstop filters).

The second command returns the positions of the zeros and poles, in rectangular form, in z and p, and the filter gain, k.

Similar commands exist for other classical filter types. For example, for Chebyshev types I and II, and elliptic filters, the syntax for the MATLAB commands is

```
[b, a] = cheby1 (N, Ap, Wc, options)
[z, p, k] = cheby1 (N, Ap, Wc, options)

[b, a] = cheby2 (N, As, Wc, options)
[z, p, k] = cheby2 (N, As, Wc, options)

[b, a] = ellip (N, Ap, As, Wc, options)
[z, p, k] = ellip (N, Ap, As, Wc, options)
```

where Ap and As are, respectively, the passband ripple and stopband attenuation, in dB.

There are a number of other useful commands that may be used to carry out intermediate tasks in the coefficient calculation process. For example, the **buttord**, **cheby1ord**, and **ellipord** commands may be used to determine the order of a suitable filter. The parameters of suitable prototype analog lowpass filters may be determined using the **butterp**, **cheby1p**, **cheby2p** and **ellipp** commands.

Example 8B.1

Design of a simple lowpass filter using impulse invariant method and MATLAB   A lowpass IIR filter with a Butterworth characteristic is required to meet the following specifications:

cutoff frequency	150 Hz
sampling frequency	1.28 kHz
filter order, N	2

(a)   Determine, using the impulse invariant method and MATLAB,

    (i)   the coefficients, poles and zeros of the frequency-scaled analog filter;

    (ii) the coefficients, poles and zeros of the IIR discrete filter. Write down its transfer function.

(b)   Plot the magnitude–frequency response and the pole–zero diagram of the discrete filter.

Solution   The MATLAB m-file for the problem is given in Program 8B.1.

(a)   (i)   Using the m-file, the coefficients, poles and zeros of the analog filter are

Coefficients:	b = 1.0e+005*[0, 0, 8.8826]
	a = 1.0e+005*[0.00001, 0.0133, 8.8826]
Poles:	1.0e+002*(−6.6643 ± 6.6643j)
Zeros:	None
Gain	8.8826e+005

Program 8B.1   MATLAB m-file for Example 8B.1.

```
%
%   Program name: EX8B1.m
%   A simple lowpass filter
%
N=2;                            %   Filter order
Fs=1280;                        %   Sampling frequency
fc=150;                         %   Cutoff frequency
WC=2*pi*fc;                     %   Cutoff frequency in radians
[b, a]=butter (N,WC,'s');       %   Create the analog filter
[z, p, k]=butter (N, WC, 's');
[bz, az]=impinvar (b, a, Fs);   %   Convert into discrete filter
subplot (2,1,1)                 %   Plot magnitude freq. response
[H, f]=freqz(bz, az, 512, Fs);
plot(f, 20*log10(abs(H)))
xlabel('Frequency (Hz)')
ylabel('Magnitude Response (dB)')
subplot(2,1,2)                  %   Plot pole-zero diagram
zplane(bz, az)
zz=roots(bz);                   %   Determine poles and zeros
pz=roots(az);
```

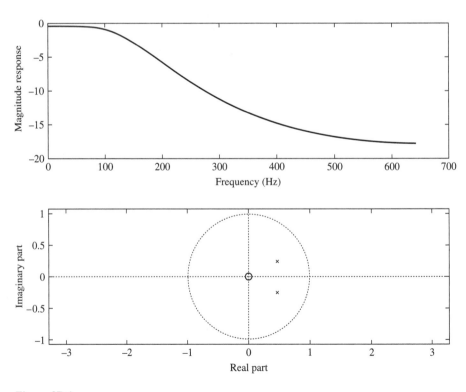

Figure 8B.1

(ii) Using the m-file, the coefficients, poles and zeros of the discrete-time filter are

Coefficients: b = [0 0.3078 0]
a = [1.0000 −1.0308 0.3530]

Poles: 0.5154 ± 0.2955j
Zero: 0

Using the coefficients, the transfer function of the filter is given by

$$H(z) = \frac{0.3078z^{-1}}{1 - 1.0308z^{-1} + 0.3553z^{-2}}$$

The results are the same as in Example 8.4.

(b) The magnitude–frequency response and the pole–zero diagram are depicted in Figure 8B.1.

Example 8B.2

Design of a simple lowpass filter using the bilinear z-transform method and MATLAB A digital IIR filter with Butterworth characteristics is required to meet the following specifications:

cutoff frequency	150 Hz
sampling frequency	1.28 kHz
filter order, N	2

(a) Determine, using the bilinear z-transform method and MATLAB, the coefficients, poles and zeros of the discrete filter.

(b) Plot the magnitude–frequency response and the pole–zero diagram of the discrete filter.

Solution

The MATLAB m-file is listed in Program 8B.2.

(a) With the aid of the m-file, the coefficient vectors (b and a), zeros and poles (z and p) of the IIR filter are

b = [0.0878, 0.1756, 0.0878]
a = [1.0000, −1.0048, 0.3561]
z = [−1, −1]
p = [0.5024 ± 0.3220j]
k = 0.0878

Using the coefficients, the transfer function of the filter is given by

$$H(z) = \frac{0.0878(1 + 2z^{-1} + z^{-2})}{1 - 1.0308z^{-1} + 0.3553z^{-2}}$$

(b) The magnitude–frequency response and the pole–zero diagram are depicted in Figure 8B.2.

Program 8B.2   MATLAB m-file for Example 8B.2.

```
%
%   Program name: EX8B2.m
%   A simple lowpass filter
%
N=2;                                    %   Filter order
Fs=1280;                                %   Sampling frequency
FN=Fs/2;
fc=150;                                 %   Cutoff frequency
Fc=fc/FN;                               %   Normalized Cutoff frequency
[b,a]=butter (N,Fc);                    %   Create and digitize analog filter.
[z,p,k]=butter (N, Fc);
subplot(2,1,1)                          %   Plot magnitude freq. response
[H, f]=freqz (b, a, 512, Fs);
plot(f, abs(H))
xlabel('Frequency (Hz)')
ylabel('Magnitude Response ')
subplot(2,1,2)                          %   Plot pole-zero diagram
zplane(b, a)
```

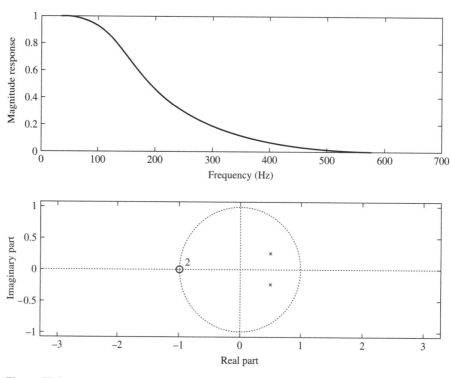

Figure 8B.2

Program 8B.3   MATLAB m-file for Example 8B.3.

```
%
%   Program name: EX8B3.m
%   A simple bandpass filter
%
Fs=2000;                              %   Sampling frequency
FN=Fs/2;
fc1=200/FN;
fc2=300/FN;
[b,a]=butter (4,[fc1, fc2]);          %   Create/digitize analog filter.
[z,p,k]=butter (4, [fc1, fc2]);
subplot(2,1,1)                        %   Plot magnitude freq. response
[H, f]=freqz (b, a, 512, Fs);
plot(f, abs(H))
xlabel ('Frequency (Hz)')
ylabel ('Magnitude Response ')
subplot (2,1,2)                       %   Plot pole-zero diagram
zplane (b, a)
```

Example 8B.3

Design of a simple bandpass filter using the bilinear z-transform method and MATLAB

(a)   Calculate the coefficients of a discrete bandpass filter with Butterworth characteristics that meets the following specifications:

passband	200–300 Hz
sampling frequency	2000 Hz
filter order	8

(b)   Plot its magnitude–frequency response and pole–zero diagram.

Solution

The MATLAB m-file is listed in Program 8B.3.

(a)   The coefficients, b and a, and the zeros, poles and gain, z, p and k respectively, are given below (note that the filter order used in the m-file is half that specified in the problem; for bandpass and bandstop filters the order is $2N$):

$b = [0.0004, 0, -0.0017, 0, 0.0025, 0, -0.0017, 0, 0.0004]$
$a = [1.0000, -5.1408, 13.1256, -20.9376, 22.6982, 17.0342, 8.6867, 2.7672, 0.4383]$
$z = [1, 1, 1, 1, -1, -1, -1, -1]$
$p = [0.5601 \pm 0.7475j, \ 0.5800 \pm 0.6286j, \ 0.6656 \pm 0.5628j, \ 0.7647 \pm 0.5648j]$
$k = 4.1660e{-}004$

(b)   The magnitude–frequency response and the pole–zero diagram are depicted in Figure 8B.3.

Example 8B.4

Design of a lowpass filter with specified pass- and stopband edge frequencies and ripples/ attenuation   A lowpass digital filter with a Butterworth characteristic is required to meet the following specifications:

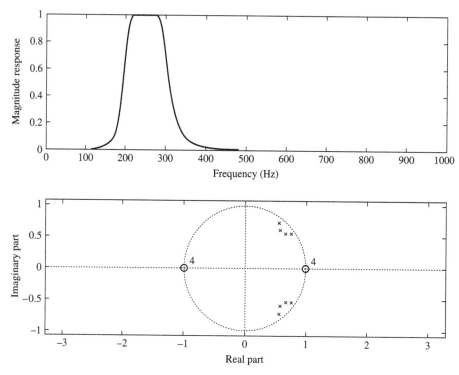

Figure 8B.3

passband	0–500 Hz
stopband	2–4 kHz
passband ripple	3 dB
stopband attenuation	20 dB
sampling frequency	8 kHz

(a)   Determine the order, N, and the coefficients of the discrete-time filter using the bilinear z-transform and MATLAB.

(b)   Plot the magnitude–frequency response and the pole–zero diagram of the filter.

Solution   The first part of the example is similar to Example 8.11 which was solved by hand. Here we will use MATLAB. It is possible to use separate MATLAB commands to determine the order of an analog filter (**buttord** command), the poles and zeros of a prototype analog lowpass filter (**buttap** command), the parameters of the transformed analog filter (**butter** command), and finally the coefficients of the discrete-time filter (**bilinear** command). However, here we will make use of the digital domain version of the butter command as this performs the intermediate steps automatically to obtain the filter coefficients.

(a)   The MATLAB implementation for the problem in the form of an m-file is given in Program 8B.4.
      The order, N, zeros and poles (zz and pz), gain (kz) and coefficients of the filter (b and a) are

Program 8B.4   MATLAB m-file for Example 8B.4.

```
%
%    Program name: EX8B4.m
%    Lowpass filter
%
Fs=8000;                                % Sampling frequency
Ap=3;
As=20;
wp=500/4000;
ws=2000/4000;
[N, wc]=buttord(wp, ws, Ap, As);        % Determine filter order
[zz, pz, kz]=butter (N,500/4000);       % Digitize filter
[b, a]=butter(N, 500/4000);
subplot (2,1,1)                         % Plot magnitude freq. response
[H, f]=freqz (b, a, 512, Fs);
plot(f, abs(H))
xlabel('Frequency (Hz)')
ylabel('Magnitude Response ')
subplot(2,1,2)                          % Plot pole-zero diagram
zplane(b, a)
```

```
N  = 2
zz = [−1, −1]
pz = [0.7271 ± 0.2130j]
kz = 0.0300
b  = [0.0300    0.0599    0.0300]
a  = [1.0000   −1.4542    0.5741]
```

The coefficients are the same as those obtained by hand in Example 8.11, and the transfer function is

$$H(z) = \frac{0.030(1 + 2z^{-1} + z^{-2})}{1 - 1.4542z^{-1} + 0.5741z^{-2}}$$

(b)   The magnitude–frequency response and the pole–zero diagram are depicted in Figure 8B.4.

Example 8B.5

Design of a highpass filter with specified pass- and stopband edge frequencies and ripples/attenuation   A highpass digital filter meeting the following specifications is required:

passband	2–4 kHz
stopband	0–500 Hz
passband ripple	3 dB
stopband attenuation	20 dB
sampling frequency	8 kHz

Determine the following:

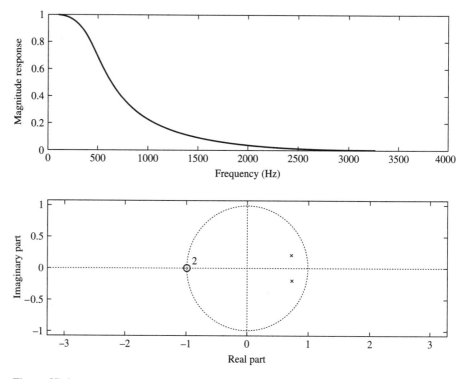

Figure 8B.4

(a) Pass- and stopband edge frequencies for a suitable analog prototype lowpass filter.
(b) Order, N, of the prototype lowpass filter.
(c) Coefficients, and hence the transfer function, of the discrete-time filter using the bilinear z-transform.

Solution

(a) The MATLAB implementation for the problem in the form of an m-file is given in Program 8B.5.
(b) The order, N, zeros and poles (zz and pz), gain (kz) and coefficients of the filter (b and a) are

$$N = 2$$
$$zz = [1,1]$$
$$pz = \pm 0.4142j$$
$$kz = 0.2929$$
$$b = [0.2929, -0.5858, 0.2929]$$
$$a = [1.0000, 0.0000, 0.1716]$$

(c) Using the coefficients, the transfer function of the filter is given by

$$H(z) = \frac{0.2929(1 - 2z^{-1} + z^{-2})}{1 + 0.1716z^{-2}}$$

The magnitude–frequency response and the pole–zero diagram are shown in Figure 8B.5.

Program 8B.5   MATLAB m-file for Example 8B.5.

```
%
%    Program name: EX8B5.m
%
Fs=8000;                                    %    Sampling frequency
Ap=3;
As=20;
wp=2000/4000;
ws=500/4000;
[N, wc]=buttord (wp, ws, Ap, As);           %    Determine filter order
[zz, pz, kz]=butter (N, 2000/4000, 'high'); %    Digitize filter
[b, a]=butter (N, 2000/4000, 'high');
subplot(2,1,1)                              %    Plot magnitude freq. response
[H, f]=freqz (b, a, 512, Fs);
plot(f, abs(H))
xlabel('Frequency (Hz)')
ylabel('Magnitude Response ')
subplot(2,1,2)                              %    Plot pole-zero diagram
zplane(b, a)
```

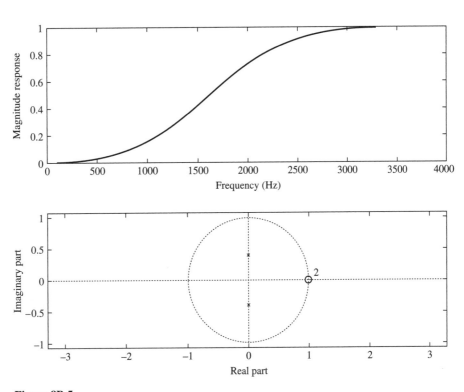

Figure 8B.5

| Example 8B.6 | *Design of a bandpass filter with specified bandedge frequencies and pass- and stopband ripples*   A requirement exists for a bandpass digital filter, with a Butterworth magnitude–frequency response, that satisfies the following specifications: |

lower passband edge frequency	200 Hz
upper passband edge frequency	300 Hz
lower stopband edge frequency	50 Hz
upper stopband edge frequency	450 Hz
passband ripple	3 dB
stopband attenuation	20 dB
sampling frequency	1 kHz

(a)   Determine, using the BZT method and MATLAB:

(i)   the order, N, of the filter;

(ii)   the poles, zeros, gain, coefficients and transfer function of the discrete-time filter.

(b)   Plot the magnitude–frequency response and the pole–zero diagram of the filter.

Solution   (a)   The MATLAB m-file is given in Program 8B.6. Using the m-file, the order, N, of the filter, the zeros and poles (zz and kz), the gain (kz) and the coefficients (b and a) of the discrete filter are

N = 2 (the order of the bandpass filter is 2 * N, i.e. 4)
zz = [1, 1, −1, −1]
pz = [−0.1884 ± 0.7791j, −0.1884 ± 0.7791j]
kz = 0.0675
b = [0.0675, 0, −0.1349, 0, 0.0675]
a = [1.0000, −0.0000, 1.1430, −0.0000, 0.4128]

Program 8B.6   MATLAB m-file for Example 8B.6.

```
%
%   Program name: EX8B6.m
%   Bandpass filter
%
Fs=1000;                           %   Sampling frequency
Ap=3;
As=20;
Wp=[200/500, 300/500];             %   Bandedge frequencies
Ws=[50/500, 450/500];
[N, Wc]=buttord (Wp, Ws, Ap, As);  %   Determine filter order
[zz, pz, kz]=butter (N,Wp);        %   Digitize filter
[b, a]=butter (N, Wp);
subplot(2,1,1)                     %   Plot magnitude freq. response
[H, f]=freqz (b, a, 512, Fs);
plot(f, abs(H))
xlabel('Frequency (Hz)')
ylabel('Magnitude Response ')
subplot(2,1,2)                     %   Plot pole-zero diagram
zplane(b, a)
```

The corresponding transfer function, $H(z)$, is

$$H(z) = \frac{0.0675(1 - 2z^{-2} + z^{-4})}{1 + 1.143z^{-2} + 0.4128z^{-4}}$$

(b) If we set $N = 1$ (as in Example 8.11), the zeros (zz), poles (pz), gain (kz) and coefficients (b and a) become

zz = [–1, 1]
pz = [0.0000 ± 0.7138j]
kz = 0.2452

b = [0.2452, 0, –0.2452]
a = [1.0000, –0.0000, 0.5095]

The corresponding transfer function, $H(z)$, is

$$H(z) = \frac{0.2452(1 - z^{-2})}{1 + 0.5095z^{-2}}$$

The magnitude–frequency response and the pole–zero diagram are shown in Figure 8B.6.

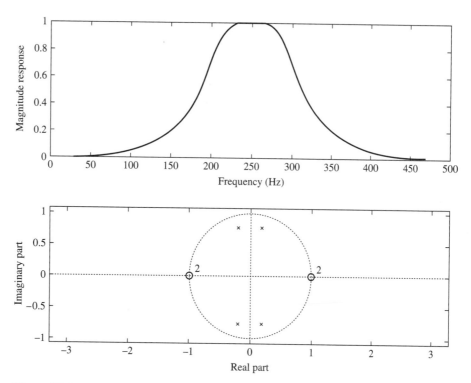

Figure 8B.6

| Example 8B.7 | *Design of a bandstop filter with specified bandedge frequencies and pass- and stopband ripples* |

A requirement exists for a bandstop digital IIR filter, with a Butterworth magnitude–frequency response, that meets the following specifications:

lower passband	0–50 Hz
upper passband	450–500 Hz
stopband	200–300 Hz
passband ripple	3 dB
stopband attenuation	20 dB
sampling frequency	1 kHz

Determine:

(a) The pass- and stopband edge frequencies of a suitable prototype lowpass filter.

(b) The order, N, of the prototype lowpass filter.

(c) The coefficients, and hence the transfer function, of the discrete-time filter using the BZT method.

Solution

The MATLAB m-file for the example is listed in Program 8B.7. Using the m-file, the filter order, N, zeros and poles (zz and pz), gain (kz) and coefficients are

```
N  = 2 (order of the bandstop filter = 2 * N, i.e. 4)
zz = two pairs of zeros at j and at –j
pz = [–0.1884 ± 0.7791j, 0.1884 ± 0.7791j]
kz = 0.6389
b = [0.6389, –0.0000, 1.2779, –0.0000, 0.6389]
a = [1.0000, –0.0000, 1.1430, –0.0000, 0.4128]
```

Program 8B.7   MATLAB m-file for Example 8B.7.

```
%
%   Program name: EX8B7.m
%   Bandstop filter
%
Fs=1000;                          %  Sampling frequency
Ap=3;
As=20;
Wp=[50/500, 450/500];             %  Bandedge frequencies
Ws=[200/500, 300/500];
[N, Wc]=buttord (Wp, Ws, Ap, As); %  Determine filter order
[zz, pz, kz]=butter (N,Ws, 'stop'); %  Digitize filter
[b, a]=butter (N, Ws, 'stop');
subplot (2,1,1)                   %  Plot magnitude freq. response
[H, f]=freqz (b, a, 512, Fs);
plot (f, abs (H))
xlabel ('Frequency (Hz)')
ylabel ('Magnitude Response')
subplot (2,1,2)                   %  Plot pole-zero diagram
zplane (b, a)
```

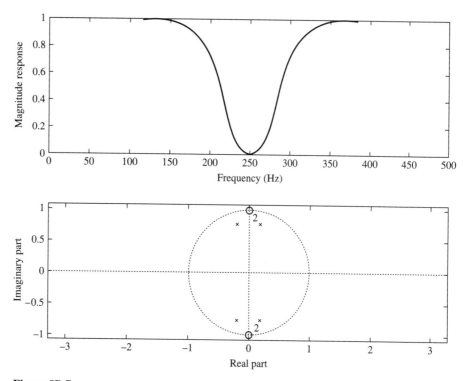

Figure 8B.7

The transfer function of the filter is

$$H(z) = \frac{0.6389(1 + 2z^{-2} + z^{-4})}{1 + 1.143z^{-2} + 0.4128z^{-4}}$$

If we set $N = 1$ (as in Example 8.12), the transfer function becomes

$$H(z) = \frac{0.7548(1 + z^{-2})}{1 + 0.5095z^{-2}}$$

The magnitude–frequency response and the pole–zero diagram are shown in Figure 8B.7.

Example 8B.8

Design example of an elliptic digital bandpass filter   A digital filter is required to satisfy the following frequency response specifications:

passband	20.5–23.5 kHz
stopband	0–19 kHz, 25–50 kHz
passband ripple	$\leq$0.25 dB
stopband attenuation	>45 dB
sampling frequency	100 kHz

Program 8B.8   MATLAB m-file for the design Example 8B.8.

```
%
%   Program Name: EX8B.m
%
Ap=0.25;
As=45;
Fs=100000;
Wp=[20500/50000, 23500/50000];          %   Bandedge frequencies
Ws=[19000/50000, 25000/50000];
[N,Wc]=ellipord(Wp, Ws, Ap, As);        %   Determine filter order
[b, a]=ellip(N, Ap, As, Wc);            %   Determine filter coeffs
[z, p, k]=ellip(N, Ap, As, Wc);         %   Determine poles and zeros
sos=zp2sos (z, p, k);                   %   Convert to second order sections
subplot(2,1,1)                          %   Plot magnitude freq. response
[H, f]=freqz(b, a, 512, Fs);
plot(f, 20*log10 (abs(H)))
xlabel('Frequency (Hz)')
ylabel('Magnitude Response (dB)')
subplot(2,1,2)                          %   Plot pole-zero diagram
zplane(b, a)
```

(a) Determine, using the BZT method and MATLAB, a suitable transfer function for the filter in the form of second-order sections suitable for cascade realization.

(b) Obtain and plot the magnitude–frequency response and pole–zero diagram of the filter.

Assume an elliptic characteristic.

Solution   (a) The MATLAB m-file for the design problem is listed in Program 8B.8. The **ellipord** command is used to determine the order of a suitable elliptic prototype filter. The **ellip** command is then used to determine the coefficients, poles and zeros of the transfer function. The **zp2sos** command converts the transfer function into second-order sections. The coefficients for the second-order sections are returned in the matrix, sos.

The filter coefficients (b and a), poles and zeros (p and z), gain (k), and the matrix of second-order sections, sos, are listed below:

```
Filter order, N = 4
b = [0.0061, -0.0083, 0.0236, -0.0221, 0.0351, -0.0221, 0.0236, -0.0083, 0.0061]
a = [1.0000, -1.4483, 4.4832 -4.2207, 6.6475, -3.9458, 3.9187, -1.1828, 0.7634]
z = [-0.0118 ± 0.9999j, 0.3737 ± 0.9275j, -0.2553 ± 0.9669j, 0.5663 ± 0.8242j]
p = [0.0872 ± 0.9793j, 0.1352 ± 0.9404j, 0.2795 ± 0.9432j, 0.2223 ± 0.9246j]
k = 0.0061
sos =
        0.1203      0.0614      0.1203      1.0000      -0.2705      0.9026
        0.2051     -0.2323      0.2051      1.0000      -0.4446      0.9042
        0.3740      0.0088      0.3740      1.0000      -0.1744      0.9665
        0.6642     -0.4965      0.6642      1.0000      -0.5589      0.9678
```

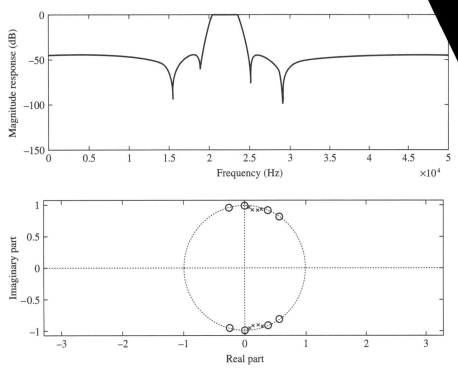

Figure 8B.8

The sos is a 4×6 matrix of second-order sections which has the following format:

$$
\begin{matrix}
b_{01} & b_{11} & b_{21} & a_{01} & a_{11} & a_{21} \\
b_{02} & b_{12} & b_{22} & a_{02} & a_{12} & a_{22} \\
b_{03} & b_{13} & b_{23} & a_{03} & a_{13} & a_{23} \\
b_{04} & b_{14} & b_{24} & a_{04} & a_{14} & a_{24}
\end{matrix}
$$

It is evident that the coefficients of the second-order sections are held in the rows, one row per second-order section.

The magnitude–frequency response and the pole–zero diagram are shown in Figure 8B.8.

8C Evaluation of complex square roots using real arithmetic

Given a complex number with rectangular coordinates $x + jy$, let its square root be $u + jv$:

$$u + jv = (x + jy)^{1/2} \tag{8C.1}$$

Squaring Equation 8.C1 we have

$$u^2 - v^2 + j2uv = x + jy$$

Equating real and imaginary parts gives

$$u^2 - v^2 = x \qquad\qquad\qquad (8C.2a)$$

$$2uv = y \qquad\qquad\qquad (8C.2b)$$

From Equation 8C.2b

$$v = y/2u \qquad\qquad\qquad (8C.2c)$$

Substituting Equation 8C.2c in 8C.2a we have

$$u^2 - y^2/4u^2 = x \qquad\qquad\qquad (8C.3)$$

and simplifying 8C.3 gives

$$4u^4 - 4xu^2 - y^2 = 0$$

This is a quadratic in u^2, which has the solution

$$u^2 = \frac{x}{2} + \frac{1}{8}(16x^2 + 16y^2)^{1/2}$$

$$= \frac{x + (x^2 + y^2)^{1/2}}{2}$$

(Note that the negative solution is not permitted since $u^2 \geqslant 0$.) Solving for u and using Equation 8C.2c, the solution for u is

$$u = \left[\frac{x + (x^2 + y^2)^{1/2}}{2} \right]^{1/2} \qquad\qquad\qquad (8C.4a)$$

$$v = y/2u \qquad\qquad\qquad (8C.4b)$$

or

$$u = -\left[\frac{x + (x^2 + y^2)^{1/2}}{2} \right]^{1/2} \qquad\qquad\qquad (8C.5a)$$

$$v = y/2u \qquad\qquad\qquad (8C.5b)$$

The algorithm for computing the square root of $x + jy$ is then

- Step 1: let $u + jv = (x + jy)^{1/2}$
- Step 2: $u = \pm\left[\dfrac{x + (x^2 + y^2)^{1/2}}{2} \right]^{1/2} = \pm\left(\dfrac{x + |x + jy|}{2} \right)^{1/2}$
- Step 3: $v = \pm y/2u$

For example

$$(-1 + j)^{1/2} = 0.455\,09 + 1.098\,68j \text{ and } -0.455\,09 - 1.098\,68j$$

$$(-1 - j)^{1/2} = 0.455\,09 - 1.098\,68j \text{ and } -0.455\,09 + 1.098\,68j$$

Multirate digital signal processing

9

This chapter discusses, from a practical standpoint, the important topic of multirate processing. The basic concepts are explained and illustrated with fully worked examples. The design of actual multirate systems is presented to enable the reader to design his/her own system. A set of programs, in the C language, are provided for the design and software implementation of multirate processing on a personal computer.

9.1 Introduction

The increasing need in modern digital systems to process data at more than one sampling rate has led to the development of a new sub-area in DSP known as multirate processing (Crochiere and Rabiner, 1975, 1976, 1979, 1981, 1983, 1988). The two primary operations in multirate processing are decimation and interpolation and they enable the data rate to be altered in an efficient manner. Decimation reduces the sampling rate (that is, the sampling frequency), effectively compressing the data, and retaining only the desired information. Interpolation on the other hand increases the sampling rate. Often, the purpose of converting data to a new rate is to make it easier (for example, computationally more efficient) to process or to achieve compatibility with another system. To take an obvious example, if we reduce the sampling rate of a signal from 100 kHz to only 10 kHz, without loss of desired information, then at a stroke we reduce the computational burden in subsequent signal processing operations by a factor of 10. As another example, if we wish to play CD (compact disc) music which has a rate of 44.1 kHz in a studio which handles data at a 48 kHz rate, then the CD data must first be increased to 48 kHz using a multirate approach.

The aim of this chapter and the associated end-of-chapter problems is to provide an applied understanding of the theory, practice, and applications of multirate signal processing. Specific learning objectives are

(1) *Principles/theory* The reader should learn the theory of sampling rate conversion, in particular:

- principles of sampling rate reduction (decimation – down sampling and digital anti-aliasing filtering);
- principles of sampling rate increase (interpolation – up sampling and digital anti-imaging filtering);
- theory of multistage sampling rate conversion;
- principles of polyphase filtering.

(2) *Practice* You should be able to design, at a block diagram level, practical multirate sampling rate converters from a given set of specifications, in particular:

- how to specify, design and analyze filters for multirate converters;
- how to determine the parameters of multirate converters;
- how to assess the computational efficiency of multirate systems;
- how to implement sample rate converters.

(3) *Applications* You will have a good appreciation of the principles and applications of multirate techniques in audio engineering, digital communication and biomedicine, in particular:

- audio signal processing – oversampling (single-bit) ADC/DAC, CD player and data acquisition;
- digital communication – transmultiplexers, communication receivers;
- biomedicine – narrow band filtering for fetal ECG and EEG.

The chapter draws from many published materials, especially those of Rabiner and Crochiere who have made considerable contributions to multirate processing. We are indebted to them.

9.1.1 Some current uses of multirate processing in industry

Advantages of multirate processing are many and have been exploited in many and in an increasing number of modern systems. High quality data acquisition and storage systems are increasingly taking advantage of multirate techniques to avoid the use of expensive anti-aliasing analog filters and to handle efficiently signals of different bandwidths which require different sampling frequencies. The basis for these applications is that, if an analog signal is sampled at a rate much higher than specified by the sampling theorem, then a much simpler analog anti-aliasing filter can be used to bandlimit it before it is digitized. Once in a digital form, the signal can be readily reduced to the desired rate using the multirate approach. A good example of such systems is the EDR8000 (Earth Data, UK) tape recorder.

In speech processing, multirate techniques are employed to reduce the storage space or the transmission rate of speech data. Estimates of speech parameters are computed at a very low sampling rate for storage or transmission. When required, the original speech is reconstructed from the low bit-rate representation at much higher rates using the multirate approach.

The need for inexpensive, high resolution analog-to-digital converters (ADCs) in digital audio has led to the use of oversampling techniques in the design of such converters rather than the traditional successive-approximation technique (Adams, 1986; Agrawal and Shenoi, 1983; Claasen et al., 1980; Matsuya et al., 1987; Welland et al., 1989). For example, by oversampling the quantization noise inherent in ADCs is spread over a wider frequency range, so that the in-band noise is made small, effectively increasing the number of ADC bits. Further, these new ranges of high performance ADCs employ delta sigma modulation because of its simplicity (for example, it requires no sample and hold amplifiers) and low cost. Most, if not all, the inexpensive, high resolution ADCs (18, 20, 24 bits) in use today employ multirate processing. Examples include devices such as CS532X (from Crystal Semiconductor) and DSP56ADCx (from Motorola).

Multirate processing has found important application in the efficient implementation of DSP functions. For example, the implementation of narrowband digital FIR filters using conventional DSP poses a serious problem because such filters require a very large number of coefficients to meet their tight frequency response specifications. The use of multirate techniques leads to very efficient implementation by allowing filtering to be performed at a much lower rate, which greatly reduces the filter order. Multirate techniques have also been used in many other applications including the well-known compact disc player. Details of the application of multirate processing in the CD player as well as many other applications are covered later in this chapter.

9.2   Concepts of multirate signal processing

A simple but naive approach to changing the sampling rate of a digital signal is to convert it back into analog and then to re-digitize it at the new rate. Errors inherent in digital–analog–digital conversion processes, such as quantization and aliasing errors, would degrade the signal. As the signal is already in a digital form, it is best to process it digitally throughout until conversion to analog is mandatory, for example when the destination is the loudspeaker. Multirate processing is basically an efficient technique for changing the sampling frequency of a signal digitally. Its main attraction is that it allows the strengths of conventional DSP to be exploited. For example, much of the anti-aliasing and anti-imaging filtering in real-time DSP systems can be performed in the digital domain, enabling both sharp magnitude frequency as well as linear phase responses to be achieved.

The processes of decimation and interpolation are the fundamental operations in multirate signal processing, and they allow the sampling frequency to be decreased or increased without significant undesirable effects of errors such as quantization and aliasing. We present details of these basic operations next.

9.2.1   Sampling rate reduction: decimation by integer factors

A block diagram showing the process of decimating a signal $x(n)$ by an integer factor M is shown in Figure 9.1(a). It consists of a digital anti-aliasing filter, $h(k)$, and a sample rate compressor, symbolized by a down-arrow and the decimation factor, M. The rate compressor reduces the sampling rate from F_s to F_s/M. To prevent aliasing at the lower rate the digital filter is used to bandlimit the input signal to less than $F_s/2M$ beforehand. Thus the signal $x(n)$ is first bandlimited (the requirements of the decimating filter are discussed in detail in Section 9.3.1). Sampling rate reduction is

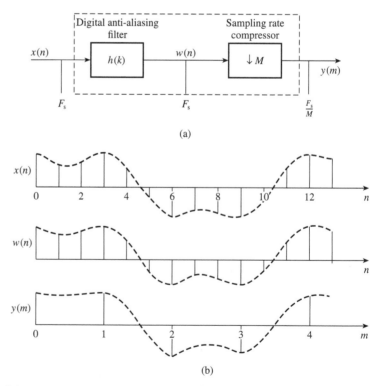

Figure 9.1   (a) Block diagram of decimation by a factor of M. (b) An illustration of decimation by a factor of $M = 3$. Note that only one out of every three samples of $w(n)$ appears at the output.

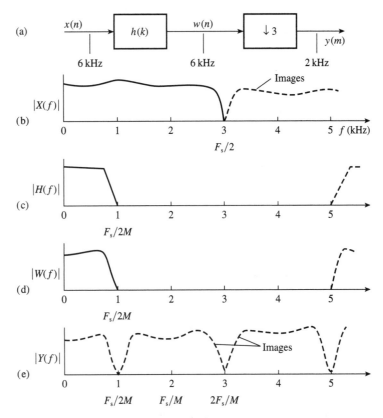

Figure 9.2   Spectral interpretation of decimation of a signal, from 6 kHz to 2 kHz. The image components would have aliased into the signal but for the digital filter.

achieved by discarding $M - 1$ samples for every M samples of the filtered signal, $w(n)$. The input–output relationship for the decimation process is

$$y(m) = w(mM) = \sum_{k=-\infty}^{\infty} h(k)x(mM - k) \qquad (9.1a)$$

where

$$w(n) = \sum_{k=-\infty}^{\infty} h(k)x(n - k) \qquad (9.1b)$$

Figure 9.1(b) illustrates the process for the simple case where $M = 3$. In this case, two samples out of every three samples of $x(n)$ are discarded. In effect, decimation is a data compression operation.

The spectral description of the decimation process is shown in Figure 9.2, where we have assumed that the input $x(n)$ is a wideband signal. The broken lines in Figure 9.2(b) indicate the image components that would have caused aliasing had the input signal $x(n)$ not been bandlimited prior to decimation.

9.2.2 Sampling rate increase: interpolation by integer factors

In many ways, interpolation is the digital equivalent of the digital-to-analog conversion process where the analog signal is recovered by interpolating the digital samples applied to the digital-to-analog converter. In the case of digital interpolation, however, the process yields specific values.

Given a signal, $x(n)$, at a sampling frequency F_s, the interpolation process increases the sampling rate by L to LF_s. Figure 9.3(a) shows the interpolator. It consists of a sample rate expander, symbolized by an up-arrow and the interpolation factor, L, which indicates the amount by which the rate is increased. For each sample of $x(n)$ the expander inserts $L - 1$ zero-valued samples to form the new signal $w(m)$ at a rate of LF_s. This signal is then lowpass filtered to remove image frequencies created by the rate increase to yield $y(m)$. The insertion of $L - 1$ zeros spreads the energy of each signal sample over L output samples, effectively attenuating each sample by a factor of L. Thus it is necessary to compensate for this, for example by multiplying each sample of $y(n)$ by L.

The interpolation process is characterized by the following input–output relationship:

$$y(m) = \sum_{k=-\infty}^{\infty} h(k)w(m - k) \tag{9.2a}$$

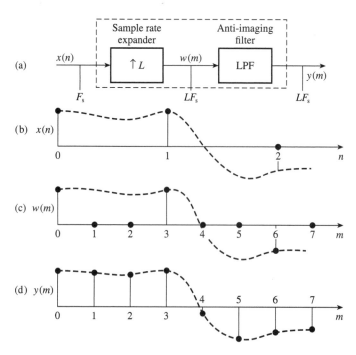

Figure 9.3 Time domain illustration of interpolation by a factor of $L = 3$. Note that for each sample of $x(n)$, three output samples $y(m)$ are obtained.

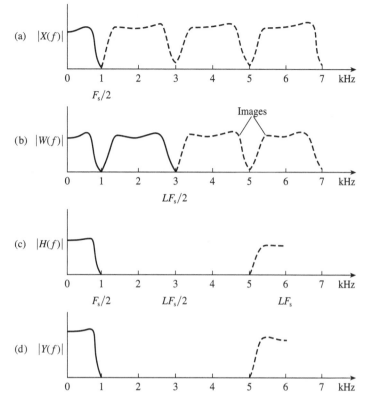

Figure 9.4 Spectral interpretation of interpolation of a signal from 2 kHz to 6 kHz.

where

$$w(m) = \begin{cases} x(m/L), & m = 0, \pm L, \pm 2L, \ldots \\ 0 \end{cases} \tag{9.2b}$$

Figures 9.3(b)–9.3(d) illustrate the interpolation process in the time domain, for the simple case of $L = 3$. Notice that, for each sample of $x(n)$, three output samples are produced. This is due to the two zero-valued samples inserted by the expander.

The frequency domain interpretation of the process is depicted in Figure 9.4. $X(f)$, $W(f)$, and $Y(f)$ are the frequency responses of the signals $x(n)$, $w(m)$ and $y(m)$, respectively. $H(f)$ is the amplitude response of the anti-imaging filter. The filter is necessary to remove the image components indicated by the broken lines in $W(f)$. It is worth pointing out at this stage, although the alert reader may have suspected this already, that decimation and interpolation processes are duals of each other, that is one is the inverse of the other. This duality property means that an interpolator can be readily derived from an equivalent decimator and vice versa.

9.2.3   Sampling rate conversion by non-integer factors

In some applications, the need often arises to change the sampling rate by a non-integer factor. An example is in digital audio applications where it may be necessary to transfer data from one storage system to another, where both systems employ different rates, possibly to discourage illegal copying of material. An example is transferring data from the compact disc system at a rate of 44.1 kHz to a digital audio tape (DAT) at 48 kHz. This can be achieved by increasing the data rate of the CD by a factor of 48/44.1, a non-integer.

In practice, such a non-integer factor is represented by a rational number, that is a ratio of two integers, say L and M, where L and M are integers such that L/M is as close to the desired factor as possible. The sampling frequency change is then achieved by first interpolating the data by L and then decimating by M (Figure 9.5(a)). It is necessary that the interpolation process precedes decimation otherwise the decimation process would remove some of the desired frequency components. In the

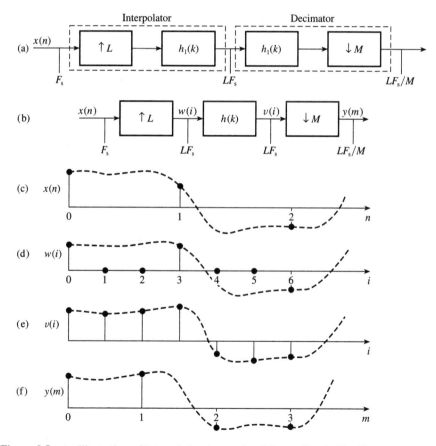

Figure 9.5   An illustration of interpolation by a rational factor ($L = 3$, $M = 2$).

CD–DAT example above, the rate conversion by 48/44.1 can be achieved by first interpolating by a factor $L = 160$ and then decimating by $M = 147$, that is we first increase the CD data rate by $L = 160$ to 7056 kHz and then reduce it by $M = 147$ to 48 kHz.

The two LPFs, $h_1(k)$ and $h_2(k)$, in Figure 9.5(a), can be combined into a single filter since they are in cascade and have a common sampling frequency to give the generalized sample rate converter in Figure 9.5(b). If $M > L$, the resulting operation is a decimation process by a non-integer, and when $M < L$ it is interpolation. If $M = 1$, the generalized system reduces to the simple integer interpolation described earlier, and if $L = 1$ it reduces to integer decimation.

Figure 9.5(c) illustrates interpolation by a factor of 3/2. The sample rate is first increased by 3, by inserting two zero-valued samples for each sample of $x(n)$, and then lowpass filtered to yield $v(i)$. The filtered data is then reduced by a factor of 2 by retaining only one sample for every two samples of $v(i)$. Figure 9.6 illustrates, in the

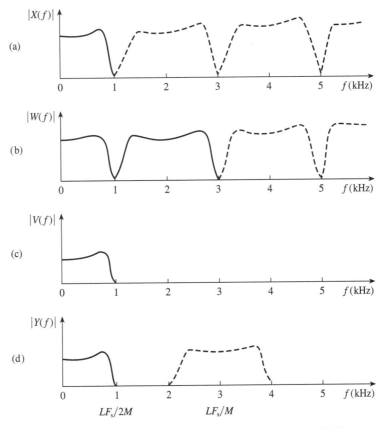

Figure 9.6 Spectral interpretation of sample rate increase of a signal at 2 kHz rate by a factor of 3/2. Signal rate is first increased by a factor of 3 to 6 kHz (a); after bandlimiting to avoid aliasing (c), signal is reduced by a factor of 2 to 3 kHz.

frequency domain, the process of interpolation by 3/2. The input signal, $x(n)$, at a rate of 2 kHz, is first increased by a factor of 3 to 6 kHz, filtered to remove the image frequencies which would otherwise cause aliasing, and then reduced by a factor of 2 to 3 kHz.

Example 9.1

A signal, $x(n)$, at a sampling frequency of 2.048 kHz is to be decimated by a factor of 32 to yield a signal at a sampling frequency of 64 Hz. The signal band of interest extends from 0 to 30 Hz. The anti-aliasing digital filter should satisfy the following specifications:

passband deviation	0.01 dB
stopband deviation	80 dB
passband	0–30 Hz
stopband	32–64 Hz

The signal components in the range from 30 to 32 Hz should be protected from aliasing. Design a suitable one-stage decimator.

Solution

The block diagram of the single-stage decimator, and the specifications for the lowpass anti-aliasing filter, are shown in Figure 9.7. From the specifications and the figure we can determine the following:

$$\Delta f = (32 - 30)/2048 = 9.766 \times 10^{-4}$$

$$\delta_p = 0.001\ 15, \quad \text{from } 20 \log (1 + \delta_p) = 0.01 \text{ dB}$$

$$\delta_s = 0.0001, \quad \text{from } -20 \log (\delta_s) = 80 \text{ dB}$$

An estimate of the number of filter coefficients for the single-stage decimator is given by (see Chapter 7)

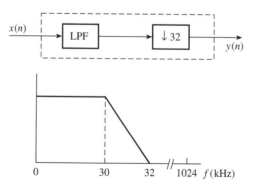

Figure 9.7   A single-stage decimator for Example 9.1.

$$N \approx \frac{D_\infty(\delta_p, \delta_s)}{\Delta f} - f(\delta_p, \delta_s) \, \Delta f + 1 \tag{9.3}$$

where Δf is the width of the transition normalized to the sampling frequency,

$$D_\infty(\delta_p, \delta_s) = (\log_{10} \delta_s)[a_1(\log_{10} \delta_p)^2 + a_2(\log_{10} \delta_p) + a_3]$$
$$+ a_4(\log_{10} \delta_p)^2 + a_5(\log_{10} \delta_p) + a_6$$
$$f(\delta_p, \delta_s) = 11.012\,17 + 0.512\,44(\log_{10} \delta_p - \log_{10} \delta_s)$$
$$a_1 = 5.309 \times 10^{-3}; \qquad a_2 = 7.114 \times 10^{-2};$$
$$a_3 = -4.761 \times 10^{-1}; \qquad a_4 = -2.66 \times 10^{-3};$$
$$a_5 = -5.941 \times 10^{-1}; \qquad a_6 = -4.278 \times 10^{-1}.$$

δ_p is the passband ripple or deviation and δ_s is the stopband ripple or deviation.

Using the values of δ_p, δ_s, and Δf in Equation 9.3, we find that $N = 3947$. It is quite obvious that N is too large. In fact, none of the available filter design methods can be used to obtain the coefficients for such a filter because approximation errors would be too large. For all practical purposes, the design of the lowpass filter for the single-stage decimator is not possible. This example makes evident the need for an alternative, more efficient method of sampling rate conversion, especially when the rate change is large. Such an approach will be discussed in the next section.

9.2.4 Multistage approach to sampling rate conversion

In the previous sections, the changes in sampling rates were achieved in one fell swoop using one decimation or interpolation factor. When large changes in the sampling rate are required it is more efficient to change the rate in two or more stages than in one single stage. In fact, most practical multirate systems employ the multistage approach because it allows a gradual reduction or increase in the sampling rate, leading to a significant relaxation in the requirements of the anti-aliasing or anti-imaging filter at each stage.

Figure 9.8 shows an I-stage decimation process. The overall decimation factor, M, is expressed as the product of smaller factors:

$$M = M_1 M_2 \ldots M_I \tag{9.4}$$

where M_I, an integer, is the decimation factor for stage I. Each stage is an independent decimator as shown in dashed boxes. If $M \gg 1$, the multistage approach leads to much reduced computational and storage requirements, a relaxation in the characteristics of the filters used in the decimators, and consequently to filters that are less sensitive to finite wordlength effects.

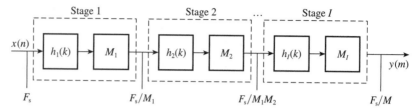

Figure 9.8 Multistage decimation process.

These advantages are achieved at the expense of increased difficulty in the design and implementation of the system. Many examples are given later to illustrate the multirate approach after we have covered the design methodology.

9.3 Design of practical sampling rate converters

The design of a practical multistage sampling rate converter can be broken down into four steps:

(1)　Specify the overall anti-aliasing or anti-imaging filter requirements.
(2)　Determine the optimum number of stages of decimation or interpolation that will yield the most efficient implementation.
(3)　Determine the decimation or interpolation factors for each stage.
(4)　Design an appropriate filter for each stage.

9.3.1 Filter specification

From previous discussions, the need for a digital filter for anti-aliasing or anti-imaging filtering in sampling rate converters should be very clear. In fact, the performance of a multirate system depends critically on the type and quality of the filter used. Either FIR or IIR filters can be used for decimation or interpolation, but the FIR is the more popular.

In multirate processing, unlike conventional DSP, the computational efficiency of FIR filters is comparable with and, in some cases, exceeds that of IIR filters (Crochiere and Rabiner, 1975, 1976, 1983). Further, FIR filters have many desirable attributes (see Chapters 6 and 7 for details), such as linear phase response and low sensitivity to finite wordlength effects, as well as being simple to implement. For these reasons, only FIR filters will be considered in this chapter. All the coefficient calculation methods described in Chapter 7 for FIR filters can be used to design filters for multirate systems. In particular, the optimal and half-band filters are widely used.

The overall filter requirements for decimation, to avoid aliasing after rate reduction, are

passband	$0 \leqslant f \leqslant f_\mathrm{p}$	(9.5a)
stopband	$F_\mathrm{s}/2M \leqslant f \leqslant F_\mathrm{s}/2$	(9.5b)
passband deviation	δ_p	(9.5c)
stopband deviation	δ_s	(9.5d)

where $f_\mathrm{p} < F_\mathrm{s}/2M$, and F_s is the original sampling frequency. Typically, f_p is the highest frequency of interest in the original signal.

In the case of interpolation, the anti-imaging filter must remove all but the useful information by bandlimiting the modified data to $F_\mathrm{s}/2$ or less. Although the highest valid frequency after raising the rate to LF_s is $LF_\mathrm{s}/2$, according to the sampling theorem, it is necessary to bandlimit to $F_\mathrm{s}/2$ as this is the highest valid frequency in $x(n)$. The overall filter requirements for interpolation are

passband	$0 \leqslant f \leqslant f_\mathrm{p}$	(9.6a)
stopband	$F_\mathrm{s}/2 \leqslant f \leqslant LF_\mathrm{s}/2$	(9.6b)
passband deviation	δ_p	(9.6c)
stopband deviation	δ_s	(9.6d)

where $f_\mathrm{p} < F_\mathrm{s}/2$. A gain of L is necessary in the passband to compensate for the amplitude reduction by the interpolation process.

9.3.2 Filter requirements for individual stages

The equiripple (optimal) filter is often used for sampling rate conversion, although filters obtained by the window method can also be used. The tolerance scheme for an equiripple lowpass filter is depicted in Figure 9.9(a).

For a multistage decimator (Figure 9.9(b)) the filter requirements for each stage to ensure that the overall filter requirements are met are (see also Figure 9.9(c))

passband	$0 \leqslant f \leqslant f_\mathrm{p}$	(9.7a)
stopband	$(F_i - F_\mathrm{s}/2M) < f < F_{i-1}/2, \quad i = 1, 2, \ldots, I$	(9.7b)
passband ripple	δ_p/I	(9.7c)
stopband ripple	δ_s	(9.7d)
filter length	$N \approx \dfrac{D_\infty(\delta_\mathrm{p}, \delta_\mathrm{s})}{\Delta f_i} - f(\delta_\mathrm{p}, \delta_\mathrm{s}) \Delta f_i + 1$	(9.7e)

where F_i, N_i and Δf_i are, respectively, the output sampling frequency, the filter length and the normalized transition width for the ith-stage decimator. The parameters $D_\infty(\delta_\mathrm{p}, \delta_\mathrm{s})$ and $f(\delta_\mathrm{p}, \delta_\mathrm{s})$ have the same meaning as in Equation 9.3. The output sampling frequency for stage i is given by

$$F_i = F_{i-1}/M_i, \quad i = 1, 2, \ldots, I \tag{9.8}$$

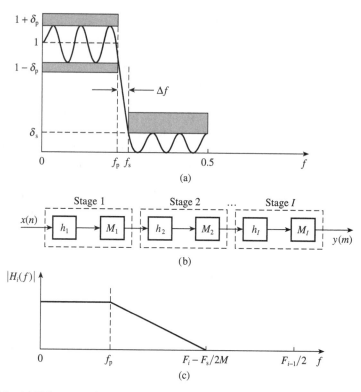

Figure 9.9   (a) Tolerance scheme for an equiripple lowpass filter; (b) multistage structure; (c) filter specifications for stage i, $i = 1, 2, \ldots, I$.

where M_i is the decimation factor for the stage. The initial and final sampling rates are F_0 and F_I, respectively. To relate to previous discussions, $F_0 = F_s$ and $F_I = F_s/M$.

For multistage decimation, a lower passband deviation is necessary for each stage to ensure that the overall passband deviation is δ_p. The stopband deviation for each stage is the same as the overall stopband deviation because as the signal goes from stage to stage the stopband components are attenuated further. If anything, the overall stopband deviation will be better than the overall stopband requirements. For a one-stage decimator, the filter requirements are the same as specified in Equation 9.5.

9.3.3   Determining the number of stages and decimation factors

The multistage design offers significant savings in computation and storage requirements over a single-stage design. The extent of this saving depends on the number of stages used and the choice of decimation factors for the individual stages. A major problem is determing the optimum number of stages, I, and the decimation factors for each stage. An optimum number of stages is one which leads to the least computational effort, for example as measured by the number of multiplications per second (MPS) or the total storage requirements (TSR) for the coefficients:

Table 9.1   Algorithm for finding optimum values of I and M_i.

- Specify overall filter parameters $(F_s, M, f_p, f_s, \delta_s, \delta_p)$.
- For each value of I $(I = 1, 2, \ldots, I_{\max})$, obtain all the possible set of integer decimation factors of M.
- For each set of decimation factors determine the filter requirements, the MPS and storage from Equation 9.9.
- For each value of I, select the decimation factors giving the most efficient design in terms of storage requirements.
- Select the most efficient or desired solution.

$$\text{MPS} = \sum_{i=1}^{I} N_i F_i \tag{9.9a}$$

$$\text{TSR} = \sum_{i=1}^{I} N_i \tag{9.9b}$$

where N_i is the number of filter coefficients for stage i, and we have ignored any symmetry in the filter coefficients.

The choice of the number of stages, I, and the decimation factors is not a trivial problem. However, in practice the number of stages, I, is rarely more than 3 or 4. Further, for a given value of M, there are only a limited set of possible integer factors. Thus a viable approach is to determine all the possible factors of M, that is all the set of M_i values, and their corresponding MPS or total storage requirements. The most efficient or preferred solution can then be chosen by inspection. The algorithm for this approach is summarized in Table 9.1. A C language implementation is provided in the disk for this book, and a spreadsheet implementation is described by DeFatta *et al.* (1988).

In general, for optimum MPS or TSR the decimation factors satisfy the following relationship (Crochiere and Rabiner, 1975, 1976):

$$M_1 > M_2 > \ldots M_I \tag{9.10}$$

where M_i are continuous. However, when the factors are integers it is not always possible to satisfy Equation 9.10 for some values of I, for example if $I = 3$ and $M = 32$ (see the comments in Example 9.3).

For $I = 2$, that is two stages of decimation, the optimal values of the decimation factors for which TSR is minimized are

$$M_{1_{\text{opt}}} = \frac{2M}{2 - \Delta f + (2M\,\Delta f)^{1/2}} \tag{9.11a}$$

$$M_{2_{\text{opt}}} = \frac{M}{M_{1_{\text{opt}}}} \tag{9.11b}$$

For $I > 2$, no simple closed-form expression exists and it becomes necessary to use a computer-aided optimization routine or the algorithm given in Table 9.1 to find the optimum decimation factors, M_i.

9.3.4   Illustrative design examples

Example 9.2

(a) The block diagram of a three-stage decimator which is used to reduce the sampling rate from 3072 kHz to 48 kHz is given in Figure 9.10. Assuming decimation factors of 8, 4 and 2, indicate the sampling rate at the output of each of the three stages.

(b) Assume that the decimator in part (a) satisfies the following overall specifications:

input sampling frequency, F_s	3072 kHz
decimation factor, M	64
passband ripple	0.01 dB
stopband ripple	60 dB
frequency band of interest	0–20 kHz

Determine the bandedge frequencies for the decimating filter at each stage.

(c) Assuming that the input and output sampling rates of a decimator are 3072 kHz and 48 kHz, respectively:

(i)   write down the overall decimation factor;

(ii) write down all the possible sets of integer decimation factors (written in descending order only), assuming two stages of decimation;

(iii) repeat (ii) but assume three stages of decimation;

(iv) repeat (ii) but assume four stages of decimation.

(d) For the decimator in part (a), calculate the total number of multiplications per second (MPS) and the total storage requirements (TSR) in terms of the filter lengths, N_1, N_2 and N_3.

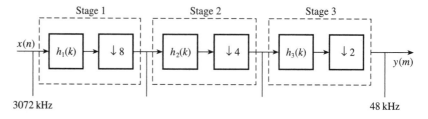

Figure 9.10   Block diagram for Example 9.2.

Solution   (a)   At the first stage, the sampling rate is reduced by a factor of 8 from 3072 kHz to 384 kHz. At the second stage the rate is further reduced by a factor of 4 from 384 kHz to 96 kHz. The third stage reduces this rate by a factor of 2 from 96 kHz to 48 kHz.

(b)   The passband edge frequency of each of the three decimating (anti-aliasing) filters is the same (namely 24 kHz) to preserve the frequency band of interest. The stopband edge frequencies are different to exploit the differences in the sampling rates. The bandstop frequency is given by the following relationship (see Figure 9.9(c) and Equation 9.7b):

$$f_{si} = F_i - \frac{F_s}{2M}, \quad 1, 2, 3$$

where

F_i = output sampling rate for the stage

F_s = the system basic sampling rate

f_{si} = stopband edge frequency for the stage.

For stage one, $f_{s1} = 384 - 3072/(2 \times 64) = 360$ kHz. Thus, the bandedge frequencies for the first stage of decimation are: 0, 20 kHz, 360 kHz and 1536 kHz (the Nyquist frequency for the stage, i.e. 3072 kHz/2).
    For stage two, $f_{s2} = 96 - 3072/(2 \times 64) = 72$ kHz and the set of bandedge frequencies are: 0, 20 kHz, 72 kHz and 192 kHz.
    For stage three, $f_{s3} = 48 - 3072/(2 \times 64) = 24$ kHz, leading to the following set of bandedge frequencies for the anti-aliasing filter: 0, 20 kHz, 24 kHz and 48 kHz.

(c)   (i)   The overall decimation factor is $3072/48 = 64$.
    (ii)   The possible sets of integer decimation factors for two stages of decimation are (in descending order):

$$32 \times 2$$
$$16 \times 4$$
$$8 \times 8$$

(iii)   The possible sets of integer decimation factors for three stages of decimation are (in descending order):

$$16 \times 2 \times 2$$
$$8 \times 4 \times 2$$

(iv)   The possible set of integer decimation factors for four stages of decimation is:

$$4 \times 4 \times 2 \times 2$$

(d) Assuming that the decimation factors are $8 \times 4 \times 2$ and that the filter lengths for stages 1, 2 and 3 are N_1, N_2 and N_3, respectively, then the total number of multiplications per second is

$$N_1 \times F_1 + N_2 \times F_2 + N_3 \times F_3 = N_1 \times 384 \times 10^3 + N_2 \times 96 \times 10^3$$
$$+ N_3 \times 48 \times 10^3$$

The total storage is

$$N_1 + N_2 + N_3$$

Example 9.3

The sampling rate of a signal $x(n)$ is to be reduced, by decimation, from 96 kHz to 1 kHz. The highest frequency of interest after decimation is 450 Hz. Assume that an optimal FIR filter is to be used, with an overall passband ripple, $\delta_p = 0.01$, and passband deviation, $\delta_s = 0.001$. Design an efficient decimator.

Solution

We will start by finding the most efficient design for each value of I, $I = 1, 2, 3, 4$. We will then compare these designs and select the best.

(1) First let us consider a one-stage design ($I = 1$). The block diagram and filter specifications for the stage are given in Figure 9.11(a).

(2) Next, we consider a two-stage design. Using the design program referred to in the text, the optimum integer decimation factors for $I = 2$ are $M_1 = 32$, $M_2 = 3$. The two-stage system, including its specifications, is shown in Figure 9.11(b). At the first stage, the sampling rate is reduced by a factor of 32 to 3 kHz, and, at the second stage, this is further reduced by 3 to 1 kHz.

(3) For the three-stage case ($I = 3$), the optimum integer decimation factors, in terms of storage, are $M_1 = 8$, $M_2 = 6$, $M_3 = 2$. The system for this is depicted in Figure 9.11(c).

(4) For the four-stage design, the optimum integer decimation factors are $M_1 = 4$, $M_2 = 4$, $M_3 = 3$, $M_4 = 2$. The system and filter specifications for this are depicted in Figure 9.11(d).

The results are summarized below:

I	N_1	N_2	N_3	N_4	M_1	M_2	M_3	M_4	MPS	TSR
1	4881	–	–	–	96	–	–	–	48 881 000	4881
2	131	167	–	–	32	3	–	–	560 000	298
3	25	34	117	–	8	6	2	–	485 000	176
4	11	13	17	120	4	4	3	2	496 000	161

It is clear that, in general, multistage designs yield very significant reductions in both computation and storage requirements compared with single-stage designs. The reductions are due to the wide transitions of filters at the early stages (even though the rates are still high), leading to small values of N (filter coefficients).

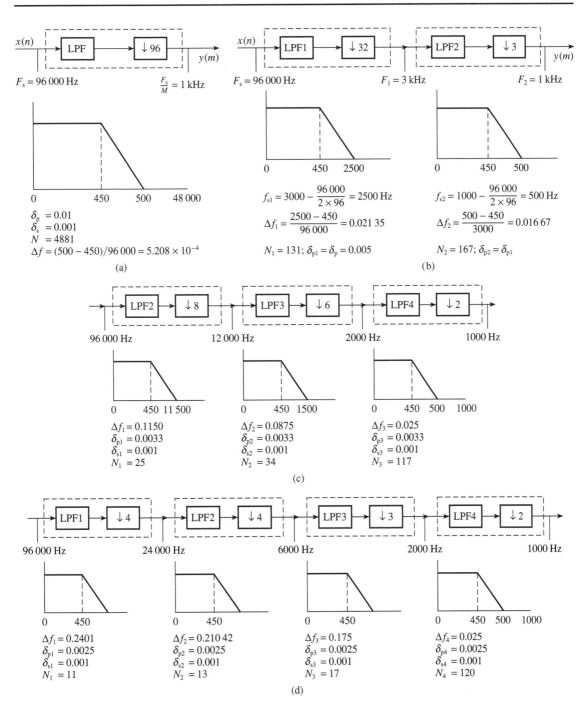

Figure 9.11 (a) Single-stage decimator. (b) Two-stage decimator. (c) Three-stage decimator. (d) Four-stage decimator (Example 9.3).

When we compare the efficiencies of the multistage designs, we find that the reduction in computation (MPS) and storage (TSR) are greatest in going from one stage to two stages. The reduction in storage requirements in going from two to three or from three to four stages is significant, but less marked. The reduction in computation between two and three stages is also significant. From three to four stages, the computational effort (MPS) actually increases. Overall, $I = 3$ would appear to be the most efficient implementation, bearing in mind that as I increases so do the implementational difficulties. In practice, account will need to be taken of the hardware and software complexity before the final choice is made.

Comments   As discussed in Crochiere and Rabiner (1975, 1976), when M_i ($i = 1$, $2, \ldots, I$) are continuous variables, the optimum M_i values satisfy the condition $M_1 > M_2 > \ldots M_I$.

Further, the values of M_i which minimize the storage requirements also minimize the MPS. However, when the values of M_i are restricted to integers these conditions are not always satisfied. For this reason our design program actually computes the solution for all possible sets of integer factors. The best solution can then be selected by inspection.

The most efficient solution, in terms of MPS or storage, may have an excessively large (impractical) value of N in one of its stages. A different set of decimation factors or an increase in the number of stages may produce the desired reduction in the value of N for these stages at the expense perhaps of increasing the filter lengths in other stages. By producing all the possible solutions, these trade-offs can easily be made by inspection.

Example 9.4

Design, at a block diagram level, a two-stage decimator that downsamples an audio signal by a factor of 30 and satisfies the specifications given in Table 9.2.

Your answer must indicate a suitable pair of decimation factors, with appropriate detailed analysis of computational and storage complexities to justify your choice. Specify the sampling frequencies at the input and output of each stage of decimation, and the following parameters for each of the decimating filters in your design:

Table 9.2   Filter specifications for the two-stage decimator

input sampling frequency, F_s	240 kHz
highest frequency of interest in the data	3.4 kHz
passband ripple, δ_p	0.05
stopband ripple, δ_s	0.01

$$\text{filter length,} \quad N = \frac{-10 \log (\delta_p \delta_s) - 13}{14.6 \Delta f} + 1$$

where

Δf = normalized transition width

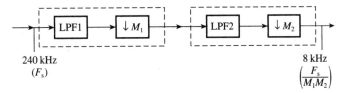

Figure 9.12   A two-stage decimator.

the bandedge frequencies
the normalized transition width
passband and stopband ripples
filter length

You may assume that the filters are direct form FIR, each with a length defined in Table 9.2.

Solution   The decimator will have the general form given in Figure 9.12.

Assume integer factors, then the possible pairs of decimation factors are (for computational efficiency, the first stage always has the largest factor)

15×2

10×3

6×5

To determine which pairs of decimation factors to use requires analysis of the computational complexities.

(i)   For $M_1 = 15$ and $M_2 = 2$, the sample rate will first be reduced by a factor of 15 to 16 kHz and then by 2 to 8 kHz. The parameters for the decimating filters in the two stages are

first stage: bandedge frequencies: 3.4 kHz and $12 \text{ kHz} \left(16 - \dfrac{240}{2 \times 30} \right)$

$$\Delta f = \frac{12 - 3.4}{240} = 0.0358$$

$$\delta_p = \frac{0.05}{2} = 0.025; \delta_s = 0.01; N_1 = 45$$

second stage: bandedge frequencies: 3.4 kHz and 4 kHz

$\Delta f = 0.0375$

$$\delta_p = \frac{0.05}{2} = 0.025; \delta_s = 0.01; N_2 = 43$$

Two measures of complexity are the number of multiplications per second (MPS) and storage requirements (TSR). For efficient implementation, the decimator structure is chosen such that at each stage the filtering operations are performed at the lower sampling rate, leading to

$$\text{MPS} = (45 \times 16 + 43 \times 8) \times 10^3 = 1064 \times 10^3; \text{ storage} = 88.$$

(ii)   For $M_1 = 10$, $M_2 = 3$, the sample rate in this case will first be reduced by 10 to 24 kHz and then by 3 to 8 kHz. The parameters for decimating filters are

first stage: bandedge frequencies: 3.4 kHz and 20 kHz $\left(24 - \dfrac{240}{2 \times 30} = 20 \text{ kHz}\right)$

$$\Delta f = \frac{20 - 3.4}{240} = 0.0691; \ \delta_p = \frac{0.05}{2} = 0.025; \ \delta_s = 0.01; \ N_1 = 23.81 \approx 24$$

second stage: bandedge frequencies: 3.4 kHz and 4 kHz $(8 - 4 = 4 \text{ kHz})$

$$\Delta f = \frac{4 - 3.4}{24} = 0.025; \ \delta_p = \frac{0.05}{2} = 0.025; \ \delta_s = 0.01; \ N_2 = 64$$

Two measures of complexity in this case are

$$\text{MPS} = (24 \times 24 + 64 \times 8) \times 10^3 = 1088 \times 10^3; \text{ storage} = 88.$$

(iii)   The final possible pair of decimation factors are $M_1 = 6$, $M_2 = 5$. A similar analysis shows that

first stage: bandedge frequencies: 3.4 kHz and 36 kHz

$$\Delta f = 0.1358; \ \delta_p = \frac{0.05}{2} = 0.025; \ \delta_s = 0.01; \ N_1 = 13$$

second stage: bandedge frequencies: 3.4 kHz and 4 kHz

$$\Delta f = 0.015; \ \delta_p = \frac{0.05}{2} = 0.025; \ \delta_s = 0.01; \ N_2 = 106$$

$$\text{MPS} = 1368 \times 10^3; \text{ storage} = 119.$$

The measures of complexity are summarized in Table 9.3 for the three cases above. Comparing the computational and storage complexities, we see that the most appropriate pair of decimation factors are $M_1 = 15$, $M_2 = 2$.

Table 9.3   Computational (MPS) and storage (TSR) complexities.

Decimation factors	*MPS*	*TSR*
$M_1 = 15; M_2 = 2$	1064×10^3	88
$M_1 = 10; M_2 = 3$	1088×10^3	88
$M_1 = 6; M_2 = 5$	1368×10^3	119

9.4   Software implementation of sampling rate converters–decimators

A simple block diagram representation of the decimator is given in Figure 9.13(a), where $h(k)$ is an anti-aliasing digital filter. Assuming a direct form implementation (that is, the use of tapped-delay lines), the output of the filter, $w(n)$, and the input, $x(n)$, are related as

$$w(n) = \sum_{k=0}^{N-1} h(k)x(n-k) \tag{9.12a}$$

where N is the number of FIR filter coefficients. The output of the decimator $y(m) = w(mM)$, which on combining with Equation 9.12a leads to the decimator equation

$$y(m) = \sum_{k=0}^{N-1} h(k)x(Mm-k) \tag{9.12b}$$

The signal flow diagram for the decimator is given in Figure 9.13(b). The input $x(n)$ is fed into the delay line one sample at a time. For every M samples of $x(n)$ applied to the delay line, one output sample $y(m)$ is computed. This involves keeping the first sample of $w(n)$, discarding the next $M-1$ samples, keeping the next sample, and discarding the next $M-1$ samples, and so on. Since for each sample that is kept, the next $M-1$ samples of $w(n)$ are discarded, it is unnecessary to perform Equation 9.12a for those samples of $w(n)$ that are discarded. Thus, the down-sampling (discarding of samples) operation can be performed before the multiplication of the input samples by the coefficients (Figure 9.14(a)). The multiplications and additions involving the filter coefficients are now performed at the lower sampling frequency F_s/M, leading to a reduction in the computational effort by a factor of M.

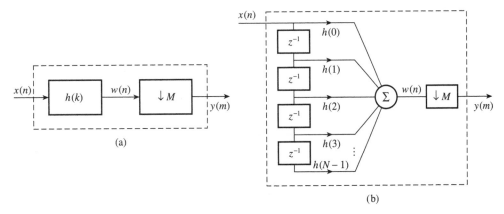

Figure 9.13   (a) A simple block diagram of the decimator; (b) signal flowgraph for the decimator.

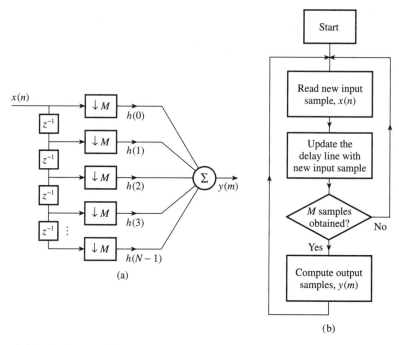

Figure 9.14   (a) A more efficient flowgraph for decimation; (b) flowchart of the decimation process.

The operation of a single-stage decimator is summarized in the flowchart of Figure 9.14(b). The flowchart for a three-stage decimation is given in Figure 9.15, which is a straightforward extension of the single-stage operation.

9.4.1   Program for multistage decimation

A self-contained, interactive C language program based on the above methods that runs on a PC (personal computer) is described in Appendix 9A. The program decimates input data using up to three decimation stages; see the flowchart of Figure 9.15. Each stage of decimation requires an integer decimation factor and a set of N-point filter coefficients representing a linear phase FIR digital filter.

The input data is read from a data file in the PC and the decimated data is written to a user-specified output data file one sample at a time. Assuming a three-stage decimation (see Figure 9.15), for every M_1 input samples fed into stage 1, one output sample is computed. For every M_2 output samples computed for stage 1, one output sample is computed at stage 2. Finally, for every M_3 output samples from stage 2 one output sample is computed from stage 3. Thus at the end of a decimation cycle for each M input samples of $x(n)$, where $M = M_1 M_2 M_3$, one output sample is computed and stored in the output data file. The process is repeated until all the input samples have been processed.

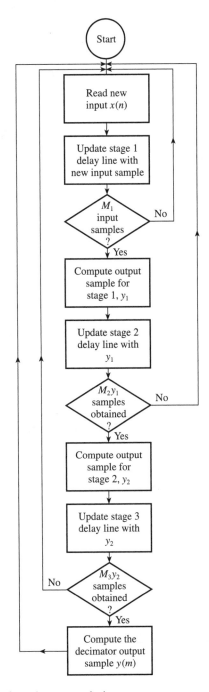

Figure 9.15 Flowchart for a three-stage decimator.

The program is self-contained. To use the program, the user must specify the number of stages, the overall decimation factor, the decimation factor for each stage, and a set of direct form FIR linear phase filter coefficients for each stage. The user specifies the names of files containing the data to be decimated, files containing filter coefficients, and the name of the output file which holds the decimated data.

9.4.2   Test example for the decimation program

The input sequence used to test the decimation process is the allpass signal derived as follows (Crochiere and Rabiner, 1979):

$$x(n) = -\alpha, \quad n = 0$$
$$= (1 - \alpha^2)\alpha^{n-1}, \quad n = 1, 2, \dots \tag{9.13}$$

where $\alpha = 0.9$. The first 29 data samples of the sequence are listed in Table 9.4.

In this example, a two-stage decimation is performed. Decimation factors of 5 and 2 are used for the first and second stages, giving an overall decimation factor of 10. Table 9.4 gives a list of the FIR filter coefficients used which are of lengths 25 and 28, respectively. The result of the decimation operation is also given.

9.4.2.1   Output delay

The output of a decimator will be delayed from the input by a certain number of samples, depending on the type of filter used at the decimator stages. Assuming that the filters used are linear phase FIR filters, the group delays for single, double, and three-stage decimation are, respectively,

$$T(1 \text{ stage}) = \frac{1}{M}[T_1 - (M - 1)] \text{ samples} \tag{9.14a}$$

$$T(2 \text{ stages}) = \frac{1}{M_1 M_2}[T_1 + M_1 T_2 - (M_1 M_2 - 1)] \text{ samples} \tag{9.14b}$$

$$T(3 \text{ stages}) = \frac{1}{M_1 M_2 M_3}[T_1 + M_1 T_2 + M_1 M_2 T_3 - (M_1 M_2 M_3 - 1)] \text{ samples}$$
$$\tag{9.14c}$$

where T_i, the delay in the ith-stage filter, is given by

$$T_i = (N_i + 1)/2 \text{ samples}$$

and N_i is the number of filter coefficients for stage i. In the two-stage test example above, the filters have delays of 14.5 and 13 samples, respectively, giving an overall delay of

Table 9.4 Data for test example on decimation.

n	x(n)	y(m)	$h_1(k)$	$h_2(k)$
0	−0.9000		−0.000 174	−0.000 303
1	0.1900		−0.002 682	0.001 807
2	0.1710		−0.006 346	0.003 120
3	0.1539		−0.011 033	−0.001 169
4	0.1385		−0.014 156	−0.009 267
5	0.1247		−0.012 024	−0.007 792
6	0.1122		−0.000 775	0.011 124
7	0.1010		0.021 904	0.027 651
8	0.0909		0.055 181	0.007 674
9	0.0818	0.000 040	0.094 397	−0.045 444
10	0.0736		0.131 836	−0.064 816
11	0.0662		0.158 866	0.022 946
12	0.0596		0.168 728	0.202 371
13	0.0537		0.158 866	0.352 610
14	0.0483		0.131 836	0.352 610
15	0.0435		0.094 397	0.202 371
16	0.0391		0.055 181	0.022 946
17	0.0352		0.021 904	−0.064 816
18	0.0317		−0.000 775	−0.045 444
19	0.0285	−0.000 286	−0.012 024	0.007 674
20	0.0257		−0.014 156	0.027 651
21	0.0231		−0.011 033	0.011 124
22	0.0208		−0.006 346	−0.007 792
23	0.0187		−0.002 682	−0.009 267
24	0.0168		−0.000 174	−0.001 169
25	0.0152			0.003 120
26	0.0136			0.001 807
27	0.0123			−0.000 303
28	0.0110			
29	0.0099	0.001 116		
30	0.0089			
31	0.0081			
32	0.0072			
33	0.0065			
34	0.0059			
35	0.0053			
36	0.0048			
37	0.0043			
38	0.0039			
39	0.0035	−0.001 659		
40	0.0031			
41	0.0028			
42	0.0025			
43	0.0023			
44	0.0020			
45	0.0018			
46	0.0017			
47	0.0015			
48	0.0013			
49	0.0012	−0.000 402		
50	0.0011			

$x(n)$ and $y(m)$ are the input and decimated data. $h_1(k)$ and $h_2(k)$ are decimating filter coefficients.

$$T(2 \text{ stages}) = [1/(5 \times 2)][14.5 + 5 \times 13 - (5 \times 2 - 1)] = 7.05 \text{ samples}$$

If an integer delay is desired then N_i should be chosen such that the value of T in the appropriate equation above is an integer. Such is the case when the input and output samples are to be compared; for example in multirate highpass filtering the output sample needs to be corrected for the delay through the decimator and interpolator.

9.5 Software implementation of interpolators

A block diagram representation of the interpolator is shown in Figure 9.16(a) and its signal flow diagram is depicted in Figure 9.16(b). For every input sample, $x(n)$, fed into the interpolator, the rate expander (the box with an up-arrow) inserts $L - 1$ zero-valued samples after the input sample. These are then filtered to yield $y(m)$. Thus for each input sample of $x(n)$ we have L samples of $y(m)$. Effectively, the input sampling frequency is increased from F_s to LF_s by the interpolator. One implication of inserting $L - 1$ zeros after each sample is that the energy of each input sample is spread across L output samples. Thus the interpolator has a gain of $1/L$. After interpolation, each output sample should be multiplied by L to restore it to its proper level.

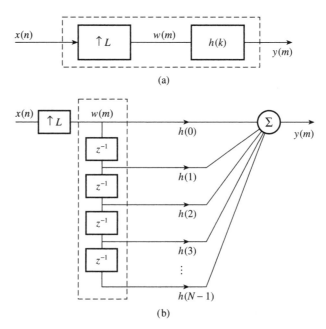

Figure 9.16 (a) Block diagram of the interpolator; (b) signal flow diagram of the interpolator.

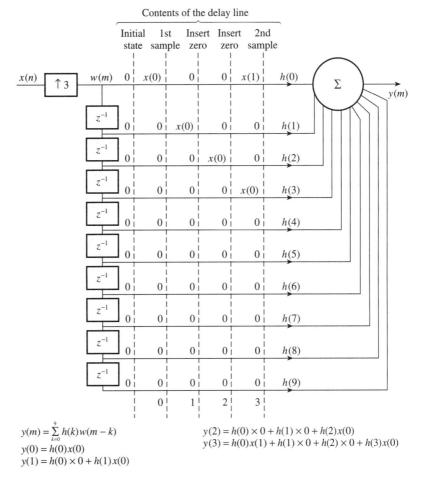

Figure 9.17   An illustration of the process of interpolation for the simple case of $L = 3$.

The interpolation equations are

$$y(m) = \sum_{k=0}^{N-1} h(k)w(m-k) \tag{9.15a}$$

$$w(m-k) = \begin{cases} x[(m-k)/L], & m-k = 0, L, 2L, \ldots \\ 0 \end{cases} \tag{9.15b}$$

Figure 9.17 illustrates the process for the simple case where $L = 3$ and the filter length is 10. The delay line is fed with an input sample followed by two zeros, then the next input sample followed by two zeros and so on. An output sample is computed for each sample (data or zero) fed into the delay line. The contents of the delay line after two samples of $x(n)$ have been fed into the interpolator are shown. We see that for

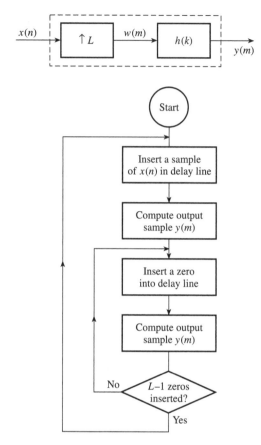

Figure 9.18   Flowchart of the interpolation process.

each input sample fed in, three samples are computed. The nonzero samples (that is the actual samples of $x(n)$) in the delay line are separated by $L - 1$ zeros (two in this example). Clearly, multiplication operations by the zero-valued samples are unnecessary.

Figure 9.18 gives the flowchart of a single-stage interpolator. In this implementation, only the nonzero-valued samples are fetched and used in the computation of the output samples. The flowchart for up to three stages of interpolation is given in Figure 9.19.

Another efficient implementation, known as polyphase filtering (Crochiere and Rabiner, 1983), exploits the fact that some of the delay line samples are zero. In this case the rate expander is removed altogether to eliminate the need to store zero-valued samples. The delay line is then shortened to N/L locations. In this approach, for each input sample fed into the delay line, the N/L delay line samples are used to compute L output samples, with each output sample computed with a different set of filter coefficients (that is, those filter coefficients that correspond to zero-valued samples

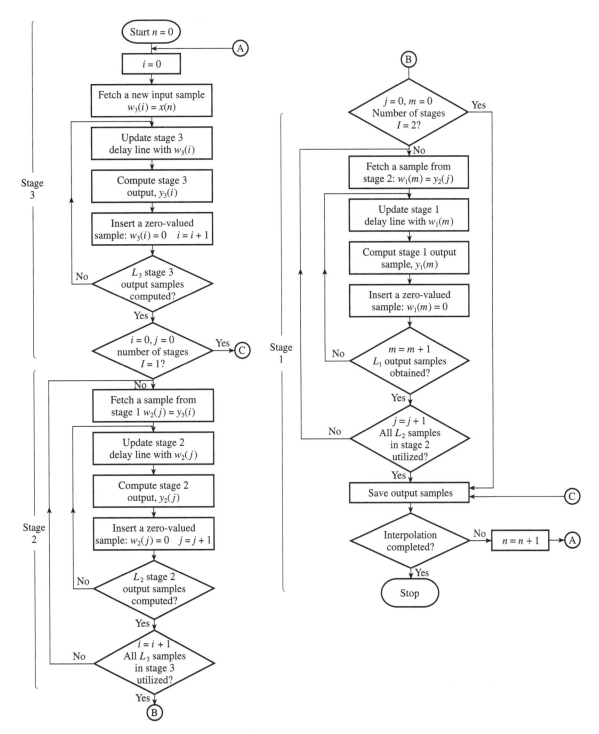

Figure 9.19   Interpolation using up to three stages. If, for example, $I = 1$ then only the sections labelled stage 3 are executed.

are skipped). The limitation of the polyphase filter approach is that the ratio of N to L must be an integer. In the next section, we describe a C language implementation of a three-stage interpolator.

9.5.1   Program for multistage interpolation

The program is based on the method described in the last section. The program interpolates input data using up to three interpolation stages; see the flowchart of Figure 9.19. Note that because of the duality between decimation and interpolation (a decimator and interpolator form a sample rate converter pair) the interpolation stages are numbered in reverse order. Each stage of interpolation requires an integer interpolation factor and a set of N-point filter coefficients representing a linear phase FIR digital filter. The program is intended to run on a personal computer, but can be readily modified to run on other hardware.

The input data is read from a data file in a computer and the interpolated data is written to an output file a sample at a time. Assuming a three-stage interpolation (see Figure 9.19), L_3 samples are computed at stage 3 of the interpolator. For each of the L_3 samples at stage 3, L_2 samples are computed in stage 2. For each of the L_2 samples obtained in stage 2, L_1 output samples are computed in stage 1. At the end of the cycle, L output samples (where $L = L_1 L_2 L_3$) are computed and stored in the output data file for each input sample, $x(n)$. The process is repeated until all the input samples have been processed. The multistage interpolation program is available on the CD in the companion handbook (see the Preface for details).

The program is self-contained. To use the program, the user must specify the number of stages, the overall interpolation factor, the interpolation factor for each stage, and a set of FIR filter coefficients for each stage. The user specifies the names of files containing the data to be interpolated, filter coefficients, and the name of the file to hold the results.

9.5.2   Test example

The input sequence used is derived in the same way as for the decimator. The first five samples of the sequence are given in Table 9.5.

In this example, two-stage interpolation is performed. Interpolation factors of 2 and 5 are used for the third and second stages, giving an overall interpolation factor of 10. Table 9.5 gives a list of the results of the interpolation. The FIR filter coefficients used are given in Table 9.5, and are of lengths 25 and 28 respectively.

9.5.2.1   Output delay

The output of the interpolator will be delayed from the input by a certain number of samples. The group delays for single, double, and three-stage decimation are, respectively,

Table 9.5 Data for test example on interpolation.

n	$x(n)$	$y(m)$	$h_2(k)$	$h_1(k)$
0	−0.9000	$-4.744\,98 \times 10^{-7}$	−0.000 303	−0.000 174
		$-7.313\,815 \times 10^{-6}$	0.001 807	−0.002 682
		$-1.730\,554 \times 10^{-5}$	0.003 120	−0.006 346
		$-3.008\,7 \times 10^{-5}$	−0.001 169	−0.011 033
		$-3.860\,341 \times 10^{-5}$	−0.009 267	−0.014 156
		$-2.995\,969 \times 10^{-5}$	−0.007 792	−0.012 024
		$4.150\,394 \times 10^{-5}$	0.011 124	−0.000 775
		$1.629\,372 \times 10^{-4}$	0.027 651	0.021 904
		$3.299\,083 \times 10^{-4}$	0.007 674	0.055 181
		$4.876\,397 \times 10^{-4}$	−0.045 444	0.094 397
1	0.1900	$5.600\,492 \times 10^{-4}$	−0.064 816	0.131 836
		$5.226\,861 \times 10^{-4}$	0.022 946	0.158 866
		$2.857\,456 \times 10^{-4}$	0.202 371	0.168 728
		$-1.480\,226 \times 10^{-4}$	0.352 610	0.158 866
		$-7.700\,115 \times 10^{-4}$	0.352 610	0.131 836
		$-0.001\,544\,5$	0.202 371	0.094 397
		$-2.448\,377 \times 10^{-3}$	0.022 946	0.055 181
		$-3.400\,52 \times 10^{-3}$	−0.064 816	0.021 904
		$-4.320\,959 \times 10^{-3}$	−0.045 444	−0.000 775
		$-5.079\,387 \times 10^{-3}$	0.007 674	−0.012 024
2	0.1710	$-5.534\,875 \times 10^{-3}$	0.027 651	−0.014 156
		$-5.728\,923 \times 10^{-3}$	0.011 124	−0.011 033
		$-5.466\,501 \times 10^{-3}$	−0.007 792	−0.006 346
		$-4.756\,987 \times 10^{-3}$	−0.009 267	−0.002 682
		$-3.522\,772 \times 10^{-3}$	−0.001 169	−0.000 174
		$-1.715\,353 \times 10^{-3}$	0.003 120	
		$5.557\,998 \times 10^{-4}$	0.001 807	
		$3.324\,822 \times 10^{-3}$	−0.000 303	
		$6.400\,164 \times 10^{-3}$		
		$9.565\,701 \times 10^{-3}$		
3	0.1539	$0.012\,597\,6$		
		$1.544\,317 \times 10^{-2}$		
		$1.774\,401 \times 10^{-2}$		
		$1.933\,819 \times 10^{-2}$		
		$1.984\,539 \times 10^{-2}$		
		$1.894\,878 \times 10^{-2}$		
		$1.681\,832 \times 10^{-2}$		
		$1.303\,225 \times 10^{-2}$		
		$7.845\,689 \times 10^{-3}$		
		$1.357\,867 \times 10^{-3}$		
4	0.1385	$-6.262\,392 \times 10^{-3}$		
		$-1.462\,789 \times 10^{-2}$		
		$-2.343\,143 \times 10^{-2}$		
		$-3.207\,272 \times 10^{-2}$		
		$-3.972\,186 \times 10^{-2}$		
		$-4.567\,938 \times 10^{-2}$		
		$-4.993\,166 \times 10^{-2}$		
		$-5.142\,782 \times 10^{-2}$		
		$-5.009\,625 \times 10^{-2}$		
		$-4.527\,419 \times 10^{-2}$		

$x(n)$ and $y(m)$ are the input and interpolated data. $h_2(k)$ and $h_1(k)$ are interpolating filters.

$$T(1 \text{ stage}) \ = T_1 \text{ samples} \tag{9.16a}$$

$$T(2 \text{ stages}) = T_1 + M_1 T_2 \text{ samples} \tag{9.16b}$$

$$T(3 \text{ stages}) = T_1 + M_1 T_2 + M_1 M_2 T_3 \text{ samples} \tag{9.16c}$$

where T_i is the delay in the ith-stage filter: $T_i = (N_i + 1)/2$ samples. N_i is the number of filter coefficients for the stage. In the test example above, the filters have delays of 13 and 14.5 samples, giving an overall delay of $13 + 2 \times 14.5 = 42$ samples.

If an integer delay is desired then N_i should be chosen such that the appropriate equation of Equations 9.16 is an integer. Thus if the input and output of the interpolator are to be compared (for example in highpass narrowband filtering where the filtering operation is performed as the inverse of a lowpass), the output signal should be corrected or adjusted for this delay.

9.6 Sample rate conversion using polyphase filter structure

Another efficient structure for implementing decimators and interpolators is the so-called polyphase structure. We will consider polyphase filtering for interpolators first as they are easier to understand.

9.6.1 Polyphase implementation of interpolators

Polyphase implementation for interpolators exploits the fact that some of the delay line samples in an interpolator are zero valued. In this case the rate expander is removed altogether to eliminate the need to store zero-valued samples. The delay line is then shortened to N/L locations (where N is the length of the anti-image filter and L is the interpolation factor). In this approach, for each input sample fed into the delay line, the N/L delay line samples are used to compute L output samples, with each sample computed with a different set of filter coefficients (i.e. those filter coefficients that correspond to zero-valued samples are skipped).

To gain some insight into polyphase implementation of interpolators, we will examine a simple 1-to-3 interpolator, see Figure 9.20. In this case, $L = 3$ and the number of filter coefficients, $N = 8$.

Referring to Figure 9.20, the delay line is fed with an input sample followed by two zeros ($L - 1$) then the next input sample followed by two zeros and so on.

The contents of the delay line after feeding four input data samples – $x(0)$, $x(1)$, $x(2)$ and $x(3)$ – each followed by two zero-valued samples are shown in Figure 9.21 with the sampling instants for the two sampling rates indicated.

An output sample is computed for every sample (data or zero) fed into the delay line. Thus, for each actual sample fed in, three samples are computed. The non-zero samples (i.e. the actual samples of $x(n)$ in the delay line) are separated by $L - 1$ zeros

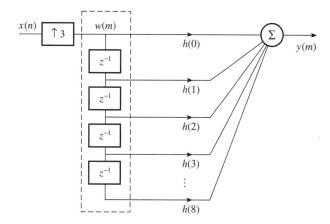

Figure 9.20 A 1-to-3 interpolator with a 9-point direct form FIR filter.

(two in this example). Clearly, multiplication operations by the zero-valued samples are unnecessary.

Successive output samples of the interpolator are given in Figure 9.21. In this implementation, only the non-zero valued input samples are used in the computation of the output samples. The following points should be noted:

(1) Each input sample leads to three output samples, where each output sample is computed using three sub-filters, each with three coefficients:

- {h(0), h(3), h(6)}, {h(1), h(4), h(7)}, {h(2), h(5), h(8)};
- the process repeats for each new input sample;
- the sub-filters operate at the lower sampling rate.

(2) The overall filter is effectively a parallel realization of the sub-filters: see Figure 9.22(a). The sub-filters, known as polyphase filters, share a common delay line leading to a reduction in the overall storage requirements by a factor of 3. For each new input sample, each polyphase filter provides an output sample at the lower sampling rate: see Figure 9.22(b).

The general structure for polyphase filter implementation of interpolation is shown in Figure 9.23. In this case, for each new input sample of $x(n)$, there are L output samples of $y(m)$, one generated for each polyphase filter. The upper polyphase filter, i.e. $\rho_0(n)$, produces the output samples $y_0(n)$, the next polyphase filter, $\rho_1(n)$, produces the output $y_1(n)$ and so on. In practice, polyphase filters are often implemented using the commutative model depicted in Figure 9.23(b). The commutator rotates in the anti-clockwise direction starting at the top.

Analysis of polyphase filters shows that they are allpass filters, but the phase shifts through them are different. This is why the filters are called polyphase filters. The computational efficiency of the polyphase filter structure comes from dividing the single N-point FIR filter into a set of smaller sub-filters of length N/L, where N is chosen to be an integer multiple of L, each of which operates at a lower sampling rate.

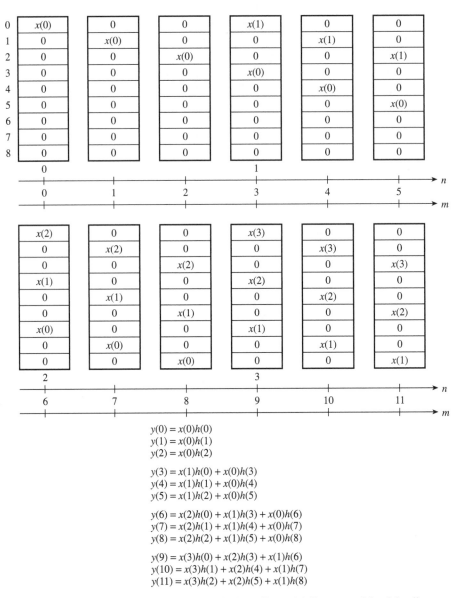

$$y(0) = x(0)h(0)$$
$$y(1) = x(0)h(1)$$
$$y(2) = x(0)h(2)$$

$$y(3) = x(1)h(0) + x(0)h(3)$$
$$y(4) = x(1)h(1) + x(0)h(4)$$
$$y(5) = x(1)h(2) + x(0)h(5)$$

$$y(6) = x(2)h(0) + x(1)h(3) + x(0)h(6)$$
$$y(7) = x(2)h(1) + x(1)h(4) + x(0)h(7)$$
$$y(8) = x(2)h(2) + x(1)h(5) + x(0)h(8)$$

$$y(9) = x(3)h(0) + x(2)h(3) + x(1)h(6)$$
$$y(10) = x(3)h(1) + x(2)h(4) + x(1)h(7)$$
$$y(11) = x(3)h(2) + x(2)h(5) + x(1)h(8)$$

Figure 9.21   A 1-to-3 interpolation using polyphase filters. (a) Contents of the delay line at the two different sampling instants, n and m. (b) Output samples of the polyphase filters.

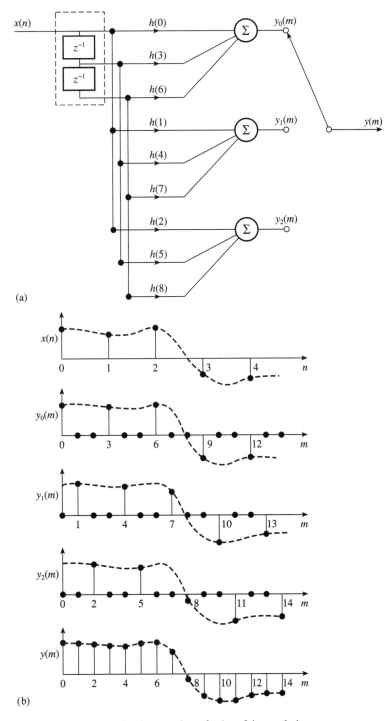

(a)

(b)

Figure 9.22   Polyphase filter implementation of a 1-to-3 interpolation.

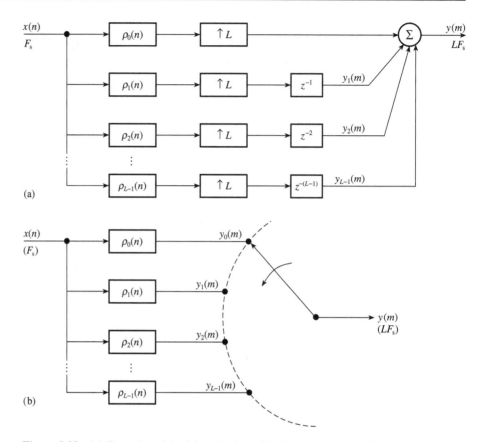

Figure 9.23   (a) General model of the polyphase filter implementation of interpolation.
(b) Commutative model of the polyphase filter implementation of interpolation.

The coefficients of the polyphase filters for interpolation are given by

$$\rho_k(n) = h(k + nL), \quad k = 0, 1, \ldots, L - 1; \quad n = 0, 1, \ldots, \frac{N}{L} - 1$$

The polyphase filter implementation of the decimator can be obtained by transposing the interpolator of Figure 9.23, leading to Figure 9.24. Note that in this case the commutator rotates in the clockwise direction starting from the top at the sampling instant, $m = 0$. The polyphase filters in this case are related to the original decimation filters as:

$$\rho_k(n) = h(k + nM), \quad k = 0, 1, \ldots, M - 1; \quad n = 0, 1, \ldots, \frac{N}{M} - 1$$

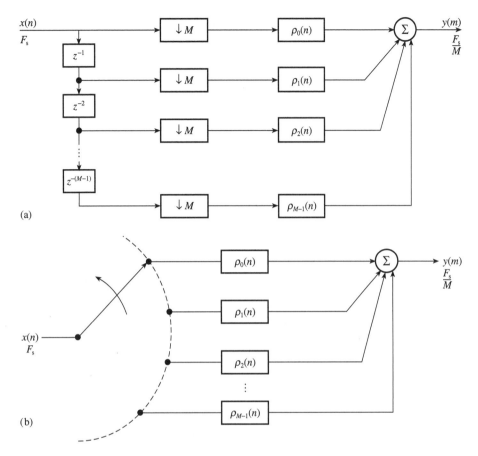

Figure 9.24 (a) General model of the polyphase filter implementation of decimation. (b) Commutative model of the polyphase filter implementation of decimation.

9.7 Application examples

Digital audio engineering is an area that has benefited significantly from multirate techniques. For example, they are used in the compact disc player to simplify the D/A conversion processes, while at the same time maintaining the quality of the reproduced sound. At the front end of digital audio systems, efforts have been directed at using delta modulation techniques, combined with multirate processing, to obtain high quality digital data from analog audio signals.

Other areas where multirate techniques have been used are in the acquisition of high quality data, high resolution spectral analysis, and the design and implementation of narrowband digital filtering.

In the next few sections we will describe a number of these applications.

9.7.1   High quality analog-to-digital conversion for digital audio

The constant demand in digital audio for better quality, higher resolution and higher speed ADC has led to the development of single-bit ADCs using delta sigma modulation techniques. This offers the possibility of eliminating altogether from the conversion process most of the analog circuitry at the front end of a digital audio system, including the analog anti-aliasing filters and sample-and-hold circuits.

A simplified block diagram of a fast single-bit ADC process is shown in Figure 9.25 (Adams, 1986; Matsuya *et al.*, 1987; Welland *et al.*, 1989). The analog audio signal is first converted into a single bit stream, using delta sigma modulation at a 3.072 MHz rate. The single bit stream is then downsampled to 48 kHz, using a multistage decimator, to yield 16-bit PCM words. Many ADCs utilizing multirate techniques are now available, off-the-shelf. Examples are 16- and 18-bit stereo ADCs by Crystal Semiconductor (CS5326, CS5327, CS5328, CS5329), and by Motorola Semiconductor (DSP56ADC16).

9.7.2   Efficient digital-to-analog conversion in compact hi-fi systems

One of the first serious applications of multirate techniques was in the reproduction of sound and music in the compact disc (CD) player.

Figure 9.26 depicts the process of reconstituting the analog audio signal from the digital signal read from the CD. After decoding and error correction the digital signals are in 16-bit words, representing the acoustic information at a 44.1 kHz sampling rate.

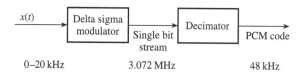

Figure 9.25   Simplified block diagram of single-bit ADC scheme.

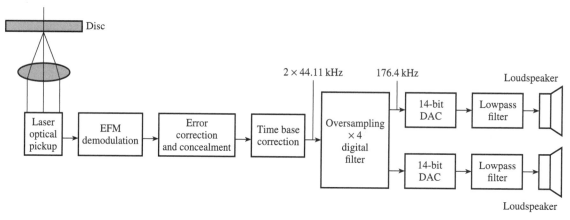

Figure 9.26   Audio signal reproduction in the compact disc system.

If these digital codes were converted directly into analog, image frequency bands centred at multiples of the sampling frequency of 44.1 kHz would be produced. Although the image frequencies would be inaudible as they are above the baseband of 0–20 kHz, they could cause overloading if passed on to the player's amplifier and loudspeaker, or they could set up intermodulation distortion. Thus the frequency components above the baseband need to be attenuated by at least 50 dB. Analog filters that can provide this level of attenuation will have to meet very tight specifications, and would require trimming to ensure that the filters for the two stereo channels are matched.

To avoid the problems with analog filters, multirate filtering is employed in the compact disc player. This is achieved by increasing, by interpolation, the sampling frequency of the data by 4 to 176 kHz (4×44.1 kHz = 176.4 kHz) before it is applied to the DAC. In the time domain, the effect of this is to give a signal that has much finer steps. In the frequency domain, the image frequencies are now pushed up to higher frequencies, making it easier to filter them out. Thus, only a relatively simple lowpass filter is required after the D/A conversion. In the actual implementation, the digital filter incorporates a $\sin x/x$ correction (see Chapter 2) to compensate for the effects of the holding circuit following the DAC. The $\sin x/x$ correction has a beneficial effect in that it attenuates the signals on either side of 174 kHz by more than 18 kB, further simplifying the analog image filtering requirements. A simple third-order Bessel filter is used after interpolation to provide additional attenuation. This has a 3 dB cutoff frequency of 30 kHz and a reasonably linear phase response in the passband.

Oversampling the data has other beneficial effects. It reduces the noise floor, as the quantization noise is now spread over a wider bandwidth, making it possible to use a DAC with fewer bits, and still to achieve an SNR performance that is equivalent to a 16-bit DAC. Thus, in Figure 9.26, the 16-bit-word interpolated data, after oversampling and noise shaping, is rounded off to 14 bits before it is fed into the 14-bit DAC.

There are other DACs on the market that exploit the oversampling concepts presented in this chapter. Examples are the bit stream DACs by Philips Components (SAA7322, SAA7323, SAA7350).

Example 9.5

A digital audio system exploits oversampling techniques to relax the requirements of the analog anti-imaging filter. The overall filter specifications for the system is given below:

baseband	0–20 kHz
input sampling frequency F_s	44.1 kHz
output sampling frequency	176.4 kHz
stopband attenuation	50 dB
passband ripple	0.5 dB
transition width	2 kHz
stopband edge frequency	22.05 kHz

Design a suitable interpolator.

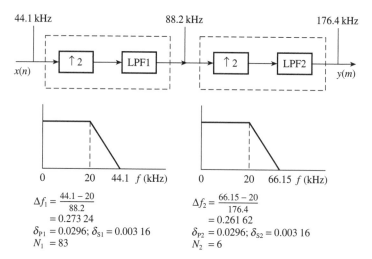

$$\Delta f_1 = \frac{44.1 - 20}{88.2}$$
$$= 0.273\,24$$
$$\delta_{P1} = 0.0296;\ \delta_{S1} = 0.003\,16$$
$$N_1 = 83$$

$$\Delta f_2 = \frac{66.15 - 20}{176.4}$$
$$= 0.261\,62$$
$$\delta_{P2} = 0.0296;\ \delta_{S2} = 0.003\,16$$
$$N_2 = 6$$

Figure 9.27   A two-stage interpolator for Example 9.5.

Solution   Using the multirate design program on the CD in the companion handbook (see the Preface for details), the interpolation factors and filter characteristics of the possible interpolators (with integer factors) are summarized below.

Number of stages	Interpolation factor, L_i	Filter length, N_i	Normalized transition width, Δf_i	Passband ripple, δ_p	Stopband ripple, δ_s
1	4	146	0.045 35	0.059 25	0.003 16
2	2	6	0.261 62	0.029 6	0.003 16
	2	83	0.273 24	0.029 6	0.003 16

For the two-stage interpolator, the system has the form shown in Figure 9.27.

9.7.3   Application in the acquisition of high quality data

In the acquisition of almost any real-life data, the need to keep aliasing low often dictates the use of a relatively complex analog anti-aliasing filter. In a multichannel system, each analog channel must be fitted with a separate anti-aliasing filter as such filters cannot be readily multiplexed. Where there is a large number of analog channels (for example, in biomedicine as many as 32 channels may be required), the use of analog anti-aliasing filters can become expensive. By using anti-aliasing digital filters, the complex analog filters on each channel can be replaced by a much simpler filter, leading to a substantial cost reduction. Further, the phase matching problems of tightly specified analog filters can be avoided. The difficulty of supporting multiple sampling frequencies (each sampling frequency would require a different cutoff frequency) when analog anti-aliasing filters are used is also overcome.

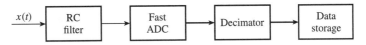

Figure 9.28 A simple multirate data acquisition system.

Figure 9.28 shows a block diagram of a multirate data acquisition system (Quarmby, 1984). The desired aliasing level is achieved by the front-end RC filters oversampling the input signal, then downsampling to the desired frequency using multirate techniques. The main price is that the ADC must operate at a faster rate.

To consolidate the materials presented here and to provide a better appreciation of the benefits of implementing the anti-aliasing filter digitally, we will discuss a real-life application by way of an example.

Example 9.6

A requirement exists for a general purpose, multichannel (up to 32 channels) data acquisition system for collecting physiological data. Each analog channel is individually configurable, by the user, to have a cutoff frequency of between 0.5 and 200 Hz, and a selectable sampling frequency in the range from 1 to 1000 Hz. The overall filter requirements for each channel are

passband ripple	≤ 0.5 dB
signal-to-aliasing ratio	≥ 45 dB (in the passband)
passband edge frequency	0.5 Hz $\leq f_p \leq 200$ Hz
stopband edge frequency	$\leq 3f_P$

Both amplitude and phase distortion should be kept as low as possible. To reduce component count and cost and the size of the PCB for the system only simple analog filters should be used at the front-end.

Solution

To provide the anti-aliasing filtering using analog filters alone would require a very high-order filter. An alternative approach is to fit each channel with a simple, identical filter, to oversample at a common, fixed rate, and then to decimate to the desired rate(s). At each stage we must ensure that the specifications are met.

A simple, one-pole RC filter could be used for each channel, but this would require a very high sampling frequency to satisfy the specification. We will use a second-order Butterworth filter, as we have found it satisfactory in our biomedical engineering work.

The amplitude response of a second-order Butterworth filter is given by

$$A(f) = \frac{1}{[1 + (f/f_c)^4]^{1/2}}$$

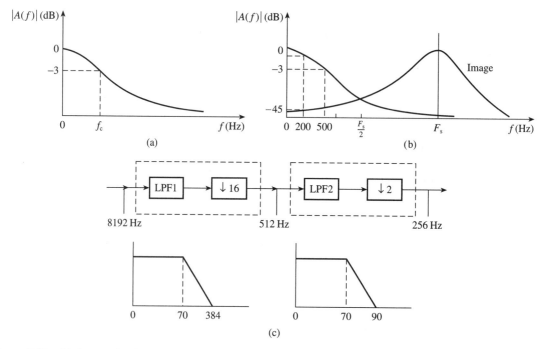

Figure 9.29   (a) A second-order Butterworth response. (b) Spectrum of bandlimited, sampled, wideband signal. (c) A two-stage decimator for Example 9.6.

This is depicted in Figure 9.29(a). It is evident from the figure that a significant amplitude error exists in the band from 0 to f_c. To keep the error within the specification, the highest frequency of interest (200 Hz in our case) should be well below f_c. A suitable value for f_c can be obtained from

$$20 \log [1 + (200/f_c)^4]^{1/2} \leq 0.5 \text{ dB}$$

Solving for f_c gives $f_c \geq 338.39$ Hz. For convenience and to allow for additional errors that would be introduced by subsequent stages $f_c = 500$ Hz would be used. An f_c of 500 Hz gives a response that is down, at 200 Hz, by about 0.11 dB.

Next, we establish a common sampling frequency for all the channels. After bandlimiting each channel with a second-order Butterworth filter and sampling, the spectrum of the signal is depicted in Figure 9.29(b) (assuming a wideband signal). Referring to the figure, it should be clear that we need a sampling frequency, F_s, such that the aliasing level is at least 45 dB down on the signal level at f_p (where $f_p = 200$ Hz):

$$20 \log \{1 + [(F_s - 200)/500]^4\}^{1/2} \geq 45 \text{ dB}$$

Solving for F_s, we have $F_s \approx 6.67$ kHz. For convenience during decimation, let $F_s = 8192$ Hz. A suitable overall specification for the general purpose decimator might look like this:

input sampling frequency	8.192 kHz
output sampling frequency	1 Hz $< F_s < 1000$ Hz
stopband attenuation	50 dB
passband ripple	0.01 dB
passband edge frequency	0.5 Hz $< f_p < 200$ Hz
decimation factor	$8.192 < M < 8192$
stopband edge frequency	$<2f_p$

For convenience, we will place a restriction on the sampling frequency, F_s, that the user may select. This is to allow us to decimate by integer factors only. If the processing capacity is available to cope with non-integer decimation factors this restriction would be unnecessary. Thus the possible sampling frequencies and their corresponding decimation factors are as follows:

M	8	16	32	64	128	256	512	1024	2048	4096	8192
F_s (Hz)	1024	512	256	128	64	32	16	8	4	2	1

Conceptually, we can think of the system as consisting of 11 multistage decimators with only one selectable for a given specification.

As an illustration, let us consider a system for collecting EEG (electro-encephalography) signals. The user specification for each channel could be

sampling frequency	256 Hz
stopband attenuation	45 dB
passband ripple	0.5 dB
passband	0–70 Hz

This specification is translated into that for a rate converter, consistent with that for the general purpose decimator above:

input sampling frequency	8192 kHz
output sampling frequency	256 Hz
decimation factor	32
stopband attenuation	50 dB
passband ripple	0.01 dB
passband	0–70 Hz
stopband	90–128 Hz

Using the design program (provided on the CD in the companion handbook), an efficient decimator (in terms of computation and system complexity) is the two-stage system depicted in Figure 9.29(c).

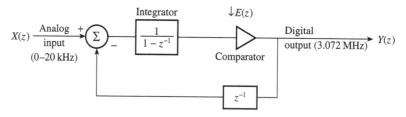

Note. The output transform, $Y(z)$, of the first order delta-sigma modulator is given by:
$Y(z) = X(z) + E(z)(1 - z^{-1})$. The variables have the usual meaning.

Figure 9.30

Example 9.7
(a) A digital signal processing system, with analog audio input signal in the range 0–20 kHz, uses oversampling techniques and a first order delta-sigma modulator to convert the analog signal into a digital bit stream at a rate of 3.072 MHz. The z-plane model of the delta sigma modulator is depicted in Figure 9.30. Determine the overall improvement in the signal-to-quantization noise ratio made possible by oversampling and noise spectrum shaping and hence estimate the effective resolution, in bits, of the converter.

(b) Design, at a block diagram level, a two-stage decimator to convert the output of the delta-sigma modulator depicted in Figure 9.30 from a single-bit stream at 3.072 MHz into a multi-bit stream at 48 kHz. The pass- and stopband ripples for the decimator are 0.001 and 0.0001, respectively.

Your answer must include the following:

∎ values for the overall decimation and interpolation factors, with a justifiable indication of how they were determined;

∎ values for the pairs of decimation and interpolation factors for the two-stage sample rate converters, with a detailed analysis of computational and storage complexities to justify your choice;

∎ a specification of the bandedge frequencies, lengths, pass- and stopband ripples for the anti-aliasing and anti-imaging filters.

Note. You may assume that the filters are direct form FIR, each with a length given by

$$\text{filter length, } N = \frac{-10 \log(\delta_p \delta_s) - 13}{14.6 \Delta f} + 1$$

where

Δf = normalized transition width

Solution   (a) The noise transfer function is given by

$$N(z) = 1 - z^{-1}$$

The magnitude response is given by

$$|N(z)|_{z=e^{j\omega T}} = |(1 - e^{-j\omega T})|$$
$$= [(1 - \cos\omega T)^2 + \sin^2\omega T]^{\frac{1}{2}}$$

At $f = 20$ kHz, $F_s = 3.072$ MHz, $\omega T = 2.3438°$ and $|N(e^{j\omega T})| = 0.0409$. This is equivalent to a reduction in quantization noise level by 27.76 dB.

The effective wordlength of the ADC is determined mainly by the combined SQNR due to oversampling and noise shaping, i.e. $18.85 + 27.76 = 46.61$ dB. This corresponds to an effective ADC resolution of about 7 bits (from SQNR = 6.02 dB + 1.77 dB).

(b) The overall decimation factor is 64, from the ratio of the input to output sampling rates. Given a decimation factor of 64, there are three possible combinations of factors in a two-stage design: 8×8, 16×4, 32×2. Measures of computational complexities are storage requirements and the number of multiplications per second (MPS).

For the 8×8 decimator, the sampling rates at the outputs of the sub-decimators are 384 kHz and 48 kHz. The bandedge frequencies for the first anti-aliasing filter are 0, 20 kHz and 360 kHz, giving a normalized transition width of 0.1106. The passband and stopband ripples are $0.001/2 = 0.0005$ and 0.0001, respectively. Using these values gives $N_1 = 38$. For the second stage, the bandedge frequencies are 0, 20 kHz and 24 kHz, giving a normalized transition width of 0.0104 and $N_2 = 396$.

Similarly, for the 16×4 structure, the rates are 192 kHz and 48 kHz. The first anti-aliasing is characterized by bandedge frequencies 0, 20 kHz and 168 kHz, and a transition width of 0.048 17, $N_2 = 198$. For the 32×2 structure, the rates are 96 kHz and 48 kHz. The bandedge frequencies for the first anti-aliasing filter are 0, 20 kHz and 72 kHz. The transition width is 0.0169 and $N_1 = 244$; the second filter has bandedge frequencies 0, 20 kHz, and 24 kHz, and a transition width of 0.041 66 and $N_2 = 100$.

The computational complexities of the structures are:

$M_1 \times M_2$	Storage	MPS
8×8	434	33.6×10^6
16×4	284	26.01×10^6
32×2	384	28.82×10^6

From the analysis, the 16×4 structure is the most efficient structure. A block diagram is given in Figure 9.31.

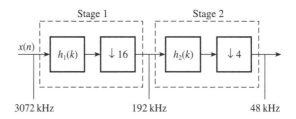

Figure 9.31

9.7.4　Multirate narrowband digital filtering

Narrowband digital filters are characterized by sharp transitions between the passband and stopbands, and by passbands which are very small compared with the sampling frequency. As a result narrowband FIR filters require a very large number of coefficients. This poses a problem in the design and implementation of such filters because they are highly susceptible to finite wordlength effects (for example roundoff noise and coefficient quantization errors). Further, large coefficient storage requirements and computational effort are required. The multirate approach overcomes these problems, and leads to FIR filters with a computational performance that is comparable with that of elliptic IIR filters.

Figure 9.32 shows a simple arrangement for multirate filtering. The sampling frequency of the input sequence is first reduced as far as possible by decimation, the desired filtering operation is then performed at the low sampling frequency, and finally the sampling frequency of the filtered data is restored back to its original rate by interpolation. The use of the same sampling rate conversion factor at the decimator and the interpolator ensures that the input signal, $x(n)$, and the output signal, $y(n)$, are at the same sampling rate.

9.7.4.1　Narrowband lowpass and bandpass filtering

For narrowband lowpass filtering, the filters $h_1(k)$ and $h_2(k)$ in Figure 9.32 would be lowpass filters and there may be no need for $h_3(k)$. A design aim would be for the overall input–output characteristics of the structure in Figure 9.32 to be equivalent to those of the desired conventional lowpass filter. The characteristics that result would,

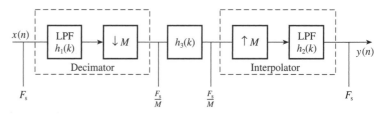

Figure 9.32　Multirate narrowband filtering.

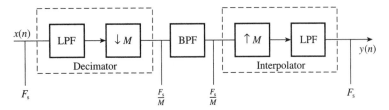

Figure 9.33   An approach to narrowband multirate bandpass filtering.

however, not be identical to those of a conventional lowpass filter because of the effects of aliasing and imaging. In practice, to guarantee that the overall filter meets the desired specifications, the filters $h_1(k)$ and $h_2(k)$ are identical, each with a passband ripple of $\delta_p/2$ and a stopband ripple of δ_s, where δ_p and δ_s are, respectively, the passband and stopband deviations of the equivalent lowpass filter.

The design of multirate bandpass filters is somewhat more involved, except when one wishes to design the so-called integer-band bandpass filters where the bandedges are exact multiples of the lowest sampling frequency in the system, that is $F_s/2M$. In these cases, the decimation/interpolation factor, M, and the filter bandedges satisfy the following conditions (Crochiere and Rabiner, 1983):

$$M = F_s/2(f_{su} - f_{sl}) \qquad (9.17a)$$

$$f_{sl} = kF_s/M, \ k \ \text{an integer}, \quad 0 < k < M - 1 \qquad (9.17b)$$

$$f_{su} = (k + 1)F_s/M \qquad (9.17c)$$

where f_{sl} and f_{su} are the lower and upper stopband edge frequencies respectively. Equation 9.17a gives the maximum decimation factor possible, and Equations 9.17b and 9.17c specify the lower stopband edge and upper stopband edge and the band number k.

A simple alternative, but less efficient, multirate bandpass filtering scheme is to decimate the data as far as possible as described previously using suitable lowpass filters, to bandpass filter the low rate signal, and then to interpolate back to the desired rate. This is illustrated in Figure 9.33. Quite clearly, care must be taken to ensure that the desired passband is protected from the effects of aliasing and imaging during decimation and interpolation.

9.7.4.2   Narrowband highpass and bandstop filters

Narrowband highpass and bandstop filters can be realized as the duals of the lowpass and bandpass filters, respectively:

$$H_{hp}(w) = 1 - H_{lp}(w) \qquad (9.18a)$$

$$H_{bs}(w) = 1 - H_{bp}(w) \qquad (9.18b)$$

The structures for highpass and bandstop filters are depicted in Figure 9.34. In the case of a highpass filter, for example, the signal, $x(n)$, is first lowpass filtered. The filtered

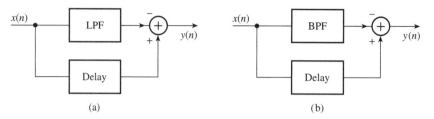

Figure 9.34   Multirate realization of highpass and bandstop filtering as duals of lowpass and bandpass filters: (a) multirate highpass filtering; (b) multirate bandstop filtering.

signal is then subtracted from the unfiltered signal to yield the desired signal. The signal $x(n)$ must be delayed by an amount equal to the delay of the lowpass filter before subtraction. Clearly, it is necessary for the delay through the lowpass filter to be an integer number of samples for this to be possible. Correct passband and stopband specifications must be used for the lowpass filter so that the desired highpass filter is obtained.

Example 9.8   In connection with a research project in fetal monitoring, a need arose to assess the effects of the measurement system on the electrical activity of the foetal heart (the ECG or electrocardiogram) (Westgate *et al.*, 1990). To achieve this required that certain features of the ECG be quantified including the mains frequency content of the signal. Because of the existence of signal energy in the neighbourhood of the mains frequency (50 Hz), it was necessary to use a very narrowband filter. The specifications of the filter are

passband	49–51 Hz
stopband edge frequencies	47 and 53 Hz
stopband attenuation	30 dB ($\delta_p = 0.031\ 62$)
passband ripple	0.1 dB ($\delta_s = 0.011\ 579$)
sampling frequency	500 Hz

Solution   If a direct design of the above filters were attempted, Equation 9.3 suggests that we would require 4018 coefficients, which is rather too long.

Using the multirate approach, a number of options exist. One option is to decimate the data to as low a rate as possible (consistent with the specifications above) (see also Problem 9.6). In this case, the lowest rate would be 125 Hz as this would still allow us to have available the band from 0 to 62.5 Hz. The overall specifications for the decimator are

passband ripple	0.05 dB ($\delta_p = 0.005\ 789\ 5$)
stopband attenuation	30 dB ($\delta_s = 0.031\ 62$)
passband	0–53 Hz
input sampling frequency	500 Hz
output sampling frequency	125 Hz

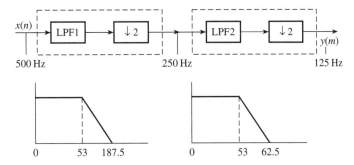

Figure 9.35 Decimator for reducing the rate of the ECG data.

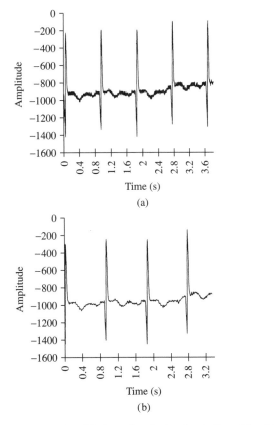

Figure 9.36 (a) Raw ECG data. (b) ECG data after decimation (adjusted for delay).

The decimation factor is 4. Using the multirate design program on the CD in the companion handbook (see the Preface for details), a two-stage decimator was designed (see Figure 9.35). The optimal method (see Chapter 7) was used to obtain the coefficients of the two filters in Figure 9.35. The ECG data were then decimated using the filters and the multistage decimation program described in Section 9.4. Examples of the data before and after decimation are shown in Figure 9.36.

A mains filter satisfying the specifications given in the problem (but at the new, reduced rate) was designed. In this case, the number of filter coefficients was 113.4 for the mains filter ($\delta_p = 0.005\ 789\ 5$, $\delta_s = 0.031\ 62$). After filtering, the data can be restored to its original rate by interpolation.

Example 9.9

Design a suitable multirate lowpass filter for extracting the baseline shifts in the fetal ECG. The filter should satisfy the following specifications:

passband	0–0.4 Hz
stopband	0.5–250 Hz
passband ripple	0.01
stopband ripple	0.001
sampling frequency	500 Hz

Solution

The approach we will use is to decimate down to a 1 Hz rate and then interpolate back up to 500 Hz. In this case, the overall filter specifications for the decimator are

passband ripple	0.01
stopband ripple	0.001
passband	0–0.4 Hz
stopband edge	0.5 Hz
sampling frequency	500 Hz
decimation factor	500

Using the multirate design program on the CD in the companion handbook (see the Preface for details), various practical decimators with up to four stages of decimation were obtained. The characteristics of the most interesting ones are summarized in Table 9.6. Taking into account implementational complexity, the three-stage decimator depicted in Figure 9.37 was selected as the best solution. Again using the optimal design program (Chapter 7) the filter coefficients of the filters of the decimator were

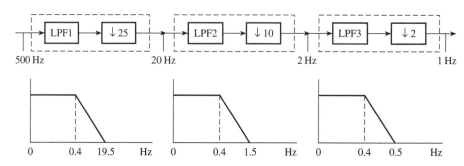

Figure 9.37 The three-stage decimator for Example 9.9.

Table 9.6   Summary of the efficient decimators.

Number of stages	MPS	TSR	M_i	Filter length N_i	Passband edge f_p (Hz)	Stopband edge (Hz)	Normalized transition width Δf_i	Passband ripple	Stopband ripple
1			500	12 707	0.4	0.5	0.002 2	0.01	0.001
2	1807	430	50	153	0.4	9.5	0.018 20	0.005	0.001
			10	277	0.4	0.5	0.01	0.005	0.001
3	1705	189	25	77	0.4	19.5	0.038 20	0.0033	0.001
			10	53	0.4	1.5	0.055	0.0033	0.001
			2	59	0.4	0.5	0.05	0.0033	0.001
4	1444	172	2	2	0.4	249.5	0.498 2	0.0025	0.001
			25	83	0.4	9.5	0.036 40	0.0025	0.001
			5	27	0.4	1.5	0.110 00	0.0025	0.001
			2	60	0.4	0.5	0.050 0	0.0025	0.001
4	1724	169	25	79	0.4	19.5	0.038 20	0.0025	0.001
			2	3	0.4	9.5	0.455 00	0.0025	0.001
			5	27	0.4	1.5	0.110 0	0.0025	0.001
			2	60	0.4	0.5	0.050 0	0.0025	0.001

computed. Using the multistage decimation program, the ECG data was decimated down to 1 Hz and then interpolated back up to 500 Hz.

9.7.5   High resolution narrowband spectral analysis

As discussed in Chapter 11, an important application of the FFT is in the estimation of the spectrum of signals. The FFT provides spectral components of a signal at uniform intervals between 0 and one-half the sampling frequency. In many applications such as sonar, seismology, radar, biomedicine and vibration analysis, the desired signal may occupy only a narrow band in the spectrum of the acquired data. In such cases, direct use of the FFT would require a significant and unnecessarily high computational effort. The multirate technique can be used to isolate and translate the frequency band of interest to a lower frequency before the FFT is applied, leading to a significant reduction in computation, as well as permitting a trade-off between resolution and computational effort.

The FFT performed on the downsampled data allows an equivalent resolution for a much reduced computation or a greater resolution for about the same amount of computation as direct FFT on the original sequence. Effectively, downsampling allows us to see the spectrum of the narrowband on an expanded scale.

Narrowband spectral analysis using multirate techniques is essentially an extension of narrowband bandpass filtering discussed earlier, and has the same restrictions. The signal is first bandpass filtered to isolate the band of interest. The sampling frequency of the filtered signal is then reduced by decimation to F_s/M, where F_s is the sampling frequency of $x(n)$. The spectrum of the much reduced sequence $y(n)$ is then computed using the FFT. A correction factor is used to compensate for the errors in the spectrum due to the ripples in the passband of $h(n)$. When the band of interest does not meet the conditions, this can be circumvented by using a slightly wider frequency band encompassing the desired band. Another approach is to use the method described in Liu and Mintzer (1978), which involves a computer search to find permissible decimation factors.

9.8 Summary

Digital systems that handle more than one sampling rate are known as multirate systems. The two key elements of a multirate system are the decimator and interpolator. The decimator allows us to reduce, efficiently, the rate of a signal by an integer factor M or a rational factor L/M ($L < M$). The interpolator allows us to increase the sampling rate by an integer factor L or a rational factor L/M ($L > M$).

In practice, sample rate changes are implemented in two or more stages for maximum computational and/or storage efficiency. The individual digital filters used in multistage designs have relaxed specifications, leading to fewer coefficients and hence to lower sensitivity to finite wordlength effects, both being directly related to the number of filter coefficients. A practical method of designing sample rate converters was described in detail.

The main strength of multirate systems lies in their ability to exploit the advantages of DSP, in particular the ability to use DSP to bandlimit a signal almost to the Nyquist frequency, with substantial attenuation, and without violating the sampling theorem requirements. The advantages have been exploited in many applications including the compact disc, digital filtering, data acquisition and high resolution data acquisition systems. Many of these systems have been described in detail and their multirate elements designed.

A set of C language programs are provided on the CD in the companion handbook (see the Preface for details) which allow the design and software implementation of multirate systems. The use of MATLAB for multirate DSP is discussed in Appendix 9B.

Problems

9.1   A one-stage decimator is characterized by the following:

decimation factor   3
anti-aliasing filter coefficients

$$h(0) = -0.06 = h(4)$$

$$h(1) = \ 0.30 = h(3)$$

$$h(2) = \ 0.62$$

Given the data, $x(n)$, with successive values $\{6, -2, -3, 8, 6, 4, -2\}$, calculate and list the filtered output, $w(n)$, and the output of the decimator, $y(m)$.

9.2   (a)   The block diagram of a three-stage decimator which is used to reduce the sampling rate from 96 kHz to 1 kHz is given in Figure 9.38. Assuming decimation factors of 8, 4 and 2, indicate the sampling rate at the output of each of the three stages.

(b)   Assume that the decimator in part (a) satisfies the following overall specifications:

input sampling frequency, F_s	96 kHz
decimation factor, M	96
passband ripple	0.01 dB
stopband ripple	60 dB
frequency band of interest	0–450 Hz

Determine the bandedge frequencies for the decimating filter at each stage.

(c)   Assuming that the input and output sampling rates of a decimator are 96 kHz and 1 kHz, respectively:

(i)   write down the overall decimator factor;

(ii)   write down all the possible sets of integer decimation factors (written in descending order only), assuming two stages of decimation;

(iii)   repeat (ii) but assume three stages of decimation;

(iv)   repeat (ii) but assume four stages of decimation.

(d)   For the decimator in part (a), calculate the total number of multiplications per second (MPS) and the total storage requirements (TSR).

9.3   (a)   Design, at a block diagram level, a two-stage decimator that downsamples an audio signal by a factor of 32 and satisfies the specifications given below. Your answer must indicate a suitable pair of decimation factors, with appropriate detailed analysis of computational and storage complexities to justify your choice. Specify the sampling frequencies at the input and output of each stage of decimation, and the following parameters for each of the decimating filters in your design:

the bandedge frequencies
the normalized transition width
passband and stopband ripples
filter length

You may assume that the filters are direct form FIR, each with a length defined as

$$\text{filter length, } N = \frac{-10 \log (\delta_p \delta_s) - 13}{14.6 \Delta f} + 1$$

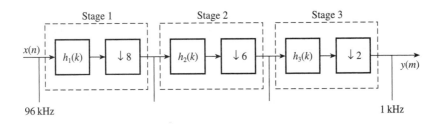

Figure 9.38

where

$$\Delta f = \text{normalized transition width}$$

(b) Show, with the aid of appropriate sketches, that the frequency band of interest (0–3.4 kHz) is protected from aliasing by the decimator.

Input sampling frequency, F_s	256 kHz
Highest frequency of interest in the data	3.4 kHz
Passband ripple, δ_p	0.05
Stopband ripple, δ_s	0.01

9.4 Design a decimator for a high quality data acquisition system with the following overall specification for the decimation filter:

audio band	0 to 20 kHz
input sampling frequency	3.072 MHz
output sampling frequency	48 kHz
passband ripple	<0.001 dB
stopband attenuation	>86 dB

9.5 A requirement exists to compute the spectrum of a narrowband signal embedded in a wideband signal. The band of interest is 49–51 Hz, but the composite signal occupies the band from 0 to 100 Hz. An N-point sequence $x(n)$ is obtained by sampling the composite signal at a rate of 1 kHz.

(1) Illustrate how the desired signal spectrum would be obtained using the multirate approach.

(2) Estimate the computational advantage of the multirate approach over the conventional FFT approach. Compare the resolution of the spectrum for both methods.

9.6 A high quality, efficient narrowband filter is required to extract and assess the mains component in a signal. The filter should satisfy the following specifications:

passband	49 to 51 Hz
stopband edge frequencies	48 and 52 Hz
stopband attenuation	60 dB
passband ripple	0.01 dB
sampling frequency	1000 Hz

Using the multirate approach, design a suitable filter.

9.7 There is a need to interpret the activity in a certain physiological signal, captured at a rate of 256 Hz. To achieve this requires the extraction and analysis of both time and frequency domain features in each band. As a first step towards this, design a suitable multirate system for splitting the signal into the following bands:

0.5–4 Hz
4–8 Hz
8–13 Hz
13–16 Hz

The multirate system should introduce no more than 0.01 dB ripple in the bands and the out-of-band signals should be attenuated by at least 50 dB.

9.8 A DSP system, with analog audio input signal in the range 0–20 kHz, uses oversampling techniques and a first order delta-sigma modulator to convert the analog signal into a digital bit stream at a rate of 3.072 MHz. The z-plane model of the delta-sigma modulator is depicted in Figure 9.39.

(a) Determine the overall improvement in the signal-to-quantization noise ratio made possible by oversampling and noise spectrum shaping and hence estimate the effective resolution, in bits, of the converter.

(b) Design, at a block diagram level, a two-stage decimator to convert the output of the delta-sigma modulator depicted in Figure 9.39 from a single-bit stream at 3.072 MHz into a multi-bit stream at 48 kHz. The pass- and stopband ripples for the decimator are 0.001 and 0.0001, respectively.

Your answer must include the following:

■ values for the overall decimation and interpolation factors, with a justifiable indication of how they were determined;

■ values for the pairs of decimation and interpolation factors for the two-stage sample rate converters, with a detailed analysis of computational and storage complexities to justify your choice;

■ a specification of the bandedge frequencies, lengths, pass- and stopband ripples for the anti-aliasing and anti-imaging filters.

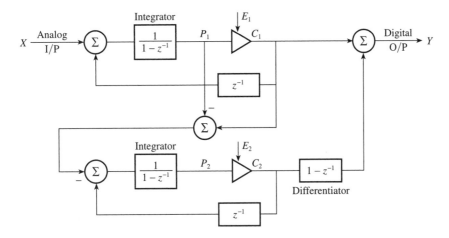

Note. The output transform, $Y(z)$, of the second-order delta-sigma modulator is given by:
$Y(z) = X(z) + E_2(z)(1 - z^{-1})^2$. The variables have the usual meaning.

Figure 9.39

You may assume that the filters are direct form FIR, each with a length given by

$$\text{filter length,} \quad N = \frac{-10 \log(\delta_p \delta_s) - 13}{14.6 \Delta f} + 1$$

where

Δf = normalized transition width

9.9 (a) (i) Explain, with the aid of a suitable block diagram, the principles of multirate lowpass digital filtering using decimator/interpolator structure. What are the main disadvantages of multirate digital filtering?

(ii) Show, from analysis of a one-stage multirate lowpass filter, that when the decimation factor is greater than 2, the multirate approach offers a computational gain over conventional fixed rate filter.

Extend your analysis to show that as the bandwidth of the lowpass filter decreases the computational gain of the multirate filter increases. Comment on the practical implications of this.

(b) A requirement exists for a lowpass filter meeting the following specifications:

passband edge frequency	4 Hz
stopband edge frequency	6.25 Hz
passband ripple	0.001
stopband ripple	0.0001
sampling rate	500 Hz

Design, at a block diagram level, an efficient multirate filter that meets the above specifications using a two-stage decimator and a two-stage interpolator. Your answer must include the following:

- values for the overall decimation and interpolation factors, with a justifiable indication of how they were determined;
- values for the pairs of decimation and interpolation factors for the two-stage sample rate converters, with a detailed analysis of computational and storage complexities to justify your choice;
- a block diagram of the multirate lowpass filter suitably labelled;
- a specification of the bandedge frequencies, lengths, pass- and stopband ripples for the anti-aliasing and anti-imaging filters.

(c)   Compare the computational complexities of the multirate lowpass filter in (b) with those of a conventional direct form FIR implementation. Comment on your answer.

You may assume that the filters are direct form FIR, each with a length given by

$$\text{filter length,} \quad N = \frac{-10\log(\delta_p\delta_s) - 13}{14.6\Delta f} + 1$$

where

$$\Delta f = \text{normalized transition width}$$

9.10 (a)   Explain clearly the roles of each of the following in a multirate processing system:

(i)   decimating filter;

(ii)   sampling rate compressor.

(b)   A requirement exists for the design of a two-stage decimator for an audio multirate system that meets the following specifications:

input sampling frequency, F_s	96 kHz
decimating factor, M	96
highest frequency of interest in the data, f_p	450 Hz
passband ripple, δ_p	0.05
stopband ripple, δ_s	0.01

A preliminary exploration of the design problem, with the aid of a software tool, produced the information summarized in Table 9.7. From the design specifications and the information in Table 9.7:

(i)   design, at a block diagram level, the two-stage decimator. Your answer should indicate a suitable pair of decimation factors, with a justification for your choice, and the sampling frequencies at the input and output of each stage of decimation;

(ii)   for each stage of the decimator, determine the following parameters for the decimating filter:

the bandedge frequencies
the normalized transition width
the stop and passband ripples
the number of filter coefficients

Table 9.7   Two-stage decimator design information

#	M_1	M_2	N_1	N_2	MPS	TSR
0	2	48	2	2651	2 747 000	2653
1	3	32	6	1768	1 960 000	1774
2	4	24	10	1326	1 566 000	1336
3	6	16	17	885	1 157 000	902
4	8	12	24	664	952 000	688
5	12	8	38	443	747 000	481
6	16	6	53	333	651 000	386
7	24	4	88	222	574 000	310
8	32	3	131	167	560 000	298
9	48	2	254	112	620 000	366

Notes

#	design option number
M_1, M_2	decimation factor pair for stages 1 and 2
N_1, N_2	number of filter coefficients for stages 1 and 2
MPS	number of multiplications per second
TSR	total storage requirements

(iii)   show, with the aid of appropriate sketches, that the frequency band of interest (0–450 Hz) is protected from aliasing by the decimator.

9.11   A requirement exists for the design of a two-stage decimator for an audio multirate system that meets the following specifications:

input sampling frequency, F_s	96 kHz
decimating factor, M	96
highest frequency of interest in the data, f_p	450 Hz
passband ripple, δ_p	0.05
stopband ripple, δ_s	0.01

A preliminary exploration of the design problem, with the aid of a software tool, produced the information summarized in Table 9.7. From the design specifications and the information in Table 9.7:

(a)   determine the optimum decimation factors for the stages using Equations 9.11;

(b)   round the decimation factors to the nearest integers to give the correct overall decimation factor;

(c)   compare the result with the result obtained in Problem 9.10(b).

MATLAB problems

9.12 A continuous-time signal is characterized by the following equation:

$$x(t) = A \cos(2\pi f_1 t) + B \cos(2\pi f_2 t)$$

The equivalent discrete-time signal, $x(nT)$, is obtained by sampling the continuous signal at a frequency

$$F_s = \frac{1}{T}$$

(a) Generate 1000 data samples, with the aid of MATLAB, from the continuous signal. Assume $F_s = 5000$ Hz, $f_1 = 50$ Hz, $f_2 = 100$ Hz, $A = 2$ and $B = 1$.

(b) Use the MATLAB function **decimate** to reduce the sampling rate by a factor of 10.

(c) Use the MATLAB function **interp** to raise the sampling rate of the decimated data by a factor of 4.

Display an appropriate number of samples of each of the discrete-time signals obtained in steps (a)–(c) using the **stem** function. Comment on your results.

9.13 Repeat Problem 9.12 using the **resample** function. Extract the internal filter used by the resample function in each case and plot its magnitude–frequency response.

9.14 Repeat Problem 9.12 using the **upfirdn** function and a suitable lowpass FIR filter designed by the optimal method. List the filter coefficients and plot the filter's magnitude–frequency response. State any reasonable assumptions made.

9.15 Repeat Problem 9.14 using the **resample** function.

9.16 (a) Design, at a block diagram level, a two-stage decimator that downsamples an audio signal by a factor of 30 and satisfies the following specifications:

input sampling frequency, F_s	240 kHz
highest frequency of interest in the data	3.4 kHz
passband ripple, δ_p	0.05
stopband ripple, δ_s	0.01

$$\text{filter length,} \quad N = \frac{-10 \log(\delta_p \delta_s) - 13}{14.6 \Delta f} + 1$$

where

$$\Delta f = \text{normalized transition width}$$

Assume decimation factors of 15 and 2 for stages 1 and 2, respectively.

(b) Calculate the coefficients and plot the magnitude–frequency responses of the anti-aliasing filters for stages 1 and 2 using the optimal method.

(c) Use suitable MATLAB functions and a discrete-time signal derived from the following continuous signal to test the decimator:

$$x(t) = 2 \sin(2\pi 100 t) + 3 \sin(2\pi 1000 t)$$
$$+ 2 \sin(2\pi 3400 t)$$

References

Adams R.W. (1986) Design and implementation of an audio 18 bit analog-to-digital converter using oversampling techniques. *J. Audio Engineering Society*, **34**(3), 153–66.

Agrawal B.P. and Shenoi K. (1983) Digital methodology for ΣΔM. *IEEE Trans. Communications*, **31**(3), 360–70.

Claasen T.A.C.M., Mecklenbrauker W.F.G., Peek J.B.H. and Van Hurck N. (1980) Signal processing method for improving the dynamic range of A/D and D/A converters. *IEEE Trans. Acoustics, Speech and Signal Processing*, **28**(5), 529–38.

Crochiere R.E. and Rabiner L.R. (1975) Optimum FIR digital filter implementations for decimation, interpolation, and narrow-band filtering. *IEEE Trans. Acoustics, Speech and Signal Processing*, **23**(5), 444–56.

Crochiere R.E. and Rabiner L.R. (1976) Further considerations in the design of decimators and interpolators. *IEEE Trans. Acoustics, Speech and Signal Processing*, **24**, 296–311.

Crochiere R.E. and Rabiner L.R. (1979) A program for multistage decimation, interpolation, and narrow band filtering. In *IEEE Programs for DSP*. Institute of Electrical and Electronics Engineers.

Crochiere R.E. and Rabiner L.R. (1981) Interpolation and decimation of digital signals – a tutorial review. *Proc. IEEE*, **69**(3), 300–31.

Crochiere R.E. and Rabiner L.R. (1983) *Multirate Digital Signal Processing*. Englewood Cliffs NJ: Prentice-Hall.

Crochiere R.E. and Rabiner L.R. (1988) Multirate processing of digital signals. In *Advanced Topics in Signal Processing* (Lim J.S. and Oppenheim A.V. (eds)). Englewood Cliffs NJ: Prentice-Hall.

DeFatta D.J., Lucas J.G. and Hodgkiss W.S. (1988) *Digital Signal Processing: A System Design Approach*. New York: Wiley.

Liu B. and Mintzer F. (1978) Calculation of narrow-band spectra by direct decimation. *IEEE Trans. Acoustics, Speech and Signal Processing*, **26**(6), 529–34.

Matsuya Y., Uchimura K., Iwata A., Kobayashi T., Ishikawa M. and Yoshitome T. (1987) A 16-bit oversampling A-to-D conversion technology using triple integration noise shaping. *IEEE J. Solid State Circuits*, **22**(6), 921–8.

Quarmby D. (ed.) (1984) *Signal Processor Chips*, Chapter 5. London: Granada.

Welland D.R., Del Signore B.P., Swanson E.J., Tanaka T., Hamashita, K., Hara S. and Takasuka K. (1989) A stereo 16-bit delta-sigma A/D converter for digital audio. *J. Audio Engineering Society*, **37**(6), 476–86.

Westgate J.A., Keith R.D.F., Gurnow J.S.K., Ifeachor E.C. and Greene K.R.G. (1990) Suitability of fetal scalp electrodes for fetal electrocardiogram during labour. *J. Clin. Physics & Physiological Measurement*, **11**(4), 297–306.

Bibliography

Analog Devices (1988) *ADSP-2100 Family Applications Handbook*, Volume 2, Chapter 3. Analog Devices, Inc.

Bellanger M.G. (1977) Computation rate and storage estimation in multirate digital filtering with halfband filters. *IEEE Trans. Acoustics, Speech and Signal Processing*, **25**, 344–6.

Bellanger M.G., Daquet J.L. and Lepagnol G.P. (1974) Interpolation, extrapolation and reduction of computation speed in digital filters. *IEEE Trans. Acoustics, Speech and Signal Processing*, **22**, 231–5.

Brown Jr J.L. (1981) Multichannel sampling of lowpass signals. *IEEE Trans. Circuits Systems*, **28**, 101–6.

Cox R.V., Bock D.E., Bauer K.B., Johnston J.D. and Snyder J.H. (1987) The analogue voice privacy system. *AT&T Technical J.*, **66**, 119–31.

Elliot D.F. (ed.) (1987) *Handbook of Digital Signal Processing*. New York: Academic Press.

Goedhart D., van der Plassche R.J. and Stikvoort E.F. (1982) Digital-to-analog conversion in playing a compact disc. *Philips Technical Rev.*, **40**(6), 174–9.

Goodman D.J. and Carey M.J. (1977) Nine digital filters for decimation and interpolation. *IEEE Trans. Acoustics, Speech and Signal Processing*, **25**(2), 121–6.

Goodman D.J. and Flanagan J.L. (1971) Direct digital conversion between linear and adaptive delta modulation formats. In *Proc. IEEE Int. Communications Conf.*, Montreal, Canada, June 1971.

Huber A., De Man E., Schiller E. and Ulbrich W. (1986) FIR lowpass filter for signal decimation with 15 MHz clock frequency. In *IEEE Int. Conf. Acoustics, Speech and Signal Processing*, Tokyo, 7–11 April, pp. 1533–6.

Jerri A.J. (1977) The Shannon sampling theorem – its various extensions and applications: a tutorial review. *Proc. IEEE*, **65**(11), 1565–96.

Linden D.A. (1959) A discussion of sampling theorems. *Proc. IRE.*, **47**, 1219–26.

Mintzer F. (1982) On half-band, third-band and Nth-band FIR filters and their design. *IEEE Trans. Acoustics, Speech and Signal Processing*, **30**, 734–8.

Mintzer F. and Liu B. (1978) The design of optimal multirate bandpass and bandstop filters. *IEEE Trans. Acoustics, Speech and Signal Processing*, **26**(6), 534–43.

Mintzer F. and Liu B. (1978) Aliasing error in the design of multirate filters. *IEEE Trans. Acoustics, Speech and Signal Processing*, **26**, 76–88.

Montijo B.A. (1988) Digital filtering in a high-speed digitizing oscilloscope. *Hewlett Packard J.*, June, 70–6.

Mou Z.J. and Duhamel P. (1987) Fast FIR filtering: algorithms and implementations. *Signal Processing*, **13**, 377–84.

Princen J.P. and Bradley A.B. (1986) Analysis/synthesis filter banks design based on time domain aliasing cancellation. *IEEE Trans. Acoustics, Speech and Signal Processing*, **23**, 1153–61.

Rabiner L.R. and Crochiere R.E. (1975) A novel implementation for narrow-band FIR digital filters. *IEEE Trans. Acoustics, Speech and Signal Processing*, **23**(5), 457–64.

Regalia P.A., Fujii N., Mitra S.K. and Neuvo Y. (1987) Active RC crossover networks with adjustable characteristics. *J. Audio Engineering Society*, January–February, **35**(1/2), 24–30.

Rorabacher D.W. (1975) Efficient FIR filter design for sample rate reduction and interpolation. In *Proc. IEEE International Symposium on Circuits and Systems*, 21–23 April, pp. 396–9.

Schafer R.W. and Rabiner R.L. (1973) A digital signal processing approach to interpolation. *Proc. IEEE*, **61**, 692–702.

Scheuermann H. and Gockler H. (1981) A comprehensive survey of digital transmultiplexing methods. *Proc. IEEE*, **69**, 1419–50.

Shannon C.E. (1949) Communications in the presence of noise. *Proc. IRE*, **37**, 10–21.

Thong T. (1989) Practical consideration for a continuous time digital spectrum analyzer. In *Proc. IEEE International Symposium on Circuits and Systems*, Portland OR, May 1989, 1047–50.

Tuffs D.W., Rorabacher D.W. and Mosier W.E. (1970) Designing simple, effective digital filters. *IEEE Trans. Audio Electro-Acoustics*, **18**, 142–58.

Vaidyanathan P.P. (1990) Multirate digital filters, filter banks, polyphase networks, and applications: a tutorial. *Proc. IEEE*, **78**(1), 56–93.

Vaidyanathan P.P. (1993) *Multirate Systems and Filter Banks*. Englewood Cliffs NJ: Prentice-Hall.

Vaidyanathan P.P. and Nguyen T.Q. (1987) A trick for the design of FIR half-band filters. *IEEE Trans. Circuits and Systems*, **34**, 297–300.

Van De Plassche R.J. and Dijkmans E.C. (1983) A monolithic 16-bit D/A conversion system for digital audio. In *Digital Audio* (Blesser B. (ed.)), pp. 54–60. Audio Engineering, Inc.

Zobel R.N. and Tang P.S. (1985) A high performance multichannel decimating FIR digital filter system for microprocessor based data acquisition. *Proc. ISCAS*, 1149–52.

Appendices

9A C programs for multirate processing and systems design

The following C language programs are available on the CD in the companion handbook, together with illustrative examples (see the Preface for details).

(1) decimate.c, which decimates the data using up to three stages of decimation;

(2) interpol.c, which interpolates the data using up to three stages of interpolation;

(3) moptimum.c, which determines the characteristics of an I-stage decimator (or interpolator) for $I = 1, 2, 3$, or 4. The characteristics include decimation factor and filter

characteristics for each stage, and the measures of efficiency (such as number of multiplications per second) for various configurations.

9B   Multirate digital signal processing with MATLAB

The MATLAB Signal Processing Toolbox may be used to carry out a variety of operations associated with sampling rate conversion and multirate processing. In this section we will highlight the key MATLAB functions for multirate processing and illustrate their use.

As discussed in the main text, the design of a practical multirate system involves the following steps:

(1)   specification of the requirements for the sampling rate converter;

(2)   determination of the parameters for the sampling rate converter (e.g. number of stages, sampling rate conversion factors and filter characteristics for the stages);

(3)   design of digital anti-aliasing and/or anti-imaging filters;

(4)   multirate filtering, upsampling and/or downsampling.

The MATLAB Signal Processing Toolbox can assist greatly in performing steps 3 and 4. The four key MATLAB functions for multirate processing are **decimate, interp, resample** and **upfirdn**. There are many other functions, especially those associated with FIR filters, that complement the multirate DSP functions.

The **decimate** function is used to reduce the sampling rate of a data sequence by a factor of M, where M is a positive integer. The function performs anti-image filtering using a lowpass filter and then resamples the signal at the lower rate by retaining one data point out of every M data points. The syntax for the **decimate** command is

```
y=decimate(x,M)
y=decimate(x,M,N)
y=decimate(x,M,'fir')
y=decimate(x, M, N, 'fir')
```

The y=decimate(x,M,N,'fir') command filters the data sequence in the vector x using an N-point, FIR lowpass filter with a normalized cutoff frequency of $1/M$ and then reduces the sampling rate by a factor M. The filter is automatically generated by the Toolbox.

If the parameter 'fir' is omitted (as in the first command above), the decimate function uses an nth order lowpass Chebyshev type I filter with a normalized cutoff frequency of $0.8/M$ and a passband ripple of 0.05 dB (the default value for N is 8). When an IIR filter is used, the function applies the filter in both the forward and reverse time directions to compensate for phase distortion.

The **interp** function is used to increase the sampling rate of a data sequence by an integer factor, L. The function first expands the data sequence by inserting zeros between data samples, designs automatically a lowpass FIR filter, and then performs anti-image filtering using the FIR filter. The syntax for the command is

```
y=interp(x,L)
y=interp(x, L, N, alpha)
[y, b] = interp(x, L, N, alpha)
```

The parameter, N, specifies the filter length (filter length $= 2^{NL+1}$, default length is 4), and alpha is the normalized cutoff frequency (default is 0.5, i.e. alpha $= f_c/f_{Nyquist}$). The last command allows the coefficients of the anti-image filter to be retrieved via the vector, b.

Example 9B.1

A continuous time signal is characterized by the following equation:

$$x(t) = A \cos (2\pi f_1 t) + B \cos (2\pi f_2 t)$$

(a) Generate, with the aid of MATLAB, a discrete-time equivalent of the signal. Assume a sampling frequency of 1 kHz, $f_1 = 50$ Hz, $f_2 = 100$ Hz, and that the ratio of the amplitudes of the frequency components, $A/B = 1.5$.

(b) Interpolate the discrete-time signal by a factor of 4 using the **interp** function.

(c) Decimate the output of the interpolator in step (b) by a factor of 4 using the **decimate** function.

(d) Plot the original, interpolated and decimated discrete-time signals.

Solution

The MATLAB m-file is listed in Program 9B.1. The original, interpolated and decimated discrete-time signals are shown in Figure 9B.1(a), (b) and (c), respectively. The reader should note the differences between Figures 9B.1(a) and (c) which are due to imperfections in the sampling rate conversion operations.

Program 9B.1   MATLAB m-file to illustrate simple interpolation and decimation operations.

```
%
%   File name: Program EX9B1.m
%   An Illustration of interpolation by a factor of 4
%
Fs=1000;                                % sampling frequency
A=1.5;                                  % relative amplitudes
B=1;
f1=50;                                  % signal frequencies
f2=100;
t=0:1/Fs:1;                             % time vector
x=A*cos(2*pi*f1*t)+B*cos(2*pi*f2*t);    % generate signal
y=interp(x,4);                          % interpolate signal by 4
stem(x(1:25))                           % plot original signal
xlabel('Discrete time, nT')
ylabel('Input signal level')
figure
stem(y(1:100))                          % plot interpolated signal.
xlabel('Discrete time, 4 x nT')
ylabel('Interpolated output signal level')
y1=decimate(y,4);
stem(y1(1:25))                          % plot decimated signal.
xlabel('Discrete time, nT')
ylabel('Decimated output signal level')
```

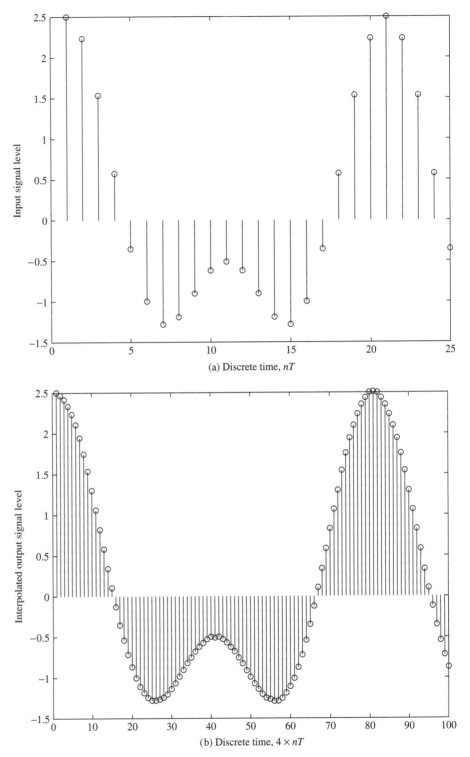

Figure 9B.1 (a) Input signal. (b) An illustration of interpolation by a factor of 4.

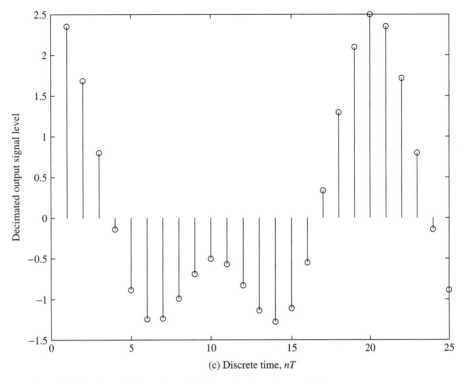

(c) Discrete time, nT

Figure 9B.1   (c) An illustration of decimation by a factor of 4.

The **decimate** and **interp** functions are useful for rapid implementation of sampling rate changes. However, they are somewhat restrictive (e.g. the anti-aliasing and anti-imaging filters used are designed automatically). The **upfirdn** and **resample** functions are more versatile as they allow greater design choice.

The **upfirdn** and **resample** functions may be used to increase and/or decrease the sampling rate of a data sequence by a rational factor L/M (L and M are positive integers). The two commands perform similar operations. The syntax for the **upfirdn** command is

```
y=upfirdn(x, h)
y=upfirdn(x, h, L)
y=upfirdn(x, h, L, M)
```

The y=upfirdn(x, h, L, M) command, for example, first increases the sampling rate by a factor of L, performs FIR filtering using the filter coefficients in the vector h, and then downsamples the sequence by a factor of M. The FIR filter may be designed using the optimal, window or frequency sampling methods. The command returns the output signal in the vector y. If both L and M are equal to 1 then the operation is a mere FIR filtering. If $L = 1$, the data is decimated (i.e. its sampling rate is reduced) by a factor of M, whereas if $M = 1$ it is interpolated (i.e. its sampling rate is raised by a factor of L).

Program 9B.2   MATLAB m-file to illustrate simple interpolation and decimation operations.

```
%
%   File name: EX9B2.m
%   An Illustration of sampling rate changes using resample by a factor of 4
%
Fs=1000;                                    % sampling frequency
A=1.5;                                       % relative amplitudes
B=1;
f1=50;                                       % signal frequencies
f2=100;
t=0:1/Fs:1;                                  % time vector
x=A*cos(2*pi*f1*t)+B*cos(2*pi*f2*t);        % generate signal
y=resample(x,4,1);                           % interpolate signal by 4
stem(x(1:25))                                % plot original signal
xlabel('Discrete time, nT')
ylabel('Input signal level')
figure
stem(y(1:100))                               % plot interpolated signal.
xlabel('Discrete time, 4 x nT')
ylabel('Interpolated output signal level')
y1=resample(y,1,4);
figure
stem(y1(1:25))                               % plot decimated signal.
xlabel('Discrete time, nT')
ylabel('Decimated output signal level')
```

The syntax for the resample command is

```
y=resample(x, L, M).
[y, b]=resample(x, L, M)
y=resample(x, L, M, b)
```

The y=resample(x, L, M) command bandlimits the data sequence in the vector x using an FIR filter designed with Kaiser window and the **fir1** command. The coefficients of the FIR filter may be retrieved by using the second command, i.e. [y, b]=resample(x, L, M). The y=resample(x, L, M, b) command uses a user-designed FIR filter.

As an illustration, we have used the resample command to solve the last problem (Example 9B.1), that is, to raise and then lower the sampling rate by a factor of 4. The MATLAB m-file for this is listed in Program 9B.2. The results are similar but not identical because of differences in implementation.

Adaptive digital filters

10

An adaptive filter is essentially a digital filter with self-adjusting characteristics. It adapts, automatically, to changes in its input signals. Adaptive filters are the central topic in the sub-area of DSP known as adaptive signal processing. This chapter describes key aspects of this important topic based on the LMS (least mean square) and RLS (recursive least squares) algorithms which are two of the most widely used algorithms in adaptive signal processing. The treatment is practical with only the essential theory included in the main text. C language implementations of a variety of LMS and RLS based adaptive filters can be found on the CD included with the companion manual – *A Practical Guide for MATLAB and C Language Implementations of DSP Algorithms* (see the Preface for details). A number of real-world applications are presented.

10.1   When to use adaptive filters and where they have been used

The contamination of a signal of interest by other unwanted, often larger, signals or noise is a problem often encountered in many applications. Where the signal and noise occupy fixed and separate frequency bands, conventional linear filters with fixed coefficients are normally used to extract the signal. However, there are many instances when it is necessary for the filter characteristics to be variable, adapted to changing signal characteristics, or to be altered intelligently. In such cases, the coefficients of the filter must vary and cannot be specified in advance. Such is the case where there is a spectral overlap between the signal and noise (see Figure 10.1) or if the band occupied by the noise is unknown or varies with time. Typical applications where fixed coefficient filters are inappropriate are the following:

(1)   Electroencephalography (EEG), where artefacts or signal contamination produced by eye movements or blinks is much larger than the genuine electrical activity of the brain and shares the same frequency band with signals of clinical interest. It is not possible to use conventional linear filters to remove the artefacts while preserving the signals of clinical interest.

(2)   Digital communication using a spread spectrum, where a large jamming signal, possibly intended to disrupt communication, could interfere with the desired signal. The interference often occupies a narrow but unknown band within the wideband spectrum, and can only be effectively dealt with adaptively.

(3)   In digital data communication over the telephone channel at a high rate. Signal distortions caused by the poor amplitude and phase response characteristics of the channel lead to pulses representing different digital codes to interfere with each other (intersymbol interference), making it difficult to detect the codes reliably at the receiving end. To compensate for the channel distortions which may be varying with time or of unknown characteristics at the receiving end, adaptive equalization is used.

An adaptive filter has the property that its frequency response is adjustable or modifiable automatically to improve its performance in accordance with some criterion,

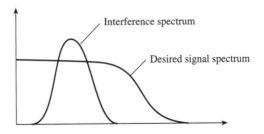

Figure 10.1   An illustration of spectral overlap between a signal and a strong interference.

allowing the filter to adapt to changes in the input signal characteristics. Because of their self-adjusting performance and in-built flexibility, adaptive filters have found use in many diverse applications such as telephone echo cancelling, radar signal processing, navigational systems, equalization of communication channels, and biomedical signal enhancement.

In summary we use adaptive filters

- when it is necessary for the filter characteristics to be variable, adapted to changing conditions;
- when there is spectral overlap between the signal and noise (see Figure 10.1); or
- if the band occupied by the noise is unknown or varies with time.

The use of conventional filters in the above cases would lead to unacceptable distortion of the desired signal. There are many other situations, apart from noise reduction, when the use of adaptive filters is appropriate (see later).

10.2 Concepts of adaptive filtering

10.2.1 Adaptive filters as a noise canceller

An adaptive filter consists of two distinct parts: a digital filter with adjustable coefficients, and an adaptive algorithm which is used to adjust or modify the coefficients of the filter (Figure 10.2). Two input signals, y_k and x_k, are applied simultaneously to the adaptive filter. The signal y_k is the contaminated signal containing both the desired signal, s_k, and the noise, n_k, assumed uncorrelated with each other. The signal, x_k, is a measure of the contaminating signal which is correlated in some way with n_k. x_k is processed by the digital filter to produce an estimate, $\hat{n}_k$, of n_k. An estimate of the desired signal is then obtained by subtracting the digital filter output, $\hat{n}_k$, from the contaminated signal, y_k:

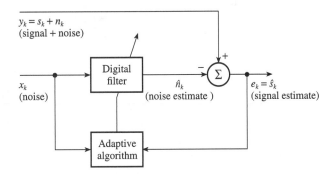

Figure 10.2 Block diagram of an adaptive filter as a noise canceller.

$$\hat{s}_k = y_k - \hat{n}_k = s_k + n_k - \hat{n}_k \tag{10.1}$$

The main objective in noise cancelling is to produce an optimum estimate of the noise in the contaminated signals and hence an optimum estimate of the desired signal. This is achieved by using $\hat{s}_k$ in a feedback arrangement to adjust the digital filter coefficients, via a suitable adaptive algorithm, to minimize the noise in $\hat{s}_k$. The output signal, $\hat{s}_k$, serves two purposes: (i) as an estimate of the desired signal and (ii) as an error signal which is used to adjust the filter coefficients.

10.2.2 Other configurations of the adaptive filter

The discussions above are based on the adaptive noise cancelling principles. It is important to keep in mind that adaptive filters can be and have been used for other purposes, such as for linear prediction, adaptive signal enhancement and adaptive control. In general, the meaning of the signals x_k, y_k, and e_k or the way they are derived are application dependent, a fact which should be borne in mind. Figure 10.3 shows different configurations of the adaptive filter.

10.2.3 Main components of the adaptive filter

In most adaptive systems, the digital filter in Figure 10.2 is realized using a transversal or finite impulse response (FIR) structure (Figure 10.4). Other forms are sometimes used, for example the infinite impulse response (IIR) or the lattice structures, but the FIR structure is the most widely used because of its simplicity and guaranteed stability. For the N-point filter depicted in Figure 10.4, the output is given by

$$\hat{n}_k = \sum_{i=0}^{N-1} w_k(i) x_{k-i} \tag{10.2}$$

where $w_k(i)$, $i = 0, 1, \ldots$, are the adjustable filter coefficients (or weights), and $x_k(i)$ and $\hat{n}_k$ are the input and output of the filter. Figure 10.4 illustrates the single-input, single-output system. In a multiple-input single-output system, the x_k may be simultaneous inputs from N different signal sources.

10.2.4 Adaptive algorithms

Adaptive algorithms are used to adjust the coefficients of the digital filter (in Figure 10.2) such that the error signal, e_k, is minimized according to some criterion, for example in the least squares sense. Common algorithms that have found widespread application are the least mean square (LMS), the recursive least squares (RLS), and the Kalman filter algorithms. In terms of computation and storage requirements, the LMS algorithm is the most efficient. Further, it does not suffer from the numerical instability problem inherent in the other two algorithms. For these reasons, the LMS

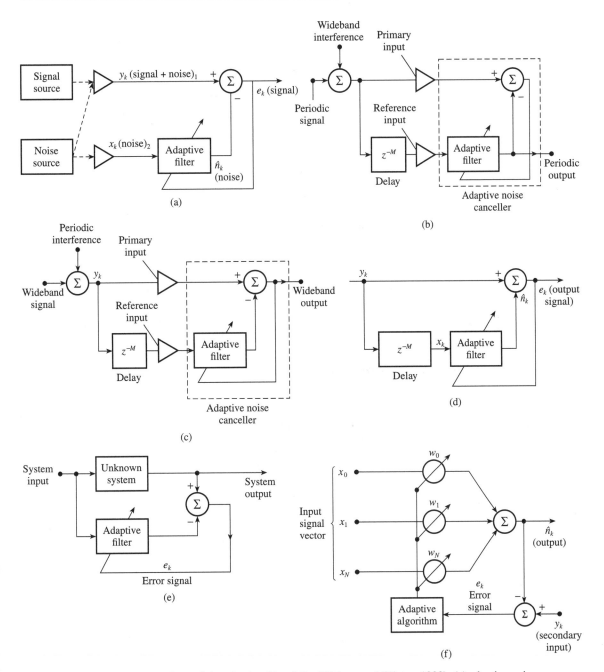

Figure 10.3   Some configurations of the adaptive filter (after Widrow and Winter, 1988): (a) adaptive noise canceller; (b) adaptive self-tuning filter; (c) cancelling periodic interference without an external reference source; (d) adaptive line enhancer; (e) system modelling; (f) linear combiner.

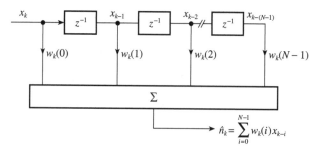

Figure 10.4   Finite impulse response filter structure.

algorithm has become the algorithm of first choice in many applications. However, the RLS algorithm has superior convergence properties.

Example 10.1   The estimate of the desired signal at the output of an adaptive noise canceller is given by (Widrow *et al.*, 1975a)

$$\hat{s}_k = y_k - \hat{n}_k = s_k + n_k - \hat{n}_k$$

Show that minimizing the total power at the output of the canceller maximizes the output signal-to-noise ratio.

Solution   The contaminated signal is given by

$$y_k = s_k + n_k \tag{10.3}$$

and the estimate of the desired signal is given by

$$\hat{s}_k = y_k - \hat{n}_k = s_k + n_k - \hat{n}_k \tag{10.4}$$

Squaring Equation 10.4 we have

$$\hat{s}_k^2 = s_k^2 + (n_k - \hat{n}_k)^2 + 2s_k(n_k - \hat{n}_k) \tag{10.5}$$

Taking the expectations of both sides of Equation 10.5,

$$E[\hat{s}_k^2] = E[s_k^2] + E[(n_k - \hat{n}_k)^2] + 2E[s_k(n_k - \hat{n}_k)] \tag{10.6}$$

Since the desired signal, s_k, is uncorrelated with n_k or with $\hat{n}_k$ the last term in Equation 10.6 is zero and we have

$$E[\hat{s}_k^2] = E[s_k^2] + E[(n_k - \hat{n}_k)^2] \tag{10.7}$$

where $E[s_k^2]$ represents the total signal power, $E[\hat{s}_k^2]$ represents the estimate of the signal power (it also represents the total output power) and $E[(n_k - \hat{n}_k)^2]$ represents the remnant noise power which may still be in s_k. It is evident in Equation 10.7 that if the estimate $\hat{n}_k$ is the exact replica of n_k, the output power will contain only the signal power. By adjusting the adaptive filter towards the optimum position, the remnant noise power and hence the total output power are minimized. The desired signal power is unaffected by this adjustment since s_k is uncorrelated with n_k. Thus

$$\min E[\hat{s}_k^2] = E[s_k^2] + \min E[(n_k - \hat{n}_k)^2] \tag{10.8}$$

It is clear in Equation 10.8 that the net effect of minimizing the total output power is to maximize the output signal-to-noise ratio. When the filter setting is such that $\hat{n}_k = n_k$, then $\hat{s}_k = s_k$. In this case, the output of the adaptive noise canceller is noise free. When the signal y_k contains no noise, that is when $n_k = 0$, the adaptive filter turns itself off (in theory at least) by setting all the weights to zero.

10.3 Basic Wiener filter theory

Many adaptive algorithms can be viewed as approximations of the discrete Wiener filter (Figure 10.5). Two signals, x_k and y_k, are applied simultaneously to the filter. Typically, y_k consists of a component that is correlated with x_k and another that is not. The Wiener filter produces an optimal estimate of the part of y_k that is correlated with x_k which is then subtracted from y_k to yield e_k.

Assuming an FIR filter structure with N coefficients (or weights – the popular term in the literature), the error, e_k, between the Wiener filter output and the primary signal, y_k, is given by

$$e_k = y_k - \hat{n}_k = y_k - \mathbf{W}^T\mathbf{X}_k = y_k - \sum_{i=0}^{N-1} w(i)x_{k-i} \tag{10.9}$$

where $\mathbf{X}_k$ and $\mathbf{W}$, the input signal vector and weight vector, respectively, are given by

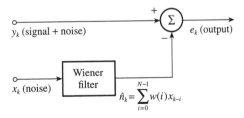

Figure 10.5   The basic Wiener filter.

$$\mathbf{X}_k = \begin{bmatrix} x_k \\ x_{k-1} \\ \vdots \\ x_{k-(N-1)} \end{bmatrix} \qquad \mathbf{W} = \begin{bmatrix} w(0) \\ w(1) \\ \vdots \\ w(N-1) \end{bmatrix} \qquad (10.10)$$

The square of the error is given as

$$e_k^2 = y_k^2 - 2y_k \mathbf{X}_k^{\mathrm{T}} \mathbf{W} + \mathbf{W}^{\mathrm{T}} \mathbf{X}_k \mathbf{X}_k^{\mathrm{T}} \mathbf{W} \qquad (10.11)$$

The mean square error (MSE), J, is obtained by taking the expectations of both sides of Equation 10.11, assuming that the input vector $\mathbf{X}_k$ and the signal y_k are jointly stationary:

$$J = E[e_k^2] = E[y_k^2] - 2E[y_k \mathbf{X}_k^{\mathrm{T}} \mathbf{W}] + E[\mathbf{W}^{\mathrm{T}} \mathbf{X}_k \mathbf{X}_k^{\mathrm{T}} \mathbf{W}]$$
$$= \sigma^2 + 2\mathbf{P}^{\mathrm{T}} \mathbf{W} + \mathbf{W}^{\mathrm{T}} \mathbf{R} \mathbf{W} \qquad (10.12)$$

where $E[\cdot]$ symbolizes expectation, $\sigma^2 = E[y_k^2]$ is the variance of y_k, $\mathbf{P} = E[y_k \mathbf{X}_k]$ is the N length cross-correlation vector and $\mathbf{R} = E[\mathbf{X}_k \mathbf{X}_k^{\mathrm{T}}]$ is the $N \times N$ autocorrelation matrix. A plot of the MSE against the filter coefficients, $\mathbf{W}$, is bowl shaped with a unique bottom (see Figure 10.6). This figure is known as the performance surface and is non-negative. The gradient of the performance surface is given by

$$\nabla = \frac{\mathrm{d}J}{\mathrm{d}\mathbf{W}} = -2\mathbf{P} + 2\mathbf{R}\mathbf{W} \qquad (10.13)$$

Each set of coefficients, $w(i)$ ($i = 0, 1, \ldots, N-1$), corresponds to a point on the surface. At the minimum point of the surface, the gradient is zero and the filter weight vector has its optimum value, $\mathbf{W}_{\mathrm{opt}}$ (see Example 10.2):

$$\mathbf{W}_{\mathrm{opt}} = \mathbf{R}^{-1} \mathbf{P} \qquad (10.14)$$

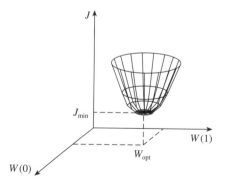

Figure 10.6 Error–performance surface.

Equation 10.14 is known as the Wiener–Hopf equation or solution. The task in adaptive filtering is to adjust the filter weights, $w(0)$, $w(1)$, ..., using a suitable algorithm, to find the optimum point on the performance surface.

The Wiener filter has a limited practical usefulness because

- it requires the autocorrelation matrix, $\mathbf{R}$, and the cross-correlation vector, $\mathbf{P}$, both of which are not known *a priori*;
- it involves matrix inversion, which is time consuming; and
- if the signals are nonstationary, then both $\mathbf{R}$ and $\mathbf{P}$ will change with time and so $\mathbf{W}_{opt}$ will have to be computed repeatedly.

For real-time application, a way of obtaining $\mathbf{W}_{opt}$ on a sample-by-sample basis is required. Adaptive algorithms are used to achieve this without having to compute $\mathbf{R}$ and $\mathbf{P}$ explicitly or performing a matrix inversion.

Example 10.2

Starting with the equation for the mean square error (Equation 10.12), derive the Wiener–Hopf equation.

Solution

The MSE is given by

$$MSE = J = \sigma^2 + 2\mathbf{P}^T\mathbf{W} + \mathbf{W}^T\mathbf{R}\mathbf{W} \tag{10.15}$$

The gradient, $\boldsymbol{\nabla}$, of the MSE is obtained by differentiating the MSE with respect to the weight vector $\mathbf{W}$, and setting the result to zero (Haykin, 1986):

$$\boldsymbol{\nabla} = \frac{dJ}{d\mathbf{W}} = \frac{d\sigma^2}{d\mathbf{W}} + \frac{d(\mathbf{P}^T\mathbf{W})}{d\mathbf{W}} + \frac{d(\mathbf{W}^T\mathbf{R}\mathbf{W})}{d\mathbf{W}} \tag{10.16}$$

Now,

$$\frac{d\sigma^2}{d\mathbf{W}} = 0$$

$$\frac{d(2\mathbf{P}^T\mathbf{W})}{d\mathbf{W}} = -2\mathbf{P}$$

$$\frac{d(\mathbf{W}^T\mathbf{R}\mathbf{W})}{d\mathbf{W}} = 2\mathbf{R}\mathbf{W}$$

Using these results, and setting $\boldsymbol{\nabla} = \mathbf{0}$, Equation 10.16 becomes

$$\nabla = \frac{dJ}{d\mathbf{W}} = -2\mathbf{P} + 2\mathbf{RW} = 0 \tag{10.17}$$

The optimum coefficient vector is then given by

$$\mathbf{W}_{opt} = \mathbf{R}^{-1}\mathbf{P} \tag{10.18}$$

10.4 The basic LMS adaptive algorithm

One of the most successful adaptive algorithms is the LMS algorithm developed by Widrow and his coworkers (Widrow *et al.*, 1975a). Instead of computing $\mathbf{W}_{opt}$ in one go as suggested by Equation 10.18, in the LMS the coefficients are adjusted from sample to sample in such a way as to minimize the MSE. This amounts to descending along the surface of Figure 10.6 towards its bottom.

The LMS is based on the steepest descent algorithm where the weight vector is updated from sample to sample as follows:

$$\mathbf{W}_{k+1} = \mathbf{W}_k - \mu\nabla_k \tag{10.19}$$

where $\mathbf{W}_k$ and ∇_k are the weight and the true gradient vectors, respectively, at the kth sampling instant. μ controls the stability and rate of convergence.

The steepest descent algorithm in Equation 10.19 still requires knowledge of $\mathbf{R}$ and $\mathbf{P}$, since ∇_k is obtained by evaluating Equation 10.17. The LMS algorithm is a practical method of obtaining estimates of the filter weights $\mathbf{W}_k$ in real time without the matrix inversion in Equation 10.18 or the direct computation of the autocorrelation and cross-correlation. The Widrow–Hopf LMS algorithm for updating the weights from sample to sample is given by

$$\mathbf{W}_{k+1} = \mathbf{W}_k + 2\mu e_k \mathbf{X}_k \tag{10.20a}$$

where

$$e_k = y_k - \mathbf{W}_k^T \mathbf{X}_k \tag{10.20b}$$

Clearly, the LMS algorithm above does not require prior knowledge of the signal statistics (that is the correlations $\mathbf{R}$ and $\mathbf{P}$), but instead uses their instantaneous estimates (see Example 10.3). The weights obtained by the LMS algorithm are only estimates, but these estimates improve gradually with time as the weights are adjusted and the filter learns the characteristics of the signals. Eventually, the weights converge. The condition for convergence is

$$0 < \mu > 1/\lambda_{max} \tag{10.21}$$

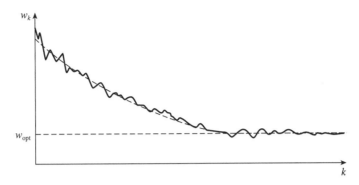

Figure 10.7 An illustration of the variations in the filter weights.

where $\lambda_{\max}$ is the maximum eigenvalue of the input data covariance matrix. In practice, $\mathbf{W}_k$ never reaches the theoretical optimum (the Wiener solution), but fluctuates about it (see Figure 10.7).

10.4.1 Implementation of the basic LMS algorithm

The computational procedure for the LMS algorithm is summarized below.

(1) Initially, set each weight $w_k(i)$, $i = 0, 1, \ldots, N-1$, to an arbitrary fixed value, such as 0.

For each subsequent sampling instant, $k = 1, 2, \ldots$, carry out steps (2) to (4) below:

(2) compute filter output

$$\hat{n}_k = \sum_{i=0}^{N-1} w_k(i)x_{k-i}$$

(3) compute the error estimate

$$e_k = y_k - \hat{n}_k$$

(4) update the next filter weights

$$w_{k+1}(i) = w_k(i) + 2\mu e_k x_{k-i}$$

The simplicity of the LMS algorithm and ease of implementation, evident from above, make it the algorithm of first choice in many real-time systems. The LMS algorithm requires approximately $2N + 1$ multiplications and $2N + 1$ additions for each new set of input and output samples. Most signal processors are suited to the mainly multiply–accumulate arithmetic operations involved, making a direct implementation of the LMS algorithm attractive.

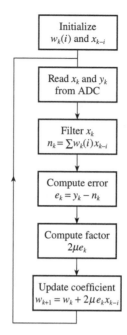

Figure 10.8   Flowchart for the LMS adaptive filter.

Inputs:	xk(i)	vector of the latest input samples
	yk	current contaminated signal sample
	wk(i)	vector of filter coefficients
Outputs:	ek	current desired output (or error) sample
	wk(i)	vector of updated filter coefficients

```
/* compute the current error estimate */

        ek = yk
        for i=1 to N do
          ek = ek – xk(i) * wk(i)
        end

/* update filter coefficients              */

        gk = 2u * ek
        for i = 1 to N do
          wk(i) = wk(i) + xk(i) * gk
        end

        return
```

Figure 10.9   Coding of the LMS adaptive filter.

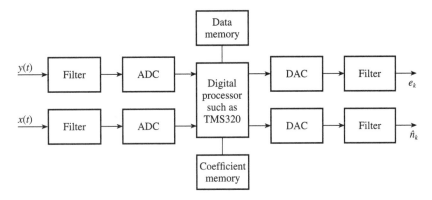

Figure 10.10   Hardware implementation for real-time LMS adaptive filtering.

The flowchart for the LMS algorithm is given in Figure 10.8. Figures 10.9 and 10.10, respectively, show a pseudo-code for the software and hardware implementations. A C language implementation of the LMS algorithm is given in the appendix.

Example 10.3

Starting with the steepest descent algorithm

$$\mathbf{W}_{k+1} = \mathbf{W}_k - \mu \mathbf{\nabla}_k$$

where $\mathbf{W}_k$ is the filter weight vector at the kth sampling instant, μ controls stability and rate of convergence and $\mathbf{\nabla}_k$ is the true gradient of the error–performance surface, derive the Widrow–Hopf LMS algorithm for adaptive noise cancelling, stating any reasonable assumptions made.

Solution

The steepest descent algorithm is given by

$$\mathbf{W}_{k+1} = \mathbf{W}_k - \mu \mathbf{\nabla}_k \tag{10.22}$$

The gradient vector, $\mathbf{\nabla}$, the cross-correlation between the primary and secondary inputs, $\mathbf{P}$, and the autocorrelation of the primary input, $\mathbf{R}$, are related as

$$\mathbf{\nabla} = -2\mathbf{P} + 2\mathbf{R}\mathbf{W} \tag{10.23}$$

In the LMS algorithm, instantaneous estimates are used for $\mathbf{\nabla}$. Thus

$$\mathbf{\nabla}_k = -2\mathbf{P}_k + 2\mathbf{R}_k\mathbf{W}_k = -2\mathbf{X}_k y_k + 2\mathbf{X}_k\mathbf{X}_k^{\mathrm{T}}\mathbf{W}_k$$
$$= -2\mathbf{X}_k(y_k - \mathbf{X}_k^{\mathrm{T}}\mathbf{W}_k) = -2e_k\mathbf{X}_k \tag{10.24}$$

where

$$e_k = y_k - \mathbf{X}_k^{\mathrm{T}} \mathbf{W}_k$$

Substituting Equation 10.24 in the equation for the steepest descent algorithm we have the basic Widrow–Hopf LMS algorithm:

$$\mathbf{W}_{k+1} = \mathbf{W}_k + 2\mu e_k \mathbf{X}_k \qquad (10.25a)$$

where

$$e_k = y_k - \mathbf{W}_k^{\mathrm{T}} \mathbf{X}_k \qquad (10.25b)$$

10.4.2 Practical limitations of the basic LMS algorithm

In practice, several practical problems are encountered when using the basic LMS algorithm, leading to a lowering of performance. Some of the more important problems are discussed here.

10.4.2.1 Effects of non-stationarity

In a stationary environment, the error performance surface of the filter has a constant shape and orientation, and the adaptive filter merely converges to and operates at or near the optimum point. If the signal statistics change after the weights have converged, the filter responds to the change by re-adjusting its weights to a new set of optimal values, provided that the change in signal statistics is sufficiently slow for the filter to converge between changes. In a non-stationary environment, however, the bottom or minimum point continually moves, and its orientation and curvature may also be changing (see Figure 10.11). Thus the algorithm in this case has the task not only of seeking the minimum point of the surface but also of tracking the changing position, leading to a significant lowering of performance. Note that a variable is said to be stationary if its statistics (such as mean, variance, autocorrelation) change with time. Such changes can result from, for example, sudden changes due to sporadic interference of short duration (Figure 10.12) or bad data, and often upset the filter weights.

A number of schemes have been developed to overcome this problem but these in general tend to increase the complexity of the basic LMS algorithm. One such scheme is the time-sequenced adaptive filter (Ferrara and Widrow, 1981).

10.4.2.2 Effects of signal component on the interference input channel

The performance of the algorithm relies on the measured interference signal, $x_k(i)$, being highly correlated with the actual interference, but weakly correlated (theoretically zero) with the desired signal. In most cases, this condition is not met. In some applications, the contaminating input may contain both the undesired interference as well as low level signal components. This leads to a cancellation of some of the

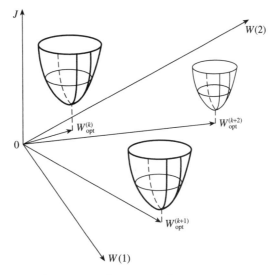

Figure 10.11   Time-varying error–performance surface.

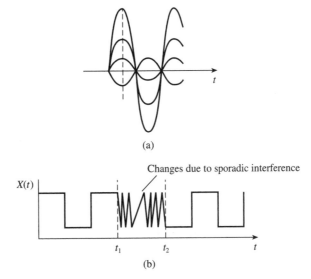

Figure 10.12   An illustration of non-stationary processes: (a) modulated waveform; (b) sporadic interference.

desired signal components. Such a situation is illustrated in Figure 10.13. It is shown in Widrow *et al.* (1975a) that the adaptive noise cancelling process still leads to a significant improvement in the desired signal-to-noise ratio in these cases but only at the expense of a small signal distortion. However, if x_k contains only signals and no noise component whatsoever, the desired signal in y_k may be completely obliterated.

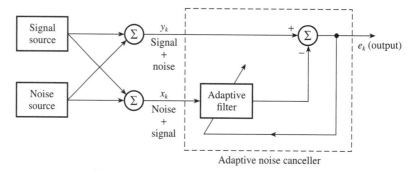

Figure 10.13   Adaptive noise cancelling with some signal components in both the desired signal and interference input channels.

Our work in biomedical signal processing confirms their results (Ifeachor *et al.*, 1986).

10.4.2.3   Computer wordlength requirements

The LMS-based FIR adaptive filter is characterized by the following equations:

for the digital filter
$$\hat{n}_k = \sum_{i=0}^{N-1} w_k(i) x_{k-i} \qquad (10.26a)$$

for the adaptive algorithm
$$\mathbf{W}_{k+1} = \mathbf{W}_k + 2\mu e_k \mathbf{X}_k \qquad (10.26b)$$

where

$$e_k = y_k - \mathbf{W}_k^{\mathrm{T}} \mathbf{X}_k$$

When adaptive filters are implemented in the real world, the filter weights, w_k, and the input variables, x_k and y_k, are of necessity represented by a finite number of bits. Similarly, the numerical operations involved are carried out using a finite precision arithmetic. The recursive nature of the LMS algorithm means that the wordlength will grow without limit and so some of the bits must be discarded before each updated weight is stored. Thus the y_k, e_k and $w_k(i)$ may differ significantly from their true values. The use of filter weights and results of arithmetic operations with limited accuracy may introduce errors into the adaptive filter whose effects may include (i) possible non-convergence of the adaptive filter to the optimal solution, leading to an inferior performance (for example, if the filter is used as an interference canceller some residual interference may remain), (ii) the filter outputs may contain noise which will cause it to fluctuate randomly, and (iii) a premature termination of the algorithm may occur. Thus a sufficient number of bits should be used to keep these errors at tolerable levels. Most adaptive systems described in the open literature represent the digital signals, x_{k-i} and y_k, as fixed point numbers of between 8 and 16 bits, with the coefficients quantized to between 16 and 24 bits. The multipliers used range from 8×8 bit to 24×16 bit, and accumulators of between 16 and 40 bits are used. It appears that for low order filters (up to about 100 coefficients) it is sufficient to store the coefficient to no more than 16-bit accuracy and to use a 16×16 bit multiplier with an accumulator of length 32 bits.

10.4.2.4   Coefficient drift

In the presence of certain types of inputs (for example narrowband signals), the filter coefficients may drift from the optimum values and grow slowly, eventually exceeding the permissible wordlength. This is an inherent problem in the LMS algorithm and leads to a long-term degradation in performance. In practice, coefficient drift is counteracted by introducing a leakage factor which gently nudges the coefficients towards zero. Two such schemes are given in Equations 10.27:

$$w_{k+1}(i) = \delta w_k(i) + 2\mu e_k x_{k-i} \qquad 0 < \delta < 1 \tag{10.27a}$$

$$w_{k+1}(i) = w_k(i) + 2\mu e_k x_{k-i} \pm \delta \qquad 0 < \delta < 1 \tag{10.27b}$$

Small δ, the leakage factor, ensures that drift is contained, but introduces bias in the error term, e_k.

The usefulness of the basic LMS algorithm has been extended by more sophisticated LMS-based algorithms as mentioned before. These include

(1)   the complex LMS algorithm which allows the handling of complex data;
(2)   the block LMS algorithm which offers substantial computational advantages and in some cases faster convergence; and
(3)   time-sequenced LMS algorithms to deal with particular types of non-stationarity.

10.4.3   Other LMS-based algorithms

10.4.3.1   Complex LMS algorithm

The complex LMS algorithm for updating the filter weights is given by (Widrow et al., 1975b)

$$\tilde{\mathbf{W}}_{k+1} = \tilde{\mathbf{W}}_k + 2\mu \tilde{e}_k \tilde{\mathbf{X}}_{k-i} \tag{10.28}$$

where the symbol ~ denotes a complex variable. The Mitel PDSP16XXX processors are ideally suited to the complex LMS algorithm as they can perform arithmetic operations directly on complex data, which is a distinct advantage over conventional processors.

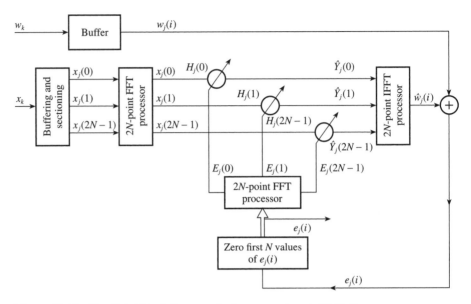

Figure 10.14 Simplified block diagram of a frequency domain LMS filter.

10.4.3.2 Fast LMS algorithms

A number of block LMS algorithms have been proposed which offer substantial computational savings especially when the number of filter coefficients is large. The computational savings result from processing the data in blocks instead of one sample at a time. Frequency domain implementations of the block LMS exploit the computational advantages of the fast Fourier transform (FFT) in performing convolutions (Mansour and Gray, 1982).

An efficient frequency domain filter is depicted in Figure 10.14.

10.5 Recursive least squares algorithm

The RLS algorithm is based on the well-known least squares method (Figure 10.15). An output signal, y_k, is measured at the discrete time, k, in response to a set of input signals, $x_k(i)$, $i = 1, 2, \ldots, n$. The input and output signals are related by the simple regression model

$$y_k = \sum_{i=0}^{n-1} w(i)x_k(i) + e_k \qquad (10.29)$$

where e_k represents measurement errors or other effects that cannot be accounted for, and $w(i)$ represents the proportion of the ith input that is contained in the primary

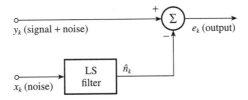

Figure 10.15   An illustration of the basic idea of the least-squares method.

signal, y_k. The problem in the LS method is, given the $x_k(i)$ and y_k above, to obtain estimates of $w(0)$ to $w(n-1)$.

Optimum estimates (in the least squares sense) of the filter weights, $w(i)$, are given by

$$\mathbf{W}_m = [\mathbf{X}_m^T\mathbf{X}_m]^{-1}\mathbf{X}_m^T\mathbf{Y}_m \tag{10.30}$$

where $\mathbf{Y}_m$, $\mathbf{W}_m$ and $\mathbf{X}_m$ are given by

$$\mathbf{Y}_m = \begin{bmatrix} y_0 \\ y_1 \\ y_2 \\ \vdots \\ y_{m-1} \end{bmatrix} \qquad \mathbf{X}_m = \begin{bmatrix} \mathbf{x}^T(0) \\ \mathbf{x}^T(1) \\ \mathbf{x}^T(2) \\ \vdots \\ \mathbf{x}^T(m-1) \end{bmatrix} \qquad \mathbf{W}_m = \begin{bmatrix} w(0) \\ w(1) \\ w(2) \\ \vdots \\ w(n-1) \end{bmatrix}$$

$$\mathbf{x}^T(k) = [x_k(0) \quad x_k(1) \quad \ldots \quad x_k(n-1)], \quad k = 0, 1, \ldots, m-1$$

The suffix m indicates that each matrix above is obtained using all m data points and T indicates transposition. Equation 10.30 gives the OLS estimate of $\mathbf{W}_m$ which can be obtained using any suitable matrix inversion technique. The filter output is then obtained as

$$\hat{n}_k = \sum_{i=0}^{n-1} \hat{w}(i)x_{k-i}, \quad k = 1, 2, \ldots, m \tag{10.31}$$

10.5.1   Recursive least squares algorithm

The computation of $\mathbf{W}_m$ in Equation 10.30 requires the time-consuming computation of the inverse matrix. Clearly, the LS method above is not suitable for real-time or on-line filtering. In practice, when continuous data is being acquired and we wish to improve our estimate of $\mathbf{W}_m$ using the new data, recursive methods are preferred. With the recursive least squares algorithm the estimates of $\mathbf{W}_m$ can be updated for each new set of data acquired without repeatedly solving the time-consuming matrix inversion directly.

A suitable RLS algorithm can be obtained by exponentially weighting the data to remove gradually the effects of old data on $\mathbf{W}_m$ and to allow the tracking of slowly varying signal characteristics. Thus

$$\mathbf{W}_k = \mathbf{W}_{k-1} + \mathbf{G}_k e_k \tag{10.32a}$$

$$\mathbf{P}_k = \frac{1}{\gamma}[\mathbf{P}_{k-1} - \mathbf{G}_k \mathbf{x}^{\mathrm{T}}(k)\mathbf{P}_{k-1}] \tag{10.32b}$$

where

$$\mathbf{G}_k = \frac{\mathbf{P}_{k-1}\mathbf{x}(k)}{\alpha_k}$$

$$e_k = y_k - \mathbf{x}^{\mathrm{T}}(k)\mathbf{W}_{k-1}$$

$$\alpha_k = \gamma + \mathbf{x}^{\mathrm{T}}(k)\mathbf{P}_{k-1}\mathbf{x}(k)$$

$\mathbf{P}_k$ is essentially a recursive way of computing the inverse matrix $[\mathbf{X}_k^{\mathrm{T}}\mathbf{X}_k]^{-1}$.

The argument k emphasizes the fact that the quantities are obtained at each sample point. γ is referred to as the forgetting factor. This weighting scheme reduces to that of the LS when $\gamma = 1$. Typically, γ is between 0.98 and 1. Smaller values assign too much weight to the more recent data, which leads to wildly fluctuating estimates. The number of previous samples that significantly contribute to the value of $\mathbf{W}_k$ at each sample point is called the asymptotic sample length (ASL) given by

$$\sum_{k=0}^{\infty} \gamma^k = \frac{1}{1 - \gamma} \tag{10.33}$$

This effectively defines the memory of the RLS filter. When $\gamma = 1$, that is when it corresponds to the LS, the filter has an infinite memory.

10.5.2 Limitations of the recursive least squares algorithm

The RLS method is very efficient and involves exactly the same number of arithmetic operations between samples as $\mathbf{W}_k$ and $\mathbf{P}_k$ in Equation 10.32 have fixed dimensions. This is an important requirement for efficient real-time filtering. There are, however, two main problems that may be encountered when the RLS algorithm is implemented directly. The first, referred to as 'blow-up', results if the signal $x_k(i)$ is zero for a long time, when the matrix $\mathbf{P}_k$ will grow exponentially as a result of division by γ (which is less than unity) at each sample point:

$$\lim_{k \to \infty} \mathbf{P}_k = \lim_{k \to \infty} \left(\frac{\mathbf{P}_{k-1}}{\gamma_{k-1}} \right) \tag{10.34}$$

The second problem with the RLS is its sensitivity to computer roundoff errors, which results in a negative definite **P** matrix and eventually to instability. For successful estimation of **W**, it is necessary that the matrix **P** be positive semi-definite which is equivalent to requiring in the LS method that the matrix $\mathbf{X}^T\mathbf{X}$ be invertible, but, because of differencing of terms in Equation 10.32b, positive definiteness of **P** cannot be guaranteed. This problem can be worse in multiparameter models, especially if the variables are linearly dependent and when the algorithm is implemented on a small system with a finite wordlength. When the algorithm has iterated for a long time the two terms in the parentheses in Equation 10.32b are very nearly equal and subtraction of such terms in a finite wordlength system may lead to errors and a negative definite $\mathbf{P}_k$ matrix.

The problem of numerical instability may be solved by suitably factorizing the matrix **P** such that the differencing of terms in Equation 10.32b is avoided. Such factorization algorithms are numerically better conditioned and have accuracies that are comparable with the RLS algorithm that uses double precision. Two such algorithms are the square root and the UD factorization algorithms. In terms of storage and computation the UD algorithm is more efficient, and is thus preferred. In fact, the UD algorithm is a square-root-free formulation of the square root algorithm and thus shares the same properties as the latter.

10.5.3   Factorization algorithms

10.5.3.1   Square root algorithm

In the square root method, the matrix $\mathbf{P}_k$ is factored as (Peterka, 1975)

$$\mathbf{P}_k = \mathbf{S}_k\mathbf{S}_k^T \tag{10.35}$$

where $\mathbf{S}_k$, an upper triangular matrix, and $\mathbf{S}_k^T$, its transpose, are square roots of $\mathbf{P}_k$. Thus if $\mathbf{S}_k$ instead of $\mathbf{P}_k$ is updated the positive definiteness of $\mathbf{P}_k$ is guaranteed, since the product of two square roots is always positive. $\mathbf{S}_k$ is updated as

$$\mathbf{S}_k = \frac{1}{\gamma^{1/2}}\, \mathbf{S}_{k-1}\mathbf{H}_{k-1} \tag{10.36}$$

where $\mathbf{H}_k$ is an upper triangular matrix.

10.5.3.2   UD factorization algorithm

In the UD method $\mathbf{P}_k$ is factored as (Bierman, 1976)

$$\mathbf{P}_k = \mathbf{U}_k\mathbf{D}_k\mathbf{U}_k^T$$

where $\mathbf{U}_k$ is a unit upper triangular matrix, $\mathbf{U}_k^T$ is its transpose and $\mathbf{D}_k$ is a diagonal matrix. Thus instead of updating $\mathbf{P}_k$ as in the RLS, its factors **U** and **D** are updated. A C language code for the UD algorithm is given on CD in the companion handbook (see the Preface for details).

10.6   Application example 1 – adaptive filtering of ocular artefacts from the human EEG

10.6.1   The physiological problem

The human electroencephalogram (EEG) is the electrical activity of the brain and contains useful diagnostic information on a variety of neurological disorders. Normal EEG signals are measured from electrodes placed on the scalp, and are often very small in amplitude, of the order of 20 μV. The EEG, like all biomedical signals, is very susceptible to a variety of large signal contaminations or artefacts which reduce its clinical usefulness. For example, blinking or moving the eyes produces large electrical potentials around the eyes called the electrooculogram (EOG). The EOG spreads across the scalp to contaminate the EEG, when it is referred to as an ocular artefact (OA). Examples of measured EOG and the corresponding contaminated EEG are given in Figure 10.16.

Ocular artefacts are a major source of difficulty in distinguishing normal brain activities from abnormal ones. In some cases, for example brain-damaged babies and patients with frontal tumours, it is difficult to distinguish between the associated pathological slow waves in the EEG and OAs. The similarity between the OAs and

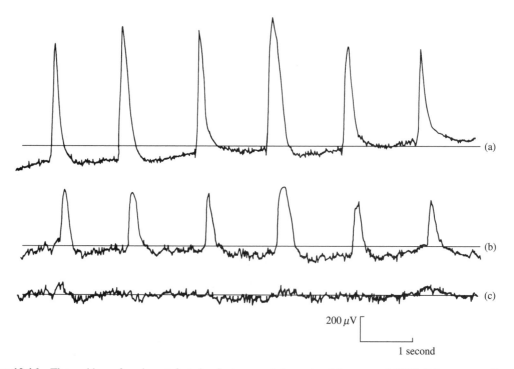

Figure 10.16   The problem of ocular artefacts in electroencephalography: (a) measured EOG, (b) corresponding contaminated EEG signal and (c) EEG signal corrected for artefact.

signals of clinical interest also makes it difficult to automate the analysis of the EEG by computer. In general, neurological disorders often manifest themselves in the EEG as slow waves which unfortunately not only appear similar to OAs but share the same frequency bands as OAs. The problem then is to remove the OAs while preserving the signals of clinical interest.

10.6.2   Artefact processing algorithm

Several methods have been proposed for processing OAs. However, factors such as the requirements of the clinical laboratory, constraints of real-time applications, costs, the random nature of OAs and the spectral overlap between OAs and some signals of cerebral origin dictate that OA processing should be adaptive and in real time.

An adaptive ocular artefact filtering scheme is depicted in Figure 10.17. In this method estimates of OAs are obtained by suitably scaling the EOGs. The OA estimates are then subtracted from the contaminated EEGs to yield 'artefact-free' EEG signals. To illustrate this, consider the simple problem of correcting a single EEG channel for ocular artefact using four EOG signals (Figure 10.17(b)). The information contained in the contaminated EEG, y_k, and the EOGs, $x_k(0)$ to $x_k(3)$, is used to obtain

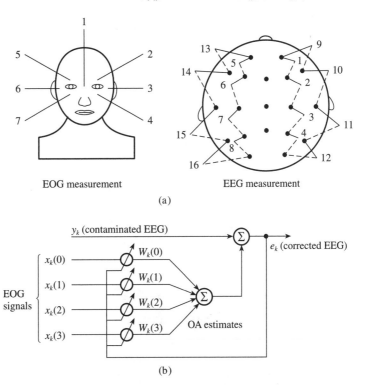

EOG measurement   EEG measurement

(a)

y_k (contaminated EEG)

e_k (corrected EEG)

EOG signals

$x_k(0)$ $W_k(0)$

$x_k(1)$ $W_k(1)$

$x_k(2)$ $W_k(2)$

$x_k(3)$ $W_k(3)$ OA estimates

(b)

Figure 10.17   Adaptive ocular artefact removal method: (a) possible electrode positions for EOG (ocular movement) and EEG measurements; (b) adaptive ocular artefact filter.

an estimate of the ocular artefact, $\sum_{i=0}^{3} w_k(i)x_k(i)$. The OA estimate is then subtracted from the contaminated EEG to yield an 'artefact-free' EEG, e_k:

$$e_k = y_k - \sum_{i=0}^{n-1} w_k(i)x_k(i) \tag{10.37}$$

where $w_k(i)$, $i = 0, 1, \ldots, n - 1$, are the coefficients of the adaptive filter and represent the fractions of the EOGs that reach the EEG as artefacts. e_k is also used to adjust the coefficients (weights) of the adaptive filter, using a numerically stable recursive least squares algorithm, so that optimal estimates of OAs are obtained. Continuous adjustment of $w_k(i)$ is necessary to account for changes in OAs due, for example, to changes in ocular movements.

The adaptive filtering algorithm used to remove the OAs is the UD algorithm described previously. This numerically stable formulation of the RLS algorithm was preferred to the LMS because of its superior convergence time, enabling it to cope better with different OAs each of which requires a different optimum set of coefficients for effective removal. An example of an EEG signal adaptively corrected for artefacts is shown in Figure 10.16(c).

10.6.3 Real-time implementation

An on-line microprocessor-based ocular artefact removal system that uses the UD algorithm described above has been developed (Ifeachor *et al.*, 1986). The system implements a variety of user-selectable models. The system has been tested on several normal and patient subjects. Good results were obtained for various categories of patient subjects.

However, it was found that when pathological waves, such as slow waves, epileptic spike and wave complexes, were picked up at both the EEG and EOG electrodes, the waves in the corrected EEG were reduced in amplitude. This is because the fraction of the EOG subtracted depends on the degree of correlation between the EOG and its component in the EEG, and the presence of slow waves of similar shape to the OA can lead to the subtraction of a fraction which depends on slow waves as well as the EOG. Thus, it is necessary to distinguish between the OA and slow waves, using a knowledge-based system, for example (Ifeachor *et al.*, 1990).

10.7 Application example 2 – adaptive telephone echo cancellation

Echoes arise primarily in communication systems when signals encounter a mismatch in impedance. Figure 10.18(a) shows a simplified long-distance telephone circuit. The hybrid circuit at the exchange converts the two-wire circuit from the customer's premises to a four-wire circuit, and provides separate paths for each direction of transmission. This is largely for economic reasons, for example to allow multiplexing, that is simultaneous transmission of many calls.

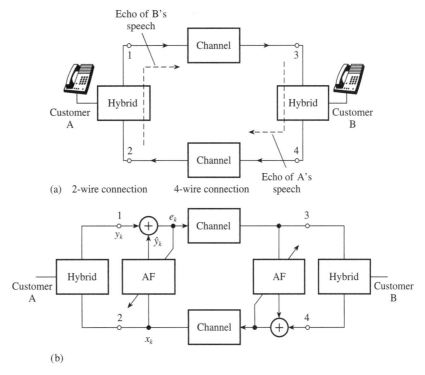

Figure 10.18   (a) Simplified long-distance telephone circuit; (b) echo cancellation in long-distance voice telephony.

Ideally, the speech signal originating from customer A travels along the upper transmission path to the hybrid on the right and from there to customer B, while that from B travels along the lower transmission path to A. The hybrid network at each end should ensure that the speech signal from the distant customer is coupled into its two-wire port and none to its output port. However, because of impedance mismatches the hybrid network allows some of the incoming signals to leak into the output path and to return to the talker as an echo. When the telephone call is made over a long distance (for example using geostationary satellites) the echo may be delayed by as much as 540 ms and represents an impairment that can be annoying to the customers. The impairment increases with distance. To overcome this problem, echo cancellers are installed in the network in pairs, as illustrated in Figure 10.18(b).

At each end of the communication system (Figure 10.18(b)), the incoming signal, x_k, is applied to both the hybrid and the adaptive filter (Duttweiller, 1978). The cancellation is achieved by making an estimate of the echo and subtracting it from the return signal, y_k. The underlying assumption here is that the echo return path (through the hybrid) is linear and time invariant. Thus the return signal at time k may be expressed as

$$y_k = \sum_{i=0}^{N-1} w_k(i)x_{k-i} + s_k \tag{10.38}$$

where x_k are samples of the incoming signal (from the far-end speaker), s_k is the near-end speaker plus any additive noise and w_k is the impulse response of the echo path. The echo canceller makes an estimate of this impulse response and produces a corresponding estimate, $\hat{y}_k = \sum w_k(i)x_{k-j}$, of the echo which is then subtracted from the normal return signal, y_k. Economic considerations place a limit on the sampling rate and the wordlengths of filter coefficients and input data which in turn limits the canceller's performance. Fundamental limits come from misadjustment in the adaptive filter and from nonlinearities in the echo path.

10.8   Other applications

10.8.1   Loudspeaking telephones

- The hybrid network is used to separate the transmit and receive paths (that is, the loudspeaker from the microphone), but there is a significant acoustic coupling between the loudspeaker and the microphone because of their proximity as well as a leakage across the imperfectly matched hybrid network (South *et al.*, 1979).

- The difficulty then is how to provide adequate gain for the receive and transmit directions without causing instability.

- The conventional solution to the problems is to use a voice-activated switch to select the transmit and receive paths, but this is not satisfactory because it does not allow full duplex communication.

- A better solution is to use adaptive filtering techniques to estimate and control the acoustic and hybrid echoes (Figure 10.19(b)). The number of filter coefficients here can be quite large, for example 512, making the use of a fast algorithm attractive.

- In teleconferencing networks (or public address systems) acoustic feedback leads to problems similar to those described above. Adaptive filters used for these may require large numbers of coefficients (250 to 1000), especially in rooms with long reverberation times, and must converge rapidly.

10.8.2   Multipath compensation

- In a type of spread spectrum system each data bit is transmitted as one of two orthogonal M-length pseudorandom sequences of bits. The sequence transmitted depends on whether the data bit is a logic 0 or 1. At the receiver two sequences identical to those used at the transmitter are cross-correlated with the received sequence to determine whether a 1 or 0 is received.

- In the presence of a multipath, the signal travels through separate paths to the receiver. Such effects could occur in mountainous or urban regions through reflection. The received signal is the sum of a number of components whose amplitudes and phases may differ (Figure 10.20). This reduces the performance of the receiver.

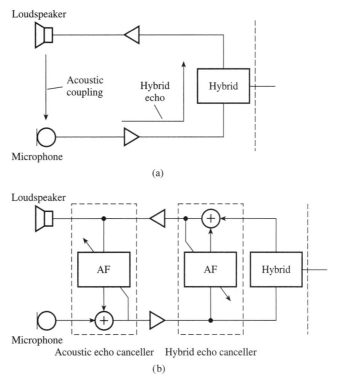

Figure 10.19 (a) Loudspeaking telephone; (b) acoustic and hybrid echo cancellation in loudspeaking telephone.

■ The adaptive filter is used to estimate the overall multipath response and to compensate for its effects.

10.8.3 Adaptive jammer suppression

■ In direct sequence spread spectrum a need often arises to suppress the effects of a jamming signal at the receiver to improve the performance of the receiver. Adaptive filtering may be used for this purpose (Figure 10.21). In such a system, use is made of the fact that the jammer is highly correlated whereas the pseudorandom code is weakly correlated. Thus the output of the filter, y_k, is an estimate of the jammer. This is subtracted from the received signal, x_k, to yield an estimate of the spread spectrum signal.

■ To enhance the performance of the system a two-stage jammer suppressor is used. The adaptive line enhancer, which is essentially another adaptive filter, counteracts the effects of finite correlation which leads to partial cancellation of the desired signal. The number of coefficients required for either filter is moderate (about 16), but the sampling frequency may be well over 400 kHz.

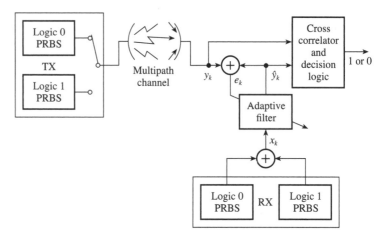

Figure 10.20   An adaptive spread spectrum communication system with multipath-effect compensation.

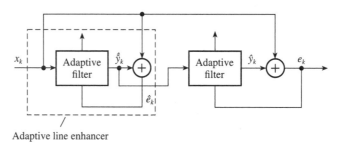

Adaptive line enhancer

Figure 10.21   Jammer suppression in direct sequence spread spectrum receiver.

10.8.4   Radar signal processing

Adaptive signal processing techniques are widely used to solve a number of problems associated with radar. For example, adaptive filters are used in monostatic radar systems to remove or cancel clutter components from the desired target signals. In HF groundwave radar, adaptive filters are used to reduce co-channel interference which is a major problem in the HF band.

10.8.5   Separation of speech signals from background noise

Acoustic background noise is a serious problem in speech processing. An adaptive filter may be used to enhance the performance of speech systems in noisy environments (for example in fighter aircraft, tanks, or cars) to improve both intelligibility and recognition of speech.

10.8.6 Fetal monitoring – cancelling of maternal ECG during labour

- Information derived from the fetal electrocardiogram (ECG), such as the fetal heart rate pattern, is valuable in assessing the condition of the baby before or during childbirth.
- The ECG derived from electrodes placed on the mother's abdomen is susceptible to contamination from much larger background noise (for example muscle activity and fetal motion) and the mother's own ECG.
- Adaptive filters have been used to derive a 'noise-free' fetal ECG. Figure 10.22 illustrates the concept.
- Four chest leads are used to detect the baby's ECG, and one or more leads to detect the combined mother and baby's ECG. A four-channel adaptive filter, with 32 coefficients per channel, is used to cancel the mother's heartbeat as shown.

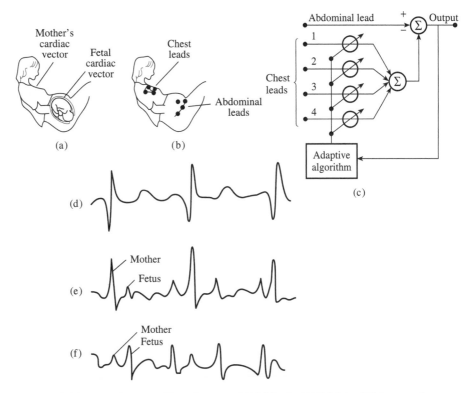

Figure 10.22 Adaptive cancelling of maternal ECG in fetal ECG (after Widrow *et al.*, 1975a): (a) cardiac electric field vectors of mother and fetus; (b) placement of leads; (c) adaptive; (d) idealized mother's ECG (chest leads); (e) idealized contaminated fetal ECG (abdominal lead); (f) output of noise canceller showing reduced mother's ECG.

Problems

10.1 Justify the use of adaptive filters instead of conventional filters in applications such as

   (1)   the removal of ocular artefacts from human EEGs,
   (2)   echo cancellation in long distance telephony, and
   (3)   suppression of jammer signal in spread spectrum communication.

   Starting with the steepest descent algorithm:

   $$\mathbf{W}_{k+1} = \mathbf{W}_k - \mu\nabla_k$$

   where $\mathbf{W}_k$ is the filter weight vector at the discrete time k, μ controls stability and rate of convergence, and ∇_k is the true gradient vector of the error–performance surface at the discrete time k, derive the Widrow–Hopf LMS algorithm for adaptive noise cancelling, stating any reasonable assumptions made. Comment on the practical significance of the LMS algorithm.

   Comment on two major practical limitations of the LMS algorithm and how they lower the performance of the algorithm. Suggest how these limitations may be overcome.

10.2 The output signal from an adaptive noise canceller is given by

   $$e_k = y_k - \mathbf{X}_k^{\mathsf{T}}\mathbf{W}_k$$

   where $\mathbf{W}_k$ is the adaptive filter weight vector and the other variables have the usual meaning. Starting with this equation, derive

   (1)   the discrete Wiener–Hopf equation and
   (2)   the basic LMS adaptive algorithm.

   State any assumptions made.

10.3 Show that the adaptive filter turns itself off when there is no correlation between the interference signal, x_k, and the contaminated signal y_k.

10.4 Explain briefly, with the aid of a block diagram, the basic concepts of adaptive noise cancelling. Discuss critically the benefits and limitations of adaptive noise cancelling in a real-time application of your choice and suggest ways of overcoming the limitations.

References

Bierman G.J. (1976) Measurement updating using the UD factorization. *Automatica*, **12**, 375–82.

Duttweiler D.L. (1978) A twelve-channel digital echo canceler. *IEEE Trans. Communications*, **26**, 647–53.

Ferrara E.R. and Widrow B. (1981) The time-sequenced adaptive filter. *IEEE Trans. Acoustics, Speech and Signal Processing*, **29**(3), 766–70.

Haykin S. (1986) *Adaptive Filter Theory*. Englewood Cliffs NJ: Prentice-Hall.

Ifeachor E.C., Jervis B.W., Morris E.L., Allen E.M. and Hudson N.R. (1986) A new microcomputer-based on-line ocular artefact removal (OAR) system. *IEE Proc.*, **133**, 291–300.

Ifeachor E.C., Hellyar M.T., Mapps D.J. and Allen E.M. (1990) Knowledge based enhancement of EEG signals. *Proc. IEE (Part F)*, **137**(5), 302–10.

Mansour D. and Gray A.H. (1982) Unconstrained frequency domain adaptive filter. *IEEE Trans. Acoustics, Speech and Signal Processing*, **30**, 726–34.

Peterka P. (1975) A square root filter for real-time multivariate regression. *Kybernetika*, **11**, 53–67.

South C.R., Hoppitt C.E. and Lewis A.V. (1979) Adaptive filters to improve loudspeaker telephone. *Electronics Lett*, **15**, 673–4.

Widrow B. and Winter R. (1988) Neural nets for adaptive filtering and adaptive pattern recognition. *IEEE Computer*, 25–30.

Widrow B., Glover J.R., McCool J.M., Kaunitz J., Williams C.S., Hearn R.H., Zeidler J.R., Dong E. and Goodlin R.C. (1975a) Adaptive noise cancelling: principles and applications. *Proc. IEEE*, **63**, 1692–716.

Widrow B., McCool J.M. and Ball M. (1975b) The complex LMS algorithm. *Proc. IEEE*, 719–20.

Bibliography

Clark G.A., Mitra S.K. and Parker S.R. (1981) Block implementation of adaptive digital filters. *IEEE Trans. Acoustics, Speech and Signal Processing*, **29**, 744–52.

Clark G.A., Parker S.R. and Mitra S.K. (1983) A unified approach to time- and frequency-domain realization of FIR adaptive digital filters. *IEEE Trans. Acoustics, Speech and Signal Processing*, **31**, 1073–83.

Cowan C.F.N. and Grant P.M. (eds) (1985) *Adaptive Filters*. Englewood Cliffs NJ: Prentice-Hall.

De Courville M. and Duhamel P. (1995) Adaptive filtering in subbands using a weighted criterion. In *Proc. ICASSP*, Detroit, MI, Vol. 2, pp. 985–8.

Dentino M., McCool J.M. and Widrow B. (1978) Adaptive filtering in the frequency domain. *Proc. IEEE*, **66**, 1658–9.

Dudek M.T. and Robinson J.M. (1981) A new adaptive circuit for spectrally efficient digital microwave-radio-relay systems. *Electronics and Power*, 397–401.

Falconer D.D. (1982) Adaptive reference echo cancellation. *IEEE Trans. Communications*, **30**, 2083–94.

Ferrara E.R. and Widrow B. (1981) Multichannel adaptive filtering for signal enhancement. *IEEE Trans. Acoustics, Speech and Signal Processing*, **29**, 766–70.

Gilloire A. and Vetterli M. (1992) Adaptive filtering in subbands with critical sampling: analysis, experiments, and applications to acoustic echo cancellation. *IEEE Trans. Circuits and Systems*, **40**, 1862–75.

Harrison W.A., Lim J.S. and Singer E. (1986) A new application of adaptive noise cancellation. *IEEE Trans. Acoustics, Speech and Signal Processing*, **34**, 21–7.

Holte N. and Stueflotten S. (1981) A new digital echo canceler for two-wire subscriber lines. *IEEE Trans. Communications*, **29**, 1573–81.

Lappage R., Clarke J., Palma G.W.R. and Huizing A.G. (1987) The Byson research radar. In *International Conf. Radar 87*, October 1987, London: IEE, 453–61.

Levin M.D. and Cowan C.F.N. (1994) The performance of eight recursive least squares adaptive filtering algorithms in a limited precision environment. In *Proc. European Signal Processing Conf.*, Edinburgh, pp. 1261–4.

Lewis A. (1992) Adaptive filtering applications in telephony. *BT Technology J.*, **10**, 49–63.

Li Y. and Ding Z. (1995) Convergence analysis of finite length blind adaptive equalizers. *IEEE Trans. Signal Processing*, **43**, 2120–9.

Macchi O. (1995) *Adaptive Processing: The LMS Approach with Applications in Transmission*. New York: Wiley.

Messerschmitt D.G. (1984) Echo cancellation in speech and data transmission. *IEEE J. Selected Areas in Communications*, **2**, 283–97.

Mikhael W.B. and Wu F.H. (1987) Fast algorithms for block FIR adaptive digital filtering. *IEEE Trans. Circuits and Systems*, **34**, 1152–60.

Mueller K.H. (1976) A new digital echo canceler for two-wire full-duplex data transmission. *IEEE Trans. Communications*, **24**, 956–62.

Ochia K., Araseki T. and Ogihara T. (1977) *IEEE Trans. Communications*, **25**, 589–94.

Ogue J.C., Saito T. and Hoshiko Y. (1983) A fast convergence frequency domain adaptive filter. *IEEE Trans. Acoustics, Speech and Signal Processing*, **31**, 1312–14.

Reed F.A., Feintuch P.L. and Bershad N.J. (1985) The application of the frequency domain LMS adaptive filter to split array bearing estimation with a sinusoidal signal. *IEEE Trans. Acoustics, Speech and Signal Processing*, **33**, 61–9.

Saulnier G.J., Das P.K. and Milstein L. (1985) An adaptive digital suppression filter for direct sequence spread spectrum communications. *IEEE J. Selected Areas in Communications*, **3**, 676–86.

Sethares W.A., Lawrence D.A., Johnson C.R. and Bitmead R.R. (1986) Parameter drift in LMS adaptive filters. *IEEE Trans. Acoustics, Speech and Signal Processing*, **34**, 868–79.

Sondhi M.M. and Berkley D.A. (1980) Silencing echoes on the telephone network. *Proc. IEEE*, **68**, 948–63.

Tao Y.G., Kolwicz K.D., Gritton C.W.K. and Duttweiller D.L. (1986) A cascadable VLSI echo canceller. *IEEE Trans. Acoustics, Speech and Signal Processing*, **34**, 297–303.

Thornton C.L. and Bierman G.J. (1978) Filtering and error analysis via the UDU covariance factorization. *IEEE Trans. Automatic Control*, **23**, 901–7.

Widrow B. (1966) *Adaptive Filters 1: Fundamentals*. Report SU-SEL-66-126, Stanford Electronics Laboratory, Stanford University, CA.

Widrow B. (1971) Adaptive filters. In *Aspects of Network and System Theory* (Kalman R. and DeClaris N. (eds)), pp. 563–87. New York: Holt, Rinehart and Winston.

Widrow B. (1976) Stationary and nonstationary learning characteristics of the LMS adaptive filter. *Proc. IEEE*, **64**, 1151–62.

Widrow B. and Stearns S.D. (1985) *Adaptive Signal Processing*. Englewood Cliffs NJ: Prentice-Hall.

Widrow B., Mantey P., Griffiths L. and Goode B. (1967) Adaptive antenna systems. *Proc. IEEE*, **55**, 2143–59.

Appendices

10A C language programs for adaptive filtering

Several adaptive algorithms have been implemented in the C language which may be used to explore further the topics covered in this chapter. These are

(1) lmsflt.c, the LMS algorithm,
(2) uduflt.c, the UD algorithm,
(3) sqrflt.c, the square root algorithm and
(4) rlsflt.c, the recursive least squares algorithm.

Only the first program is listed here to limit the size of the book (see Program 10A.1). However, all the programs are available on the CD in the companion handbook *A Practical Guide for MATLAB and C Language Implementations of DSP Algorithms* (see the Preface for details).

Program 10A.1 C language implementation of the LMS algorithm (lmsflt.c).

```
/* ------------------------------------------------------------------- */
/*      implementation of the LMS algorithm                            */
/*                                                                     */
/*      manny 6.11.92                                                  */
/*                                                                     */
/*      inputs:                                                        */
/*      x[]       input data vector                                    */
/*      dk        latest input data value                             */
/*      w[]       coefficient vector                                   */
/*                                                                     */
/*      outputs:                                                       */
/*      ek        error value                                         */
/*      yk        digital filter output                                */
/*      w[]       updated coefficient vector                           */
/* ------------------------------------------------------------------- */
```

```
double    lmsflt()
{
          int     i;
          double  uek,yk;

          yk = 0;
          for(i=0; i<N; ++i){                    /* digital filtering */
                    yk=yk+w[i]*x[i];
          }
          ek=dk-yk;                              /* compute output error */

          uek=2*mu*ek;                           /* update the weights */
          for(i=0; i<N; i++){
                    w[i]=w[i]+uek*x[i];
          }
          return(yk);
}
```

To illustrate how to implement adaptive filters, we will use the program listed in Program 10A.1 to detect a tone in broadband noise.

10A.1 Adaptive enhancement of narrowband signals buried in noise

Adaptive filters are often used to detect or enhance narrowband signals buried in wideband noise. The structure that is commonly used for this purpose is depicted in Figure 10A.1. It consists of a delay element, symbolized by z^{-M}, and an adaptive predictor. The delay element removes any correlation that may exist between the samples of the noise component. The adaptive predictor is essentially an FIR filter with adjustable coefficients and its output, y_k, gives the enhanced narrowband signal. In some applications, the second output of the adaptive filter, e_k, and not y_k is the desired output. The prediction coefficients, $w_k(i)$, are optimized by a suitable adaptive algorithm, which in our case is an LMS algorithm (see Section 10.4 for details).

In the case of the LMS algorithm, the adaptive filter is characterized by the following equations:

$$y_k = \sum_{i=0}^{N-1} w_k(i)x_k(i) \tag{10A.1}$$

$$e_k = d_k - y_k \tag{10A.2}$$

$$w_{k+1}(i+1) = w_k(i) + 2\mu e_k x_k(i), \quad i = 0, 1, \ldots, N-1 \tag{10A.3}$$

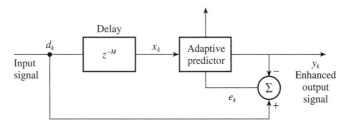

Figure 10A.1 Adaptive signal enhancement.

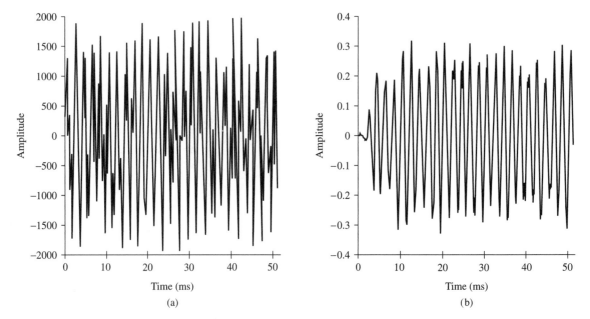

Figure 10A.2   Adaptive enhancement of a narrowband signal: (a) noisy signal; (b) enhanced signal.

where d_k is the noisy narrowband signal sample, $x_k(i)$ is the input data vector, derived from delayed values of d_k, $w_k(i)$ is the prediction coefficient vector at the kth sampling instant, μ is the stability factor, and y_k is the enhanced, narrowband signal.

The function, lmsflt.c, in Program 10A.1 is a C language implementation of the above equations. The program listed in Program 10A.2 illustrates how to use the function lmsflt.c for signal enhancement. To simulate the problem a broadband noise was added to a 500 Hz sine wave signal, and the composite data stored in ASCII format in a file din.dat. The noisy sine wave was then applied to the adaptive filter. To simulate real-time adaptive filtering, the input data is read from the file and applied to the adaptive filter one sample at a time. For very long lengths of data the user may need to read the data in blocks for efficiency. Figure 10A.2 shows the results for the LMS based filters.

As may be evident from Figure 10A.2, to use the adaptive algorithms, the user needs to specify the parameters of the adaptive filters, for example the length of the FIR filter, N, the delay factor, M, and the stability factor, μ. Attention must also be paid to the format of the input data. For example, in some applications the input data may come from a multichannel source, with each element of $x_k(i)$ representing the data value from a channel. In this case, the input data array to the adaptive algorithm will need to be suitably modified.

Program 10A.2   Program for adaptive signal enhancement.

```
/* ----------------------------------------------------------------------------------- */
/*                                                                                      */
/*         program to illustrate adaptive filtering using                               */
/*         the LMS algorithms                                                            */
/*                                                                                      */
/*         program name: adfilter.c                                                     */
/*                                                                                      */
```

```
/*      manny, 7.11.92                                                    */
/* --------------------------------------------------------------------- */

#include        <stdio.h>
#include        <math.h>
#include        <dos.h>

/* constant definitions */

#define N       30              /* filter length */
#define M       1               /* delay */
#define w0      0               /* initial value for adaptive filter coefficients */
#define npt     N+M
#define SF      2048            /* factor for reducing the data samples – 11 bit ADC assumed */
#define mu      0.04

double      lmsflt();
void        initlms();
void        update__data__buffers();
void        initfiles();
float       x[npt], d[npt], dk, ek;
double      w[npt];
FILE        *in,*out,*fopen();
char        din[30];

main()
{
        double yk, yk1;

        initfiles();
                                                        /* lms-based adaptive filter */
        initlms();
        while(fscanf(in,"%f",&dk)!=EOF){
            dk=dk/SF;
            update__data__buffers();
            yk=lmsflt();
            yk1 =SF.yk;
            fprintf(out,"%lf \n",yk1);
        }
        fcloseall();
}


/* --------------------------------------------------------------------- */
void        initfiles()
{
        clrscr();
        printf("enter name of file holding data to be filtered \n");
        scanf("%s",din);
        printf("\n");
        printf("the filtered data will be stored in dout.dat \n");

        if((in=fopen(din,"r"))= =NULL){
            printf("cannot open input data file \n");
            exit(1);
        }
```

```
                    if ((out = fopen ("dout.dat", "w"))= =NULL){
                        printf("cannot open output data file \n");
                        exit(1);
                    }
                    return;
    }
    /* ---------------------------------------------------------------------------------------------- */

    void         update__data__buffers()
    {
                    long    j, k;
                    for(j=1; j<N; + +j){                          /*update x-data buffer*/
                        k=N–j;
                        x[k]=x[k–1];
                    }
                    x[0]=dk;
                    if(M>0)
                        x[0]=d[M–1];
                    for(j=1; j<M; + +j){                          /*update d-data buffer*/
                        k=M–j;
                        d[k]=d[k–1];
                    }
                    d[0]=dk;
    }


    /* ---------------------------------------------------------------------------------------------- */
    void         initlms()
    {
                    long    i;
                    for(i=0; i<npt; + +i){
                        x[i]=0;
                        d[i]=0;
                        w[i] = w0;
                    }
    }
    /* ---------------------------------------------------------------------------------------------- */

    #include      "lmsflt.c";
```

10B   MATLAB programs for adaptive filtering

MATLAB does not explicitly support adaptive signal processing. However, we have developed MATLAB programs for the two basic adaptive algorithms, namely, the LMS and the RLS algorithms:

> lmsadf.m – function for performing LMS based adaptive filtering
> rlsadf.m – function for performing RLS based adaptive filtering

The programs are available on the web. Illustrative examples of the use of the programs can be found in the companion manual, *A Practical Guide for MATLAB and C Language Implementations of DSP Algorithms*, published by Pearson Education (see the Preface for details).

Spectrum estimation and analysis

The following topics are covered in this chapter: the basic concepts of spectrum analysis, the pitfalls to be avoided especially in the nonparametric methods, the properties of data windows, pre-processing of the data and choice of the data window function. Nonparametric methods described and compared include the modified periodogram, the Blackman–Tukey method based on fast Fourier transformation of the autocorrelation function of the data and the fast correlation method. The most commonly used method of parametric spectrum estimation based upon auto-regressive modelling is described, together with a brief review of other parametric methods. Applications described are the processing of electrical waveforms of the brain to differentiate between patients with brain diseases and normal subjects, and the analysis of human electroencephalograms based on autoregressive modelling.

11.1   Introduction

The transformation of data from the time domain to the frequency domain was introduced in Section 3.1. The principles and practice of the estimation and analysis of spectra in the frequency domain are the subject of this chapter. Plots of harmonic amplitude or phase versus frequency often result in a more comprehensible presentation of the data or waveform, particularly when the latter is of a random nature. By selecting certain harmonics on some suitable criterion and rejecting others, significant data reduction may be possible. Spectrum analysis has found various applications such as the study of communication engineering signals, of event-related or stimulated responses of the human electroencephalogram (EEG) in the diagnosis of brain diseases (Jervis *et al.*, 1993), of other biological signals, of meteorological data, and in industrial process control and the measurement of noise spectra for the design of optimal linear filters.

The spectrum estimation techniques available may be categorized as nonparametric and parametric. The nonparametric methods include the periodogram, the Bartlett and Welch modified periodogram, and the Blackman–Tukey methods. All these methods have the advantage of possible implementation using the fast Fourier transform, but with the disadvantage in the case of short data lengths of limited frequency resolution. Also, considerable care has to be exercised to obtain meaningful results. Parametric methods on the other hand can provide high resolution in addition to being computationally efficient. However, it is necessary to form a sufficiently accurate model of the process from which to estimate the spectrum. The most common parametric approach is to derive the spectrum from the parameters of an autoregressive model of the signal. Autoregressive spectrum estimation is described in Section 11.5.

A number of pitfalls have to be avoided in performing nonparametric spectral analysis and the associated topics of aliasing, scalloping loss, finite data length, spectral leakage, and spectral smearing should be understood and are discussed in Section 11.3.

The deleterious effects of spectral leakage and smearing may be minimized by windowing the data by a suitable window function. The sampled data values are multiplied point by point by the sampled values of the selected window function. The topic of windowing is discussed in Section 11.3.2. The equivalent noise bandwidth, processing gain, worst-case processing loss, and minimum resolution bandwidth are all properties of window functions which are considered when choosing a suitable window (see Section 11.3.2.1). In the section on overlap correlation it is shown that averaging the spectra of a number of windowed sections of the data instead of computing the spectrum of the windowed data directly leads to a significantly improved estimate of the spectrum. The part of Section 11.3.2.1 on the biasing effect of data windows shows how the loss of signal energy and the dc biasing effect of data windows may be overcome by pre-processing the data prior to windowing.

Judgements of the quality of spectral estimates are based on estimation theory, and so some basic concepts of that theory are now presented. Statistical estimation involves determining the expected values of statistical quantities derived from samples of the population. However, in time series analysis, discrete data obtained as a function of time is usually available rather than samples of the population taken simultaneously. This difficulty is commonly avoided by assuming that the process is ergodic, that is the properties of the time series data are the same as would be those of the hypothetical samples. Some statistical definitions are now in order.

The mean value of a time series with data values $x(n)$, $n = 0, 1, \ldots, N-1$, is the expected value of $x(n)$, $E[x(n)]$, given by

$$E[x(n)] = \frac{1}{N} \sum_{n=0}^{N-1} x(n) \qquad (11.1)$$

where E denotes expectation. The variance of the same time series is

$$\text{var}\,[x(n)] = E\{[x(n) - \bar{x}(n)]^2\} \qquad (11.2)$$

The autocovariance of $x(n)$ is given by

$$c_{xx}(m) = E\{[x(n) - \bar{x}(n)][x(n + m) - \bar{x}(n)]\} \qquad (11.3)$$

where m denotes the lag in the data points and $\bar{x}(n)$ denotes $E[x(n)]$. The power spectral density estimated from the finite realization is

$$P_E(\omega) = \sum_{m=-\infty}^{\infty} c_{xx}(m) \exp(-j\omega m) \qquad (11.4)$$

Note that if finite duration waveforms as opposed to infinitely long stochastic processes are being analyzed it would more appropriate to use the energy spectral density. The power spectral density has units of $V^2\,Hz^{-1}$. If the statistical attribute α is being estimated then the bias of the estimate is defined to be the difference between the true (population) value and the estimate:

$$\text{bias} = \alpha - E[\alpha] \qquad (11.5)$$

If the bias is zero then the estimate gives the true value; if the bias is not zero it represents the error in α and the estimated value of α is said to be biased. Clearly good estimators will not be biased. The variance of α is a measure of the width of the peak of the probability density distribution function of α. Small variances correspond to narrow peaks, and as the variance tends to zero the estimated value approaches the population (true) value if the estimate is unbiased. If the variance tends to zero as the number of data, N, increases, the estimate is said to be consistent. If the estimate is inconsistent then the estimates will fluctuate more and more wildly from realization to realization as the number of data is increased. It is therefore desirable that statistical estimates be both unbiased and consistent.

The properties of the periodogram methods of spectrum estimation are discussed in Sections 11.3.3 and 11.3.4. It is shown that the spectrum estimates derived as periodograms are inconsistent, that is successive realizations yield fluctuating estimates and are only unbiased for a large number of data. Stable and more accurate estimates are obtained by windowing sections of the data and averaging their spectra. The associated methods are known as the Bartlett and Welch modified periodogram methods. The final nonparametric method described is known as the Blackman–Tukey method. The windowed autocorrelation function of the data is first computed and the energy spectrum is then obtained from its FFT. The Blackman–Tukey spectrum estimate is characterized by a larger quality factor than the other periodogram methods.

It is also desirable to identify a quality factor for the estimates of power spectral density to allow a comparison of the different estimations. The ratio of the square of the mean of the power spectral density to its variance has been proposed as a suitable quality factor (Proakis and Manolakis, 1989):

$$Q = \frac{\{E[P_\mathrm{E}(f)]\}^2}{\mathrm{var}\,[P_\mathrm{E}(f)]} \tag{11.6}$$

11.2   Principles of spectrum estimation

In this section a voltage waveform, plotted against time, will be considered initially. The shape of the waveform may provide useful information. For example, it may be a sine wave which can obviously be characterized by its amplitude, frequency, and phase angle. To be more specific, the waveform would be describable as consisting of a single component with a certain amplitude and phase at the known frequency. As an alternative to representing the waveform as a plot of voltage versus time, it could be represented by two plots: one of amplitude versus frequency and the other of phase versus frequency. Because the sine wave only has one amplitude, one phase, and one frequency the amplitude and phase plots would each consist of a single point only. It can be shown by Fourier analysis (see Chapter 3) that all waveforms can be represented mathematically as the summation of a number of sinusoidal waveforms, each with a specific amplitude and phase at its specific frequency. Thus any waveform can be alternatively represented by a plot of amplitude versus frequency together with a plot of phase versus frequency. These plots are known as the amplitude and phase spectra. These spectra are important because they provide a complementary way of representing the waveform which more clearly reveals information about the frequency content of the waveform. The observed shapes of the spectra and changes in them are often helpful in the understanding and interpretation of the waveforms. Amplitude and phase spectra very often provide more useful information than the waveforms. The topic of transformation from the time to the frequency domain, and the converse, was described in Chapter 3. The transformation of periodic waveforms to the frequency domain by the Fourier series and the complex Fourier series was reviewed. It was shown that the frequencies of the sinusoidal frequency components of the periodic waveform, known as the Fourier components, are harmonically related to each other, i.e. each is an integral multiple of the first harmonic frequency, f, where

$$f = 1/T_p$$

where T_p is the repetition period of the waveform. It follows that adjacent harmonic components are all equally spaced by the frequency interval $f = 1/T_p$, which in this sense of separation is known as the frequency resolution. The amplitude spectrum has an amplitude measured in volts. As an example Figure 3.1(a) shows a periodic voltage pulse waveform while Figures 3.1(b) and 3.1(c) display the amplitude spectrum and the phase spectrum of this waveform respectively. Various uses of the amplitude and phase spectra are mentioned in the introduction to Chapter 3.

Non-periodic but continuous signals can be transformed from the time to the frequency domain using the Fourier transform described in Section 3.1.2. The 'amplitude' of this transform was shown to have the dimension of $V\,Hz^{-1}$ and when plotted against frequency therefore represents the amplitude spectral density. Thus the area under the curve between two frequencies yields the 'average' voltage of the waveform for those frequency components lying between the two frequencies. Squaring the calculated Fourier transform 'amplitudes' gives the energy spectral density of the voltage waveform in $J\,Hz^{-1}$. The term 'spectrum' is often used to refer to plots of the energy spectral density versus frequency. Figures 3.2(a) and 3.2(b) show the amplitude and energy spectral densities of a rectangular pulse respectively. It was also shown in Chapter 3 how the spectra of sampled and nonperiodic voltage waveforms may be calculated using the discrete Fourier transform (DFT). The DFT components were shown to be harmonically related with the first harmonic angular frequency being $\Omega = 2\pi/(N-1)T$ so that the first harmonic frequency is f where

$$f = 1/(N-1)T \tag{11.7a}$$

where N is the number of data, and T is the sampling interval. Since $(N-1)T = T_p$, the duration of the sampled waveform, the first harmonic frequency is expressible as

$$f = 1/T_p \tag{11.7b}$$

Again, owing to the harmonic relationship the frequency resolution of the spectra is also given by $1/T_p$. Thus the longer the waveform, the greater will be the frequency resolution of the spectra.

An example calculation of the DFT of a data sequence $\{1, 0, 0, 1\}$ is given in Section 3.2. The DFT of the data sequence was shown to be $\{2, 1+j, 0, 1-j\}$. Thus, the second harmonic component, $1 + j$, is of magnitude $\sqrt{(1^2 + 1^2)} = \sqrt{2}$. If the data sequence represented sampled voltages the amplitude of this second harmonic would be $\sqrt{2}$ V, and its energy would be $(\sqrt{2}\,V)^2$, i.e. 2 joules. The corresponding phase angle is given by $\tan^{-1}$ (imaginary component/real component) $= \tan^{-1} 1 = 45°$. The data sequence is plotted in Figure 3.3(a) and its amplitude and phase spectra in Figures 3.3(b) and 3.3(c) respectively. The amplitude spectrum has the dimension of volts. The DFT and the Fourier transform were shown in Section 3.2 to be related by $F(j\omega) = TX(k)$. The decimation-in-time fast Fourier transform (FFT) algorithm for accelerating the computation of the DFT was introduced in Section 3.5 and an example of its use given in Section 3.5.1 for calculating the DFT of the above sequence $\{1, 0, 0, 1\}$.

If the waveform under investigation is long compared with the time interval over which it may be regarded as having constant statistical moments, then the spectrum estimate is likely to be inaccurate. This will also be the case when there is a large noise component in the waveform. It is then desirable to smooth the estimated spectrum to obtain an improved estimate. The purpose of spectrum smoothing is essential to remove randomness. The signal-to-noise ratio in random waveforms can be improved by averaging the waveforms, the signal-to-noise ratio being improved by a factor of $\sqrt{K}$ if K waveforms are averaged. Thus one method of improving the accuracy of the estimated spectrum consists of dividing the data into K equal length sections, determining the spectrum of each section, and then averaging the spectra. In this way the average amplitude and average phase of each harmonic frequency component of the K spectra are obtained and plotted to yield the average amplitude and phase spectra. The accuracy of the spectra may be obtained in terms of their variance. For example the smaller the variance of the power spectral density the more accurate its estimate. It is therefore important to know what is the effect of a spectrum estimation method on the variance of the spectrum. Spectrum estimation by averaging is discussed in Section 11.3.2.1 in which it is explained that spectra estimated as the averages of the spectra of K data sections have a lower variance than those computed directly, with the variance decreasing in proportion to the number of sections taken. Even if the noise content of the waveform is low, that is high signal-to-noise ratio, the averaged result from K sections still results in a significant improvement in accuracy by this modified periodogram method. However, sectioning the data results in fewer data per FFT and consequently a coarser spectrum. This disadvantage may be overcome by the addition of augmenting zeros (Section 11.3.1.2). It is therefore always necessary to bear in mind the opposing claims of the accuracy of the estimate and the required spectral smoothness, and to design for the best compromise. Another approach to smoothing the periodogram is to calculate it from the DFT of the windowed auto-correlation function of the data. This is the Blackman–Tukey method as described in Section 11.3.5. Since the autocorrelation function of the data consists of the average of the sums of products of the data by themselves at different lags (Section 5.2), the signal-to-noise ratio is improved (Section 5.2.2). The Blackman–Tukey method results in a spectrum with a larger quality factor than the modified periodogram methods.

Data windows also exert a smoothing effect on spectra. In particular, windows with small side lobes in the frequency domain filter out noise which falls outside the main lobe and will offer improved smoothing. In fact this type of spectral smoothing is sometimes performed by convolving the spectrum of the data with that of the chosen window function.

Parametric and more recent methods of spectrum estimation are discussed in Section 11.4. These methods are less of an art than the nonparametric methods and may be automated. The concepts of forming models of the data in terms of model parameters and of obtaining the spectra from the frequency response functions of linear systems in terms of these models are outlined as are more recent methods.

11.3   Traditional methods

In Section 11.3.1 various pitfalls found in spectral analysis are detailed and explanations of how they may be overcome are provided. The technique of windowing and the properties of data windows are described in Section 11.3.2. Some properties of spectra and spectral smoothing are described in Section 11.3.3.

11.3.1   Pitfalls

11.3.1.1   Sampling rate and aliasing

Before the spectral analysis can be carried out the first procedure must be to pass the analog signals through an anti-aliasing filter, the function of which is to prevent aliasing of the sampled signal after the following stage of analog-to-digital conversion. Aliasing refers to distortion of the signal spectrum by the introduction of spurious low frequency components owing to a combination of an inadequate anti-aliasing filter and too low a sampling rate. This topic is fully discussed and explained in Section 2.2.

11.3.1.2   Scalloping loss or picket-fence effect

As explained in Section 3.2, the discrete Fourier transform (DFT) consists of harmonic amplitude and phase components regularly spaced in frequency. The spacing of the spectral lines decreases with the length of the sampled waveform. If, therefore, there is a signal component which falls between two adjacent harmonic frequency components in the spectrum then it cannot be properly represented. Its energy will be shared between neighbouring harmonics and the nearby spectral 'amplitudes' will be distorted. The amplitude density spectrum for a uniform spectral density signal is shown in Figure 11.1. Note the finite width of the main lobes occurring at the harmonic frequencies and that a signal component at a nonharmonic frequency such as f_{nh} cannot be properly represented. A solution to this difficulty lies in arranging for the harmonic components to be more closely spaced and coincident with the signal frequencies. This may be achieved by adding additional data in the form of zeros to the true data. The added zeros are termed augmenting zeros and serve to increase the fidelity of the estimated spectrum to the true spectrum without adding additional information. Sufficient augmenting zeros, N', are added to the N data to satisfy the simultaneous requirements that

$$N + N' = 2^m \tag{11.8}$$

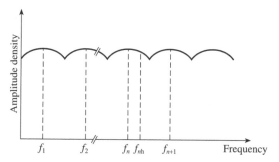

Figure 11.1   The amplitude density spectrum for a signal of uniform spectral density.

for a radix-2 fast Fourier transform (FFT) algorithm, where m is an integer, and that harmonics of the frequency $1/(N + N' - 1)T$ in which T represents the sampling interval coincide with the signal frequencies.

The scalloping loss, SL, is defined to represent the maximum reduction in processing gain which occurs mid-way between the harmonically related frequencies (Harris, 1978):

$$SL = \frac{|W(\omega_s/2N)|}{W(0)} = \frac{\left| \sum_{n=0}^{N-1} w(nT) \exp(-j\pi n/N) \right|}{\sum_{n=0}^{N-1} w(nT)} \qquad (11.9)$$

where W represents the DFT of the window function (Section 11.3.2), $\omega_s = 2\pi f_s$ is the angular sampling frequency, f_s is the sampling frequency, N is the number of data, n is the datum number, and $w(nT)$ is the time domain sampled window function.

As already stated, the effect of finite data length is to limit the achievable frequency resolution to $1/(N - 1)T$ (Hz). This results in a coarse spectrum which may be rendered smooth and continuous by supplementing the data with augmenting zeros. The process is simply one of interpolation of the spectral curve between adjacent harmonics. A real improvement in resolution can only be achieved if a longer realization is available. The frequency interval between the lines of the spectrum becomes $1/(N + N' - 1)T$ (Hz) on addition of N' augmenting zeros.

11.3.1.3   Trend removal

Any trends in the data must be removed prior to computation of the spectrum because error terms owing to addition of the trend to the data will be integrated and may produce large errors in the estimated spectrum.

11.3.1.4   Spectral leakage and spectral smearing

The FFT of a set of sampled data is not the true FFT of the process from which the data was obtained. This is because the process is continuous whereas the data is the sampled values of a realization which is truncated at its beginning and end. The data which represents a length T_s (s) of the signal is effectively obtained by multiplying all the sampled values in the interval T_s by unity, while all values outside this interval are multiplied by zero. This is equivalent to multiplying, or windowing, the signal by a rectangular pulse, or window, of width T_s and height 1. In this case the sampled data values $v(n)$ are given by the product of the data values, $s(n)$, and the values of the window function, $w(n)$:

$$v(n) = w(n)s(n) \tag{11.10}$$

This time-domain product is equivalent to a convolution in the frequency domain; see Sections 3.3 and 5.3. Thus the value of the FFT for the nth harmonic is

$$V(\omega_n) = \sum_{k=-N}^{N} W(\omega_n - \omega_k)S(\omega_k) \tag{11.11}$$

where ω_n is the angular frequency of the nth harmonic, $V(\omega_n)$ is the complex DFT component at frequency ω_n, $W(\omega_n)$ is the DFT of the window at frequency ω_n, and $S(\omega_k)$ is the true DFT component of the signal at frequency ω_k.

Equation 11.11 shows that the computed spectrum consists of the true spectrum of the data convolved with that of the window function. The amplitude spectrum of the rectangular pulse $S_R(\omega_n)$ is given by the following expression, known as a Dirichlet kernel:

$$S_R(\omega_n) = \frac{T_s \sin(\omega_n T_s/2)}{\omega_n T_s/2} = \text{Sa}\left(\frac{\omega_n T_s}{2}\right) \tag{11.12}$$

$\text{Sa}(\omega_n T_s/2)$ is the sampling function of $\omega_n/2$ (see Sections 3.1.1 and 3.1.2) and was illustrated in Figure 3.2(a). It is seen to consist of a main lobe and an infinite number of side lobes peaking at 0 Hz and $(n + 0.5)/T_s$ Hz respectively. Now the amplitude spectrum of a single sine wave component of the signal at frequency f_n comprises two impulses at frequencies $\pm f_n$. Convolution by the sampling function produces the spectrum of Figure 11.2. The two impulses have been transformed into two

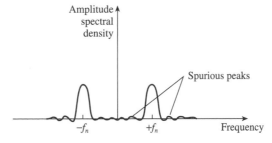

Figure 11.2 Amplitude spectral density of a sinusoidal signal convolved with the sampling function.

overlapping sampling functions. The effect of the rectangular window has been to introduce spurious peaks into the computed spectrum owing to the effect of the side lobes. This will be true of each frequency component of the signal, and so the amplitude spectrum of the signal will be distorted by the addition and subtraction of the large number of window side lobes and main lobes. The effect may be to introduce spurious peaks, or to conceal true peaks in the spectrum, the phenomenon being known as spectral leakage. To avoid this it is necessary to modify the data by multiplying them by a window shaped to reduce the side lobe effect. Suitable windows have a value of 1 at the mid-data point and are tapered to zero at points $n = 0$ and $n = N - 1$. At least 23 such windows have been developed and their relative suitabilities have been investigated by Harris (1978).

In order to minimize spectral leakage the window shape is chosen to minimize its side lobe levels. Unfortunately, this has the effect of increasing the main lobe width, causing it to spread into the adjacent side lobes. This therefore aliases the side lobes. This is repeated at each harmonic frequency, the overall result being an aliasing of the signal spectrum, known as smearing. Windows and their parameters therefore have to be carefully chosen in order to strike the optimum balance between frequency resolution and the statistical accuracy of the spectral estimate.

11.3.2   Windowing

Various properties of windows are described in this section, basically in the time domain. However, it should again be noted that windowing may be performed either in the time domain (data windows) or in the frequency domain (frequency windows) because of the equivalence between multiplication in the time domain and convolution in the frequency domain (Section 11.3.1.4). Frequency domain windowing may thus be carried out by convolving the frequency domain window with the signal spectrum. This procedure may be achieved by applying Equation 5.104.

11.3.2.1   Properties of windows

Equivalent noise bandwidth

It was seen in Section 11.3.1.4 that owing to the phenomenon of spectral leakage the impulse functions which theoretically represent the amplitude spectral densities become sampling functions. That is, the infinitely narrow spectral components are replaced by the broader bandwidth sampling function. The side lobes of these functions which bias the signal components may be regarded as contributing unwanted noise to the signal, and the frequency window may be regarded as a broadband filter. It is therefore desirable from this point of view to design the windows to have a low noise bandwidth by reducing the side lobe amplitudes. This noise bandwidth is measured for comparison purposes between different windows by the equivalent noise bandwidth. This is defined to be the bandwidth of an ideal rectangular filter which passes the same amount of applied white noise as the spectral filter in question

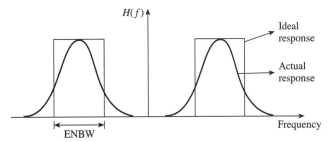

Area under ideal response curves = area under actual response curves

Figure 11.3   Equivalent noise bandwidth of a filter.

(see Figure 11.3). With this definition it is possible to compare the side lobe properties of different windows by comparing their equivalent noise bandwidths. Thus equivalent noise bandwidth is an important window parameter. The smaller it is then the better the window from this point of view.

The equivalent noise bandwidth is given by

$$\text{ENBW} = \frac{\displaystyle\sum_{n=0}^{n=N-1} w^2(nT)}{\left[\displaystyle\sum_{n=0}^{n=N-1} w(nT)\right]^2} \tag{11.13}$$

Overlap correlation

When data is windowed the ends of the data sequence are tapered to zero. This represents a loss of information. In particular, short duration events occurring in the tapered region may be missed. One approach to solving this problem is to partition the data sequence into overlapping sections and to window and transform each section separately. If the overlap is about 50% to 75%, then most features of the data will be included in the sequences. The resulting spectra are then averaged to obtain an estimate of the true spectrum. The partitioning of the data is illustrated in Figure 11.4. This procedure is called redundancy or overlap processing. Generally 50–75% overlap processing provides 90% of the possible performance improvement for most weighting functions (DeFatta *et al.*, 1988). By averaging the spectra of the sections the variance of the spectrum is reduced. For K statistically identical but independent measurements the variance of the average is $1/K$ times the variance of the individual spectrum. However, this is not true when the spectra of overlapped sections are averaged because of the correlation between them. For the cases of 50% and 75% overlap the variance of the averaged spectrum to the individual spectrum is given in Harris (1978), from which it can be shown that, for example, the averaging of four spectra reduces the variance to 25% of that of the original spectrum. This represents a significant improvement in the estimate of the spectrum.

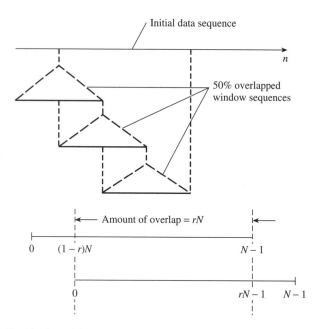

Figure 11.4   Partitioning of the data for overlap processing.

Processing gain

The processing gain, PG, is defined as the ratio of the output signal-to-noise ratio subsequent to windowing to the input signal-to-noise ratio prior to windowing:

$$PG = \frac{(S/N)_{\mathrm{O/P}}}{(S/N)_{\mathrm{I/P}}} \tag{11.14}$$

The processing gain depends on the shape of the window since this determines its equivalent noise bandwidth (see the previous section on this subject). The taper of the window reduces the signal power causing a processing loss, while the side lobes enhance the noise bandwidth.

Worst-case processing loss

The worst-case processing loss (WCPL) is defined to be the sum (in decibels) of the maximum scalloping loss of a window and its processing loss (PL). It represents the reduction in output signal-to-noise ratio owing to windowing and worst-case frequency location. It always lies between 3.0 and 4.3 dB. Windows for which it exceeds 3.8 dB should be avoided. These include the rectangular, Poisson ($\alpha = 4$), Hanning–Poisson ($\alpha = 2.0$), Cauchy ($\alpha > 4$), and minimum four-sample Blackman–Harris windows.

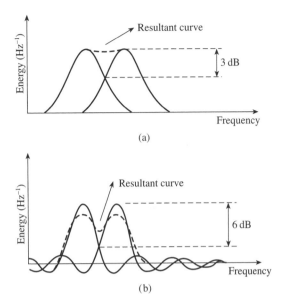

Figure 11.5   Minimum resolution bandwidth. (a) Resolution of spectral peaks determined by
3 dB bandwidths. (b) Spectral resolution of DFT peaks determined by 6 dB
bandwidths.

Minimum resolution bandwidth

Normally when two identical spectral peaks overlap they may be resolved provided
that they do not overlap beyond their 3 dB points (Figure 11.5(a)). However, in
the case of DFT components the adjacent spectral components are weighted by the
window and then summed coherently, that is the side lobes are included in the
summation. The gain of each component at cross-over must not exceed 0.5. This
means that the spectral resolution is determined by the 6 dB bandwidth of the
components rather than the 3 dB bandwidth (Figure 11.5(b)).

Biasing effect of data windows

Multiplication of the data by a tapered window reduces the sample magnitudes at
the taper and hence the total signal power. It can be shown that each frequency
component is affected identically by the window and that the multiplying factor is
proportional to the square root of the coherent power gain, which represents the
normalized power of the data window regarded as a voltage waveform. Thus the
reduction in signal power may be restored without distorting the power density
spectrum. The window also adjusts the mean level of the data, thereby increasing the
apparent energy of the lower frequency components in the spectrum. Some means of
compensating for this is required but the straightforward subtraction of the mean of
the windowed data results in pronounced high frequency side lobes.

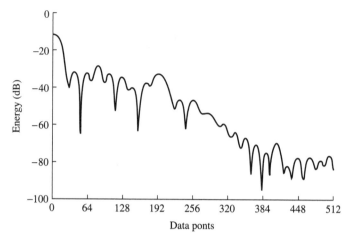

Figure 11.6   Energy spectrum of 1 s interstimulus interval (ISI) CNV, Kaiser–Bessel window. 64 data points, 1024-point FFT.

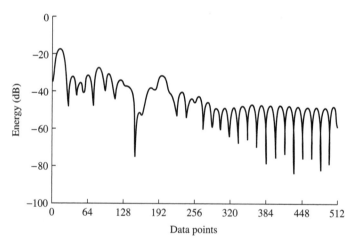

Figure 11.7   Energy spectrum of 1 s ISI CNV, Kaiser–Bessel window. 64 data points, 1024-point FFT. Mean level of windowed data removed by subtraction of mean from data.

Figure 11.6 shows the energy density spectrum of 64 data points from which the mean level of the data had first been subtracted after which the data had then been multiplied by the corresponding values of a Kaiser–Bessel window. It can be seen that the spectrum contains low frequency components introduced by the window-ing process. Figure 11.7 shows the spectrum obtained when the mean level of the windowed data had first been subtracted. The low frequency components are seen to have been removed but at high frequencies a marked side lobe structure has become apparent. The following example shows how the deleterious effects of windowing may be overcome.

Example 11.1   Show that the twin effects of the window in reducing the signal energy and in introducing low frequency components into the spectrum may be overcome by windowing a linear function of the data rather than the data (Jervis *et al.*, 1989a).

Solution   Assume the initial data $s(n)$ has zero mean. Let the mean level introduced into the data by windowing be removed by subtracting a constant k_1 from $s(n)$. The new windowed data values are now given by $s^1(n)$ where

$$s^1(n) = w(n)[s(n) - k_1] \tag{11.15}$$

in which the $w(n)$ are the window function values or weights. The reduction in signal power caused by windowing may now be restored by multiplying each value $s^1(n)$ by a carefully selected constant k_2. Thus the data points are transformed to become

$$S(n) = k_2 w(n)[s(n) - k_1] \tag{11.16}$$

The necessary values of k_1 may be found from the condition that the average value of $S(n)$ must be zero. Therefore

$$\sum_{n=0}^{N-1} S(n) = 0$$

Hence

$$k_2 \left[\sum_{n=0}^{N-1} w(n)s(n) - \sum_{n=0}^{N-1} w(n)k_1 \right] = 0$$

Therefore

$$k_1 = \frac{\displaystyle\sum_{n=0}^{N-1} w(n)s(n)}{\displaystyle\sum_{n=0}^{N-1} w(n)} \tag{11.17}$$

The normalized ac power of the data prior to windowing is

$$E[(s(n) - k_1)^2] = \sigma_{sN}^2 \tag{11.18}$$

where E denotes the expected value and σ_{sN}^2 is the variance of $s(n)$ with mean k_1. The normalized ac power of the windowed data is

$$E\{k_2^2 w^2(n)[s(n) - k_1]^2\}$$

where $w(n)$ and $s(n)$ are mutually independent. Now

$$E\{k_2^2 w^2(n)[s(n) - k_1]^2\} = E[k_2^2 w^2(n)]E\{[s(n) - k_1]^2\}$$
$$= k_2^2 E[w^2(n)]\sigma_{sN}^2 \qquad (11.19)$$

The value of k_2 required to make the powers of the windowed and unwindowed data the same may be obtained by equating Equations 11.18 and 11.19:

$$\sigma_{sN}^2 = k_2 E[w^2(n)]\sigma_{sN}^2$$

so

$$k_2^2 = \frac{1}{E[w^2(n)]} = \frac{1}{(1/N)\sum\limits_{n=0}^{N-1} w^2(n)} = \frac{N}{\sum\limits_{n=0}^{N-1} w^2(n)}$$

Therefore

$$k_2 = \left[\frac{N}{\sum\limits_{n=0}^{N-1} w^2(n)}\right]^{1/2} \qquad (11.20)$$

Substituting Equations 11.17 and 11.20 into Equation 11.16 finally gives

$$S(n) = w(n)\left[s(n) - \frac{\sum\limits_{n=0}^{N-1} w(n)s(n)}{\sum\limits_{n=0}^{N-1} w(n)}\right]\left[\frac{N}{\sum\limits_{n=0}^{N-1} w^2(n)}\right]^{1/2} \qquad (11.21)$$

Figure 11.8 shows the resulting energy spectrum when the mean of the windowed data is removed and the signal energy restored using Equation 11.21. It can be seen that the dc level owing to windowing and the side lobe effect have both been removed.

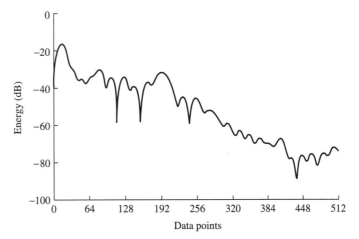

Figure 11.8   Energy spectrum of 1 s ISI CNV, Kaiser–Bessel window. 64 data points, 1024-point FFT. Data pre-processed to remove windowed mean level and to maintain mean power.

The recommended procedure is therefore to modify the data $s(n)$ according to Equation 11.21 prior to computation of its DFT. This is equivalent to first subtracting k_1 from the data and then multiplying the difference by k_2 prior to windowing.

11.3.2.2   Window choice

Harris (1978) has considered in detail the effects of the different window characteristics on window performance and concluded that the major influences on window quality are the highest side lobe level and the worst-case processing loss. The preferred windows are then the Blackman–Harris, Dolph–Chebyshev, and Kaiser–Bessel windows. The well-known Tukey (cosine-tapered), Poisson, Hanning, and Hamming windows all have inferior performance.

The taper on many windows and their shape may in many cases be adjusted by selecting the value of a parameter, α. The effect of this is to adjust the main lobe width and the side lobe level. Part of the art of windowing is to select by trial and error the value of α which optimizes the results in a particular application.

Example 11.2

The effect of different windows on an amplitude spectrum   Figure 11.9(a) shows the DFT components of two sine waves, differing in amplitude by 40 dB, and of frequencies $100f$ and $120f$, where f is the first harmonic frequency corresponding to a record length of T_s (s), and which were obtained without windowing, that is with a rectangular window. When the signal is harmonically related to the window length in this way the signal appears periodic and infinite and is faithfully reproduced even by a rectangular window. This periodic relationship was broken to yield the results of Figure 11.9(b) by changing the frequency of the larger signal to $102.5f$ which is not a

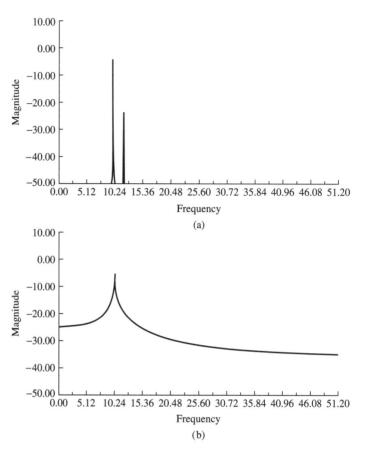

Figure 11.9   Amplitude spectra of two sine waves differing in amplitude by 40 dB. (a)
Window length is an integer multiple of both periods. (b) Window length is
not an integer multiple of one of the periods (the larger period).

harmonic frequency. It can be seen that the result has been a big increase in side lobe
level which almost masks out the smaller signal. This effect may be largely overcome
by the choice of a suitable window. Figures 11.10(a) and 11.10(b) show tapered
cosine (Tukey) windows with $\alpha = 0.1$ and $\alpha = 0.5$ respectively while Figures 11.11(a)
and 11.11(b) show the respective spectra. Figures 11.12(a) and 11.12(b) show a
Hamming window ($\alpha = 0.54$) and the resulting spectrum respectively. Figures
11.13(a)–11.13(c) show Kaiser–Bessel windows with $\alpha = 2.0$, 3.0, and 4.0 respectively while Figure 11.14 illustrates the corresponding spectra. The result nearest the
true spectrum is given by the Kaiser–Bessel window with $\alpha = 4.0$. However, it should
be noted that as α is increased to decrease the side lobe level the main lobe width
increases. A compromise value of α must be sought. Of course in a real situation all
the harmonics will be affected so that window selection with respect to multitone
performance is very important if spurious results are to be avoided.

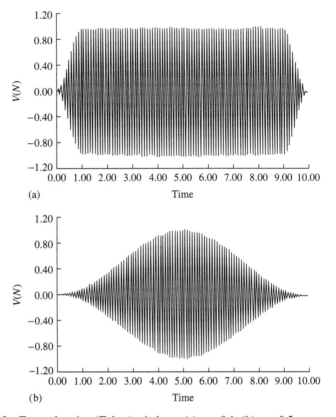

Figure 11.10   Tapered cosine (Tukey) windows. (a) $\alpha = 0.1$. (b) $\alpha = 0.5$.

The Dolph–Chebyshev window gives the best results with regard to low side lobes and worst-case processing loss but the coherent addition of its side lobes detracts from its multitone detection performance. Also, its side lobe structure exhibits extreme sensitivity to coefficient errors. Therefore the Blackman–Harris or Kaiser–Bessel windows are to be preferred. The Kaiser–Bessel window has the advantages that its coefficients are easier to generate and the side lobe level versus main lobe width compromise is easier to adjust by varying α. The expression for the Kaiser–Bessel window is (Kuo and Kaiser, 1966)

$$w(n_{\text{KB}}) = I_0 \left\{ \pi\alpha \left[1.0 - \left(\frac{n_{\text{KB}}}{N/2} \right)^2 \right]^{1/2} \right\} \Big/ I_0(\pi\alpha), \quad 0 \leqslant |n_{\text{KB}}| \leqslant N/2 \qquad (11.22)$$

in which n_{KB} is the window function sample number, α is a numerical parameter which may be adjusted to select the best side lobe level versus main lobe width compromise, N is the number of sample points in the window, and

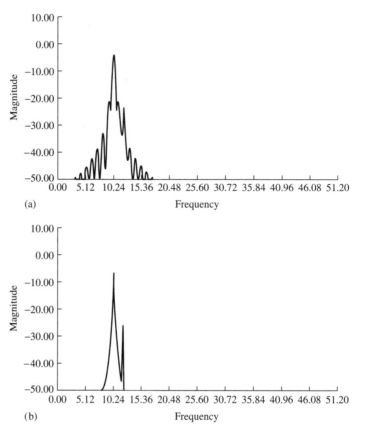

Figure 11.11 Amplitude spectra of the two sine waves multiplied by the tapered cosine windows.

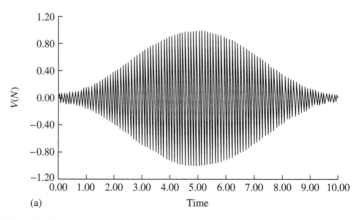

Figure 11.12 (a) Hamming window, $\alpha = 0.54$.

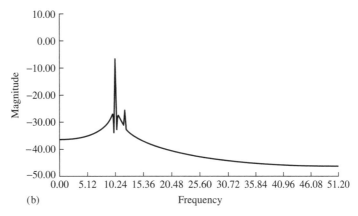

(b)

Figure 11.12   (b) The corresponding amplitude spectrum of the two sine waves.

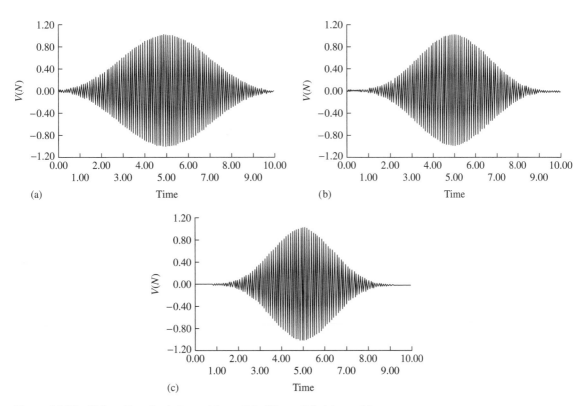

Figure 11.13   Kaiser–Bessel windows. (a) $\alpha = 2.0$. (b) $\alpha = 3.0$. (c) $\alpha = 4.0$.

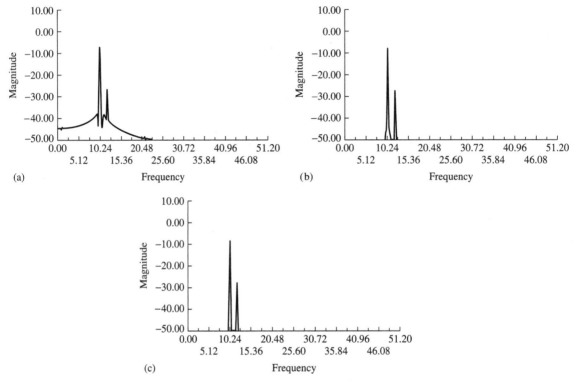

(a)

(b)

(c)

Figure 11.14 The amplitude spectra of the two sine waves calculated using the corresponding Kaiser–Bessel windows of Figure 11.13. (a) $\alpha = 2.0$. (b) $\alpha = 3.0$. (c) $\alpha = 4.0$.

$$I_0(x) = \sum_{k=0}^{K} \left[\frac{(x/2)^k}{k!} \right]^2 \tag{11.23}$$

is a zero-order modified Bessel function of the first kind, and K is theoretically infinite but, because the magnitude of the Bessel function decreases rapidly with k, it is usually adequate to set $K = 32$.

Equation 11.22 defines the window between the points $-N/2$ and $N/2 - 1$. The DFT is normally required to extend from $n_{\mathrm{DFT}} = 0$ to $n_{\mathrm{DFT}} = N - 1$ where n_{DFT} is the DFT datum number. Therefore for use with the DFT the Kaiser–Bessel window has to be shifted right $N/2$ places to satisfy

$$n_{\mathrm{DFT}} = n_{\mathrm{KB}} + N/2$$

or $\hfill (11.24)$

$$n_{\mathrm{KB}} = n_{\mathrm{DFT}} - N/2$$

Hence Equation 11.22 must be modified to

$$w(n_{\text{DFT}}) = I_0 \left\{ \pi \alpha \left[1.0 - \left(\frac{n_{\text{DFT}} - N/2}{N/2} \right)^2 \right]^{1/2} \right\} \Big/ I_0(\pi \alpha), \quad 0 \leqslant n_{\text{DFT}} \leqslant N - 1$$

(11.25)

11.3.3 The periodogram method and periodogram properties

The square of the modulus of the Fourier transform, $|F(\text{j}\omega)|^2$, is the estimated power spectral density, $E[P(f)]$, and is also known as the periodogram. $E[P(f)]$ can be shown (Proakis and Manolakis, 1989; DeFatta *et al.*, 1988) to be given by

$$E[P(f)] = \sum_{m=-\infty}^{\infty} w_{\text{B}}(m) c_{xx}(m) \exp(-\text{j}2\pi fm)$$

(11.26)

in which $c_{xx}(m)$ is the autocovariance of $x(n)$ evaluated at lag m, f is the frequency, and $w_{\text{B}}(m)$ is the triangular (Bartlett) window defined by

$$w_{\text{B}}(m) = 1 - \frac{|m|}{N} \quad |m| \leqslant N - 1$$

(11.27)

By comparison the true power spectral density, $P(f)$, of $x(n)$ is

$$P(f) = \sum_{m=-\infty}^{\infty} c_{xx}(m) \exp(-\text{j}2\pi fm)$$

(11.28)

The power spectral density given by the periodogram is therefore biased with the bias given by

$$P(f) - E[P(f)] = \sum_{m=-\infty}^{\infty} [1 - w_{\text{B}}(m)] c_{xx}(m) \exp(-\text{j}2\pi fm)$$

$$= \frac{|m|}{N} P(f)$$

(11.29)

For $N \gg |m|$ the bias becomes small and $E[P(f)] \rightarrow P(f)$, that is the periodogram is asymptotically unbiased. Also, for large N the variance of the periodogram becomes

$$\text{var}[P(f)] \approx FP^2(f)$$

(11.30)

where F depends on the window function used, that is the variance depends on the square of the power spectral density and does not converge towards zero with increasing N. This means that power spectral density estimates obtained from the periodogram are inconsistent and yield fluctuating estimates of $P(f)$ from successive realizations.

Note that the autocovariance is usually determined by averaging over N terms rather than the alternative $N - |m|$ terms. Both estimates are consistent and asymptotically unbiased, but the former has a smaller variance which is why it is preferred.

Furthermore, when the DFT is applied to obtain the spectrum, the corresponding periodogram is defined as $(1/N)|X(k)|^2$ and has the dimension of normalized energy although the reader may note that some authors still refer to the function $(1/N)|X(k)|^2$ as the power spectral density.

11.3.4   Modified periodogram methods

Welch (1967) suggested that the inconsistency of the periodogram method could be overcome by averaging a number of modified periodograms. Each of the latter consists of a section of the data. These sections may be sequential (the Bartlett method) or overlapped (the Welch method). The methods also reduce the variance of the power spectral density estimates. This is demonstrated for the Bartlett method in Example 11.3.

11.3.4.1   The Welch method

An advantage of the Welch method over the Bartlett method is that the variance of the power spectral density is further reduced. However, this is at the expense of a further decrease in spectral resolution. In the Welch method L data sections of length M are overlapped and the periodograms are computed from the L windowed data sections. Also, the periodograms are normalized by the factor U to compensate for the loss of signal energy owing to the windowing procedure. In fact U equates to $1/k_2^{1/2}$ where k_2 is the factor derived in part of Section 11.3.2.1 on the biasing effect of data windows as necessary to compensate for this reduction in signal energy. Thus,

$$U = \frac{1}{M} \sum_{n=0}^{M-1} w^2(n) \tag{11.31}$$

The Welch power density spectral estimate, $P_{\text{WE}}(f)$, is therefore

$$P_{\text{WE}}(f) = \frac{1}{L} \sum_{j=0}^{L-1} P_j(f) \tag{11.32}$$

The expected value of the Welch estimate is

$$E[P_{\text{WE}}(f)] = \frac{1}{L} \sum_{j=0}^{L-1} E[P_j(f)] = E[P_j(f)] \tag{11.33}$$

that is the same as the expected value of the modified periodogram. It can be shown (Proakis and Manolakis, 1989) that as $N \to \infty$ and $M \to \infty$ the value converges to that of the true power spectral density, $P(f)$. Thus for large N and M the Welch power spectral density estimate is unbiased. Under the same conditions the variance of the Welch estimate converges to zero, that is the estimate is consistent. Welch showed that for the case of no overlap $(L = K)$

$$\text{var}\,[P_{\text{WE}}(f)] \approx (1/K)P^2(f)$$

which is equal to that of the Bartlett variance under the same conditions. For 50% overlap $(L = 2K)$

$$\text{var}\,[P_{\text{WE}}(f)] \approx (9/8L)P^2(f)$$

which is less than that for the Bartlett case by the factor $9/16 = 0.56$.

11.3.5 The Blackman–Tukey method

It was established in Chapter 3 that the power density spectrum is given by the DFT of the autocorrelation function of the data. One might query the usefulness of this approach knowing that the periodogram can be computed directly from the data as the square of the DFT. Well, first it is noteworthy that the Blackman–Tukey method was introduced in 1958 (Blackman and Tukey, 1958) while the FFT algorithm for fast computation of the DFT was not published by Cooley and Tukey until 1965 (Cooley and Tukey, 1965). Second, it is possible that the Blackman–Tukey approach might contain some advantage over the periodogram method. Indeed, it will be shown in the next section that the Blackman–Tukey method is characterized by a larger quality factor. In addition autocorrelation functions may now be computed using DFTs by the fast correlation method (Section 11.3.6). The Blackman–Tukey procedure then is

(1) to calculate the autocorrelation function of the data,
(2) to apply a suitable window function to the data, and
(3) to compute the FFT of the resulting data to obtain the power density spectrum.

By comparison with the periodogram method we see that the smoothing is achieved by the averaging effect of the autocorrelation process rather than by the averaging of several periodograms.

The autocorrelation function is windowed to taper it towards its extremes because at larger lags fewer data points enter the computation so these estimates are less accurate. Tapering has the effect of attaching less weight to these estimates.

The Blackman–Tukey estimate, $P_{\text{BTE}}(f)$, is

$$P_{\text{BTE}}(f) = \sum_{m=-(M-1)}^{M-1} r_{xx}(m)w(m)\exp\left(-j2\pi fm\right) \tag{11.34}$$

where $r_{xx}(m)$ is the autocorrelation function of the data and $w(n)$ is the window function of length $2M - 1$ and is zero for $|m| \geqslant M$.

In order to obtain real estimates $w(n)$ must be symmetrical about $m = 0$, and for the estimates to be positive its transform must be positive. Not all windows satisfy these criteria. The Hanning and Hamming windows are examples of two which do not.

It may be shown that the expected value of the Blackman–Tukey estimate is

$$E[P_{\text{BTE}}(f)] = \sum_{m=-(M-1)}^{M-1} c_{xx}(m) w_{\text{B}}(m) \exp(-\text{j}2\pi fm) \tag{11.35}$$

in which $w_{\text{B}}(m)$ is the triangular Bartlett window.

The condition $M < N$ has to be satisfied to obtain additional smoothing of the spectrum. If $N \gg m$ the estimate will be asymptotically unbiased. Also, if $W(k)$, the DFT of the window, $w(n)$, is narrower than $P(f)$, the true power-density spectrum, then

$$\text{var}\,[P_{\text{BTE}}(f)] \approx P^2(f)\left[\frac{1}{N} \sum_{m=-(M-1)}^{M-1} w^2(m)\right] \tag{11.36}$$

and as $N/M \to \infty$, var $[P_{\text{BTE}}(f)] \to 0$, showing that under these conditions the Blackman–Tukey estimate is consistent.

11.3.6   The fast correlation method

It was pointed out in Section 5.3.7 that if in excess of 128 data are to be correlated the computation is quicker if use is made of the correlation theorem (Equation 5.63) to implement the calculations using FFTs. For example, this yields a tenfold speed increase if $N = 1024$. In addition, if large amounts of input data are involved such as may exceed the memory capacity of the system then the overlap–add or overlap–save sectioning techniques may be applied (Sections 5.3.8–5.3.10). When the autocorrelation in the Blackman–Tukey method is computed using FFTs in these ways the method is known as the fast correlation method for spectral estimation.

11.3.7   Comparison of the power spectral density estimation methods

A quality factor for estimates of power spectral density was given in Equation 11.6. It can be shown (Proakis and Manolakis, 1989) that the quality factors for the four nonparametric spectral analysis methods are as given in Table 11.1 where f is the 3 dB main lobe width of the associated windows. It is seen that the Blackman–Tukey method is superior for quality, and that, with the exception of the periodogram method, the quality may be maintained as the frequency resolution is increased (decrease in f) by increasing N.

Table 11.1   Quality factors Q for power spectral density estimates.

Estimation method	Conditions	Q	Comments
Periodogram	$N \to \infty$	1	Inconsistent, independent of N
Bartlett	$N, M \to \infty$	$1.11Nf$	Quality improves with data length
Welch	$N, M \to \infty$, 50% overlap	$1.39Nf$	Quality improves with data length
Blackman–Tukey	$N, M \to \infty$, triangular window	$2.34Nf$	Quality improves with data length

Great care and some trial computations are required to ensure satisfactory results. On balance the Blackman–Tukey method would seem to be marginally the best, but other considerations may lead to a preference for one of the other methods.

11.4   Modern parametric estimation methods

The nonparametric methods described in the previous sections of this chapter which utilize periodograms and FFTs are subject to the aforementioned limitations of low spectral resolution in the case of short records and the requirement for windowing to reduce the spectral leakage. These difficulties may be overcome by parametric methods (Burg, 1968; Nuttall, 1976; Ulrych and Clayton, 1976; Marple, 1980; Cadzow, 1979, 1982; Graupe et al., 1975; Kay, 1980; Friedlander, 1982). The price to be paid is an extensive investigation of an appropriate model for each process, a determination of the necessary order of the chosen model for adequate representation of the data (Whittle, 1965; Jenkins and Watts, 1968; Box and Jenkins, 1976; Chatfield, 1979; Akaike, 1969, 1973, 1974, 1978, 1979; Shibata, 1976; Rissanen, 1983), and computation of the model parameters (Proakis and Manolakis, 1989; Makhoul, 1975; Levinson, 1947; Durbin, 1959; Priestley, 1981; Wold, 1954; Chatfield, 1984). The advantages gained are increased spectral resolution, applicability to short data lengths, and avoidance of spectral leakage, scalloping loss, spectral smearing, and window biasing effects. Because of the importance of these parametric techniques, the most commonly used method of autoregressive modelling is presented. Nevertheless, although an improvement over the nonparametric methods described, these parametric methods do have some disadvantages which may be avoided by alternative modern approaches such as the sequential or adaptive (Friedlander, 1982; Kalouptsidis and Theodoridis, 1987) and the maximum likelihood methods (Capon, 1969; Lacoss, 1971).

To summarize, the parametric approach calls for parametric modelling of the data, a well-established branch of time series analysis (Jenkins and Watts, 1968; Box and Jenkins, 1976; Priestley, 1981), combined with an interpretation of the data as being the output of a linear system excited by white noise. This system is represented by a polynomial transfer function expressed in terms of the model parameters. The spectrum of the data is computed from this transfer function.

11.5   Autoregressive spectrum estimation

In this method the digitized signal is modelled as an autoregressive (AR) time series plus a white noise error term. The spectrum is then obtained from the AR model parameters and the variance of the error term. The model parameters are found by solving a set of linear equations obtained by minimizing the mean squared error term (the white noise power) over all the data. There are a number of ways of solving the equations, which are described below. An important consideration is the choice of the number of terms in the AR model. This is known as its order. If the order is too low the power density estimate will be excessively smoothed, so some peaks may be obscured. If the order is too high, spurious peaks may be introduced. Hence, it is important to determine the appropriate model order for each set of data, and so this topic is discussed. The method is applicable to signals from the power density spectra which exhibit sharp peaks. Other models such as the moving average or the autoregressive moving average models would have to be used if the signals did not fulfil this criterion. Because the autoregressive approach results in solvable equations it is preferred whenever possible. The references provide more detailed information (Pardey, Roberts and Tarassenko, 1996; Kay, 1988; Candy, 1989; Proakis and Manolakis, 1989; Marple, 1987; Clarkson, 1993).

11.5.1   Autoregressive model and filter

In an AR model of a time series the current value of the series, $x(n)$, is expressed as a linear function of previous values plus an error term, $e(n)$, thus:

$$x(n) = -a(1)x(n-1) - a(2)x(n-2) - \ldots - a(k)x(n-k) - \ldots$$
$$- a(p)x(n-p) + e(n) \tag{11.37}$$

This equation incorporates p previous terms and represents a model of order p. It is more compactly written as

$$x(n) = -\sum_{k=1}^{p} a(k)x(n-k) + e(n) = -\sum_{k=1}^{p} a(k)z^{-k}x(n) + e(n) \tag{11.38}$$

where z^{-k} is the back-shift operator which denotes a delay of k sampling intervals. Rewriting 11.38:

$$x(n) + \sum_{k=1}^{p} a(k)z^{-k}x(n) = \left(1 + \sum_{k=1}^{p} a(k)z^{-k}\right)x(n) = e(n) \tag{11.39}$$

whence

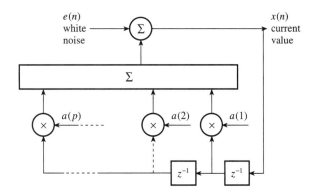

Figure 11.15   AR filter.

$$x(n) = \frac{e(n)}{1 + \displaystyle\sum_{k=1}^{p} a(k)z^{-k}} \tag{11.40}$$

By taking the ratio $x(n)/e(n)$ we have

$$\frac{x(n)}{e(n)} = \frac{1}{1 + \displaystyle\sum_{k=1}^{p} a(k)z^{-k}} = H(z) \tag{11.41}$$

Here, $H(z)$ is interpretable as the z-transform of an all-pole IIR digital filter with coefficients, $a(k)$. This filter is called an autoregressive (AR) filter, and is illustrated in Figure 11.15. From Equation 11.41 the $x(n)$ may be regarded as the outputs of this filter caused by the random inputs $e(n)$. $e(n)$ represents the error between the value predicted by the model, $\hat{x}(n)$, and the true datum value, $x(n)$. $e(n)$ is usually assumed to have the properties of white noise, i.e., it is assumed to have a Gaussian probability density distribution and a uniform power density spectrum. Thus $x(n)$ may be regarded as having been generated by the AR filter from a white noise source. The frequency response of the filter, $H(f)$, is obtained by substituting $z = e^{j\omega T}$ in Equation 11.39 (see Section 4.4), where ω is the angular frequency and T is the sampling period. Hence

$$H(f) = \frac{1}{1 + \displaystyle\sum_{k=1}^{p} a(k)e^{-jk\omega T}} \tag{11.42}$$

11.5.2 Power spectrum density of AR series

The power spectrum density, $P_x(f)$, of the AR series $x(n)$ is required. This is related to the power spectrum density of the white noise error signal, $P_e(f)$, which is its variance, $\sigma_e^2(n)$, by

$$P_x(f) = |H(f)|^2 P_e(f) = |H(f)|^2 \sigma_e^2(n) = \frac{\sigma_e^2(n)}{\left| 1 + \sum_{k=1}^{p} a(k)e^{-jk\omega T} \right|^2} \tag{11.43}$$

The variance of the white noise is its mean square value, which is the mean squared value of $e(n)$, subsequently referred to as E. The model parameters can be determined, and $\sigma_e^2(n)$ (or E) may be derived in terms of these parameters. Hence the power density spectrum may be found.

11.5.3   Computation of model parameters – Yule–Walker equations

The optimum model parameters will be those which minimize the errors, $e(n)$, for each sampled point, $x(n)$, represented by an equation like Equation 11.38. These errors are given by re-ordering 11.38 to

$$e(n) = x(n) + \sum_{k=1}^{p} a(k)x(n-k) \tag{11.44}$$

A measure of the total error over all samples, N $(1 \leqslant n \leqslant N)$, is required. Each error, $e(n)$, may be positive or negative, and so for a large number of sampled points the mean error tends to be small. Thus the mean error is an unsuitable measure of the accuracy of the model and so it is usual to quantify it in terms of the mean squared error, when all of the squared error terms are positive. The mean squared error is given by:

$$E = \frac{1}{N}\sum_{n=1}^{N} e^2(n) = \frac{1}{N}\sum_{n=1}^{N} \left(x(n) + \sum_{k=1}^{p} a(k)x(n-k) \right)^2 \tag{11.45}$$

We saw above that $E = \sigma_e^2(n)$. The requirement now is to select the model parameters to minimize E. The optimum value of each parameter is obtained by setting the partial derivative of Equation 11.45 with respect to the model parameter to zero. Thus, for the kth parameter, we have:

$$\frac{\partial E}{\partial a(k)} = \frac{2}{N}\sum_{n=1}^{N} \left(x(n) + \sum_{k=1}^{p} a(k)x(n-k) \right) \frac{\partial}{\partial a(k)} \sum_{k=1}^{p} a(k)x(n-k) = 0, \, 1 \leqslant k \leqslant p \tag{11.46}$$

Now

$$\frac{\partial}{\partial a(k)} \sum_{k=1}^{p} a(k)x(n-k) = x(n-k)$$

and so Equation 11.46 simplifies to

$$\frac{\partial E}{\partial a(k)} = \frac{2}{N}\sum_{n=1}^{N}\left(x(n) + \sum_{k=1}^{p} a(k)x(n-k)\right)x(n-k) = 0 \qquad (11.47)$$

giving for the kth parameter:

$$\frac{1}{N}\sum_{n=1}^{N}\left(\sum_{k=1}^{p} a(k)x(n-k)\right)x(n-k) = -\frac{1}{N}\sum_{n=1}^{N}x(n)x(n-k) \qquad (11.48)$$

Writing out the left-hand side of Equation 11.48 for the example case of $k = 1$ gives

$$\frac{1}{N}\sum_{n=1}^{N}(a(1)x(n-1) + a(2)x(n-2) + \ldots + a(p)x(n-p))x(n-1)$$

$$= \frac{1}{N}(a(1)x(0)x(0) + a(2)x(-1)x(0) + \ldots + a(p)x(1-p)x(0))$$

$$+ \frac{1}{N}(a(1)x(1)x(1) + a(2)x(0)x(1) + \ldots + a(p)x(2-p)x(1)) + \ldots$$

$$+ \frac{1}{N}(a(1)x(N-1)x(N-1) + a(2)x(N-2)x(N-1) + \ldots + a(p)x(N-p)x(N-1))$$

Inspection of the first terms on each line shows that they sum to the zero-lag autocorrelation function, $R_{xx}(0)$, of the series $x(n)$ multiplied by $a(1)$. Similarly the second terms sum to the lag-one autocorrelation function, $R_{xx}(-1)$, times $a(2)$, and the summed pth terms equate to $R_{xx}(-(p-1))$ by $a(p)$. Since in the case of autocorrelation functions $R_{xx}(-j) = R_{xx}(j)$, the expression may be written as

$$R_{xx}(0)a(1) + R_{xx}(1)a(2) + \ldots + R_{xx}(k-1)a(k) + \ldots + R_{xx}(p-1)a(p)$$

The right-hand side of Equation 11.48 is equal to $-R_{xx}(1)$. Equating the left- and right-hand sides gives

$$R_{xx}(0)a(1) + R_{xx}(1)a(2) + \ldots + R_{xx}(k-1)a(k) + \ldots + R_{xx}(p-1)a(p) = -R_{xx}(1)$$

$$(11.49)$$

For each value of k, $1 \leqslant k \leqslant p$, a similar equation may be written. These equations may be written in matrix form as

$$\begin{pmatrix} R_{xx}(0) & R_{xx}(1) & \cdots & R_{xx}(p-1) \\ R_{xx}(1) & R_{xx}(0) & \cdots & R_{xx}(p-2) \\ \vdots & \vdots & \vdots & \vdots \\ R_{xx}(p-1) & R_{xx}(p-2) & \cdots & R_{xx}(0) \end{pmatrix} \begin{pmatrix} a(1) \\ a(2) \\ \vdots \\ a(p) \end{pmatrix} = - \begin{pmatrix} R_{xx}(1) \\ R_{xx}(2) \\ \vdots \\ R_{xx}(p) \end{pmatrix} \qquad (11.50)$$

The model parameters, $a(k)$, may now be obtained from this set of equations which are known as the Yule–Walker (YW) equations. In matrix notation 11.50 may be written

$$\mathbf{R}_{xx}(k - j)\mathbf{a}(k) = -\mathbf{R}_{xx}(k) \tag{11.51}$$

Hence, in principle,

$$\mathbf{a}(k) = -\mathbf{R}_{xx}^{T}(k - j)\mathbf{R}_{xx}(k) \tag{11.52}$$

It is seen that $\mathbf{R}_{xx}(k - j)$ is symmetrical, and because each element of the leading diagonal is the same (equalling $R_{xx}(0)$) it is said to be Toeplitz. It is also positive definite, provided that $x(n)$ does not consist purely of sinusoids. There are several ways of solving Equation 11.50 for the $a(k)$.

Equation 11.45 allows calculation of E, but another expression in terms of the autocorrelation functions and the $a(k)$ may be found as follows. Assuming the $a(k)$ and the $x(n)$ are real and expanding Equation 11.45 gives

$$\begin{aligned}
E &= \frac{1}{N}\sum_{n=1}^{N}\left\{x(n) + \sum_{k=1}^{p}a(k)x(n-k)\right\}\left\{x(n) + \sum_{k=1}^{p}a(k)x(n-k)\right\} \\
&= \frac{1}{N}\sum_{n=1}^{N}\Bigg[\left\{x(n) + \sum_{k=1}^{p}a(k)x(n-k)\right\}x(n) \\
&\quad + \left\{x(n) + \sum_{k=1}^{p}a(k)x(n-k)\right\}\sum_{k=1}^{p}a(k)x(n-k)\Bigg]
\end{aligned} \tag{11.53}$$

From Equation 11.47, which is true for all k, it is seen that in 11.53

$$\frac{1}{N}\sum_{n=1}^{N}\left\{x(n) + \sum_{k=1}^{p}a(k)x(n-k)\right\}\sum_{k=1}^{p}a(k)x(n-k) = 0$$

hence 11.53 simplifies to

$$\begin{aligned}
E &= \frac{1}{N}\sum_{n=1}^{N}\left\{x(n) + \sum_{k=1}^{p}a(k)x(n-k)\right\}x(n) \\
&= \frac{1}{N}\sum_{n=1}^{N}x^{2}(n) + \frac{1}{N}\sum_{n=1}^{N}\left(\sum_{k=1}^{p}a(k)x(n-k)\right)x(n) \\
&= R_{xx}(0) + \sum_{k=1}^{p}\frac{1}{N}\sum_{n=1}^{N}a(k)x(n)x(n-k)
\end{aligned}$$

so that finally

$$E = R_{xx}(0) + \sum_{k=1}^{p} a(k) R_{xx}(k) \tag{11.54}$$

Equations 11.54 or 11.45 and the model parameters from 11.52 may now be inserted in 11.43 to obtain the autoregressive power density spectrum. However, the possible ways of solving 11.50 for the $a(k)$ and the choice of the model order, p, must first be described.

11.5.4 Solution of the Yule–Walker equations

The mean squared error, E, given by Equation 11.45, is computed using the available sampled values, $x(n)$, for $n = 1$ to $n = N$. Previous or succeeding values of $x(n)$ are effectively set to zero. As already explained, this is equivalent to windowing the data, and in the case of the non-parametric methods of spectrum estimation leads to spectral smearing by side lobes and reduced resolution. However, this is not the case for the autoregressive method. It can be shown (Kay, 1988) that this method implicitly estimates the autocorrelation functions for lags greater than p even when there are no corresponding values of $x(n)$. Hence autoregressive methods offer improved spectral resolution. It is possible, however, to improve the spectrum estimations by, for example, basing the method only upon the available data. In this approach, values of $n, 0 \geqslant n < N$, are not required, and consequently do not have to be set to zero. These methods are the subject of this section. A more complete specialized treatment is given in texts such as Kay (1988) and Candy (1989). Programs for the algorithms are given in Kay (1988) and other texts, and in software packages.

11.5.4.1 The autocorrelation method

The autocorrelation method is based upon the mean squared error expression in Equation 11.45. The Levinson–Durbin algorithm (Kay, 1988; Pardey, Roberts, and Tarassenko, 1996) provides a computationally efficient way of solving the YW equations of 11.50 for the model parameters. This method gives poorer frequency resolution than the others to be described, and is therefore less suitable for shorter data records.

11.5.4.2 The covariance method

In this method the limits of summation in Equation 11.45 are modified to run from $n = p$ to $n = N$. This means that only available values of $x(n)$ are required for the autocorrelation function calculations. Also, the average is calculated over $N - p$ terms rather than N. Thus, 11.45 becomes

$$E = \frac{1}{N - p} \sum_{n=p}^{N} \left(x(n) + \sum_{k=1}^{p} a(k) x(n - k) \right)^2 \tag{11.55}$$

The equivalent of Equation 11.50 is

$$\begin{pmatrix} C_{xx}(1,1) & C_{xx}(1,2) & \cdots & C_{xx}(1,p) \\ C_{xx}(2,1) & C_{xx}(2,2) & \cdots & C_{xx}(2,p) \\ \vdots & \vdots & \vdots & \vdots \\ C_{xx}(p,1) & C_{xx}(p,2) & \cdots & C_{xx}(p,p) \end{pmatrix} \begin{pmatrix} a(1) \\ a(2) \\ \vdots \\ a(p) \end{pmatrix} = - \begin{pmatrix} C_{xx}(1,0) \\ C_{xx}(2,0) \\ \vdots \\ C_{xx}(p,0) \end{pmatrix} \tag{11.56}$$

where

$$C_{xx}(j,k) = \frac{1}{N-p} \sum_{n=p}^{N} x(n-j)x(n-k) \tag{11.57}$$

E is given by

$$E = C_{xx}(0,0) + \sum_{k=1}^{p} a(k)C_{xx}(0,k) \tag{11.58}$$

The $p \times p$ matrix $C_{xx}(j,k)$ is Hermitian and positive semi-definite. Equation 11.56 may be solved using the Cholensky decomposition method (Lawson and Hanson, 1974). Only $N - p$ lagged components are summed, so for short data lengths there could be some end effects. The covariance method results in better spectral resolution than the autocorrelation method.

11.5.4.3   The modified covariance method

In the modified covariance method the average of the estimated forward and backward prediction errors is minimized (Kay, 1988; Candy, 1989). Equations 11.56 and 11.58 still apply, but Equation 11.57 is modified to

$$C_{xx}(j,k) = \frac{1}{2(N-p)} \left\{ \sum_{n=p}^{N} x(n-j)x(n-k) + \sum_{n=1}^{N-p} x(n+j)x(n+k) \right\} \tag{11.59}$$

The $p \times p$ matrix $C_{xx}(j,k)$ is again Hermitian and positive semi-definite, and Equation 11.56 may be solved using the Cholensky decomposition method (Lawson and Hanson, 1974). The method does not guarantee a stable all-pole filter, but this usually results. It yields statistically stable spectral estimates of high resolution.

11.5.4.4   The Burg method

This method relies upon aspects beyond the present scope. It produces accurate spectral estimates for AR data.

11.5.5   Model order

The order of the autoregressive model which best fits the data has to be carefully chosen for each set of data, since it depends upon the statistical properties of the data. An example of this is to be found in EEG data, for which different data segments require different orders of model (Pardey, Roberts and Tarassenko, 1996). Models of low order are preferred on the grounds that fewer parameters have to be fitted. However, if the order is too small the spectrum estimate will be too smoothed. On the other hand too large an order results in spurious peaks and spectral instability. Two of the most commonly used order estimation parameters were developed by Akaike. These are the final prediction error, $FPE(p)$ (Akaike, 1969), given by

$$FPE(p) = \frac{N + p}{N - p} E(p) \tag{11.60}$$

and the Akaike information criterion, $AIC(p)$ (Akaike, 1974), which is

$$AIC(p) = N \ln E(p) + 2p \tag{11.61}$$

and is particularly recommended for short data records, while $FPE(p)$ is recommended for longer data records. A practical approach is to attempt to select p to minimize both $FPE(p)$ and $AIC(p)$.

11.6 Comparison of estimation methods

Of the nonparametric methods the Blackman–Tukey method has the larger quality factor and is therefore to be preferred, although for convenience one of the other approaches may be employed.

The parametric methods give greater frequency resolution and avoid the use of window functions. The Capon, maximum likelihood, method yields a minimum variance unbiased estimate with a spectral resolution intermediate between the Burg or the unconstrained least squares methods, and the non-parametric methods. The adaptive filtering methods emphasize the more recent data and are suitable for nonstationary data.

11.7 Application examples

11.7.1 Use of spectral analysis by a DFT for differentiating between brain diseases

Amplitude and phase spectra derived using an FFT have been utilized in a procedure to differentiate between Huntington's disease, schizophrenia, Parkinson's disease, and normal subjects by analyzing selected harmonics in the spectrum of the contingent negative variation (CNV) in the subjects' electroencephalogram (EEG) (Jervis *et al.*, 1993).

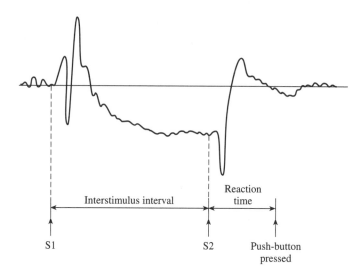

Figure 11.16   Schematic CNV waveform.

The CNV is an event-related potential (ERP) manifesting as a negative electrical potential shift on the scalp on elicitation by a suitable auditory stimulus paradigm. A schematic drawing of a CNV waveform is shown in Figure 11.16. The CNV is the negative waveform found between the points of onset of the auditory stimuli at S_1 and S_2. There are reasons to believe that the CNV waveform is affected in people who suffer from one of the above-mentioned diseases. By determining the spectra of appropriate parts of the waveform, and treating them statistically, it is possible to differentiate between the subject categories.

Several CNVs were recorded from each subject using a purpose-designed signal processing instrumentation system (Jervis and Saatchi, 1990; Saatchi and Jervis, 1991). The data was then pre-processed to reduce the effect of the background EEG and of ocular artefacts on the CNV waveform. The mean level of the signals was removed so that a comparison over time could be made and to ensure that the ocular artefact removal algorithm functioned properly. The mean level removal caused a positive shift of the pre- and post-stimulus baseline. Therefore the baseline was corrected by subtracting the means of the different sections of the response from the corresponding sections. Digital lowpass filtering was then applied to filter out the unwanted high frequency components of the EEG. An FIR filter was used rather than an IIR filter because it would not distort the waveform. An ocular artefact removal algorithm was then applied to remove ocular artefacts by the method of proportional subtraction (Jervis *et al.*, 1989b). The averages of eight pre-processed CNV waveforms for one each of the subject categories are shown in Figures 11.17–11.20. These are the averaged CNVs of a normal, a Huntington's disease, a schizophrenic, and a Parkinson's disease subject respectively.

Two 512 ms (64 sample) segments of each individual CNV waveform were then windowed using a Kaiser–Bessel window. Experiments indicated that a value of the

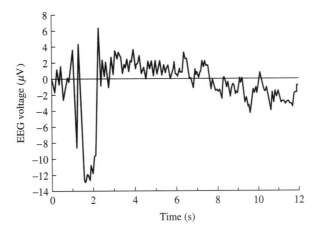

Figure 11.17 Pre-processed and averaged CNV waveform of a normal subject.

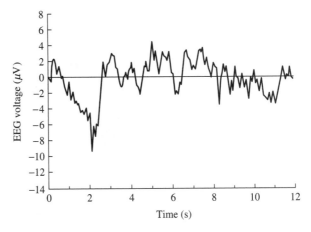

Figure 11.18 Pre-processed and averaged CNV waveform of a Huntington's disease patient.

window parameter, α, of 0.75 offered a satisfactory compromise between side lobe level and main lobe width. 960 augmenting zeros were added to the 64 data samples to reduce the scalloping loss. The DFTs of the sets of 1024 data were then calculated. Four statistical tests were applied to the first 96 harmonic components of the spectra thus produced. These tests are described in Jervis *et al.* (1983). The tests were entitled the nearest and furthest mean amplitude test, the pre- and post-stimulus mean amplitude difference test, the Rayleigh test of circular variance, and the modified Rayleigh test of circular variance. These tests resulted in a number of test statistics.

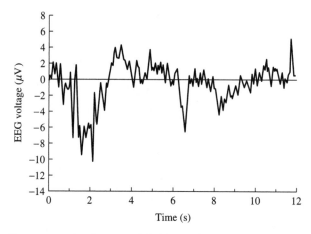

Figure 11.19   Pre-processed and averaged CNV waveform of a schizophrenic patient.

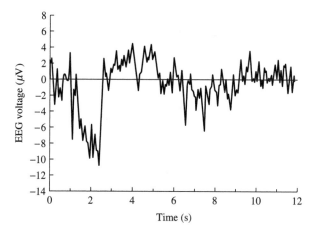

Figure 11.20   Pre-processed and averaged CNV waveform of a Parkinson's disease patient.

In order to reduce the number of test statistics by selecting the more discriminatory ones they were subjected to the univariate test, the t-test, and stepwise discriminant analysis. This procedure was described in Jervis *et al.* (1993) and was implemented by means of the statistical programs package SAS (SAS, 1982).

The classification of the individuals was now carried out using discriminant analysis (Morrison, 1976). Again, further details are given in Jervis *et al.* (1993) and the SAS package was used for implementation. The results are summarized in Table 11.2.

These results show the usefulness of spectral analysis applied to the CNV waveform when combined with statistical techniques in successfully differentiating to a

Table 11.2 Summary of brain disease differentiation results.

Subject types		Differentiation success rates (%) in test domain	
Type 1	Type 2	Type 1	Type 2
HD	control	100	100
schizophrenic	control	95	100
PD	control	93.8	87.5
HD	schizophrenic	100	90.9
HD	PD	90.9	81.8
schizophrenic	PD	81.3	93.8

HD, Huntington's disease subject; PD, Parkinson's disease subject; control, age- and sex-matched control subject.

high degree of accuracy between Huntington's disease, schizophrenic, Parkinson's disease, and normal subjects.

11.7.2 Spectral analysis of EEGs using autoregressive modelling

A recorded event-related potential (ERP) signal was simulated by adding together a section of measured electroencephalogram (EEG) signal and a simulated ERP. The simulated recorded ERP and the simulated ERP are shown in Figure 11.21(a). In Figure 11.21(b) the power density spectrum of the simulated EP recording is illustrated, as determined using the Fast Fourier Transform method. The method of Section 11.5 for determining spectra autoregressively was then applied to obtain the spectra of the signals. The covariance method, Section 11.5.4.2, was used to calculate the model parameters. The AR spectrum of the simulated recorded ERP is shown in Figure 11.21(c) for an autoregressive model of order 50. This is smoother than that derived using the FFT. The AR spectrum of the measured EEG is shown for comparison in Figure 11.21(d). This contains less energy below 2 Hz than that of the simulated ERP, as indeed may be expected, since the energy of the ERP is concentrated at low frequencies. The AR spectrum of the ERP is shown in Figure 11.21(e). The energy is seen to be confined to a bandwidth of 2 Hz, confirming the previous observation. Comparison of Figures 11.21(d) and 11.21(e) reveals that most of the energy below 2 Hz is associated with the ERP and not so much with the EEG. Finally the AR spectrum of the ERP is also plotted in Figure 11.21(f), but with a model order of 6 instead of 50. These spectra are identical, showing that a low order model is adequate for the determination of the spectrum of the narrow bandwidth ERP, whilst a higher order model is necessary to find the spectrum of the EEG, which is of wider bandwidth.

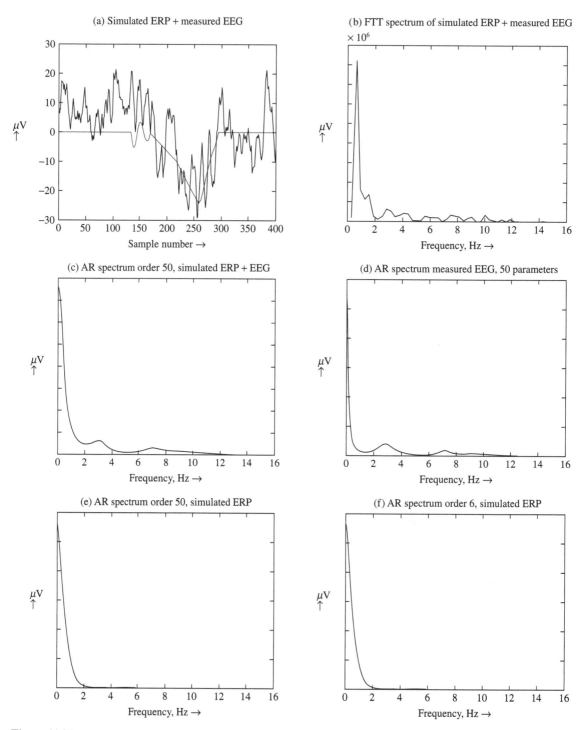

Figure 11.21   An illustration of autoregressively derived spectra.

11.8 Summary

It has been suggested in this chapter that parametric methods of spectrum estimation may yield more reliable results. Since they may be automated they are probably preferable to the non-parametric periodogram-based methods. The latter are less reliable and are more of an art in that they require the application of significant expertise to secure meaningful results. However, their usefulness was illustrated particularly well by the clinical application to the differentiation of brain diseases.

11.9 Worked example

Example 11.3

Show that the Bartlett estimate of the power spectral density is asymptotically unbiased, that the variance of the estimate decreases with the number of data sections, and that the spectrum estimates are consistent. What effect does the modified periodogram method have on the frequency resolution?

Solution

In this method the periodograms of K non-overlapping sections of M data are computed. Given a total of N data this means $K = N/M$. The average of the K periodograms then gives the Bartlett power density spectral estimate, $P_{BE}(f)$, where

$$P_{BE}(f) = \frac{1}{K} \sum_{j=0}^{K-1} P_j(f) \tag{11.62}$$

and j denotes the sequence number and $P_j(f)$ is the corresponding jth estimate of the power spectral density.

Examining the expected values of $P_{BE}(f)$ gives

$$E[P_{BE}(f)] = \frac{1}{K} \sum_{j=0}^{K-1} E[P_j(f)]$$

$$= E[P_j(f)]$$

$$= \sum_{m=-(M-1)}^{M-1} \left(1 - \frac{|m|}{M}\right) c_{xx}(m) \exp(-j2\pi fm) \tag{11.63}$$

For $M \gg |m|$, the Bartlett window term $1 - |m|/M$ approaches unity and $E[P_{BE}(f)]$ becomes the true power spectral density, $P(f)$. The Bartlett estimate of $P(f)$ is therefore asymptotically unbiased.

Turning now to the variance,

$$\text{var}\,[P_{\text{BE}}(f)] = \frac{1}{K^2}\sum_{j=0}^{K-1}\text{var}\,[P_j(f)] = \frac{1}{K}\text{var}\,[P_j(f)] \qquad (11.64)$$

Thus the variance decreases in inverse proportion to the number of sections, K, into which the N data sequence has been divided. Because the variance decreases with K the Bartlett estimate of $P(f)$ is consistent. However, because the number of data from which the periodograms are calculated has been reduced by the factor K from N to $M = N/K$ the spectral resolution has also been reduced by the same factor. Thus the main lobe width obtained by the Bartlett method is K times that obtained using the full N data set.

Problems

11.1 (1) A waveform is sampled at the rate of 30 kHz. The first 524 288 samples are fast Fourier transformed. Calculate the frequency of the first harmonic and the frequency resolution of the spectrum.

(2) If the true spectrum of the waveform contained a sinusoidal component at 5.7505 Hz how would you modify the data to ensure this component was properly represented in the estimated spectrum?

11.2 A waveform is sampled at 8 kHz, the sampled voltages being 1.0, 1.0, 1.0, 1.0, 1.0, 1.0, 1.0, 1.0 V. The data is then windowed by a window function, the corresponding sampled values of which are 0, 0.5, 1.0, 1.0, 1.0, 1.0, 0.5, 0.0. Determine the DFT and hence the Fourier transform of the windowed data.

11.3 Calculate the scalloping loss and equivalent noise bandwidth associated with the data of Problem 11.2.

11.4 By referring to Figures 11.10, 11.12 and 11.13 estimate the equivalent noise bandwidth and processing gain of the following windows by using eight sampled values in each case:

(1) rectangular;
(2) Tukey (tapered cosine), $\alpha = 0.1$;
(3) Tukey, $\alpha = 0.5$;
(4) Hamming, $\alpha = 0.54$;

(5) Kaiser–Bessel, $\alpha = 2.0$;
(6) Kaiser–Bessel, $\alpha = 4.0$.

11.5 A telecommunications pulse waveform is sampled every 0.167 μs giving the values 0, 0, 1, 1, 1, 1, 1, 0 V. The nonzero values are windowed with a Kaiser–Bessel window with $\alpha = 4.0$. Compute the energy spectrum of the windowed pulse using an eight-point FFT (or DFT). (Obtain estimated window data from Figure 11.13(c).)

11.6 Determine the scalloping loss, the processing loss, and the worst-case processing loss of the window in Problem 11.5.

11.7 For the data of Problem 11.5 determine the worst-case processing loss and the amplitude of the first side lobe for the Kaiser–Bessel window with $\alpha = 4.0$, the Hamming window with $\alpha = 0.54$, the Tukey window with $\alpha = 0.5$, and the rectangular window. Tabulate the results and select the most suitable window. (Use Figures 11.13(c), 11.12(a) and 11.10(b) to obtain estimates of the window functions.)

11.8 The sampled voltages obtained from a waveform sampled at 8 kHz are 0, 4.0, 2,4, 1.0, −1.0, −3.8, −1.3, 0 V. Compute and plot the energy spectra:

(1) applying a Kaiser–Bessel window, $\alpha = 2.0$, to the data;
(2) after modification of the data according to the formula of Equation 11.21,

$$S(n) = w(n) \left[s(n) - \frac{\sum\limits_{n=0}^{N-1} w(n)s(n)}{\sum\limits_{n=0}^{N-1} w(n)} \right]$$

$$\times \left[\frac{N}{\sum\limits_{n=0}^{N-1} w^2(n)} \right]^{1/2}$$

where the $w(n)$ are the sampled values of the Kaiser–Bessel window with $\alpha = 2.0$;

(3) explain the reasons for any differences between the results obtained in parts (1) and (2).

11.9 Obtain the energy spectrum of the sampled data sequence $\{0, 1, 0, 1, 0, 1, 0, 1\}$ and compare it with what you would expect for a square waveform.

11.10 Apply the Bartlett modified periodogram method to obtain the energy spectrum of the data of Problem 11.9 by sectioning the data into the two non-overlapping sequences $\{0, 1, 0, 1\}$ and $\{0, 1, 0, 1\}$.

11.11 Apply the Welch modified periodogram method to obtain the energy spectrum of the data of Problem 11.9 by subdividing the data sequence into three equal length sections with 50% overlap.

11.12 The data sequence of Problem 11.9 is now assumed to contain a random noise component such that the new data sequence becomes $\{0.763, 1.656, 0.424, 1.939, 0.133, 1.881, 0.328, 1.348\}$. Calculate the energy spectrum of this noisy data and compare it with that of the original data. Estimate the signal-to-noise ratio from the sampled data.

11.13 Now repeat the Bartlett modified periodogram method of obtaining the energy spectrum by using the two non-overlapping sequences $\{0.763, 1.656, 0.424, 1.939\}$ and $\{0.133, 1.881, 0.328, 1.348\}$.

11.14 Now repeat Problem 11.11 using the noisy data sequence of Problem 11.12.

11.15 Now assume that the original data sequence of Problem 11.9 is more highly contaminated with noise, so that the noisy data sequence is $\{6.03, 6.18, 3.35, 8.42, 1.05, 7.96, 2.59, 3.75\}$. Calculate

the energy spectrum of this data and estimate the signal-to-noise ratio from the sampled data.

11.16 Obtain an improved estimate of the energy spectrum of the data of Problem 11.15 by the Bartlett method using two equal length sections.

11.17 Enhance the quality of the estimated energy spectrum of the data of Problem 11.15 by using the Welch modified periodogram method with three equal sections with 50% overlap.

11.18 Draw up a table to compare the results of Problems 11.9–11.17 for the different methods of spectrum estimation and for the different signal-to-noise ratios. Discuss the effects of the signal-to-noise ratio on the different methods and select your preferred method.

11.19 Obtain the power density spectrum of the data of Problem 11.9, that is $\{0, 1, 0, 1, 0, 1, 0, 1\}$ using the Blackman–Tukey method and compare the result with that obtained for Problem 11.9. Use the window function $\{0, 0.5, 1, 1, 1, 1, 0.5, 0\}$.

11.20 Repeat Problem 11.19 using the noisy data of Problem 11.12.

11.21 Repeat Problem 11.19 using the noisy data of Problem 11.15.

11.22 Now compare the results obtained as answers to Problems 11.12–11.17 and ascertain whether a periodogram method or the Blackman–Tukey method gives the best result in the presence of noise.

11.23 Determine the energy spectrum of a square wave of period $1.0\ \mu s$ which alternates between 0 V and 5 V in amplitude by using the Blackman–Tukey method. Compare your result with the theoretical one.

11.24 Write computer programs to generate waveforms of interest to yourself and design suitable spectrum analysis procedures using the techniques described in the chapter to evaluate the energy and phase spectra.

11.25 Devise some amplitude and phase spectra which show some interesting features which would challenge the spectrum estimation techniques, such as close peaks. Transform them to the time domain and then add noise to achieve a low, a unity, and a high signal-to-noise ratio. Now determine and criticize the amplitude and phase spectra.

References

Akaike H. (1969) Fitting autoregressive models for prediction. *Ann. Institute of Statistical Mathematics*, **21**, 243–7.

Akaike H. (1973) Information theory and an extension of the maximum likelihood principle. In *2nd International Symposium on Information Theory* (Petrov B.N. and Csaki F. (eds)), pp. 267–81. Budapest: Akademiai Kiade.

Akaike H. (1974) A new look at the statistical model identification. *IEEE Trans. Automatic Control*, **19**, 716–22.

Akaike H. (1978) A Bayesian analysis of the minimum AIC procedure. *Ann. Institute of Statistical Mathematics*, **30A**, 9–14.

Akaike H. (1979) A Bayesian extension of the minimum AIC procedure of autoregressive model fitting. *Biometrika*, **66**, 237–42.

Blackman R.B. and Tukey J.W. (1958) *The Measurement of Power Spectra*. New York: Dover.

Box G.E.P. and Jenkins G.M. (1976) *Time-Series Analysis, Forecasting, and Control*. San Francisco CA: Holden-Day.

Burg J.P. (1967) Maximum entropy spectral analysis. In *Proc. 37th Meeting Society Exploration Geophysicists*, Oklahoma City, October. Reprinted in Childers D.G. (ed.) (1968) *Modern Spectrum Analysis*. New York: IEEE Press.

Burg J.P. (1968) A new analysis technique for time series data. In *NATO Advanced Study Institute on Signal Processing with Emphasis on Underwater Acoustics*, 12–23 August. Reprinted in Childers D.G. (ed.) (1968) *Modern Spectrum Analysis*. New York: IEEE Press.

Cadzow J.A. (1979) ARMA spectral estimation: an efficient closed-form procedure. In *Proc. RADC Spectrum Estimation Workshop*, Rome NY, October 1979, pp. 81–97.

Cadzow J.A. (1982) Spectral estimation: an overdetermined rational model equation approach. *Proc. IEEE*, **70**, 907–38.

Candy J.V. (1989) *Signal Processing: The Modern Approach*. New York: McGraw-Hill.

Capon J. (1969) High-resolution frequency–wavenumber spectrum analysis. *Proc. IEEE*, **57**, 1408–18.

Chatfield C. (1979) Inverse autocorrelation. *J. Royal Statistical Society A*, **142**, 363–77.

Chatfield C. (1984) *The Analysis of Time Series*, 3rd edn. London: Chapman and Hall.

Clarkson P.M. (1993) *Optimal and Adaptive Signal Processing*. Boca Raton FL: CRC Press.

Cooley J.W. and Tukey J.W. (1965) An algorithm for the machine calculation of complex Fourier series. *Mathematics of Computation*, **19**, 297–301.

DeFatta D.J., Lucas J.G. and Hodgkiss W.S. (1988) *Digital Signal Processing: A System Design Approach*, Section 6.6.5, p. 263. New York: Wiley.

Durbin J. (1959) Efficient estimation of parameters in moving-average models. *Biometrika*, **46**, 306–16.

Friedlander B. (1982) Lattice methods for spectral estimation. *Proc. IEEE*, **70**, 990–1017.

Gersch W. (1970) Spectral analysis of EEGs by autoregressive decomposition of time series. *Mathematical Biosciences*, **7**, 205–22.

Graupe D., Krause D.J. and Moore J.B. (1975) Identification of autoregressive-moving average parameters of time series. *IEEE Trans. Automatic Control*, **20**, 104–7.

Harris F.J. (1978) On the use of windows for harmonic analysis with the discrete Fourier transform. *Proc. IEEE*, **66**(1), 51–84.

Jenkins G.M. and Watts D.G. (1968) *Spectral Analysis and its Applications*. San Francisco CA: Holden-Day.

Jervis B.W. and Saatchi M.R. (1990) An integrated system for process control and the acquisition, storage and processing of data. In *IEE Colloq. on PC-Based Instrumentation*, 31 January, IEE Digest No. 1990/025.

Jervis B.W., Nichols M.J., Johnson T.E., Allen E.M. and Hudson N.R. (1983) A fundamental investigation of the composition of auditory evoked potentials. *IEEE Trans. Biomedical Engineering*, **30**(1), 43–50.

Jervis B.W., Coelho M. and Morgan G.W. (1989a) Spectral analysis of EEG responses. *Medical and Biological Engineering and Computing*, **27**, 230–8.

Jervis B.W., Coelho M. and Morgan G.W. (1989b) Effect on EEG responses of removing ocular artefacts by proportional EOG subtraction. *Medical and Biological Engineering and Computing*, **27**, 484–90.

Jervis B.W., Saatchi M.R., Allen E.M., Hudson N.R., Oke S. and Grimsley M. (1993) A pilot study of computerised differentiation of Huntington's disease, schizophrenia, and Parkinson's disease patients using the contingent negative variation. *Medical and Biological Engineering and Computing*, **31** (January), 31–8.

Kalouptsidis N. and Theodoridis S. (1987) Fast adaptive least-squares algorithms for power spectral estimation. *IEEE Trans. Acoustics, Speech and Signal Processing*, **35**, 661–70.

Kay S.M. (1980) A new ARMA spectral estimator. *IEEE Trans. Acoustics, Speech and Signal Processing*, **28**, 585–8.

Kay S.M. (1988) *Modern Spectral Estimation: Theory and Application*. Englewood Cliffs NJ: Prentice-Hall.

Kuo F.F. and Kaiser J.F. (1966) *System Analysis by Digital Computer*, Chapter 7, pp. 232–8. New York: Wiley.

Lacoss R.T. (1971) Data adaptive spectral analysis methods. *Geophysics*, **36**, 661–75.

Lawson C.L. and Hanson R.J. (1974) *Solving Least Squares Problems*. Englewood Cliffs NJ: Prentice-Hall.

Levinson N. (1947) The Weiner RMS error criterion in filter design and prediction. *J. Mathematical Physics*, **25**, 261–78.

Makhoul J. (1975) Linear prediction: a tutorial review. *Proc. IEEE*, **63**, 561–80.

Marple S.L. (1980) A new autoregressive spectrum analysis algorithm. *IEEE Trans. Acoustics, Speech and Signal Processing*, **28**, 441–54.

Marple S.L. (1987) *Digital Spectral Analysis*. Englewood Cliffs NJ: Prentice-Hall.

Morrison D.F. (1976) *Multivariate Statistical Methods*, 2nd edn. New York: McGraw-Hill.

Nuttall A.H. (1976) *Spectral Analysis of a Univariate Process with Bad Data Points via Maximum Entropy and Linear Predictive Techniques*. NUSC Technical Report TR-5303, New London CN.

Pardey J., Roberts S. and Tarassenko L. (1996) A review of parametric modelling techniques for EEG analysis. *Medical Engineering Physics*, **18**, 2–11.

Priestley M.B. (1981) *Spectral Analysis and Time Series*, Volume 1, *Univariate Series*, Chapters 6 and 7. New York: Academic Press.

Proakis J.G. and Manolakis D.G. (1989) *Introduction to Digital Signal Processing*, Sections 1.3.2, 11.2.4, 11.3 and 11.3.4 and Appendix 6A. Basingstoke: Macmillan.

Rissanen J. (1983) A universal prior for the integers and estimation by minimum description length. *Ann. Statistics*, **11**, 417–31.

Saatchi M.R. and Jervis B.W. (1991) PC-based integrated system developed to diagnose specific brain disorders. *Computing and Control Engineering J.*, **2**(2), 61–8.

SAS (1982) *SAS User Guide*. SAS Institute.

Shibata R. (1976) Selection of the order of an autoregressive model by Akaike's information criterion. *Biometrika*, **63**, 117–26.

Ulrych T.J. and Clayton R.W. (1976) Time series modelling and maximum entropy. *Physics Earth and Planetary Interiors*, **12**, 188–200.

Welch P.D. (1967) The use of fast Fourier transform for the estimation of power spectra. *IEEE Trans. Audio and Electroacoustics*, **15**, 70–3.

Whittle P. (1965) *Prediction and Regulation*. London: English Universities Press.

Wold H. (1954) *A Study of the Analysis of Stationary Time Series*. Stockholm: Almquist and Wichsells.

Appendix

11A MATLAB programs for spectrum estimation and analysis

The MATLAB Signal Processing Toolbox contains useful functions for spectrum estimation and analysis using parametric and non-parametric methods. We have developed the following simple programs to illustrate the use of the functions in practice:

welchm.m	program that illustrates power spectrum estimation and analysis using Welch's method;
burgm.m	program that illustrates power spectrum estimation and analysis using Burg's method;
yulewalkm.m	program that illustrates power spectrum estimation and analysis using Yule–Walker's method.

The programs are available on the web. Illustrative examples of the use of the programs can be found in the companion manual, *A Practical Guide for MATLAB and C Language Implementations of DSP Algorithms*, published by Pearson Education (see the Preface for details).

12

General- and special-purpose digital signal processors

The main objectives of this chapter are to provide an understanding of the key issues underlying general- and special-purpose processors for DSP, the impact of DSP algorithms on the hardware and software architectures of these processors, and how key DSP algorithms are implemented for real-time execution on general-purpose digital signal processors or realized as a piece of special-purpose hardware.

Real-time often implies 'as soon as possible' but within specified time limits. Real-time processing may be divided into two broad categories (although further subdivision is possible): stream processing, for example digital filtering, where data is processed one sample at a time, and block processing, for example FFT and correlation, where fixed blocks of data points are processed at a time. The implementation of DSP algorithms in real time requires both hardware and software. The hardware may be an array of processors, standard microprocessors, DSP chips or microprogrammed special-purpose devices. The software may be low level assembly language codes or microcodes native to the DSP hardware, and/or codes in an efficient high level language, such as C or C++. The use of high level languages is now widespread, especially with newer DSP processors which are very complex and sophisticated.

Since their introduction in the early 1980s, DSP processors have grown substantially in complexity and sophistication to enhance their capability and range of applicability. This has also led to a substantial increase in the number of DSP processors available. To reflect this, features of successive generations of fixed- and floating-point DSP processors and factors that affect choice of DSP processors are covered.

12.1 Introduction

For convenience, DSP processors can be divided into two broad categories: general purpose and special purpose. DSP processors include fixed-point devices such as Texas Instruments TMS320C54x, and Motorola DSP563x processors, and floating-point processors such as Texas Instruments TMS320C4x and Analog Devices ADSP21xxx SHARC processors.

There are two types of special-purpose hardware.

(1)   Hardware designed for efficient execution of specific DSP algorithms such as digital filters, fast Fourier transform. This type of special-purpose hardware is sometimes called an algorithm-specific digital signal processor.

(2)   Hardware designed for specific applications, for example telecommunications, digital audio, or control applications. This type of hardware is sometimes called an application-specific digital signal processor.

In most cases application-specific digital signal processors execute specific algorithms, such as PCM encoding/decoding, but are also required to perform other application-specific operations. Examples of special-purpose DSP processors are Cirrus's processor for digital audio sampling rate converters (CS8420), Mitel's multi-channel telephony voice echo canceller (MT9300), FFT processor (PDSP16515A) and programmable FIR filter (VPDSP16256).

Both general-purpose and special-purpose processors can be designed with single chips or with individual blocks of multipliers, ALUs, memories, and so on.

First, we will discuss the architectural features of digital signal processors that have made real-time DSP in many areas possible.

12.2 Computer architectures for signal processing

Most general purpose processors available today are based on the von Neumann concepts, where operations are performed sequentially. Figure 12.1 shows a simplified architecture for a standard von Neumann processor. When an instruction is processed in such a processor, units of the processor not involved at each instruction phase wait idly until control is passed on to them. Increase in processor speed is

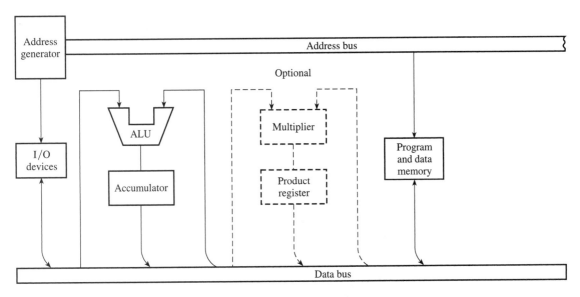

Figure 12.1   A simplified architecture for standard microprocessors.

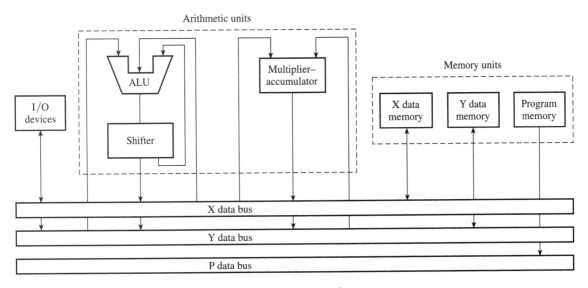

Figure 12.2   Basic generic hardware architecture for signal processing.

achieved by making the individual units operate faster, but there is a limit on how fast they can be made to operate.

If it is to operate in real time, a DSP processor must have its architecture optimized for executing DSP functions. Figure 12.2 shows a generic hardware architecture suitable for real-time DSP. It is characterized by the following:

- Multiple bus structure with separate memory space for data and program instructions. Typically the data memories hold input data, intermediate data values and output samples, as well as fixed coefficients for, for example, digital filters or FFTs. The program instructions are stored in the program memory.

- The I/O port provides a means of passing data to and from external devices such as the ADC and DAC or for passing digital data to other processors. Direct memory access (DMA), if available, allows for rapid transfer of blocks of data directly to or from data RAM, typically under external control.

- Arithmetic units for logical and arithmetic operations, which include an ALU, a hardware multiplier and shifters (or multiplier–accumulator).

Why is such an architecture necessary? Most DSP algorithms (such as filtering, correlation and fast Fourier transform) involve repetitive arithmetic operations such as multiply, add, memory accesses, and heavy data flow through the CPU. The architecture of standard microprocessors is not suited to this type of activity. An important goal in DSP hardware design is to optimize both the hardware architecture and the instruction set for DSP operations. In digital signal processors, this is achieved by making extensive use of the concepts of parallelism. In particular, the following techniques are used:

- Harvard architecture;
- pipelining;
- fast, dedicated hardware multiplier/accumulator;
- special instructions dedicated to DSP;
- replication;
- on-chip memory/cache;
- extended parallelism – SIMD, VLIW and static superscalar processing.

For successful DSP design, it is important to understand these key architectural features.

12.2.1   Harvard architecture

The principal feature of the Harvard architecture is that the program and data memories lie in two separate spaces, permitting a full overlap of instruction fetch and execution. Standard microprocessors, such as the Intel 6502, are characterized by a single bus structure for both data and instructions, as shown in Figure 12.1.

Suppose that in a standard microprocessor we wish to read a value op1 at address ADR1 in memory into the accumulator and then store it at two other addresses, ADR2 and ADR3. The instructions could be

LDA	ADR1	load the operand op1 into the accumulator from ADR1
STA	ADR2	store op1 in address ADR2
STA	ADR3	store op1 in address ADR3

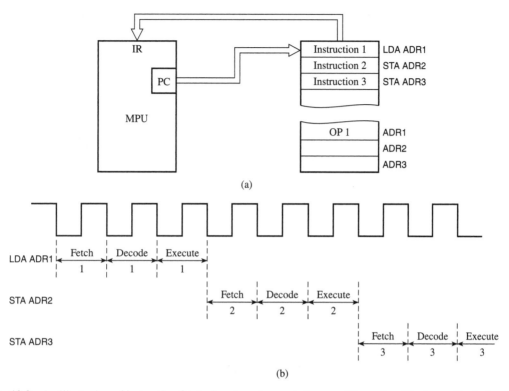

Figure 12.3   An illustration of instruction fetch, decode and execute in a non-Harvard architecture with single memory space: (a) instruction fetch from memory; (b) timing diagram.

Typically, each of these instructions would involve three distinct steps:

■ instruction fetch;
■ instruction decode;
■ instruction execute.

In our case, the instruction fetch involves fetching the next instruction from memory, and instruction execute involves either reading or writing data into memory. In a standard processor, without Harvard architecture, the program instructions (that is, the program code) and the data (operands) are held in one memory space; see Figure 12.3. Thus the fetching of the next instruction while the current one is executing is not allowed, because the fetch and execution phases each require memory access.

In a Harvard architecture (Figure 12.4), since the program instructions and data lie in separate memory spaces, the fetching of the next instruction can overlap the execution of the current instruction; see Figure 12.5. Normally, the program memory holds the program code, while the data memory stores variables such as the input data samples.

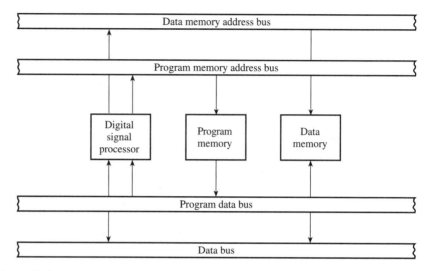

Figure 12.4 Basic Harvard architecture with separate data and program memory spaces. Data and program instruction fetches can be overlapped as two independent memories are used.

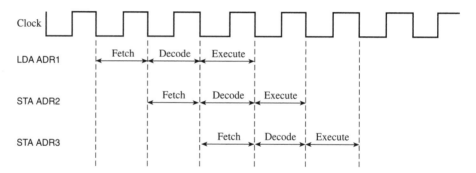

Figure 12.5 An illustration of instruction overlap made possible by the Harvard architecture.

Strict Harvard architecture is used by some digital signal processors (for example, Motorola DSP56000), but most use a modified Harvard architecture (for example, the TMS320 family of processors). In the modified architecture used by the TMS320, for example, separate program and data memory spaces are still maintained, but communication between the two memory spaces is permissible, unlike in the strict Harvard architecture.

12.2.2   Pipelining

Pipelining is a technique which allows two or more operations to overlap during execution. In pipelining, a task is broken down into a number of distinct subtasks

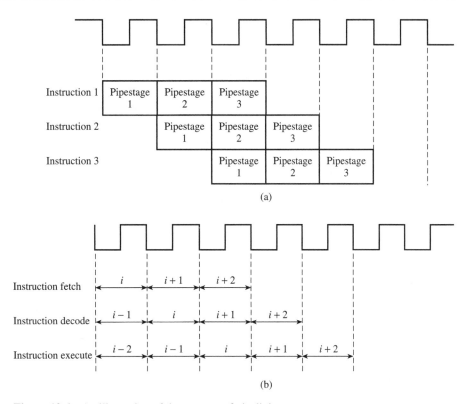

Figure 12.6   An illustration of the concept of pipelining.

which are then overlapped during execution. It is used extensively in digital signal processors to increase speed. A pipeline is akin to a typical production line in a factory, such as a car or television assembly plant. As in the production line, the task is broken down into small, independent subtasks called pipe stages. The pipe stages are connected in series to form a pipe and the stages executed sequentially.

As we have seen in the last section, an instruction can be broken down into three steps. Each step in the instruction can be regarded as a stage in a pipeline and so can be overlapped. By overlapping the instructions, a new instruction is started at the start of each clock cycle (Figure 12.6(a)).

Figure 12.6(b) gives the timing diagram for a three-stage pipeline, drawn to highlight the instruction steps. Typically, each step in the pipeline takes one machine cycle. Thus during a given cycle up to three different instructions may be active at the same time, although each will be at a different stage of completion. The key to an instruction pipeline is that the three parts of the instruction (that is, fetch, decode and execute) are independent and so the execution of multiple instructions can be overlapped. In Figure 12.6(b), it is seen that, at the ith cycle, the processor could be simultaneously fetching the ith instruction, decoding the $(i-1)$th instruction and at the same time executing the $(i-2)$th instruction.

The three-stage pipelining discussed above is based on the technique used in the Texas Instruments TMS320 processors. As in other applications of pipelining, in the TMS320 a number of registers are used to achieve the pipeline: a prefetch counter holds the address of the next instruction to be fetched, an instruction register holds the instruction to be executed, and a queue instruction register stores the instructions to be executed if the current instruction is still executing. The program counter contains the address of the next instruction to execute.

By exploiting the inherent parallelism in the instruction stream, pipelining leads to a significant reduction, on average, of the execution time per instruction. The throughput of a pipeline machine is determined by the number of instructions through the pipe per unit time. As in a production line, all the stages in the pipeline must be synchronized. The time for moving an instruction from one step to another within the pipe (see Figure 12.6(a)) is one cycle and depends on the slowest stage in the pipeline. In a perfect pipeline, the average time per instruction is given by (Hennessy and Patterson, 1990)

$$\frac{\text{time per instruction (nonpipeline)}}{\text{number of pipe stages}} \tag{12.1}$$

In the ideal case, the speed increase is equal to the number of pipe stages. In practice, the speed increase will be less because of the overheads in setting up the pipeline, and delays in the pipeline registers, and so on.

Example 12.1   In a nonpipeline machine, the instruction fetch, decode and execute take 35 ns, 25 ns, and 40 ns, respectively. Determine the increase in throughput if the instruction steps were pipelined. Assume a 5 ns pipeline overhead at each stage, and ignore other delays.

Solution   In the nonpipeline machine, the average instruction time is simply the sum of the execution time of all the steps: 35 + 25 + 40 ns = 100 ns. However, if we assume that the processor has a fixed machine cycle with the instruction steps synchronized to the system clock, then each instruction would take three machine cycles to complete: 40 ns × 3 = 120 ns. This corresponds to a throughput of 8.3×10^6 instructions s^{-1}.

In the pipeline machine, the clock speed is determined by the speed of the slowest stage plus overheads. In our case, the machine cycle is 40 + 5 = 45 ns. This places a limit on the average instruction execution time. The throughput (when the pipeline is full) is 22.2×10^6 instructions s^{-1}. Then

$$\text{speedup} = \frac{\text{average instruction time (nonpipeline)}}{\text{average instruction time (pipeline)}} \tag{12.2}$$

$$= 120/45$$

$$= 2.67 \text{ times (assuming nonpipeline executes in three cycles)}$$

In the pipeline machine, each instruction still takes three clock cycles, but at each cycle the processor is executing up to three different instructions. Pipelining increases the system throughput, but not the execution time of each instruction on its own. Typically, there is a slight increase in the execution time of each instruction because of the pipeline overhead.

Pipelining has a major impact on the system memory. The number of memory accesses in a pipeline machine increases, essentially by the number of stages. In DSP the use of Harvard architecture, where data and instructions lie in separate memory spaces, promotes pipelining.

When a slow unit, such as a data memory, and an arithmetic element are connected in series, the arithmetic unit often waits idly for a good deal of the time for data. Pipelining may be used in such cases to allow a better utilization of the arithmetic unit. The next example illustrates the concept.

Example 12.2

Most DSP algorithms are characterized by multiply-and-accumulate operations typified by the following equation:

$$a_0x(n) + a_1x(n-1) + a_2x(n-2) + \ldots + a_{N-1}x(n-(N-1))$$

Figure 12.7 shows a nonpipeline configuration for an arithmetic element for executing the above equation. Assume a transport delay of 200 ns, 100 ns, and 100 ns, respectively, for the memory, multiplier and accumulator.

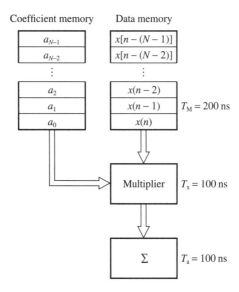

Figure 12.7   Non-pipelined MAC configuration. Products are clocked into the accumulator every 400 ns.

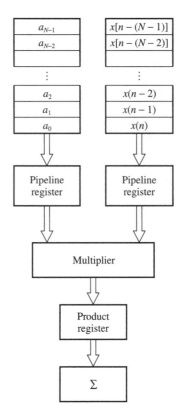

Figure 12.8   Pipelined MAC configuration. The pipeline registers serve as temporary store for coefficient and data sample pair. The product register also serves as a temporary store for the product.

(1)   What is the system throughput?

(2)   Reconfigure the system with pipelining to give a speed increase of 2 : 1. Illustrate the operation of the new configuration with a timing diagram.

Solution   (1)   The coefficients, a_k, and the data arrays are stored in memory as shown in Figure 12.7(a). In the nonpipelined mode, the coefficients and data are accessed sequentially and applied to the multiplier. The products are summed in the accumulator. Successive multiplication–accumulation (MAC) will be performed once every 400 ns (200 + 100 + 100), giving a throughput of 2.5×10^6 operations s^{-1}.

(2)   The arithmetic operations involved can be broken up into three distinct steps: memory read, multiply, and accumulate. To improve speed these steps can be overlapped. A speed improvement of 2 : 1 can be achieved by inserting pipeline registers between the memory and multiplier and between the multiplier and accumulator as shown in Figure 12.8. The timing diagram for

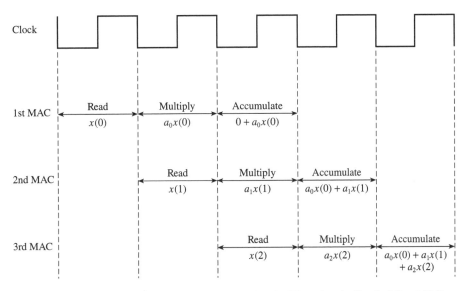

Figure 12.9   Timing diagram for a pipelined MAC unit. When the pipeline is full, a MAC operation is performed every clock cycle (200 ns).

the pipeline configuration is shown in Figure 12.9. As is evident in the timing diagram, the MAC is performed once every 200 ns. The limiting factor is the basic transport delay through the slowest element, in this case the memory. Pipeline overheads have been ignored.

DSP algorithms are often repetitive but highly structured, making them well suited to multilevel pipelining. For example, FFT requires the continuous calculation of butterflies. Although each butterfly requires different data and coefficients the basic butterfly arithmetic operations are identical. Thus arithmetic units such as FFT processors can be tailored to take advantage of this. Pipelining ensures a steady flow of instructions to the CPU, and in general leads to a significant increase in system throughput. However, on occasions pipelining may cause problems. For example, in some digital signal processors, pipelining may cause an unwanted instruction to be executed, especially near branch instructions, and the designer should be aware of this possibility.

12.2.3 Hardware multiplier–accumulator

The basic numerical operations in DSP are multiplications and additions. Multiplication, in software, is notoriously time consuming. Additions are even more time consuming if floating point arithmetic is used. To make real-time DSP possible a fast, dedicated hardware multiplier–accumulator (MAC) using fixed or floating point arithmetic is mandatory. Fixed or floating hardware MAC is now standard in all digital signal processors. In a fixed point processor, the hardware multiplier typically

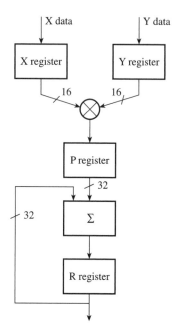

Figure 12.10   A typical MAC configuration in DSPs.

accepts two 16-bit 2's complement fractional numbers and computes a 32-bit product in a single cycle (25 ns typically). The average MAC instruction time can be significantly reduced through the use of special repeat instructions.

A typical DSP hardware MAC configuration is depicted in Figure 12.10. In this configuration, the multiplier has a pair of input registers that hold the inputs to the multiplier, and a 32-bit product register which holds the result of a multiplication. The output of the P (product) register is connected to a double-precision accumulator, where the products are accumulated.

The principle is very much the same for hardware floating-point multiplier–accumulators, except that the inputs and products are normalized floating-point numbers. Floating-point MACs allow fast computation of DSP results with minimal errors. As discussed in Chapters 7 and 8 DSP algorithms such as FIR and IIR filtering suffer from the effects of finite wordlength (coefficient quantization and arithmetic errors). Floating point offers a wide dynamic range and reduced arithmetic errors, although for many applications the dynamic range provided by the fixed-point representation is adequate.

12.2.4   Special instructions

Digital signal processors provide special instructions optimized for DSP. The benefits of these special instructions are twofold: they lead to a more compact code which takes up less space in memory (nearly as compact as a code written in a high level language such as C) and they lead to an increase in the speed of execution of DSP algorithms. Special instructions provided by DSP chips include (i) instructions that support basic DSP operations, (ii) instructions that reduce the overhead in instruction loops and (iii) application-oriented instructions.

Many key algorithms in DSP, such as digital filtering and correlation, require data shifts or delays to make room for new data samples. Digital signal processors provide special instructions that allow a data sample to be copied to the next higher memory address as it is being fetched from memory or operated on, all in one cycle. For example, in the second generation TMS320 family of DSP processors, the instruction pair, LTD and MPY, permits simultaneous loading of data into the temporary register for the multiplier, data shifting (to implement the unit delay symbolized by z^{-1}), and accumulation of products. Special instructions that support multiply and accumulate (MAC) operations with data shifts are now standard in modern DSP processors.

Special instructions are also provided to speed up DSP operations that are often repeated. For example, in the second generation TMS320 family of DSP processors, a repeat instruction is provided that allows the next instruction to be repeated a specified number of times. As the repeat instruction requires only a single instruction fetch, a piece of code that would normally require a multi-cycle loop effectively becomes a single-cycle instruction. Repeat instructions are especially useful in DSP for inner loop computation, e.g. in FIR and adaptive filtering, and data move for bit reversal in FFTs.

Recall that FIR filters are characterized by the following equation:

$$y(n) = \sum_{k=0}^{N-1} h(k)x(n-k)$$

where N is the filter length.

In the TMS320C50, for example, the FIR equation can be efficiently implemented using the instruction pair:

```
RPT     NM1
MACD    HNM1, XNM1
```

The first instruction, RPT NM1, loads the filter length minus 1 ($N-1$) into the repeat instruction counter, and causes the multiply–accumulate with data move (MACD) instruction following it to be repeated N times. The MACD instruction performs a number of operations in one cycle:

(1)   multiplies the data sample, $x(n-k)$, in the data memory by the coefficient, $h(k)$, in the program memory;

(2)   adds previous product to the accumulator;

(3)   implements the unit delay, symbolized by z^{-1}, by shifting the data sample, $x(n-k)$, up to update the tapped delay line.

In the Motorola DSP56000 DSP processor family, as in the TMS320 family, the MAC instruction, together with the repeat instruction (REP) may be used to implement an FIR filter efficiently:

```
REP    #N−1
MAC    X0, Y0, A    X: (R0)+, X0    Y: (R4)+, Y0
```

Here the repeat instruction is used with the MAC instruction to perform sustained multiplication and sums of product operations. Again, notice the ability to perform multiple operations with one instruction, made possible by having multiple data paths. The contents of the registers X0 and Y0 are multiplied together and the product added to the accumulator. At the same time, the next data sample and corresponding coefficient are fetched from the X and Y memories for multiplication.

In most modern DSP processors, the concept of instruction repeat has been taken further by providing instructions that allow a block of code, not just a single instruction, to be repeated a specified number of times. In the TMS320 family (e.g. TMS320C50, TMS320C54 and TMS320C30), the format for repeat execution of a block of instructions, with a zero-overhead loop, is:

```
RPTB loop
     :
     :
loop    (last instruction)
```

Repeat instructions provided by some DSP processors have high-level language features. In the Motorola DSP56000 and DSP56300 families zero-overhead DO loops are provided which may also be nested. The example below illustrates a nested DO loop in which the outer loop is executed N times and the inner loop NM times.

```
        DO #N, LOOP1
        :
        DO #M, LOOP2
        :
        :
LOOP2 (last instruction is placed here)
        :
LOOP1 (last instruction in the outer loop is placed here)
```

Nested loops are useful for efficient implementation of DSP functions such as FFT algorithms and two-dimensional signal processing.

Analog Devices DSP processors (e.g. ADSP-2115 and SHARC processors) also have nested-looping capability. The ADSP-2115 supports up to four levels of nested loops. The format for looping is

```
        CNTR = N
        DO LOOP UNTIL CE
        :
        :
LOOP: (last instruction in the loop)
```

The loop is repeated until the counter expires. The loop can contain a large block of instructions, not just a single instruction. The format for nested looping is essentially the same as for the DSP56000 family.

Adaptive filtering is another key generic function in modern signal processing and special instructions are being provided to support it. In adaptive filtering, coefficient update is an important step (see Chapter 10) and involves computing a new set of coefficient values based on previous values. For example, in LMS-based adaptive filters, the coefficient update task is characterized by the following equation:

$$h_{k+1}(i) = h_k(i) + 2\mu e_k x(k - i) \tag{12.3a}$$

where

$$e_k = y_k - \hat{y}_k \tag{12.3b}$$

and $\hat{y}_k$ is the output of the adaptive filter,

$$\hat{y}_k = \sum_{i=0}^{N-1} h_k(i) x(k - i) \tag{12.3c}$$

Typically, the FIR filtering part of the coefficient update task (Equation 12.3c) is implemented using special MAC instructions as described previously. The coefficient update task (Equation 12.3a) may be performed using a repeat block instruction with zero-overhead looping. In the TMS320C54, the LMS-based adaptive filter instruction, LMS, combined with the multiply instruction, ST//MPY, and repeat block instructions, RPTBD, can be used to calculate the adaptive filter output and update the filter coefficients. This can lead to reductions in execution times and code size of adaptive filters.

In fast Fourier transformation, another key DSP function, there is always a need for scrambling the input data sequence before FFT or unscrambling the data after FFT to ensure that the data points appear in the correct sequence. All high performance general purpose DSP processors provide special instructions for bit-reversed addressing to perform the required scrambling/unscrambling at the same time as a data sample is being moved or fetched.

For example, the bit-reversed addressing of the TMS320 can be used to perform bit reversal whilst storing an N-point complex input data sequence:

```
RPT     N2
BLDD    #XN, *BR0+
```

In this case, the input data is complex and so each data sample consists of a real and an imaginary data value. Thus, each data sample is stored in two data locations.

Modern DSP processors also feature application-oriented instructions for applications such as speech coding (e.g. those for codebook search), digital audio (e.g. those for surround sound) and telecommunications (e.g. those for Viterbi decoding).

12.2.5 Replication

In DSP, replication involves using two or more basic units, for example using more than one ALU, multiplier or memory unit. Often the units are arranged to work simultaneously. In DSP, the norm is to have one CPU, with one or more arithmetic elements replicated.

However, full-blown parallel processing concepts where for example a number of independent processors work on a given task, or several processors under one control unit work simultaneously on a single problem, are now being extended to DSP. A number of parallel DSP processors, e.g. TMS320C40 and ADSP-21060 SHARC, are now available.

12.2.6   On-chip memory/cache

In most cases, DSP chips operate so fast that slow inexpensive memories are unable to keep up. The common practice is to slow the processor down by adding wait states. In some processors, wait states are software programmable, but in others a piece of external hardware is necessary to slow the processor down. Wait states mean of course that the processor cannot operate at full speed.

To alleviate this problem many DSP chips contain fast on-chip data RAMs and/or ROMs. In such processors, slow external memories may be used to hold the program code. At initialization, the code may be transferred to the fast, internal memory for full-speed execution. Fast on-chip EPROMs are useful for real-time development and for final prototyping. Some chips provide an on-chip program cache which may be used to hold often repeated sections of a program. Execution of codes in the cache avoids further memory fetches and speeds up program execution.

Provision of an on-chip memory is now a norm.

12.2.7   Extended parallelism – SIMD, VLIW and static superscalar processing

The trend in DSP processor architecture design is to increase both the number of instructions executed in each cycle and the number of operations performed per instruction to enhance performance (Hacker, 1999; Texas Instruments, 1999; Berkeley Design Technology, 1999; Levy, 1999; Blalock, 1997).

In newer DSP processor architectures, parallel processing techniques are extensively used to achieve increased computational performance. The three techniques that are used, often in combination, are single instruction, multiple data (SIMD), very-large-instruction-word (VLIW) and superscalar processing (Hacker, 1999; Texas Instruments, 1999; Hayes, 1998).

SIMD processing is used to increase the number of operations performed per instruction. Typically, in DSP processors with SIMD architectures the processor has multiple data paths and multiple execution units. Thus, a single instruction may be issued to the multiple execution units to process blocks of data simultaneously and in this way the number of operations performed in one cycle is increased (see Figure 12.11, for example). In the case of Figure 12.11, the DSP processor has two execution units, each with its own ALU, MAC and shifter. The processor is able to perform two separate arithmetic operations (e.g. additions and MACs) simultaneously with a single instruction. Examples of DSP processors with SIMD architectures and dual execution units include Lucent DSP16000, Texas Instruments TMS320C62x and Analog Devices TigerSHARC, ADSP-TS001.

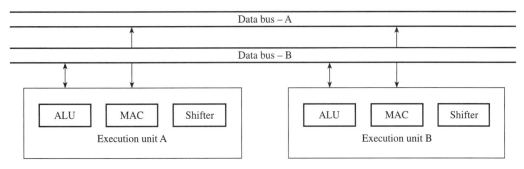

Figure 12.11 Dual arithmetic units with dual data paths for SIMD processing.

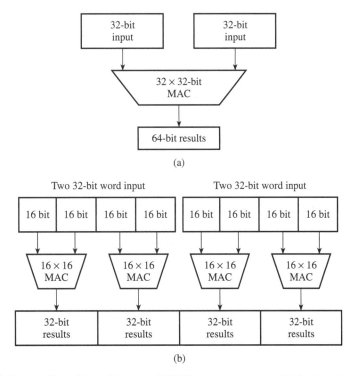

Figure 12.12 An illustration of the use of SIMD processing and multiple data size capability to extend the number of multiplier/accumulators (MACs) from one to four in a TigerSHARC processing element.

An attractive feature of SIMD architectures, especially those that support multiple data sizes, is the ability to effectively increase the number of execution units available and hence the number of operations performed per cycle by partitioning the units. For example, in the TigerSHARC, the two multipliers can each be partitioned to perform four 16-bit × 16-bit multiply–accumulate operations concurrently instead of one 32-bit × 32-bit multiply–accumulate operation (see Figure 12.12).

Clearly, in applications where data can be processed in parallel, SIMD processing can substantially enhance processor performance. However, in applications with sequential data, the scope for increased computational performance by SIMD processing is less. It is for this reason that the next generation DSP processors that are aimed at multi-channel applications, such as the third generation mobile communications systems, tend to have SIMD capability (Hacker, 1999; Texas Instruments, 1999; Levy, 1999).

Very-long-instruction-word (VLIW) processing is an important approach for substantially increasing the number of instructions that are processed per cycle (Texas Instruments, 1999). A *very-long-instruction word* is essentially a concatenation of several *short* instructions and requires multiple execution units, running in parallel, to carry out the instructions in a single cycle.

The principles of VLIW architecture and data flow for the TMS320C62x family of advanced, fixed-point DSP processors is illustrated in Figure 12.13. The CPU contains two data paths and eight independent execution units, organized in two sets – (L1, S1, M1 and D1) and (L2, S2, M2 and D2). In this case, each *short* instruction is 32 bits wide and eight of these are linked together to form a *very long instruction* word packet which may be executed in parallel.

The VLIW processing starts when the CPU fetches an instruction packet (eight 32-bit instructions) from the on-chip program memory. The eight instructions in the

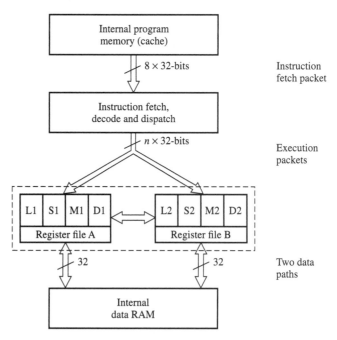

Figure 12.13   Principles of very long instruction word (VLIW) architecture and data flow in the TMS320C62x. Note: L1, L2 are logic units; S1, S2 are shifter/logic units; M1, M2 are multipliers; D1, D2 are data address units.

fetch packet are formed into an *execute* packet, if they can be executed in parallel, and then dispatched to the eight execution units as appropriate. The next 256-bit instruction packet is fetched from the program memory while the execute packet is decoded and executed. If the eight instructions in a fetch packet are not executable in parallel (for example if the eight instructions were all multiply–accumulate instructions, then only two can be performed in a cycle because there are only two multipliers available), then several execute packets will be formed and dispatched to the execution units, one at a time. A fetch packet is always 256 bits wide (eight instructions), but an execute packet may vary between one and eight instructions.

The VLIW architecture is clearly designed to support instruction level parallelism. This architecture, together with fast clock speeds (typically, 200 MHz), leads to very high performance DSP processors. In the TMS320C62x, the instruction parallelism is scheduled at compile time. However, the computational efficiency of such processors falls if the instructions cannot be executed in parallel.

Superscalar processing is another technique for increasing the instruction rate of a DSP processor (the number of instructions processed in a cycle) by exploiting instruction-level parallelism. Traditionally, the term *superscalar* refers to computer architectures that enable multiple instructions to be executed in one cycle (Hayes, 1998). Such architectures are widely used in general purpose processors, such as PowerPC and Pentium processors. In *superscalar* DSP processors, multiple execution units are provided and several instructions may be issued to the units for concurrent execution. Extensive use is also made of pipelining techniques to increase performance further.

The best known superscalar DSP processor is the Analog Devices TigerSHARC (Hacker, 1999): see Figure 12.14. The TigerSHARC is described as a static superscalar DSP processor because parallelism in the instructions is determined before run-time. In fact, the TigerSHARC processor combines SIMD, VLIW and superscalar concepts. This advanced, DSP processor has multiple data paths and two sets of independent execution units, each with a multiplier, ALU, a 64-bit shifter and a register file (see Figure 12.14). TigerSHARC is a floating point processor, but it supports fixed arithmetic with multiple data types (8-, 16-, and 32-bit numbers).

The instruction width is not fixed in the TigerSHARC processor. In each cycle, up to four 32-bit instructions are fetched from the internal program memory and issued to the two sets of execution units in parallel. An instruction may be issued to both units in parallel (Single Instruction, Multiple Data – SIMD – instructions) or to each execution unit independently. Each execution unit (ALU, multiplier or shifter) takes its inputs from and returns its results to the register file. The register files are connected to the three data paths and so can simultaneously read two inputs and write an output to memory in a cycle. This load/store architecture is suited to basic DSP operations which often take two inputs and compute an output.

As discussed earlier, because the processor can work on several data sizes (8-bit, 16-bit, 32-bit and 64-bit), the execution units allow further levels of parallel computation. Thus, in each cycle the TigerSHARC can execute up to eight addition/ subtract operations and eight multiply–accumulate operations with 16-bit inputs, instead of two multiply–accumulate operations with 32-bit inputs. The capability for handling mixed data types and for breaking large instruction words into separate

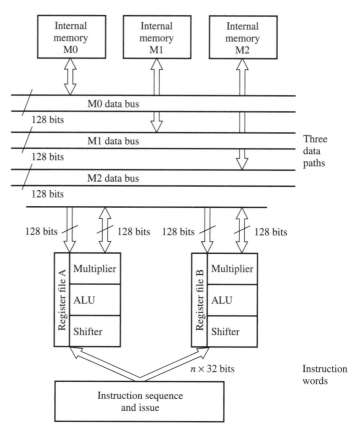

Figure 12.14   Principles of superscalar architecture and data flow in the TigerSHARC DSP processor.

instructions for the execution units enables the processor to exploit extensively instruction-level parallelism.

It is worth mentioning that in DSP processors using advanced architectures, such as VLIW and superscalar, some form of static scheduling of instructions prior to runtime is necessary to make an efficient use of the parallel execution units. There is also a need to avoid problems associated with data dependencies (e.g. results needed before they are ready) and control dependencies (e.g. branch instructions).

12.3   General-purpose digital signal processors

General-purpose digital signal processors are basically high speed microprocessors, with hardware architectures and instruction sets optimized for DSP operations. These processors make extensive use of parallelism, Harvard architecture, pipelining and dedicated hardware whenever possible to perform time-consuming operations, such as shifting/scaling, multiplication, and so on.

General-purpose DSPs have evolved substantially over the last decade as a result of the never-ending quest to find better ways to perform DSP operations, in terms of computational efficiency, ease of implementation, cost, power consumption, size, and application-specific needs (Levy, 1998, 1999; Berkeley Design Technology, 1999). The insatiable appetite for improved computational efficiency has led to substantial reductions in instruction cycle times and, more importantly, to increasing sophistication in the hardware and software architectures. It is now common to have dedicated, on-chip arithmetic hardware units (e.g. to support fast multiply/accumulate operations), large on-chip memory with multiple access and special instructions for efficient execution of inner core computations in DSP. We have also seen a trend towards increased data word sizes (e.g. to maintain signal quality) and increased parallelism (to increase both the number of instructions executed in one cycle and the number of operations performed per instruction). Thus, we find that in newer general-purpose DSP processors increasing use is made of multiple data paths/arithmetic to support parallel operations. DSP processors based on SIMD (Single Instruction, Multiple Data), VLIW (Very Large Instruction Word) and superscalar architectures are being introduced to support efficient parallel processing. In some DSPs, performance is enhanced further by using specialized, on-chip co-processors to speed up specific DSP algorithms such as FIR filtering and Viterbi decoding. The explosive growth in communications and digital audio technologies has had a major influence in the evolution of DSPs, as has growth in embedded DSP processor applications.

In the next two sections, we will describe briefly the architectural features of several generations of fixed-point and floating-point DSP processors.

12.3.1 Fixed-point digital signal processors

Fixed-point DSP processors available today differ in their detailed architecture and the onboard resources provided. A summary of key features of four generations of fixed-point DSP processors from four leading semiconductor manufacturers is given in Table 12.1. The classification of DSP processors into the four generations is partly based on historical reasons, architectural features, and computational performance.

The basic architecture of the *first generation* fixed-point DSP processor family (TMS320C1x), first introduced in 1982 by Texas Instruments, is depicted in Figure 12.15. Key features of the TMS320C1x are the dedicated arithmetic units which include a multiplier and an accumulator. The processor family has a modified Harvard architecture with two separate memory spaces for programs and data. It has an on-chip memory and special instructions for execution of basic DSP algorithms, although these are limited.

Second generation fixed-point DSPs have substantially enhanced features compared to the first generation. In most cases, these include much larger on-chip memories and more special instructions to support efficient execution of DSP

Table 12.1 Features of general-purpose fixed-point DSPs from Texas Instruments, Motorola and Analog Devices.

Gener-ation	Fixed-point DSP	Data path width (bits)	No. of data paths	Data wordlength (bits)	Accum. wordlength (bits)	Instruction width (bits)	On-chip RAM size (words)	Instruction cache size (no. of inst.)	No. of multipliers	Performance index*
1	Texas Instruments TMS320C10	16	1	16	32	16	144		1	
2	Texas Instruments TMS320C50	16	2	16	32	16	10 K		1	10 @ 50 MHz
	Motorola DSP56002	24	2	24	56	24	1 K		1	13 @
	Analog Devices DSP-2100	16	2	16	40	24	32 K	16	1	13 @ 52 MHz
	Lucent Technologies 1600	16	2	16	36	16		15	1	22 @ 120 MHz
3	Texas Instruments TMS320C54	16	3	16	40	16	32 K		1	25 @ 100 MHz
	Motorola DSP56300	24	3	24	56	24		3 K	1	25 @ 100 MHz
	Lucent Technologies 16000	32	2	32	40	32	127 K	31	2	36 @ 100 MHz
4	Texas Instruments TMS320C6200		2		40	256	17 K	64 K	2	86 @ 133.6 MHz

* Performance index is based on execution speed of benchmark DSP kernels/algorithms (Levy, 1998; Berkeley Design Technology, 1999).

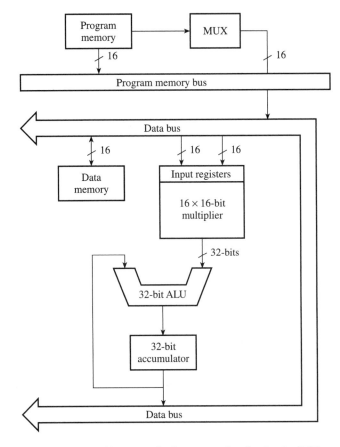

Figure 12.15 A simplified architecture of a first generation fixed-point DSP processor (Texas Instruments TMS320C10).

algorithms. As a result, the computational performance of second generation DSP processors is four to six times that of the first generation.

Typical second generation DSP processors include Texas Instruments TMS320C5x, Motorola DSP5600x, Analog Devices ADSP21xx and Lucent Technologies DSP16xx families. Texas Instruments first and second generation DSPs have a lot in common, architecturally, but second generation DSPs have more features and increased speed (see Table 12.1). The internal architecture that typifies the TMS320C5x family of processors is shown in Figure 12.16 in a simplified form to emphasize the dual internal memory spaces which are characteristic of the Harvard architecture. Special instructions for DSP operations include a multiply and accumulate with data move instruction which, for example, can be combined with a repeat instruction to execute an FIR filter with considerable time savings. Its bit-reversed addressing capability is useful in FFTs. Unlike the first generation fixed-point processor family, C1x, which has a very limited internal memory, the C5x provides more on-chip memory.

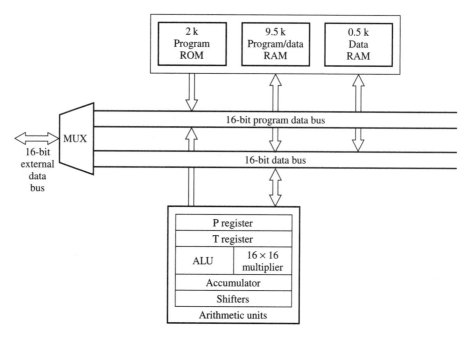

Figure 12.16 A simplified architecture of a second generation fixed-point DSP
(Texas Instruments TMS320C50).

The Motorola DSP5600x processor is a high-precision fixed-point digital signal
processor. Its architecture is depicted in Figure 12.17. Internally, it has two independent data memory spaces, the X-data and Y-data memory spaces, and one program
memory space. Having two separate data memory spaces allows a natural partitioning
of data for DSP operations and facilitates the execution of the algorithm. For example,
in graphics applications data can be stored as X and Y data, in FIR filtering as
coefficients and data, and in FFT as real and imaginary. During program execution,
pairs of data samples can be fetched or stored in internal memory simultaneously in
one cycle. Externally, the two data spaces are multiplexed into a single data bus,
reducing somewhat the benefits of the dual internal data memory. The arithmetic
units consist of two 56-bit accumulators and a single cycle, fixed-point hardware
multiplier–accumulator (MAC). The MAC accepts 24-bit inputs and produces a
56-bit product. The 24-bit wordlength provides sufficient accuracy for representing
most DSP variables while the 56-bit accumulator (including eight guard bits) prevents
arithmetic overflows. These wordlengths are adequate for most applications, including digital audio, which impose stringent requirements. The 5600x processors provide
special instructions that allow zero-overhead looping and bit-reversed addressing
capability for scrambling input data before FFT or unscrambling the fast Fourier
transformed data.

Analog Devices ADSP21xx is another family of second generation fixed-point DSP
processors with two separate external memory spaces – one holds data only, and the
other holds program code as well as data. A simplified block diagram of the internal
architecture of the ADSP21xx is depicted in Figure 12.18. The main components are

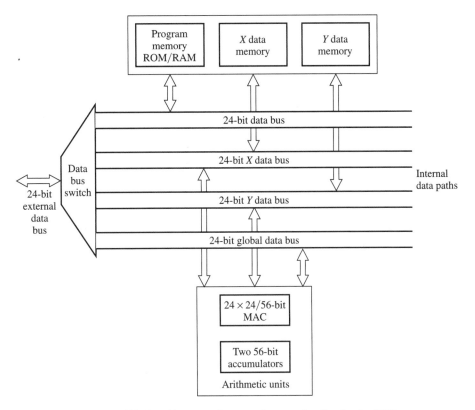

Figure 12.17   A simplified architecture of a second generation fixed-point DSP (Motorola DSP56002).

the ALU, multiplier–accumulator, and shifters. The MAC accepts 16×16-bit inputs and produces a 32-bit product in one cycle. The accumulator of the ADSP21xx has eight guard bits which may be used for extended precision. The ADSP21xx departs from the strict Harvard architecture, as it allows the storage of both data and program instructions in the program memory. A signal line (data access signal) is used to indicate when data and not program instructions are being fetched from the program memory. Storage of data in the program memory inhibits a steady data flow through the CPU as data and instruction fetches cannot occur simultaneously. To avoid a bottleneck, the ADSP21xx family has an on-chip program memory cache which holds the last 16 instructions executed. This eliminates the need, especially when executing program loops, for repeated instruction fetches from program memory. The ADSP21xx provides special instructions for zero-overhead looping and supports a bit-reversing addressing facility for FFT. The processor family has a large on-chip memory (up to 64 Kbytes of internal RAM are provided for increased data transfer). The processor has an excellent support for DMA. External devices can transfer data and instructions to or from the DSP processor RAM without processor intervention.

Lucent Technologies' DSP16xx family of fixed-point DSPs (see Figure 12.19) is targeted at the telecommunications and modem market. In terms of computational

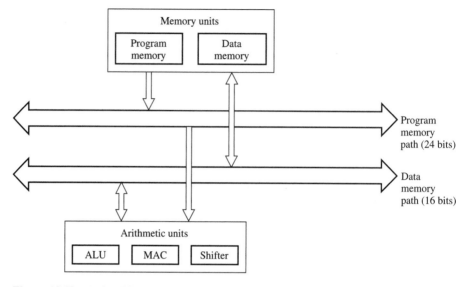

Figure 12.18   A simplified architecture of a second generation fixed-point DSP (Analog Devices ADSP2100).

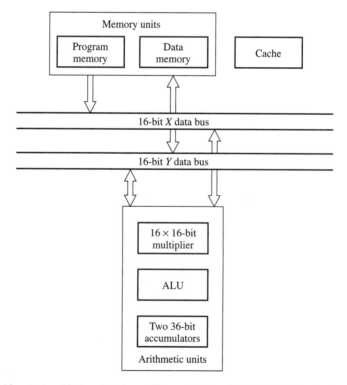

Figure 12.19   A simplified architecture of Lucent Technologies' DSP16xx fixed-point DSP.

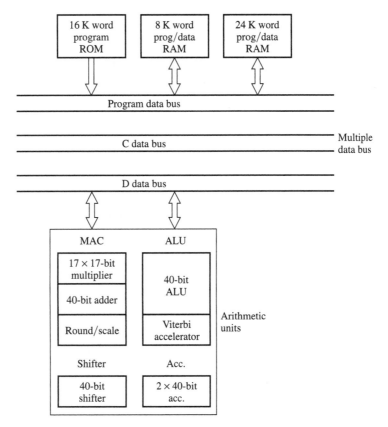

Figure 12.20   A simplified architecture of a third generation fixed-point DSP (Texas Instruments TMS320C54x).

performance, it is one of the most powerful second generation processors. The processor has a Harvard architecture, and like most of the other second generation processors, it has two data paths, the *X* and *Y* data paths. Its data arithmetic units include a dedicated 16 × 16-bit multiplier, a 36-bit ALU/shifter (which includes four guard bits) and dual accumulators. Special instructions such as those for zero-overhead single and block instruction looping are provided.

Third generation fixed-point DSPs are essentially enhancements of second generation DSPs. In general, performance enhancements are achieved by increasing and/or making more effective use of available on-chip resources. Compared to the second generation DSPs, features of the third generation DSPs include more data paths (typically three compared to two in the second generation), wider data paths, larger on-chip memory and instruction cache and in some cases a dual MAC. As a result, the performance of third generation DSPs is typically two or three times superior to that of the second generation DSP processors of the same family (Levy, 1998; Berkeley Design Technology, 1999). Simplified architectures of three third generation DSP processors, TMS320C54x, DSP563x and DSP16000, are depicted in Figures 12.20,

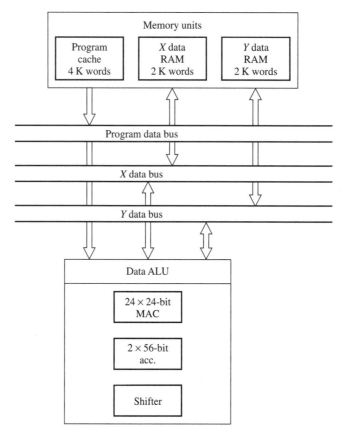

Figure 12.21   A simplified architecture of a third generation fixed-point DSP (Motorola DSP56300).

12.21 and 12.22. Most of the third generation fixed-point DSP processors are aimed at applications in digital communications and digital audio, reflecting the enormous growth and influence of these application areas on DSP processor development. Thus, we find features in some of the processors that support these applications. The TMS320C54x, for example, includes special instructions for adaptive filtering (which is often used for echo cancellation and adaptive equalization in telecommunications) and to support Viterbi decoding. In the third generation processors, semiconductor manufacturers have also taken the issue of power consumption seriously (because of its importance in portable and hand-held devices such as the mobile phone). Most of the third generation DSP processors are low power and have a power management facility.

Fourth generation fixed-point DSP processors with their new architectures are primarily aimed at large and/or emerging multichannel applications, such as digital subscriber loops, remote access server modems, wireless base stations, third generation mobile systems and medical imaging. The new fixed-point architecture that has

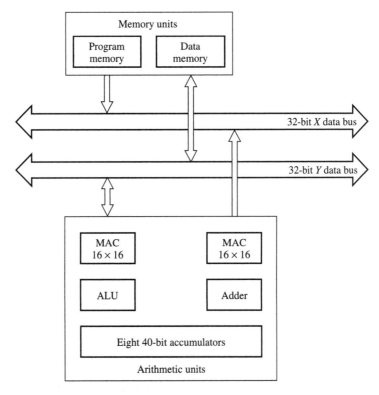

Figure 12.22   A simplified architecture of a third generation fixed-point DSP (Lucent Technologies DSP16000).

attracted a great deal of attention in the DSP community is the *very long instruction word* (VLIW) (see Section 12.2 for more details). The new architecture makes extensive use of parallelism whilst retaining some of the good features of previous DSP processors. Compared to previous generations, fourth generation fixed point DSP processors, in general, have wider instruction words, wider data paths, more registers, larger instruction cache and multiple arithmetic units, enabling them to execute many more instructions and operations per cycle.

Texas Instruments' TMS320C62x family of fixed-point DSP processors is based on the VLIW architecture: see Figure 12.23. The core processor has two independent arithmetic paths, each with four execution units – a logic unit (Li), a shifter/logic unit (Si), a multiplier (Mi) and a data address unit (Di). Typically, the core processor fetches eight 32-bit instructions at a time, giving an instruction width of 256 bits (and hence the term *very long instruction word*). With a total of eight execution units, four in each data path, the TMS320C62x can execute up to eight instructions in parallel in one cycle. The processor has a large program and data cache memories (typically, 4 Kbyte of level 1 program/data caches and 64 Kbyte of level 2 program/data cache). Each data path has its own register file (sixteen 32-bit registers), but can also access registers on the other data path. Advantages of VLIW architectures

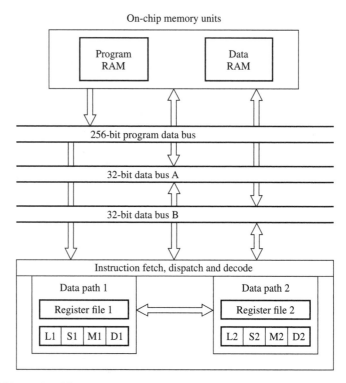

Figure 12.23   A simplified architecture of a fourth generation fixed-point, very long instruction word, DSP processor (Texas Instruments TMS320C62x). Note the two independent arithmetic data paths, each with four execution units – L1, S1, M1 and D1; L2, S2, M2 and D2.

include simplicity and high computational performance. Disadvantages include increased program memory usage (organization of codes to match the inherent parallelism of the processor may lead to inefficient use of memory). Further, optimum processor performance can only be achieved when all the execution units are busy which is not always possible because of data dependencies, instruction delays and restrictions in the use of the execution units. However, sophisticated programming tools are available for code packing, instruction scheduling, resource assignment and in general to exploit the vast potential of the processor.

12.3.2   Floating-point digital signal processors

The ability of DSP processors to perform high speed, high precision DSP operations using floating point arithmetic has been a welcome development. This minimizes finite word length effects such as overflows, roundoff errors, and coefficient quantization errors inherent in DSP. It also facilitates algorithm development, as a designer can develop an algorithm on a large computer in a high level language and then port it to a DSP device more readily than with a fixed point.

Floating-point DSP processors retain key features of fixed-point processors such as special instructions for DSP operations and multiple data paths for multiple operations. As in the case of fixed-point DSP processors, floating-point DSP processors available are significantly different architecturally. Some of the key features of the three generations of floating-point DSP processors from Texas Instruments and Analog Devices are summarized in Table 12.2.

The TMS320C3x is perhaps the best known family of first generation general-purpose floating-point DSPs. The C3x family are 32-bit single chip digital signal processors and support both integer and floating-point arithmetic operations. They have a large memory space and are equipped with many on-chip peripheral facilities to simplify system design. These include a program cache to improve the execution of commonly used codes, and on-chip dual access memories. The large memory spaces cater for memory intensive applications, for example graphics and image processing. In the TMS320C30, a floating-point multiplication requires 32-bit operands and produces a 40-bit normalized floating-point product. Integer multiplication requires 24-bit inputs and yields 32-bit results. Three floating-point formats are supported. The first is a 16-bit short floating-point format, with 4-bit exponents, 1 sign bit and 11 bits for mantissa. This format is for immediate floating-point operations. The second is a single-precision format with an 8-bit exponent, 1 sign bit and 23-bit fractions (32 bits in total). The third is a 40-bit extended precision format which has an 8-bit exponent, 1 sign bit and 31-bit fractions. The floating-point representation differs from that of standard IEEE, but facilities are provided to allow conversion between the two formats. The TMS320C3x combines the features of Harvard architecture (separate buses for program instructions, data and I/O) and von Neumann processor (unified address space).

The emphasis in the second generation, general-purpose floating-point DSPs is on multiprocessing and multiprocessor support. Key issues in multiprocessor support include inter-processor communication, DMA transfers and global memory sharing. The best known second generation floating-point DSP families are Texas Instruments TMS320C4x and Analog Devices ADSP-2106x SHARC (Super Harvard Architecture Computer). The C4x shares some of the architectural features of the C3x, but it was designed for multiprocessing. The C40x family has good I/O capabilities – it has six COMM ports for inter-processor communication and six 32-bit wide DMA channels for rapid data transfers. The architecture allows multiple operations to be performed in parallel in one instruction cycle. The C4x family supports both floating- and fixed-point arithmetic. The native floating-point data format in the C40 differs from the IEEE 754/854 standard, although conversion between them can be readily accomplished.

Analog Devices ADSP-2106x SHARC DSP processors are also 32-bit floating-point devices. They have large internal memory and impressive I/O capability – 10 DMA channels to allow access to internal memory without intervention and six Link ports for inter-processor communications at high speed. The architecture allows shared global memory, making it possible for up to six SHARC processors to access each other's internal RAM at up to full data rate. The ADSP-2106x family supports both fixed-point and floating-point arithmetic. Its single precision floating-point

Table 12.2   Features of general-purpose floating-point DSP processors from Texas Instruments and Analog Devices.

Gener- ation	Floating-point DSPs	Data path width (bits)	No. of data paths	Data wordlength (bits)	Accum. wordlength (bits)	Instruction width (bits)	On-chip RAM size (words)	Instruction cache size (no. of inst.)	No. of multipliers	Performance index*
1	Texas Instruments TMS320C30	16	1	32	40	32	2 K	64	1	7 @ 30 MHz
2	Texas Instruments TMS320C40	16	2	32	40	32	2 K	128	1	7 @ 30 MHz
	Analog Devices ADSP-21060	24	2	32	80	48	128 K	32	1	14 @ 50 MHz
3	Texas Instruments TMS320C67x	16	3	16	40		17 K		2	
	Analog Devices TigerSHARC	128	3	32	40/80	128	192 K		2	

* Performance index is based on execution speed of benchmark DSP kernels/algorithms (Levy, 1998; Berkeley Design Technology, 1999).

format complies with the single precision IEEE 754/854 floating-point standard (24-bit mantissa and 8-bit exponent). The architecture also supports multiple operations per cycle.

Third generation floating-point DSP processors take the concepts of parallelism much further to increase both the number of instructions and the number of operations in a cycle to meet the challenges of multichannel and computationally intensive applications. This is achieved by the use of new architectures, the VLIW (very long instruction word) and superscalar architectures in particular. The two leading third generation floating-point DSP processor families are the Texas Instruments TMS320C67x and Analog Devices ADSP-TS001. The TMS320C67x family has the same VLIW architecture as the advanced, fourth generation fixed-point DSP processors, TMS320C62x (see Figure 12.23 and Section 12.3.1 for more details).

The TigerSHARC DSP family supports mixed arithmetic types (fixed and floating point arithmetic) and data types (8-, 16-, and 32-bit numbers). This flexibility makes it possible to use the arithmetic and data type most appropriate for a given application to enhance performance. As with the TMS320C67x, the TigerSHARC is aimed at large-scale, multi-channel applications, such as the third generation mobile systems (3G wireless), digital subscriber lines (xDSL) and remote, multiple access server modems for Internet services. TigerSHARC, with its static superscalar architecture (see Figure 12.14 and Section 12.3.1 for more details), combines the good features of VLIW architecture, conventional DSP architecture, and RISC computers. The processor has two computation blocks, each with a multiplier, ALU and 64-bit shifter. The processor can execute up to eight MAC operations per cycle with 16-bit inputs and 40-bit accumulation, two 40-bit MACs on 16-bit complex data or two 80-bit MACs with 32-bit data. With 8-bit data, TigerSHARC can issue up to 16 operations in a cycle. TigerSHARC has a wide memory bandwidth, with its memory organized in three 128-bit wide banks. Access to data can be in variable data sizes – normal 32-bit words, long 64-bit words or quad 128-bit words. Up to four 32-bit instructions can be issued in one cycle. To avoid the use of large NOPs (which is a disadvantage of VLIW designs), the large instruction words may be broken down into separate short instructions which are issued to each unit independently.

12.4 Selecting digital signal processors

The choice of a DSP processor for a given application has become an important issue in recent years because of the wide range of processors available (Levy, 1999; Berkeley Design Technology, 1996, 1999). Specific factors that may be considered when selecting a DSP processor for an application include architectural features, execution speed, type of arithmetic and wordlength.

(1) *Architectural features*   Most DSP processors available today have good architectural features, but these may not be adequate for a specific application.

Key features of interest include size of on-chip memory, special instructions and I/O capability. On-chip memory is an essential requirement in most real-time DSP applications for fast access to data and rapid program execution. For memory hungry applications (e.g. digital audio – Dolby AC-2, FAX/Modem, MPEG coding/decoding), the size of internal RAM may become an important distinguishing factor. Where internal memory is insufficient this can be augmented by high speed, off-chip memory, although this may add to system costs. For applications that require fast and efficient communication or data flow with the outside world, I/O features such interface to ADC and DACs, DMA capability and support for multiprocessing may be important. Depending on the application, a rich set of special instructions to support DSP operations are important, e.g. zero-overhead looping capability, dedicated DSP instructions, and circular addressing.

(2) *Execution speed*   The speed of DSP processors is an important measure of performance because of the time-critical nature of most DSP tasks. Traditionally, the two main units of measurement for this are the clock speed of the processor, in MHz, and the number of instructions performed, in millions of instructions per second (MIPS) or, in the case of floating-point DSP processors, in millions of floating-point operations per second (MFLOPS). However, such measures may be inappropriate in some cases because of significant differences in the way different DSP processors operate, with most able to perform multiple operations in one machine instruction. For example, the C62x family of processors can execute as many as eight instructions in a cycle. The number of operations performed in each cycle also differs from processor to processor. Thus, comparison of execution speed of processors based on such measures may not be meaningful. An alternative measure is based on the execution speed of benchmark algorithms (Levy, 1998; Berkeley Design Technology, 1999) – e.g. DSP kernels such as FFT, FIR and IIR filters. In Tables 12.1 and 12.2, performance indices based on such benchmarks give an indication of the relative performance of a number of popular DSP processors.

(3) *Type of arithmetic*   The two most common types of arithmetic used in modern DSP processors are fixed- and floating-point arithmetic. Floating arithmetic is the natural choice for applications with wide and variable dynamic range requirements (dynamic range may be defined as the difference between the largest and smallest signal levels that can be represented, or the difference between the largest signal and the noise floor, measured in decibels). Fixed-point processors are favoured in low cost, high volume applications (e.g. cellular phones and computer disk drives). The use of fixed-point arithmetic raises issues associated with dynamic range constraints which the designer must address (see Chapter 13 for more details). In general, floating processors are more expensive than fixed-point processors, although the cost difference has fallen significantly in recent years. Most floating-point DSP processors available today also support fixed-point arithmetic.

(4) *Wordlength*   Processor data wordlength is an important parameter in DSP as it can have a significant impact on signal quality. It determines how accurately parameters and results of DSP operations can be represented (see Chapter 13 for more details). In general, the longer the data word the lower the errors that are introduced by digital signal processing. In fixed-point audio processing, for example, a processor wordlength of at least 24 bits is required to keep the smallest signal level sufficiently above the noise floor generated by signal processing to maintain CD quality. A variety of processor wordlengths are used in fixed-point DSP processors, depending on application (see Table 12.1). Fixed-point DSP processors aimed at telecommunications markets tend to use a 16-bit wordlength (e.g. TMS320C54x), whereas those aimed at high quality audio applications tend to use 24 bits (e.g. DSP56300). In recent years, we have seen a trend towards the use of more bits for the ADC and DAC (e.g. Cirrus 24-bit audio codec, CS4228) as the cost of these devices falls to meet the insatiable demand for increased quality. Thus, we are likely to see an increased demand for larger processor wordlengths for audio processing. In fixed-point processors, it may also be necessary to provide guard bits (typically 1 to 8 bits) in the accumulators to prevent arithmetic overflows during extended multiply and accumulate operations. The extra bits effectively extend the dynamic range available in the DSP processor. In most floating-point DSP processors, a 32-bit data size (24-bit mantissa and 8-bit exponent) is used for single-precision arithmetic. This size is also compatible with the IEEE floating-point format (IEEE 754). Most floating-point DSP processors also have fixed-point arithmetic capability, and often support variable data size, fixed-point arithmetic.

In practice, factors such as experience/familiarity with a particular DSP processor family, ease of use, time to market and costs may be the overriding factors in selecting a given processor.

12.5   Implementation of DSP algorithms on general-purpose digital signal processors

12.5.1   FIR digital filtering

Nonrecursive *N*-point FIR filters, with the structure given in Figure 12.24(a), are characterized by the following difference equation (see Chapter 7 for details):

$$y(n) = \sum_{k=0}^{N-1} h(k)x(n-k) \tag{12.4}$$

A fragment of a C language implementation of the general FIR filter is given in Program 12.1. For real-time FIR filtering, the data and coefficients are stored in

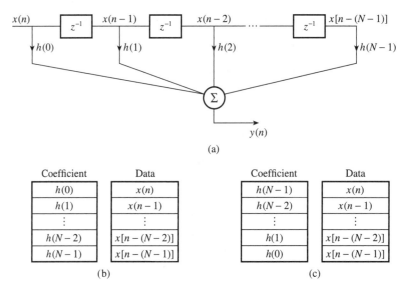

Figure 12.24 Implementation of FIR filter: (a) filter structure; (b) coefficient and data memory maps; (c) alternative memory map.

Program 12.1 A C language pseudo-code for FIR filtering.

```
nm1=N−1;
yn=0;
for(k=0; k<nm1;++k){               /* shift data to make room for new sample */
        x[nm1−k]=x[nm1−k−1];
        x[0]=xn;
}
for(k=0; k<N;++k){
        yn=yn+h[k]*x[k];           /* filter data and compute output sample */
}
return(yn);                        /* filter output sample */
```

---

memory, conceptually, as shown in Figure 12.24(b). To appreciate how the FIR filter works, consider the simple case of $N = 3$, with the following difference equation:

$$y(n) = h(0)x(n) + h(1)x(n-1) + h(2)x(n-2) \tag{12.5}$$

$x(n)$ represents the latest input sample, $x(n-1)$ the last sample, and $x(n-2)$ the sample before that.

Suppose the three-coefficient digital filter is fed from an ADC. The first thing to do is to allocate two sets of contiguous memory locations (in RAM), one for storing the input data $(x(n), x(n-1), x(n-2))$ and the other for the filter coefficients $(h(0), h(1), h(2))$ as depicted below:

Data	Coefficient
RAM	memory
0	$h(0)$
0	$h(1)$
0	$h(2)$

At initialization, the RAM locations where the data samples are to be stored are set to zero since we always start with no data. The following operations are then performed.

(1) *First sampling instant*   Read data sample from the ADC, shift data RAM one place (to make room for the new data), save the new input sample, compute output sample from Equation 12.5 and then send the computed output sample to the DAC:

Data	Coefficient	
RAM	memory	
$\rightarrow x(1)$	$h(0)$	$y(1) = h(0)x(1) + h(1)x(0) + h(2)x(-1)$
0	$h(1)$	
0	$h(2)$	

(2) *Second sampling instant*   Repeat the above operation and work out the new output sample and send to the DAC:

Data	Coefficient	
RAM	memory	
$\rightarrow x(2)$	$h(0)$	$y(2) = h(0)x(2) + h(1)x(1) + h(2)x(0)$
$x(1)$	$h(1)$	
0	$h(2)$	

(3) *Third sampling instant*   Repeat the above operation and work out the new output sample and send to the DAC:

Data	Coefficient	
RAM	memory	
$\rightarrow x(3)$	$h(0)$	$y(3) = h(0)x(3) + h(1)x(2) + h(2)x(1)$
$x(2)$	$h(1)$	
$x(1)$	$h(2)$	

(4) *Fourth sampling instant*   Repeat the above operation and work out the new output sample and send to the DAC:

Data	Coefficient	
RAM	memory	
$\rightarrow x(4)$	$h(0)$	$y(4) = h(0)x(4) + h(1)x(3) + h(2)x(2)$
$x(3)$	$h(1)$	
$x(2)$	$h(2)$	

Note that the oldest data sample has now fallen off the end.

Program 12.2   Straight-line code for three-point FIR filter.

```
NXTPT    IN      XN, ADC
         ZAC
         LT      XNM2
         MPY     H2          ;h(2)x(n–2)

         LTD     XNM1        ;0+h(2)x(n–2); x(n–2)=x(n–1)
         MPY     H1          ;h(1)x(n–1); x(n–1)=x(n–2)

         LTD     XN          ;h(2)x(n–2)+h(1)x(n–1); x(n–1)=x(n)
         MPY     H0          ;h(0)x(n)

         APAC                ;h(2)x(n–2)+h(1)x(n–1)+h(0)x(n)

         SACH    YN,1        ;save output sample
         OUT     YN,DAC      ;output sample to DAC

         B       NXTPT
```

(5)   *nth sampling instant*   Repeat the above operation and work out the new
output sample and send to the DAC:

$$
\begin{array}{ll}
\text{Data} & \text{Coefficient} \\
\text{RAM} & \text{memory} \\
\rightarrow x(n) & h(0) \qquad\qquad y(n) = h(0)x(n) + h(1)x(n-1) + h(2)x(n-2) \\
x(n-1) & h(1) \\
x(n-2) & h(2)
\end{array}
$$

An implementation of the three-point FIR filter in a first generation fixed-point DSP
(the TMS320C10) is given in Program 12.2. In this case, the computation of the products starts at the bottom of the data and coefficients to exploit the TMS320C10 data
move instructions. The instruction pair LTD and MPY are central to the TMS320C10-
based FIR filter implementation. For example, the instruction pair below performs the
shift implied in Equation 12.4 or represented by z^{-1} in Figure 12.24(a), adds the
previous product to the accumulator and calculates the next product, $h(k)x(n-k)$.

```
LTD    XNM1
MPY    H1
```

Specifically, the instruction LTD XNM1 loads the T (temporary) register with the
data sample $x(n-1)$ (held in data RAM address XNM1), adds the previous product,
$h(2)x(n-2)$, which is still in the P (product) register, to the accumulator, and shifts
$x(n-1)$ up to the next address, that is $x(n-2) = x(n-1)$. The second instruction
MPY multiplies the contents of the T register with $h(1)$ and leaves the result in the
product register. The shifting scheme ensures that the input data samples are in the
right locations when the next sample is to be computed.

Straight-line coding of the FIR filter, such as Program 12.2, leads to a fast
implementation, but is not general purpose, and for large N-point filters will not yield
a compact program. In particular, a general purpose FIR filter is implemented by

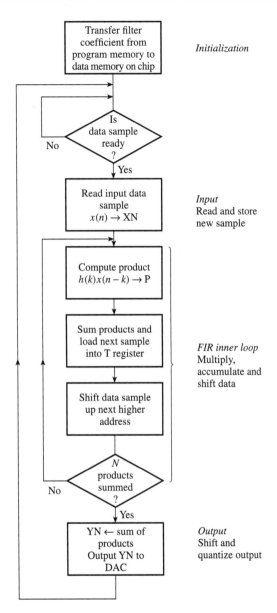

Figure 12.25   Flowchart for the FIR filter. The FIR inner loop executes the convolution sum in Equation 12.4.

setting up an inner loop to execute the FIR equation and calculate the filter output as specified in Equation 12.4.

The flowchart for an N-point FIR showing the inner loop is given in Figure 12.25. In the first generation DSP processor, the inner loop of the FIR filter may be executed by the following instructions:

```
LOOP  LTD      *, AR0      ; shift/update delay line and accumulate products
      MPY      *-, AR1     ; multiply next coefficient and data value
      BANZ     LOOP
```

In this case, the auxiliary registers, AR0 and AR1, are used to point to the data value and coefficient to be multiplied. The auxiliary register, AR1, contains the filter length and acts as a loop counter. The branch on register not zero instruction, BANZ, together with AR1, is used to control the loop. FIR filter implementation in the first generation DSP processor is not efficient because of the overhead associated with loop control.

The second generation fixed-point DSP processors, such as TMS320C50 and Motorola DSP56000, have zero-overhead looping capability and special multiply and accumulate instructions which help to cut down the time to execute the FIR inner loop. In the TMS320C50, the inner loop of an N-point FIR filter which is shown in Figure 12.25 can be efficiently executed using the following instructions:

```
RPT     NM1
MACD    HNM1, XNM1
```

The instruction RPT NM1 loads the filter length minus 1 ($N - 1$) into the repeat register and causes the multiply and accumulate with data shift instruction, MACD, to be repeated $N - 1$ times with zero overhead. The MACD combines the instruction pair LTD MPY into a single instruction, enabling faster execution. The instruction pair RPT and MACD is a good example of time-saving special instructions available in DSP processors.

An alternative approach to implementing the N-point FIR filters in second and later generations of DSP processors is to use circular buffers. It is evident that in FIR filtering, the content of the coefficient memory is static, but the data memory changes when each new input data sample arrives. Effectively, successive new data samples are fed into a sliding window whilst the oldest data samples drop off. A circular buffer may be used to handle the changes in the block of input data samples that are used for FIR filtering without having to shift the data as in linear data buffers.

Conceptually, a circular buffer is the same as a linear buffer if we consider the two ends of the linear buffer to be adjacent, i.e. the latest and oldest data samples, $x(n)$ and $x[n - (N - 1)]$ are adjacent: see Figure 12.26(a). In the circular buffer in Figure 12.26, the data pointer (symbolized by the arrow) points to the memory location of the newest input sample, $x(n)$, and previous input data samples, $x(n - 1), x(n - 2), \ldots x(n - 7)$ are stored in successive locations, clockwise. The FIR inner loop is executed at each sampling period, as before, by multiplying each data sample by the corresponding filter coefficient, $h(k)$ and accumulating the products. The only difference is that the data samples are not shifted. After the inner loop computation, the pointer is located at the $x(n - 7)$, the oldest data sample, which is then overwritten by the next input sample, $x(n)$. Figure 12.26(a)–(c) illustrates how the circular buffer works for three successive data samples.

In practice, circular addressing is achieved by using modulo arithmetic to create an automatic wraparound when the address pointers fall outside the buffer boundary. Typically, we need to specify the start address of the circular buffer and the buffer size

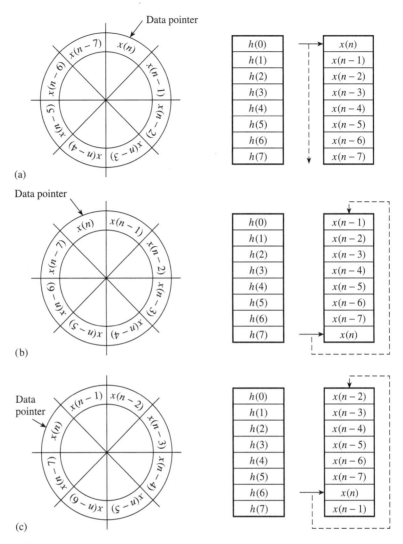

Figure 12.26   An illustration of the principles of circular buffer-based FIR implementation.

(or modulo size). A DSP56000 implementation of the inner loop of an N-point FIR using circular addressing is illustrated below.

```
MOVE     #XDATA, R0
MOVE     #COEFF, R4
MOVE     #N-1, M0                        ; buffer/modulo size
MOVEP    X: INPUT, X: (R0)               ; read and store input sample
CLR      A                               ; clear the accumulator
REP      #N-1                            ; execute FIR inner loop
MAC      X0, Y0, A      X:(R0)+, X0    Y:(R4)+, Y0
MACR     X0, X0, A      (R0)-
```

In this case, circular buffers are used to store both the data and coefficients. The circular data buffer performs the time shift implicitly as described above. However, the circular coefficient buffer is used here for convenience for automatic wraparound of the coefficient pointer. The first four instructions above set up the address pointers, R0 and R4. The inner FIR loop is executed by the instruction-pair REP and MAC. The repeat instruction, REP, repeats the next instruction $N - 1$ times. The next instruction line exploits the multi-path architecture and parallelism of the DSP56000 to perform a set of multiple operations – it multiplies the data value and coefficient which are in X0 and Y0, adds the product to the accumulator and fetches the next data value and coefficient pair to be multiplied from X and Y memories and updates the pointers.

Apart from FIR filtering, circular addressing is useful in the efficient implementation of a number of DSP functions that require time shifts or FIFO queues, e.g. correlation, multirate filters (decimation and interpolation filters) and periodic waveform generation. Its use eliminates the need to move data or for constant checking/resetting of address pointers. Later generations of DSP processors have enhanced circular addressing capability.

| Example 12.3 | A digital FIR notch filter satisfying the specifications given below is to be implemented on the second-generation fixed-point DSP processor, TMS320C50. |

Notch frequency	1.875 kHz
Attenuation at notch frequency	60 dB
Passband edge frequencies	1.575 kHz and 2.175 kHz
Passband ripple	0.01 dB
Sampling frequency	7.5 kHz

A 61-point, optimal FIR filter satisfies the above specifications. The design of this filter was discussed in detail in Section 7.6.5. Here, we will concentrate only on the implementation. The coefficients of the filter are quantized to 16 bits (Q15 format), by multiplying each coefficient by 2^{15}, and then rounding to the nearest integer. The quantized and unquantized coefficients are listed in Table 12.3. As shown in the flowchart in Figure 12.25, the complete FIR filter has at least four essential parts:

(1) *Initialization*   Initialize system; this may include setting up a coefficient table.

(2) *Input section*   This may include reading of the input sample, $x(n)$, e.g. from an ADC via a serial port.

(3) *Inner loop computation*   The execution of the FIR equation to obtain $y(n)$.

(4) *Output section*   This may include shifting/rounding of the result of the inner loop computation and sending this, e.g. to the DAC via a serial port.

As much of steps 1, 2 and 4 are system dependent, we will concentrate on the inner loop computation here. The FIR inner loop may be implemented with the following instructions in the TMS320C50:

```
SACL    XN           ; store newest sample, x(n), in data memory
LAR     AR1, #XNM1   ; point to location of oldest data sample, x[n–(N–1)]
ZAP                  ; clear the accumulator and product register
```

Table 12.3   Filter coefficients for Example 12.3.

	Quantized coefficients
FILTER LENGTH = 61	
***** IMPULSE RESPONSE *****	
H(1) = 0.12743640E−02 = H(61)	42
H(2) = 0.26730640E−05 = H(60)	0
H(3) = −0.23681110E−02 = H(59)	−78
H(4) = −0.17416350E−05 = H(58)	0
H(5) = 0.43428480E−02 = H(57)	142
H(6) = 0.53579250E−05 = H(56)	0
H(7) = −0.71570240E−02 = H(55)	−235
H(8) = −0.49028620E−05 = H(54)	0
H(9) = 0.10897540E−01 = H(53)	357
H(10) = 0.89629280E−05 = H(52)	0
H(11) = −0.15605960E−01 = H(51)	−511
H(12) = −0.85508990E−05 = H(50)	0
H(13) = 0.21226410E−01 = H(49)	695
H(14) = 0.12250150E−04 = H(48)	0
H(15) = −0.27630130E−01 = H(47)	−905
H(16) = −0.11091200E−04 = H(46)	0
H(17) = 0.34579770E−01 = H(45)	1133
H(18) = 0.13800660E−04 = H(44)	0
H(19) = −0.41774130E−01 = H(43)	−1369
H(20) = −0.11560390E−04 = H(42)	0
H(21) = 0.48832790E−01 = H(41)	1600
H(22) = 0.12787590E−04 = H(40)	0
H(23) = −0.55359840E−01 = H(39)	−1814
H(24) = −0.90065860E−05 = H(38)	0
H(25) = 0.60944450E−01 = H(37)	1997
H(26) = 0.88997300E−05 = H(36)	0
H(27) = −0.65232190E−01 = H(35)	−2137
H(28) = −0.38167120E−05 = H(34)	0
H(29) = 0.67925720E−01 = H(33)	2226
H(30) = 0.27041150E−05 = H(32)	0
H(31) = 0.93115220E+00 = H(31)	30512

```
MAR     *, AR1        ; make AR1 current auxiliary register
RPT     #60           ; execute FIR inner loop
MACD    #COEFF, *−    ; multiply and accumulate with data shift
APAC                  ; add last product
```

In this case, the coefficient and data memories are organized as shown in Figure 12.24(c). The auxiliary register AR1 is used for indirect addressing in the inner loop computation (the MACD instruction) and initially points to the oldest data sample, XNM1, in the data memory. In the inner loop, the MACD instruction does the following:

■ adds the previous product to the accumulator – initially, the product is zero;

■ multiplies the coefficient, $h(k)$, by the data pointed to by AR1 – initially, $h(k) = h(N − 1)$, and the auxiliary register points to $x[n − (N − 1)]$;

- copies data pointed to by AR1 to the next higher location – initially, $x[n-(N-1)]$ is copied to $x(n-N)$; i.e. the oldest data sample drops off. The last MACD instruction copies $x(n)$ into $x(n-1)$ to make room for the next input sample;

- decrements AR1 by 1 (i.e. points to the next sample in data memory) – initially AR1 points to $x[n-(N-1)]$, and then, successively, points to $x[n-(N-2)]$, $x[n-(N-3)], \ldots, x(n)$ as we go round the loop;

- increments the COEFF address by 1 – successive addresses are $h(N-1)$, $h(N-2), \ldots, h(0)$.

12.5.2   IIR digital filtering

12.5.2.1   The basic building blocks for IIR filters

Second-order IIR filter sections form the basic building blocks for digital IIR filters. The two most widely used second-order structures are the canonic section (Figure 12.27) and the direct form (Figure 12.28). The canonic second-order section is characterized by the following equations:

$$w(n) = SF_1 x(n) - a_1 w(n-1) - a_2 w(n-2) \tag{12.6a}$$

$$y(n) = b_0 w(n) + b_1 w(n-1) + b_2 w(n-2) \tag{12.6b}$$

where $x(n)$ represents the input data, $w(n)$ represents the internal node, $y(n)$ is the filter output sample and SF_1 is a scale factor, equal to $1/s_1$. The difference equation for the direct form second-order IIR filter (Figure 12.28(b)) section is given by

$$y(n) = b_0 x(n) + b_1 x(n-1) + b_2 x(n-2) - a_1 y(n-1) - a_2 y(n-2) \tag{12.7}$$

where $x(n-k)$ are the input data sequence and $y(n-k)$ are the output data sequence. The data and coefficient storage for the direct structure is depicted in Figure 12.28(b).

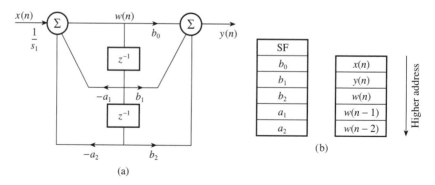

Figure 12.27   (a) Second-order canonic section; (b) coefficient–data storage.

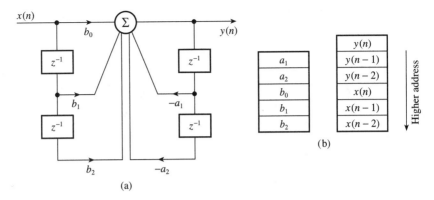

Figure 12.28 Implementation of the direct form second-order section; (a) realization diagram; (b) data and coefficient storage.

The direct form filter is simpler to program and can lead to a somewhat faster implementation than the canonic section because of the simpler indexing involved: compare, for example, Equations 12.6 and 12.7.

Higher-order IIR filters are realized as either a cascade or a parallel combination of the second-order filter sections (see Chapter 8 for more details).

Cascade realization

The transfer function, $H(z)$, of an Nth-order IIR filter, using second-order sections in cascade, is given by

$$H(z) = \prod_{k=1}^{N/2} \frac{b_{0k} + b_{1k}z^{-1} + b_{2k}z^{-2}}{1 - a_{1k}z^{-1} - a_{2k}z^{-2}} \tag{12.8}$$

The cascade realization of a fourth-order ($N = 4$) IIR filter using second-order canonic sections is shown in Figure 12.29(a). The storage of the filter variables (data and coefficients) is shown in Figure 12.29(b). The set of difference equations for the fourth-order IIR filter, using canonic sections, is given by

$$w_1(n) = \text{SF}_1 x(n) - a_{11}w_1(n-1) - a_{21}w_1(n-2) \tag{12.9a}$$

$$y_1(n) = b_{01}w_1(n) + b_{11}w_1(n-1) + b_{21}w_1(n-2) \tag{12.9b}$$

$$w_2(n) = y_1(n) - a_{12}w_1(n-1) - a_{22}w_2(n-2) \tag{12.9c}$$

$$y_2(n) = b_{02}w_2(n) + b_{12}w_2(n-1) + b_{22}w_2(n-2) \tag{12.9d}$$

A C language pseudo-code for an IIR filter realized as a cascade of second-order canonic sections is given in Program 12.3.

Fragments of TMS320C50 and DSP56000 implementations of an Nth order IIR filter which consists of M biquadratic sections (where $M = N/2$) in cascade are given in Programs 12.4 and 12.5, respectively.

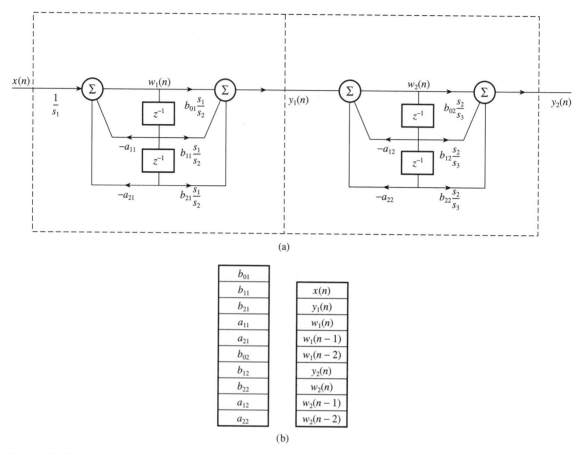

(a)

(b)

Figure 12.29 Cascade realization of an IIR filter: (a) realization diagram; (b) data and coefficient storage.

Program 12.3 C language pseudo-code for a cascade IIR filter.

```
for(n=0; n<(Nsamples–1); ++n){                    /* Nsamples no of data samples */
        xn=x[n];
        for(k=1; k<N; ++k){
                wk=sk[k]*xn–a1[k]*w1[k]–a2[k]*w2[k];
                yk=(b0[k]*wk+b1[k]*w1[k]+b2[k]*w2[k]);
                                                  /* output of 1st section */
                w2[k]=w1[k];
                                                  /* shift and save delay node data */
                w1[k]=wk;
                xn=yk;                            /* kth section feeds next section */
        }
        y[n]=yk;                                  /* nth output sample */
}
```

Program 12.4   TMS320C50 implementation of an *M* second-order canonic filter section in cascade.

```
        SPLK    #M−1, BRCR      ; no. of biquadratic sections
        RPTB    M_IIR           ; compute M biquads in cascade
        LT      *−, AR2         ; load wk(n−2)
        MPYA    +, AR1          ; compute wk(n−2)*ak(2)
        MPY     +               ; wk(n−1)*ak(1)
        LTA     *+, AR1         ; compute and save wk(n)=x(n)+
                                ; wk(n−1)*ak(1)+wk(n−2)*ak(2)

        SACH    *0+, 1
        MPY     *−              ; compute wk(n−2)*bk(2)
        LACL    #0              ;
        LTD     *−, AR2         ; shift data wk(n−2) = wk(n−1)
        MPY     *+, AR1         ; yk=yk+wk(n−2)*bk(2), wk(n−1)*bk(1)
        LTD     *−, AR2         ; shift data wk(n−1) = wk(n)
        MPY     *+, AR1         ;compute wk(n−2)*bk(2) +
                                ; wk(n−1)*bk(1), wk(n)*bk(0)
M_IIR   :
        LTA     *, AR4          ; add last product
        SACH    *, 1            ; quantize and save output sample
```

Program 12.5   DSP56000 implementation of an *M* second-order canonic filter section in cascade.

```
        DO      #M, M_IIR                                    ; compute M biquads
        MAC     −X0, Y0, A   X: (R0)−, X1   Y: (R4)+, Y0    ;
        MACR    −X1, Y0, A   X1, X: (R0)+   Y: (R4)+, Y0    ; shift data
                                                            ; w(n−2)=w(n−1)
        MAC     X0, Y0, A    A, X: (R0)+    Y: (R4)+, Y0
        MAC     X1, Y0, A    X: (R0)+, X0   Y: (R4)+, Y0
M_IIR
        RND
        MOVEP A, Y: OUTPUT
```

Example 12.4

(1) Design and implement a lowpass IIR digital filter using the TMS320C50 fixed-point DSP processor to meet the following specifications:

sampling frequency	15 kHz
passband	0–3 kHz
transition width	450 Hz
passband ripple	0.5 dB
stopband attenuation	45 dB

(2) Repeat (1) using the TMS320C54 fixed-point DSP processor.

(3) Repeat (1) and (2) using the DSP56000 and DSP56300, respectively.

Table 12.4   Filter coefficients before and after quantization to 16 bits.

	Coefficient	Scaled	Quantized
b_{02}	1	0.999 969 5	32 767
b_{12}	0.675 718	0.675 718	22 142
b_{22}	1	0.999 969 5	32 767
a_{12}	−0.495 935	−0.495 935	−16 251
a_{22}	0.761 864	0.761 864	24 965
b_{01}	1	0.131 113 6	4 296
b_{11}	1.649 656	0.216 292 4	7 087
b_{21}	1	0.131 113 6	4 296
a_{11}	−0.829 328	−0.829 328	−27 175
a_{21}	0.307 046	0.307 046	10 061

$s_1 = 2.479\ 158\ (L_1)$; SF = 0.403 362 7; $s_2 = 18.908\ 47\ (L_1)$.

Solution   A detailed design of this filter was given in Chapter 8 (Section 8.8). It was shown there that a fourth-order elliptic filter with the following transfer function would meet the specifications:

$$H(z) = \frac{1 + 0.675\ 718z^{-1} + z^{-2}}{1 - 0.495\ 935z^{-1} + 0.761\ 864z^{-2}} \times \frac{1 + 1.649\ 656z^{-1} + z^{-2}}{1 - 0.829\ 328z^{-1} + 0.307\ 046z^{-2}}$$

The difference equations for the cascade realization using canonic sections are the same as Equation 12.9.

The coefficients, scaled to avoid overflow and quantized to 16 bits, are listed in Table 12.4. The TMS320C50 and TMS320C54 codes for the fourth-order filter using the canonic form filters are not listed here for lack of space, but are available on the CD and in the companion handbook (see the Preface for details). For the DSP56000 and DSP56300 implementation, the coefficients are quantized to 24 bits, the processor wordlength. The codes for these are also available on the CD.

Parallel realization

The transfer function of an Nth-order IIR filter for parallel realization is given by

$$H(z) = \prod_{k=1}^{N/2} \frac{b_{0k} + b_{1k}z^{-1}}{1 + a_{1k}z^{-1} + a_{2k}z^{-2}} + C \tag{12.10}$$

The realization diagram, using second-order canonic sections, for $N = 4$ is given in Figure 12.30. For the canonic section, the difference equation is given by

$$w_1(n) = SF_1 x(n) - a_{11}w_1(n-1) - a_{21}w_1(n-2)$$
$$y_1(n) = b_{01}w_1(n) + b_{11}w_1(n-1)$$

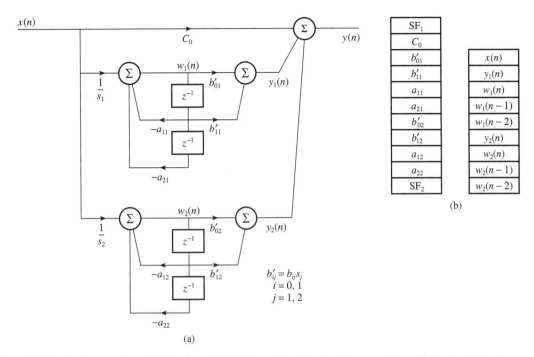

Figure 12.30   Implementation of a fourth-order IIR filter: (a) realization diagram; (b) coefficient and data storage.

Program 12.6   C language pseudo-code for parallel realization.

```
for(n=0; n<(Nsamples–1); ++n){                 /* Nsamples no of data samples */
       y[n] = c*x[n];                          /* output through constant path */
       for(k=1; k<N; ++k){
              wk=sk[k]*x[n]–a1[k]*w1[k]–a2[k]*w2[k];
              yk=(b0[k]*wk+b1[k]*w1[k])/sk[k]; /* output of 1st section */
              w2[k]=w1[k];                     /* shift and save delay node data */
              w1[k]=wk;
              y[n]=yk+y[n];
       }
}
```

$$w_2(n) = SF_2 x(n) - a_{12} w_2(n-1) - a_{22} w_1(n-2)$$

$$y_2(n) = b_{02} w_2(n) + b_{12} w_2(n-1)$$

$$y(n) = c_0 x(n) + y_1(n) + y_2(n)$$

A simple C language code for an IIR filter realized as a parallel combination of second-order canonic sections is given in Program 12.6.

Example 12.5

Represent the transfer function of Example 12.4 in a parallel form using second-order canonic sections as building blocks. Implement the filter using the same hardware as the last example.

Solution

Using the partial fraction expansion program discussed in Chapter 4, the coefficients for parallel realization were obtained from those of the cascade realization. The transfer function becomes

$$H(z) = \frac{-0.132\,922\,5 - 0.180\,523\,2z^{-1}}{1 - 0.028\,994z^{-1} + 0.044\,541\,6z^{-2}}$$

$$+ \frac{-0.058\,534 + 0.508\,420z^{-1}}{1 - 0.048\,489\,9z^{-1} + 0.017\,951\,1z^{-2}} + 0.249\,923\,79$$

$$s_1 = 5.524\,484\,4, \quad s_2 = 2.4794$$

Table 12.5 gives the coefficient values before and after quantization to 16 bits. The TMS320C50, TMS320C54, DSP56000 and DSP56300 codes for the filter are available on the CD and in the companion handbook (see the Preface for details). For the DSP56000 and DSP56300, the coefficients were quantized to 24 bits.

Table 12.5   Implementation of fourth-order IIR filter of Example 12.5: filter coefficients before and after quantization to 16 bits.

	Unquantized coefficients	*Quantized coefficients*
SF_1	0.181 00	5 931
C_0	0.249 923 79	8 190
b_{01}	−0.132 922 5	−24 063
b_{11}	−0.180 523 2	−32 670
a_{11}	0.028 994	16 251
a_{21}	−0.044 541 6	−24 965
b_{02}	−0.058 534	−4 756
b_{12}	0.508 420 5	20 653
a_{12}	0.048 489 9	27 178
a_{22}	−0.017 951	−10 061
SF_2	0.403 32	13 216

Extension of the implementation techniques discussed above for both the cascade and parallel structures to higher-order IIR filters is relatively easy. However, a more compact code may be obtained by implementing the second-order building block as a subroutine.

12.5.3 FFT processing

The discrete Fourier transform (DFT) of a finite data sequence, $x(n)$, is defined as

$$X(k) = \sum_{n=0}^{N-1} x(n) W_N^{nk}$$

where W_N, often called the twiddle factor, is a set of complex coefficients.

Direct computation of the DFT coefficients, $X(k)$, is time consuming when N is large. FFT algorithms provide efficient ways of computing $X(k)$ with significant reduction in computation time. As discussed in Chapter 3, the butterfly and twiddle factor are central to FFT algorithms.

12.5.3.1 Implementation of the butterfly

Figures 12.31(a) and 12.31(b) depict the two types of butterflies used in the radix-2 FFT. FFTs based on these butterflies lead to the same result. For the decimation in time (Figure 12.31(a)) the butterfly takes a pair of input data, A and B, and produces a pair of outputs:

$$A' = A + BW_N^k \tag{12.11a}$$

$$B' = A - BW_N^k \tag{12.11b}$$

In general the input and output data samples as well as the twiddle factors are all complex and can be expressed as

$$A = A_r + jA_i \tag{12.12a}$$

$$B = B_r + jB_i \tag{12.12b}$$

$$W_N^k = e^{-j2\pi k/N} = \cos(2\pi k/N) - j\sin(2\pi k/N) \tag{12.12c}$$

where the suffix r indicates the real part and i the imaginary part of the data. The butterfly operation in Equations 12.11 involves complex arithmetic, but in practice it is often carried out using real arithmetic. To express the operation in a form suitable for real arithmetic, we note that the product of B and W in Equations 12.11 has the form:

$$BW_N^k = B_r \cos(X) + B_i \sin(X) + j[B_i \cos(X) - B_r \sin(X)] \tag{12.13}$$

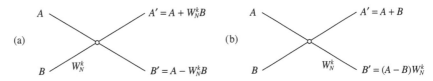

Figure 12.31   The two types of butterflies used in radix-2 FFT algorithms: (a) butterfly for the decimation in time radix-2 FFT; (b) butterfly for the decimation in frequency radix-2 FFT.

Program 12.7   A C language pseudo-code for pre-calculating the twiddle factor values.

```
pi=6.28315307179586/N;
for(k=0; k<N/2; ++k){
      X=k*pi;
      w.real[k]=cos[X];
      w.imag[k]=sin[X];
}
```

Program 12.8   A C language pseudo-code for the butterfly.

```
t.real=br*w.real[k]+bi*w.imag[k];
t.imag=bi*w.real[k]–br*w.imag[k];
b.real[j]=a.real–t.real;
b.imag[j]=a.imag–t.imag;
a.real[j]=a.real+t.real;
a.imag[j]=a.imag+t.imag;
```

where $X = 2\pi k/N$. Using Equations 12.12 and 12.13 in Equations 12.11a and 12.11b we have

$$A' = A_r + [B_r \cos(X) + B_i \sin(X)] + j\{A_i + [B_i \cos(X) - B_r \sin(X)]\} \qquad (12.14a)$$

$$B' = A_r - [B_r \cos(X) + B_i \sin(X)] + j\{A_i - [B_i \cos(X) - B_r \sin(X)]\} \qquad (12.14b)$$

The outputs of the butterfly, A' and B', are now in the desired form. Thus, given a pair of complex data points, A and B, in rectangular form, Equations 12.14a and 12.14b are used to compute the output of the butterfly using real arithmetic.

The computation of the sine and cosine terms in Equations 12.14 is time consuming. In real-time FFT, a more efficient approach is to pre-calculate the real and imaginary parts of the twiddle factor (Equation 12.12c), and to store these values in a look-up table. The C language pseudo-code in Program 12.7 illustrates how the twiddle factor values are pre-calculated.

A C language pseudo-code for the radix-2 butterfly, with twiddle factor values pre-calculated and stored in a look-up table, is given in Program 12.8.

A TMS320C25 pseudo-code is shown in Program 12.9. The pre-calculated twiddle factor values are stored in Q15 format. The input data is assumed complex, with the real and imaginary parts stored in consecutive locations in data RAM. For complex inputs, the values of A' or B' can attain a maximum value of 2.414 42, and 2 for real data input. In fixed-point arithmetic this will cause overflow. To avoid overflow the input data to a butterfly should be scaled. In the C50 implementation, the scaling is dynamic, advantage being taken of the fact that the product of two fixed-point numbers produces an extra sign bit. The extra sign bit is normally removed by a left shift, but by leaving it the result is effectively scaled by 2.

Program 12.9   TMS320C50 code for the butterfly.

```
*
*        compute terms common to the two butterfly outputs, A' and B'
*
*
         LT        BR          ;compute 1/2*[b.real*cos(X)+b.imag*sin(X)]
         MPY       WREAL       ;1/2*b.real*cos(X)
         LTP       BI
         MPY       WIMAG       ;1/2*b.imag*sin(X)
         APAC                  ;1/2[b.real*cos(X)+b.imag*sin(X)]
         MPY       WREAL       ;1/2*b.imag*cos(X)
         LT        BR
         SACH      BR          ;1/2[b.real*cos(X)+b.imag*sin(X)]

         PAC                   ;compute [q.imag*cos(X)–q.real*sin(X)]
         MPY       WIMAG
         SPAC
         SACH      BI
*
*        compute and save the butterfly outputs
*
         LACC      AR, 14      ;compute and save the real parts of the output
         ADD       BR, 15
         SACH      AR, 1       ;save a.real
         SUB       BR
         SACH      BR, 1       ;save b.real

         LAC       AI,14       ;compute and save the imaginary parts of the
                               ;output
         ADD       BI, 15
         SACH      AI, 1       ;save a.imag
         SUB       BI
         SACH      BI, 1       ;save b.imag
```

12.5.3.2   In-place computation and constant geometry

The signal flowgraph of an eight-point FFT is shown in Figure 12.32. From the figure, it is evident that to obtain the DFT coefficients, $X(k)$, shown on the right-hand side, given the input, a series of butterfly computations will have to be performed. The radix-2 FFT algorithm is a method of carrying out the series of butterfly computations in an orderly manner. In the flowgraph, the data flows from left to right. Thus, once the outputs of a butterfly, A' and B', are computed the inputs, A and B, are no longer required and so can be overwritten by the outputs. This is the basis of the concept of in-place computation.

The in-place algorithm makes efficient use of available memory as the transformed data overwrites the input data. In the past, when memory was very expensive, this was an important consideration. However, with in-place computation the indexing required to determine where in memory to fetch the input data to each butterfly is quite complex. For example, in Figure 12.32 the top butterfly in stage 1 takes its inputs

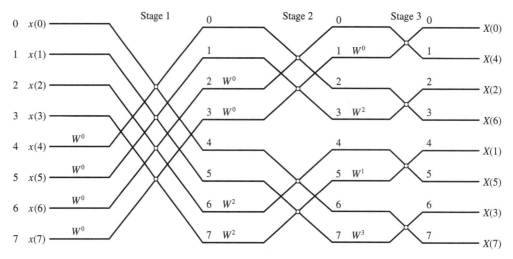

Figure 12.32 An in-place DIT FFT flowgraph, with input in natural order but output in bit-reversed order.

from addresses 0 and 4 and writes the outputs back to the same addresses. On the other hand, the top butterfly in stage 2 takes its inputs from addresses 0 and 2. In general, in the in-place FFT the input/output addresses vary from stage to stage. Further, for high speed FFTs, the use of the same memory for input and output slows down computations because of long memory access (except, for example, if dual-port RAMs are used). As both memory and multipliers are cheap, the trend now is to optimize the whole FFT processor to push speed up.

An alternative FFT implementation, known as the non-in-place or constant geometry, reads the input data to a butterfly from a pair of addresses and stores the output in another pair of addresses as shown in Figure 12.33. Unlike the in-place FFT, where the input/output addresses for each butterfly vary from stage to stage, in this case the addressing for each butterfly is fixed and much simpler. For an N-point FFT, the inputs of the nth butterfly at each stage are $2n$ and $2n + 1$, $n = 0, 1, \ldots, N/2 - 1$. The outputs of the nth butterfly are stored at addresses n and $N/2 + n$. For example, at the second stage of Figure 12.33 the top butterfly takes its inputs from addresses 0 and 1 and stores its outputs at addresses 0 and 4. Clearly, for the non-in-place FFT to work two separate memories or arrays are required; one holds the inputs and the other the outputs. After each stage the roles of the memories are reversed.

12.5.3.3 Data scrambling and bit reversal

In the DIT (decimation-in-time) FFT, if the input data sequence is applied to the FFT processor in natural order, the output of the FFT appears scrambled (see Figure 12.32). To ensure the output appears in the correct order (that is, as $X(0)$, $X(1), \ldots, X(N - 1)$), we either scramble the input data sequence before taking the FFT (see Figures 12.33 and 12.34) or unscramble the output after taking the FFT (see Figure 12.32).

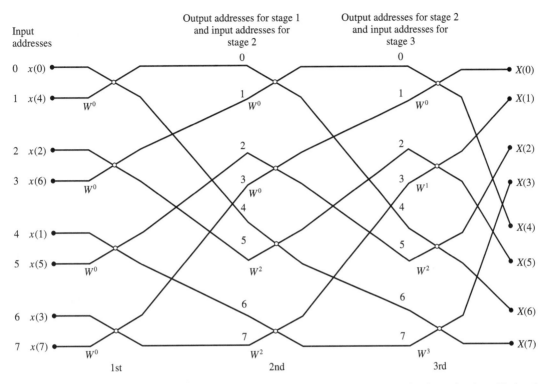

Figure 12.33   Constant geometry radix-2 FFT. In the constant geometry, the computation is not in place. Notice that for each butterfly the input and output span is constant.

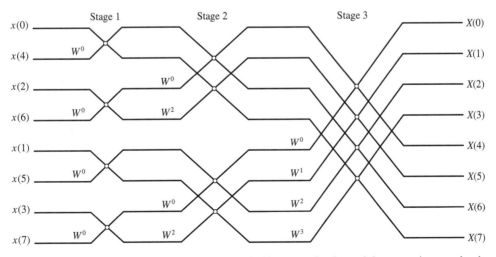

Figure 12.34   An in-place DIT FFT flowgraph with input in bit-reversed order and the output in natural order.

Table 12.6 Data for eight-point FFT showing the concepts of bit reversing.

Input sequence, natural order	Binary code of sequence	Input sequence, bit reversed	Binary code of sequence (bit reversed)
$x(0)$	000	$x(0)$	000
$x(1)$	001	$x(4)$	100
$x(2)$	010	$x(2)$	010
$x(3)$	011	$x(6)$	110
$x(4)$	100	$x(1)$	001
$x(5)$	101	$x(5)$	101
$x(6)$	110	$x(3)$	011
$x(7)$	111	$x(7)$	111

For radix-2 FFT, input data scrambling is achieved by storing the input sequence in bit-reversed order. Assuming the input data has already been stored in natural order (that is, as $x(0)$, $x(1)$, $x(2)$, ..., $x(N - 1)$), the bit-reversed order is achieved by representing the indices of the input data in binary as shown in the second column of Table 12.6 for an eight-point FFT, and then swapping the bits about the centre (fourth column in Table 12.6). Notice, for example, that in the table the index of data sample $x(3)$ has a binary representation of 011. Swapping the first and third bits about the middle bit gives 110 (that is, bit 1 which is 0 becomes 1 and bit 3 which is 1 becomes 0 while the middle bit about which we perform the operation remains unchanged). The bit-reversed code, 110, is decimal number 6. To effect scrambling, we swap the locations of the data samples $x(3)$ and $x(6)$. Applying the same principle to the remainder of the inputs, we obtain the bit-reversed sequence given in Table 12.6, third column. Notice that, after scrambling, the locations of the first and last data samples remain unchanged. This is because bit reversal of the indices 000 and 111 does not have any effect. In general, the first and last data points are not affected by scrambling in radix-2 FFT. The alert reader will have spotted other samples of the input sequence immune to bit reversal.

When the input data is held in a memory or an array, scrambling the input data involves identifying pairs of input data locations and interchanging or swapping the data in those locations. A bit reversal algorithm due to Rader (Rabiner and Gold, 1975) is the most widely used to determine the indices of the memory locations to swap. A C language pseudo-code for the bit reversing algorithm is given in Program 12.10.

Advanced DSP chips now provide instructions to perform bit reversal on the input data samples as they are fetched from memory in readiness for FFT or on the transformed data as they are being stored in memory after FFT.

12.5.4 Multirate processing

As discussed in Chapter 9, multirate processing involves performing DSP operations at more than one sampling rate. The two fundamental operations in multirate

Program 12.10   In-place bit reversal for radix-2 FFT.

```
/* perform in-place bit reversal */

j=1;
for(i=1; i<N; ++i){
        if(i<j){
                tr=x.real[j];      /* swap x[j] and x[i] */
                ti=x.imag[j];
                x.real[j]=x.real[i];
                x.imag[j]=x.imag[i];
                x.real[i]=tr;
                x.imag[i]=ti;
                k=N/2;
                while(k<j){
                        j=j-k;
                        k=k/2;
                }
        }
        else  {
                k=N/2;
                while(k<j){
                        j=j-k;
                        k=k/2;
                }
        }
j=j+k;
}
```

processing are decimation (sample rate reduction) and interpolation (sampling rate increase). We will illustrate the implementation of a real-time decimator by an example.

Example 12.6   The sampling rate of a signal is to be reduced by a three-stage decimation process from 30 kHz to 1 kHz. Assume the highest frequency of interest after decimation is 400 Hz, an in-band ripple of 0.08 dB and stopband rejection of 50 dB. The decimator is to be implemented on the TMS320C50.

Solution   Using the multirate design program (available on the CD and in the companion handbook – see the Preface for details), parameters of the three-stage decimator were obtained; see Figure 12.35. The coefficients of the three filters, using the optimal FIR design program, are given in Table 12.7. The filter lengths are slightly longer than the estimates predicted by the decimator design program (12, 13, 48 instead of 13, 12, 46) to ensure the specifications are met.

The flowchart for a general three-stage decimator is given in Figure 9.15. The coefficient and data storage map for the TMS320C50-based decimator is shown in Figure 12.36. The filter coefficients are quantized to 16 bits by multiplying each

Table 12.7   Filter coefficients for the three-stage decimator.

FILTER LENGTH = 12

***** IMPULSE RESPONSE *****

H(1) =	0.73075550E–02 = H(12)	239
H(2) =	0.27123260E–01 = H(11)	889
H(3) =	0.59286430E–01 = H(10)	1943
H(4) =	0.10198970E+00 = H(9)	3342
H(5) =	0.14187870E+00 = H(8)	4649
H(6) =	0.16675770E+00 = H(7)	5464

***** IMPULSE RESPONSE *****

H(1) =	–0.86768190E–02 = H(13)	–284
H(2) =	–0.25476870E–01 = H(12)	–835
H(3) =	–0.25468170E–01 = H(11)	–834
H(4) =	0.24184320E–01 = H(10)	792
H(5) =	0.13238570E+00 = H(9)	4338
H(6) =	0.24907950E+00 = H(8)	8162
H(7) =	0.30075170E+00 = H(7)	9855

FILTER LENGTH = 48

***** IMPULSE RESPONSE *****

H(1) =	0.17780220E–02 = H(48)	585
H(2) =	–0.17396640E–02 = H(47)	–57
H(3) =	–0.49461790E–02 = H(46)	–162
H(4) =	–0.25451430E–02 = H(45)	–83
H(5) =	0.40843330E–02 = H(44)	134
H(6) =	0.42773070E–02 = H(43)	140
H(7) =	–0.45042640E–02 = H(42)	–148
H(8) =	–0.80385180E–02 = H(41)	–263
H(9) =	0.29002500E–02 = H(40)	95
H(10) =	0.12193670E–01 = H(39)	400
H(11) =	0.92281120E–03 = H(38)	30
H(12) =	–0.16199860E–01 = H(37)	–531
H(13) =	–0.76966970E–02 = H(36)	–252
H(14) =	0.18898710E–01 = H(35)	619
H(15) =	0.17966280E–01 = H(34)	589
H(16) =	–0.18756490E–01 = H(33)	–615
H(17) =	–0.32451860E–01 = H(32)	–1063
H(18) =	0.13458800E–01 = H(31)	441
H(19) =	0.52945520E–01 = H(30)	1735
H(20) =	0.17620600E–02 = H(29)	58
H(21) =	–0.86433440E–01 = H(28)	–2832
H(22) =	–0.44585360E–01 = H(27)	–1461
H(23) =	0.18176500E+00 = H(26)	5956
H(24) =	0.41039480E+00 = H(25)	13448

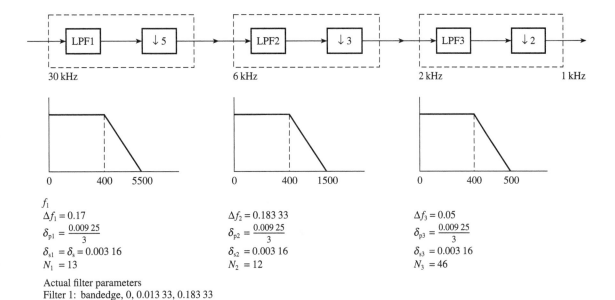

f_1

$\Delta f_1 = 0.17$

$\delta_{p1} = \dfrac{0.009\,25}{3}$

$\delta_{s1} = \delta_s = 0.003\,16$

$N_1 = 13$

$\Delta f_2 = 0.183\,33$

$\delta_{p2} = \dfrac{0.009\,25}{3}$

$\delta_{s2} = 0.003\,16$

$N_2 = 12$

$\Delta f_3 = 0.05$

$\delta_{p3} = \dfrac{0.009\,25}{3}$

$\delta_{s3} = 0.003\,16$

$N_3 = 46$

Actual filter parameters
Filter 1: bandedge, 0, 0.013 33, 0.183 33
        filter length, 12; weight, 1, 3
Filter 2: bandedge, 0, 0.066 66, 0.25, 0.5
        filter length, 13; weight, 1, 3
Filter 3: bandedge, 0, 0.2, 0.25, 0.5
        filter length, 48; weight, 1, 3

Figure 12.35   Parameters of the three-stage decimator.

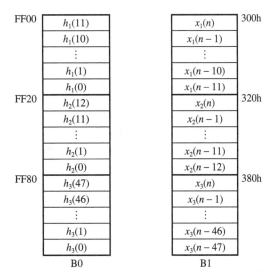

Figure 12.36   Coefficient and data storage map for a three-stage decimator.

coefficient by 2^{15} and then rounding up to the nearest integer. The decimation program, in TMS320C50 assembly language, is general purpose, and can be modified to implement one-stage, two-stage or three-stage decimation by replacing the coefficients, specifying the filter lengths, number of decimation stages and decimation factors.

12.5.5   Adaptive filtering

The general structure for an adaptive filter is depicted in Figure 12.37. As discussed in Chapter 10, adaptive filtering involves two processes.

(1) *Digital filtering*   The coefficients of the digital filter in Figure 12.37 are used to extract appropriate information from the input signal, $x(n)$, to yield $y(n)$. Assuming a transversal FIR structure, the filter is given by

$$y(n) = \sum_{k=0}^{N-1} w_n(k)x(n - k) \tag{12.15}$$

where $w_n(k)$, $k = 0, 1, \ldots, N - 1$, are the digital filter coefficients (often called weights) and $x(n - k)$, $k = 0, 1, \ldots, N - 1$, is a sequence of the input data.

The implementation of the digital filter given in Equation 12.15 is very similar to that of a standard FIR filter discussed earlier. Thus a C language implementation of the filter would have the familiar form

```
y[n]=0;
for(k=0; k<N; k++){
      y[n]=y[n]+wn[k]*xn[k];
}
```

(2) *Adaptive process*   This process involves updating, that is adjusting the filter coefficients towards their optimal values. When the basic LMS algorithm is used, the coefficients are updated as follows:

$$w_{n+1}(k) = w_n(k) + 2\mu e(n)x(n - k), \quad k = 0, 1, 2, \ldots, N - 1 \tag{12.16}$$

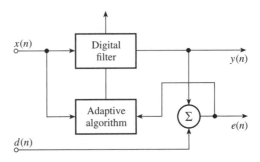

Figure 12.37   General structure of an adaptive filter: a pair of inputs and a pair of outputs.

Program 12.11   C pseudo-code for the LMS adaptive filter coefficient update.

```
uen=2*u*e[n]
for(k=0; k<N; k++){
      wn[k]=wn[k]+uen*xn[k];
}
```

Program 12.12   TMS320C50 pseudo-code for the LMS adaptive filter coefficient update.

```
        LT      ERR
        MPY     U               ;compute μ * e(n)
        PAC
        ADD     ONE, 15
        SACH    ERRF
        LACC    #N-1            ; specify filter length
        SAMM    BRCR
        LAR     AR2, #WNM1      ; point to the last coefficient, wk(N–1)
        LAR     AR3, #XNM1      ; point to x(n–(N–1))
        LT      ERRF
        MPY     –, AR2          ; compute μ e(n) + x(n–k)
        RPTB    LMS–1           ; update coefficients, wk+1(n)
        ZALR    *, AR3
        MPYA    *–, AR2
        SACH    *+
LMS     ZALR    *, AR3
        APAC
        SACH    *+
```

where $w_n(k)$ is the kth coefficient of the digital filter at the nth sampling instants, μ is the stability factor, and $x(n - k)$ is the kth input data sample at the kth delay line. A C language implementation of the basic LMS update equation is given by Program 12.11. The term 2ue[n] is a scalar and is the same for all coefficients and so it is computed once and placed outside the loop. A TMS320C50 implementation of the adaptation process is given in Program 12.12.

12.6 Special-purpose DSP hardware

Why special purpose?

Digital signal processing operations are computationally intensive. In wide bandwidth applications where the input/output data rates are high, most general-purpose digital signal processors (DSPs) cannot perform the required computations fast enough. This is of course the reason why general-purpose DSPs are often found in audio frequency applications. Further, for a given application most general-purpose DSPs contain many on-chip resources that are either redundant or underutilized, for example addressing modes, instruction set and I/O peripherals. In special-purpose DSPs, the hardware is optimized to execute a specific algorithm or to perform certain functions in a specific application. This leads to greater utilization of on-chip resources and increased speed of operation.

Special-purpose hardware can be implemented as a single-chip product or realized as blocks of individual components. The building block approach using individual components is more flexible and leads to increased speed, but the hardware is difficult to develop and may be more expensive. Single-chip DSPs, if they exist for the task, have lower chip counts, do not require knowledge of an obscure assembly language, and do not have problems of software debugging.

Basic requirements of special-purpose DSPs

The most common arithmetic operation in DSP algorithms, such as digital filtering, correlation and transformations, is the sum of products:

$$y = \sum_{k=0}^{N-1} a_k x_k \tag{12.17}$$

where a_k is a set of coefficients or variables and x_k a data sequence.

The characteristic Equation 12.17 can be written in a recursive form to allow for a more efficient computation of the sums of products:

$$y_k = a_k x_k + y_{k-1}, \quad k = 0, 1, \ldots, N-1 \tag{12.18}$$

where

$$y_{-1} = 0$$

$$y = y_{N-1}$$

In special-purpose DSPs Equation 12.18 is computed with a multiplier–accumulator (MAC) at a very fast rate, for example 40 ns per MAC.

Like general-purpose DSPs, the architecture of a special-purpose DSP includes data memory, RAM and/or ROM, for storing data and variables (such as filter or FFT coefficients), fast hardware multiplier–accumulator, and temporary registers to store data or intermediate results. Extensive use is made of parallelism, multiplexing and pipelining to achieve maximum speed.

In the next few sections, we will discuss some basic issues involved in the design of special-purpose hardware for DSP.

12.6.1   Hardware digital filters

12.6.1.1   FIR digital filters

The direct form FIR filter is characterized by the following equation:

$$y(n) = \sum_{k=0}^{N-1} h(k)x(n - k)$$

Figure 12.38 shows a basic architecture for an FIR digital filter using blocks of individual components. The main components are coefficient and data memories, analog input/output units (ADC and DAC), multiplier–accumulator (MAC), and a controller (not shown). The components of the FIR filter can be implemented with fast, off-the-shelf products.

At each sampling instant, a new data sample, $x(n)$, is read from the ADC and saved in the data memory. Each input data sample and the corresponding coefficient are fetched from the memory simultaneously and applied to the multiplier. The products are then accumulated to yield the output sample. The computation of each output sample, $y(n)$, would require N data–coefficient fetches from memory and N MACs.

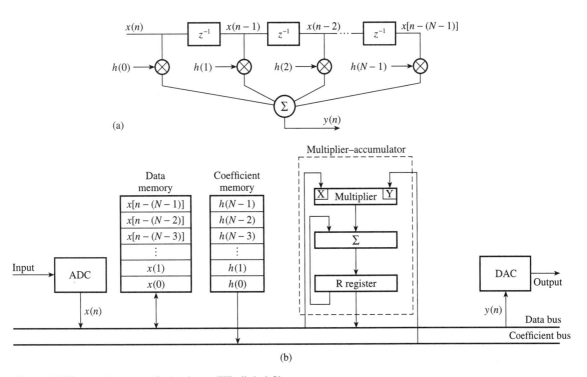

Figure 12.38   Architecture of a hardware FIR digital filter.

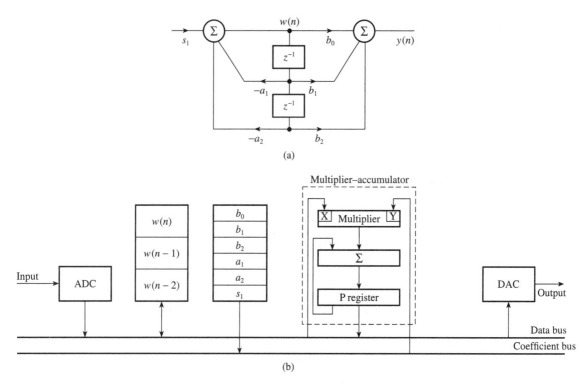

Figure 12.39   (a) IIR filter structure. (b) Hardware architecture for an IIR filter biquadratic section.

The FIR filtering operation is regular and well structured, and can be readily implemented in a single IC. Special-purpose single-chip FIR filters are now available such as Mitel's PDSP16256 preprogrammable FIR filter.

12.6.1.2   IIR digital filter

The architecture for a second-order canonic IIR filter is shown in Figure 12.39. In this case the data memory holds the internal node data $w(n)$. The standard second-order canonic section in Figure 12.39(a) is characterized by the following equations:

$$w(n) = s_1 x(n) - a_1 w(n-1) - a_2 w(n-2)$$

$$y(n) = b_0 w(n) + b_1 w(n-1) + b_2 w(n-2)$$

where $x(n)$ represents the input data, $w(n)$ represents the internal node, $y(n)$ is the filter output sample and s_1 is a scale factor.

12.6.2   Hardware FFT processors

The DFT takes a set of N time domain samples and transforms these into a set of N frequency domain samples, $X(k)$. The FFT is an efficient way of computing the DFT

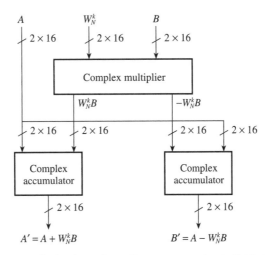

Figure 12.40   Concepts of a hardware butterfly processor using individual blocks of complex arithmetic units.

coefficients, $X(k)$. Butterfly arithmetic is the basic operation in the FFT. A butterfly is characterized by the following equations:

$$A' = A + W_N^k B$$

$$B' = A - W_N^k B$$

where A and B are a pair of complex-valued input data samples to the butterfly, A' and B' are the outputs of the butterfly, and W_N is the twiddle factor, also complex valued.

Each butterfly operation requires a complex multiplication, that is $W_N^k B$, a complex addition and a complex subtraction. Figure 12.40 shows a direct hardware implementation of a butterfly processor using individual blocks of complex arithmetic units: a complex multiplier and a pair of complex accumulators. The complex multiplier calculates the common terms, $W_N^k B$. The two complex accumulators calculate the two outputs of the butterfly, A' and B'.

A 50 ns, single-cycle butterfly processor can be implemented using the Mitel complex multiplier (PDSP16112A) and a pair of complex accumulators (PDSP16318A). With standard real arithmetic units, an equivalent butterfly processor would consist of four multipliers and six adders. The use of complex arithmetic units clearly leads to a lower chip count and possibly to enhanced system performance. A hardware FFT processor built around the butterfly processor is shown in Figure 12.41. Standalone, high-speed FFT processors, such as Mitel's PDSP16510 device, are also available.

A real-time, double-buffered, FFT configuration is depicted in Figure 12.42. N-point FFT is performed alternately from each of the two buffers. FFT is performed on the N-point data in buffer A, while buffer B is being filled with new data. The double buffering allows real-time continuous FFT without loss of data. The maximum time to complete the N-point FFT is the interval $T_f = NT$ (s).

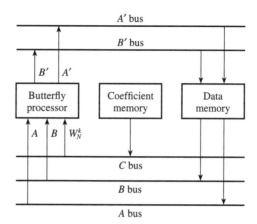

Figure 12.41   A simplified architecture of a hardware FFT processor. Controller and address generator are not shown.

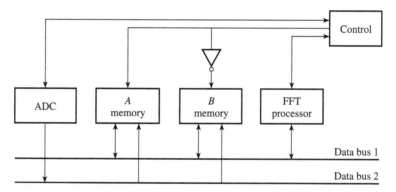

Figure 12.42   Double buffering in real-time FFT.

12.7 Summary

DSP algorithms involve extensive arithmetic operations, in particular multiplications and additions with heavy data flow through the CPU. Efficient execution of such algorithms in real time requires a hardware architecture and instruction set radically different from those of standard microprocessors. In digital signal processors, this is achieved by making use of the concepts of Harvard architecture, pipelining and dedicated hardware, for example fast hardware multiplier–accumulator and shifters, and by providing fast internal memories and many DSP-oriented instructions.

To meet challenges of multichannel, computation-intensive, applications, such as modern remote access server modems, 3G (third generation) mobile communications and multimedia information processing, new architectures are being introduced. In particular, the *very long instruction word* (VLIW) and *static superscalar* architectures are used in the latest generation of DSP processors. Both architectures have multiple data paths and arithmetic units and exploit parallelism at instruction level to increase performance.

There are two types of digital signal processors: general-purpose processors (which are akin to standard microprocessors, except that they have architectures and instruction sets tailored for DSP operations), and special-purpose processors. The latter are used to perform specific DSP algorithms, for example digital FIR filtering (algorithm-specific digital signal processors), or for efficient execution of some application-dependent operations (application-specific digital signal processors). Compared with general-purpose digital signal processors, special-purpose DSPs offer speed, but are less flexible.

The basic ideas underlying DSP hardware and the impact of DSP algorithms on the architecture of DSP processors were discussed in detail. The implementation of some key DSP algorithms using general-purpose digital signal processors as well as special-purpose DSP processors was discussed to illustrate the issues involved.

Problems

12.1   Write short critical notes on each of the following concepts, using diagrams where appropriate to illustrate your answer:

- Harvard architecture;
- pipelining;
- multiplier–accumulator;
- special instructions;
- data and program memory.

Explain how Harvard architecture as used by the TMS320 family differs from the strict Harvard architecture. Compare this with the architecture of a standard von Neumann processor.

12.2   (1)   A multiplier–accumulator, with three pipe stages, is required for a digital signal processor. Sketch a block diagram of a suitable configuration for the MAC. Explain, briefly, and with the aid of a timing diagram, how your MAC works.

(2)   Assume a memory access time of 150 ns, multiplication time of 100 ns, addition time of 100 ns, and overhead of 5 ns at each pipe stage. Determine the throughput of the MAC. Comment on your answer.

(3)   The DSP system is required to execute the following algorithm in real time:

$$y(n) = a_0 x(n) + a_1 x(n-1)$$
$$+ a_2 x(n-2) + \ldots$$
$$+ a_{N-1} x[n - (N-1)]$$

How long will it take the MAC to compute each output sample?

12.3   M.J. Flynn in his paper (Flynn, 1966) divides high speed computers into the following four categories:

(1)   single instruction stream, single data stream (SISD);

(2)   single instruction stream, multiple data stream (SIMD);

(3)   multiple instruction stream, single data stream (MISD);

(4)   multiple instruction stream, multiple data stream (MIMD).

where an instruction stream is the sequence of program instructions executed by the computer, and a data stream is the sequence of data required by the computer to execute the instructions.

Determine, with justification, the appropriate category for each of the following processors:

■ Motorola 68000;

■ Motorola DSP56000;

■ Analog Devices ADSP2100;

■ Texas Instruments TMS320C50;

■ Texas Instruments TMS320C30;

■ Texas Instruments TMS320C40;

■ Texas Instruments TMS320C62X;

■ Analog Devices TS001.

12.4 (a) Explain why traditional measures such as processor clock speed, MIPS and MFLOPS may not be suitable for comparing the execution performance of DSP processors. Suggest, with justifications, an alternative method of comparing execution performance.

(b) State and then discuss four key factors, apart from execution speed, that should be considered in choosing a DSP processor for each of the following applications:

(i) high fidelity digital audio;
(ii) voice over Internet Protocol telephony;
(iii) physiological signal processing for diagnosis in biomedicine.

12.5 (a) Compare the computational performance of the TMS320C50, DSP56000 and ADSP2100 fixed point processors based on the execution of the inner loop of an N-point FIR filter. State any assumptions made.

(b) Repeat (a) for an Nth order IIR filter with M second-order canonic filter sections in cascade. Assume that N is even.

(c) Repeat (a) for the LMS adaptive filter defined in Equation 12.16.

12.6 In relation to DSP processors, write brief explanatory notes, with the aid of sketches where appropriate, for each of the following techniques:

(1) circular addressing;

(2) SIMD (single input multiple data);

(3) superscalar architecture;

(4) very long instruction word architecture;

(5) zero-overhead looping.

In each case, clearly point out the advantages and disadvantages of the technique in signal processing.

12.7 Given in Figure 12.43 is the signal flowgraph for a 16-point DIT FFT. Construct an equivalent constant geometry signal flowgraph. Comment on the relative advantages of the two flowgraphs.

12.8 The signal flowgraph of an 8-point DIT FFT is shown in Figure 12.32 with the output scrambled. Show that the output can be obtained in natural order by bit reversing $X(k)$ and hence show that, in a DIT FFT, the final output will appear in the correct order by either scrambling the input data sequence before taking the FFT or unscrambling the output after FFT.

12.9 The input data sequence for a 16-point FFT is given in Table 12.8 together with the binary representation of its indices. Determine the input sequence in bit-reversed order and hence complete the table.

12.10 Design an efficient special-purpose hardware, using individual blocks of arithmetic elements, for a real-time Nth-order IIR digital filter realized as a cascade of second-order sections. Assume an ADC/DAC resolution of 12 bits, and a coefficient wordlength of 16 bits. A sampling frequency of 100 kHz is required. State any assumptions made.

12.11 (a) In relation to DSP-based systems, write a brief explanatory note on each of the following:

(i) dynamic range;
(ii) fixed- and floating-point arithmetic.

Refer to specific applications in audio, communications and biomedicine to illustrate your answer.

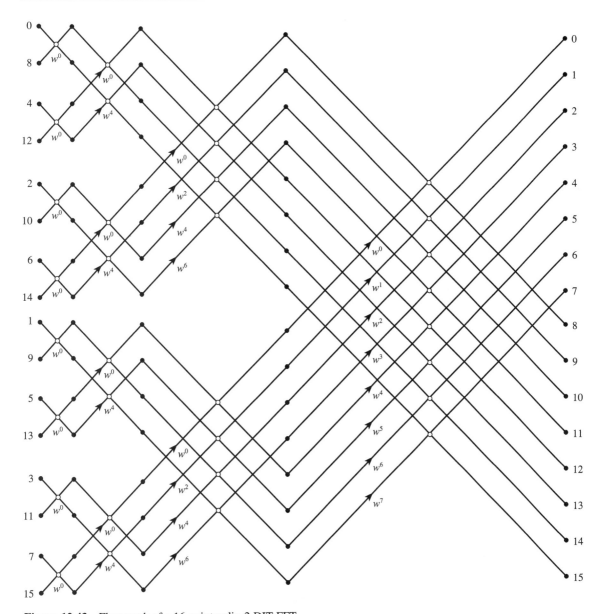

Figure 12.43 Flowgraph of a 16-point radix-2 DIT FFT.

(b) (i) A DSP-based system uses a fixed point digital signal processor with 16-bit wordlength. Estimate the dynamic range provided by the system.

   (ii) Repeat (i) if the wordlength is 24 bits.

(c) Repeat (b) if the multiplier–accumulator provides one guard bit to prevent overflow.

(d) Repeat (c) if eight guard bits are used.

Estimate and compare the additional dynamic range gained for each of the above cases.

Table 12.8   Input data sequence for Problem 12.9.

Input sequence, natural order	Binary code of sequence	Input sequence, bit reversed	Binary code of sequence (bit reversed)
$x(0)$	0000		
$x(1)$	0001		
$x(3)$	0011		
$x(5)$	0101		
$x(6)$	0110		
$x(7)$	0111		
$x(8)$	1000		
$x(9)$	1001		
$x(10)$	1010		
$x(11)$	1011		
$x(12)$	1100		
$x(13)$	1101		
$x(14)$	1110		
$x(15)$	1111		

References

Berkeley Design Technology (1999) *Buyer's Guide to DSP Processors*. Fremont CA: Berkeley Design Technology Inc. Details available at *www.BDTI.com*

Blalock G. (1997) General-purpose μPs for DSP applications: consider the trade-offs. *EDN*, 23 October, pp. 165–72.

Flynn M.J. (1966) Very high-speed computing systems. *Proc. IEEE*, **54**(12), 1901–9.

Hacker S. (1999) Static superscalar design: a new architecture for the TigerSHARC DSP processor. Analog Devices Whitepaper. *www.analog.com/publications/whitepapers/products/sharc.html*

Hayes J.P. (1998) *Computer Architecture and Organization*, 3rd edn. Boston MA: McGraw-Hill.

Hennessy J.L. and Patterson D.A. (1990) *Computer Architecture: A Quantitative Approach*. San Mateo CA: Morgan Kaufmann.

Levy M. (1998) DSP architecture directory. *EDN*, 23 April, pp. 40–110.

Levy M. (1999) DSP architecture directory. *EDN*, 15 April, pp. 67–102.

Rabiner L.R. and Gold B. (1975) *Theory and Applications of Digital Signal Processing*. Englewood Cliffs NJ: Prentice-Hall.

Texas Instruments (1988) *TMS320C2x Software Development System User's Guide*. Dallas TX: Texas Instruments.

Texas Instruments (1999) TMS320C6000 Technical Brief, Literature No. SPRU197D, Texas Instruments, Austin TX. Also available at *www.ti.com*

Bibliography

Casey P.E. and Simmers L. (1986) Digital signal processing IC helps to shed new light on image processing applications. *Electronic Design*, 20 March, 135.

Chassaing R. and Horning D.W. (1990) *Digital Signal Processing with the TMS320C25*. New York: Wiley.

Cragon H. (1980) The elements of single-chip microcomputer architecture. *Computing Mag.*, **13**(10), 27–41.

Croisier A., Estaban D.J., Levilion M.E. and Rizo W. (1973) *Digital Filter for PCM Encoded Signal*. US Patent 3777130, 3 December, 1973.

Dahnoun N. (2000) *Digital Signal Processing Implementation Using the TMS320C6000 DSP Platform*. Englewood Cliffs NJ: Prentice-Hall.

De Roberts R.B. and Rifaat R. (1999) DSPs enhance flexible third-generation base-station design. *Wireless System Design*, November, *www.wsdmag.com*

DSP-architecture directory, 1997, *EDN*, 8 May, 43–107.

DSP56300 Family Manual, Motorola, *www.motorola.com/SPS/DSP/documentation*

DSP56000 Digital Signal Processor Family Manual, Motorola, Austin TX, 1995.

Fine B. and McGuire G. Considerations for selecting a DSP processor – ADSP-2101 vs TMS320C50, AN-233 Analog Devices Application Note, Digital Signal Processing Products 9-77–9-92.

Gallant J. (1990) Plug-in DSP boards. *EDN*, **35**, 142–60.

Ganesan S. (1991) A dual-DSP microprocessor system for real-time digital correlation. *Microprocessor and Microsystems*, **15**(7), 379–84.

Gore A.E. (1986) Cascadable digital signal processor. *New Electronics*, 14 October, 39–40.

Jouppi P. and Wall D.W. (1989) Available instruction-level parallelism for superscalar and superpipelined machines. In *Proc. Third Conf. on Architectural Support for Programming Languages and Operating Systems*. IEEE/ACM (April), Boston MA, pp. 272–82.

Kloker K.L. (1986) The Motorola DSP56000 Digital Signal Processor, *IEEE Micro*, December, pp. 29–48.

Kogge P.M. (1981) *The Architecture of Pipelined Computers*. New York: McGraw-Hill.

Leary K. and Morgan D. (1986) Fast and accurate analysis with LPC gives a DSP chip speech-processing power. *Electronic Design*, 17 April, 153.

Levy M. (1996) DSP-chip directory, *EDN*, March, 42–103.

Levy M. (1997) C compilers for DSPs flex their muscles. *EDN*, 7 June, 93–107.

Lin K.S. (ed.) (1988) *Digital Signal Processing Applications with the TMS320 Family*, Vol. 1. Englewood Cliffs NJ: Prentice-Hall.

Lin K.S., Frantz G.A. and Simar R. (1987) The TMS320 family of digital signal processors. *Proc. IEEE*, **75**(9), 1143–59.

McKee D. (1990) TMS32010 routine finds phase. *EDN*, **35**, 148.

Mennen P. (1991) DSP chips can produce random numbers using proven algorithm. *EDN*, **36**, 141–6.

Messer D.D. (1991) Convolutional encoding and viterbi decoding using the DSP56001. *Microprocessors and Microsystems*, **15**(1), 54–62.

Motorola (1986) *DSP56000 Digital Signal Processor User's Manual*. Motorola.

Nath N.S.M. (1999) C compilers and development tools simplify DSP assembly-language programming. *EDN*, 21 January, 103–10.

Papamichalis P. (1990) *Digital Signal Processing Applications with the TMS320 Family. Theory, Algorithms, and Implementations*, Vol. 3. Dallas TX: Texas Instruments.

Papamichalis P. and Simar R. (1988) The TMS320C30 floating-point digital signal processor. *IEEE Micro Mag.*, **8**(6), 10–28.

Roesgen J. (1986) Fast modem designs benefit from DSP chip's versatility. *Electronic Design*, 12 June.

Roesgen J. and Tung S. (1986) Moving memory off chip, DSP μP squeezes in more computational power. *Electronic Design*, 20 February, 131.

Rosen S. (1969) Electronic computers: a historical survey. *Computer Survey*, **1**(1), 7–36.

Schmalzel J., Hein D. and Ahmed N. (1980) Some pedagogical considerations of digital filter hardware implementation. *IEEE Circuits and Systems Mag.*, **2**(1), 4–13.

So J. (1983) TMS 320 – step forward in digital signal processing. *Microprocessors and Microsystems*, **7**(10), 451–60.

Stokes J. and Sohie G.R.L. (1991) Implementation of PID controllers on the Motorola DSP56000/DSP56001. Part 1. *Microprocessors and Microsystems*, **15**(6), 321–31.

Stokes J. and Sohie G.R.L. (1991) Implementation of PID controllers on the Motorola DSP56000/DSP56001. Part 2. *Microprocessors and Microsystems*, **15**(7), 385–92.

Texas Instruments (1989) *Second-Generation TMS320 User's Guide*. Dallas, TX: Texas Instruments.

TMS320C5x User's Guide, Texas Instruments, 1995.

TMS320C54x DSP Reference Set, Volume 4: Applications Guide. Texas Instruments, October 1996. *www.ti.com*

Tomarakos J. and Ledger D. (1998) Using the Low-cost, High Performance ADSP-21065L Digital Signal Processor for Digital Audio Applications. Analog Devices DSP Application. Details are available at *www.analog.com*

Zolzer U. (1997) *Digital Audio Signal Processing*. Wiley.

Other useful web addresses

Special purpose hardware, e.g. MT9300 multi-channel voice echo canceller, PDSP16256 programmable FIR filters. Mitel Semiconductor, *www.mitelsemi.com*

Audio codecs and processors, e.g. CS4228 24 bit 96 kHz surround sound codec, digital audio sample rate converters. Cirrus, *www.cirrus.com*

Appendix

12A   TMS320 assembly language programs for real-time signal processing and a C language program for constant geometry radix-2 FFT

The following TMS320C10/C25 programs are available on the CD included with the companion manual – *A Practical Guide for MATLAB and C Language Implementations of DSP Algorithms* (see the Preface for details).

(1)   a TMS320C10-based FIR digital notch filter;

(2)   a TMS320C10 implementation of an FIR digital bandpass filter;

(3)   a TMS320C25-based FIR digital notch filter;

(4)   a TMS320C10 fourth-order digital IIR filter realized with second-order sections in cascade;

(5)   a TMS320C25 fourth-order digital IIR filter realized with second-order sections in cascade;

(6)   a TMS320C25 fourth-order digital IIR filter realized with second-order sections in parallel;

(7)   a C language program for constant geometry radix-2 FFT;

(8)   a TMS320C25-based radix-2 FFT algorithm;

(9)   a TMS320C25 adaptive filter.

Only programs (2) and (5) are listed in this appendix because of lack of space.

Program 12A.1   A TMS320C10-based implementation of an FIR digital bandpass filter.

METAi Assembler 4.00 ©1988 Crash Barrier Thu Nov 19 00:37:40 1992
Page 1 Assembler

targbpf.asm

```
00000000                        1    c:\metai\32010.tab/
                                2
00000000                        3         .ctrl              27, 15
00000000                        4         SEGMENT            word at 0000 'ram'
00000000                        5    ;;;;;;;;;;;;;;;;;;;;;;;;;;;;;;;;;;;;;;;;;;;;;;;;;;;;;;;;;;;;;;;;;;;;;;;;;;;;;;;;;;;;;;;;;;;
00000000                        6    ;                                                                                     ;
00000000                        7    ;      FIR BANDPASS FILTER                                                           ;
00000000                        8    ;                                                                                     ;
00000000                        9    ;      Filter specification:                                                         ;
00000000                       10    ;                                                                                     ;
00000000                       11    ;      filter type                        bandpass filter                           ;
00000000                       12    ;      sampling frequency                 15 kHz                                    ;
00000000                       13    ;      passband                           900–1100 Hz                               ;
00000000                       14    ;      transition width                   450 Hz                                    ;
00000000                       15    ;      passband ripple                    <0.87 dB                                  ;
00000000                       16    ;      stopband attenuation               >30 dB                                    ;
00000000                       17    ;      filter length                      41                                        ;
00000000                       18    ;                                                                                     ;
00000000                       19    ;      Hardware: TMS320C10 Target board with 8-bit ADC/DAC    ;
00000000                       20    ;                                                                                     ;
00000000                       21    ;;;;;;;;;;;;;;;;;;;;;;;;;;;;;;;;;;;;;;;;;;;;;;;;;;;;;;;;;;;;;;;;;;;;;;;;;;;;;;;;;;;;;;;;;;;
00000000                       22    ;
                               23
00000000    F900002B           24         B        START
                               25
00000028                       26    NM1      EQU    40             ;N–1
00000000                       27    XN       EQU    0              ;CURRENT I/P SAMPLE
00000028                       28    XNM1     EQU    NM1
00000029                       29    H0       EQU    NM1+1
00000051                       30    HNM1     EQU    H0+NM1
0000007B                       31    YN       EQU    123
0000007C                       32    ONE      EQU    124
00000000                       33    PA0      EQU    0              ;address of I/O for D/A IN TARGET BOARD
00000001                       34    PA1      EQU    1
00000002                       35    PA2      EQU    2
00000003                       36    PA3      EQU    3
00000002                       37    COEFF    EQU    2              ;START ADDRESS OF COEFFS.
00000000                       38    R0       EQU    0
00000001                       39    R1       EQU    1
00000002                       40    ;
00000002                       41    ;TABLE OF COEFFS. THESE ARE INITIALLY
00000002                       42    ;STORED IN PROGRAM MEMORY.
                               43
                               44
00000002    FE09FFFEFF5B019F   45
00000002    02B3038E03D9       45         DC.W     –503,–2,–165,415,691,910,985
00000009    035001 D9FF97FCE8  46
```

```
00000009   FA57F881            46              DC.W   848,473,-105,-792,-1449,-1919
0000000F   F7EAF8DDFB52FEE8     47
0000000F   02F406A8            47              DC.W   -2070,-1827,-1198,-280,756,1704
00000015   093F0A2E093F06A8     48
00000015   02F4FEE8            48              DC.W   2367,2606,2367,1704,756,-280
0000001B   FB52F8DDF7EAF881     49
0000001B   FA57                49              DC.W   -1198,-1827,-2070,-1919,-1449
00000020   FCE8FF9701D90350     50
00000020   03D9038E02B3        50              DC.W   -792,-105,473,848,985,910,691
00000027   019F00A5FFFEFE09     51              DC.W   415,165,-2,-503
```

METAi Assembler 4.00 © 1988 Crash Barrier   Thu Nov 19 00:37:40 1992
Page 2 Assembler

                                      targbpf.asm

```
                               52
                               53
0000002B                       54   ;=============== START OF MAIN PROGRAM ===============
0000002B                       55   ;
0000002B                       56   ;INITIALIZATION
0000002B                       57   ;
0000002B   7E01                58   START   LACK   1
0000002C   507C                59           SACL   ONE
0000002D   6E00                60           LDPK   0          ;POINT TO PAGE ZERO OF DATA
                                                              ;MEMORY
0000002E                       61   ;
0000002E                       62   ;TRANSFER COEFFICIENTS TO DATA MEMORY FROM PROGRAM
                                     ;MEMORY IN EPROM SPACE
0000002E                       63   ;
0000002E   7E02                64           LACK   COEFF      ;LOAD COEFF ADDRESS INTO
                                                              ;ACCUMULATOR
0000002F   7028                65           LARK   AR0,NM1    ;NO OF COEFF INTO AUXILIARY
                                                              ;REGISTER 0
00000030   7129                66           LARK   AR1,H0     ;AND DATA MEMORY ADDRESS OF
                                                              ;COEFFICIENTS INTO AR1
00000031   6881                67   LOAD    LARP   R1         ;SELECT AR1 AND BEGIN TO TRANSFER
                                                              ;COEFF.
00000032   67A0                68           TBLR   *+,R0      ;INTO DATA MEMORY, THEN INCREMENT
                                                              ;THE CONTENTS OF AR1
00000033   007C                69           ADD    ONE        ;INCREMENT THE ACCUMULATOR
00000034   F4000031            70           BANZ   LOAD       ;DEC AR0, AND BRANCH IF NOT ZERO
00000036                       71   ;
00000036                       72   ;WAIT FOR NEW INPUT SAMPLE
00000036                       73   ;
00000036   6880                74           LARP   R0
00000037   F600003B            75   WAIT    BIOZ   NXTPT      ;SEE IF SAMPLE IS RDY
00000039   F9000037            76           B      WAIT       ;IF NOT GO WAIT
                               77
0000003B   4000                78   NXTPT   IN     XN,PA0     ;IF READY THEN READ SAMPLE . . . was
                                                              ;PA2 for EVM
0000003C                       79   ;
0000003C                       80   ;CALCULATE FILTER OUTPUT IN YN AND OUTPUT TO DAC
0000003C                       81   ;
0000003C   7028                82   skip    LARK   AR0,XNM1
```

0000003D	7151	83		LARK	AR1,HNM1	
0000003E	7F89	84		ZAC		
		85				
0000003F	6A91	86		LT	*-,R1	;LOAD XN(N-1) SAMPLE
00000040	6D90	87		MPY	*-,R0	;COMPUTE H(N-1)*XN(N-1)
00000041	6B81	88	LOOP	LTD	*,R1	;COMPUTE SIG[H(K)*X(N-K)]
00000042	6D90	89		MPY	*-,R0	
00000043	F4000041	90		BANZ	LOOP	
00000045	7F8F	91		APAC		;ADD H(N-1)*X(N-1)
00000046	597B	92		SACH	YN,1	;OUTPUT SAMPLE
00000047	487B	93		OUT	YN,PA0	;was PA3 for EVM
		94				
00000048	F6000048	95	onhi	BIOZ	onhi	;wait here until BIO line goes high before ;going for next
		96				
0000004A	F9000037	97		B	WAIT	
		98				
0000004C		99		end		

No errors on assembly of 'targbpf.asm'

Program 12A.2   A TMS320C25 fourth-order digital IIR filter realized with second-order sections in cascade.

```
;;;;;;;;;;;;;;;;;;;;;;;;;;;;;;;;;;;;;;;;;;;;;;;;;;;;;;;;;;;;;;;;;;
; Fourth order Elliptic filter, connected        ;
; as a cascade of 2 biquad canonic sections      ;
; Manny Ifeachor, Jan., 1992                     ;
;                                                ;
;         FILTER SPECIFICATIONS:                 ;
;                                                ;
;     Filter Type              lowpass           ;
;     Sampling frequency       15 kHz            ;
;     Passband                 0–3 kHz           ;
;     transition width         450 Hz            ;
;     Passband ripple          0.5 dB            ;
;     Stopband attenuation     45 dB             ;
;                                                ;
;     Hardware: TMS320C25 SWDS with AIB          ;
;     1-bit ADC/DAC (filter B)                   ;
; -------------------------------------------------;
;
XN       .set     0
YN1      .set     1
W1N      .set     2
W1NM1    .set     3
W1NM2    .set     4
YN2      .set     5
W2N      .set     6
W2NM1    .set     7
W2NM2    .set     8
SF1      .set     9
A01      .set     10
```

```
A11        .set    11
A21        .set    12
B11        .set    13
B21        .set    14
A02        .set    15
A12        .set    16
A22        .set    17
B12        .set    18
B22        .set    19
SF2        .set    20
ONE        .set    21
RATED      .set    22
MODED      .set    23
WONE       .set    24
TEMP       .set    25
PBM1       .set    0300 h
;
           .sect   "IRUPTS"
START      B       INIT
;
           .text
COEFFS     .word   13217      ; SF1 = 0.4033627 IN Q15 FORMAT
                              ;(SCALE FACTOR)
;
           .word   4296       ;a01=0.1311136
           .word   7087       ;a11=0.2162924
           .word   4296       ;a21=0.1311136
           .word   27175      ;−b11=0.829328
           .word   −10061     ;−b21=−0.307046
;
           .word   32767      ;a02=0.9999695 (largest + ve number)
           .word   22142      ;a12=0.675718
           .word   32767      ;a22=0.9999695
           .word   16251      ;−b12=0.495935
           .word   −24965     ;−b22=−0.761864
;
           .word   29769      ;SF2=0.90847
;
MODEP      .word   0Ah
RATEP      .word   0299h
;
*
**         INITIALIZE THE AIB **
*
INIT       LDPK    6
           SSXM
           LACK    MODEP
           TBLR    MODED
           OUT     MODED,PA0
           LACK    RATEP      ;SET UP AIB SAMPLING FREQ
           TBLR    RATED
           OUT     RATED,PA1
           OUT     RATED,PA3  ;LET GO AIB
*
**         TRANSFER COEFFS FROM PROG MEMORY TO DATA MEMORY
```

```
*              LARP    AR0
               LRLK    AR0,PMB1+SF1
               RPTK    11
               BLKP    COEFFS,*+
*
**             INTIALIZE DMA FOR INTERNAL NODE DATA
*
INITWN         ZAC
               SACL    W1N
               SACL    W1NM1
               SACL    W1NM2
               SACL    W2N
               SACL    W2NM1
               SACL    W2NM2
*
**             WAIT FOR NEW DATA SAMPLE TO BE READY
*
RDATA          BIOZ    NXTPT              ;FETCH THE NEW SAMPLE
               B       RDATA
NXTPT          IN      XN,PA2
               LT      XN
*
**             START OF FILTER BLOCK 1
*
BLOCK1         MPY     SF1                ;SCALE INPUT DATA SAMPLE: SF*X(N)
               PAC
               LT      W1NM1              ;LOAD T-REGISTER WITH W(N-1)
               MPY     B11                ;B11W1(N-1)
               LTA     W1NM2              ;SFX(N)+B11W1(N-1)
               MPY     B21                ;B21W1(N-2)
               APAC                       ;SFX(N)+B11W1(N-1)+B21W1(N-2)
               SACH    W1N
               ZAC
               MPY     A21                ;A21W1(N-2)
               LTD     W1NM1              ;SUM = A21W(N-2); W1(N-2)=W1(N-1)
               MPY     A11                ;A11W1(N-1)
               LTD     W1N                ;SUM=A21W1(N-2) + A11W1(N-1); W1(N-1)=W1(N)
               MPY     A01                ;A01W1(N)
*
**             START OF FILTER BLOCK 2
*
BLOCK2         LTA     W2NM1              ;Y1(N)=A21W(N-2)+A11W1(N-1)+A01W1(N)
               MPY     B12                ;B12W2(N-1)
               LTA     W2NM2              ;Y1(N)+B12W2(N-1)
               MPY     B22                ;B22*W2(N-2)
               APAC                       ;Y1(N)+B12*W2(N-1)+B22*W2(N-2)
;                                         ;SUM=Y1(N)+B12*W2(N-1)+2*B22*W2(N-2)
               SACH    W2N                ;STORE HIGH 16 BIT WORD OF SUM IN W2(N)
               MPY     A22                ;A22*W2(N-2)
               ZAC                        ;SUM=SUM+A22*W2(N-2)
               LTD     W2NM1              ;W2(N-2)=W2(N-1); SUM = A22*W2(N-2)
               MPY     A12                ;A12*W2(N-1)
               APAC                       ;SUM=A22*W2(N-2)+A12*W2(N-1)
               LTD     W2N                ;W2(N-1)=W2(N)
```

```
        MPY    A02              ;A02*W2(N)
        APAC
        SACH   YN2              ;
*
**      SCALE OUTPUT SAMPLE AND SEND IT TO DAC
*
        LT     YN2              ;SCALE OUTPUT BACKUP
        MPY    SF2              ;SF2*YN2
        PAC                     ;
        APAC
        APAC
        APAC
        APAC
        APAC
        APAC
        APAC
        APAC
        APAC
        APAC
        APAC
        APAC
        APAC
        APAC
        APAC
        APAC
        APAC
        APAC
        APAC
        SACH   YN2              ;21*SF2*YN2
;
        OUT    YN2,PA2
        B      RDATA
;
        END
;@\\\000300AB1800001100AB18;
```

```
/* This is the link command file */
MEMORY
{                                                                   /* Program Memory */
    PAGE 0: VECTORS:origin=0h, length=01Fh
            CODE:origin=20h, length=0F90h

                                                                    /* Data Memory */
    PAGE 1: RAMB2:origin=60h, length=020h
            RAMB0:origin=200h, length=0FFh
            RAMB1:origin=300h, length=0FFh
}
SECTIONS
{   IRUPTS    :{}    > VECTORS    PAGE 0
    .text     :{}    > CODE       PAGE 0
    .data     :{}    > RAMB2      PAGE 1
    .bss      :{}    > RAMB0      PAGE 1
}
```

Analysis of finite wordlength effects in fixed-point DSP systems

The purpose of this chapter is to provide an understanding of the errors that arise in practical DSP systems due to quantization and use of finite-wordlength arithmetic units to perform DSP operations. The effects of the errors on signal quality and techniques for combating them are also discussed. An understanding of the issues covered in the chapter should assist in the design of DSP systems with predictable performance.

The emphasis in the chapter is on fixed-point DSP systems because they are more prevalent.

13.1 Introduction

In most cases, the final objective in a DSP design problem is to implement a DSP function, e.g. filtering or FFT, in a digital processor. In practice, a finite number of bits is used to represent variables and to perform arithmetic operations. Typical wordlengths in modern DSP processors are 16 bits (e.g. TMS320C54), 24 bits (e.g. DSP56300) and 32 bits (e.g. ADSP-21065). The use of finite wordlengths introduces errors which can affect the performance of DSP systems. Before implementing a DSP function, the designer should ascertain the extent to which its performance will be degraded by errors due to the effects of finite wordlength and to find a remedy if necessary.

The main errors in DSP are:

(1)   ADC quantization error – this results from representing the input data by a limited number of bits.

(2)   Coefficient quantization error – this is caused by representing the coefficients or DSP parameters by a finite number of bits. The coefficients, a_k and b_k, from stage 2 of filter design, for example, are normally of very high precision but in a DSP processor they must be quantized, typically to the processor wordlength.

(3)   Overflow error – this is caused by the addition of two large numbers of the same sign which produces a result that exceeds permissible wordlength.

(4)   Roundoff error – this is caused when the result of a multiplication is rounded (or truncated) to the nearest discrete value or permissible wordlength.

The effects of the errors introduced by signal processing depend on a number of factors, including the type of arithmetic used, the quality of the input signal, and the type of DSP function or algorithm implemented. In the discussions, we will use IIR digital filters as the main vehicle to study the effects of finite wordlength on DSP system performance because they embody most of the problems that are commonly encountered in practice.

13.2   DSP arithmetic

We have seen in previous chapters that the basic operations in DSP are multiplications, additions and delays (or shifts). For example, in digital FIR filtering the coefficients $h(k)$, $(k = 0, 1, \ldots, N-1)$, and the input data samples, $x(n)$, $(n = 0, 1, \ldots)$ are multiplied and the products added as follows:

$$y(n) = \sum_{k=0}^{N-1} h(k)x(n - k) = h(0)x(n) + h(1)x(n - 1) + \ldots$$
$$+ h(N - 1)x[n - (N - 1)] \tag{13.1}$$

In practice, the arithmetic operations involved in DSP (such as those indicated above) are often carried out using fixed-point or floating-point arithmetic (Rabiner and Gold, 1975). Other types of arithmetic are sometimes used, for example block floating point arithmetic, which seeks to combine the benefits of the two above. Fixed-point arithmetic is the most prevalent in DSP work because it leads to a fast and inexpensive implementation, but it is limited in the range of numbers that can be represented, and is susceptible to problems of overflow which may occur when the result of an addition exceeds the permissible number range (for example large-scale limit cycles in IIR filters and overload in high order FFTs). To prevent the results of arithmetic operations going outside the permissible number range, the operands are scaled. Such scaling degrades the performance of DSP systems, that is it reduces the signal-to-noise ratio achievable.

Floating point arithmetic is preferred where the magnitudes of variables or system coefficients vary widely (Flores, 1963). It allows a much wider dynamic range, and virtually eliminates overflow problems. Further, floating-point processing simplifies programming. DSP algorithms developed on large machines, for example on personal or mainframe computers, in a high level language can be implemented directly in DSP hardware with little change to the core algorithms. However, floating-point arithmetic is more expensive and often slower, although high speed digital signal processors with a built-in floating-point processor (such as the Texas Instruments TMS320C30) are becoming widely available. Efficient floating-point software routines are also available in the open literature (Texas Instruments, 1986). Thus the disparity in price and speed between fixed- and floating-point has diminished significantly.

Increasingly, DSP techniques are being used in applications where both wide dynamic range and high precision are required. Fixed-point digital signal processors with large wordlengths (24 bits or more) may satisfy such requirements, but floating point processing provides a simpler and a more natural way of catering for these cases.

Application areas where wide dynamic range and high precision are required and the use of floating-point arithmetic is necessary or desirable are many. An example is in real-time parametric equalization of digital audio signals using digital filters, where the values of filter coefficients vary widely as the user sweeps through the audio band and adjusts the equalizer parameters. In certain signal processing tasks (for example spectrum analysis in radar and sonar, seismology or biomedicine) a need often arises to resolve very low level components in a signal of wide dynamic range. In such cases both wide dynamic range and high precision are required. Other application examples include high resolution graphics stations and general engineering computations. Desirable maximum dynamic range and accuracy requirements in a number of applications are summarized in Table 13.1 (Weitek, 1984).

Table 13.1 Dynamic range and accuracy requirements.

	Dynamic range (bits)	Accuracy (bits)
Noise cancelling	32	20
Radar processing	32	20
Broadcast quality picture processing	20	20
Image processing	30	20
Medical spectrum analysis	20	20
Seismic data processing	70	20

The key lesson is that the type of arithmetic used is a major factor in the performance of DSP systems. In the next few sections, we will discuss the basic concepts of the two types of arithmetic (that is fixed- and floating-point).

13.2.1   Fixed-point arithmetic

13.2.1.1   Fixed-point representation

In DSP, variables are often represented as fixed-point, 2's complement fractions; see, for example, Table 13.2. In this representation, the binary point is to the right of the MSB (most significant bit) which is also the sign bit. Each number lies in the range from -1 to $1 - 2^{-(B-1)}$, where B is the number of bits used to represent the number. A common representation in DSP is the so-called Q15 format which uses 16 bits (1 sign bit and 15 fractional bits):

0110 0000 0000 0000
| binary point

Two's complement positive numbers are in natural binary form; see Table 13.2. A negative number is formed from the corresponding positive number by complementing all the bits of the positive number and then adding 1 LSB. For example,

Table 13.2   A comparison of 2's complement and offset binary number systems for a 4-bit wordlength.

Number	Decimal fractions	Two's complement	Offset binary
7	7/8	0111	1111
6	6/8	0110	1110
5	5/8	0101	1101
4	4/8	0100	1100
3	3/8	0011	1011
2	2/8	0010	1010
1	1/8	0001	1001
0	0	0000	1000
−1	−1/8	1111	0111
−2	−2/8	1110	0110
−3	−3/8	1101	0101
−4	−4/8	1100	0100
−5	−5/8	1011	0011
−6	−6/8	1010	0010
−7	−7/8	1001	0001
−8	−1	1000	0000

the 2's complement representation of $-3/8$ is obtained from $3/8$ (that is, 0011) as $1100 + 0001 = 1101$.

When the input to a DSP system is from an ADC (analog-to-digital converter) the data fed to the digital processor may be in offset binary form. Similarly, the output of the DSP system may need to be converted to offset binary if it feeds a DAC (digital-to-analog converter). Conversion from offset binary to 2's complement representation is trivially achieved by complementing the MSB of the offset binary code. For example, in Table 13.2, the offset binary code 1111, that is $7/8$, is easily converted to 2's complement code (0111) by complementing the MSB. In practice, the DSP chip bus is often wider than the ADC resolution. In this case, after conversion to 2's complement, the sign bit is extended to fill the remaining spaces to the left. For example, the code (1111 1101) which has been 2's complemented becomes (1111 1111 1111 1101) after sign extension.

In fixed-point, 2's complement representation, if each number is represented by B bits then a maximum of 2^B different numbers can be represented, with adjacent numbers separated by approximately 2^{-B}. It is useful to know the accuracy to which we can represent each number compared with decimal representation.

Given a decimal fraction, X, consisting of d digits, its accuracy is $\pm 0.5 \times 10^{-d}$. If we represent the same number in binary with B bits, its accuracy now becomes $\pm 0.5 \times 2^{-B}$. To retain the same accuracy for the two representations requires

$$0.5 \times 10^{-d} = 0.5 \times 2^{-B}, \text{ that is } \quad B = d \log_2 10 \simeq 3.3d \text{ bits} \tag{13.2}$$

For example, suppose the decimal number 0.234 56 is to be represented in binary; then we require $3.3 \times 5 = 17$ bits to represent it as accurately as before. Table 13.3 summarizes the relationship between the number of bits in a binary system and their accuracies in decimal digits or places.

Table 13.3   Relationship between the number of bits and accuracies in decimal digits.

Number of bits	Accuracy (number of decimal digits)
7	2.1
8	2.4
10	3
12	3.6
14	4.2
15	4.5
16	4.8
18	5.4
20	6.1
23	7.0
24	7.3
64	19.4

Example 13.1	Represent the decimal number, 0.956 24, as

(1)   a Q3 number, and

(2)   a Q4 number.

Compare the errors in the two cases.

(3)   Estimate the number of bits required to represent the decimal number above to retain the same accuracy.

Solution   (1)   A Q3 number is a 2's complement number with 1 sign bit and 3 fractional bits. To convert the decimal number to a Q3 format we simply multiply it by 2^3 and then round the product to the nearest permissible integer: $0.956\,24 \times 2^3 = 7.649\,92$, which is rounded to $7 = 0111$ (the highest permissible number).

(2)   In this case, we represent the number using 1 sign bit and 4 fractional bits: $0.956\,24 \times 2^4 = 15.299\,84$, which is rounded to $15 = 01111$.

The errors in representing the number in parts (1) and (2) are, respectively, $0.649\,92/8 = 0.081\,24$ and $0.299\,84/16 = 0.018\,74$. The error in representing the number is often referred to as the coefficient quantization error.

(3)   From Equation 13.2, we require $3.3 \times 5 = 16.5$ bits $\simeq 17$ bits.

13.2.1.2   Fixed-point multiplication

In fixed-point multiplication, use is made of the fact that the product of two fractions is also a fraction and that of an integer is also an integer. We will illustrate with an example.

Example 13.2	Find the square of 0.5625 using fixed-point 2's complement arithmetic. Assume a Q4 format.

Solution

```
        0 1001 = 0.5625
        0 1001 = 0.5625
     0 0000
      0100 1
       000 00
        00 000
         0 1001
     00 0101 0001
     ↑
 binary point        this bit is lost when product is quantized
```

After shifting left, to remove the extra sign bit, and rounding, we obtain the final answer: 0 0101 = 0.25 + 2^{-4} = 0.3125, instead of 0.316 406 25.

Example 13.2 shows that an additional sign bit is created after multiplication and that the product of two 5-bit numbers is 10 bits long, so that the result should be truncated or rounded to 5 bits before it can be saved in memory. In general, the product of two B-bit numbers is $2B$ bits long. For example, the 10-bit result is shifted left once to remove the extra sign bit and then rounded to 0.0101. The ability to round (or truncate) results as we have just done is one of the major attractions of 2's complement fractional arithmetic as it means that overflow never occurs in multiplications. Such rounding (or truncation), however, introduces errors into the signal which can lead to instability or some undesirable side-effects in DSP systems with feedback.

13.2.1.3   Fixed-point addition

Addition of two fixed-point fractions is more difficult than multiplication. This is because the operands to be added must be in the same Q format and attention must be paid to the possibility of an overflow.

Example 13.3   Find the sum of the following 2's complement numbers: 0001 1001 and 0110 1101 0111 1101.

Solution   The operands are first expressed in the same Q format and then added:

> 0110 1101 0111 1101
> 0001 1001 0000 0000
> 1000 0110 0111 1101
> | overflow

One way to correct for the overflow is to shift the result one place to the right and then set an exponent flag. Thus, the answer becomes

> 1˙ 0100 0011 0011 1110
> ↑
> exponent
> flag

Another alternative is to represent the result using double precision or to provide sufficient headroom to allow for growth due to overflow.

Example 13.4   Assuming a register length of 4 bits (1 sign bit and 3 data bits), find the sum of

(1) −0.25 and 0.75, and

(2) 0.5, 0.75, −0.5.

Solution   From Table 13.2, $0.25_{10} = 0.010_2$, $0.5_{10} = 0.100_2$ and $0.75_{10} = 0.110_2$.

(1)   0.75   0.110
       −0.25   1.110
        0.50   1 0.100

Answer: 0.100

(2)   0.5   0.100
      +0.75   0.110
        1.25   1.010 ← partial sum
      −0.50   1.100
        0.75   1 0.110 ← final sum

Answer: 0.110

13.2.2   Floating-point arithmetic

13.2.2.1   Floating-point representation

A binary floating-point number X is represented as the product of two signed numbers, the mantissa M and the exponent, E:

$$X = M \cdot 2^E \tag{13.3}$$

where 2 is the base of the binary system.

The exponent determines the range of numbers that can be represented, the mantissa the accuracy of the numbers. For example, if the exponent and the mantissa are represented by 8 and 16 bits, respectively, the range of the floating-point numbers that can be represented in this simple case is from 0.5×2^{-128} to $1 - (2^{-15}) \times 2^{128}$. Of the 16 bits used to represent the mantissa, one bit is the sign bit and the least significant bit may be of a doubtful accuracy owing to rounding effects. Thus the accuracy of the floating point numbers is 1 in 2^{14} (0.61×10^4), that is about 4 decimal digits.

13.2.2.2   IEEE floating point format

One of the most widely used binary floating-point systems is the IEEE 754 Standard (Patterson and Hennessy, 1990; IEEE, 1985). The format for single precision is shown

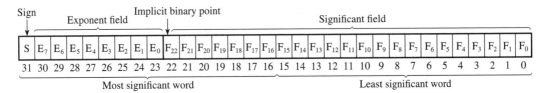

Figure 13.1   Floating-point representation (IEEE single precision).

in Figure 13.1. In this case, floating-point (FP) numbers with exponents in the range $0 < E < 255$ are said to be normalized.

The decimal equivalent, X, of a normalized IEEE FP number is given by

$$X = (-1)^s(1 \cdot F)2^{E-127}$$

where

- F is the mantissa in 2's complement binary fraction represented by bits 0 to 22,
- E is the exponent in excess 127 form, and
- $s = 0$ for positive numbers, $s = 1$ for negative numbers.

Two important features of the IEEE floating-point format are the assumed 1 preceding the mantissa and the biased exponent.

Example 13.5   (1)   For the FP number

sign
0 1000 0011.1100 . . . 00000
|←—exponent—→|←——mantissa——→|

the exponent is 1000 0011 = 131, the mantissa is 0.1100 . . . = 0.75, and $s = 0$. Therefore $X = (1.75)2^{131-127} = 1 \times 1.75 \times 2^4$.

(2)   For the FP number

sign
1 0000 1111.0110 . . . 00000
|←—exponent—→|←——mantissa——→|

the exponent is 0000 1111 = 15, the mantissa is 0.0110 . . . = 0.375, and $s = 1$. Therefore $X = (1.375)2^{15-127} = -1 \times 1.375 \times 2^{-112}$.

13.2.2.3   Floating-point additions and multiplications

If X_1 and X_2 are two floating point numbers to be added, where $X_1 = M_1 \times 2^{E_1}$ and $X_2 = M_2 \times 2^{E_2}$, then their sum X is given by

$$X = M \times 2^E$$

where

$$M = M_1 + M_2 \times 2^{E_1 - E_2}; \; E = E_1, \text{ assuming that } X_1 > X_2 \qquad (13.4)$$

Before two floating-point numbers are added, their exponents must be made equal. This is called alignment, and involves shifting right the mantissa of the smaller operand and incrementing its exponent until it equals that of the larger operand.

If X_1 and X_2 are two properly normalized FP numbers to be multiplied, where

$$X_1 = M_1 \times 2^{E_1}, \; X_2 = M_2 \times 2^{E_2} \qquad (13.5)$$

then their product X is given by

$$X = M \times 2^E$$

where

$$M = M_1 \times M_2, \; E = E_1 + E_2$$

Thus the mantissas are multiplied and their exponents added. Since M_1 and M_2 are both normalized then their product, M, will be in the range $0.25 < M < 1$. Thus the product, M, cannot overflow but may not be properly normalized (mantissa underflow).

| Example 13.6 | (1) Find the sum of the two numbers A and B, where $A = 9.985 \times 10^4$ and $B = 5.6756 \times 10^2$.

(2) Find the product of the two numbers A and B, where $A = 2.75 \times 10^{-16}$ and $B = 4.5 \times 10^{10}$.

Solution   (1) First, the exponents of the two numbers are compared and if they are not equal, the mantissa of the operand with the smaller exponent is then shifted so that the two exponents are equal:

$$5.6756 \times 10^2 = 0.056\,756 \times 10^4$$

The mantissas are then added to give

$$M = 9.985 + 0.056\,756 = 10.041\,756; \; E = 10^4$$

and the sum is $10.041\ 756 \times 10^4$. The sum is next normalized by moving the decimal point of the mantissa and then adjusting the exponent (if necessary) to give the final answer:

$$1.004\ 175\ 6 \times 10^5$$

(2) The mantissas are multiplied and the exponents added:

$$M = 2.75 \times 4.5 = 12.375;\ E = -16 + 10 = -6$$

giving the product $A \times B = 12.375 \times 10^{-6}$. The product is then normalized, where we have assumed that a properly normalized floating point number is less than 10, by adjusting the exponent and moving the position of the decimal point in the mantissa to obtain

$$A \times B = 1.2375 \times 10^{-5}$$

In floating-point arithmetic, roundoff errors occur in both addition and multiplication, whereas in fixed-point roundoff errors are only possible in multiplication. However, unlike fixed-point arithmetic overflow is very unlikely in floating-point addition because of the very wide dynamic range that is available: the more bits in the exponent the wider the dynamic range. Floating-point DSP processors and routines are now widely available (see Chapter 12).

13.3 ADC quantization noise and signal quality

The ADC quantizes the analog input signal into a finite number of bits, typically 8, 12 or 16, which gives rise to quantization noise.

As discussed in Chapter 2, the quantization noise power (or variance, σ_{ADC}^2) is given by

$$\sigma_A^2 = \frac{q^2}{12} = \frac{2^{-2B}}{3} \tag{13.6}$$

where q is the quantization step size and B is the number of ADC bits. Clearly, the level of the noise can be readily reduced by increasing the number of ADC bits. It is also possible to reduce it using multirate techniques (see Chapter 9). In general for values of B above 12 bits, the noise due to quantization error is insignificant, except for applications such as professional audio where at least 16 bits are required for acceptable performance.

The noise due to ADC quantization is fed into the DSP system as an irreversible error. The noise power at the output of the DSP system, due to the ADC, is given by

$$\sigma_{oA}^2 = \sigma_A^2 \left[\frac{1}{2\pi j} \oint_c H(z)H(z^{-1}) \frac{dz}{z} \right] \tag{13.7a}$$

$$= \sigma_A^2 \sum_{k=0}^{\infty} h^2(k) \tag{13.7b}$$

where

σ_{oA}^2 = ADC quantization noise at system output
$\oint_c$ = contour integral
$h(k)$ = impulse response of the system

The term in the square brackets can be viewed as the 'system power gain', which amplifies (or alters) the ADC noise, depending on the characteristics of the DSP system.

Example 13.7   An 8-bit ADC feeds a DSP system characterized by the following transfer function:

$$H(z) = \frac{1}{z + 0.5}$$

Estimate the steady state quantization noise power at the output of the system.

Solution   The noise power, due to the ADC, at the system input is given by (Equation 13.6)

$$\sigma_A^2 = \frac{2^{-16}}{3}$$

The output noise, due to the ADC, is given by

$$\sigma_{oA}^2 = \sigma_A^2 \left[\frac{1}{2\pi j} \oint_c H(z)H(z^{-1}) \frac{dz}{z} \right] = \sigma_A^2 \sum_{k=0}^{\infty} h^2(k) \tag{13.8}$$

where $\oint_c$ indicates a contour integral and $h(k)$ is the impulse response of the system (see Chapter 4). The term in the square brackets can be viewed as the system power gain, serving to amplify (or alternate) the ADC noise.

For the simple transfer function in this example, the term in the square brackets can be readily obtained using the residue method discussed in Chapter 4. Using the results in Chapter 4 and the unit circle as the contour, the term inside the square brackets is simply equal to 4/3. Thus, the quantization noise power is

$$\sigma_{oA}^2 = \frac{2^{-16}}{3} \times \frac{4}{3}$$

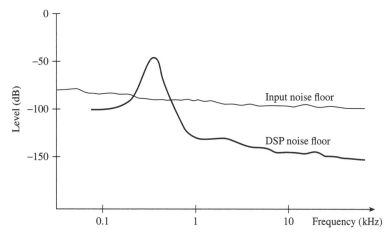

Figure 13.2   An illustration of the effects of roundoff noise in signal processing on system noise floor (after Wilson, 1993).

The ADC noise, together with the noise inherent in the signal, establishes the input noise floor. To maintain signal quality, the level of distortion introduced by subsequent digital signal processing should be below the noise floor. In Figure 13.2, we give an example of a DSP system in which the noise introduced by DSP (in this case IIR filtering) has risen above the input noise floor (Wilson, 1993). In such a case, a remedial action would be required to push the noise to below the noise floor (e.g. by using a roundoff error feedback scheme or a processor with a sufficiently long wordlength).

In practice, the number of bits used to represent the input data gives an indication of the dynamic range and signal quality. For example, in consumer CD audio systems the audio input is represented as 16-bit samples which corresponds to a theoretical input signal-to-noise ratio (SNR) of 96 dB. To maintain CD audio quality, DSP operations should be performed with a wordlength of at least 24 bits. A processor wordlength of 16 bits gives a dynamic range of 96 dB, but allowing for computational errors the effective dynamic range would be less than this. Thus, in carrying out finite wordlength error analysis for a specific system an important factor to consider is the input signal quality.

13.4   Finite wordlength effects in IIR digital filters

The coefficients, a_k and b_k, obtained from stage 2 of IIR filter design are of infinite or very high precision, typically six decimal places. When an IIR digital filter is implemented in a small system, such as an 8-bit microcomputer, errors arise in representing the filter coefficients and in performing the arithmetic operations indicated by the difference equation. These errors degrade the performance of the filter and in extreme cases lead to instability.

Before implementing an IIR filter, it is important to ascertain the extent to which its performance will be degraded by finite wordlength effects and to find a remedy if the degradation is not acceptable. In general, the effects of these errors can be reduced to acceptable levels by using more bits but this may be at the expense of increased cost.

The main errors in digital IIR filters are as follows:

- ADC quantization noise, which results from representing the samples of the input data, $x(n)$, by only a small number of bits;

- coefficient quantization errors, caused by representing the IIR filter coefficients by a finite number of bits;

- overflow errors, which result from the additions or accumulation of partial results in a limited register length;

- product roundoff errors, caused when the output, $y(n)$, and results of internal arithmetic operations are rounded (or truncated) to the permissible wordlength.

The extent of filter degradation depends on (i) the wordlength and type of arithmetic used to perform the filtering operation, (ii) the method used to quantize filter coefficients and variables, and (iii) the filter structure. From a knowledge of these factors, the designer can assess the effects of finite wordlength on the filter performance and take remedial action if necessary. Depending on how the filter is to be implemented some of the effects may be insignificant. For example, when implemented as a high level language program on most large computers, coefficient quantization and roundoff errors are not important. For real-time processing, finite wordlengths (typically 8 bits, 12 bits, and 16 bits) are used to represent the input and output signals, filter coefficients and the results of arithmetic operations. In these cases, it is nearly always necessary to analyze the effects of quantization on the filter performance.

The effects of finite wordlength on performance are more difficult to analyze in IIR filters than in FIR filters because of their feedback arrangements. However, the use of the PC-based program on the CD in the companion handbook (see the Preface) allows practical solutions to be obtained for specific filters. The effects of each of the four sources of errors listed above will be discussed, in turn, in the next few sections.

13.4.1 Influence of filter structure on finite wordlength effects

As the reader may already be aware, IIR filters may be represented by a variety of structures which are theoretically equivalent. However, when implemented in fixed- or floating-point DSP processors the behaviour of the filters may be significantly different.

In practice, IIR filters are often implemented using second-order direct form I and direct form II (or canonic) structures (Figures 13.1 and 13.2). The direct form I is characterized by the following features:

$$y(n) = \sum_{i=0}^{2} b_i x(n-1) - \sum_{i=1}^{2} a_i y(n-i) \qquad \text{– difference equation}$$

$$H(z) = \frac{b_0 + b_1 z^{-1} + b_2 z^{-2}}{1 + a_1 z^{-1} + a_2 z^{-2}} \qquad \text{– transfer function}$$

- 5 filter coefficients
- 4 delay elements
- 1 adder (4 additions)
- 1 quantization point for sum of products
- 1 multiplier (5 multiplications)
- 9 memory locations required to store data and coefficients

The canonic section is characterized by the following features:

$$\left. \begin{array}{l} y(n) = \displaystyle\sum_{i=0}^{2} b_i w(n-i) \\[4mm] w(n) = x(n) - \displaystyle\sum_{i=1}^{2} a_i w(n-i) \end{array} \right\} \quad \text{– two-step, difference equation}$$

$$H(z) = \frac{b_0 + b_1 z^{-1} + b_2 z^{-2}}{1 + a_1 z^{-1} + a_2 z^{-2}} \qquad \text{– transfer function}$$

- 5 filter coefficients
- 2 delay elements
- 2 adders (4 additions)
- 2 quantization points for sum of products
- 1 multiplier (5 multiplications)
- 7 memory locations required for data and coefficients

We note that although the two transfer functions are theoretically the same, there are important differences between them. For example, in the direct form I structure (Figure 13.3(a)), the feedforward terms (associated with the zeros) precede the feedback terms (associated with the poles). The opposite is the case for the canonic section (Figure 13.3(b)). The implication of this in practice is that the poles of the canonic section tend to amplify noise generated during computation more.

The differences between the two structures are more evident if we compare Figures 13.3(a) and 13.3(b) directly. The direct form I, Figure 13.3(a), has only 1 adder and 1 quantization point for sum of products whereas the canonic section, Figure 13.3(b), has 2 adders and 2 quantization points. The output of the left adder in Figure 13.3(b) is an internal node, $w(n)$. Thus the canonic section is susceptible to

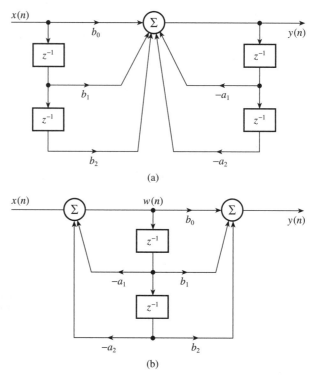

Figure 13.3   Basic building blocks for IIR filters: (a) direct form I second-order filter section; (b) canonic second-order filter section.

internal self-sustaining overflow. The direct form I does not have an internal node and overflow at its output during additions for 2's complement arithmetic either is self-correcting or can be handled readily. Further, its input, $x(n)$, is scaled by the coefficient, b_0, unlike in the canonic section where the input is not constrained in any way. Thus, we see that the structure used to implement a DSP function, in this case an IIR filter, can have a profound influence on the performance of the resulting DSP system.

In practice, higher order IIR filters are realized as cascade or parallel combinations of the second-order building blocks: see Figure 13.4. Three difficulties arise in connection with the cascade realization:

■ how to pair the numerator factors with the denominator factors to determine the coefficients of the second-order filter sections,

■ the order in which the second-order filter sections should be connected, and

■ the need to scale signal levels at various points within the composite filter to keep the levels within the permissible wordlength.

As discussed in Chapter 8, pairing and ordering of the filter sections are intimately linked to the effects of finite wordlength. Depending on the order of the filter, there may be a large number of possible equivalent filters resulting from pairing and

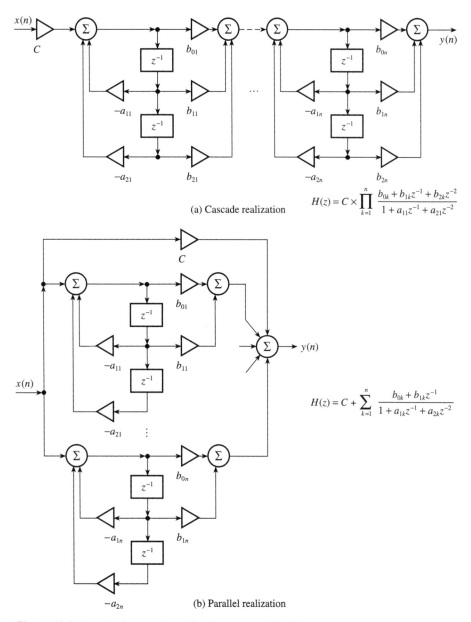

$$H(z) = C \times \prod_{k=1}^{n} \frac{b_{0k} + b_{1k}z^{-1} + b_{2k}z^{-2}}{1 + a_{1k}z^{-1} + a_{21}z^{-2}}$$

(a) Cascade realization

$$H(z) = C + \sum_{k=1}^{n} \frac{b_{0k} + b_{1k}z^{-1}}{1 + a_{1k}z^{-1} + a_{2k}z^{-2}}$$

(b) Parallel realization

Figure 13.4 Realization structures for higher-order IIR filters.

ordering of second-order sections, but these are not affected in the same way by finite wordlength errors. The finite wordlength analysis program in the companion handbook (Ifeachor, 2001) may be used to determine a suitable filter configuration.

For parallel realization, the order in which the sections is connected is not important.

13.4.2   Coefficient quantization errors in IIR digital filters

The IIR filter is characterized by the following equation:

$$H(z) = \frac{\sum_{k=0}^{N} b_k z^{-k}}{1 + \sum_{k=1}^{M} a_k z^{-k}}$$

$$[H(z)]_q = \frac{\sum_{k=0}^{N} [b_k]_q z^{-k}}{1 + \sum_{k=1}^{M} [a_k]_q z^{-k}}$$

where

$$[a_k]_q = a_k + \Delta a_k; \quad [b_k]_q = b_k + \Delta b_k$$

Δa_k, Δb_k are changes in the coefficients, a_k and b_k, respectively

q denotes a quantized quantity

The primary effect of quantizing filter coefficients into a finite number of bits is to alter the positions of the poles and zeros of $H(z)$ in the z-plane. For narrowband filters, for example, the poles will be close to the unit circle so that any significant deviation in their positions could make the filter unstable. The fewer the number of bits used to represent the coefficients the more will be the deviation in the pole and zero positions.

As well as potential instability, deviations in the locations of the poles and zeros also lead to deviations in the frequency response. The quantized filter should be analyzed to ensure that its wordlength is sufficient for both stability and satisfactory frequency response. Thus, essentially, at this stage the designer should determine the number of bits that are required to represent the filter coefficients for stability and for the desired frequency response to be satisfied.

Example 13.8   A bandpass digital IIR filter to be used in digital clock recovery for a 4.8 kbit/s modem is characterized by the following transfer function:

$$H(z) = \frac{1}{1 + a_1 z^{-1} + a_2 z^{-2}}$$

where

$$a_1 = -1.957\,558, \quad a_2 = 0.995\,813$$

Assuming a sampling frequency of 153.6 kHz, assess the effects of quantizing the coefficients to 8 bits on the pole positions and hence on the centre frequency.

Solution   First we find the pole positions of the unquantized filter. The pole position is (see Equation 13.10)

$$r = \sqrt{0.995\,913} = 0.997\,95, \quad \theta = \cos^{-1}\left(\frac{1.957\,558}{2r}\right) = 11.25°$$

This corresponds to a centre frequency of 4.7999 kHz ($153.6 \times 10^3 \times 11.25/360$).

Next we quantize the coefficients to 8 bits. As one of the coefficients is greater than unity, we will assign 1 bit as the sign bit, 1 bit for the integer and 6 bits for the fractional part of the coefficients. Quantizing the coefficients we have

$$a_1' = -1.957\,558 \times 2^6 = -125 \equiv 10000011$$

$$a_2' = 0.995\,913 \times 2^6 = 63 \equiv 00111111$$

In fractional notation, the quantized coefficient values are

$$a_1' = -\frac{125}{64} = -1.953\,125; \quad a_2' = \frac{63}{64} = 0.984\,375$$

The new pole position becomes

$$r' = 0.992\,156; \quad \theta' = 10.171\,853°$$

and the centre frequency now becomes

$$f_\text{o} = \left(\frac{10.171\,853}{360}\right) \times 153.6 \times 10^3 = 4.3399 \text{ kHz}$$

13.4.3 Coefficient wordlength requirements for stability and desired frequency response

Our stability discussions will be restricted to second-order filter sections since these are the basic building blocks of any filter. Consider a second-order section characterized by the familiar equations

$$H(z) = \frac{b_0 + b_1 z^{-1} + b_2 z^{-2}}{1 + a_1 z^{-1} + a_2 z^{-2}}$$

$$y(n) = \sum_{k=0}^{2} b_k x(n - k) - \sum_{k=1}^{2} a_k y(n - k)$$

The poles (or the roots of the denominator) are located at

$$p_1 = \tfrac{1}{2}[-a_1 + (a_1^2 - 4a_2)^{1/2}] \tag{13.9a}$$

$$p_2 = \tfrac{1}{2}[-a_1 - (a_1^2 - 4a_2)^{1/2}] \tag{13.9b}$$

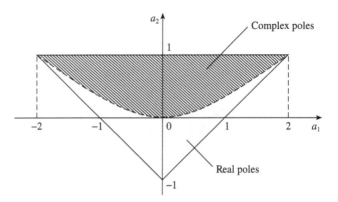

Figure 13.5   Stability triangle showing values of filter coefficients, a_1 and a_2, for which the filter is stable.

For each second-order section, three types of poles arise: complex conjugate poles, poles that are real and unequal, and real and equal (multiple-order) poles. Complex conjugate poles are the most common and occur if $a_1^2 < 4a_2$. For this case, the poles are each located at a radius, r, from the origin and at an angle θ given by

$$p_1 = r \angle \theta, \, p_2 = r \angle -\theta \tag{13.10}$$

where

$$r = a_2^{1/2}, \, \theta = \cos^{-1}\left(-\frac{a_1}{2r}\right)$$

Small changes in the coefficients a_1 and a_2, due to coefficient quantization, will lead to changes in both r and θ. For stability, the filter coefficients must lie inside the stability triangle (Figure 13.5) bounded by

$$0 \leq |a_2| < 1 \tag{13.11a}$$

$$|a_1| \leq 1 + a_2 \tag{13.11b}$$

The first bound specifies that the poles must lie inside the unit circle, since the pole radius is given by Equation 13.10. From Equations 13.10 and 13.11, a number of simple formulas can be derived for estimating the number of bits required to maintain stability, but these are applicable to a restricted set of cases. An alternative way of estimating suitable coefficient wordlength for stability is to analyze individual second-order blocks for various values of coefficient wordlengths (see Example 13.9).

The number of bits required for stability may not guarantee a satisfactory response. The effect of representing the coefficients by too few bits is to alter the frequency

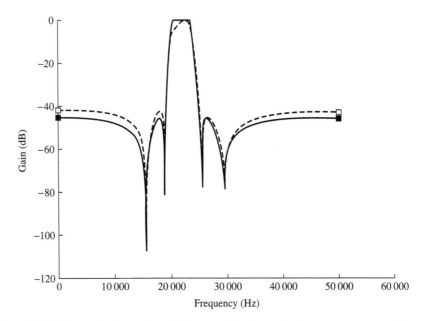

Figure 13.6   Practical effects of coefficient quantization on the frequency response: ■, unquantized; □, quantized, 5 bits.

response in the passband and stopband (see Figure 13.6). The changes in the passband are caused primarily by changes in the positions of the poles, and those in the stopband by changes in the locations of the zeros.

A procedure for determining a suitable coefficient wordlength for satisfactory frequency response is to find the minimum coefficient wordlength for which the passband and stopband requirements are satisfied. Although this approach may involve a lot of computation, the widespread availability of PCs and programs for finite wordlength analysis (FWA) makes it relatively straightforward to determine the wordlength for specific filters (an example program is on the CD in the companion handbook). An alternative approach employs a statistical method and is said to yield reasonably accurate estimates (Antoniou, 1979).

A smaller number of bits may be used to represent the filter coefficients if an optimization approach is used to quantize the filter coefficients, or by increasing the filter order, all other parameters remaining unchanged. This offers a trade-off between the coefficient wordlength and the filter order (Rabiner and Gold, 1975). For example, in a particular problem the designer may wish to use an existing processor or system with wordlengths already predetermined. If the designer finds that the required wordlength is greater than the processor wordlength, then it may be preferable to increase the order of the filter to bring the coefficient wordlength down to match that of the processor. However, the use of a higher-order filter would require more computational effort, which has some speed implications, and may be more susceptible to roundoff noise. The designer needs to consider the trade-off carefully.

Example 13.9   A digital filter is required to satisfy the following frequency response specifications:

passband	20.5–23.5 kHz
stopband	0–19 kHz, 25–50 kHz
passband ripple	≤0.25 dB
stopband attenuation	>45 dB
sampling frequency	100 kHz

(1)   Determine a suitable transfer function for the filter.

(2)   Determine a suitable coefficient wordlength

   (a)   to maintain stability and

   (b)   to satisfy the frequency response specifications.

(3)   Obtain and plot the frequency response of the unquantized filter and those of quantized filters corresponding to part (2).

Solution   (1)   Using the design program (available on the CD in the companion handbook – see the Preface for details) it was found that an elliptic filter characterized by the following transfer function is suitable:

$$H(z) = H_1(z)H_2(z)H_3(z)H_4(z)$$

where

$$H_1(z) = \frac{1 + 0.0339z^{-1} + z^{-2}}{1 - 0.1743z^{-1} + 0.9662z^{-2}}$$

$$H_2(z) = \frac{1 - 0.7563z^{-1} + z^{-2}}{1 - 0.5588z^{-1} + 0.9675z^{-2}}$$

$$H_3(z) = \frac{1 + 0.5331z^{-1} + z^{-2}}{1 - 0.2711z^{-1} + 0.9028z^{-2}}$$

$$H_4(z) = \frac{1 - 1.1489z^{-1} + z^{-2}}{1 - 0.4441z^{-1} + 0.9045z^{-2}}$$

(2)   (a)   The denominator coefficients of each of the second-order sections were each quantized by rounding to B bits ($B = 2, 3, \ldots, 29$) including the sign bits. For each value of B, the quantized coefficients and the pole location in polar form were computed. To illustrate, consider the first second-order filter section, $H_1(z)$. For $B = 8$ bits, the denominator coefficients are quantized by rounding as follows:

$$a_1 = -(0.1743 \times 2^7 + 0.5) = -22.8104 = -22$$
$$a_2 = 0.9662 \times 2^7 + 0.5 = 124.1736 = 124$$

In fractional notation, the coefficients are

$$a_1 = -22/128 = -0.171\,875$$

$$a_2 = 124/128 = 0.968\,75$$

From Equation 13.10, the pole radius and angle for the section for $B = 8$ bits are given by

$$r = \sqrt{0.968\,75} = 0.9843,$$

$$\theta = \cos^{-1}\left(-\frac{b_1}{2r}\right) = \cos^{-1}(0.087\,308) = 84.99°$$

All the quantized coefficients and polar coordinates were computed using an analysis program. If, for any coefficient wordlength, the pole radial distance of a filter section is equal to or greater than unity then there is potential instability. It was found that for all the filter sections as few as $B = 5$ bits are required to maintain stability. In general, if the pole of an unquantized second-order section is at a radius $r < 0.9$, instability is unlikely if a coefficient wordlength of 8 bits or more is used.

(b)   The coefficients of the second-order sections were each quantized to various wordlengths as described above. For each wordlength, the quantized coefficients were then combined to yield an overall quantized transfer function in direct form. Examples for wordlengths of 5 and 16 bits are given in Table 13.4. The passband ripples and stopband attenuation of the quantized filter for the various coefficient wordlengths were obtained. It was found that, to satisfy the frequency response specifications in both passband and stopband, 16 or more bits are required. We note that this is more than the wordlength required for stability.

Table 13.4   Coefficients for Example 13.9, showing wordlength effects.

	$B(k)$			$A(k)$		
k	*Ideal*	*5 bits*	*16 bits*	*Ideal*	*5 bits*	*16 bits*
0	1.000 000	1.000 000	1.000 000	1.000 000	1.000 000	1.000 000
1	−1.338 200	−1.250 000	−1.338 165	−1.448 300	−1.437 500	−1.448 273
2	3.806 737	3.707 031	3.806 700	4.483 108	4.355 469 0	4.483 071
3	−3.556 357	−3.288 574	−3.556 255	−4.220 527	−4.060 791	−4.220 431
4	5.629 177	5.443 726	5.629 105	6.647 162	6.261 536	6.647 087
5	−3.556 357	−3.288 574	−3.556 255	−3.945 450	−3.677 216	−3.945 354
6	3.806 737	3.707 031	3.806 700	3.918 398 1	3.573 486	3.918 352
7	−1.338 200	−1.250 000	−1.338 165	−1.182 602 0	−1.067 047	−1.182 575
8	1.000 000	1.000 000	1.000 000	0.763 340 2	0.672 912	0.763 338

(3)   The frequency responses, scaled to have a maximum of 0 dB, for the unquantized and quantized ($B = 5$ bits) filters are depicted in Figure 13.6. Visually, the response for the 16-bit quantized filter was the same as that of the unquantized filter and is therefore not shown.

13.4.4   Addition overflow errors and their effects

In 2's complement arithmetic, the addition of two large numbers of a similar sign may produce an overflow, that is a result that exceeds the permissible wordlength, which would cause a change in the sign of the output sample. Thus a very large positive number becomes a very large negative number and vice versa (Figure 13.7). Consider the canonic section in Figure 13.8. Because of the recursive nature of the IIR filter, an

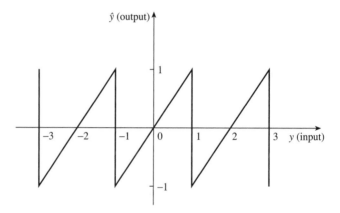

Figure 13.7   Overflow characteristics in 2's complement arithmetic. At instants when the input exceeds the permissible range $(-1, 1)$ overflow occurs.

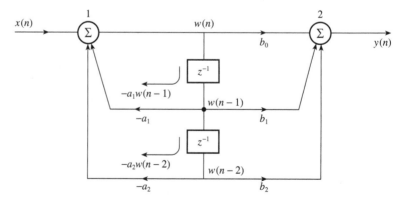

Figure 13.8   An illustration of the effects of addition overflow. Large inputs of the same sign at adder 1 will cause $w(n)$ to become too large. As $w(n)$ is fed back, the effect is self-sustaining.

overflow at $w(n)$ is fed back and used to compute the next output where it can cause further overflow, creating undesirable self-sustaining oscillations. Large-scale overflow limit cycles, as they are called, are difficult to stop once they start and may only be stopped by reinitializing the filter.

Large-scale overflow occurs at the outputs of the adders and may be prevented by scaling the inputs to the adders in such a way that the outputs are kept low, but this is at the expense of reduced signal-to-noise ratio (SNR). Thus it is important to select scale factors to prevent overflow while at the same time maintaining the largest possible SNR.

13.4.5   Principles of scaling

13.4.5.1   Canonic section

Consider the second-order canonic section in Figure 13.9(a). The scaling factor, s_1, at the filter input is chosen to avoid or reduce the possibility of overflow at the output of the left adder. To keep the overall filter gain the same, the numerator coefficients are multiplied by s_1.

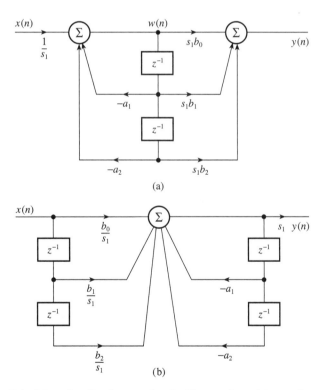

Figure 13.9   Principles of scaling in second-order filter sections: (a) canonic; (b) direct form.

There are three common methods of determining suitable scale factors for a filter. In method 1, often called the L_1 norm, the scale factor is chosen as follows:

$$s_1 = \sum_{k=0}^{\infty} |f(k)| \tag{13.12}$$

where $f(k)$ is the impulse response from the input to the output of the first adder, that is $w(n)$. This scale factor s_1 ensures that the overall gain of the filter from the input to $w(n)$ is unity and so there cannot be an overflow at $w(n)$. The impulse response, $f(k)$, may be obtained by first determining the corresponding transfer function $F(z)$ and then inverse z-transforming.

In the second method, often called the L_2 norm, the scale factor, s_1, is obtained as

$$s_1 = \left[\sum_{k=0}^{\infty} f^2(k) \right]^{1/2} \tag{13.13}$$

Alternatively, the L_2 norm scale factor may be obtained using contour integration via the relationship

$$\sum_{k=0}^{\infty} f^2(k) = \frac{1}{2\pi j} \oint F(z)F(z^{-1}) \frac{dz}{z} \tag{13.14}$$

where $F(z)$ is the z-transform of $f(k)$ and $\oint$ indicates a contour integral around the unit circle $|z| = 1$. Evaluating this, we have (see Appendix 13B)

$$s_1^2 = \sum_{k=0}^{\infty} f^2(k) = \frac{1}{2\pi j} \oint \frac{1}{1 + a_1 z^{-1} + a_2 z^{-2}} \frac{1}{1 + a_1 z + a_2 z^2} \frac{dz}{z}$$
$$= \frac{1}{1 - a_2^2 - a_1^2(1 - a_2)/(1 + a_2)} \tag{13.15}$$

The use of Equation 13.15 avoids the evaluation of the infinite summation of Equation 13.13. In practice, however, $f(k)$ has only a finite number of significant terms and is readily evaluated with a suitable finite wordlength analysis program.

In method 3, known as the L_∞ norm, the scale factor is obtained as

$$s_1 = \max |F(w)| \tag{13.16}$$

where $F(w)$ is the peak amplitude of the frequency response between the input and $w(n)$.

The underlying assumption in method 1 is that the input is bounded, that is $|x(n)| < 1$. The scaling scheme is such that regardless of the type of input there will be no overflow. This is a somewhat drastic scaling scheme, as it caters for situations which are unlikely to happen in normal, real-world situations. The L_2 norm corresponds to placing an energy constraint on both the input and the transfer function. Its main attraction is that finite wordlength effect analysis requires the evaluation of L_2 norms (compare for example Equation 13.8 and Equation 13.14). It is also possible to derive closed form expressions for a variety of filter structures. Method 3 ensures that the filter does not overflow when a sine wave is applied and offers the best compromise. It is the scaling scheme often preferred, especially as it allows the effects of scaling to be verified experimentally using sine waves.

A compact way of expressing the ith scale factor is:

$$s_i = \|F\|_p$$

where the symbol $\|\cdot\|$ indicates the norm, and $p = 1, 2, \infty$ denotes the type of norm. Scale factors obtained by the three methods satisfy the following relationship:

$$L_2 < L_\infty < L_1$$

13.4.5.2   Direct structure

Consider the direct structure in Figure 13.9(b). Since the filter has one accumulator, internal overflows are not a problem, and so input scaling is not strictly necessary. This is one of the attractions of the direct structure. Intermediate overflows may occur in the output of the adder in the course of computing $y(n)$. Provided that the final output does not overflow, they do not matter. The scaling arrangement in Figure 13.9 may be used if scaling is required.

Example 13.10   Determine a suitable scale factor to prevent or reduce the possibility of overflow in an IIR lowpass filter characterized by the following transfer function:

$$H(z) = \frac{1 + 2z^{-1} + z^{-2}}{1 - 1.058\,135\,9z^{-1} + 0.338\,544z^{-2}}$$

Solution   The block diagram representation of the filter, using a second-order canonic section, is shown in Figure 13.10. Using the FWA program (available on the CD in the companion handbook – see the Preface for details) to evaluate Equations 13.12, 13.13 and 13.16, the scale factors for the three methods were computed. These are summarized below:

	L_1	L_2	L_∞
s_1	3.7112	1.7352	3.5663

Just as an illustration, we will also compute the L_2 norm using Equation 13.15:

$$s_1^2 = \frac{1}{1 - (0.3385)^2 - (1.058)^2[(1 - 0.3385)/(1 + 0.3385)]}$$

$$= 1/0.3322 = 3.01$$

$$s_1 = 1.7350$$

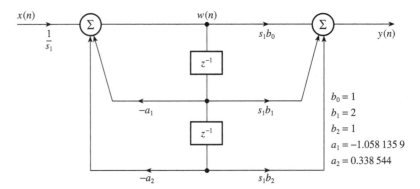

Figure 13.10   Block diagram representation for Example 13.9.

13.4.6   Scaling in cascade realization

In practice, filters are realized as cascades or parallel combinations of second–first-order sections. A scaling scheme for a sixth-order cascade realization is shown in Figure 13.11. As before, the scale factors s_i, $i = 1, 2, 3$, are chosen to avoid or minimize overflows in the filter sections at the nodes labelled $w_i(n)$. The scaling scheme for each second-order section is essentially the same as for the single section considered before. The scale factors are obtained as

$$s_i = \left\| F_i(z) \right\|_p \tag{13.17}$$

where p denotes the type of norm: $p = 1, 2, \infty$. $F_i(z)$ is the transfer function from the input to the node $w_i(n)$ and is given by

$$F_i(z) = \frac{\displaystyle\prod_{k=1}^{i-1} H_k(z)}{1 + a_{1i}z^{-1} + a_{2i}z^{-2}}, \quad i = 1, 2, 3$$

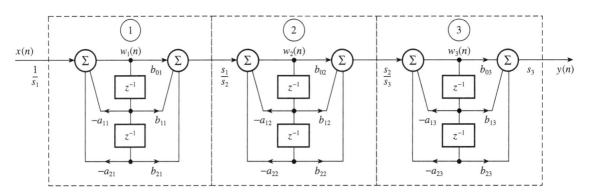

Figure 13.11   Scaling in a cascade realization of a sixth-order IIR filter.

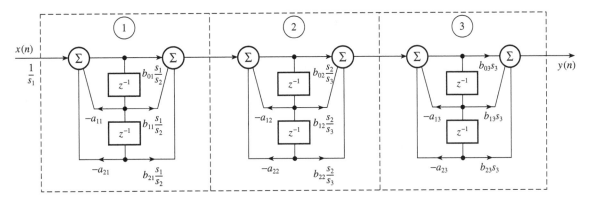

Figure 13.12   Scaling in a cascade realization of a sixth-order IIR filter (scaling factors absorbed into numerator, that is feedforward coefficients).

For the cascade realization, it is common practice to absorb the input scaling factor s_1/s_2 into the numerator of the first stage, s_2/s_3 into that of the second, and so on. Thus the scaling factors in Figure 13.11 can be rearranged as shown in Figure 13.12. It should be noted that the transfer function of the filter after scaling as discussed above is the same as that of the unscaled filter (theoretically at least).

Example 13.11   Compare the scale factors using the three methods above for the filter with the following transfer function, assuming cascade realization with second-order sections:

$$H(z) = H_1(z)H_2(z)H_3(z)$$

where

$$H_1(z) = \frac{1 + 0.2189z^{-1} + z^{-2}}{1 - 0.0127z^{-1} + 0.9443z^{-2}}$$

$$H_2(z) = \frac{1 - 0.5291z^{-1} + z^{-2}}{1 - 0.1731z^{-1} + 0.7252z^{-2}}$$

$$H_3(z) = \frac{1 + 1.5947z^{-1} + z^{-2}}{1 - 0.6152z^{-1} + 0.2581z^{-2}}$$

Solution   Using the FWA program, the scale factors s_1 to s_3 for the three methods were obtained. These are summarized below.

	L_1	L_2	L_∞
s_1	20.9608	3.0388	13.4098
s_2	19.0361	2.5358	10.1366
s_3	14.4467	2.9146	6.4087

As pointed out before, the L_1 norm is always the largest and the L_2 norm the smallest.

13.4.7 Scaling in parallel realization

Figure 13.13 depicts the scaling scheme for a parallel realization of a sixth-order, IIR filter. It is seen that, in this case, the second-order sections are individually scaled as discussed previously. The scaling factor, s_i, at the input of each filter section ensures there is no overflow at the corresponding node $w_i(n)$. To keep the gain of the section the same as before scaling, the feedforward coefficients, b_{ki}, are multiplied by s_i as shown in the diagram. The scaling factors are given by

$$s_i = \|F_i(z)\|_p$$

where $F_i(z)$, the transfer function from the input $x(n)$ to the node $w_i(n)$, is given by

$$F_i(z) = \frac{1}{1 + a_{1i}z^{-1} + a_{2i}z^{-2}}, \quad i = 1, 2, 3$$

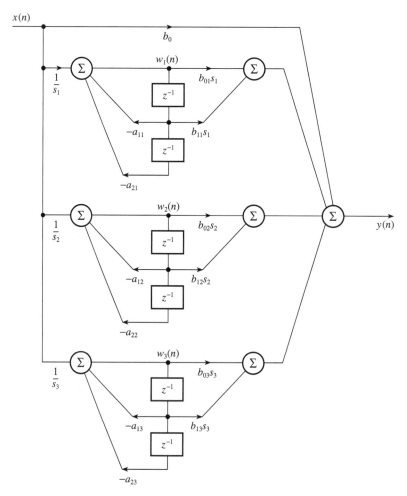

Figure 13.13 Scaling in parallel realization of a sixth-order IIR filter.

Example 13.12   Compare the scale factors for the filter with the transfer function given below using the three methods of scaling:

$$H(z) = \frac{1.2916 - 0.084\,07z^{-1}}{1 - 0.131z^{-1} + 0.3355z^{-2}} + \frac{7.5268}{1 - 0.049z^{-1}} - 8.6788$$

Solution   Using an FWA routine, the scale factors for the three methods were computed and are summarized below.

	L_1	L_2	L_∞
s_1	1.7345	1.0667	1.5126
s_2	1.0515	1.0012	1.0515

13.4.8   Output overflow detection and prevention

When the L_2 and L_∞ norms are used, then output overflow is possible, albeit occasionally. In these cases, it is common practice to employ saturation arithmetic at the filter output. Essentially, when the output overflows, it is set to the maximum permissible positive or negative value depending on the sign of the true data sample. This approach has been shown to be effective in dealing with overflow in the final output. If the output is not saturated, then it will be wrong and may lead to undesirable effects, for example unpleasant sound in digital audio. In the case of the canonic section, that is all there is. In the case of the direct method, if the final output overflows without correction, this is also fed back into the multipliers and will affect subsequent output samples.

Figures 13.14(a) and 13.14(b) depict the overflow detection characteristics of 2's complement and saturation arithmetic, respectively. In the figure, y is the correct

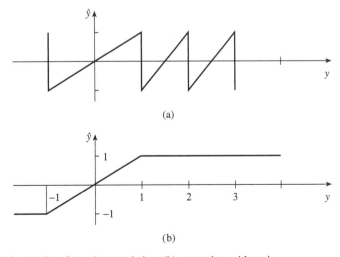

(a)

(b)

Figure 13.14   (a) Overflow characteristics; (b) saturation arithmetic.

output and $\hat{y}$ is the overflow output. A trend in modern DSP processors is to provide extra guard bits in the accumulator to prevent or reduce the possibility of overflow errors. For example, in the DSP56300, a 56-bit accumulator includes 8 guard bits. Thus, it would take 256 overflows in the calculation for an overflow error to occur.

13.4.9 Product roundoff errors in IIR digital filters

Product roundoff error analysis is an extensive topic. Our presentation here will be brief and aims to make you aware of the nature of the errors, their effects and how to reduce them if necessary.

The basic operations in IIR filtering are defined by the familiar second-order difference equation:

$$y(n) = \sum_{k=0}^{2} b_k x(n-k) - \sum_{k=1}^{2} a_k y(n-k) \tag{13.18}$$

where $x(n-k)$ and $y(n-k)$ are the input and output data samples, and b_k and a_k are the filter coefficients. In practice these variables are often represented as fixed point numbers. Typically, each of the products $b_k x(n-k)$ or $a_k y(n-k)$ would require more bits to represent than any of the operands. For example, the product of a B-bit data and a B-bit coefficient is $2B$ bits long. For recursive systems, if this result is not reduced subsequent computations will cause the number of bits to grow without limit.

Truncation or rounding is used to quantize the products back to the permissible wordlength. Quantizing the products leads to errors, popularly known as roundoff errors, in the output data and hence a reduction in the SNR. These errors can also lead to small-scale oscillations in the output of the digital filter, even when there is no input to the filter.

Figure 13.15(a) shows a block diagram representation of the product quantization process, and Figure 13.15(b) a linear model of the effect of product quantization. The model consists of an ideal multiplier, with infinite precision, in series with an adder fed by a noise sample, $e(n)$, representing the error in the quantized product, where we have assumed, for simplicity, that $x(n)$, $y(n)$, and K are each represented by B bits. Thus

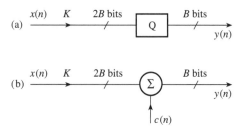

Figure 13.15   Representation of the product quantization error: (a) a block diagram representation of the quantization process; (b) a linear model of the quantization process.

$$y(n) = Kx(n) + e(n) \tag{13.19}$$

The noise power, due to each product quantization, is given by

$$\sigma_r^2 = \frac{q^2}{12}$$

where r symbolizes the roundoff error and q is the quantization step defined by the wordlength to which the product is quantized. The roundoff noise is assumed to be a random variable with zero mean and a constant variance. Although this assumption may not always be valid (for example in the presence of narrowband, low level signals), it is useful in assessing the performance of the filter.

The roundoff noise may be fed into subsequent sections or stages of a DSP system where it is amplified, attenuated or modified in some way. The total output noise due to roundoff errors depends on the realization structure. When the filter is realized using a cascade structure, the noise produced by one section is fed into subsequent sections. Thus the sections should be arranged such that the total noise due to roundoff error is minimized.

13.4.10 Effects of roundoff errors on filter performance

The effects of roundoff noise on filter performance depend on the type of filter structure used and the point at which the results are quantized. Figure 13.16(a) shows the quantization noise model for the direct form building block described earlier. It

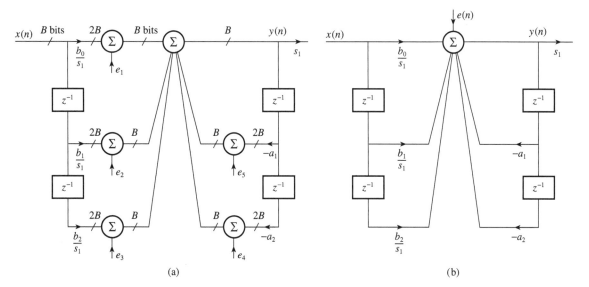

(a)   (b)

Figure 13.16 Product quantization noise model for the direct form filter section. All the noise sources in (a) have been combined in (b) as they feed to the same point.

is assumed in the figure that the input data, $x(n)$, the output, $y(n)$, and the filter coefficients are represented as B-bit numbers (including the sign bit). The products are quantized back to B bits after multiplication by rounding (or truncation).

Since all five noise sources, e_1 to e_5 in Figure 13.16(a), feed to the same point (that is into the middle adder), the total output noise power is the sum of the individual noise powers (Figure 13.16(b)):

$$\sigma_{\text{or}}^2 = \frac{5q^2}{12} \left[\frac{1}{2\pi j} \oint_c F(z)F(z^{-1}) \frac{dz}{z} \right] s_1^2$$

$$= \frac{5q^2}{12} \left[\sum_{k=0}^{\infty} f^2(k) \right] s_1^2 \tag{13.20}$$

$$= \frac{5q^2}{12} \| F(z) \|_2^2 s_1^2 \tag{13.21}$$

where

$$F(z) = \frac{1}{1 + a_1 z^{-1} + a_2 z^{-2}}$$

$$f(k) = Z^{-1}[F(z)]$$

is the inverse z-transform of $F(z)$, which is also the impulse response from each noise source to the filter output, $\| \cdot \|_2^2$ is the L_2 norm squared and $q^2/12$ is the intrinsic product roundoff noise power. The total noise power at the filter output is the sum of the product roundoff noise and the ADC quantization noise (Equations 13.8 and 13.20):

$$\sigma_o^2 = \sigma_{oA}^2 + \sigma_{\text{or}}^2$$

$$= \frac{q^2}{12} \left[\sum_{k=0}^{\infty} h^2(k) + 5s_1^2 \sum_{k=0}^{\infty} f^2(k) \right]$$

$$= \frac{q^2}{12} [\| H(z) \|_2^2 + 5s_1^2 \| F(z) \|_2^2] \tag{13.22}$$

For the canonic section, Figure 13.17(a), the noise model again includes a scale factor as this generates a roundoff noise of its own. The noise sources $e_1(n)$ to $e_3(n)$ all feed to the left adder, whilst the noise sources $e_4(n)$ to $e_6(n)$ feed directly into the filter output. Combination of the noise sources feeding to the same point leads to the noise model of Figure 13.17(b). Assuming uncorrelated noise sources, the total noise contribution is simply the sum of the individual noise contributions:

$$\sigma_{\text{or}}^2 = \frac{3q^2}{12} \sum_{k=0}^{\infty} f^2(k) + \frac{3q^2}{12} = \frac{3q^2}{12} [\| F(z) \|_2^2 + 1] \tag{13.23}$$

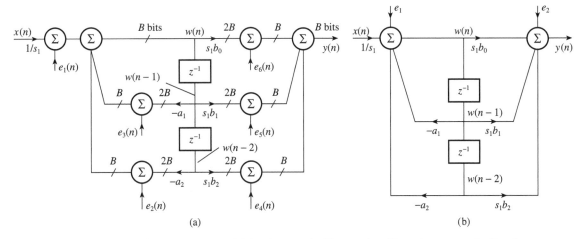

Figure 13.17   Product quantization noise model for the canonic filter section. The noise sources feeding the same point in (a) have been combined in (b).

where $f(k)$ is the impulse response from the noise source e_1 to the filter output, and $F(z)$ the corresponding transfer function given by

$$F(z) = s_1 \frac{b_0 + b_1 z^{-1} + b_2 z^{-2}}{1 + a_1 z^{-1} + a_2 z^{-2}} = s_1 H(z) \tag{13.24}$$

The total noise (ADC + roundoff noises) at the filter output is given by

$$\sigma_o^2 = \sigma_{oA}^2 + \sigma_{or}^2$$

$$= \frac{q^2}{12} \left\{ 3\left[1 + s_1^2 \sum_{k=0}^{\infty} h^2(k)\right] + \sum_{k=0}^{\infty} h^2(k) \right\}$$

$$= \frac{q^2}{12} \{3[1 + s_1^2 \|H(z)\|_2^2] + \|H(z)\|_2^2\} \tag{13.25}$$

Equation 13.25 should be compared with Equation 13.22. Note also that the inclusion of the scale factor increases the output noise.

Example 13.13   A filter is characterized by the following transfer function:

$$H(z) = \frac{0.1436 + 0.2872z^{-1} + 0.1436z^{-2}}{1 - 1.8353z^{-1} + 0.9747z^{-2}}$$

The filter is to be implemented in an 8-bit system with the input data, $x(n)$, the output data, $y(n)$, and filter coefficients represented as 8-bit fractional fixed point, 2's complement numbers.

Assuming that a second-order canonic section is used to realize the filter, and that each product (represented by 16 bits) is quantized to 8 bits immediately after multiplication,

(1) (a) sketch the realization diagram showing the sources of roundoff errors within the filter and determine a suitable scale factor for the system,
   (b) estimate the total steady state output noise power and the degradation in the output SNR, in decibels, caused by roundoff errors, and

(2) repeat part (1) if direct form structure is used to realize the filter.

Solution   (1) (a) The realization diagram of the system with the noise sources is given in Figure 13.17(b). The noise source e_1 represents the sum of the errors resulting from rounding (or truncating) each of the products $a_1 w(n - 1)$, $a_2 w(n - 2)$ and $x(n)/s_1$ from 16 bits to 8 bits. The noise source e_2 is the sum of the three 16-bit inputs to the right adder, quantized to 8 bits.

Using the FWA program the scale factors for the three norms are $s_1 = 133.899\,66$ (L$_1$), $s_1 = 12.1395$ (L$_2$) and $s_1 = 102.088$ (L$_\infty$).

(b) The output noise power due to roundoff error is obtained with the aid of the finite wordlength analysis program, based on Equation 13.23:

$$\sigma_{or}^2 = 1668.03 q^2$$

where we have assumed L$_2$ scaling.

An estimate of the output noise due to the ADC, using Equation 13.8, is $3.7724 q^2$. Thus total output noise power is

$$\sigma_o^2 = (1668.03 + 3.7724)q^2 = 1671.8024 q^2$$

The output signal power, assuming a random signal input, is given by

$$\sigma_x^2 = \tfrac{1}{3}\| H(z)\|_2^2 = 15.0896$$

The SNR (without roundoff error) is

$$\frac{15.0896}{3.7724 q^2} = \frac{4}{q^2}$$

The SNR (with roundoff error) is

$$\frac{15.0896}{1671.8024 q^2}$$

The degradation in SNR due to roundoff error is

$$10\log\left(\frac{4/q^2}{9.0260 \times 10^{-3}/q^2}\right) = 26.47\,\text{dB}$$

(2) For the direct realization, the total output noise power due to roundoff error is $9048.82q^2$. The degradation in SNR due to roundoff error is 33.8 dB, with L_2 scaling. Without scaling, the degradation in SNR is about 1.11 dB. This low degradation in SNR when there is no scaling is why, in some applications, the direct realization is preferred as it offers the possibility of avoiding scaling (Dattorro, 1988).

13.4.11   Roundoff noise in cascade and parallel realizations

13.4.11.1   Cascade

Figure 13.18 shows the cascade realizations for a sixth-order IIR system using the second-order canonic section, where noise sources feeding the same point have been combined as suggested above and renumbered for simplicity. Thus e_1 is the sum of three noise sources, derived from the three multipliers feeding the leftmost adder. The composite noise source e_1 goes through the three filter sections $H_1(z)$, $H_2(z)$ and $H_3(z)$. The composite noise source e_2 goes through the transfer functions $H_2(z)$ and $H_3(z)$, and so on.

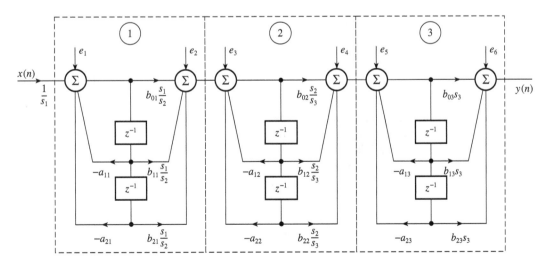

Figure 13.18   Noise model of a cascade realization of a sixth-order IIR filter.

The total output noise due to roundoff error is the sum of all six noise sources:

$$\sigma_{or}^2 = \frac{3q^2}{12}\sum_{k=0}^{\infty} f_1^2(k) + \frac{3q^2}{12}\sum_{k=0}^{\infty} f_2^2(k) + \frac{2q^2}{12}\sum_{k=0}^{\infty} f_3^2(k) + \frac{3q^2}{12}\sum_{k=0}^{\infty} f_4^2(k)$$

$$+ \frac{2q^2}{12}\sum_{k=0}^{\infty} f_5^2(k) + \frac{3q^2}{12}$$

$$= \frac{q^2}{12}\left[3\sum_{k=0}^{\infty} f_1^2(k) + 5\sum_{k=0}^{\infty} f_3^2(k) + 5\sum_{k=0}^{\infty} f_5^2(k) + 3\right]$$

$$= \frac{q^2}{12}[3\|F_1(z)\|_2^2 + 5\|F_3(z)\|_2^2 + 5\|F_5(z)\|_2^2 + 3] \tag{13.26}$$

where $f_i(k)$ is the impulse response between the noise source e_i and the output. The noise components due to e_2 and e_3 (Figure 13.18) each go through the same filter sections, that is through $H_2(z)$ and $H_3(z)$, and so their contributions at the output have been combined. The same is true of the contributions of the noise components e_4 and e_5.

13.4.11.2   Parallel

The roundoff noise model for parallel realization of a sixth-order filter is given in Figure 13.19. As before the noise sources due to individual product quantization have been combined. The noise sources e_1 to e_3 each go through a filter section to reach the output whereas the remaining noise sources feed directly to the output. The contribution of each of the noise sources, e_1 to e_3, to the output noise is

$$\sigma_{r,i}^2 = \frac{3q^2}{12}\sum_{k=0}^{\infty} f_i^2(k) = \frac{3q^2}{12}\|F_i(z)\|_2^2, \quad i = 1, 2, 3$$

$$= \frac{3q^2}{12}s_i^2\sum_{k=0}^{\infty} h_i^2(k) = \frac{3q^2}{12}s_i^2\|H_i(z)\|_2^2 \tag{13.27}$$

where $H_i(z)$ and $h_i(k)$ are, respectively, the transfer function of filter section i and its corresponding impulse response. $F_i(z)$ and $f_i(k)$ are, respectively, the transfer function seen by the noise source i and the corresponding impulse response.

The noise sources e_i, $i = 4, 5, 6$, all feed directly into the output as does e_7. Thus the total output noise power is given by

$$\sigma_{or}^2 = \frac{q^2}{12}\left\{7 + 3\sum_{i=1}^{3}\left[s_i^2\sum_{k=0}^{\infty} h_i^2(k)\right]\right\}$$

$$= \frac{q^2}{12}\left[7 + 3\sum_{i=1}^{3} s_i^2\|H_i(z)\|_2^2\right] \tag{13.28}$$

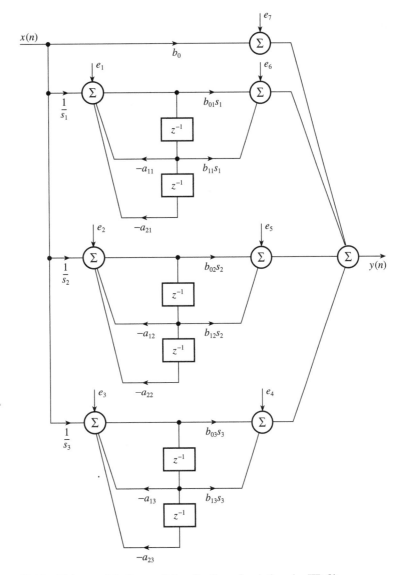

Figure 13.19 Noise model of a parallel realization of a sixth-order IIR filter.

In general, for a parallel realization with L sections, the output power due to roundoff error is given by

$$\sigma_{or}^2 = \frac{q^2}{12}\left[2L + 1 + 3\sum_{i=1}^{L} s_i^2 \|H_i(z)\|_2^2\right] \qquad (13.29)$$

Estimates of all the equations above for roundoff noise power can be readily obtained using a suitable computer program (for example, the finite wordlength analysis program described earlier).

Example 13.14

Given the following transfer function representing a fourth-order IIR filter (Mitra *et al.*, 1974),

$$H(z) = \frac{1 - 2z^{-1} + z^{-2}}{1 + 0.777z^{-1} + 0.3434z^{-2}} \frac{1 - 0.707z^{-1} + z^{-2}}{1 + 0.018\,77z^{-1} + 0.801z^{-2}} 0.093\,226$$

estimate the reduction in SNR, in decibels, if the filter is realized as

(1)   a cascade of two second-order sections paired and ordered as given in $H(z)$, and

(2)   parallel combinations of two second-order sections.

Assume in each case that the products are quantized to 8 bits before they are summed.

Solution

The noise models for the cascade and parallel realization structure are identical to Figures 13.18 and 13.19 respectively, if we ignore the third filter stage in each figure.

(1)   For the cascade realization, the L_1 scale factors (using the PC-based programs) are

$$s_1 = 2.395\,746$$

$$s_2 = 15.703\,627$$

The roundoff noise power, the ADC noise power, and the signal power at the filter output are

$$\sigma_{or}^2 = 345.0391q^2$$

$$\sigma_{oA}^2 = 0.039\,76q^2$$

$$\sigma_y^2 = 0.1591$$

The SNR (without roundoff error) is

$$\frac{0.1591}{0.0397q^2}$$

and the SNR (with roundoff error) is

$$\frac{0.1591}{345.0789q^2}$$

Degradation in SNR due to roundoff error is

$$10 \log (345.0789/0.039\,76) \approx 39.4 \text{ dB}$$

(2) For the parallel realization, using a partial fraction expansion (Mitra *et al.*, 1974), the transfer function becomes

$$H(z) = 0.093\,326\left(1 + \frac{-5.162 + 0.7867z^{-1}}{1 + 0.777z^{-1} + 0.3434z^{-2}} + \frac{1.657\,36 + 0.2759z^{-1}}{1 + 0.018\,77z^{-1} + 0.801z^{-2}}\right)$$

The roundoff noise power at the filter output is given by

$$\sigma_{or}^2 = \frac{q^2}{12}\left[5 + 3\sum_{i=1}^{2} s_i^2 \|H_i(z)\|_2^2\right]$$

Using the FWA program we obtain the L_2 scale factors as $s_1 = 2.395\,746$, $s_2 = 5.450\,612$, $\|H_1(z)\|_2^2 = (7.378\,492)^2$, $\|H_2(z)\|_2^2 = (2.801\,937)^2$. Thus we have

$$\sigma_{or}^2 = \frac{q^2}{12}\{5 + 3[(2.395\,746)^2(7.378\,492)^2 + (5.450\,612)^2(2.801\,937)^2]\}$$

$$= 136.85q^2$$

Thus we have at the filter output, assuming the same output signal and ADC noise, an SNR (with roundoff error) of

$$0.1591/(136.85 + 0.039\,76)q^2 = 1.162\,28 \times 10^{-3}/q^2$$

The degradation in SNR due to roundoff error then becomes

$$10 \log\left[(136.85 + 0.039\,76)/0.039\,76\right] = 35.37 \text{ dB}$$

13.4.12 Effects of product roundoff noise in modern DSP systems

Early work on the effects of product roundoff errors on filter performance was based on a single, fixed internal wordlength, where it was mandatory to quantize each $2B$-bit product (strictly speaking, $2B - 1$ bits) back to B bits before summation. With modern DSP processors such constraints do not exist because they support double-wordlength accumulation. All modern DSP processors feature, at least, a built-in 16×16-bit multiplier and a 32-bit product register, and allow for the products to be accumulated as 32 bit numbers – the so-called 16/32 bit architecture.

Figure 13.20(a) shows the noise model for the direct second-order section when quantization is carried out after the products have been added. In the figure, the $2B$-bit sum of the products, $y'(n)$, is quantized to B bits. To distinguish this from the case where each product is separately quantized, we will refer to this as the post-accumulation quantization. It is clear that, in this case, there is only one noise source following the quantization. The output noise power in this case is given by

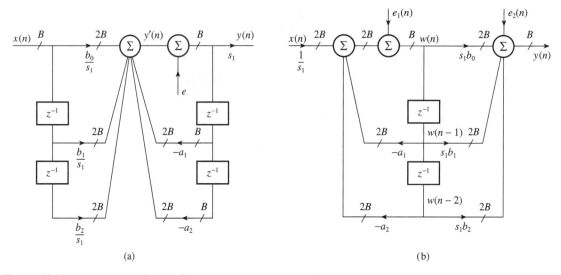

Figure 13.20   Noise models for IIR filter sections in a modern DSP system. The wordlengths at various points within the filter are shown. It is assumed that the input data and filter coefficients are each B bits long.

$$\sigma_{or}^2 = \frac{q^2}{12} \| F(z) \|_2^2 \tag{13.30}$$

where

$$F(z) = \frac{1}{1 + a_1 z^{-1} + a_2 z^{-2}}$$

In the case of the canonic second-order section (Figure 13.20(b)) the output noise due to the noise sources, and the corresponding SNR, is given by

$$\sigma_{or}^2 = \frac{q^2}{12} [s_1^2 \| H(z) \|_2^2 + 1] \tag{13.31}$$

It is evident that rounding after accumulation of the products (Equations 13.30 and 13.31) leads to a significant reduction in the roundoff noise compared with rounding after each product.

13.4.13   Roundoff noise reduction schemes

In practice, rounding or truncation at some point within the filter is necessary to satisfy the wordlength requirements of the multipliers, data memory and those of the interfaces to the outside world. Product roundoff errors due to rounding or truncation of products may lead to a noticeable distortion at the filter output with low level input signals and for high fidelity systems should be minimized. A number of schemes have been devised for reducing or eliminating the effects of roundoff errors in IIR filters. Effectively, these schemes shape the noise spectrum in such a way as to reduce or cancel their effects over certain bands of the filter. The schemes have been collectively called error spectral shaping (ESS).

We will introduce the basic principles of roundoff noise reduction using the direct form I second-order filter section, Figure 13.21(a). In the figure, filter parameters (coefficients and data) are represented as B bit fixed-point numbers and the accumulator is $2B$ bits wide, leading to a $B/2B$ bit implementation. In modern DSP processors, B would be typically 16 bits or 24 bits. When the output of the accumulator is quantized to B bits roundoff noise is generated. It can be shown that the transform of the quantized output is given by

$$Y(z) = \frac{b_0 + b_1 z^{-1} + b_2 z^{-2}}{1 + a_1 z^{-1} + a_2 z^{-2}} X(z) - \frac{1}{1 + a_1 z^{-1} + a_2 z^{-2}} E(z) \qquad (13.32)$$

We see that the spectrum of the quantized output is equal to the ideal output spectrum plus a scaled error spectrum. The error spectrum is amplified by the poles of the filter, $1 + a_1 z^{-1} + a_2 z^{-2}$, regardless of the characteristics of the filter. Depending on the type of filter the noise may be boosted at the low, mid or high end of the frequency range.

A first-order error feedback scheme for reducing the effects of roundoff noise is shown in Figure 13.21(b). It can be shown (see Example 13.16) that in this case the output transform of the filter section with error feedback is given by

$$Y(z) = \frac{b_0 + b_1 z^{-1} + b_2 z^{-2}}{1 + a_1 z^{-1} + a_2 z^{-2}} X(z) - \frac{1 - k z^{-1}}{1 + a_1 z^{-1} + a_2 z^{-2}} E(z) \qquad (13.33)$$

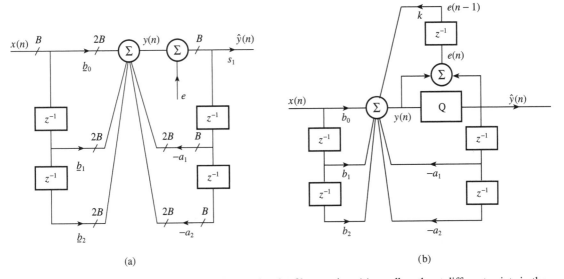

(a) (b)

Figure 13.21 Roundoff noise in direct form I second-order filter section: (a) wordlengths at different points in the filter; (b) a first-order scheme for reducing roundoff noise.

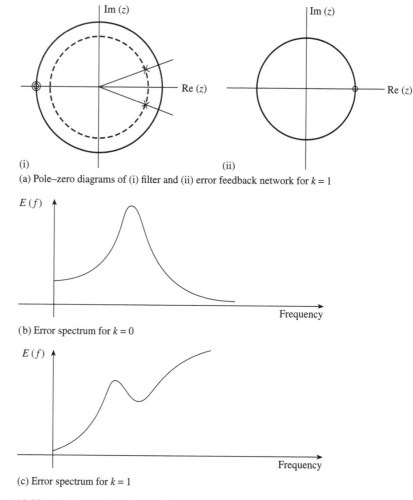

(a) Pole–zero diagrams of (i) filter and (ii) error feedback network for $k = 1$

(b) Error spectrum for $k = 0$

(c) Error spectrum for $k = 1$

Figure 13.22   An illustration of the effects of the error feedback coefficient. Notice that the position of the zero of the error feedback network is close to the position of the poles of the filter.

The feedback coefficient, k, has effectively introduced a zero in the path of the error spectrum to counteract the effects of the poles of the filter. Figure 13.22 illustrates the effects of the zero of the error feedback on the error spectrum. It is evident that without the error feedback, the error spectrum is amplified by the poles of the filter. With an error feedback coefficient a zero is introduced in the path of the error leading to a reduction in the noise spectrum. An appropriate strategy is to place the zero of the feedback network as close as possible to the pole frequency to counteract the effects of the poles.

A second-order noise reduction may be used to achieve greater reduction in the noise spectrum: see Figure 13.23.

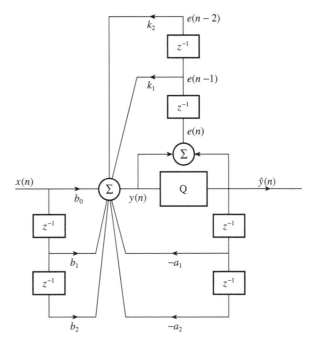

Figure 13.23 A second-order noise reduction scheme.

Generalized noise reduction strategies for the direct and canonic filter sections are depicted in Figures 13.24(a) and 13.24(b), respectively. In both figures the feedback and feedforward coefficients, a'_i and b'_i, are used to modify the transfer function seen by the roundoff errors such that the output roundoff noise is minimized without affecting the desired signal. In Figure 13.24(a), quantization of the sum of products at the output of the left adder generates an error, $e_1(n)$, which is the lower half of the double precision variable, $y(n)$. The products $a'_1 e_1(n-1)$ and $a'_2 e_2(n-2)$ do not have the same weight as the other inputs to the adder although they are all $2B+1$ bits long and so must be re-aligned or quantized before summing with the other inputs. The quantization error in this case is denoted by $e_2(n)$. Similarly, the terms $b'_0 e(n)/s_1$, $b'_1 e(n-1)/s_1$, and $b'_2 e(n-2)/s_1$ need to be quantized before being summed with the other inputs to the right adder, giving rise to the error $e_3(n)$. Finally, the output of the right adder will be quantized from $2B+1$ bits to $B+1$ bits giving rise to the error $e_4(n)$. Similar considerations apply to the canonic section in Figure 13.24(b).

For the direct form realization with ESS (Figure 13.24(a)) the output noise is given by

$$\sigma_{\text{or}}^2 = \frac{q^2}{12} \left[\sum_{k=0}^{\infty} f_1^2(k) + \sum_{k=0}^{\infty} f_2^2(k) + 2s_1^2 \right] \tag{13.34}$$

where $f_1(k)$ and $f_2(k)$ are, respectively, the impulse responses from the noise sources 1 and 2 to the output of the filter.

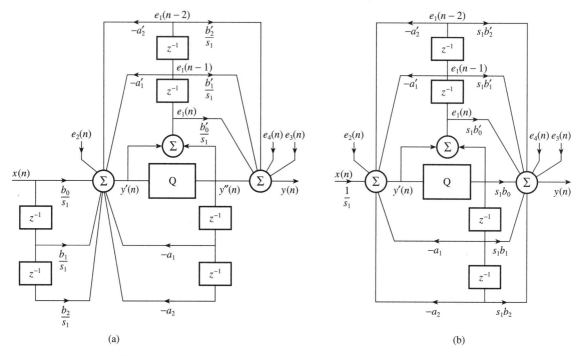

Figure 13.24 Generalized noise reduction schemes: (a) direct form filter; (b) canonic filter.

For the canonic section (Figure 13.24(b)) the output noise power due to roundoff is given by

$$\sigma_{\text{or}}^2 = \frac{q^2}{12}\left[\sum_{k=0}^{\infty} f_1^2(k) + \sum_{k=0}^{\infty} f_2^2(k) + 2\right]$$

$$= \frac{q^2}{12}[\|F_1(z)\|_2^2 + \|F_2(z)\|_2^2 + 2] \tag{13.35}$$

where

$$F_1(z) = (b_0' + b_1'z^{-1} + b_2'z^{-2})s_1 + \frac{(1 + a_1'z^{-1} + a_2'z^{-2})(b_0 + b_1z^{-1} + b_2z^{-2})s_1}{1 + a_1z^{-1} + a_2z^{-2}}$$

$$F_2(z) = \frac{(b_0 + b_1z^{-1} + b_2z^{-2})s_1}{1 + a_1z^{-1} + a_2z^{-2}}$$

The choice of the error spectral shaping (ESS) coefficients, a_i' and b_i', determines the effectiveness of the scheme in reducing noise. In practice, the ESS coefficients are often constrained to be integers to avoid further quantization processes. In practice, both first- and second-order ESS are normally used.

In first-order noise reduction schemes, the error feedforward coefficients are set to zero, and one of the error feedback coefficients is set to an integer. First-order noise reduction schemes are particularly effective in reducing roundoff noise in narrow-band lowpass and highpass filters as they allow a single zero to be placed in the path of the roundoff error at either the low frequency or the high frequency end. The scheme is attractive as it involves only a modest increase in the computational complexity of the filter.

Optimum ESS can be achieved using second-order schemes in which the effect of roundoff noise at the filter output is completely nullified. For the direct form section (Figure 13.24(a)) the optimal ESS is obtained by setting

$$b_i' = 0, \, i = 0, \, 1, \, 2; \quad a_i' = a_i, \, i = 1, \, 2 \tag{13.36}$$

In this case, the output roundoff noise reduces to largely the intrinsic roundoff noise. The output roundoff noise power reduces to

$$\sigma_{or}^2 = \frac{q^2}{12} \left[1 + \sum_{k=0}^{\infty} f^2(k) \right]$$

$$= \frac{q^2}{12} [1 + \| F(z) \|_2^2] \tag{13.37}$$

where

$$F(z) = \frac{s_1}{1 + a_1 z^{-1} + a_2 z^{-2}}$$

For the canonic section an optimal solution is obtained by setting

$$b_i' = -b_i, \, i = 0, \, 1, \, 2; \quad a_i' = a_i, \, i = 1, \, 2 \tag{13.38}$$

The output noise power is given by

$$\sigma_{or}^2 = \frac{q^4}{12} \sum_{k=0}^{\infty} f^2(k) = \frac{q^4}{12} \| F(z) \|_2^2$$

where

$$F(z) = \frac{(b_0 + b_1 z^{-1} + b_2 z^{-2}) s_1}{1 + a_1 z^{-1} + a_2 z^{-2}}$$

The optimal solution is computationally more expensive and, as pointed out by Mullis and Roberts (1982), effectively involves a double precision representation of the internal filter variables. A number of other suboptimal solutions, besides the integer ones considered previously, are available (for example in Higgins and Munson, 1982).

Example 13.15

Compare the roundoff noise performance of a second-order IIR filter which is characterized by the following transfer function:

$$H(z) = \frac{0.1436(1 + 2z^{-1} + z^{-2})}{1 - 1.8353z^{-1} + 0.9748z^{-2}}$$

if the filter is realized using (a) a canonic section or (b) a direct filter section for the following cases:

(1) $a_i' = 0$, $i = 1, 2$ (no error feedback)
(2) $a_1' = -1$, $a_2' = 0$
(3) $a_1' = -2$, $a_2' = 0$
(4) $a_1' = -1$, $a_2' = 1$
(5) $a_1' = -2$, $a_2' = 1$

Assume that all the feedforward error coefficients are zero in each case.

Solution

The realization structures, with ESS, are shown in Figure 13.25. The roundoff noise output power for the canonic and direct realization structures are, respectively,

$$\sigma_{or}^2 = \frac{q^2}{12}[\|F_1(z)\|_2^2 + 1]$$

and

$$\sigma_{or}^2 = \frac{q^2}{12}[\|F_2(z)\|_2^2 + 1]$$

where

$$F_1(z) = (1 + a_1'z^{-1} + a_2'z^{-2})\frac{(b_0 + b_1z^{-1} + b_2z^{-2})s_1}{1 + a_1z^{-1} + a_2z^{-2}}, \quad s_1 = 12.1395 \text{ (}L_2 \text{ scaling)}$$

$$F_2(z) = \frac{(1 + a_1'z^{-1} + a_2'z^{-2})s_1}{1 + a_1z^{-1} + a_2z^{-2}}, \quad s_1 = 6.7282 \text{ (}L_2 \text{ scaling)}$$

Using the FWA program, the output noise power for each case was obtained and is summarized in Table 13.5. Note that for case 3 ($a_1' = -2$, $a_2' = 0$) the noise output has actually gone up instead of improving, emphasizing the importance of the choice of the feedback coefficients. The first order scheme, case 2, is quite effective in reducing the output noise.

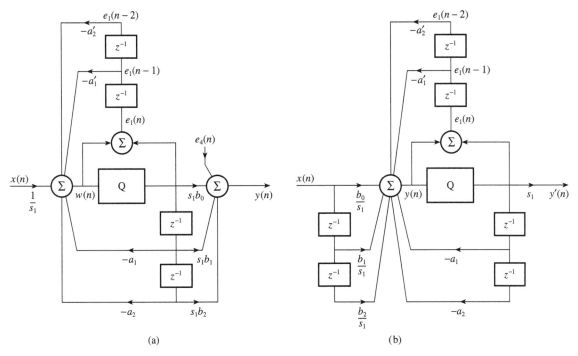

Figure 13.25 Two different realization structures of a second-order IIR filter.

Table 13.5 Output of the FWA program for Example 13.15.

	Noise power	
Case	Canonic	Direct
1	$556.0108q^2$	$556.0108q^2$
2	$77.1261q^2$	$78.6247q^2$
3	$710.0842q^2$	$713.0933q^2$
4	$414.1014q^2$	$413.845q^2$
5	$12.2659q^2$	$14.9999q^2$

13.4.14 Determining practical values for error feedback coefficients

It is evident from the discussions so far that the error spectrum, $E(z)$, is affected by the poles of the filter. Essentially, the error spectrum is amplified by the poles of the filter. If we assume that the error has a flat spectrum, then at the filter output the noise will be amplified near the pole frequency. The error feedback coefficients counteract the amplification of the error spectrum by introducing one or more zeros in the path of the noise. For a first-order error feedback network, a single zero is introduced in the

Table 13.6   Integer error feedback coefficients and the corresponding zero locations.

Case no.	k_1 values	k_2 values	Location of zeros
1	0	1	A pair of zeros at 0 and 180° (i.e. at dc and $F_s/2$)
2	0	−1	A pair of complex conjugate zeros at ±90° (i.e. at ±$F_s/4$)
3	1	0	A single zero at 0° (i.e. at dc)
4	1	−1	A pair of complex conjugate zeros at ±60° (i.e. at ±$F_s/6$)
5	2	−1	A double zero at 0° (i.e. at dc)
6	−2	−1	A double zero at 180° (i.e. at $F_s/2$)
7	−1	−1	A pair of complex conjugate zeros at ±120° (i.e. at ±$F_s/3$)
8	−1	0	A single zero at 180° (i.e. at $F_s/2$)

numerator of the error transfer function. For a second-order feedback network, the feedback coefficients introduce two zeros in the noise transfer function. In both cases, the zeros do not affect the input of the filter.

A simple, but effective strategy is to place the feedback zeros as close to the pole frequencies as possible to counteract the effects of the poles: see Figure 13.21. In practice, factors that affect the choice of the error feedback coefficients include the desire to avoid further roundoff errors, the use of double precision, the desire to simplify multipliers, and the need to position the zeros of the error feedback network as close as possible to the poles of the filter to counteract their effects on the roundoff noise. For these reasons, the values of the error feedback coefficients are often constrained to be simple integers with values for k_1 and k_2 of 0, ±1, ±2.

For the feedback coefficients to be simple integers, the zeros of the feedback network should be located at 0, ±60°, ±90°, ±120°, ±180°, depending on the values of k_1 and k_2. The possible values of the error feedback coefficients and the locations of the corresponding zeros are summarized in Table 13.6.

The choice of the error feedback coefficients depends on the type of the filter. For example, a lowpass filter will have poles at or near dc. Thus, we see from the table that the choice of values of k_1 and k_2 is limited to entries 1, 3, 4 or 5 in the table as these are the only cases that would produce zeros closest to the poles of the filter. On the other hand, a highpass filter will have poles near the high frequency end of the available spectrum (i.e. near $F_s/2$). A possible choice for k_1 and k_2 is 1, 6, 7 or 8.

Example 13.16

(a)   Discuss, with the aid of diagrams where appropriate, the problems of roundoff noise in fixed-point digital IIR filters. Your answer should cover the following points:

(i)   how roundoff noise arises in IIR filters;

(ii)   the effects of roundoff noise on the performance of IIR filters.

(b)   The structure of a second-order filter section, with an error feedback scheme, is shown in Figure 13.26. Assume the filter section is to be implemented using 2's complement, fixed-point arithmetic, with quantization taking place after addition of products.

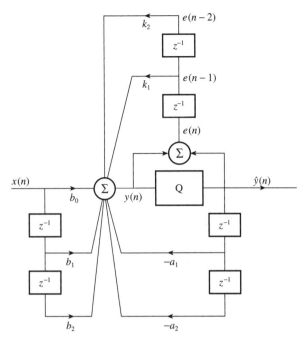

Figure 13.26   A second-order noise reduction scheme – choice of error feedback coefficients.

(i)   Derive an expression for the transform of the quantized output, $\hat{Y}(z)$, in terms of the input transform, $X(z)$, and the quantization error, $E(z)$, and hence show that the error feedback network has no adverse effect on the input signal.

(ii)   Deduce the expression for the error feedback function.

(iii)   What are the key factors that influence the choice of values of the error feedback coefficients in practice?

(c)   Deduce, from analysis of pole and zero positions as appropriate, suitable pairs of integer values for the error feedback coefficients to minimize the level of roundoff noise at the output of each of the following filters:

(i) $H(z) = \dfrac{1 + 2z^{-1} + z^{-2}}{1 - 1.75z^{-1} + 0.81z^{-2}}$

(ii) $H(z) = \dfrac{1 - 2z^{-1} + z^{-2}}{1 + 1.75z^{-1} + 0.81z^{-2}}$

(iii) $H(z) = \dfrac{1 - z^{-2}}{1 + 0.81z^{-2}}$

Assume that each filter is implemented using the structure depicted in Figure 13.26 and that the roots of the second-order polynomial

$$1 + d_1 z^{-1} + d_2 z^{-2}$$

are $r \angle \theta$ and $r \angle -\theta$, where

$$r = \sqrt{d_2} \quad \text{and} \quad \theta = \cos^{-1}\left(\frac{-d_1}{2r}\right)$$

Solution　(a)　Roundoff noise arises from product quantization – rounding or truncation – and/or quantization of sum of products which is necessary in recursive realizations to keep variables within permissible system wordlength. For example, multiplication of two numbers each represented by B bits gives a result with $2B$ bits. If the result is not quantized to B bits (say), the wordlength of subsequent results will grow without limit. The quantization errors are amplified by the poles of the filter and appear at the output as noise. This noise raises the overall system noise floor. This distorts low level input signals and in applications requiring high fidelity reproduction this may not be acceptable. Roundoff errors can also lead to small-scale oscillations in the output of the filter, even when there is no input. Diagrams of filter topologies (e.g. a second-order canonic section) and models of roundoff noise may be used to illustrate the answer.

(b)　(i)　From Figure 13.26, the roundoff error, $e(n)$, and the quantized and unquantized outputs of the filter, $y'(n)$ and $y(n)$, respectively are relayed as

$$e(n) = y(n) - \hat{y}(n) \tag{13.39a}$$

$$y(n) = \sum_{i=0}^{2} b_i x(n-i) - \sum_{i=1}^{2} a_i \hat{y}(n-i) + \sum_{i=1}^{2} k_i e(n-i) \tag{13.39b}$$

Using Equations (13.39a) and (13.39b), taking z-transforms and simplifying leads to the desired equation:

$$\hat{Y}(z) = \left(\frac{\sum_{i=0}^{2} b_i z^{-i}}{1 + \sum_{i=1}^{2} a_i z^{-i}} \right) X(z) - \left(\frac{1 - \sum_{i=1}^{2} k_i z^{-i}}{1 + \sum_{i=1}^{2} a_i z^{-i}} \right) E(z) \tag{13.39c}$$

(ii)　The noise transfer function can be obtained by setting the input to zero and from Equation 13.39c above is given as

$$H_e(z) = \frac{1 - \sum_{i=1}^{2} k_i z^{-i}}{1 + \sum_{i=1}^{2} a_i z^{-i}} = \frac{1 - k_1 z^{-1} - k_2 z^{-2}}{1 + a_1 z^{-1} + a_2 z^{-2}}$$

(iii)   The main factors are the need to avoid further roundoff errors or the use of double precision, the need to avoid the use of another multiplier, the need for a minimum phase system, and the need to position the zeros of the error feedback network as close as possible to the poles of the filter to counteract their effects on the roundoff noise. For these reasons, the values of the error feedback coefficients are constrained to be simple integers with values for k_1 and k_2 of 0, ±1, ±2.

(c)   (i)   Using the equation for the roots of the polynomial, we find that the poles of the filter are located in the z-plane at a radius of $r = \sqrt{0.81} = 0.9$ and angles $\theta = \pm\cos^{-1}(1.75/2 \times 0.9) = \pm13.5°$; a double zero exists at a radius $r = 1$ and an angle $\theta = 180°$ (i.e. at $F_s/2$). Thus, the filter is a lowpass filter. The error feedback coefficients should have a double zero at $r = 1$ and $\theta = 0$ to counteract the effects of the poles on roundoff noise, leading to the following values: $k_1 = 2$ and $k_2 = -1$. We could also use first-order error feedback coefficients with values $k_1 = 1$ and $k_2 = 0$ (to give a zero at dc).

     (ii)   For the second filter, the poles are located at a radius and angles $r = 0.9$ and $\theta = \pm\cos^{-1}(-1.75/2 \times 0.9) = \pm166.4°$, with a double zero at dc, and so the filter is clearly a highpass filter. The nearest feedback network zeros to the poles of the filter correspond to feedback coefficient values of $k_1 = -2$ and $k_2 = -1$ (and these give a double zero located at $r = 1$ and an angle of $\theta = 180°$).

    (iii)   For the third filter, there are two complex conjugate poles located at a radius $r = 0.9$ and angles $\theta = \pm90°$ in the z-plane and a pair of zeros at $0°$ and $180°$. To nullify the effects of the poles as much as possible, the best choice of integer feedback coefficients is $k_1 = 0$ and $k_2 = -1$. As can be seen from Table 13.6, this choice would produce a pair of complex conjugate zeros at a radius of 1 and angles ±90°.

13.4.15   Limit cycles due to product roundoff errors

In addition to the degradation in SNR, the error due to roundoff can cause an oscillation at the filter output or the output to remain stuck at a fixed nonzero value, even when there is no input. This effect is known as low level limit cycle. We will illustrate with an example.

Example 13.17   A first-order IIR filter is characterized by the difference equation

$$y(n) = x(n) + \alpha y(n-1) \quad n > 0$$

Given the initial condition $y(0) = 6$ and a zero input, that is $x(n) = 0$, $n = 0, 1, \ldots,$

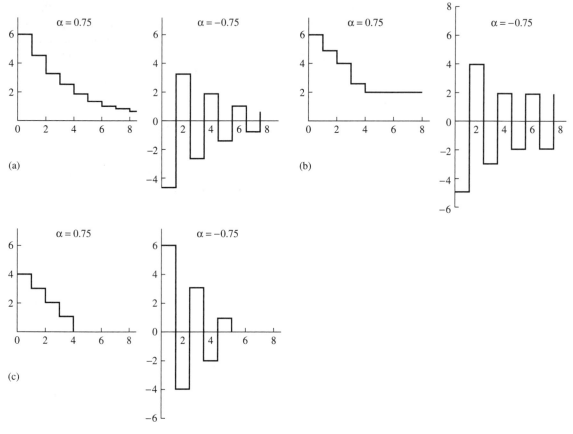

Figure 13.27 An illustration of low level limit cycle due to product quantization in a first-order IIR filter: (a) infinite precision; (b) quantization by rounding; (c) quantization by truncation.

(1) obtain and plot, assuming infinite precision, the first 10 output values for (i) $\alpha = -0.75$ and (ii) $\alpha = 0.75$;

(2) repeat part (1), but assume that the data and register lengths are each four bits long (that is 3 data bits and a sign bit) and that the products are rounded; and

(3) repeat parts (1) and (2), but assume truncation of products immediately after multiplication.

Solution The values of the output samples for the three cases above are depicted in Figure 13.27 and listed in Table 13.7.

It is seen that if the input $x(n)$ is zero indefinitely the output $y(n)$, using infinite precision, decays exponentially to zero, regardless of the sign of α. However, if finite precision arithmetic is used, with the output rounded to the nearest integer, then for positive α the output remains fixed at a given level. The range of output levels over which the output is confined is known as the deadband. In this example, the deadband

Table 13.7 Results for Example 13.17.

	$y(n)$, ($\alpha = 0.75$)			$y(n)$, ($\alpha = -0.75$)		
n	Infinite	Rounding	Truncation	Infinite	Rounding	Truncation
0	6	6	6	6	6	6
1	4.5	5	4	−4.5	−5	−4
2	3.38	4	3	3.38	4	3
3	2.53	3	2	−2.53	−3	−2
4	1.90	2	1	1.90	2	1
5	1.42	2	0	−1.42	−2	0
6	1.07	2	0	1.07	2	0
7	0.80	2	0	−0.80	−2	0
8	0.60	2	0	0.60	2	0
9	0.45	2	0	−0.45	−2	0

is the interval $[-2, 2]$. For the first-order filter the deadband interval is given by (Jackson, 1986)

$$k = \text{int}\left[\frac{0.5}{1 - \|\alpha\|}\right]$$

where int$[\cdot]$ is the integer part of the quantity in the square bracket. If α is negative, the output oscillates at a frequency of $F_s/2$ between two fixed levels of alternating sign. This results from the fact that when the input is removed the filter output decays to less than the quantization level, but then it is rounded up to the next level and the process repeats, creating a low level oscillation. These low level oscillations are undesirable in some applications. For example, they produce unpleasant noise in a telephone system during idle channel conditions when the speakers are silent. A way to reduce limit cycles is to increase the processor wordlength or to add a dither signal to the output before it is rounded. ESS, already discussed, has also been shown to reduce the amplitude of limit cycles and in some cases to eliminate them completely.

In general roundoff limit cycles will not exist in a second-order filter if the coefficients lie inside the hatched area of the stability triangle.

13.4.16 Other nonlinear phenomena

As well as overflow and product limit cycles, other nonlinear effects that may influence the behaviour of an IIR digital filter include the following:

(1) *Jump phenomenon*   When the filter is fed by a sine wave, two possible output levels may exist for the same input signal. A small change in the amplitude or

frequency of the input signal causes a jump from one output level to another. Several regions inside the stability triangle where such a phenomenon may exist have been identified. In these regions, the filter coefficients satisfy the condition $|a_1|a_2 < -1$. ESS has been shown to reduce the consequences of such nonlinear effects.

(2) *Subharmonic response*   For a sine wave input the output may contain subharmonics of the input (Claasen, 1974). Thus, for the same input signal but different initial conditions we can have outputs which are quite different. These effects are more serious for filters with poles close to the unit circle.

13.5 Finite wordlength effects in FFT algorithms

As in most DSP algorithms, the main errors arising from implementing FFT algorithms using fixed-point arithmetic are

- roundoff errors, which are produced when the product W^kB is truncated or rounded to the system wordlength;
- overflow errors, which result when the output of a butterfly exceeds the permissible wordlength;
- coefficient quantization errors, which result from representing the twiddle factors using a limited number of bits.

We will consider the effects of these errors on the FFT output for radix-2 FFT.

13.5.1 Roundoff errors in FFT

The basic operation in any FFT algorithm is the butterfly computation, which for radix-2 DIT FFT is characterized by

$$A' = A + W^kB$$

$$B' = A - W^kB$$

where A and B are the inputs to the butterfly, A' and B' the outputs. In the general case, W^k, the twiddle factor, as well as the inputs and outputs, are all complex valued. In fixed-point implementation, the butterfly computation is carried out using real arithmetic, thus we need to express A' and B' in rectangular form (see Chapter 12):

$$A' = A_r + B_r \cos(X) + B_i \sin(X) + j[A_i + B_i \cos(X) - B_r \sin(X)]$$

$$= A_r + B_rW_r + B_iW_i + j[A_i + B_iW_r - B_rW_i] \tag{13.40a}$$

$$B' = A_r - [B_r \cos(X) + B_i \sin(X)] + j[A_i - \{B_i \cos(X) - B_r \sin(X)\}]$$

$$= A_r - (B_rW_r + B_iW_i)_i + j[A_i - (B_iW_r - B_rW_i)] \tag{13.40b}$$

where the subscript r denotes the real part and the subscript i the imaginary part of the variable, and $X = 2\pi k/N$. Thus the butterfly computation requires four multiplications and six real additions (we regard subtractions as being the same as additions). In a fixed-point implementation, each of the products above will require approximately twice as many bits to represent as the operands themselves. For example, if the variables B_r, B_i, W_r and W_i are each represented as 16-bit numbers, then after multiplication each product will require 32 bits to represent. Truncating or rounding each product back to 16 bits produces an error, the familiar roundoff error.

Thus, we can associate four roundoff noise sources to each butterfly, one for each product, and so the roundoff noise power (i.e. variance) at the output of each butterfly is given by

$$\sigma_B^2 = 4 \times \frac{q^2}{12} \tag{13.41}$$

where $q = 2^{-(B-1)}$ and the system wordlength is B bits.

The noise generated by a butterfly at one stage is fed into subsequent stages. If we examine the flowgraph of an FFT, e.g. Figure 13.28, we find that each FFT output, $X(k)$, can be traced back to $(N-1)$ butterflies. Figure 13.29 shows the butterflies that contribute to the output $X(2)$, for $N = 7$. In general, each FFT output is connected to all $N/2$ butterflies in stage 1, $N/4$ butterflies in stage 2, $N/8$ in stage 3 and so on. Assuming that each butterfly generates identical but uncorrelated errors, the maximum noise power at each FFT output, $X(k)$, is simply (Oppenheim and Weinstein, 1972)

$$\sigma_0^2 = (N-1)\sigma_B^2 \approx N\sigma_B^2 = \frac{N}{3}2^{-2(B-1)} \text{ (when } N \text{ is large)}$$

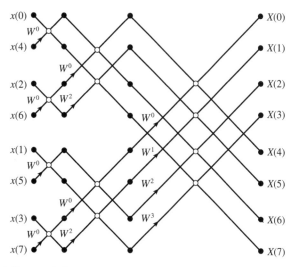

Figure 13.28 A flowgraph for an 8-point, radix-2, decimation-in-time FFT algorithm.

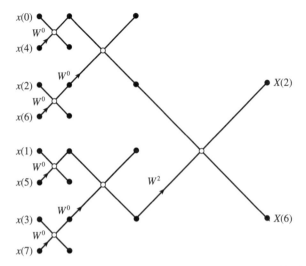

Figure 13.29   A flowgraph showing butterflies that contribute to roundoff noise at the outputs $X(2)$ and $X(6)$.

Thus, the noise power is directly proportional to the transform size. Doubling N, which is equivalent to adding a stage to the FFT, doubles the noise power. To retain the same noise power, we increase the wordlength by 1 bit since the noise power is proportional to N and to $2^{-2(B-1)}$.

The signal-to-noise ratio in this case is approximately equal to

$$SNR = \frac{1/3}{N2^{-2(B-1)}/3} = \frac{2^{2(B-1)}}{N}$$

If we consider only the noise contributions of butterflies that have non-trivial twiddle factors (butterflies with twiddle factors, $W^k = \pm1$ or $\pm j$, will have products that are exact and no errors are produced), a lower noise power will be produced as a result of roundoff error. In fact, if we make use of this information, we will find that some of the FFT outputs generate no errors at all. Clearly, the expressions above are upper bounds.

13.5.2   Overflow errors and scaling in FFT

Scaling is necessary in FFT computation to avoid overflow errors (after the additions in Equations 13.40a and b) because the data size tends to grow after each butterfly computation. There are a number of ways of scaling the data to avoid overflow during FFT computations. A popular scaling scheme is based on the observation that the output of each butterfly satisfies the relationship (Oppenheim and Weinstein, 1972)

$$\max[\,|A'|, |B'|\,] \leqslant 2 \max[\,|A|, |B|\,] \tag{13.42}$$

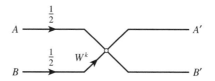

Figure 13.30   A scaling scheme to reduce overflow at each butterfly.

This implies that the maximum modulus of the butterfly output increases from stage to stage by a factor of 2. Thus, if the inputs of the butterfly are each scaled by 0.5 the outputs should not overflow provided the magnitude of the input data is within the permissible wordlength: see Figure 13.30.

In fact, in some cases scaling by 0.5 is not sufficient to avoid overflow even if the input is less than unity. To illustrate, consider the explicit expressions given in Equations 13.40:

$$A' = A_r + B_r \cos(X) + B_i \sin(X) + j[A_i + B_i \cos(X) - B_r \sin(X)]$$

$$B' = A_r - [B_r \cos(X) + B_i \sin(X)] + j[A_i - \{B_i \cos(X) - B_r \sin(X)\}]$$

If $X = 2\pi k/N = 45°$, then $\cos(45°) = \sin(45°) = \sqrt{2}/2$. Without scaling and with real and imaginary parts of inputs each set to 1 (the limiting case), we have, from the equations above,

$$A' = 2.4142 + j; \quad B' = -0.4142 + j$$

With each input scaled by 0.5, we have

$$A' = 1.2071 + 0.5j; \quad B' = -0.2071 + 0.5j$$

Clearly, the real part of A' will still cause an overflow even with scaling by 0.5 because its magnitude is greater than unity.

Despite the possibility of overflow most implementations still employ a scale factor of 0.5 because it is easy to implement – a simple shift of one place to the right (or if advantage is taken of the double sign bits normally produced after a fixed-point multiplication, we do nothing). To avoid overflow in all cases, the input should be scaled by 1.2071 (2.414 21/2) first, followed by scaling at each stage by 0.5. After the FFT the output is scaled back to the correct value. For most real data, the additional input scaling may not be necessary as the maximum cannot be attained.

Scaling the inputs to the butterflies alters the roundoff noise characteristics of the FFT. The output signal-to-noise ratio is approximately given by

$$SNR = \frac{1}{2N} 2^{2(B-1)} \tag{13.43}$$

Example 13.18   A hardware FFT processor uses fixed-point arithmetic in its butterfly computations. Estimate the maximum wordlength required to perform a 1024 point FFT with an output SNR of 40 dB. Assume that the input to each butterfly is scaled by 0.5 throughout the FFT.

Solution

$$40 = 10 \log \left(\frac{1}{2N} 2^{2(B-1)} \right); \quad 10^{\frac{40}{10}} = \frac{1}{2N} 2^{2(B-1)}$$

$$B - 1 = \tfrac{1}{2} \log_2 (2N \times 10^4)/\log(2) = 12.14 = 13 \text{ bits (approximately)}$$

System wordlength, $B = 14$ bits.

13.5.3   Coefficient quantization in FFT

In many hardware FFT implementations, the real and imaginary parts of the twiddle factor W^k are normally pre-computed, quantized to B bits and stored in a lookup table, where B bits is the system wordlength. This gives rise to the familiar quantization errors.

13.6   Summary

The performance of a DSP system is limited by the number of bits used in its implementation. The four common sources of errors are those due to (1) input quantization, (2) coefficient quantization, (3) product roundoff and (4) addition overflow. Techniques for analyzing their effects on DSP system performance and, where possible, for eliminating or minimizing them have been presented. An IIR filter was used as the main vehicle for the presentation. Coefficient wordlength must be adequate to minimize the effects of coefficient quantization on the frequency response and to prevent the possibility of instability. The stability of an IIR filter is always of concern. An IIR filter that is otherwise stable when implemented with infinite precision may become unstable if implemented with finite precision. In high fidelity audio work, for example, 24-bit coefficients are said to be necessary for processing low frequency audio signals. In most other cases, representing the coefficients with 16 or more bits and carrying out the arithmetic operations with double-length accumulators are sufficient to minimize the effects of finite wordlength.

Truncation or roundoff errors due to finite precision arithmetic operations create a nonlinear effect in the filter, such as limit cycles whereby the filter output oscillates even in the absence of input or with constant input. The effects of roundoff error on filter performance can be quantified in terms of SNRs at the filter output. Reduction in the SNR due to roundoff error can be offset by the use of the error spectral shaping (ESS) scheme. The primary effect of such schemes is to nullify the 'amplifying' effect of the poles of the filter on the roundoff errors. The price paid for this is an increase in the number of multiplications and additions, although first-order ESS with integer coefficients is computationally efficient.

Design programs are provided on the CD in the companion
designers to calculate the filter coefficients and to analyze some
wordlength on filter performance (see the Preface for details).

Problems

13.1 Figure 13.31 shows a standard second-order filter section.

   (1) Explain why overflow is permissible at nodes 1 and 3 but not at node 2.

   (2) Find suitable scale factors to reduce the possibility of overflow at node 2.

   (3) Assuming that the filter is implemented with pre-accumulator product quantization in an 8-bit system, determine the number of additional bits that will be required to achieve a reduction of roundoff noise by at least 20 dB.

13.2 An IIR filter has the following transfer function:

$$H(z) = \frac{0.1436 + 0.2872z^{-1} + 0.1436z^{-2}}{1 - 1.8353z^{-1} + 0.9748z^{-2}}$$

   (1) Determine the positions of the poles and zeros and sketch the pole–zero diagram.

   (2) Determine the radial distance of the pole from the origin.

   (3) Estimate the number of bits required to represent each coefficient

      (a) to maintain stability, and

      (b) so that the amplitude response in the passband does not change by more than 1%.

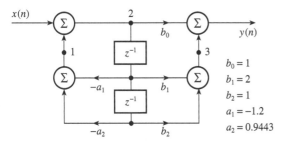

Figure 13.31 Standard second-order filter section for Problem 13.1.

13.3 The following transfer function is given:

$$H(z) = \frac{1 - 0.9631z^{-1} + z^{-2}}{1 - 1.5763z^{-1} + 0.9413z^{-2}}$$

   (1) Find a suitable scale factor to avoid overflow if a canonic second-order section is used to realize it.

   (2) Determine the minimum wordlengths required to achieve an output signal-to-noise ratio of 60 dB. State any assumptions made.

13.4 The following poles and zeros of an eighth-order IIR filter are given:

poles	zeros
$0.2870 \pm 0.9075j$	$0.0553 \pm 0.9985j$
$0.7882 \pm 0.5658j$	$0.8828 \pm 0.4698j$
$0.4089 \pm 0.7447j$	$-0.4816 \pm 0.8764j$
$0.6479 \pm 0.5975j$	$0.9617 \pm 0.2740j$

   (1) Sketch the pole–zero diagram and pair the poles and zeros, justifying your pairing scheme.

   (2) Write down the transfer function of the filter from the diagram. Assuming the filter is to be realized in cascade form, decide on a suitable ordering scheme.

   (3) Determine suitable scaling factors for the filter sections using the finite wordlength analysis program on the CD in the companion handbook.

   (4) Assume that the input data is digitized to 8 bits and that the degradation in SNR due to roundoff error is to be no more than 0.5 dB. Determine suitable wordlengths for the internal data, coefficients and data variables.

13.5 The following transfer function is given:

$$H(z) = \frac{1 - 1.4890z^{-1} + z^{-2}}{1 - 0.3724z^{-1} + 0.5119z^{-2}}$$
$$\times \frac{1 - 1.9020z^{-1} + z^{-2}}{1 - 0.3779z^{-1} + 0.0851z^{-2}}$$

(1) Determine and plot the locations of the poles and zeros.

(2) Plot the magnitude and phase responses of the filter using a sampling frequency of 48 kHz.

(3) Write down the 2's complement fixed point representation of the filter coefficients using 8 bits (including the sign bit).

(4) Repeat parts (1) and (2) for the quantized filter and compare the two sets of results.

13.6 The following transfer function is given:

$$H(z) = 0.1436 \frac{1 + 2z^{-1} + z^{-2}}{1 - 0.679\,93z^{-1} + 0.491\,33z^{-2}}$$

(1) Determine a suitable scale factor to avoid overflow at the output of adder 1 and an output scale factor for the overall gain to be unity.

(2) Encode the filter coefficients and scale factors using 8-bit fixed-point arithmetic.

(3) Determine the total roundoff noise.

13.7 A lowpass IIR filter is required in digital telephony to bandlimit the voice signal. The filter is required to satisfy the following specifications:

passband	0–3300 Hz
stopband	4.6–16 kHz
passband ripple	<0.1 dB
stopband attenuation	>30 dB
ADC	12 bits
coefficient wordlength	16 bits

Assuming a sampling frequency of 32 kHz, determine

(1) a suitable transfer for the filter, assuming the filter is realized in cascade form with second- and/or first-order sections,

(2) scale factors for each filter section,

(3) the change in passband and stopband ripples due to coefficient quantization, and

(4) the degradation in SNR due to roundoff error, assuming that post-accumulation quantization is employed.

13.8 Scaling is employed in IIR filters to prevent the adders overflowing. One scheme is to attenuate the input to each IIR filter section such that the scale factor is given by

$$s_1^2 = \frac{1}{2\pi j} \oint \frac{z^{-1}dz}{D(z)D(z^{-1})}$$

where $\oint$ indicates integration around the unit circle, that is $|z| = 1$, and

$$D(z) = 1 + a_1 z^{-1} + a_2 z^{-2}$$

(1) Find a general expression for s_1^2.

(2) Obtain the scale factor, s_1, for the filter with the following transfer function:

$$H(z) = \frac{1 + 1.2173z^{-1} + z^{-2}}{1 + 0.9140z^{-1} + 0.8793z^{-2}}$$

13.9 Design a Chebyshev lowpass IIR digital filter meeting the following specifications:

passband edge	12 kHz
stopband edge	16 kHz
passband ripples	0.5 dB
stopband attenuation	60 dB
sampling frequency	48 kHz

Assume that the filter will be implemented on a TMS320C54-based system with 12-bit ADC and DAC.

13.10 Design a Chebyshev highpass IIR digital filter meeting the following specifications:

passband edge	12 kHz
stopband edge	8 kHz
passband ripples	0.5 dB
stopband attenuation	60 dB
sampling frequency	48 kHz

Assume that the filter will be implemented on a DSP56300-based system with 16-bit ADC and DAC.

13.11 A requirement exists for a digital notch filter to reduce the effects of mains interference. The filter is to meet the following specifications:

notch frequency	50 Hz
width of notch (3 dB)	±2 Hz
sampling frequency	500 Hz
filter order	2

(a) Determine the transfer function of a suitable digital notch filter using the pole–zero placement method. With the aid of your transfer function explain why the amplitude response of the filter will be essentially flat except near the notch frequency.

(b) Find a scale factor for the filter in (a), based on the frequency response, to reduce the possibility of internal overflow.

Assume that the filter is to be realized using a second-order canonic section.

(c) Determine the change in the notch frequency if the coefficients are quantized to 8 bits.

13.12 (a) Discuss, with the aid of diagrams where appropriate, the problems of roundoff noise in fixed-point digital IIR filters. Your answer should cover the following points:

- how roundoff noise arises in IIR filters;
- the effects of roundoff noise on the performance of IIR filters.

(b) The structure of a second-order filter section, with an error feedback scheme, is shown in Figure 13.32. Assume the filter section is to be implemented using 2's complement, fixed-point arithmetic, with quantization taking place after addition of products. Derive an expression for the transform of the quantized output, $\hat{Y}(z)$, in terms of the input transform, $X(z)$, and the quantization error, $E(z)$, and hence show that the error feedback network has no adverse effect on the input signal.

(i) Deduce the expression for the error feedback function.

(ii) Explain, with the aid of sketches of appropriate frequency responses, the effect of the error feedback network on the roundoff noise at the output of the filter.

(iii) What are the key factors that influence the choice of values of the error feedback coefficients in practice?

(iv) Deduce, from analysis of pole and zero positions as appropriate, suitable integer values for the error feedback coefficients to minimize the level of roundoff noise at the output of each of the following filters:

(1) $H(z) = \dfrac{1 + 2z^{-1} + z^{-2}}{1 - 1.25z^{-1} + 0.81z^{-2}}$

(2) $H(z) = \dfrac{1 + 2z^{-1} + z^{-2}}{1 + 1.40z^{-1} + 0.53z^{-2}}$

(3) $H(z) = \dfrac{1 - 2z^{-1} + z^{-2}}{1 - 1.4z^{-1} + 0.53z^{-2}}$

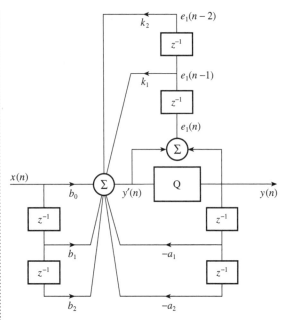

Figure 13.32 A second-order roundoff noise reduction scheme for Problem 13.12.

13.13 Figure 13.33 shows a simple first-order IIR filter with an error spectral shaping scheme to minimize product roundoff noise at the filter output. Determine, analytically,

(1) a suitable L_2 scale factor to reduce the possibility of overflow, and

(2) the output noise power due to roundoff error for each of the following cases:

(a) no error feedback, that is $k' = 0$;

(b) the error feedback coefficient $k' = 1$.

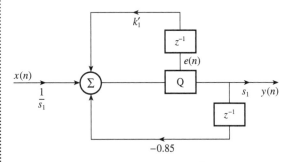

Figure 13.33 First-order IIR filter with error spectral shaping.

References

Abu-el-Haija A. and Al-Ibrahim M.M. (1986) Improving performance of digital sinusoidal oscillators by means of error feedback circuits. *IEEE Trans. Circuits and Systems*, **33**(4), 373–80.

Antoniou A. (1979) *Digital Filters Analysis and Design*. New York: McGraw-Hill.

Chen W. (1996) Performance of cascade and parallel IIR filters. *J. Audio Eng. Soc.*, **44**(3), 148–58.

Claasen T. (1974) Improvement of overflow behaviour of 2nd-order digital filters by means of error feedback. *Electronics Lett.*, **10**(12), 240–1.

Dattorro J. (1988) The implementation of recursive digital filters for high-fidelity audio. *J. Audio Eng. Soc.*, **36**(11), 851–78.

Flores I. (1963) *The Logic of Computer Arithmetic*. Englewood Cliffs NJ: Prentice-Hall.

Higgins W.E. and Munson D. (1982) Noise reduction strategies for digital filters: error spectrum shaping versus the optimal linear state-space formulation. *IEEE Trans. Acoustics, Speech and Signal Processing*, **30**(6), 963–73.

IEEE (1979) *Programs for Digital Signal Processing*. New York: IEEE Press.

IEEE (1985) IEEE Standard for Binary Floating Point Arithmetic. *SIGPLAN Notices*, **22**(2), 9–25.

Ifeachor E.C. (2001) *A Practical Guide for MATLAB and C Language Implementation of DSP Algorithms*. Harlow: Pearson Education.

Jackson L.B. (1986) *Digital Filters and Signal Processing*. Boston MA: Kluwer.

Mitra S.K., Hirano K. and Sakaguchi H. (1974) A simple method of computing the input quantization and multiplication roundoff errors in a digital filter. *IEEE Trans. Acoustics, Speech and Signal Processing*, **22**(5), 326–9.

Mullis C.T. and Roberts R.A. (1982) An interpretation of error spectrum shaping in digital filters. *IEEE Trans. Acoustics, Speech and Signal Processing*, **30**(6), 1013–15.

Oppenheim A.V. and Weinstein C.J. (1972) Effects of finite register length in digital filtering and the fast Fourier transform. *Proc. IEEE*, **60**, 957–76.

Patterson D.A. and Hennessy J.L. (1990) *Computer Architecture: A Quantitative Approach*. San Mateo CA: Morgan Kaufmann.

Rabiner L.R. and Gold B. (1975) *Theory and Applications of Digital Signal Processing*. Englewood Cliffs NJ: Prentice-Hall.

Rader C.M. and Gold B. (1967) Effects of parameter quantization on the poles of a digital filter. *Proc. IEEE*, **55**, 688–9.

Texas Instruments (1986) *Digital Signal Processing Applications with the TMS320 Family: Theory, Algorithms and Implementations*. Texas Instruments.

Tomarakos J. and Ledger D. (1998) Using the Low-cost, High-performance ADSP-21065L Digital Signal Processor for Digital Audio Applications. Analog Devices DSP Application. Details are available at *www.analog.com*

Weitek (1984) *High Speed Digital Arithmetic VLS Application Seminar Notes*. Sunnyvale CA: Weitek.

Wilson R. (1993) Filter topologies. *J. Audio Eng. Soc.*, **41**(9), 667–78.

Bibliography

Abu-el-Haija A.I. and Peterson A.M. (1979) An approach to eliminate roundoff errors in digital filters. *IEEE Trans. Acoustics, Speech and Signal Processing*, **27**, 195–8.

Ahmed N. and Natarajan T. (1983) *Discrete-time Signals and Systems*. Reston VA: Reston Publishing Inc.

Avenhaus E. (1972) Filters with coefficients of limited wordlength. *IEEE Trans. Audio Electroacoustics*, **20**, 206–12.

Barnes C.W., Tran B.N. and Leung S.H. (1985) On the statistics of fixed-point roundoff error. *IEEE Trans. Acoustics, Speech and Signal Processing*, **33**, 595–606.

Chang T.L. (1978) A low roundoff noise digital filter structure. In *Proc. IEEE Int. Symp. on Circuits and Systems*, May 1978, pp. 1004–8.

Chang T.L. (1979) Error-feedback digital filters. *Electronics Lett.* 348–9.

Chang T.L. (1980) Comments on 'An approach to eliminate roundoff errors in digital filters'. *IEEE Trans. Acoustics, Speech and Signal Processing*, **28**(2), 244–5.

Chang T.L. (1981) Suppression of limit cycles in digital filters designed with one magnitude-truncation quantizer. *IEEE Trans. Circuits and Systems*, **28**(2), 107–11.

Chang T.L. (1981) On low-roundoff noise and low-sensitivity digital filter structures. *IEEE Trans. Acoustics, Speech and Signal Processing*, **29**(5), 1077–80.

Chang T.L. and White S.A. (1981) An error cancellation digital-filter structure and its distributed-arithmetic implementation. *IEEE Trans. Circuits and Systems*, **28**(4), 339–42.

Charalambous C. and Best M.J. (1974) Optimization of recursive digital filters with finite wordlengths. *IEEE Trans. Acoustics, Speech and Signal Processing*, **22**(6), 424–31.

Claasen T.A.C.M. and Kristiansson L.O.G. (1975) Necessary and sufficient conditions for the absence of overflow phenomena in a second order recursive digital filter. *IEEE Trans. Acoustics, Speech and Signal Processing*, **23**(6), 509–15.

Claasen T.A.C.M., Mecklenbrauker W.F.G. and Peek J.B.H. (1973) Second-order digital filter with only one magnitude-truncation quantiser and having practically no limit cycles. *Electronics Lett.*, **9**, 531–2.

Claasen T.A.C.M., Mecklenbrauker W.F.G. and Peek J.B.H. (1973) Some remarks on the classification of limit cycles in digital filters. *Philips Research Rep.*, **28**, 297–305.

Claasen T., Mecklenbrauker W.F.G. and Peek J.B.H. (1975) Frequency domain criteria for the absence of zero-input limit cycles in nonlinear discrete-time systems, with applications to digital filters. *IEEE Trans. Circuits and Systems*, **22**, 232–9.

Claasen T.A.C.M., Mecklenbrauker W.F.G. and Peek J.B.H. (1976) Effects of quantization and overflow in recursive digital filters. *IEEE Trans. Acoustics, Speech and Signal Processing*, **24**(6), 517–28.

Crochiere R.E. (1975) A new statistical approach to the coefficient wordlength problem for digital filters. *IEEE Trans. Circuits and Systems*, **22**, 190–6.

Crochiere R.E. and Oppenheim A.V. (1975) Analysis of linear digital networks. *Proc. IEEE*, **63**(4), 581–94.

Diniz P.S.R. and Antoniou A. (1985) Low-sensitivity digital filter structures which are amenable to error-spectrum shaping. *IEEE Trans. Circuits and Systems*, **32**(10), 1000–7.

Elliot D.F. (ed.) (1987) *Handbook of Digital Signal Processing*. London: Academic Press.

IEEE (1978) *Digital Signal Processing II*. Institute of Electrical and Electronics Engineers.

Jackson L.B. (1970) On the interaction of roundoff noise and dynamic range in digital filters. *BSTJ*, **49**(2), 159–84.

Jackson L.B. (1976) Roundoff noise bounds derived from coefficient sensitivities for digital filters. *IEEE Trans. Circuits and Systems*, **23**(8), 481–5.

Knowles J.B. and Olcayto E.M. (1968) Coefficient accuracy and digital filter response. *IEEE Trans. Circuit Theory*, **15**, 31–41.

Liu B. (1971) Effect of finite wordlength on the accuracy of digital filters – a review. *IEEE Trans. Circuit Theory*, **18**, 670–7.

Liu B. and Kaneko T. (1969) Error analysis of digital filters realized with floating-point arithmetic. *Proc. IEEE*, **57**(10), 1735–47.

Markel J.D. and Gray A.H. (1975) Fixed-point implementation algorithms for a class of orthogonal polynomial filter structures. *IEEE Trans. Acoustics, Speech and Signal Processing*, **23**(5), 486–94.

Markel J.D. and Gray A.H. (1975) Roundoff noise characteristics of a class of orthogonal polynomial structures. *IEEE Trans. Acoustics, Speech and Signal Processing*, **23**(5), 473–86.

Motorola (1988) *Digital Stereo 10-band Graphic Equalizer Using the DSP56001*. Motorola Application Note.

Mullis C.T. and Roberts R.A. (1976) Round-off noise in digital filters: frequency transformations and invariants. *IEEE Trans. Acoustics, Speech and Signal Processing*, **24**(6), 538–50.

Munson D.C. and Liu B. (1980) Low-noise realization for narrow-band recursive digital filters. *IEEE Trans. Acoustics, Speech and Signal Processing*, **28**, 41–54.

Nagle H.T. and Nelson V.P. (1981) Digital filter implementation on 16 bit microcomputers. *IEEE Micro*, **1**, 23–41.

Oppenheim A.V. and Schafer R.W. (1975) *Digital Signal Processing*. Englewood Cliffs NJ: Prentice-Hall.

Peled A., Liu B. and Steiglitz K. (1974) A new hardware realization of digital filters. *IEEE Trans. Acoustics, Speech and Signal Processing*, **22**, 456–62.

Peled A., Liu B. and Steiglitz K. (1975) A note on implementation of digital filters. *IEEE Trans. Acoustics, Speech and Signal Processing*, **23**, 387–9.

Rabiner L.R., Cooley J.W., Helms H.D., Jackson L.B., Kaiser L.F., Rader C.M., Schafer R.W., Steiglitz K. and Weinstein C.J. (1972) Terminology in digital signal processing. *IEEE Trans. Audio and Electroacoustics*, **20**, 322–37.

Sandberg I.W. and Kaiser J.F. (1972) A bound on limit cycles in fixed-point implementations of digital filters. *IEEE Trans. Audio and Electroacoustics*, **20**, 110–12.

Sim P.K. and Pang K.K. (1985) Effects of input-scaling on the asymptotic overflow-stability properties of second recursive digital filters. *IEEE Trans. Circuits and Systems*, **32**(10), 1008–15.

Steiglitz K. (1971) Designing short-word recursive digital. *Proc. 9th Ann. Allerton Conf. on Circuit and System Theory*, 6–8 October, pp. 778–88.

Steiglitz K., Bede L. and Liu B. (1976) An improved algorithm for ordering poles and zeros of fixed-point recursive digital filters. *IEEE Trans. Acoustics, Speech and Signal Processing*, **24**, 341–3.

Taylor F.J. (1983) *Digital Filter Design Handbook*. New York: Marcel Dekker.

Thong T. (1976) Finite wordlength effects in the ROM digital filter. *IEEE Trans. Acoustics, Speech and Signal Processing*, **24**, 436–7.

Thong T. and Liu B. (1977) Error spectrum shaping in narrowband recursive digital filters. *IEEE Trans. Acoustics, Speech and Signal Processing*, **25**, 200–3.

Williamson D. and Sridharan S. (1985) An approach to coefficient wordlength reduction in digital filters. *IEEE Trans. Circuits and Systems*, **32**(9), 893–903.

Appendices

13A Finite wordlength analysis program for IIR filters

C language programs for finite wordlength analysis of IIR filters, together with illustrative examples, are available in the companion book, Ifeachor E.C. (2001) *A Practical Guide for MATLAB and C Language Implementations of DSP Algorithms*, Pearson Education (see the Preface for details).

13B L_2 scaling factor equations

The standard canonic filter section is depicted in Figure 13B.1. The transfer function of the filter section is given by

$$H(z) = \frac{b_0 + b_1 z^{-1} + b_2 z^{-2}}{1 + a_1 z^{-1} + a_2 z^{-2}} \tag{13B.1}$$

The L_2 scale factor to reduce the possibility of overflow at the node labelled $w(n)$ is given by

$$s_1^2 = \frac{1}{2\pi \mathrm{j}} \oint \frac{z^{-1}\,\mathrm{d}z}{D(z)D(z^{-1})} = \frac{1}{2\pi \mathrm{j}} \oint F(z)\,\mathrm{d}z \tag{13B.2}$$

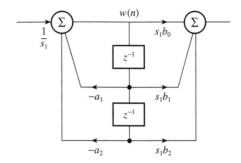

Figure 13B.1   Standard canonic filter section.

where

$$D(z) = 1 + a_1 z^{-1} + a_2 z^{-2}$$

$$F(z) = z^{-1}/D(z)D(z^{-1})$$

and $\oint$ is the contour integral around the circle $|z| = 1$.

Using the value of $D(z)$ in Equation 13B.2 we have

$$s_1^2 = \frac{1}{2\pi j} \oint \frac{z^{-1} \, dz}{(1 + a_1 z^{-1} + a_2 z^{-2})(1 + a_1 z + a_2 z^2)}$$

$$= \frac{1}{2\pi j} \oint \frac{z \, dz}{(z^2 + a_1 z + a_2)(1 + a_1 z + a_2 z^2)}$$

The poles z_1 and z_2 of the integrand inside the unit circle are given by

$$z^2 + a_1 z + a_2 = (z - z_1)(z - z_2) = 0 \tag{13B.3}$$

From the calculus of residues, s_1^2 is the sum of the residues of $F(z)$:

$$s_1^2 = \lim_{z \to z_1} \frac{(z - z_1)z}{(z^2 + a_1 z + a_2)(1 + a_1 z + a_2 z^2)} + \lim_{z \to z_2} \frac{(z - z_2)z}{(z^2 + a_1 z + a_2)(1 + a_1 z + a_2 z^2)}$$

$$= \frac{z_1}{(z_1 - z_2)(1 + a_1 z_1 + a_2 z_1^2)} - \frac{z_2}{(z_1 - z_2)(1 + a_1 z_2 + a_2 z_2^2)}$$

$$= \frac{z_1(1 + a_1 z_2 + a_2 z_2^2) - z_2(1 + a_1 z_1 + a_2 z_1^2)}{(z_1 - z_2)(1 + a_1 z_1 + a_2 z_1^2)(1 + a_1 z_2 + a_2 z_2^2)}$$

$$= \frac{1 - a_2 z_1 z_2}{(1 + a_1 z_1 + a_2 z_1^2)(1 + a_1 z_2 + a_2 z_2^2)}$$

$$= \frac{1 - a_2 z_1 z_2}{1 + a_1(z_1 + z_2) + a_2(z_1^2 + z_2^2) + a_1^2 z_1 z_2 + a_1 a_2 z_1 z_2(z_1 + z_2) + a_2^2 z_1^2 z_2^2}$$

$$= \frac{1 - a_2 z_1 z_2}{1 + a_1(z_1 + z_2) + a_2[(z_1 + z_2)^2 - 2z_1 z_2] + a_1^2 z_1 z_2 + a_1 a_2 z_1 z_2(z_1 + z_2) + a_2^2(z_1 z_2)^2}$$

$$\tag{13B.4}$$

Now, from Equation 13B.3,

$$z^2 + a_1 z + a_2 = (z - z_1)(z - z_2) = z^2 - (z_1 + z_2)z + z_1 z_2$$

Thus

$$a_1 = -(z_1 + z_2)$$

$$a_2 = z_1 z_2$$

Using the values of a_1 and a_2 in Equation 13B.4 we have

$$s_1^2 = \frac{1 - a_2^2}{1 - a_1^2 + a_2(a_1^2 - 2a_2) + a_1^2 a_2 - a_1^2 a_2^2 + a_2^4}$$

$$= \frac{1 - a_2^2}{1 - a_1^2 - 2a_2^2 + 2a_1^2 a_2 - a_1^2 a_2^2 + a_2^4}$$

$$= \frac{1 - a_2^2}{(1 - a_2^2)^2 - a_1^2(1 - 2a_2 + a_2^2)}$$

$$= \frac{1 - a_2^2}{(1 - a_2^2)^2 - a_1^2(1 - a_2)^2}$$

$$= \frac{1}{(1 - a_2^2) - a_1^2(1 - a_2)/(1 + a_2)}$$

Thus

$$s_1^2 = \frac{1}{(1 - a_2^2) - a_1^2(1 - a_2)/(1 + a_2)} \tag{13B.5}$$

14.1 Evaluation boards for real-time signal processing

14.1.1 Background

As in other areas of engineering, practical experience of designing and implementing DSP algorithms is necessary for gaining a proper appreciation of the issues involved in DSP. Engineering students with a background in only analog signal processing have genuine difficulty getting to grips with techniques involved in DSP, especially if they do not have the necessary mathematical background to understand the concepts from a theoretical point of view. Often they are puzzled by, for example, how the numerical operations used in FIR or IIR filters can lead to filtering. They are reasonably comfortable with analog filters and the concepts of how the frequency-dependent characteristics of capacitors and inductors combined with resistance achieve filtering. In DSP, there are no obvious frequency-dependent parameters. How does a digital filter actually work, some will ask.

We became convinced that what we needed was a simple, standalone piece of hardware which students could use to design and implement simple DSP functions. We also wanted to demonstrate some of the practical issues involved in real-time DSP, such as the concepts of aliasing, imaging, $\sin x/x$, overflows and so on. This led to the development of a number of simple target boards for first and second generation fixed-point DSP processors. These boards still remain useful and inexpensive platforms for demonstrating DSP algorithms in real time. Today, a number of low-cost evaluation boards for second and third generation fixed-point DSP processors are available commercially. In the next three subsections, we will describe some of the boards.

14.1.2 TMS320C10 target board

The TMS320C10 target board was our first DSP board. It still serves the useful purpose of demonstrating simple real-time DSP algorithms in standalone mode. The main features of the board are:

- standalone board capable of executing simple DSP algorithms in real time;
- single analog input/output digitized to 8 bits;
- allows codes for DSP algorithms to be modified easily;
- capable of operating at two different sampling rates;
- allows the study of aliasing and imaging.

The system block diagram, shown in Figure 14.1, consists of four main units, a first generation TMS320C10 digital signal processor which is the heart of the system, an 8-bit ADC/DAC unit, the timing circuit, and memory units. For high fidelity systems 8-bit resolution is not adequate, but for demonstrating the principles of DSP we found

Applications and design studies

The objectives of this chapter are fourfold. The first is to describe some of the low-cost boards that may be used to implement the DSP algorithms described in previous chapters. Two low-cost boards for first and second generation fixed-point DSP processors that we use to demonstrate the principles of DSP to students are described. An overview of a number of low-cost boards for third-generation fixed-point DSP processors is provided.

The second objective is to describe a number of real-world applications of DSP in the form of case studies. Applications described here include real-time audio signal processing, adaptive filtering of artefacts from the human electroencephalogram (electrical activity of the brain), and the detection of fetal heartbeats from the fetal electrocardiogram (electrical activity of the heart) which is necessary for assessing the condition of the baby during childbirth. The presentation draws on many DSP concepts discussed in earlier chapters.

The third objective is to present a number of challenging, practical problems which may be carried out in groups as design studies. The final objective is to present a set of multiple choice questions to assist the reader in gaining deeper insight into various aspects of DSP.

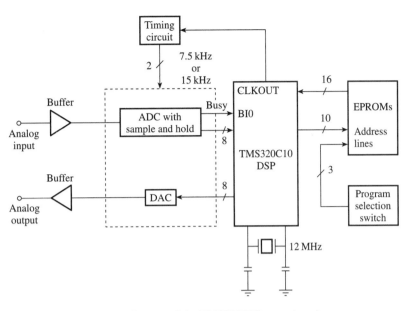

Figure 14.1 A simple block diagram of the TMS320C10 target board.

Table 14.1 Program memory selection.

Block	Address	DSP programs
0	0000–03FF	Input–output loop
1	0400–07FF	Noise generator
2	0800–0BFF	Square wave generator
3	0C00–0FFF	41-point bandpass FIR filter
4	1000–13FF	61-point FIR notch filter
5	1400–17FF	Fourth-order IIR lowpass filter in cascade
6	1800–1BFF	Fourth-order IIR lowpass filter in parallel
7	1C00–1FFF	Fourth-order IIR bandpass filter in cascade

it sufficient. The memory unit consists of a program selection switch and a pair of EPROMs, mounted on ZIF (zero insertion force) sockets for easy use. The EPROM is partitioned into eight blocks each of 1 k words, selectable through the program selection switch. This allows up to eight different programs to be held in the EPROMs. For standalone operation, the use of EPROMs is necessary. Two user-selectable sampling frequencies are available, one at 7.5 kHz and the other at 15 kHz.

Many DSP algorithms have been implemented on the target board. These include FIR and IIR filters, noise and square wave generators. Examples are listed in Table 14.1.

14.1.3   DSP56002 evaluation module for real-time DSP

The TMS320C10 board is useful for demonstrating simple DSP functions in a standalone mode, but it is limited in serious design tasks. The Motorola DSP56002EVM is a low-cost evaluation module (EVM) which is useful for rapid design and demonstration of real-time DSP systems. For the past six years we have used the EVM in one of our DSP courses, partly because of the emphasis on real audio signal processing for which the DSP56002 is well suited. Features of the EVM include:

- a 24-bit fixed-point DSP56002 processor;
- 32 k words of SRAM and optional 32 k bytes of flash EEPROM for standalone operation;
- CD quality audio CODEC (16-bit stereo A/D and D/A);
- sampling rates of 48, 32, 16, 9.6 or 8 kHz;
- assembler and debugger.

The DSP processor has two 48-bit X and Y registers which may also be used as four 24-bit registers ($X0$, $X1$, $Y0$ and $Y1$), two 56-bit accumulators and a hardware multiplier which are invaluable in signal processing. The debugger allows changes to be made to the registers and source code via simple on-screen editing. In the section on design studies, the use of the DSP56002 board will be illustrated by using a filter design problem as a vehicle.

14.1.4   TMS320C54 and DSP56300 evaluation boards

Sophisticated software and hardware development tools for new generations of fixed- and floating-point DSP processors now exist (e.g. Texas Instruments Code Composer Studio) and details of these can be found on the websites of major manufacturers. In this section, we will describe briefly two low-cost evaluation modules which are well suited to learning DSP concepts and for developing fairly advanced DSP systems.

The TMS320C54x evaluation module (Texas Instruments, 1995) is a PC-based, plug-in card that may be used to implement DSP algorithms in real time. The main features of the EVM are:

- a TMS320C541 16-bit, fixed-point DSP processor, with 5 k bytes of on-chip program/data RAM and 28 k bytes of on-chip ROM;
- a graphical, Windows-based, debugger;
- embedded emulation support for the C source debugger;
- an analog I/O interface.

The analog I/O interface supports programmable anti-aliasing and anti-image filtering, amplitude control and sampling rates (up to 43.2 kHz). A single channel 14-bit analog-to-digital/digital-to-analog conversion is provided.

The DSP56302 EVM (Motorola, 1996) is an excellent, low-cost standalone, PC-linked platform for developing DSP systems. User-developed software can be downloaded from the PC onto the on-chip memory for execution and debugging. The main features of the DSP56302 EVM include:

- a DSP56302 24-bit fixed-point DSP processor;
- onboard 32 k words of program/cache and data RAM;
- two channels of CD-quality audio codecs (16-bit ADC/DAC);
- cross-assembler and Windows-based debugger.

14.2 DSP applications

14.2.1 Detection of fetal heartbeats during labour[†]

Worldwide, the standard method of monitoring the fetus during labour is the display of continuous fetal heart rate (FHR) and the uterine activity which together constitute the cardiotocogram (CTG) (Figure 14.2). By analysis and appropriate interpretation of

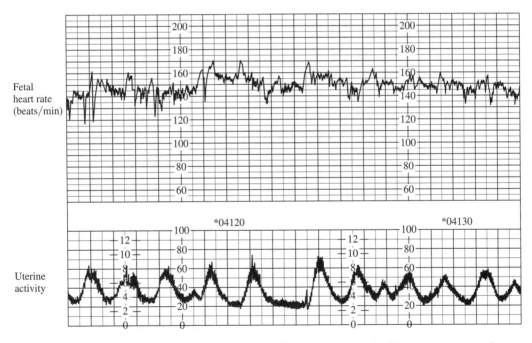

Figure 14.2   An example of a cardiotocogram (CTG). The CTG consists of the fetal heart rate pattern and the uterine activity.

† This section is based, in part, on the final year project of one of our previous students, Mr Iead Rezek.

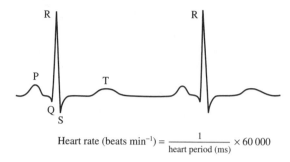

$$\text{Heart rate (beats min}^{-1}) = \frac{1}{\text{heart period (ms)}} \times 60\,000$$

Figure 14.3 The electrocardiogram.

changes in the CTG obstetricians hope to prevent the delivery of dead or impaired babies who had suffered as a result of a lack of oxygen during labour and delivery.

14.2.1.1 The fetal electrocardiogram

The fetal heart rate is routinely obtained during labour from the electrocardiogram (ECG), the electrical activity of the heart (see Figure 14.3) or ultrasound. Like the adult ECG, normal fetal ECG is characterized by five peaks and valleys labelled with successive letters of the alphabet P, Q, R, S, and T (Greene, 1987). Thus, the ECG is said to consist of the P wave, QRS complex and T wave (Greene, 1987).

As shown in Figure 14.3, the reciprocal of the heart period, that is the time interval between the R-to-R peaks (in milliseconds), multiplied by 60 000 gives the instantaneous heart rate. The FHR pattern in the upper half of Figure 14.2 is a plot of successive instantaneous heart rates.

In practice, to measure the fetal heart rate a suitable DSP algorithm is employed to detect, in hardware or software, successive QRS complexes and from these to calculate the R-to-R intervals and the corresponding FHR. Most QRS detection methods assume that the shape of the fetal QRS complex is known *a priori*, but that its time of occurrence is unknown. This assumption is reasonable, although not always valid as the shape of the QRS complex may change from patient to patient and indeed within the same patient. Thus by comparing the ECG signal against a known, representative QRS template the locations of the QRS complexes in the ECG can be determined based on some measure of similarity, for example a high value of cross-correlation.

A fundamental problem is the reliable detection of the QRS complexes. Signal degradation due, for example, to baseline wander, mains interference, uterine contractions, ADC saturation, and movements of the baby or mother will lead to false detection or missed QRS complexes. The aim in this case study is to investigate and compare two QRS detection methods that may be of practical value in real-time fetal monitoring. This work is a small part of an ongoing research initiative with local hospitals to develop intelligent systems to assist the busy clinicians in managing labour (Ifeachor *et al.*, 1991).

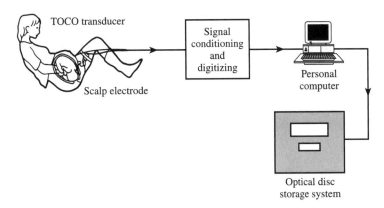

Figure 14.4   Measurement of fetal electrocardiogram.

The fetal ECG data used in the case study was taken from our fetal research database. The ECG signal was obtained by measuring differentially between an electrode on the fetal scalp and a standard skin electrode placed on the maternal thigh, and a second maternal electrode was used as earth (Figure 14.4). This lead system has a sensitivity vector in the longitudinal plane of the fetus compared with the sagittal plane of the standard fetal scalp electrode connection and should reduce ECG vector change resulting from rotation of the fetus (Lindecrantz *et al.*, 1988). The FECG was fed through a patient, isolation box amplified, analog bandpass filtered (passband 0.07–100 Hz), and digitized at 500 samples s^{-1} with a resolution of 8 bits.

Examples of measured fetal ECGs are shown in Figures 14.5(a) to 14.5(c). From visual examination, it is seen that the data in Figure 14.5(a) has a relatively high SNR, compared with Figures 14.5(b) and 14.5(c), with large amplitude R waves (seen as spikes). The data in Figure 14.5(b), on the other hand, has a relatively high noise content and significant baseline shifts, although the R waves are still discernible. The data in Figure 14.5(c) contains ADC errors, seen as large amplitude swings between the maximum and minimum ADC values at the start of the record, due perhaps to ADC saturation, as well as severe baseline shifts and high frequency noise including mains contamination. The three data sets in Figures 14.5(a), 14.5(b) and 14.5(c) have been subjectively classified as grade 1 (good), grade 2 (average) and grade 3 (poor) ECGs, respectively.

14.2.1.2   Fetal ECG signal pre-processing

For grade 2 and 3 data, noise levels and baseline shifts make detection of QRS complexes from the raw ECG more difficult. For a reliable QRS detection, it is necessary to pre-process the raw ECG to minimize the influence of these sources of signal degradation before attempting to detect the QRS complexes. It is known that significant frequency components of the QRS complex lie between about 4 and 45 Hz. Baseline shifts in the ECG are normally of a low frequency, typically less than about 3 Hz, although, for grade 3 data, baseline frequency may extend to 15 Hz or more.

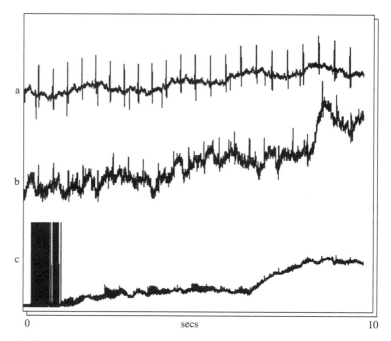

Figure 14.5   Examples of grades of ECG data: (a) grade 1 (good); (b) grade 2 (average); (c) grade 3 (poor).

An FIR or IIR bandpass digital filter may be used to pre-process the raw ECG before QRS detection. We prefer the use of FIR as an IIR filter of a high order, for example eighth order, sometimes rings when excited by the narrow width QRS complexes which may complicate the precise location of the R wave. The filter specifications used in the study are as follows:

filter length	75
sampling frequency	500 Hz
stopbands	0–1 Hz, 47–250 Hz
passband	9–39 Hz
passband ripple	0.5 dB
stopband attenuation	30 dB

The filter coefficients were obtained using the optimal method described in Chapter 7. Figures 14.6(a) to 14.6(c) show the filtered ECG data. Compared with the corresponding unfiltered raw data, Figures 14.6(a) to 14.6(c), the baseline shifts as well as the high frequency noise have been reduced in the filtered data (ignoring the initial transients in the filtered data). In the grade 3 data, the ADC error appears as a burst which no doubt will confound most QRS detection algorithms; see Figure 14.6(c).

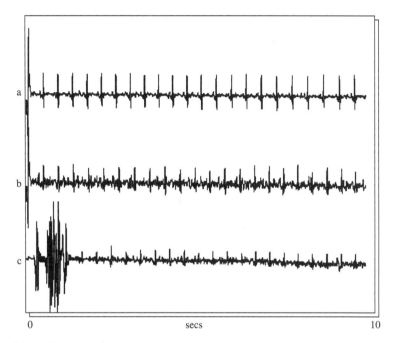

Figure 14.6   Filtered ECG data: (a) grade 1 (good); (b) grade 2 (average); (c) grade 3 (poor).

14.2.1.3   QRS template

Most QRS detection methods rely on the availability of a representative QRS template against which the incoming ECG signal is compared. The template may be generated from raw ECG data by detecting and averaging several good QRS complexes. This may be done automatically or semimanually by visually examining a grade 1 ECG record and identifying good, unambiguous ECG complexes. The R waves are then synchronized and the QRS complexes averaged. A fixed QRS template may be used to detect QRS complexes or a new one may be generated at the start of each ECG record. An example of a QRS template obtained by averaging 69 QRS complexes in grade 1 data and then taking 31 samples (15 samples on either side of the R wave) of the averaged QRS complexes is shown in Figure 14.7.

In the study, templates of various lengths were tried. Typically, the length of the template, N, was between 11 and 31 samples, that is a width of between about 20 ms and 60 ms at a sampling rate of 500 samples s^{-1}. In this report, we will give results obtained with two templates of lengths 11 and 31 samples.

14.2.1.4   QRS detection methods

A general block diagram of the QRS detection process is given in Figure 14.8. The raw ECG data is first pre-processed to reduce the effects of noise. The pre-processed data samples are fed into a buffer one data point at a time. For each new data point fed

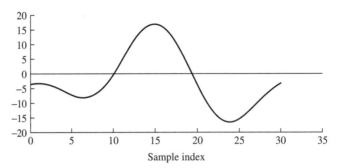

Figure 14.7 An example of a QRS complex template. This is obtained by averaging over 69 QRS complexes in a grade 1 ECG, with the R-waves synchronized. The QRS complexes were detected with a threshold level of 13.

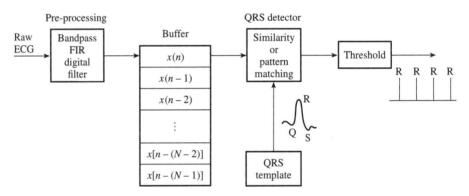

Figure 14.8 Concepts of QRS complex detection from raw ECG.

into the buffer, the oldest data point is removed and the content of the buffer compared with a QRS template in the QRS detector. The output of the QRS detector is then thresholded. If this output exceeds the threshold value then a QRS is said to have occurred. Two conventional QRS detection methods are compared in this study, selected on the basis that they are either in practical use or potentially of practical use. Many other QRS detection methods exist.

The methods are

(1) average magnitude cross-difference (AMCD) (Lindecrantz *et al.*, 1988), currently used in a new fetal monitor described in Lindecrantz *et al.* (1988), and

(2) matched filtering, which is a common QRS detection method and has been investigated by a number of workers (Azevedo and Longini, 1980; Favret, 1968): it is closely related to the correlation method.

Average magnitude cross-difference

In this method, blocks of pre-processed fetal ECG data are compared against a template QRS complex as described above. The differences between corresponding samples in the ECG and the template are computed by waveform subtraction. The sum, $y(i)$, of the absolute values of the differences is then computed:

$$y(i) = \sum_{k=0}^{N-1} |x_t(k) - x_t - [x(k+i) - x_i]|, \quad i = 0, 1, \ldots \quad (14.1)$$

where $x_t(k)$ are samples of the template QRS complex, $x(k+i)$ are samples of the ECG signal, N is the length of the template, and i is the time shift parameter. x_t is the mean value of the QRS template, and x_i the mean value of the ith data block for the ECG signal given by

$$x_t = \frac{1}{N} \sum_{k=0}^{N-1} x_t(i)$$

$$x_i = \frac{1}{N} \sum_{k=0}^{N-1} x(k+i)$$

When the ECG signal and QRS template are very similar in shape, that is in the neighbourhood of a QRS complex, the AMCD value, $y(i)$, becomes a minimum (theoretically zero).

Digital matched filtering

Matched filtering is commonly used to detect time recurring signals buried in noise. The main underlying assumptions in this method are that the signal is time limited and has a known waveshape. The problem then is to determine its time of occurrence. The impulse response of a digital matched filter, $h(k)$, is the time-reversed replica of the signal to be detected. Thus in our case, if $x_t(k)$ is the QRS template then the coefficients of the matched filter are given by

$$h(k) = x_t(N - k - 1), \quad k = 0, 1, \ldots, N - 1 \quad (14.2)$$

The digital matched filter can be represented as an FIR filter with the usual transverse structure, with the output and the input of the filter related as

$$y(i) = \sum_{k=0}^{N-1} h(k)x(i-k)$$

$$= \sum_{k=0}^{N-1} x_t(N - k - 1)x(i - k)$$

where $x(i)$ are the samples of the input ECG signal, $x_t(k)$ are the samples of the QRS template, N is the filter length, $h(k)$ are matched filter coefficients, and i is the time shift index. It is evident that when the template and the QRS complex coincide, the output of the matched filter will be a maximum. Thus by searching the output of the matched filter for a value above a threshold the occurrence of the QRS can be tested.

14.2.1.5   Performance measure for QRS detection

To evaluate and compare the algorithms requires a measure of performance. Following Azevedo and Longini (1980) we define the performance measure as

$$\frac{(\text{total number of R waves} - \text{number of misses} - \text{number of false detections}) \times 100\%}{\text{total number of fetal R waves}}$$

(14.3)

For a given ECG record, the performance measure attains a value of 100% only if all the R waves in the record are correctly detected, that is no misses (undetected R waves) and no false detections (false alarms). For a given QRS detection method, the number of misses or false detections can be determined by comparing, visually, the output of the detector and the pre-processed ECG. Another alternative is the so-called 28-beat rule employed by clinicians to distinguish between a genuine change in the fetal heart rate and a false change due, for example, to an instrumentation error. According to this rule, an R wave causing a change in the frequency of the fetal heartbeat of more than ±28 beats per minute is indicative of either a missed or a false QRS complex. A baseline may be fitted to the FHR pattern to aid the application of the rule.

14.2.1.6   Results

Figures 14.9 and 14.10 show the performances of the AMCD and matched filtering methods, respectively, plotted against threshold levels for grades 1 and 2 data.

The performances of both methods are dependent on the threshold level used (each threshold was expressed as a fraction of the maximum input signal value) and the length of the QRS template. The best performance was achieved with a threshold level of about 50% for both methods. The wider template tends to perform better than the narrow template when the quality of data is good, but the main difference between them seems to be in their sensitivity to threshold levels.

Overall, there was little to choose between the AMCD and matched filtering methods in terms of their performance. With suitable threshold levels, both methods attained the following performances:

■ 100% detection for all grade 1 ECG;
■ >90% detection for grade 2 data;
■ >60% detection for grade 3 data.

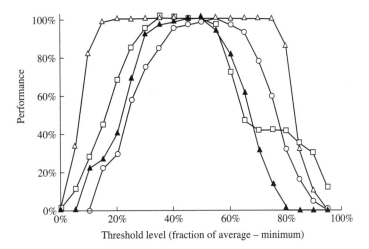

Figure 14.9   Performance of the average magnitude cross-difference method for grades 1 and 2 data, and templates of lengths 11 and 31: –□– , grade 1, T-11; –△– , grade 1, T-31; –○– , grade 2, T-11; –▲– , grade 2, T-31.

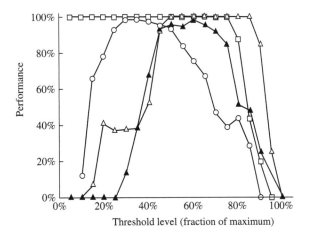

Figure 14.10   Performance of the matched filtering method for grades 1 and 2 data, and for templates of lengths 11 and 31: –□– , grade 1, T-11; –△–, grade 1, T-31; –○– , grade 2, T-11; –▲–, grade 2, T-31.

14.2.2   Adaptive removal of ocular artefacts from human EEGs†

14.2.2.1   Introduction

The work described in this section is concerned with the on-line removal of ocular artefacts from the human electroencephalogram (EEG). The EEG is widely used in

† Adapted from Ifeachor *et al.*, 1986.

clinical and psychological situations, but it is often seriously contaminated by ocular artefacts (OAs) resulting from movements in the ocular systems (eyeball, eyelids, and so on). In some cases, for example brain-damaged babies and patients with frontal tumours, it is difficult to distinguish between associated pathological slow waves in the EEG and ocular artefacts. The similarity between the OAs and signals of interest also makes it difficult to automate the analysis of the EEG by computer. A stimulus-related response, known as the contingent negative variation (CNV), which has diagnostic usefulness for patients with Huntingdon's chorea (Jervis *et al.*, 1984), is very vulnerable to ocular artefacts. It is therefore necessary to remove the OA from the EEG so that the true EEG record can be analyzed.

Although satisfactory OA removal is now possible offline, online OA removal has hitherto been unsatisfactory. The previously reported online methods required the cooperation of subjects which cannot always be guaranteed, involved time-consuming manual calibration, and at best can only deal with one type of OA as they assume a constant correction factor. In this section a new online system for removing OAs from the EEG signals is described which overcomes these disadvantages and offers additional advantages, such as flexibility. The system is based on the Motorola 68000 microprocessor and uses the numerically stable UD factorization algorithm which allows continuous adaptive OA removal. A description of the on-line algorithm, and of the hardware and software of the OAR system, is presented.

Methods for the removal and control of ocular artefacts

The problem of removing the OAs from the EEG is complicated by the similarity between them and some cerebral waves of interest, and by the spectral overlap between them. Of the various methods that have been proposed for removing or controlling the OA in the EEG signals, the electro-oculogram (EOG) subtraction methods are probably the best. In this chapter, the term EOG refers to the electric potential due to ocular movements measured between two skin electrodes placed close to the eyes. However, the various EOG subtraction techniques reported to date do not completely solve the problem, and new approaches are continually being developed. All the techniques are based on the principle that the OA is additive to the background EEG. Thus, in discrete form

$$y(i) = \sum_{j}^{n} \theta_j x_j(i) + e(i) = \mathbf{x}^{\mathrm{T}}(i)\boldsymbol{\theta} + e(i) \tag{14.4}$$

where

$$\mathbf{x}^{\mathrm{T}}(i) = [x_1(i) \quad x_2(i) \quad \ldots \quad x_n(i)]$$
$$\boldsymbol{\theta} = [\theta_1 \quad \theta_2 \quad \ldots \quad \theta_n]^{\mathrm{T}}$$

$y(i)$ and $x_j(i)$ are the samples of the measured EEG and the EOGs respectively, $e(i)$ is the 'true' EEG which may be regarded as an error term, and i is the sample number. θ_j

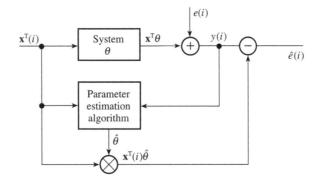

Figure 14.11 Block diagram representation of ocular artefact removal.

are constants of proportionality which will be called the ocular artefact parameters, and n is the number of parameters in the model. The θ_j have also been called the transmission coefficients. $\mathbf{x}^{\mathrm{T}}(i)$ and $\boldsymbol{\theta}$ are the vectors of the EOGs and ocular artefact parameters, respectively, and ^{T} indicates transposition. If θ_j can be estimated then an estimate of $e(i)$ can be obtained as

$$\hat{e}(i) = y(i) - \sum_{j}^{n} \hat{\theta}_j x_j(i), \quad i = 1, 2, \ldots, m \tag{14.5}$$

where $\hat{\theta}_j$ are the estimates of θ_j and $\hat{e}(i)$ is the estimate of $e(i)$, and m is the number of samples used in the estimation. The problem then is one of estimating θ_j. This problem is illustrated in Figure 14.11. For a given type of ocular movement, the $\hat{\theta}_j$ are fairly constant but differ significantly between the different types of OAs, although there is evidence to show that the $\hat{\theta}_j$ vary slowly, at least, even for a given type of OA. In general, there is no way of knowing the type of OA that will occur at a given time, and, because in many cases more than one type of OA occur simultaneously, $\hat{\theta}_j$ cannot be assumed constant so that it is best to estimate θ_j adaptively. The term 'adaptive' is used here to signify that the OA parameter estimates, and hence the OA removal, should be automatically adjusted to changes in the OA. The various EOG subtraction techniques differ primarily in the way the θ_j are estimated, in the number of EOG signals that are used and the way these are measured.

The EOG subtraction method can be carried out either online, that is as the data is being acquired, or offline. The main advantage of the offline methods over previously reported online methods is that more sophisticated removal techniques can be employed. However, in applications requiring real-time processing and analysis the delay involved when offline methods are employed is unacceptable. The trend in EEG signal processing is clearly towards real-time processing, and it is then necessary to remove the artefacts online.

In the offline methods, estimates of $\boldsymbol{\theta}$ are obtained by minimizing J, the sum of squares of the error term, that is $J = \sum_{i=1}^{m} e^2(i)$. This minimum leads to

$$\hat{\boldsymbol{\theta}}_m = [\mathbf{X}_m^{\mathrm{T}} \mathbf{X}_m]^{-1} \mathbf{X}_m^{\mathrm{T}} \mathbf{Y}_m \tag{14.6}$$

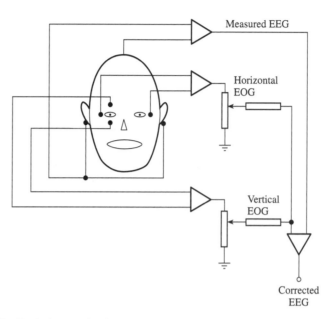

Figure 14.12   Typical example of previously reported online methods of removing artefacts (Girton and Kamiya, 1973).

where

$$\mathbf{Y}_m = [y(1) \quad y(2) \quad \ldots \quad y(m)]^T, \mathbf{X}_m = [\mathbf{x}^T(1) \quad \mathbf{x}^T(2) \quad \ldots \quad \mathbf{x}^T(m)]^T$$
$$\hat{\boldsymbol{\theta}}_m = [\hat{\theta}_1 \quad \hat{\theta}_2 \quad \ldots \quad \hat{\theta}_m]^T, \mathbf{E}_m = [e(1) \quad e(2) \quad \ldots \quad e(m)]^T$$

This equation gives the ordinary least-squares (OLS) estimate of $\boldsymbol{\theta}$ which can be obtained using any suitable matrix inversion technique and forms the basis of the offline OA removal methods. Having obtained $\hat{\boldsymbol{\theta}}_m$, estimates of the OA and hence the background EEG, $\hat{e}(i)$, can be obtained from Equation 14.5.

The OLS method described here can be extended to the multichannel case, where there is more than one EEG signal being corrected. Thus a system, with n EOG inputs and q measured EEG outputs, can be treated as q individual single-output subsystems, and the overall system identified in q separate ways.

A typical example of the previously reported online methods is depicted in Figure 14.12. In this method, due to Girton and Kamiya (1973), an initial calibration was made by adjusting the potentiometers while the subject moved his eyes repetitively in the vertical or horizontal plane until there was minimal amount of OA in the EEG trace. The device was then left at this setting during recording. A number of workers have used the method and found it unwieldy and inefficient in removing OA; see, for example, Gotman et al. (1975).

The previously reported online methods of removing ocular artefacts employed analog methods (Barlow and Rémond, 1981; Girton and Kamiya, 1973), requiring the cooperation of subjects which cannot always be guaranteed. They were, in general, inferior to the offline methods in removing OAs (Jervis *et al.*, 1985; Gotman *et al.*, 1975), required the user to be familiar with the technique, and involved a time-consuming manual calibration of each EEG channel. In addition, these methods could be adjusted for only one type of OA.

In this section, an online system for removing ocular artefacts which overcomes the disadvantages mentioned will be described. It is a microcomputer-based system using a numerically stable online algorithm based on the efficient recursive least-squares technique.

14.2.2.2   Online removal algorithms used in the OAR system

The computation of $\hat{\theta}_m$ in Equation 14.6 requires the time-consuming computation of the inverse matrix. Clearly, the OLS approach is not suitable for real-time or online estimation for which online algorithms involving a fixed and restricted number of arithmetic operations and no direct matrix inversions are preferred. With the online algorithms, $\hat{\theta}$ is updated at each sample point so that changes in ocular movements can be reflected in $\hat{\theta}$ and the OA removal in this case can be viewed as adaptive filtering of OAs from the EEGs (Widrow *et al.*, 1975).

The two parameter estimation methods that are suitable for estimating θ on line are the least-mean-square (LMS) (Widrow *et al.*, 1975) and recursive least-squares (RLS) algorithms (Peterka, 1975; Clarke, 1981). In terms of computation and storage, the LMS algorithm is more efficient than the RLS algorithm. In addition, it does not suffer from the numerical instability problem inherent in the RLS algorithm (see later). However, the RLS algorithm has superior convergence properties to those of the LMS algorithm and, for this reason, it is to be preferred.

The form of the LMS algorithm commonly used is

$$\hat{\theta}(m+1) = \hat{\theta}(m) + 2\mu\mathbf{x}(m+1)[y(m+1) - \mathbf{x}^{\mathrm{T}}(m+1)\hat{\theta}(m)] \tag{14.7}$$

where $\hat{\theta}(m)$ and $\hat{\theta}(m+1)$ are the estimates of θ at the mth and $(m+1)$th sample points, respectively, and μ is a constant that controls the rate of convergence and stability of the algorithm. For convergence μ should be within the limits

$$0 < \mu < 1/\lambda_{\max}$$

where $\lambda_{\max}$ is the maximum eigenvalue of the matrix $(\mathbf{X}_m^{\mathrm{T}}\mathbf{X}_m)$ of Equation 14.6. However, the convergence time of the algorithm is directly proportional to the ratio of the maximum to minimum eigenvalues of $(\mathbf{X}_m^{\mathrm{T}}\mathbf{X}_m)$, which can be very large when there is strong collinearity among the input variables, as is the case with the EOGs. However, the LMS algorithm has been widely used in biomedical applications to reduce noise or artefacts mainly because of its simplicity.

A suitable recursive least-squares (RLS) algorithm is obtained by exponentially weighting the data to remove gradually the effects of old data on the estimates. Thus

$$J = \sum_{i=1}^{m} \gamma^{m-i} e^2(i), \quad 0 < \gamma < 1 \tag{14.8}$$

Minimization of J with respect to the values of $\boldsymbol{\theta}$ leads to the following recursive least-squares algorithm:

$$\hat{\boldsymbol{\theta}}(m+1) = \hat{\boldsymbol{\theta}}(m) + \mathbf{G}[y(m+1) - \mathbf{x}^{\mathrm{T}}(m+1)\hat{\boldsymbol{\theta}}(m)] \tag{14.9a}$$

$$\mathbf{P}(m+1) = \frac{1}{\gamma}\left[\mathbf{P}(m) - \frac{1}{\alpha}\mathbf{P}(m)\mathbf{x}(m+1)\mathbf{x}^{\mathrm{T}}(m+1)\mathbf{P}(m)\right] \tag{14.9b}$$

where

$$\alpha = \gamma + \mathbf{x}^{\mathrm{T}}(m+1)\mathbf{P}(m)\mathbf{x}(m+1)$$

$$\mathbf{x}^{\mathrm{T}} = [x_1(m+1) \quad x_2(m+1) \quad \dots \quad x_n(m+1)]$$

$$\mathbf{G} = \mathbf{P}(m+1)\mathbf{x}(m+1) = \mathbf{P}(m)\mathbf{x}(m+1)/\alpha$$

and the argument m is used to emphasize the fact that the quantities are obtained at each sample point. γ is referred to as the forgetting factor and prevents the matrix $\mathbf{P}(m+1)$ from tending to zero (and $\hat{\boldsymbol{\theta}}(m+1)$ to a constant) with increased m, thus allowing the tracking of a slowly varying parameter. Typically, γ is between 0.98 and 1. Smaller values assign too much weight to the more recent data which leads to wildly fluctuating estimates.

There are, however, two main problems that may be encountered when the RLS algorithm is implemented directly. The first, referred to as 'blow-up', results if the signal is not 'persistently exciting' as, for example, when there is no ocular movement, leading to an exponential increase in the elements of $\mathbf{P}$ in Equation 14.9b. Thus

$$\lim_{m \to \infty}[P_{ij}(m+1)] = \lim_{m \to \infty}\left[\frac{P_{ij}(m)}{\gamma^m}\right] \to \infty \tag{14.10}$$

However, because of miniature ocular movements (which are always present), and other activities usually picked up in the EOG channels, this problem is not so serious in OA removal.

The second problem with the RLS is its sensitivity to computer roundoff errors, which results in a negative definite $\mathbf{P}$ matrix and eventually instability. For successful estimation, it is necessary that the matrix $\mathbf{P}$ be positive semi-definite which is equivalent to requiring in the offline case that the matrix $(\mathbf{X}_m^{\mathrm{T}}\mathbf{X}_m)$ be invertible, but, because of differencing of terms in Equation 14.9b, positive definiteness of $\mathbf{P}$ cannot be guaranteed (Peterka, 1975; Clarke, 1981; Bierman, 1976). This problem is worse in multiparameter models, especially if the variables (EOGs in this case) are linearly dependent (Peterka, 1975) and when the algorithm is implemented on a small system with finite wordlength (Clarke, 1981).

The problem of numerical instability may be solved by suitably factorizing the matrix **P** such that the differencing of terms in Equation 14.9b is avoided. Such factorization algorithms are numerically better conditioned and have accuracies that are comparable with the RLS algorithms that use double precision (Bierman, 1976, 1977). Two such algorithms are the square root and the UD factorization algorithms. However, in terms of storage and computation the UD algorithm is more efficient, and is thus preferred. In fact, the UD algorithm is a square-root-free arrangement of the conventional square-root algorithm and thus shares the same properties as the latter.

In the UD method, **P**$(m + 1)$ is factored as

$$\mathbf{P}(m + 1) = \mathbf{U}(m + 1)\mathbf{D}(m + 1)\mathbf{U}^{\mathrm{T}}(m + 1) \tag{14.11}$$

where **U**$(m + 1)$ is a unit upper triangular matrix, **U**$^{\mathrm{T}}(m + 1)$ is its transpose and **D**$(m + 1)$ is a diagonal matrix. Thus, instead of updating **P**, its factors **U** and **D** are updated.

Using Equation 14.11, Equation 14.9b may be written as

$$\mathbf{P}(m + 1) = \frac{1}{\gamma}\mathbf{U}(m)\left[\mathbf{D}(m) - \frac{1}{\alpha}\boldsymbol{v}\boldsymbol{v}^{\mathrm{T}}\right]\mathbf{U}^{\mathrm{T}}(m) \tag{14.12}$$

where

$$\boldsymbol{v} = \mathbf{D}(m)\mathbf{U}^{\mathrm{T}}(m)\mathbf{x}(m + 1)$$

If the term in the square brackets is further factored into an upper triangular and diagonal matrices, such that

$$\bar{\mathbf{U}}(m)\bar{\mathbf{D}}(m)\bar{\mathbf{U}}^{\mathrm{T}}(m) = \mathbf{D}(m) - \frac{1}{\alpha}\boldsymbol{v}\boldsymbol{v}^{\mathrm{T}} \tag{14.13}$$

where the bar is used to distinguish the **U** and **D** factors of **D**$(m) - (1/\alpha)\boldsymbol{v}\boldsymbol{v}^{\mathrm{T}}$ from those of **P**, then

$$\mathbf{P}(m + 1) = \frac{1}{\gamma}\mathbf{U}(m)\bar{\mathbf{U}}(m)\bar{\mathbf{D}}(m)\bar{\mathbf{U}}^{\mathrm{T}}(m)\mathbf{U}^{\mathrm{T}}(m) \tag{14.14}$$

Comparing Equations 14.11 and 14.14, and noting that the product of upper triangular matrices is itself upper triangular and the symmetry in Equation 14.14, then

$$\mathbf{U}(m + 1) = \mathbf{U}(m)\bar{\mathbf{U}}(m) \tag{14.15a}$$

$$\mathbf{D}(m + 1) = \frac{1}{\gamma}\bar{\mathbf{D}}(m) \tag{14.15b}$$

Thus, the problem of updating $U(m + 1)$ and $D(m + 1)$ depends on finding appropriate recursive formulas for $U(m)$ and $D(m)$. Bierman (1976) has given an algorithm for updating $U(m + 1)$ and $D(m + 1)$ recursively for the Kalman filter, which uses the variance of the error term $e(i)$ but not γ. This algorithm has been trivially modified for the OA problem to incorporate γ instead, as has the presentation given above. The modified algorithm is given in Appendix 14A.

The gain vector $\mathbf{G}$ obtained at step 10 of Appendix 14A is used to update the parameter estimates, as indicated in Equation 14.9a. Thus, although $\mathbf{P}(m + 1)$ can be obtained from the updated UD elements as in Equation 14.11, it is unnecessary to compute $\mathbf{P}(m + 1)$ explicitly.

Some properties of the UD algorithm

To gain some insight into the UD algorithm, it is useful to write out the algorithm explicitly. Thus, for a two-parameter model, the algorithm of the appendix becomes

- ▪ step 1: $v_1 = x_1(m + 1)$; $v_2 = x_2(m + 1) + U_{12}(m)x_1(m + 1)$
- ▪ step 2: $b_1 = d_1(m)x_1(m + 1)$; $b_2 = d_2(m)v_2$
- ▪ step 3: $\alpha_1 = \gamma + b_1 v_1 = \gamma + d_1(m)x_1^2(m + 1)$
- ▪ step 4: $d_1(m + 1) = d_1(m)/\alpha_1$
- ▪ step 5: $\alpha_2 = \alpha_1 + b_2 v_2 = \alpha_1 + d_2(m)v_2^2$
- ▪ step 6: $U_{12}(m + 1) = U_{12}(m + 1) - b_1 v_2/\alpha_1$
- ▪ step 7: $b_1 = b_1 + b_2 U_{12}(m) = d_1(m)x_1(m + 1) + d_2(m)v_2 U_{12}(m)$
- ▪ step 8: $d_2(m + 1) = d_2\alpha_1/\gamma\alpha_2$

It is seen from step 3 that, provided that the starting values for the diagonal elements (that is $d_1(0)$ and $d_2(0)$) are positive, α_1 and hence $d_1(m + 1)$ will always be positive. The same is true of α_2 (which is the same as α in Equation 14.9) and $d_2(m + 1)$.

(1) *Positive definiteness of* $\mathbf{P}$ A matrix $\mathbf{P}$ is positive definite if and only if $\mathbf{x}^T\mathbf{P}\mathbf{x} > 0$, except when $x_1 = x_2 = \ldots = x_n = 0$ (Bajpai *et al.*, 1973).
From Equations 14.9 and 14.11, $\alpha = \gamma + \mathbf{x}^T\mathbf{P}\mathbf{x} = \gamma + \mathbf{x}^T\mathbf{U}\mathbf{D}\mathbf{U}^T\mathbf{x}$, which for a two-parameter model becomes (see steps 3 and 5 above)

$$\alpha_2 = \alpha_1 + d_2 v_2^2 = \gamma + d_1(m)x_1^2(m + 1) + d_2(m)v_2^2$$

so that, in this case,

$$\mathbf{x}^T\mathbf{P}\mathbf{x} = d_1(m)x_1^2(m + 1) + d_2(m)v_2^2$$

Thus, it is seen that the sign of $\mathbf{x}^T\mathbf{P}\mathbf{x}$ depends on the signs of the diagonal elements (that is d_1 and d_2) which are always positive, as stated earlier, so that the matrix $\mathbf{P}$ satisfies the positive definiteness property. Thus, the UD factorization algorithm guarantees the positive definiteness property of $\mathbf{P}$.

(2) *Blow-up problem*   If there is no data, that is $x_1 = x_2 = \ldots = x_n = 0$, then $\alpha_1 = \alpha_2 = \gamma$ (steps 3 and 5), so that both d_1 and d_2 will be continuously scaled by γ, which is less than unity, leading to an exponential increase in the diagonal elements. Thus it appears that the problem of blow-up is not eliminated by the matrix factorization as sometimes suggested. Other RLS schemes may be used to reduce the effects of blow-up (Goodwin and Sin, 1984), but, in the OA work, miniature ocular movements and other inherent systems noise ensure that the values of x never become zero indefinitely.

Online OA removal using the UD and square-root algorithms has been simulated on a mainframe computer and these were found to give similar results to their offline equivalents.

14.2.2.3   Hardware for the online ocular artefact removal system

In this section and the following, the online ocular artefact removal (OAR) system which uses the UD algorithm is described. First, target specifications for the system are set out. Next, a suitable system is described at system and block diagram level.
The OAR system was designed with the following requirements in mind:

(1)   compatibility with standard EEG machines;

(2)   ability to provide continuous real-time OA removal in multichannel EEG signals (the OA removal should now be based on subjective criteria and should be adaptive);

(3)   ability to output the corrected EEGs and/or the uncorrected EEGs and EOGs to the EEG machine, to allow instant comparison of the corrected and uncorrected EEGs;

(4)   ability to avoid saturation, which would reduce the corrector's effectiveness, and the system should have some autoranging facility;

(5)   the instrument should be suitable for use by unskilled persons.

These requirements can be met by a suitable microprocessor-based system that implements the UD algorithm. The use of a microprocessor-based instrument also offers the following advantages.

(1)   Software-controlled design yields a very flexible system. Several OAR algorithms and models can be implemented on one system, and the models used in any application specified by the user.

(2)   New models or ideas can be investigated by mere software modifications, without having to build a new instrument. Thus a software-controlled OAR system could be an excellent research aid.

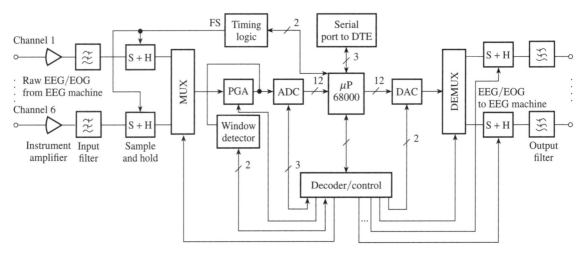

Figure 14.13   Block diagram of the microprocessor-based ocular artefact removal system.

(3)   A programmed instrument allows the provision of housekeeping routines for self-checking, automatic calibration, reduction in overload problems, and so on. Data processing on the removal system may include digital filtering of the EOGs to reduce the effects of secondary artefacts (Hamer *et al.*, 1985).

At the present, the OAR system can only process six channels of EEG and EOG signals, but is expandable to 20 channels (see later). The block diagram of the system is given in Figure 14.13. Each EEG/EOG signal from the auxiliary output of the EEG machine is first amplified and then bandlimited to 30 Hz by a lowpass filter which feeds a sample-and-hold circuit. The EEG/EOG signals are then simultaneously sampled at the positive transition of the sampling signal FS. Simultaneous sampling is employed to avoid the introduction of delays between corresponding time points. The negative transition of the signal FS then interrupts the processor, and signals the beginning of a cycle during which the 20 samples are sequentially selected by the multiplexer (MUX) and digitized by the analog converter (ADC) under the control of the microprocessor (μP).

The programmable gain amplifier (PGA) and the window detector are used to extend the dynamic range of the ADC to avoid overloading. ADC overload is a problem in the OA work and may lead to false parameter estimates. To avoid this, it is common practice to utilize only a fraction of the dynamic range of the ADC so that overloading is infrequent, and to discard all data in the region where overload occurred, but this can lead to wastage of data. The use of the PGA and window detector allows the gain to be changed dynamically. Thus, when the output of the PGA exceeds a predefined window limit, the gain of the PGA is set to a lower value which automatically halves the sample value and brings it to within the dynamic range of the ADC and thus avoids saturation. Account is taken of this before the digitized sample is saved in the memory.

The digitized samples are then processed by the OAR algorithm to obtain the corrected EEGs. The corrected EEG samples, together with the raw EEGs/EOGs if desired, are output to the auxiliary inputs of the EEG machine via the digital/analog converter (DAC) and the associated network.

The multiplexer and the demultipliers are required to allow resource sharing by the input and output channels. An alternative approach would be to provide separate ADC and DAC for each channel. This approach reduces the system noise due to cross-talk between channels to a minimum, but was considered rather too expensive.

Each channel has a separate output sample-and-hold with a separate sampling signal line. The sample-and-hold is used to hold the analog samples until the next analog sample is obtained. This stretches the sample pulses and increases the signal power, but introduces aperture distortion, which is considered small in this case.

14.2.2.4 Software for the on-line ocular artefact removal system

The OAR system software consists of data acquisition and distribution routines, an on-line OA removal routine, software floating point arithmetic routines, and a supervising main program. The whole software occupies 3 kbytes of memory. The heart of the OAR system software, which is written in the 68000 microprocessor assembly language, is the UD algorithm described in Section 14.2.2.2. An overview of the software will be given here.

The OAR system is interrupt driven. The interrupt signal is derived from the programmable timer module (PTM) on board the system controller, and has a frequency of 128 Hz (a frequency of 95 Hz was sometimes used to allow more time for computation, and the change in frequency was easily carried out in software). On interrupt, the OAR system software is used to acquire the EOG/EEG data, remove OAs from the EEG samples using the UD algorithm and output the corrected EEGs, and/or the raw data, to the EEG machine so that a paper chart record can be produced. The system operation is summarized in the flowchart of Figure 14.14.

During the initialization phase, the user is invited by means of a visual display unit (VDU) to specify the various system constants, namely the number of EEG channels to be corrected for artefacts, the number of model parameters and hence the model that should be used in the removal algorithm, and the number of corrected EEG and/or raw EEG signals to be output to the EEG machine. Some EOG signals and parameter estimates may also be output to the EEG machine. These constants are checked and, if valid, are used to initialize the system. A default value is used for any constant that is invalid. After initialization, the program loops around endlessly until valid data is available (Figure 14.14(a)). The procedure described here applies to the prototype OAR system only. In future OAR systems, the user will not require a VDU, any selections would be made by push buttons, and the 'background' program would be replaced by a more useful housekeeping routine (see Section 14.2.2.3). A flag (data flag) is set in the interrupt service routine (see Figure 14.14(b)) each time the interrupt occurs, and this is the indication to the main program that valid data is now available. After the data has been acquired, the elements of the UD algorithm are updated, the OAs are removed from the EEGs, and the corrected EEGs and/or raw data are output

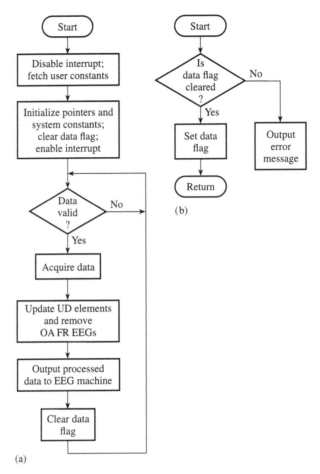

Figure 14.14 Ocular artefact removal system software: (a) main program; (b) interrupt service routine.

to the EEG machine. Finally, the data flag is cleared to indicate that the current data samples have been successfully processed.

An error message is output to the VDU and the program halted if an interrupt occurs before the previous one has been serviced. This will normally occur if more parameters and/or EEG channels are specified than the software can process within the sampling interval of 8 ms, and prevents the accumulation of unserviceable interrupts and the eventual system failure.

Arithmetic operations in the OAR system are performed using the floating point (FP) format to take advantage of the increase in the dynamic range of the numbers that it affords, and to avoid the scaling problem associated with the fixed-point approach.

As speed is vital in this application, hardware floating point was considered the best approach, but it was found that hardware floating-point devices available at the time were both expensive and too slow. Therefore, the number of EEG channels to correct was scaled down so that software FP could be used until fast FP devices became available.

14.2.2.5   System testing and experimental results

A preclinical test on the OAR system was carried out at Freedom Fields Hospital, Plymouth. In the first phase, extensive tests of the reliability of the OAR system were made using six normal subjects. This phase was also a 'learning' phase, in which the behaviour of the system was understood and minor faults were identified. In the second phase, one uncooperative subject, whose EEGs contained spike and wave discharges, and two subjects, one of whom was uncooperative, whose EEGs contained slow waves, were used to assess how the OA removal process affected the spikes and slow waves. In all the tests carried out, the subjects were asked to perform ocular movement exercises which included repetitive and random blinking, vertical eye movements (VEMs) and horizontal eye movements (HEMs).

Nine EEG signals were derived from several electrodes placed as shown in Figure 14.15. These were FP2–F4, F4–C4, C4–P4, FP1–F7, F3–C3, C3–P3, Fz–Cz, Cz referred to the right ear lobe or A2, and Cz–Pz. The EOG signals were derived from electrodes placed near the eyes, as shown in Figure 14.15(b).

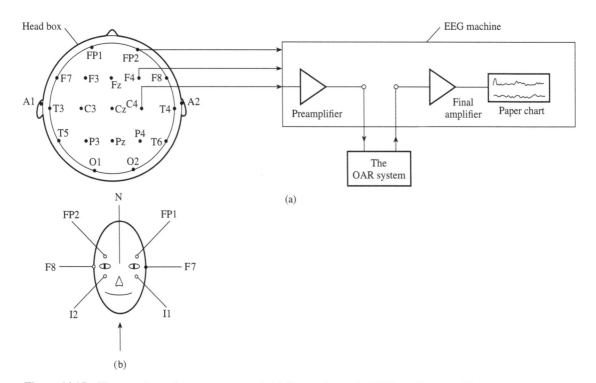

Figure 14.15 The experimental arrangement used. (a) Connection to the EEG machine; (b) EOG placements.

The EEG and EOG signals were fed into an eight-channel EEG machine via the head box, and after amplification they were fed into the OAR system via a 37-way D-type connector. After removing the OA from the EEG signals, both the corrected and raw EEGs (with the means removed) and/or the EOGs were fed into the final amplifiers of the EEG machine, and thence to the paper chart for examination (see Figure 14.15(a)).

Several models were used in the tests, but only the three that gave the best results will be described here. Two of these utilized the EOGs derived from the electrode placement of Figure 14.15(b) and were also found to give the best OA removal in a previous study. These two models are as follows:

$$3D \quad y(i) = \theta_1 VR(i) + \theta_2 HR(i) + \theta_3 HL(i) + e(i) \tag{14.16}$$

$$4D \quad y(i) = \theta_1 VR(i) + \theta_2 HR(i) + \theta_3 HL(i) + \theta_4 HL(i)HR(i) + e(i)$$

The third model, which will be called model 2H in line with earlier nomenclature, used the EOGs derived from the electrode placement pairs FP1–F7 and FP2–F8:

$$2H \quad y(i) = \theta_1 EOGR(i) + \theta_2 EOGL(i) + e(i) \tag{14.17}$$

It should also be mentioned that the selection switch on the EEG machine can be used to 'force' the OAR system to implement a variety of models by selecting the appropriate pairs of EOG electrodes to feed channels 1 to 4 (which are reserved for the EOGs).

The OAR system was used to remove OAs from a number of other EEG electrodes (see Figure 14.15(a)) using model 2H (and occasionally models 3D and 4D). It was found that in all the cases studied the OA was satisfactorily removed, and this included all the frontal EEG channels where the OA was largest. Figure 14.16 gives results for four different electrodes (Fz–Cz, Cz–A2, F4–C4 and F3–C3) for blink experiments. In both Figures 14.16(a) and 14.16(b) two different EEG signals were simultaneously corrected for OA. Comparison of the corrected and uncorrected EEGs in both sets of figures showed that the system had satisfactorily removed the OAs (compare traces (v) and (vi) with traces (iii) and (iv) in each set of figures).

There was little OA contamination at the more posteriorly placed EEG electrodes, for example Cz–Pz and C4–P4. In these cases all the models performed equally well.

Figure 14.17 shows an EEG record containing epileptic spike and wave discharges as well as OAs obtained from an uncooperative subject. (In this and all other cases where the EEG contained abnormal waves, the raw EEG was fed direct to the final amplifier of the EEG machine as well as to the OAR system as described earlier, to allow an unbiased analysis of the records.) Comparison of the corrected EEG and the raw EEG (Figures 14.17(d) and 14.17(e)) showed that the OAs have been removed, but not the spike and waves.

Good results were also obtained from an uncooperative mental patient whose EEGs contained only low amplitude slow waves. It is noteworthy that previous online methods would have been unsuitable for the cooperative subjects.

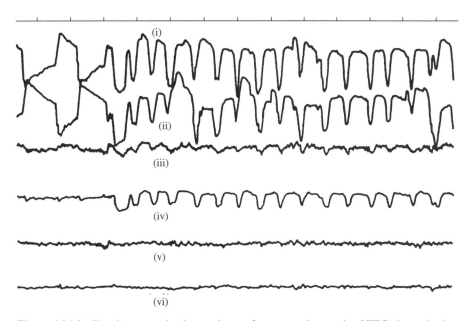

Figure 14.16   Simultaneous adaptive ocular artefact removal at a pair of EEG electrodes in a blink experiment. (i) and (ii) measured EOG signal for the right and left eyes; (iii) and (iv) measured EEGs at Cz–A2 and Fz–Cz electrodes, (v) and (vi) corresponding EEGs with artefacts removed.

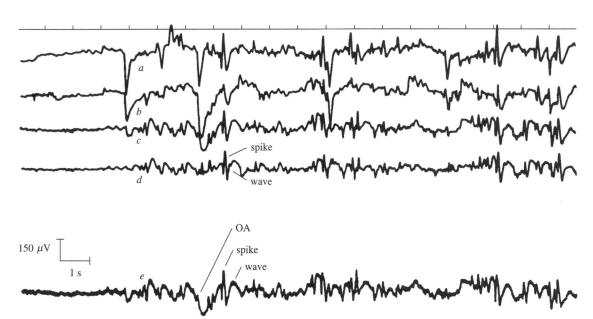

Figure 14.17   Ocular artefact removal in the presence of epileptic spike and wave discharges. (a) and (b) measured EOGs for the right and left eyes; (c) measured EEG; (d) EEG with artefacts removed; (e) raw EEG.

14.2.2.6 Discussion

The tests, using different types of ocular movements (blinks, VEM, HEM), showed that it is possible to achieve satisfactory removal of OAs due to these causes, at all EEG sites, using the numerically stable UD algorithm. It was found that, in the more posterior EEG sites, very little OA contamination of the EEG was observed and in these cases all the models performed well. It was also found that, although satisfactory OA removal was obtained during vertical eye movement, OA due to rider artefact was not completely removed. Similar results were obtained in a previous study, and this result corroborates those. The inability of the OAR system to remove completely the OA due to rider artefact is probably because the models used only take into account simultaneous changes in the EOG and EEG. A dynamic model of the form (for a single input or EOG)

$$y(m) = \sum_{n=0}^{N-1} h(n)x(m - n) + e(n)$$

may be used where it is desired to obtain improved results.

The results for the patient subjects showed that, when pathological slow waves and spikes occurred in the absence of OA, they were in general not significantly affected by the OA removal process. However, when they occurred simultaneously with the OA, they might be reduced in amplitude but not removed completely. The reduction in amplitude occurred mainly at the frontal EEG channels. Thus it may be necessary to distinguish between the OA and slow waves. This is an area that deserves further investigation.

A limitation of the OAR system at the present is that, even with the simplest model (model 2H), it can only remove the OA from a maximum of four EEG signals simultaneously, owing to the slow speed of the floating-point arithmetic routines, which take typically 70 μs to perform an arithmetic operation. A solution to this problem is to use fast hardware floating-point arithmetic devices, capable of performing an arithmetic operation in 1 or 2 μs.

14.2.2.7 Conclusions

Preliminary results obtained with both normal and patient subjects showed that the OAR system gives satisfactory OA removal for blinks, vertical and horizontal eye movements and a bipolar EEG electrode montage. The use of the UD factorization algorithm and a software-controlled system enabled us to overcome the disadvantages of the previous online OA removal methods. Thus the OAR system is able to deal with multiple artefacts, does not need the cooperation of the subjects in a preliminary calibration and bases the removal criterion on a purely objective method. This system, which is the first of its kind, is compatible with standard EEG machines, so that it could be manufactured and sold as an accessory. However, the usefulness of the instrument can only be fully assessed after extensive clinical tests.

Although the OAR system was designed specifically for removing the OA from the EEG, it could be used as a general-purpose artefact (or noise) removal system in most physiological situations where both the contaminating and the contaminated signals can be separately measured. An example is the problem of measuring the fetal electrocardiogram (ECG) in the presence of large contaminating maternal ECG (Widrow *et al.*, 1975). Another example is the case where it is necessary to remove both the OA and the ECG artefacts from the EEG (Fortgens and De Bruin, 1983). In both applications the OAR system could be used to remove the artefacts, after possible minor modifications to the software and hardware. The OAR system, suitably programmed, could also be used in other signal processing applications, for example digital filtering.

Since the design of the OAR system about 10 years ago considerable changes have taken place in DSP, especially in the area of DSP hardware. It is likely that, if the system was implemented now, a good DSP chip would be used. Today, a floating-point digital signal processor such as the TMS320C30 or TMS320C40 would be appropriate for a time-critical system such as this. Despite the changes in DSP, the design principles and issues remain pertinent. We have emphasized these and hope the reader will benefit from our experience.

14.2.3   Equalization of digital audio signals†

Equalization of audio signals is an important functional requirement of mixing consoles used in many professional and semi-professional audio applications, for example in studio recording, sound reinforcement in public address systems, and broadcasting. An audio equalizer is basically a set of filters with adjustable frequency responses which is used to shape the spectrum of the audio signals in a desired manner. In a traditional mixing console, equalization of audio signals is achieved by analog filtering, but the trend is towards an all-digital mixing console because this offers improved sound quality and potential reduction in production costs in the future. In an all-digital mixer, the analog filters would be replaced by equivalent real-time digital filters. We describe here a real-time semi-parametric equalizer for audio signals implemented using a high speed floating-point digital signal processor, the TMS320C30.

A standard parametric equalizer allows the user to sweep to a specific frequency in the audio band and to adjust the level of the audio signal at a single frequency or a range of frequencies. Three basic filter types are used.

■ *Bell filter*   This allows the user to boost or attenuate a particular frequency in the audio band. The bell filter is basically a bandpass filter with adjustable gain, Q factor, and centre frequency. The centre frequency may vary in the range 20 Hz–16 kHz, the Q value between 0.5 and 3, and the gain in the interval ±15 dB. Figure 14.18(a) illustrates the bell characteristic.

† This section is based on the final year project of one of our former students, Mr Robin Clark.

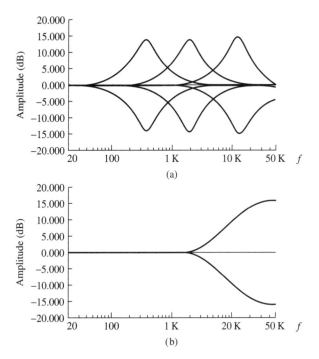

Figure 14.18 (a) Amplitude response of the bell filter for various centre frequencies and bandwidths. (b) Amplitude for high frequency shelving.

■ *Shelf filter*   This allows adjustment of the gain and cutoff frequency of the equalizer over a range of frequencies at either the low or the high frequency end of the audio band. Low frequency shelf filters are used to boost or attenuate a band of low frequencies, for example between 20 and 500 Hz, whereas high frequency shelf filters are used to boost or attenuate a band of high frequencies, for example between 1.6 kHz and 16 kHz. The familiar treble or bass controls in the home hi-fi systems are essentially shelf filters with fixed responses. Figure 14.18(b) illustrates typical responses for a shelf filter.

■ *Pass filters*   These are essentially lowpass and highpass filters with fixed cutoff frequencies and are used to remove low and/or high frequency noise from the audio signal.

Full equalization is achieved by the combined effects of the basic filters. In analog equalizers, the characteristics of the filters (gain, centre frequencies, Q factor, and so on) are adjusted by the user via interactive controls (variable resistors). In a digital implementation, the adjustment is achieved by altering the digital filter coefficients in real time in response to changes in equalizer parameters.

Analysis of a typical analog parametric equalizer showed that each filter type described above can be viewed as a Butterworth filter, with s-plane transfer functions of the following forms: for the bell filter (Clark *et al.*, 2000),

$$H(s) = \frac{s^2 + A\omega_n s + \omega_n^2}{s^2 + B\omega_n s + \omega_n^2} \qquad (14.18a)$$

for the low frequency shelf filter,

$$H(s) = \frac{s^2 + 2A\omega_n s + A^2\omega_n^2}{s^2 + 2B\omega_n s + B^2\omega_n^2} \qquad (14.18b)$$

and for the high frequency shelf filter,

$$H(s) = \frac{A^2 s^2 + 2A\omega_n s + \omega_n^2}{B^2 s^2 + 2B\omega_n s + \omega_n^2} \qquad (14.18c)$$

where

$$A = C/Q$$

$$B = 1/Q$$

To achieve a performance similar to the traditional analog equalizer, the analog filters above are replaced by equivalent digital filters, by transforming each of the s-plane transfer functions given above using the bilinear z-transform technique (see Chapter 8). Each of the resulting z-transfer functions has the form

$$H(z) = \frac{az^2 + bz + c}{dz^2 + ez + f} \qquad (14.19)$$

where

$$a = P^2 + AP + \omega_n^2, \, b = 2\omega_n^2 - 2P^2, \, c = P^2 - AP + \omega_n^2$$

$$d = P^2 + BP + \omega_n^2, \, e = 2\omega_n^2 - 2P^2, \, f = P^2 - BP + \omega_n^2$$

$$P = \frac{\omega_n}{\omega_p}, \quad \omega_p = \frac{2}{T}\tan\left(\frac{\omega_n T}{2}\right)$$

Simulation studies showed that floating-point arithmetic should be used to cater for the large variations in the range of values of the filter coefficients as the equalizer parameters are adjusted over the audio range. Floating-point arithmetic is also attractive in this application to make it easier to recompute the filter coefficients, via the BZT, in 'real time' to permit online adjustment of the equalizer characteristics.

The equalizer was implemented in the PC-based TMS320C30 evaluation module (EVM), which contains a TMS320C30 processor, a 14-bit ADC/DAC module and a software development package. A simplified block diagram of the TMS320C30-based parametric equalizer is depicted in Figure 14.19. The analog audio signal (for

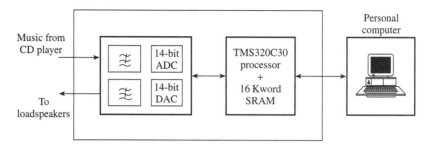

Figure 14.19 A simplified block diagram of the TMS320C30 EVM.

example, from a compact disc player) is digitized to 14 bits at a rate of about 18.9 kHz and passed to the C30 where it is digitally filtered by a bell, shelf and/or pass filter. The keyboard is used to adjust the parameters of the equalizer (gain, frequency, type of filter used). The VDU (visual display unit) displays dynamically the frequency response of the equalizer.

Several processes take place within the EVM and the PC. The C30 processor executes the filtering required to achieve equalization, and recomputes the filter coefficients for the equalizer when the user adjusts the equalizer parameters. The coefficient recalculation occurs in the background whereas the filtering is interrupt driven. The programs are written in ANSI C language, and the arithmetic operations are carried out using floating-point (24-bit mantissa and 8-bit exponent). Fragments of the C codes for the filter and coefficient calculation are available in the companion handbook (see the Preface for details).

The performance of the equalizer was evaluated for various parameter settings using music from a CD player and was found to be quite effective. A restriction at present is that the EVM allows only a maximum of about 18.9 kHz sampling rate, making it impossible at this stage to use the equalizer at the professional audio rate of 44.1 kHz.

14.3 Design studies

In most DSP courses, it is often beneficial to set one or more major, practical tasks that are challenging which may be carried out in groups or individually. Such tasks are intended to provide the student with the opportunity to gain deeper knowledge of DSP in a directed way. They draw on several aspects of DSP concepts and may be set near the end of the course or earlier (in the latter case, the concepts will be taught/learnt just as they are needed).

In this section, we present a number of such major problems that we have found useful in some of our courses. The learning objectives and the sub-tasks that are expected are provided.

(1) Roundoff noise reduction schemes for fixed-point IIR digital filters

Figure 14.20 shows the structure of a second-order, IIR digital filter with an error-feedback scheme for reducing product roundoff errors. The filter is to be implemented using 2's complement arithmetic, in a DSP processor with a $B/2B$-bit architecture. The coefficients and other variables are stored as B-bit words and the products are quantized after summation. The error-feedback coefficients can each take on only one of the following values: 0, ±0.25, ±0.5, ±0.75, ±1, ±1.25, ±1.5, ±1.75, ±2.

(a) Derive general expressions for:

   (i)   the transfer function between the roundoff noise source, $e(n)$, and the filter output, $y(n)$;

   (ii)   the output noise power due to roundoff errors;

   (iii) the signal-to-noise ratio (SNR) at the output of the filter. Assume that the input signal, $x(n)$, is a sine wave with a known frequency and quantized to B bits;

   (iv)   the positions of the zeros of the feedback network.

(b) Develop the following:

   (i)   an algorithm for computing the output noise power due to roundoff error, SNR at the output of the filter and the positions of the zeros of the feedback network, given the values of the unquantized filter coefficients, feedback coefficients (from the permissible set), and the system wordlength, B;

   (ii)   a MATLAB program that implements the algorithm in (i).

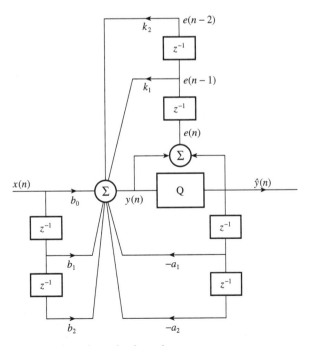

Figure 14.20   Second-order noise reduction scheme.

(c) Use your MATLAB program, together with other utilities in MATLAB, to investigate the effects of the error-feedback scheme on the level of product roundoff noise for each of the following filters (assume $B = 16$ bits and then repeat for $B = 24$ bits) (Dattorro, 1983):

(i) A 'cut' filter for audio signal processing characterized by

Filter coefficients	$b_0 = 0.996\ 450\ 761\ 790$, $b_1 = -1.992\ 821\ 454\ 486$, $b_2 = 0.996\ 443\ 656\ 208$
	$a_1 = -1.992\ 821\ 454\ 486$, $a_2 = 0.992\ 894\ 417\ 998$
Sampling frequency	48 kHz

(ii) A 'boost' filter for audio signal processing characterized by

Filter coefficients	$b_0 = 0.996\ 450\ 761\ 790$, $b_1 = -1.992\ 821\ 454\ 486$, $b_2 = 0.996\ 443\ 656\ 208$
	$a_1 = 1.992\ 821\ 454\ 486$, $a_2 = -0.992\ 894\ 417\ 998$
Sampling frequency	48 kHz
Centre frequency	15 kHz

(iii) A highpass filter characterized by

Filter coefficients	$b_0 = 0.292\ 893$, $b_1 = -0.585\ 786$, $b_2 = 0.292\ 893$
	$a_1 = 0$, $a_2 = 0.171\ 572\ 8$
Sampling frequency	8 kHz
Stopband edge frequency	500 Hz
Passband edge frequency	2 kHz

(d) The investigation above should include:

(i) for each filter and wordlength, B, the computation and tabulation of the output noise power, the SNR and the locations of the zeros of the feedback network, for each of the feedback coefficient pair in Table 14.2;

(ii) the determination of the error-feedback coefficient pair that gives the lowest output noise power (from all the possible pairs and permissible feedback coefficient values specified in section (a));

Table 14.2 Values of error feedback coefficients.

Case	k1′	k2′
1	0	−1
2	0	1
3	−1	0
4	−1	−1
5	−1	1
6	1	0
7	1	−1
8	−2	−1
9	2	−1
10	0	0

(iii) plots of the frequency response that correspond to the noise transfer function for each of the feedback coefficient pair in Table 14.2 and for the best coefficient pair found in (ii);

(iv) Write a short report (3–4 pages maximum, excluding diagrams, program listings) about the investigation and your findings. Your report should include the following:

- an introduction to the problems of product roundoff noise in DSP systems;
- details of your answer to sections (a)–(d) above;
- a critical discussion of your results and recommendation of how to choose error-feedback coefficients in practice;
- a critical discussion of the limitations of error-feedback schemes as a means of reducing roundoff noise;
- a discussion of other benefits of error-feedback schemes, besides roundoff noise reduction;
- a summary of your own assessment of specific things you believe you have learnt from this assignment;
- a listing of the MATLAB program.

(2) Multirate filter design study

Background

This design study is based on one of the MATLAB-based filtering topics developed by one of our ex-students, Dr Nick Outram (Outram *et al.*, 1995), for our DSP course. The learning objectives here include

- reinforcement of the concepts of multirate signal processing;
- demonstration of some of the limitations of FIR filters when the desired magnitude response is very sharp;
- provision of a hands-on experience of designing and applying multirate filters.

In this design study, we will use the fetal ECG (electrocardiogram) to facilitate the learning process. The shape of the fetal ECG is of clinical interest, but it is susceptible to distortion by a low frequency artefact known as baseline shift: see Figure 14.21. Baseline shifts hinder accurate measurement and analysis of features of the fetal ECG. The difficulty is that signal enhancement should not distort low frequency information of clinical interest. Unfortunately, much of the energy of baseline shift lies below 3 Hz.

In the study, two data sets are provided, good.ecg, containing a high quality fetal ECG recording with almost no baseline shift, and poor.ecg, containing a fetal ECG with noticeable baseline shifts. The fetal ECG data consists of 12-bit data values sampled at 500 Hz.

The problem

The primary problem is to develop a signal processing scheme to remove the baseline shift without distorting the ECG shape using linear phase FIR filters.

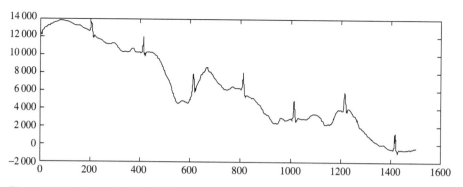

Figure 14.21 Acute baseline shifts.

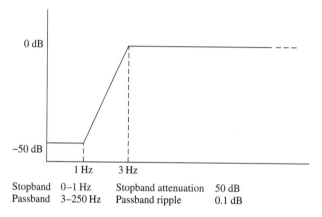

Stopband 0–1 Hz   Stopband attenuation 50 dB
Passband 3–250 Hz   Passband ripple   0.1 dB

Figure 14.22 Filter specification for removing baseline shifts.

Specific tasks are

(a) to write a MATLAB program to read the ECG data files into two separate arrays and display the first 4000 samples of each of the waveforms on the same axes;

(b) to estimate, using the MATLAB **remezord** function, the number of filter coefficients required to meet the specifications depicted in Figure 14.22. Comment on your result and the problem of designing narrowband FIR filters;

(c) to design and implement a multirate filter in MATLAB using decimation and interpolation techniques as appropriate, to meet the specifications in Figure 14.22;

(d) to test the filter using the data, check for any distortion of the ECG waveform and determine the delay through the multirate system;

(e) to write a short report about the design study which should include the design and test of your multirate filter.

(3) Digital filter design and implementation with MATLAB and DSP56002 processor

Background

The aims of this design problem are to enable students to gain hands-on experience of the design and implementation of FIR filters using MATLAB and a fixed-point DSP processor. The design problem is based on the laboratory exercise developed by our research students, Brahim Hamadicharef, Robin Clark and Eddie Riddington for one of our DSP courses.

Specific learning objectives are

(1) to use MATLAB

- to calculate the coefficients and plot the frequency response of FIR filters for a given set of specifications;
- to investigate the effects of coefficient quantization on frequency response;

(2) to implement FIR filters using a fixed-point DSP processor

- to gain familiarity with DSP development tools;
- to develop a simple assembly language program for a fixed-point FIR filter;
- to test and demonstrate real-time filtering.

The problem

A requirement exists for a linear phase FIR filter to remove a 1 kHz interference from an audio signal. The filter should meet the following specifications:

passband ripple	0.5 dB
stopband attenuation	25 dB
passband	900–1100 Hz
stopband edge frequencies	990 Hz and 1010 Hz
sampling frequency	8 kHz

You are required to design and implement the filter using MATLAB and the Motorola DSP56002 fixed-point DSP processor. Specific tasks are as follows.

(a) To calculate the coefficients of a suitable FIR filter using MATLAB and the optimal method. *Note.* You should use the MATLAB function **remezord** to estimate the length of the filter, N, calculate filter coefficients using the function **remez**, and plot the frequency response of the filter using the function **freqz**. The following equations may be useful to determine the ripple parameters:

$$R_p = -20 \log_{10} \frac{1 - \delta_p}{1 + \delta_p} \qquad \text{– passband ripple in dB}$$

$$R_s = -20 \log_{10} \frac{\delta_s}{1 + \delta_s} \qquad \text{– stopband attenuation in dB}$$

where

δ_p	– passband ripple
δ_s	– stopband ripple

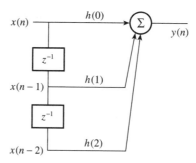

Figure 14.23   A simple three-point FIR filter.

Table 14.3   Initial memory map.

X memory		Y memory	
$h(2)$	← r0	$x(n-2)$	← r4
$h(1)$		$x(n-1)$	
$h(0)$		$x(n)$	

(b) To determine a suitable coefficient wordlength for the filter by carrying out an analysis of the effects of quantizing the coefficients to 4, 8, 16, and 24 bits on the frequency response of the filter. *Note.* To quantize the filter coefficients to B-bit, 2's complement, fixed-point numbers, we simply multiply each coefficient by 2^{B-1} and then round to the nearest integer. The difference between the actual coefficients and the fixed-point representation gives a measure of coefficient quantization error.

(c) To implement the filter on the DSP56002 evaluation module and verify that the filter is operating correctly. *Note.* As an introduction to FIR digital filtering using the DSP56002, consider the simple three-point FIR filter depicted in Figure 14.23. The FIR filter is characterized by the following equation:

$$y(n) = h(0)x(n) + h(1)x(n-1) + h(2)x(n-2)$$

Table 14.3 gives the initial memory map for the coefficient and data for the filter. Registers r0 and r4 are used as pointers to the coefficient and data memories, respectively. Circular addressing is used. Table 14.3 gives the DSP56002 code for implementing the simple filter. The comment column explains the operation of the program, Program 14.1.

Program 14.1   DSP56002 code for a simple 3-point FIR filter.

```
move   x: (r0)+, x0           ; copy h(2) into register x0 and increment pointer r0.
move   y: (r4)+, y0           ; copy x(n–1) into register y0 and increment pointer r4.
```

```
mpy    x0, y0, a x: (r0)+, x0 y: (r4)+, y0    ; multiply h(2) and x(n–2) and store in accumulator a.
                                               ; copy h(1) into register x0 and increment r0.
                                               ; copy x(n–1) into register y0 and increment r4.
mac    x0, y0, a x: (r0)+, x0 y: (r4)+, y0    ; multiply h(1) and x(n–1) and add to the accumulator a.
                                               ; Copy a0 into register x0 and increment r0.
                                               ; Copy x(n) into register y0 and increment r4.
mac    x0, y0, a y1, y: (r4)+                  ; multiply h(0) and x(n) and add to the accumulator a.
                                               ; Accumulator a contains the output sample y(n).
                                               ; x(n–2), x(n–1) and x(n) shall now be referred to as
                                               ; x(n–3), x(n–2) and x(n–1). Overwrite x(n–3) with the
                                               ; next input sample stored in y(1).
```

For an FIR filter with many coefficients, it is more efficient to use the DSP56002 repeat instruction together with the mac instruction. A fragment of DSP56000 code is given below:

```
; FIR filtering using the repeat instruction

clr     a           x(r0)+, x0    y(r4)+, y0
rep     #N–1
mac     x0, y0, a   x(r0)+, x0    y(r4)+, y0
macr    x0, y0, a   (r0)–
move    a, y:Output
```

(d) Write a short report giving details of your filter design and implementation, including listings of the MATLAB and DSP56002 programs.

14.4 Computer-based multiple choice DSP questions

In this section, we describe an approach we have found useful for rapid assessment of large groups of students and for identifying areas that require remedial action. An important feature of the approach is that it enables students to revise DSP materials by testing their understanding of specific issues.

Essentially, we have developed two banks of multiple choice questions covering basic and advanced/applied DSP topics. The banks of questions are updated and enlarged each year. The bank of questions on basic DSP topics is used to test students in the first semester and the bank of questions on advanced/applied DSP topics to test them in the second semester. In each case, about two weeks before the test students are issued with hard copies of all the possible questions in the test, but not the answers. No direct assistance is given with the questions. The software Questionmark is used to conduct and automatically mark the test. Each test is computer-timed to last for half-an-hour (students are automatically logged out) and consists of about 30 questions drawn at random from the data bank of questions.

All questions are either multiple choice or multiple response and the current marking scheme is as follows (though this can be easily changed):

Multiple choice	Correct answer	+1
	Incorrect answer	−1
	No answer	0

Multiple response	Correct answer	+1
	Incorrect answer	−1
	No answer	0

Examples of the multiple choice questions are given below. Further details of the multiple choice questions, including lists of the current data bank of questions, can be found in the companion handbook, *A Practical Guide for MATLAB and C Language Implementation of DSP Algorithms* (see the Preface for details).

(1)   Sample questions on basic DSP topics

Question 1

An analog signal to be spectrum analyzed is sampled at 125 Hz to produce 1024 data values. What is the frequency spacing between the spectral samples?

(a)   0.041 42 Hz

(b)   0.122 07 Hz

(c)   0.000 98 Hz

(d)   0.008 0 Hz

Question 2

A digital filter is said to be an IIR if

(a)   All its poles lie outside the unit circle

(b)   One or more denominator coefficients is nonzero

(c)   Current output depends on previous output

(d)   It oscillates

Question 3

A discrete time filter transfer function $H(z)$ is given by

$$H(z) = 1/(1 + 0.454\ 456z^{-1} + 0.269\ 259z^{-2})$$

Find the poles of the filter:

(a)   Poles are: $0.45 + 0.75j$ and $0.45 - 0.75j$

(b)   Poles are: $-0.2272 + 0.4664j$ and $-0.2272 - 0.4664j$

(c)   Poles are: $-0.9452 + 0.5j$ and $-0.9452 - 0.5j$

Question 4

A bandpass digital filter may be designed by transforming a suitable prototype lowpass analog filter into a bandpass filter. This transformation

(a) Halves the order of the analog lowpass filter
(b) Doubles the order of the analog lowpass filter
(c) Maps the lowpass filter cutoff F_c and $-F_c$ to the bandpass cutoffs F_{c2} and F_{c1}
(d) Maps the lowpass filter cutoff F_c and $-F_c$ to the bandpass centre frequency

Question 5

A discrete filter has poles located at $z = j0.75$ and $z = -j0.75$ and zeros located at $z = -1$ and $z = 1$. What is the amplitude response at a quarter of the sampling frequency?

(a) 1.52
(b) 4.57
(c) 7.15

Question 6

Quantizing the coefficient -0.1743 to 8 bits gives

(a) -22
(b) -23
(c) $+22$
(d) 23×10^{-7}

Question 7

Which of the following apply in DSP?

(a) The Nyquist frequency is the same as half the sampling frequency
(b) The Nyquist frequency is the same as the sampling frequency
(c) The Nyquist frequency is the same as the folding frequency
(d) The Nyquist frequency is half the oversampling rate

Question 8

A discrete-time filter transfer function is given by

$$H(z) = (1 - 1.6z^{-1} + z^{-2})/(1 - 1.5z^{-1} + 0.8z^{-2})$$

What is the amplitude response at dc?

(a) 0

(b) 1.33

(c) 1

(d) 1.6

Question 9

A discrete-time filter transfer function is given by

$$H(z) = (1 - 1.6z^{-1} + z^{-2})/(1 - 1.5z^{-1} + 0.8z^{-2})$$

What are the radii of the poles and zeros?

(a) 0.9, 1

(b) 0.81, 1

(c) 1, 0.81

(d) 1, 1

(e) 0.5, 0.5

Question 10

In relation to the bilinear z-transform (BZT) method of IIR filter design, which of the following apply?

(a) The warping effect of the BZT method gets worse as we get closer to the Nyquist frequency

(b) Pre-warping compensates for the distortion caused by the BZT over the entire frequency response

(c) Pre-warping compensates for the distortion caused by the BZT at only specific frequencies

(d) Frequency scaling eliminates the effects of warping in the frequency response

Question 11

A requirement exists for a highpass IIR digital filter with the following specifications:

stopband edge frequency	1 kHz
passband edge frequency	2 kHz
sampling frequency	16 kHz
passband ripple	3 dB
stopband attenuation	30 dB

What are the passband and stopband frequencies for a suitable prototype lowpass filter? Assume that the bilinear z-transform design method is used.

(a)   0.0414, 0.0198

(b)   1, 2

(c)   1, 2.0813

(d)   2, 1

(e)   0.0198, 0.0414

Question 12

The analog input signal to a DSP system is bandlimited by a fourth-order Butterworth filter with a cutoff frequency of 10 kHz and then sampled at a rate of 50 kHz. Assume a wideband input signal. What is the aliasing error level at 10 kHz?

(a)   0.156

(b)   0.026

(c)   0.04

(d)   0.707

Question 13

The analog input signal to a DSP system is bandlimited by a fourth-order Butterworth filter with a cutoff frequency of 10 kHz and then sampled. Assume a wideband input signal. Determine the minimum sampling frequency to give a signal-to-aliasing error level of 20 dB at 10 kHz.

(a)   10 kHz

(b)   20 kHz

(c)   29.39 kHz

(d)   19.39 kHz

(e)   50 kHz

Question 14

An analog audio signal has frequency components that extend from 0 to 24 kHz. The signal is sampled at the rate of 3.072 MHz. What is the oversampling ratio?

(a)   32

(b)   64

(c)   128

(d)   256

Question 15

An analog bandpass signal is to be sampled in accordance with the bandpass sampling theorem. Assuming the signal occupies the band 48 kHz–52 kHz, what is the minimum theoretical sampling rate to avoid aliasing?

(a) 104 kHz

(b) 96 kHz

(c) 8 kHz

(d) 16 kHz

Question 16

A data sequence consisting of 32 samples is stored in memory in natural order. If the data sequence is subsequently scrambled, using a standard bit reversal algorithm, what will be the bit reversed indices for the data samples $x(7)$ and $x(13)$? Assume a 32-point FFT processor is to be used.

(a) $x(11)$ and $x(14)$

(b) $x(22)$ and $x(28)$

(c) $x(7)$ and $x(13)$

(d) none of the above

Question 17

Given the following DFT coefficients for a 4-point DFT:

(a) $X(0) = 1.5$

(b) $X(1) = 1 - 0.5j$

(c) $X(2) = 0.5$

(d) $X(3) = 1 + 0.5j$

What is the equivalent discrete-time sequence?

(a) $x(0) = 1, \quad x(1) = 0, \quad x(2) = 1, x(3) = 0$

(b) $x(0) = 0.5, x(1) = 1, \quad x(2) = 0, x(3) = 0$

(c) $x(0) = 1, \quad x(1) = 0.5, x(2) = 0, x(3) = 0$

(d) $x(0) = 0.5, x(1) = 0.5, x(2) = 0, x(3) = 0.5$

(e) None of the above

Question 18

The first four outputs of an 8-point, radix-2 FFT are

$X(0) = 27$
$X(1) = -4 + 3j$
$X(2) = 4 + j$
$X(3) = 0 - 5j$

Which of the following statements are true?

(a)   The value of $X(7) = 0 + 5j$

(b)   The value of $X(7) = -4 - j$

(c)   The value of $X(7) = -4 - 3j$

(d)   The dc value of the output sequence is 27

(e)   None of the above

(2)   Sample questions on advanced/applied DSP topics

Question 1

Which of the following statements are true? In digital communication, marginally stable IIR filters are used for clock recovery because

(a)   their poles and zeros are close to the unit circle

(b)   they have impulse responses which decay slowly with time

(c)   this ensures that there is a clock signal even when there are no transitions in the input data stream for a relatively long period

(d)   they are not susceptible to drift with age and temperature

(e)   they are suitable for applications that involve burst transmission

Question 2

A digital communication system has a data rate of 56 kbaud and a sampling rate of 448 k samples per second. The clock recovery IIR filter for the system has a bandwidth of 25 Hz. Which of the following are true of the filter?

(a)   The pole positions are: $r = 0.9998$, ±45 degrees

(b)   The pole positions are: $r = 0.9998$, 45 degrees

(c)   The pole positions are: $r = 0.9998$, ±0.785 degrees

(d)   The coefficients associated with the poles are: $a_1 = -1.413\ 965$, $a_2 = 0.999\ 825$

(e)   The coefficients associated with the poles are: $a_1 = -1.413\ 965$, $a_2 = 0.999\ 649$

(f)   The clock recovery filter is a bandpass filter

(g)   The clock recovery filter is a lowpass filter

Question 3

A digital communication system has a data rate of 28 kbaud and a sampling rate of 224 k samples per second. The clock recovery IIR filter has a bandwidth of 100 Hz, and denominator coefficients $a_1 = -1.412\ 230$, $a_2 = 0.997\ 196$. Which of the following statements are true if the filter coefficients are quantized to 8-bit fixed-point numbers?

(a)   The fixed-point coefficients are: $a_1' = -90$, $a_2' = 63$

(b)   The fixed-point coefficients are: $a_1' = -181$, $a_2' = 127$

(c)   In fractional notation, the quantized coefficients are: $a'_1 = -1.406\ 25$, $a'_2 = 0.984\ 375$

(d)   In fractional notation, the quantized coefficients are: $a'_1 = -1.414\ 062$, $a'_2 = 0.992\ 187$

(e)   None of the above

Question 4

In relation to fixed-point IIR filters realized using second-order, canonic sections, identify the correct statements:

(a)   Overflow errors are caused by the result of addition exceeding the permissible wordlength

(b)   Overflow errors can cause a self-sustaining oscillation

(c)   In general, frequency domain scaling (Chebyshev norm) eliminates overflow errors completely

(d)   Error feedback or noise shaping techniques can be used to reduce overflow errors

(e)   Scaling for overflow improves the signal-to-noise ratio at the filter output

Question 5

An IIR digital notch filter, with a notch frequency a quarter of the sampling frequency, has the following coefficients: $b_0 = 1$, $b_1 = 0$, $b_2 = 1$; $a_1 = 0$, $a_2 = 0.81$. What should be the frequency domain scale factor at the input of the filter? Assume that it is realized using a second-order canonic section.

(a)   1.81

(b)   0.55

(c)   5.26

(d)   0.19

(e)   None of the above

Question 6

Which of the following statements are correct? In relation to DTMF decoding, only one real coefficient is required for each Goertzel filter because

(a)   only the magnitudes of the tone frequencies are required to decode DTMF signals

(b)   phase information is not required in DTMF decoding

(c)   the Goertzel filter processes each data sample as it arrives and does not wait for a set of N samples as in the FFT

(d)   the Goertzel filter, adapted for DTMF decoding, executes quickly and takes up little memory space

(e)   the Goertzel algorithm consists of a series of second-order IIR filters

Question 7

A DTMF tone detection scheme for a digital telephone uses a series of second-order Goertzel filters to extract the DTMF tones and their second harmonics. Assume that the DTMF tones for digit '0' are 941 Hz and 1336 Hz. Determine the values of the coefficients in the feedback path of the two Goertzel filters for the low frequency tone, 941 Hz, if the sequence lengths $N = 205$ and 210 are used for the fundamental and second harmonic, respectively, with corresponding discrete frequency bins of 24 and 47.

(a) For the low frequency tone, i.e. 941 Hz, the values of the feedback coefficients are
$a_1 = 0.999\ 71,\ a_2 = -1$

(b) For the low frequency tone, i.e. 941 Hz, the values of the feedback coefficients are
$a_1 = 1.482\ 867,\ a_2 = -1$

(c) For the low frequency tone, i.e. 941 Hz, the values of the feedback coefficients are
$a_1 = 1.345\ 621,\ a_2 = -1$

(d) For the second harmonic of the low frequency tone, i.e. 2×941 Hz, the values of the feedback coefficients are $a_1 = 0.463\ 812,\ a_2 = -1$

(e) For the second harmonic of the low frequency tone, i.e. 2×941 Hz, the values of the feedback coefficients are $a_1 = 0.488\ 851,\ a_2 = -1$

(f) For the second harmonic of the low frequency tone, i.e. 2×941 Hz, the values of the feedback coefficients are $a_1 = 0.327\ 635,\ a_2 = -1$

Question 8

Which of the following statements are true? In DTMF decoding:

(a) We need to know the magnitudes of the second harmonics of the DTMF frequencies as well as those of the fundamental frequencies to discriminate between speech and DTMF tones

(b) We do not need to know the magnitudes of the second harmonics of the DTMF frequencies to discriminate between speech and DTMF tones

(c) The frequency response of the telephone system is such that the DTMF high frequency tones are attenuated more than the low frequency tones

(d) The frequency response of the telephone system is such that the DTMF low frequency tones are attenuated more than the high frequency tones

(e) Speech has significant even-order harmonics, but DTMF signals do not

(f) DTMF signals have significant even-order harmonics, but speech does not

Question 9

A two-stage decimator is used to reduce the sample rate in a system from 240 k samples per second to 8 k samples per second. The decimation factors for the stages are $M_1 = 15$ and $M_2 = 2$, and the filter lengths are $N_1 = 45$ and $N_2 = 43$. Which of the following statements are correct?

(a) The sample rates at the outputs of the two stages are 16 k and 8 k samples per second, respectively

(b) The sample rates at the outputs of the two stages are 8 k and 16 k samples per second, respectively

(c) The measures of complexity for the decimator are: MPS = 1 064 000, TSR = 88

(d) The measures of complexity for the decimator are: MPS = 1064, TSR = 88

(e) None of the above

Question 10

A two-stage decimator is to be used to reduce the sampling rate from 500 Hz to 12.5 Hz. The decimation factors for the stages are 10 and 4 and the associated filter lengths are $N_1 = 55$, $N_2 = 97$, respectively. What are the efficiency indices for the decimator?

(a) MPS = 32 350.5, TSR = 152

(b) MPS = 3962.5, TSR = 152

(c) MPS = 500, TSR = 55

(d) MPS = 28 712.5, TSR = 152

(e) None of the above

Question 11

A two-stage decimator is used to reduce the sampling rate from 500 Hz to 12.5 Hz. The decimation factors for the first and second stages of decimation are 10 and 4, respectively. The frequency band of interest is 0–4 Hz. What are the bandedge frequencies for the anti-aliasing filter in the first stage?

(a) 0, 4, 6.25 Hz

(b) 0, 4, 43.75 Hz

(c) 0, 4, 12.5 Hz

(d) 0, 4, 50, 6.25 Hz

(e) None of the above

14.5 Summary

In this chapter, a number of design and development boards that may be used to implement some of the DSP algorithms described in the book have been presented. A number of real-world applications of DSP are described in the form of case studies to give the reader some idea of practical design issues. We have also presented in the problems section a number of challenging design studies and a method, based on multiple choice questions, that we have found useful for rapid assessment of DSP topics. Both should serve as a means of gaining deeper insight into DSP.

Problems

14.1 Show, stating any assumptions made, that the autocorrelation function (ACF) of a signal contaminated by a random noise is the same as the ACF of the signal alone. Explain how this result may be used to detect hidden periodicities.

14.2 Prove, stating any reasonable assumptions, that the maximum signal-to-noise ratio at the output of a digital matched filter is independent of the waveshape of the input signal.

14.3 A recurring signal, buried in noise, is to be detected by digital matched filtering. Given below are the successive sample values of the noise-free signal and the noisy signal:

noise-free signal {−0.51, −0.35, −0.29, −0.25, −0.29, −0.39, −0.47}

noisy signal {−0.18, −0.06, 0.27, 0.69, −0.50, −0.44, −0.20, −1.46, −0.93, −1.46, −0.91, −0.39, −1.70}

Determine

(a) the coefficients of the digital matched filter,

(b) the output of the digital matched filter, and

(c) the improvement in signal-to-noise ratio, expressed in decibels, achievable by matched filtering.

Note: the variance, σ_0^2, of the noise at the filter output is given by

$$\sigma_0^2 = \sigma^2 \sum_{m=0}^{\infty} h^2(m)$$

where σ^2 is the variance of the noise at filter input, and the $\{h(m)\}$ are the filter coefficients.

References

Azevedo S. and Longini R.L. (1980) Abdominal-lead fetal electrocardiographic R-wave enhancement for heart rate determination. *IEEE Trans. Biomedical Engineering*, **27**(5), 255–60.

Bajpai, A.C., Calus I.M. and Fairley J.A. (1973) *Mathematics for Engineers and Scientists*, Volume 2. New York: Wiley.

Barlow J.S. and Rémond A. (1981) Eye movement artifact nulling in EEGs by multichannel on-line EOG subtraction. *Electroencephalography and Clinical Neurophysiology*, **52**, 418–23.

Bierman G.J. (1976) Measurement updating using the *U-D* factorization. *Automatica*, **12**, 375–82.

Bierman G.J. (1977) *Factorization Methods for Discrete Sequential Estimation*. New York: Academic Press.

Clark R.J., Ifeachor E.C., Van Eetvelt P.W.J. and Rogers G.M. (2000) Techniques for generating digital equalizer coefficients. *J. Audio Eng. Soc.*, **48**(4), April, 281–98.

Clarke D.W. (1981) Implementation of self-tuning controllers. In *Self-Tuning and Adaptive Control* (Harris C.J. and Billings S.A. (eds)), pp. 144–65. Stevenage, UK: Peter Peregrinus.

Cowan C.F.N. and Grant P.M. (1984) Adaptive processing – an overview. In *IEE Colloq. Adaptive Processing and Biomedical Applications*, October 1984, Paper 1.

Dattorro J. (1988) The implementation of recursive digital filters for high-fidelity audio. *J. Audio Eng. Soc.*, **36**(11), 851–78.

Favret A.G. (1968) Computer matched filter location of fetal R-waves. *Medical and Biological Engineering*, **6**, 467–75.

Flores I. (1963) *The Logic of Computer Arithmetic*. Englewood Cliffs NJ: Prentice-Hall.

Fortgens C. and De Bruin M.P. (1983) Removal of eye movement and ECG artifacts from the non-cephalic reference EEG. *Electroencephalography and Clinical Neurophysiology*, **56**, 90–6.

Girton D.G. and Kamiya A.J. (1973) A simple on-line technique for removing eye movement artifacts from the EEG. *Electroencephalography and Clinical Neurophysiology*, **34**, 212–8.

Goodman G.C. and Sin K.S. (1984) *Adaptive Filtering, Prediction and Control*. Englewood Cliffs NJ: Prentice-Hall.

Gotman J., Gloor P. and Ray W.F. (1975) A quantitative comparison of traditional reading of the EEG and interpretation of computer-extracted features in patients with supratentorial brain lesions. *Electroencephalography and Clinical Neurophysiology*, **38**, 623–39.

Greene K.R. (1987) The ECG waveform. In *Balliere's Clinical Obstetrics and Gynaecology* (Whittle M. (ed.)), Volume 1, pp. 131–55.

Hamer C.F., Ifeachor E.C. and Jervis B.W. (1985) Digital filtering of physiological signals with minimal distortion. *Medical and Biological Engineering and Computation*, **23**, 274–8.

Harris C.J. (1983) Brainwaves appear on T.V. in real-time. *Electronics*, February, 47–8.

IEEE (1985) IEEE Standard for Binary Floating Point Arithmetic. *SIGPLAN Notices*, **22**(2), 9–25.

Ifeachor E.C. (2001) *A Practical Guide for MATLAB and C Language Implementations of DSP Algorithms*. Harlow: Pearson Education.

Ifeachor E.C., Jervis B.W., Morris E.L., Allen E.M. and Hudson N.R. (1986) A new microcomputer-based online ocular artefact removal (OAR) system. *Proc. IEE*, **133**(5), 291–300.

Ifeachor E.C., Keith R.D.F., Westgate J. and Greene K.R. (1991) An expert system to assist in the management of labour. In *Proc. World Congress on Expert Systems* (Liebowitz J. (ed.)), Volume 4, pp. 2615–22. New York: Pergamon.

Jervis B.W., Allen E., Johnson T.E., Nichols M.J. and Hudson N.R. (1984) The application of pattern recognition techniques to the contingent negative variation for the differentiation of subject categories. *IEEE Trans. Biomedical Engineering*, **31**, 342–9.

Jervis B.W., Nichols M.J., Allen E., Hudson N.R. and Johnson T.E. (1985) The assessment of two methods for removing eye movement artefact from the EEG. *Electroencephalography and Clinical Neurophysiology*, **61**, 444–52.

Lindecrantz K.G., Lilja H. and Rosen K.G. (1988) New software QRS detector algorithm suitable for realtime applications with low signal to noise ratios. *J. Biomedical Engineering*, **10**, 280–3.

Motorola (1995) DSP56000 Digital Signal Processor Family Manual. Austin TX: Motorola.

Motorola (1996) DSP56302 Evaluation Module. Motorola Inc. *www1.motoroladsp.com/docs/docs.html*

Outram N.J., Ifeachor E.C., Van Eetvelt P.W.J. and Curnow J.S.H. (1995) Techniques for optimal enhancement and feature extraction of the fetal electrocardiogram. *IEE Proc.-Sci. Meas. Technol.*, **142**(6), 482–9.

Patterson D.A. and Hennessy J.L. (1990) *Computer Architecture: A Quantitative Approach*. San Mateo CA: Morgan Kaufmann.

Peterka V. (1975) A square root filter for real-time multivariate regression. *Kybernetika*, **11**, 53–67.

Quilter P.M., Macgillivray B.B. and Wadbrook D.G. (1977) The removal of eye movement artefact from EEG signals using correlation techniques. *IEE Conf. Publ.*, **159**, 93–100.

Rabiner L.R. and Gold B. (1975) *Theory and Application of Digital Signal Processing*. Englewood Cliffs NJ: Prentice-Hall.

Takeda H. and Hata S. (1985) Development of micro-computerized topographic EEG analyzer and its application to real time display. *Electroencephalography and Clinical Neurophysiology*, **61**, 98.

Texas Instruments (1986) *Digital Signal Processing Applications with the TMS320 Family: Theory, Algorithms and Implementations*. Texas Instruments.

Texas Instruments (1995) TMS320C54x evaluation module technical reference. Texas Instruments. *www.ti.com/sc/docs/psheets/abstract/apps/spru135.html*

Tomé A.M., Principe J.C. and Da Silva A.M. (1985) Micro analysis of spike and wave bursts in children's EEG. *Electroencephalography and Clinical Neurophysiology*, **61**, 113.

Weitek (1984) *High Speed Digital Arithmetic VLS Application Seminar Notes*. Sunnyvale CA: Weitek.

Widrow B., Glover J.R., McCool J.M., Kaunitz J., Williams C.S., Hearn R.H., Zeidler J.R., Dong E. and Goodlin R.C. (1975) Adaptive noise cancelling: principles and applications. *Proc. IEEE*, **63**, 1692–716.

Young P. (1974) Recursive approaches to time series analysis. *Bull. IMA*, **10**, 209–24.

Bibliography

Clarke D.W. (1980) Some implementation considerations of self-tuning controllers. In *Numerical Techniques for Stochastic Systems* (Archetti F. and Cugiani M. (eds)), pp. 81–101. Amsterdam: North-Holland.

Clarke D.W., Cope S.N. and Gawthrop P.J. (1975) *Feasibility Study of the Application of Microprocessors to Self-tuning Controllers*. OUEL Report 1137/75.

Dattorro J. (1988) The implementation of recursive digital filters for high fidelity audio. *J. Audio Eng. Soc.*, **36**(11), 851–78.

Kay S.M. (1987) *Modern Spectrum Estimation*. Englewood Cliffs NJ: Prentice-Hall.

Marple S.L. Jr (1987) *Digital Spectral Analysis with Applications*. Englewood Cliffs, NJ: Prentice-Hall.

Motorola (1980) *16-bit Microprocessor User's Manual*. Austin TX: Motorola Semiconductor.

Otnes R.K. and Enochson L. (1978) *Applied Time Series Analysis*, Volume 1. New York: Wiley.

Rosen K.G. and Lindecrantz K.G. (1989) STAN, the Gothenburg model for fetal surveillance during labour by ST analysis of the fetal electrocardiogram. *Clinical Physiology and Physiological Measurement, Suppl. B*, **10**, 51–6.

Stanley W.D., Dougherty G.R. and Dougherty R. (1984) *Digital Signal Processing*, 2nd edn. Reston VA: Reston Publications.

Verleger R., Gasser T. and Möcks J. (1982) Correction of EOG artifacts in event-related potentials of the EEG: aspects of reliability and validity. *Psychophysiology*, **19**, 472–80.

Appendix

14A   The modified UD factorization algorithm

- Step 1: $\mathbf{v} = \mathbf{U}^{\mathrm{T}}(m)\mathbf{x}$.
- Step 2: $b_i = d_i(m)v_i, \quad i = 1, \ldots, n$.
- Step 3: $\alpha_1 = \gamma + b_1 v_1$.
- Step 4: $d_1(m + 1) = d_1(m)/\alpha_1$.

For $j = 2, \ldots, n$, recursively evaluate Equations 14.9–14.13.

- Step 5: $\alpha_j = \alpha_{j-1} + v_j b_j$.
- Step 6: $\rho_j = -v_j/\alpha_{j-1}$.

For $k = 1, 2, \ldots, j - 1$, recursively evaluate Equation 14.11.

- Step 7: $U_{kj}(m + 1) = U_{kj}(m) + b_k \rho_j$.
- Step 8: $b_k = b_k + b_j U_{kj}(m)$.

- Step 9: $d_j(m + 1) = d_j(m) \alpha_{j-1}/\alpha_j \gamma$.
- Step 10: $G = b/\alpha_n$ ($g_i = b_i/\alpha_n$, $i = 1, \ldots, n$).

The following points should be noted.

(1) α_n in step 10 is the value of α_j (step 5) after the nth iteration. This is also equal to α in Equation 14.9 of the RLS algorithm.

(2)   The elements of **D** (that is the d_i in steps 4 and 9) can be stored along the diagonals of **U**, since $U_{jj} = 1$. In addition, to save storage and for ease of programming, the elements of **U** (including those of **D**) can be stored as a vector, even though the subscripts (k, j) indicate that **U** is a two-dimensional array.

Index

COMING SOON
A Practical Guide for MATLAB and C Language Implementations of DSP Algorithms

This CD and user guide is a practical companion to the book Ifeachor & Jervis: *Digital Signal Processing*. It focuses on application and implementation of the programs covered in the main text.

CD includes:
*MATLAB m-files, C-language programs and assembly language codes referred to in the main text
*further illustrative examples of the use of the MATLAB m-files and C programs
*tutorials and suggestions for use of the code

User Guide & CD   ISBN: 0130433020   Price: £50.00/$70.00

SPECIAL OFFER: Purchase before May 31st 2002 and receive a **50%** discount
(only valid with an original order form)

If you would like to order a copy of the companion handbook & CD please fill in the order form below and send this whole page to:

Pearson Education, Customer Services (orders), PO Box 88,
Edinburgh Gate, Harlow, Essex CM19 5AA, U.K.
or fax +44 (0)1279 623627

Your details:
Name: ....
Address: ....
....
.... Post code/Zip code: ....
Email: .... Telephone: ....

Purchase Order Form:
I would like .... copies of *A Practical Guide for MATLAB and C Language Implementations of DSP Algorithms* for a total value of £........./$......... (£25/$35 per unit up to and including 31/05/02, £50/$70 after 31/05/02) + postage and packing at £........./$.........

N.B. For details of postage and packing please fax your order to the number above or call +44 (0)1279 623928.

❑ I enclose a cheque for £........./$......... made payable to **Pearson Education Ltd**.

❑ Please charge £........./$......... to my credit card: MasterCard/VISA/Diners Club
Card Number: ....
Issue from: .... Expiry Date: ....
Name: ....…...
*Address: ....
....
Signature: .... Date: ....…....
*Please ensure that the address supplied is that to which the card is registered.